Geologic Evolution
of
Atlantic Continental Rises

Geologic Evolution
of
Atlantic Continental Rises

Edited by

C. Wylie Poag
and
Pierre Charles de Graciansky

VNR SPRINGER SCIENCE+BUSINESS MEDIA, LLC

Additional material to this book can be downloaded from http://extras.springer.com.

Copyright © 1992 by Springer Science+Business Media New York
Originally published by Van Nostrand Reinhold Inc. in 1992
Softcover reprint of the hardcover 1st edition 1992
Library of Congress Catalog Card Number: 91-45498
ISBN 978-1-4684-6502-0   ISBN 978-1-4684-6500-6 (eBook)
DOI 10.1007/978-1-4684-6500-6

16 15 14 13 12 11 10 9 8 7 6 5 4 3 2 1

Library of Congress Cataloging-in-Publication Data
Geologic evolution of Atlantic continental rises/edited by C. Wylie
   Poag and Pierre Charles de Graciansky.
      p.   cm.
   Includes bibliographical references and index.
   ISBN 978-1-4684-6502-0

      1. Geology—Atlantic Ocean.   2. Continental margins—Atlantic
   Ocean.   I. Poag, C. Wylie.   II. Graciansky, Pierre Charles.
   III. Title: Continental rises.
   QE350.5.G45   1992
   551.46′08′093—dc20   91-45498
                                                                  CIP

To the Deep Sea Drilling Project, whose expeditions fundamentally altered conventional concepts of continental-rise geology, and under whose auspices our collaboration and friendship began.

Contents

Part I. Prerift and Synrift Evolution

Part II. Early Postrift Evolution

List of Foldouts

Contributors

Juan Acosta
Instituto Español de Oceanografica
Corazon de Maria, 8, Primer Piso
Madrid 28002
Spain

Christophe Basile
Laboratoire de Geodynamique Sous-Marine
B.P. 48
06230 Villefranche/Mer
France

P. Charpentier
ELF Aquitaine
Centre Scientifique et Technique Jean Feger
64018 Pau Cedex
France

Richard V. Dingle
South African Museum
P.O. Box 61
Cape Town 8000
South Africa

Max R. Dobson
University of Wales
Institute of Earth Sciences
Aberystwyth, SY23 3DB
United Kingdom

Olav Eldholm
Department of Geology
University of Oslo
P.O. Box 1047
N-0316 Blindern
Oslo 3
Norway

J. I. Faleide
Department of Geology
University of Norway
P.O. Box 1047
N-0316 Blindern
Oslo 3
Norway

Roger D. Flood
Department of Marine Science
State University of New York at Stony Brook
Stony Brook, NY 11794-5000

James V. Gardner
U.S. Geological Survey
345 Middlefield Road
Menlo Park, CA 94025

S. T. Gudlaugsson
Department of Geology
University of Oslo
P.O. Box 1047
N-0316 Blindern
Oslo 3
Norway

Pierre Charles de Graciansky
Ecole Nationale Supérieure de Mines, Paris
60 Boulevard Saint-Michel
75272 Cedex 06
Paris
France

Dennis E. Hayes
Lamont-Doherty Geological Observatory of Columbia
    University and Department of Geology, Columbia
    University
Palisades, NY 10964

P. Herranz
Instituto Español de Oceanografia
Malaga
Spain

Quintin J. Huggett
Institute of Oceanographic Sciences, Deacon
    Laboratory
Brook Road
Wormley, Godalming
Surrey GU8 5UB
United Kingdom

John E. Hughes Clarke
Department of Survey Engineering
University of New Brunswick
Fredericton, N.B.
Canada

Robert D. Jacobi
Department of Geology
State University of New York at Buffalo
4240 Ridge Lea Road
Buffalo, NY 14260

Robert B. Kidd
Department of Geology
University College of Wales, Cardiff
P.O. Box 914
Cathys Park
Cardiff CF1 3YE
United Kingdom

Allen Lowrie
Consultant Geologist
RFD Route 1
Box 164—F.Z. Goss Road
Picayune, MS 39522-5001

Jean Mascle
Laboratoire de Géodynamique Sous-Marine
B.P. 48
06230 Villefranche/Mer
France

Douglas G. Masson
Institute of Oceanographic Sciences, Deacon
    Laboratory
Brook Road
Wormley, Godalming
Surrey GU8 5UB
United Kingdom

Gilberto A. Mello
Lamont-Doherty Geological Observatory of Columbia
    University
Palisades, NY 10964

Annik M. Myhre
Department of Geology
University of Oslo
P.O. Box 1047
N-0316 Blindern
Oslo 3
Norway

Dennis W. O'Leary
U.S. Geological Survey
Branch of Atlantic Marine Geology
Woods Hole, MA 02543

Thomas H. Orsi
Department of Oceanography
Texas A&M University
College Station, TX 77843

David J. W. Piper
Geological Survey of Canada
Atlantic Geoscience Centre
Bedford Institute of Oceanography
Box 1006
Dartmouth, Nova Scotia
Canada B2Y 4A2

S. Planke
Department of Geology
University of Oslo
P.O. Box 1047
N-0316 Blindern
Oslo 3
Norway

C. Wylie Poag
U.S. Geological Survey
Branch of Atlantic Marine Geology
Woods Hole, MA 02543

J. P. Richert
ELF Aquitaine
Centre Scientifique et Technique Jean Feger
64018 Pau Cedex
France

Simon H. Robson
EPI Consultancy
12A Foregate Square
Harbour Road
Cape Town
South Africa

Sergio Rossi
Instituto di Geologia Marina
Via Zamboni
Bologna, Italy

J. L. Sanz
Instituto Español de Oceanografia
Malaga
Spain

Jean-Claude Sibuet
IFREMER Centre de Brest
B.P. 70
29280 Plouzane
France

J. Skogseid
Department of Geology
University of Oslo
P.O. Box 1047
N-0316 Blindern
Oslo 3
Norway

L. M. Steuvold
Department of Geology
University of Oslo
P.O. Box 1047

N-0316 Blindern
Oslo 3
Norway

Elazar Uchupi
Department of Geology and Geophysics
Woods Hole Oceanographic Institution
Woods Hole, MA 02543

E. Vågnes
Department of Geology
University of Oslo
P.O. Box 1047
N-0316 Blindern
Oslo 3
Norway

Frédéric Walgenwitz
ELF Aquitaine

Centre Scientifique et Technique Jean Feger
64018 Pau Cedex
France

Philip P. E. Weaver
Institute of Oceanographic Sciences, Deacon
    Laboratory
Brook Road
Wormley, Godalming
Surrey GU8 5UB
United Kingdom

Frances Westall
Sciences de la Terre
Universite de Nantes
France

Preface

Continental rises comprise approximately 10% of the earth's surface and contain 20% of its total sediment volume ($\sim$ 100 million cubic kilometers). However, their great depth (2,000–6,000 m) below the sea surface and distance from shore-based population centers, have kept them relatively insulated from man's geological explorations. During the last 10–15 years, however, driven by an ever-increasing thirst for knowledge of our planet's nature and origin, by a rising imperative to find new reserves of fossil fuels, and by a demand to find safe haven for hazardous wastes, marine scientists have intensified their surveys of this vast province.

Voluminous geological and geophysical data from multichannel and single-channel seismic-reflection profiles, high-resolution seafloor images (SEABEAM, Sea MARC, GLORIA), and various gravity, piston, and rotary cores (Deep Sea Drilling Project, Ocean Drilling Program), record the evolution of this expansive deep-water depositional regime. To date, the results of myriad individual studies have been published, but no comprehensive regional or circumoceanic syntheses have yet been attempted. The purpose of this volume is to bring together an international group of experts on continental-rise geology, who present selected regional syntheses and detailed local studies of the surficial and subsurface stratigraphic framework and the structural/depositional history of Atlantic continental rises. We focus on the Atlantic because its continental rises are larger and more clearly defined than those of any other ocean basin, and because it is the most thoroughly explored major ocean basin; it has been examined extensively by all three exploration methods (seismic profiling, seafloor imaging, and coring). We address a wide variety of structural and sedimentary settings, including sites off all the Atlantic-facing continents (North and South America, Europe, Africa, and Antarctica).

Most published volumes dealing with continental-rise deposits emphasize near-bottom lithofacies and sedimentary processes by which modern and submodern submarine fans and related features are constructed. To provide a broader perspective of continental-rise evolution, we have organized this volume into four parts:

1. Part I addresses prerift and synrift tectonic and depositional aspects that account for the location, geometry, and thickness of subsequent continental rises.
2. Part II embraces early postrift aspects, such as subsurface structure, stratigraphy, and accumulation rates; relationships to continental slope and shelf deposition and to continental source terrains; and the relative effects of local and global regulating agents, such as tectonism, eustacy, and paleoclimate. This accounts for most of the $\sim$ 187 million-year evolution of Atlantic continental rises.
3. Part III is devoted to late postrift aspects, especially Pleistocene and Holocene episodes of deposition and erosion, which have shaped the modern Atlantic continental rises.
4. Part IV summarizes our circum-Atlantic perspective of continental-rise evolution.

This volume is aimed primarily at university students and professional geologists and oceanographers; it will appeal to a wide range of industry (particularly petroleum-related), academic, and government earth scientists and environmental specialists. It is not a comprehensive textbook (though it would be an excellent nucleus for a graduate seminar), but constitutes a partial synthesis (with selected case studies) of our current knowledge of the subject, a general guide to the application of contemporary survey methods and analytical techniques, a pathway to the pertinent literature, and a reminder that the evolution of Atlantic continental rises has been intricately complex and by no means uniform. Many vital questions remain unanswered.

Acknowledgments

We are grateful to Mel Peterson, Terry Edgar, Yves Lancelot, and Matt Salisbury of the Deep Sea Drilling Project, who encouraged our interest in deep-ocean sedimentary regimes and provided a singular means of exploring them. We also thank P. Robin Brett of the USGS, who suggested this topic originally as a poster session for the 28th International Geological Congress. Thanks from C.W.P. go to John Behrendt, Bill Dillon, Dave Folger, Brad Butman, Joe Hazel, Bill Sliter, Dick Poore, and John Pojeta, who, as successive chiefs of the USGS Branch of Atlantic Marine Geology and the Branch of Paleontology and Stratigraphy, have generously supported his research on the geology of Atlantic continental margins. We owe particular thanks to Chuck Hutchinson, Mark Licker, and Marjorie Spencer for their indispensible help in initiating and consumating this publishing effort. Our sincerest gratitude goes to all the contributing authors for their dedication to this project and for their patience during its prolonged incubation.

Geologic Evolution
of
Atlantic Continental Rises

I
Prerift and Synrift Evolution

1

Galicia Continental Margin: Constraints on Formation of Nonvolcanic Passive Margins

Jean-Claude Sibuet

Figure 1-1 shows two interpretative sections across the North American margin (Baltimore Canyon Trough; Grow, 1981) and the European margin (western Galicia margin; Sibuet et al., 1987). The two margins differ considerably because a huge volume of synrift and postrift sediments exists on the American margin (up to 15 km thick close to the ocean–continent boundary; see Chapter 6), as opposed to 1–2 km on the Iberian margin. The geometry, provenance, and style of deposition of the North American sedimentary pile are described and presented in Chapters 6, 9, and 10. The North American continental rise has been formed mainly by construction of superposed deep-sea fans and other sediment–gravity flows and mass movements since the initial disruption of the Pangean continents. Three main tectonic factors have controlled the location and size of the basin in which the rise sediments accumulated:

1. The mechanism of formation of continental margins during the rifting phase, which tightly controlled the resulting depth and morphology of the basement;
2. Subsequent thermal cooling and subsidence of the basement during the postrift phase, which is determined by the mechanism of continental thinning.
3. Isostatic readjustment of basement level due to sedimentary load.

Because the shelfbreak migrated oceanward through time, and is presently above the oceanic crust in some areas (Fig. 1-1A), construction of the thick sedimentary column in the presence of a massive detrital input was tectonically controlled by the mechanism of continental thinning. Where the total subsidence curves and resulting tectonic subsidence curves are available from pa-

leobathymetric estimations deduced from drill holes (Watts and Steckler, 1979), some constraints can be placed on the thinning mechanisms. However, the continental basement and its overlying prerift sequence are so deep on the North American margin that we have only imprecise ideas of their morphology, no information on their internal structure, and consequently no information on the style of deformation during rifting.

On the European side, in contrast, sedimentary accumulations are relatively thin, except in a few structural basins. The North Spanish continental rise is such an example; there, Tertiary products of the erosion of the Pyrenees have been laterally conveyed through the Gouf de Capbreton (located in the southeastern corner of the Bay of Biscay) to reach the lower continental margin and adjacent oceanic crust. The Interior Basin, located between the Galicia Bank and Iberia, also could be considered a sedimentary trap, which has prevented any sedimentary transit in the direction of the lower western Galicia slope. For that reason, only minor detrital input is available to the western Galicia margin, and that small volume cannot overflow the structural system of parallel ridges to build a significant continental rise. The continental rise is physiographically well defined on the eastern U.S. margin, but is poorly developed or absent on starved passive margins, and in particular on the western Galicia margin. In this chapter, therefore, I use the expression "upper continental margin" for continental slope, and "lower continental margin" for continental rise. The definitions of such features by structural geologists and geophysicists may not necessarily correspond to the definitions offered by sedimentologists.

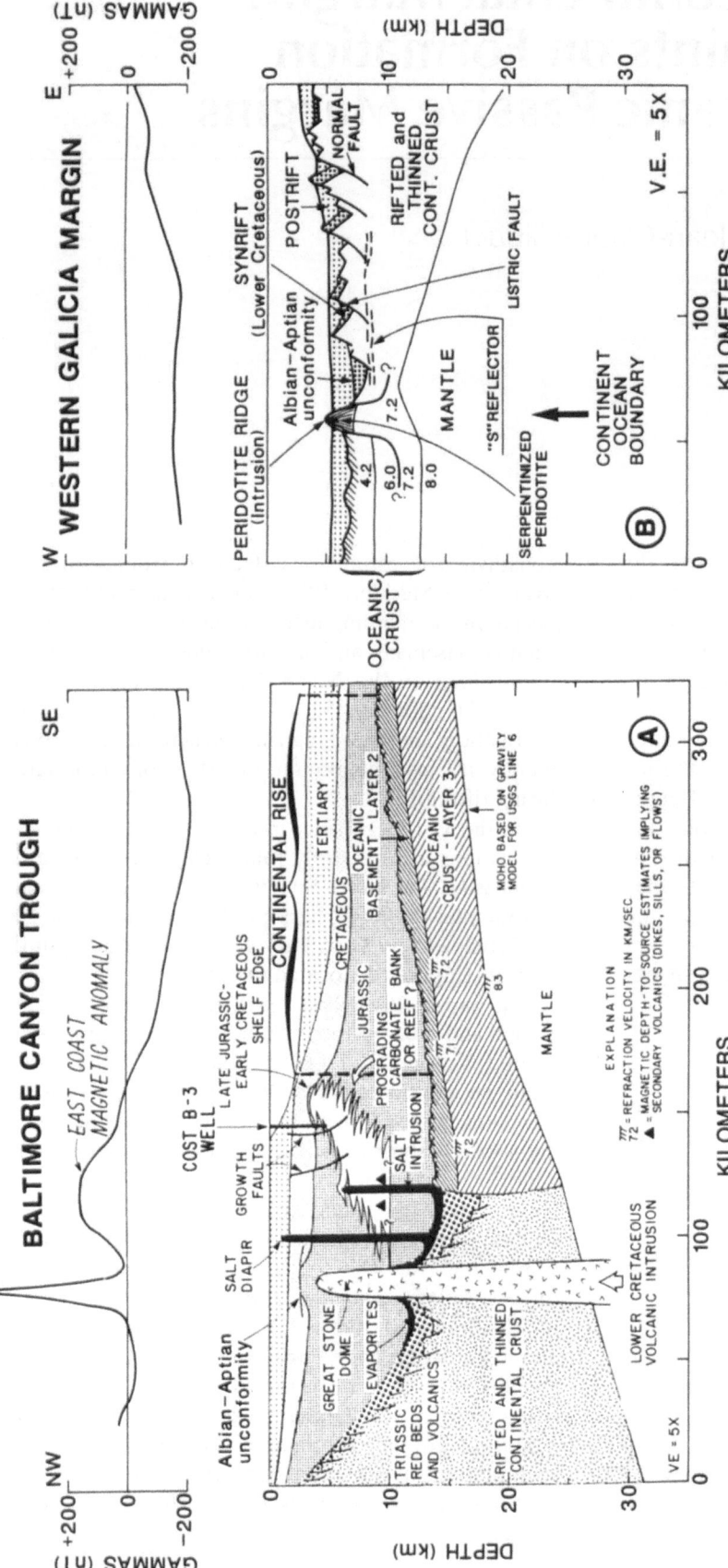

FIGURE 1-1. Interpretative sections across the North American margin. (A, Baltimore Canyon Trough; Grow, 1981) and the European margin (B, western Galicia margin; adapted from Sibuet et al., 1987, and Recq, Whitmarsh, and Sibuet, 1991) showing contrast between thickly sedimented passive margin with the building up of a distinct continental rise (A) and a sediment-starved passive margin without the construction of a distinct rise (B).

The nature, provenance, and evolution of synrift and postrift sedimentary deposits on the Galicia margin have been thoroughly studied by scientists of the Ocean Drilling Program Leg 103 (Boillot, Winterer, et al., 1988b). Therefore, I will not reexamine these aspects in this chapter. However, because favorable circumstances exist on the Galicia margin to image the basement features and their internal crustal structures, I focus mainly on them. The dislocation of the margin under tensional processes can be unravelled from these detailed structural observations, which are not accessible on the thickly sedimented American margin. Thus, western Galicia is probably one of the most appropriate margins to study rifting mechanisms; the results then can be applied to the evolution of heavily sedimented margins.

Pure and simple shear are the two main mechanisms that describe deformation of the lithosphere during continental rifting. Models using other mechanisms are either derived from these two or do not significantly contribute to understanding rifting processes. In the simple shear, or detachment, model (Bally, 1981; Wernicke and Burchfield, 1982), a major low-angle detachment zone cuts across the crust and upper mantle (Wernicke, 1985), or only the brittle portion of the continental crust, with a flattening, which soles out at the brittle–ductile interface (Le Pichon and Barbier, 1987; Radel and Melosh, 1987). In the pure shear, or stretching model (McKenzie, 1978), deformation of the whole lithosphere involves formation of a more or less symmetrical surface depression (e.g., Le Pichon and Sibuet, 1981).

Royden and Keen (1980) have developed the depth-dependent model, a two-layer model with two different stretching factors. Below the basins and margins, this model gives results similar to that of the thermomechanical model (Moretti and Froidevaux, 1986), which supposes a weak extension in the upper crust and a large scale material displacement in the ductile lower crust, following the appearance of a thermal anomaly at the base of the lithosphere. Nevertheless, outside the basins, these two models differ

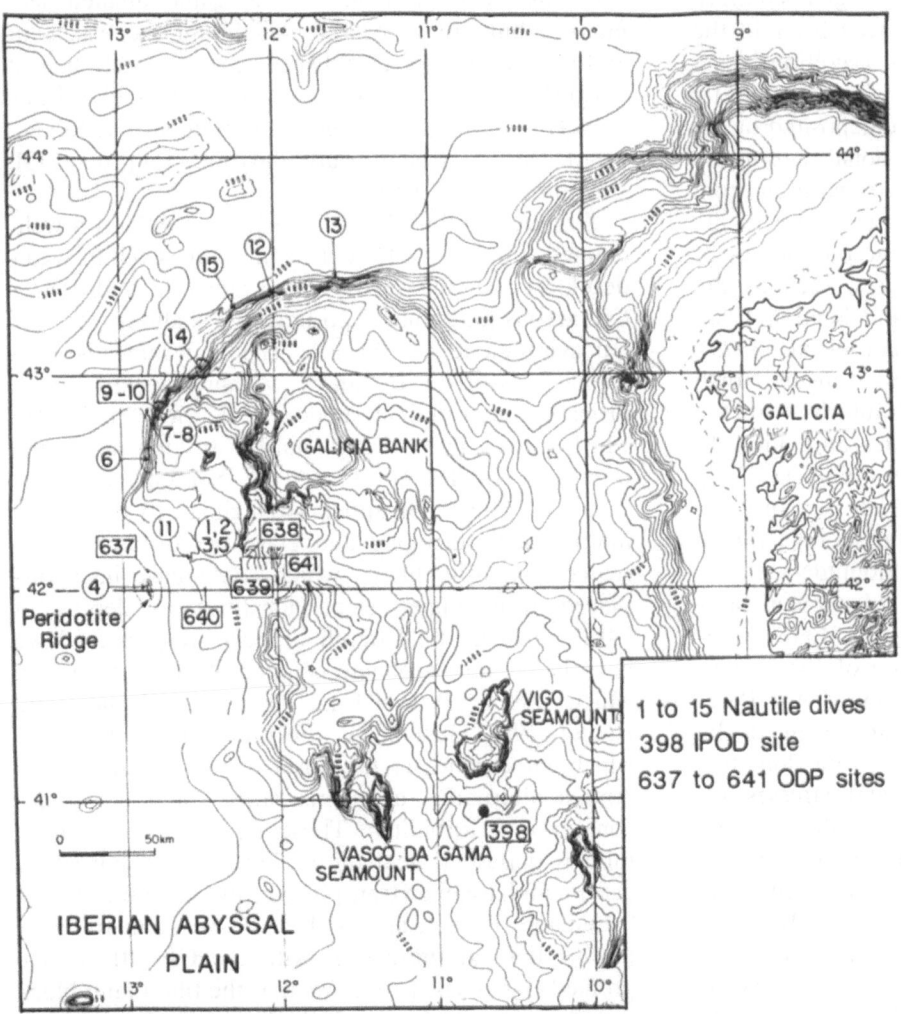

FIGURE 1-2.   Bathymetric map of the Galicia Bank area (from Lallemand and Sibuet, 1986) with locations of IPOD (International Program of Ocean Drilling) and ODP (Ocean Drilling Program) drill sites (398, 637 to 641) and *Nautile* dives (from Boillot et al., 1988).

considerably, because of the increased thickness of the lower crust due to lateral flow in the thermomechanical model.

Depending on the authors and areas of investigation, many models or combinations of models have been proposed. The aim of this chapter is to interpret new data acquired from the Galicia continental margin, in order to constrain rifting mechanisms and, thereby, to better understand the construction and sedimented margins and their continental rises.

TECTONIC SETTING

Extensional Phases

The continental margin of the western Galicia Bank was formed during several phases of extension since the Trias, and was affected later on by Early Tertiary and Miocene phases of compression. Triassic to Liassic evaporites or sandstones recovered in the deep basins of Iberia and its margins (such as the Algarve Basin) are interpreted as products of the first extensional phase recorded in northwest Europe (Ziegler, 1982). In fact, several extensional phases occurred prior to the late Dogger (Mougenot, 1988), when accretion began in the Central Atlantic. During the Middle and early Late Jurassic, marine sediments were deposited in the depressions, which suggests that regional subsidence took place around Iberia during its separation from North America, Eurasia, and Africa (Savostin et al., 1986). The Late Jurassic–Early Cretaceous phases of extension reactivated previous features, such as the Parentis Basin (Curnelle, Dubois, and Seguin, 1980) and the Lusitanian Basin (Mougenot et al., 1979). Then, the onset of spreading progressed in the North Atlantic from south to north: late Hauterivian in the Iberian Abyssal Plain (Whitmarsh, Miles, and Mauffret, 1990), late Aptian for the southern Galicia Bank (Sibuet, Ryan, et al., 1979), early Albian west of Galicia Bank and in the Bay of Biscay (Boillot et al., 1987b; Montadert, Roberts, et al., 1979; Montadert et al., 1979), and late-early Albian west of Goban Spur (Graciansky, Poag, et al., 1985). From the Late Cretaceous to late Eocene, limited rotation of Iberia with respect to Eurasia resulted in formation of the Pyrenees and the north Spanish marginal trough at the plate boundary. After relative motion ceased along this plate boundary, north–south horizontal stresses increased, and were transmitted over long distances by the lithosphere, resulting in compressive features, which appeared along previous lines of weakness from the Açores—Gibraltar region to the North Sea. Less important Miocene compressive episodes mostly affected Tore–Madeira Rise, Gorringe Bank, and the adjacent continental margins, as well as the Betic and Rif systems on land.

Using experimental and analytical techniques, Withjack and Jamison (1986) have described the strain state and fault pattern produced by oblique rifting. Applying these techniques to the northeast Atlantic continental margins, Sibuet (1988, 1989) has shown that the Galicia margin can be considered as a pure tensional rifted margin.

From Goban Spur to Gibraltar, the overall tectonic style of continental margin deformation acquired during the Late Jurassic–Early Cretaceous rifting phase is characterized by a series of tilted fault blocks bounded by normal or listric faults, which delineate half-grabens. The tilted fault-block geometry, of which the most spectacular example is the western Galicia margin (Fig. 1-2), developed within both the Jurassic platform and the underlying continental basement.

Timing and Geometry of Tilted Fault Blocks

North–south trending tilted fault blocks can be delineated on the detailed SEABEAM bathymetric map of the western Galicia margin (Fig. 1-3; Sibuet et al., 1987) as steep gradients or elongated aligned seamounts, but are especially evident on the basement and structural maps established from a closely spaced network of single-channel and multichannel seismic lines (Figs. 1-4, 1-5; Thommeret, Boillot, and Sibuet, 1988). The rift structures are illustrated on profiles GP 101 and 102 (Fig. 1-6).

Drilling at ODP Sites 638, 639, and 641 on a well-defined tilted block (Fig. 1-6) in about 4,700-m water depth, recovered synrift Upper Jurassic and Lower Cretaceous limestone deposited in a shallow-water paleoenvironment (< 100 m). The limestone is conformably overlain by lower Valanginian marlstone, probably deposited at moderate depth (< 1,000 m), which passes upward into marl with thin sandstone interbeds (Boillot et al., 1987b). Subsequent synrift sediments consist of upper Valanginian to Barremian interbedded sandstone and claystone turbidites with scarce composite fauna, which could have been displaced from upper slope localities to depths of 2,000–3,000 m; final depth of deposition might have been below the carbonate compensation depth (CCD, which was at ~ 2,500 m; Moullade, Brunet, and Boillot, 1988). Just below the breakup unconformity, which corresponds to the onset of seafloor spreading, Aptian alternating clays, marlstones, and limestones contain thin turbidites. These strata include faunas initially deposited near the CCD (2,200 m), but which also might have been transported from shallower depths (Moullade, Brunet, and Boillot, 1988). Comparison of drilling data at these sites with the GP 101 seismic profile (Figs. 1-1B, 1-6) shows that the tilted fault block includes not only the prerift Tithonian carbonate platform, but also the Valanginian to lower Hauterivian

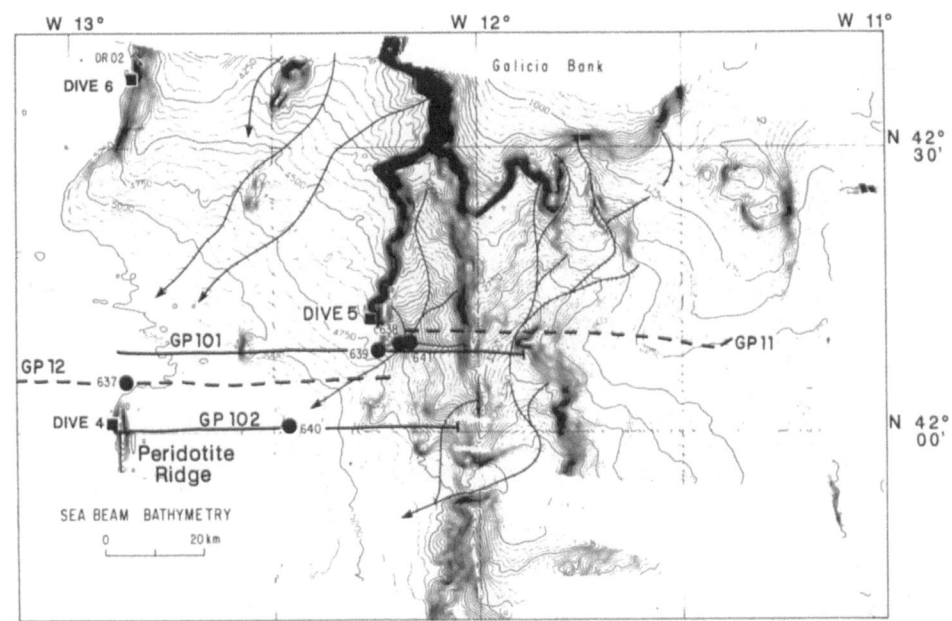

FIGURE 1-3.   Detailed SEABEAM bathymetric map of Galicia margin (from Sibuet et al., 1987) with location of ODP drill sites (637 to 641), *Nautile* dives, and seismic profiles (GP) discussed in text. Main canyons and tributaries are highlighted (arrows).

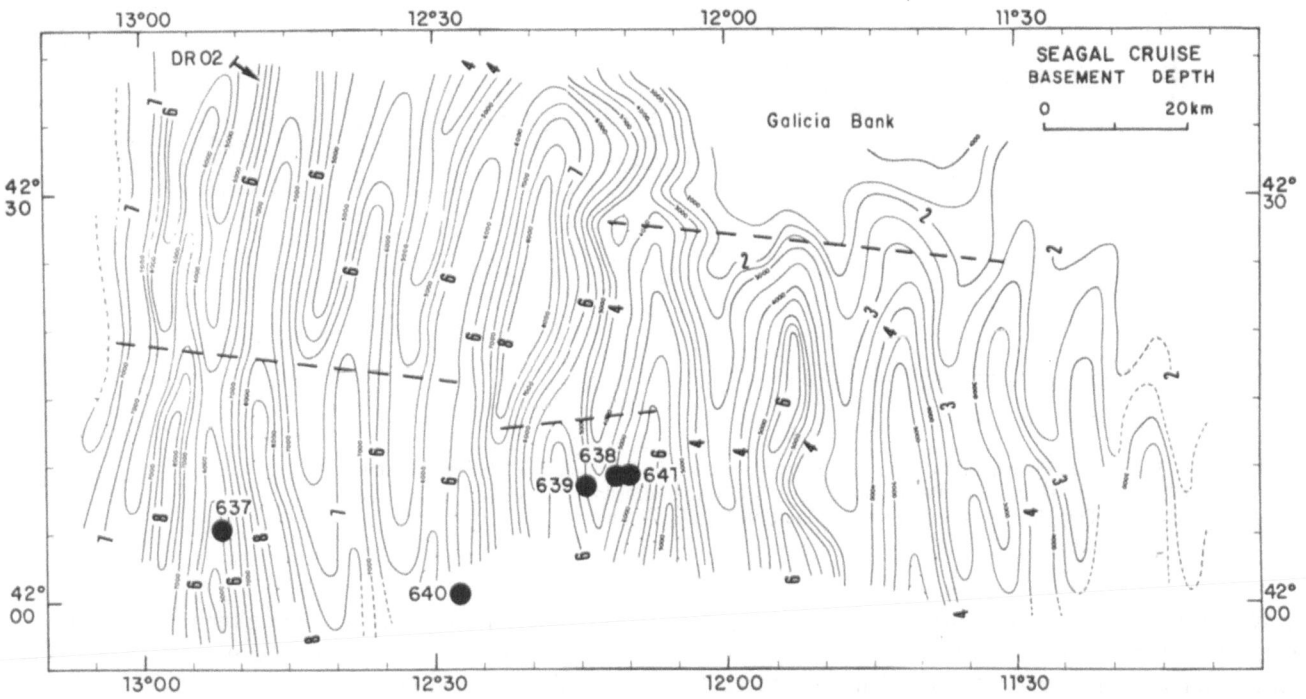

FIGURE 1-4.   Topographic map of surface of tilted blocks, in kilometers below sea level, established from all available seismic data (about 5,000 km). Contour spacing is 0.5 km; for ease of identification, contours are labelled both in meters (small numbers) and kilometers (large numbers). Numbered dots, 637 to 641, are locations of ODP Leg 103 drill sites. Serpentinized peridotites have been sampled at Site 637 and at DR02 dredge station (Sibuet et al., 1987). Dashed lines are possible transverse faults, southern edge of Galicia Bank *sensu stricto*, and location of slight 10° changes in trend of tilted fault blocks.

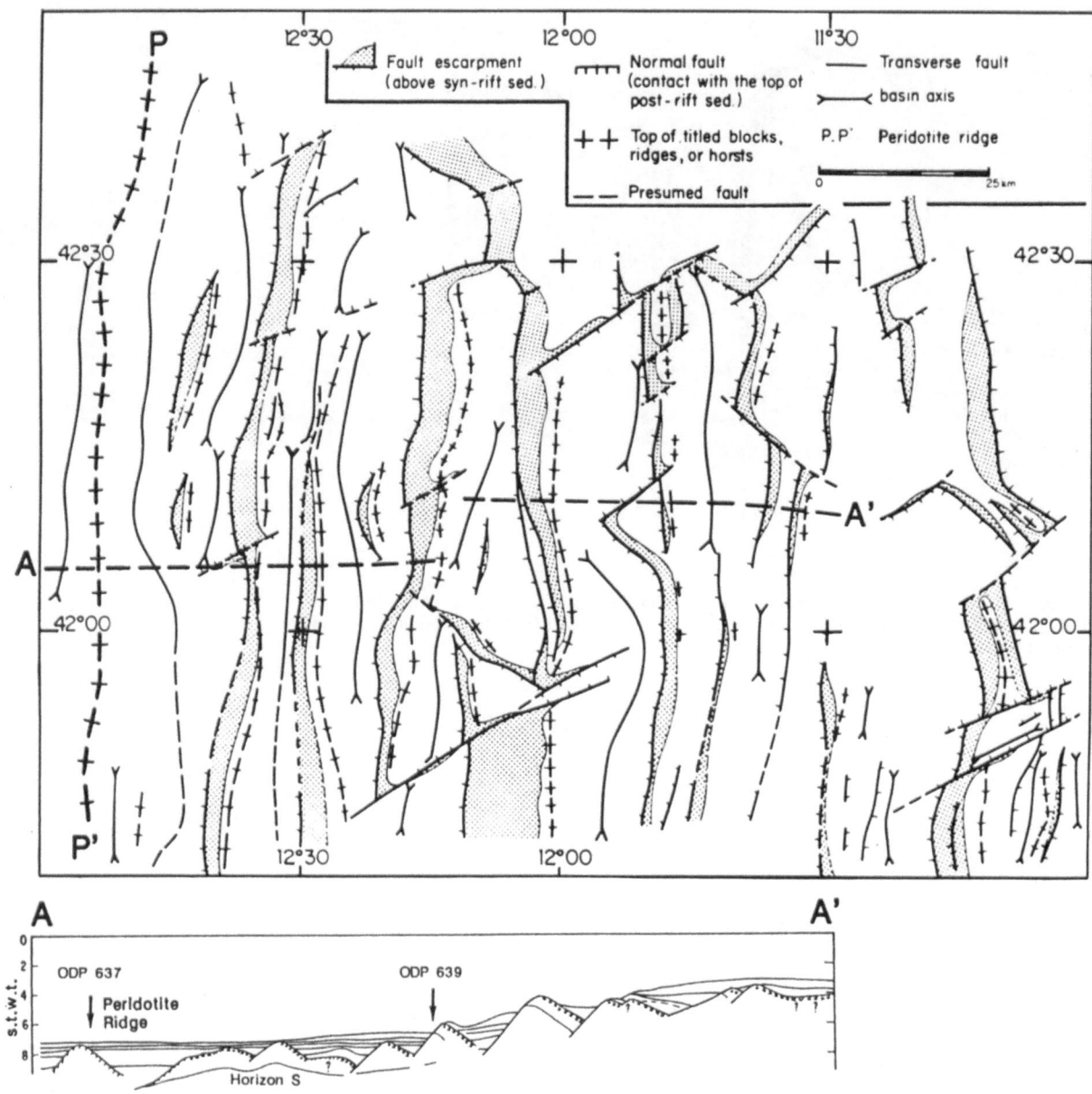

FIGURE 1-5. Structural map of the deep Galicia margin established from all available seismic data (about 5,000 km; after Thommeret, Boillot, and Sibuet, 1988). Line drawing of time section acquired along profile *AA'* shows lateral extension of tilted fault blocks throughout area.

synrift turbidities; overlying synrift sediments form a significant angular unconformity with the underlying synrift sediments. Drilling data also demonstrate that the block subsided continuously during the entire rifting phase (Boillot, Girardeau, and Kornprobst, 1988).

Site 640 is located about 20 km west of the previous sites, on a buried tilted fault block (Fig. 1-6). The upper part of the block consists of synrift Hauterivian sandy turbidites. Above the block, unconformable synrift deposits extend up to the lowermost Albian. The lithology and stratigraphy of the two drilled blocks are similar. On the seismic records of Figure 1-6, a prominent oblique reflector, located 0.4 sec (two-way traveltime) below the top of the block, was identified at Site

641 as the top of the Tithonian dolomite or limestone. This reflector is not observed at Site 640, where the internal acoustic structure is more disturbed. However, the beginning of an acoustic reflector located 0.7 sec below the crest of the block may correspond to the top of the Tithonian carbonate. Because this block is located close to the rift axis, the thickness of synrift sediments is greater, and the internal acoustic structure is more incoherent. Below the two drilled blocks, the so-called S reflector has been identified at the same depth (1.5 sec) below the crest of the blocks. Consequently, the two blocks seem to have been deformed in the same tectonic and sedimentary settings. During the initial Late Jurassic to early Hauterivian rifting phase,

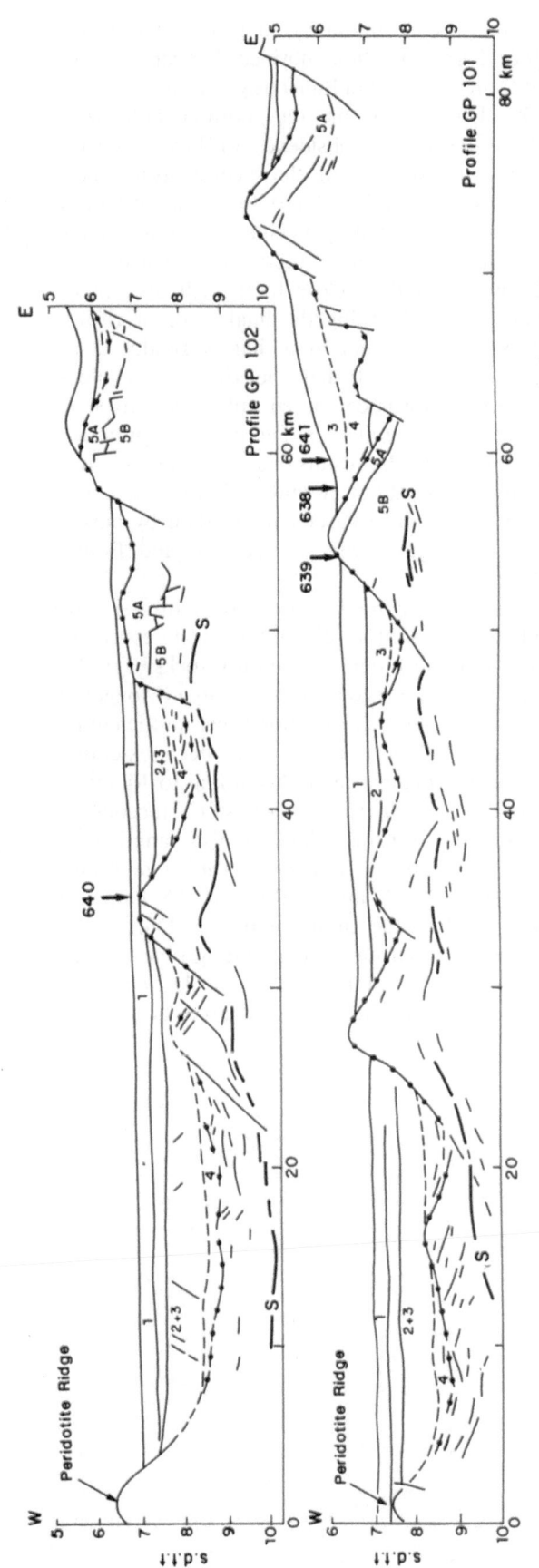

FIGURE 1-6. Line drawing of MCS lines GP 101 and GP 102 located in Fig. 1-2, with positions of ODP Leg 103 drill sites. S = "S" Reflector; 1-Recent to Oligocene section; 2 = Eocene to Santonian section; 3 = Cenomanian to Aptian section; 4 = late synrift sequences (Aptian to Hauterivian); 5A = early synrift sequences (Hauterivian to Valanginian) separated by the continuous lines with dots; 5B = prerift sequence (Berriasian to Tithonian carbonate platform).

synrift sediments accumulated conformably over the top of the blocks and then subsided 2–3 km (without any significant tilting) simultaneously with the underlying blocks. This means that only normal faults were involved in the initial subsidence of blocks located close to the rift axis. During the second rifting phase (late Hauterivian to early Albian), the tilting of blocks, including the early synrift sediments, occurred simultaneously with the deposition of the late synrift sediments. This means that, close to the rift axis, listric faults were involved during the final stage of rifting. This interpretation differs from that of Boillot et al. (1987b), who concluded that the tilted fault block drilled at Site 640 was composed only of synrift sediments down to the S reflector. The new interpretation is in agreement with the results of several *Nautile* dives on the same block, about 20 and 70 km to the north. There, the granodiorite basement, overlain by barren sandstone and conglomerate, has been sampled (Boillot et al., 1988a).

According to dredging data, the Upper Jurassic carbonate platform lies either directly upon the continental basement or on older sedimentary rocks; Triassic sandstones have been recovered in this area (Mougenot et al., 1985). On the basis of numerous multichannel seismic data, ODP observations have been extended across the whole Galicia area. According to Mauffret and Montadert (1987), the lowermost Cretaceous to lower Hauterivian synrift sediments also sampled by the *Nautile* submersible during the *Galinaute* cruise (Boillot et al., 1988a), are as thick as 2.5 km, and everywhere predate the tilting of blocks. During the late Hauterivian to earliest Albian rifting phase, older sediments, tilted with their respective blocks, may have been deeply eroded at the crests of the blocks and may

presently crop out along the west-facing scarp of Galicia Bank. Sediments partly filled the half-grabens, and sometimes formed fan-shaped prisms.

Data from ODP Leg 103 drilling indicate that the boundary between the two rifting phases is early Hauterivian. However, present data are insufficient to determine whether this discontinuity is of the same age along the whole margin, or whether it might increase in age oceanward, in concert with the progression of rifting.

Figures 1-4 and 1-5 are the basement and structural maps established from 5,000 km of available single- and multichannel seismic profiles (Mauffret and Montadert, 1987; Sibuet et al., 1987; Thommeret, Boillot, and Sibuet, 1988). The basement used there is the early Hauterivian unconformity, which corresponds to the end of the first rifting phase. The tilted blocks are continuous over large distances (up to 50 km along strike) and trend N003° ± 6°. A few discontinuous and generally faint transverse faults, oriented N058° ± 3° and N069° ±3°, slightly offset the tilted blocks. These transverse faults probably represent late Hercynian onshore strike-slip faults (Parga, 1969), which were reactivated during the rifting, as indicated by submersible observations on the western portion of the Spanish margin (Témine, 1984). They cannot be considered as transfer or accommodation faults in the sense of Le Pichon and Sibuet (1981) and Gibbs (1984), as they do not separate clear sections of tilted fault blocks, and as their orientation differs by about 30° from the expected orientation of transfer faults, according to Withjack and Jamison (1986). Spacing between consecutive fault-block crests varies from 7 to 18 km (with means values of 16 km) east of long. 12°–12°15'W and 12 km west of that longitude. The

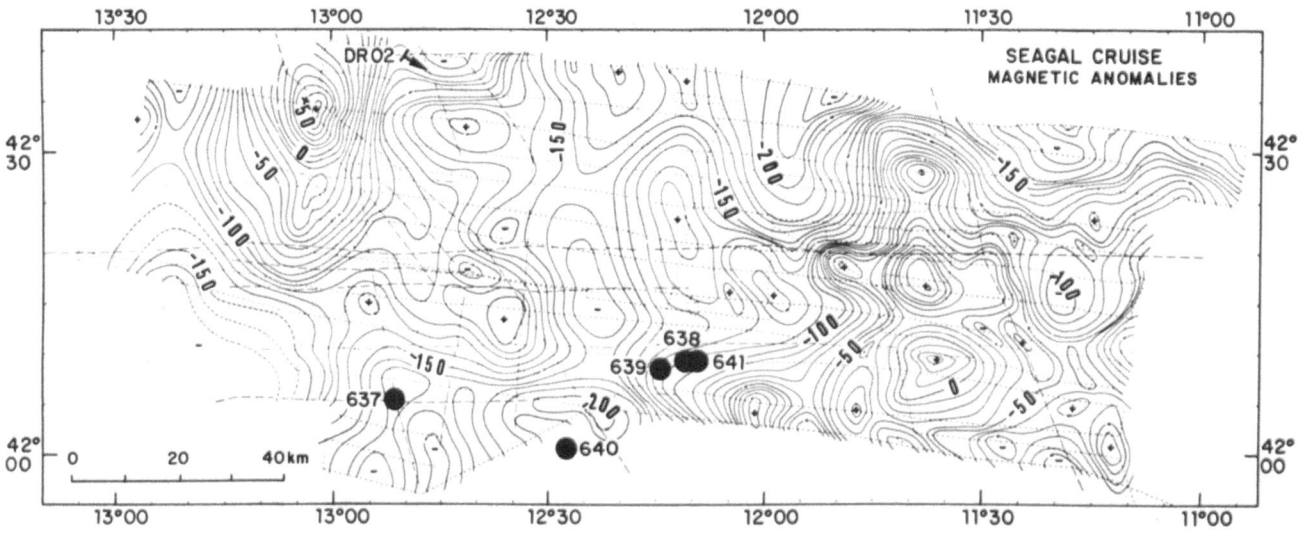

FIGURE 1-7.   Magnetic anomaly map contoured every 10 nT, with location of ODP Leg 103 drill sites (numbered dots). The meridian 12°W separates margin into domains characterized by magnetic anomalies with different amplitudes and gradients. DR02 = dredge station.

horizontal projection of the fault escarpments appears on Figure 1-5. A clear limit exists around long. 12°–12° 15′ W; the horizontal projection of vertical normal or listric faults is at least double in the western portion of the map. This limit also appears on the detailed magnetic anomaly map (Fig. 1-7; Sibuet et al., 1987). The maximum amplitude of anomalies is about 100 nT peak to peak, and their E–W wavelengths correspond to the spacing between tilted blocks. The weak amplitude of magnetic anomalies confirms that no volcanic extrusion or intrusion appears on the whole continental margin. East of long. 12°–12° 15′ W, assuming the morphology of blocks controls the trends of linear anomalies, quasicircular anomalies are superimposed on these linear anomalies. West of long. 12°–12° 15′ W, on the lower part of the margin, the amplitude and gradient of magnetic anomalies decrease by a factor of 2. This could be largely explained by greater depths to basement (6 km instead of 4 km), resulting in a ratio of about half for linear anomalies, but part of the residual anomalies could be linked to the rifting processes. Thus, both the structural and the magnetic anomaly maps of the west Galicia margin emphasize the existence of two domains, separated by the 12°–12° 15′ W meridian.

Continent–Ocean Boundary and Peridotite Ridge

Near the continent–ocean boundary (COB), a rather symmetrical ridge extends over several tens of kilometers in the N–S direction (Figs. 1-3 to 1-6; Sibuet et al., 1987; Mauffret and Montadert, 1987). Generally, the ridge does not display coherent internal reflections. This ridge is buried except near the peridotite ridge (Fig. 1-3) where a dredge haul, the drill hole at ODP Site 637, and five Nautile dives (Boillot et al., 1980; 1988; Boillot et al., 1987b) recovered serpentinized peridotites. Seismic data show that the ridge disappears about 20 km to the south, but could extend 50 km to the north in direction of the west Galicia Bank upper continental margin (Fig. 1-3), where similar peridotites have been dredged (Sibuet et al., 1987) and sampled during Nautile dives (Boillot et al., 1988). Thus, it is tempting to suggest that the peridotite ridge should extend northward and merge with the base of the continental slope. However, the ridge has a prominent saddle along its axial trend at lat. 42° 15′ N (Fig. 1-4), which could suggest that its extent is limited; the northward prolongation may be another ridge of unknown nature. Mauffret and Montadert (1987) also noticed that the shape of the ridge changes from symmetrical relief at the level of the peridotite ridge to asymmetrical relief north of lat 42° 15′ N, which resembles a tilted fault block with a west-facing scarp and indications of tilted sedimentary layers.

The dredged and drilled peridotites recovered from the peridotite ridge and the western Galicia margin (Fig. 1-3) consist of serpentinized harzburgites and lherzolites with calcite veins. Observed brown amphiboles correspond to a recrystallized mineral assemblage forming at high temperatures (> 750°C). These amphiboles have undergone the same ductile deformation as the surrounding rocks. Dates derived from ^{39}Ar-^{40}Ar analysis provide an age of 122 Ma, which corresponds to the last stage of rifting of the Galicia margin (Féraud et al., 1988). These data indicate that the peridotites must have been emplaced in the axial portion of an already well-established continental rift.

From the lithologic and structural analyses of all peridotitic samples, I conclude that a limited melt extraction of 5–9% occurred in asthenospheric conditions and probably resulted in the primary foliation. The mylonitic texture observed on some samples was acquired later during a ductile deformation event, which occurred under lithospheric conditions, during the vertical ascent of the material. Both the primary and mylonitic foliations are crosscut by ultramylonitic bands in which all previous phases are totally recrystallized. These bands are the location of a brittle deformation along shear zones, which probably act as normal faults oriented N–E west of the peridotite ridge, E–SE at ODP Site 637 (located east of the peridotite ridge), and N–W (but with a strike-slip component) in the northern Nautile dive sites 6 and 10 (Fig. 1-3). Orientation of the ductile mylonitic deformation varies between N110° at Site 637, N050° to N005° at site 5, N150° at sites 6, 10, and 14, and N060° for one sample collected at site 10 (Beslier, Girardeau, and Boillot, 1988; 1990; Girardeau, Evans, and Beslier, 1988). Though the origin of the ductile mylonitic deformation of the peridotites remains uncertain, the dispersion of values suggests that the measured samples came from the edge of mantle diapirs or ridges rising through the cold lithosphere, rather than along a large detachment fault dipping eastward across the whole lithosphere. In the same way, the dispersed orientations of the shear zones and their variable dips also support a diapiric ascent. Bulk density values measured along the 74-m peridotite column drilled at ODP Site 637, vary from 2.4 to 2.9 g/cm^3, with a mean value of 2.6 g/cm^3 (Boillot et al., 1987b). This also could suggest a vertical ascent of the peridotites through the whole continental lithosphere.

Using ocean bottom seismometers, detailed measurements were carried out across the peridotite ridge during the Réframarge cruise (1987). Preliminary results (Recq, Whitmarsh, and Sibuet, 1991) show that a major change in crustal velocities occurs on either side of the ridge. East of the ridge, the crust resembles thinned continental crust, as in the northern Bay of Biscay (Whitmarsh, Avedik, and Saunders, 1986; Le Pichon and Barbier, 1987). West of the ridge, the crust resembles oceanic crust, but without velocities characterizing the upper part of layer 3. Velocities of the upper portion of the ridge are heterogeneous, but rather low (~ 5.0 km/sec). This could be linked to a high degree of serpentinization. Two kilometers below the ridge crest, high-velocity material (> 7 km/sec) would correspond to peridotite. This suggests that the peridotite ridge may have originated by uplift of mantle material (Recq, Whitmarsh, and Sibuet, 1991).

Thus, as the present continental crust is extremely thinned (2–4 km without the synrift sediments), local ascents and outcrops of serpentinized peridotite diapirs or ridges across the remaining thin veil of continental crust are understandable. For the moment, there are no data that clearly support the tectonic denudation of the upper mantle along N–S trends as proposed by Boillot et al. (1987a).

Geometry and Nature of the S Reflector

The S reflector, initially described in the northern Bay of Biscay and Galicia margins (Charpal et al., 1978), corresponds to the lower limit of tilted fault blocks. It has been interpreted as the brittle-ductile boundary within the continental crust, the level at which listric normal faults sole, either in parallel (Montadert et al., 1979) or with a significant angle (Le Pichon and Sibuet, 1981). Where the S reflector is imaged by a single reflector (Montadert et al., 1979), it generally corresponds to a 0.2–0.4-sec-thick zone (about 1 km; Fig. 1-6).

Below the Galicia margin, the S reflector disappears about 10 km east of the axis of the peridotite ridge, and its E–W extension is only 35 km in contrast with its 90-km extension below the Biscay margin. In spite of the numerous multichannel seismic profiles, it is impossible to determine whether the S reflector merges with or intersects the peridotite ridge. Thus, nothing suggests that the S reflector is the top of the upper petrological mantle (Boillot, Girardeau, and Kornprobst, 1988). Toward the east, at the base of the upper continental margin, the S reflector seems to stop abruptly, without showing any significant increase in the thickness of the tilted blocks and sedimentary

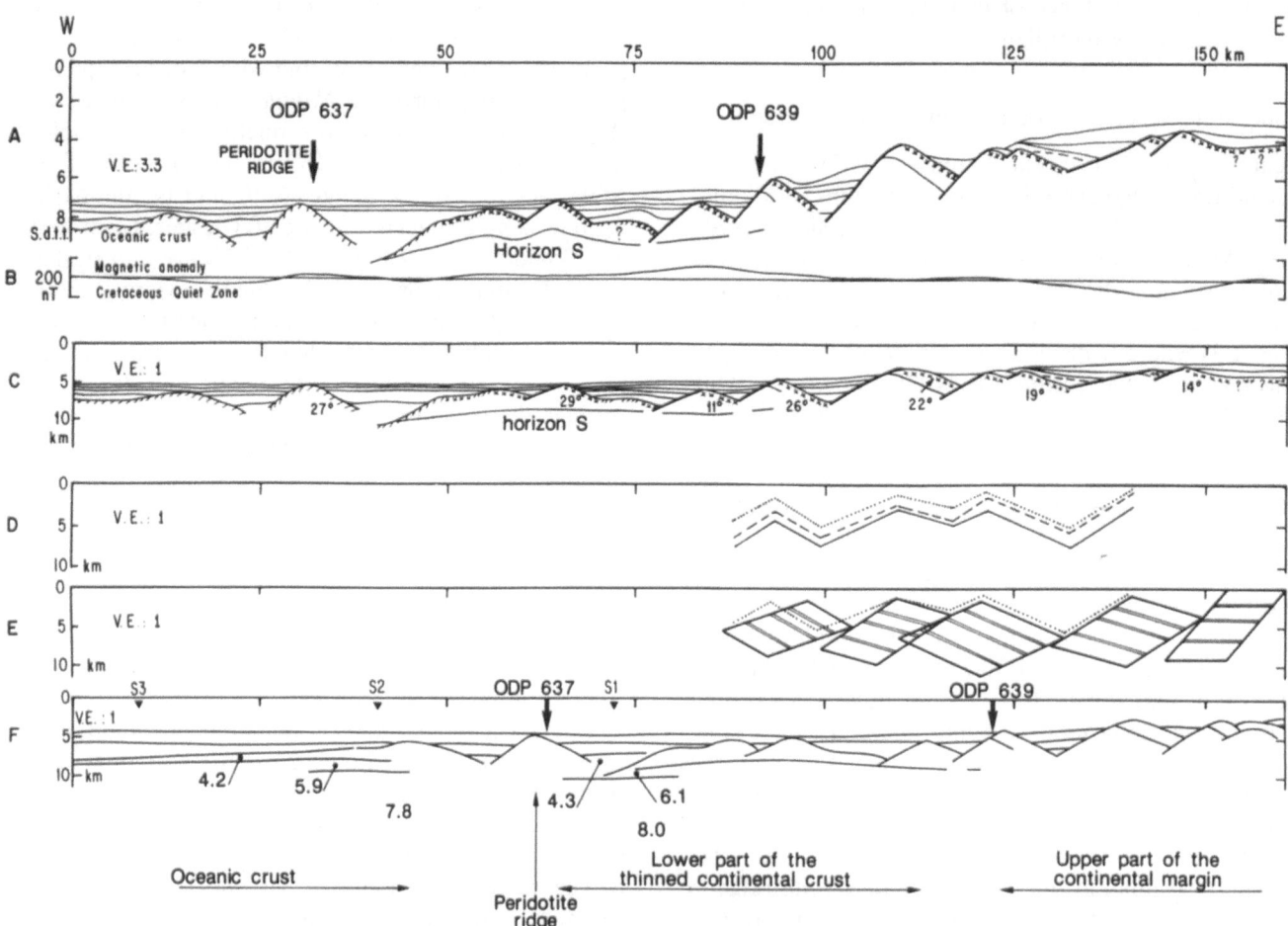

FIGURE 1-8. Interpretation of composite seismic profile (GP 11 and GP 12, located in Fig. 1-1) perpendicular to Galicia margin (time and depth sections with velocity indications from sonobuoys S1, S2, and S3; Sibuet et al., 1987).

layers. Thus, its disappearance does not seem to be linked with attenuation of the reflected signal with depth.

On time sections, the S reflector displays pull-up effects below the tilted blocks (Fig. 1-6) due to lateral velocity contrast. The presently poor knowledge of velocities (until the final *Réframarge* results are available) does not allow us to correctly restore the geometry of the S reflector. Nevertheless, if we assume a 2.9-km/sec synrift sediment velocity and a 4.3-km/sec tilted-block velocity from the few sonobuoy data acquired in this area (Sibuet et al., 1987), the S reflector is approximately a plane; below some normal faults bounding the blocks, however, it could represent vertical offsets interpreted as ramp effects by Mauffret and Montadert (1987). The reflector dips toward the ocean at a gradient at 1.6 sec over 35 km on profile 101 and 1.9 sec over 32 km on profile 102 (Fig. 1-6). Taking into account a local isostatic readjustment due to the synrift and postrift sedimentary load, the present dip is about 2° in direction of the ocean and about 1° just after the rifting episode (this takes into account the thermal cooling of the margin, whatever its mode of formation). For the northern Bay of Biscay margin, the present dip also is 2°, but in the direction of the upper continental margin, which sloped at 3° just after the rifting episode. Landward, this slope angle increases below the upper continental margin. Thus, because of this small angle, which can even be in the wrong direction, tilted blocks cannot have been formed by gravity gliding along the S reflector as proposed by Brun and Choukroune (1983).

As in the northern Bay of Biscay, the thickness of blocks, including synrift sediments of the first rifting phase, varies from 0.9 to 1.4 sec in the lower part of the margin (Fig. 1-6). Sonobuoy refraction data obtained on the thinned continental crust near the peridotite ridge suggest that 4.3-km/sec sediments overlie a 6.1-km/sec layer, 1.5 km thick, which could be some upper crustal material (Fig. 1-8, S1 sonobuoy). More recently, Recq, Whitmarsh, and Sibuet (1991) have been able to show that the S reflector is at a depth of at least 2.5 km into the crust where the velocity exceeds 6 km/sec.

THINNING AND EXTENSION OF CONTINENTAL CRUST

To a first approximation, the superficial extension measured from the geometry of tilted blocks on the Biscay and Galicia margins and the thickness of the thinned continental crust are compatible with the pure shear model (Le Pichon and Sibuet, 1981). Thus, Le Pichon and Sibuet (1981) and Le Pichon, Angelier, and Sibuet (1983) proposed the pack-of-cards model, in which the tilting angle increases toward the ocean. Heat flow measurements also are compatible with the pure shear model, which calls for a very low crustal radiogenic contribution (Foucher and Sibuet, 1980; Louden, Sibuet, and Foucher, 1991). However, Chenet et al. (1983) have discussed the validity of this method to define the superficial extension; they illustrated their concepts at that time on the only published migrated

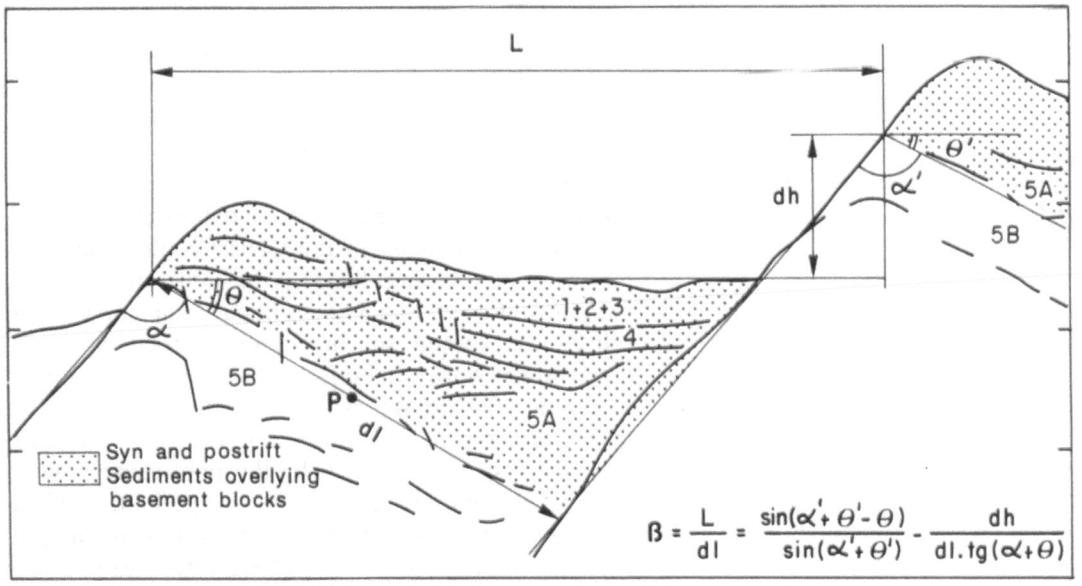

FIGURE 1-9. Computation of the superficial extension β on an example of tilted fault block belonging to the Galicia margin. L, distance between two consecutive blocks; dl, initial width of the block; θ and θ', tilting of the blocks; α and α', supplement of the angle of the fault plane with respect to the bedding place; dh, difference in height between two consecutive blocks.

section of a tilted fault block, which was located on the Meriadzek Terrace, at the base of the northern Bay of Biscay margin. The tail of this block displays faults antithetic to the main normal fault limiting the block, which obscures the problem. To avoid this difficulty, most tilted fault blocks used in this paper to compute superficial extension, do not present these complexities.

Extension Measured in the Brittle Crust

On the Galicia margin, the tails of blocks are less deformed than on the Biscay margin, but the summits of tilted blocks are often eroded. The extension measured in the upper and lower parts of the blocks differ only by a few percent instead of 20% as seen in the Meriadzek block (northern Bay of Biscay margin; Le Pichon, Angelier, and Sibuet, 1983). Extension is thus computed from the less deformed reflector located within the block (Fig. 1-9) and assigned to point P, located in the middle of the block and locally readjusted for the synrift and postrift sedimentary load. Figure 1-10 gives the depth of tilted blocks as a function of $1 - 1/\beta$ from the upper part of the margin located just south of Galicia Bank s.s. to the peridotite ridge by using a series of 12 parallel seismic profiles oriented E–W, 150 km long and spaced 5 km apart (Sibuet et al., 1987). The error bars correspond to plate isostatic readjustments using a reasonable range of lithospheric elastic thicknesses (Diament, Sibuet, and

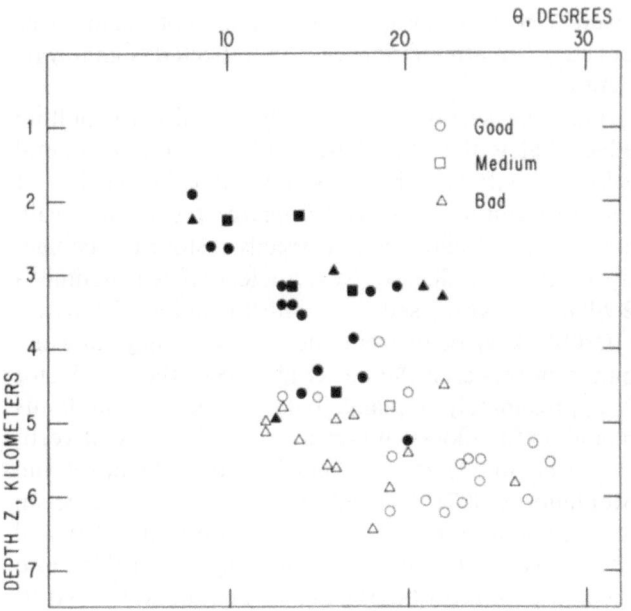

FIGURE 1-11.   Tilting θ of blocks in function of depth of tilted blocks isostatically adjusted for weight of synrift and postrift sediments. Black symbols correspond to blocks of upper part of margin; empty symbols correspond to blocks of lower part of margin located above S reflector.

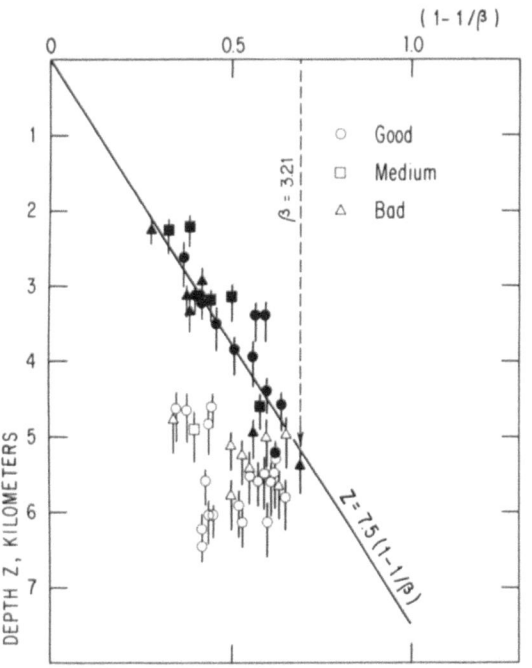

FIGURE 1-10. β extension values in function of depth of tilted blocks isostatically readjusted for weight of synrift and postrift sediments. Black symbols correspond to blocks of upper part of margin; empty symbols correspond to blocks of lower part of margin located above S reflector.

Hadaoui, 1986). Values corresponding to the five upper tilted blocks are represented with black symbols and fit well with the straight line $Z = 7.5 (1 - 1/\beta)$. Values that correspond to the five lower blocks, including the northern portion of the peridotite ridge (which could be a tilted block) are represented by open symbols. Their β values are systematically below the straight line. The limit between these two groups of values is located at long. 12° 15′ W. Figure 1-11 shows the tilting of blocks as a function of depth. Though the dispersion of data points increases with depth and reaches $\pm 6°$, the tilting angle increases with depth and seems to attain a maximum value of about 20–25°.

Consequently, the long. 12°–12° 15′ W limit, also defined on the seismic and magnetic data (Figs. 1-5, 1-7), separates two domains of the Galicia continental margin. East of this meridian, in the upper part of the margin, and down to 5.3 km (without sediments), the S reflector is absent; spacing between tilted blocks is about 16 km, horizontal offset of normal or listric faults is about 2 km above synrift deposits, amplitude of magnetic anomalies is around 200 nT, and measured extension can be explained by the pure shear model. West of this meridian, as far as the peridotite ridge, the S reflector is present; the lower part of the margin is characterized by less well-defined tilted blocks (with a mean 12-km spacing and mainly created within early synrift deposits), horizontal offset of normal or listric faults above synrift deposits is about 4 km, amplitude of magnetic anomalies is around 100 nT, and the measured extension cannot be explained by the pure shear model. The disappearance of the S reflector at the long. 12°–12° 15′W limit seems to indicate that the S reflector is a detachment zone below the lower continental margin, and does not exist below the upper margin.

Subsidence Curve ODP Site 638

Paleobathymetric estimates of the depositional environment at ODP Site 638 have been proposed for seven periods of the Galicia margin evolution (Moullade, Brunet, and Boillot, 1988) including four estimates for the rifting period (Fig. 1-12). Using the pure shear model initially proposed by McKenzie (1978), these authors have computed a subsidence curve. The thickness of the lithosphere and crust is, respectively, 125 and 30 km. The temperature at the base of the lithosphere is 1,333° C. As the prerift surface was close to sea level, and assuming that the synrift and postrift sediments at ODP Site 638 are representative of the area, a β value of 3.21 has been

assumed. Decompaction of sediments and eustatic sea-level variations have been introduced. If a continuous extension is assumed during the 25-m.y. rifting period (Berriasian to late Aptian), the resulting subsidence curve (Fig. 1-12) is, to a first approximation, close to the paleobathymetric estimates. Faunal data suggest at least a 1,500-m deepening at ODP Site 638 during a very short period of time during the Valanginian (132 to 136 Ma). This observation suggests that the initial subsidence of the margin was not regularly distributed over the whole rifting period, but could occur in several steps, as suggested by Mauffret and Montadert (1987) on the basis of structural data, and by Boillot, Winterer, et al. (1988b) from drilling observations. Due to the relative imprecision of microfaunal paleobathymetric estimates, it is not possible to quantify these steps of rifting. Nevertheless, the pure stretching model, which calls for a rapid Valanginian falling step, fits the faunal data.

ODP Site 638 is close to the boundary of the two previously identified domains of the upper and lower Galicia margin (Figs. 1-3, 1-4, 1-5, 1-7). The 3.21 β value is associated with the last tilted block of the upper domain, which follows the pure shear model (Fig. 1-10). Good agreement between the β values deduced from two independent sets of data (superficial extension and subsidence curve) confirms that the pure shear model is a possible mechanism to explain the formation of the upper part of the margin.

Two main hypotheses can explain the low β values (Fig. 1-10) of the lower part of the margin:

1. If the tilted blocks contain mainly Valanginian synrift sediments mobilized during the second rifting episode (Boillot, Winterer, et al., 1988b), the measured superficial extension would be only part of the whole extension.
2. The presence of the S reflector also could be interpreted as a zone of "décollement", which favors the simple shear mechanism (Wernicke, 1985).

Boillot et al. (1987a) have pushed this latter model to an extreme and have interpreted the peridotite ridge and the S reflector as relics of mantle denudation. In the northern Bay of Biscay, Le Pichon and Barbier (1987) also have interpreted the S reflector as a zone of "décollement", but suggested that it merges at depth with the brittle-ductile interface. Thus, complementary information on conjugate margins is needed to better constrain the mode of formation of these margins.

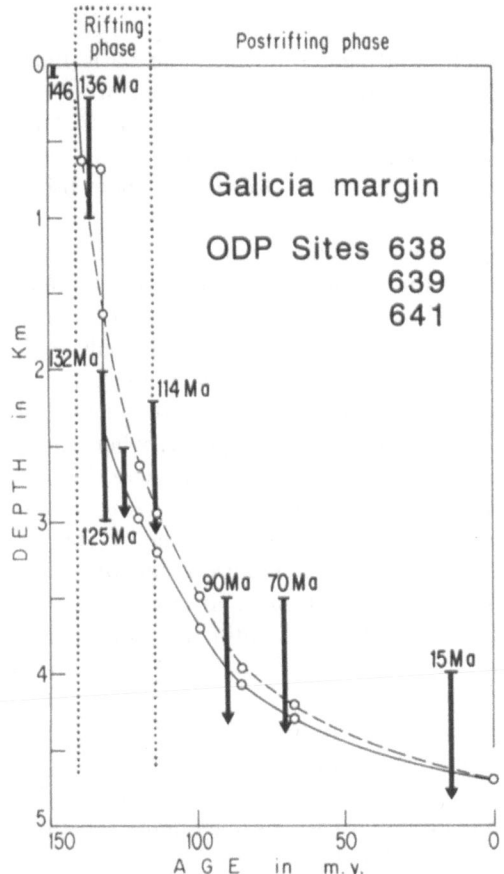

FIGURE 1-12. Subsidence curve at ODP Sites 638, 639, and 641, established from paleobathymetric data and represented by heavy arrows (Moullade, Brunet, and Boillot, 1988). Continuous thin line represents two instantaneous stretching phases of rifting followed by thermal subsidence; dashed thin line represents continuous stretching during rifting phase, following by thermal cooling. Pure stretching model, with eventually a rapid Valanginian falling step, fits faunal data in area of ODP Sites 638, 639, and 641.

SUMMARY AND CONCLUSION

I presently favor a composite pure and simple shear model for the formation of the Galicia margin: pure shear for the whole lithosphere of the upper Galicia

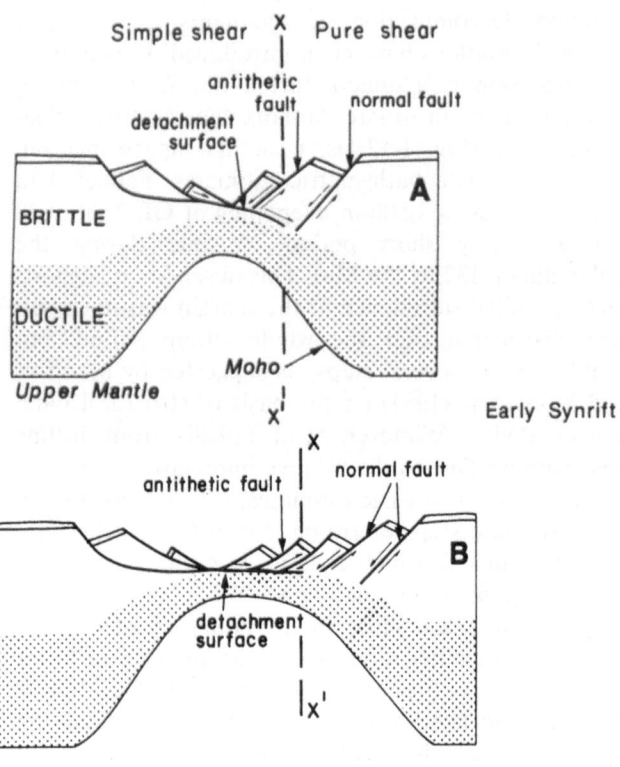

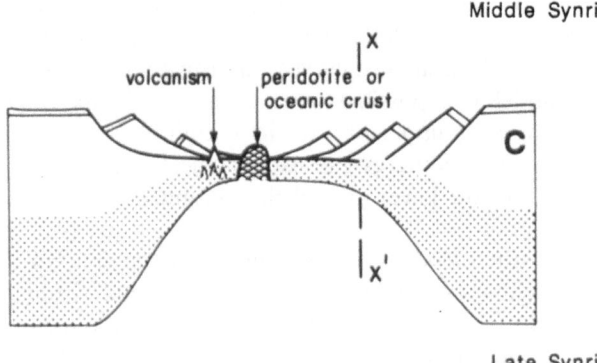

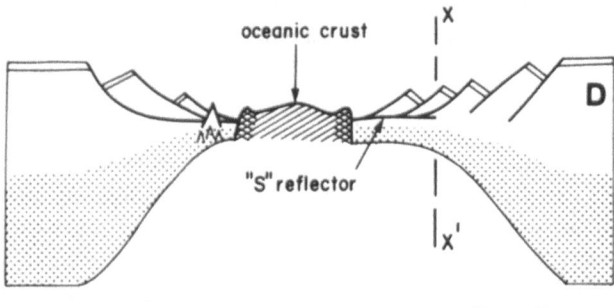

FIGURE 1-13.   Proposed four-step (A–D) model of formation of western Galicia margin: xx' dashed line separates system of conjugate margins into two asymmetrical portions. On right-hand side of xx', pure shear mechanism affects whole lithosphere. On left-hand side of xx', simple shear mechanism affects upper brittle portion of crust, and pure shear model affects remaining lower portion of lithosphere. At end of rifting phase (C), final breakup occurs where lithosphere is extremely thinned and is associated with intrusion of upper mantle material.

margin and simple shear for the upper crust of the lower Galicia margin. In this model, a detachment surface (Fig. 1-13A) has been acting since the beginning of rifting and has propagated eastward from the Canadian side (Enachescu, 1987; Tankard and Welsink, 1990) to the Iberian side (Wernicke, 1985) as extension increased. This detachment surface, which crops out on the Canadian side, affects the upper portion of the continental crust and probably merges at depth with the brittle-ductile interface (Radel and Melosh, 1987) or within the lower crust. Below the Galicia margin, this surface corresponds to the S reflector. In this hypothesis, the lateral offset along the detachment surface decreases eastward from the Canadian side to zero below the Galicia margin, where the S reflector disappears (Fig. 1-13A, B).

Below the upper portion of the Galicia margin, there is no lateral motion along the dashed line that arbitrarily prolongates the detachment surface at the brittle-ductile interface. The xx' vertical line, which intersects the detachment surface at its point of disappearance, separates the system of conjugate margins into two parts:

1. the left-hand side for which the basic mechanism within the brittle crust is simple shear.
2. The right-hand side where I suggest that the pure shear model is applied not only to the brittle portion of the crust but also to the whole lithosphere.

More generally, I suggest that the main mechanism that affects the whole lithosphere would be pure shear (Keen et al., 1989; Etheridge, Symonds, and Lister, 1989; Kuznir and Egan, 1989); it is only within the upper part of the crust that the fundamental mechanism of rifting would be either simple or pure shear.

At the end of the rifting phase, when the crust is particularly thinned, upper mantle material composed of peridotites serpentinized by seawater circulation, and less dense that the surrounding crustal material, can be uplifted and "fossilized" at the continent–ocean boundary (Fig. 1-13C). After creation of oceanic crust (Fig. 1-13D), the continental lithosphere cools and the brittle-ductile interface sinks, leaving the former brittle-ductile interface within the present-day brittle upper crust. As only a minor detrital input has been available to the distal western Galicia margin since rifting has occurred, the construction of a well-defined continental rise has been limited (contrast with Chapters 4, 5, 6, 11).

ACKNOWLEDGMENTS

I thank P.-C. de Graciansky, who asked me to propose a contribution on the geodynamic context of the formation of continental margins on which continental rises may develop through time. I also thank P.-C. de Graciansky, C. W. Poag, and R. B. Whitmarsh for their very helpful comments on the manuscript.

REFERENCES

Bally, A. W. 1981. Atlantic-type margins. Geology of Passive Continental Margins: History, Structure and Sedimentologic Record (With Special Emphasis on the Atlantic Margin). *American Association of Petroleum Geologists, Education Course Note Series 19*, 1: 1–48.

Beslier, M.-O., Girardeau, J., and Boillot, G. 1988. Lithologie et structure des péridotites à plagioclase bordant la marge continentale passive de la Galice (Espagne). *Comptes Rendus Académie des Sciences Paris* 306: 373–380.

Beslier, M.-O., Girardeau, J., and Boillot, G. 1990. Kinematics of peridotite emplacement during North Atlantic continental rifting, Galicia, northwestern Spain. *Tectonophysics* 184: 321–343.

Boillot, G., Comas, M. C., Girardeau, J., Kornprobst, J., Loreau, J.-P., Malod, J., Mougenot, D., and Moullade, M. 1988a. Preliminary results of the *Galinaute* cruise: dives of the submersible *Nautile* on the western Galicia margin, Spain. *In* Proceedings of the Ocean Drilling Program, Scientific Results, Volume 103, G. Boillot, E. L. Winterer, et al.: College Station, TX: Ocean Drilling Program, 37–51.

Boillot, G., Girardeau, J., and Kornprobst, J. 1988. Rifting of the Galicia margin: crustal thinning and emplacement of mantle rocks on the seafloor. *In* Proceedings of the Ocean Drilling Program, Scientific Results, Volume 103, G. Boillot, E. L. Winterer, et al.: College Station, TX: Ocean Drilling Program, 741–756.

Boillot, G., Grimaud, S., Mauffret, A., Mougenot, D., Kornprobst, J., Mergoil-Daniel, J., and Torrent, G. 1980. Ocean–continent transition off the Iberian margin: A serpentinite diapir west of Galicia Bank. *Earth and Planetary Science Letters* 48: 23–34.

Boillot, G., Recq, M., Winterer, E. L., Meyer, A. W., Applegate, J., Baltuck, M., Bergen, J. A., Comas, M. C., Davies, T. A., Dunham, K., Evans, C. A., Girardeau, J., Goldberg, G., Haggerty, J., Jansa, L. F., Johnson, J. A., Kasahara, J., Loreau, J.-P. Luna-Sierra, E., Moullade, M., Ogg, J., Sarti, M., Thurow, J., and Williamson, M. 1987a. Tectonic denudation of the upper mantle along passive margins: a model based on drilling results (ODP Leg 103, western Galicia margin, Spain). *Tectonophysics* 132: 335–342.

Boillot, G., Winterer, E. L., Meyer, A. W., Applegate, J., Baltuck, M., Bergen, J. A., Comas, M. C., Davies, T. A., Dunham, K., Evans, C. A., Girardeau, J., Goldberg, G., Haggerty, J., Jansa, L. F., Johnson, J. A., Kasahara, J., Loreau, J.-P., Luna-Sierra, E., Moullade, M. Ogg, J., Sarti, M., Thurow, J., and Williamson, M. 1987b. Introduction, objectives, and principal results. *In* Proceedings of the Ocean Drilling Program, Initial Reports, Volume 103, G. Boillot, E. L. Winterer, et al.: College Station, TX: Ocean Drilling Program, 3–17.

Boillot, G., Winterer, E. L., Meyer, A. W., Applegate, J., Baltuck, M., Bergen, J. A., Comas, M. C., Davies, T. A.,

Dunham, K., Evans, C. A., Girardeau, J., Goldberg, G., Haggerty, J., Jansa, L. F., Johnson, J. A., Kasahara, J., Loreau, J.-P., Luna-Sierra, E., Moullade, M., Ogg, J., Sarti, M., Thurow, J., and Williamson, M. 1988b. Proceedings of the Ocean Drilling Program, Scientific Results, Volume 103: College Station, TX: Ocean Drilling Program.

Brun, J.-P., and Choukroune, P. 1983. Normal faulting, block tilting, and decollement in a stretched crust. *Tectonics* 2: 345–556.

Charpal, O. de, Guennoc, P., Montadert, L., and Roberts, D. G. 1978. Rifting, crustal attenuation and subsidence in the Bay of Biscay. *Nature* 275: 707–711.

Chenet, P.-Y., Montadert, L., Gairaud, H., and Roberts, D. G. 1983. Extension ratio measurements on the Galicia, Portugal and northern Biscay continental margins: implications for evolutionary models of passive continental margins. *In* Studies in Continental Margin Geology, J. S. Watkins and C. L. Drake, Eds., *American Association of Petroleum Geologists Memoir 34*, 703–715.

Curnelle, E., Dubois, P., and Sequin, J.-C. 1980. Le bassin d'Aquitaine; substratum antetertiaire et bordures mésozoïques. *Bulletin du Centre de Recherche, d'Exploration et de Production Elf Aquitaine, Memoir 3*, 47–58.

Diament, M., Sibuet, J.-C., and Hadaoui, A. 1986. Isostasy of the northern bay of Biscay continental margin. *Geophysical Journal of the Royal Astronomical Society* 86: 893–907.

Enachescu, M. E. 1987. Tectonic and structural framework of the northeast Newfoundland continental margin. *In* Sedimentary Basins and Basin-Forming Mechanisms, C. Beaumont and A. J. Tankard, Eds., *Canadian Society of Petroleum Geologists Memoir 12*, 117–146.

Etheridge, M. A., Symonds, P. A., and Lister, G. S. 1989. Application of the detachment model to reconstruction of conjugate passive margins. *In* Extensional Tectonics and Stratigraphy of the North Atlantic Margins, A. J. Tankard and H. R. Balkwill, Eds., *American Association of Petroleum Geologists Memoir 46*, 23–40.

Féraud, G., Girardeau, J., Beslier, M.-O., and Boillot, G. 1988. Datation ^{39}Ar-^{40}Ar de la mise en place des péridotites bordant la marge de la Galice (Espagne). *Comptes Rendus Académie des Sciences Paris* 307: 49–55.

Foucher, J.-P., and Sibuet, J.-C. 1980. Thermal regime of the northern Bay of Biscay continental margin in the vicinity of the DSDP Sites 400–402. *Philosophical Transactions of the Royal Society of London* A294: 157–167.

Gibbs, A. D. 1984. Structural evolution of extensional basin margins. *Journal of the Geological Society of London* 141: 609–620.

Girardeau, J., Evans, C. A., and Beslier, M.-O. 1988. Structural analysis of plagioclase-bearing peridotites emplaced at the end of continental rifting: Hole 637A, ODP Leg 103 on the Galicia margin. *In* Proceedings of the Ocean Drilling Program, Scientific Results, Volume 103, G. Boillot, E. L. Winterer, et al.: College Station, TX: Ocean Drilling Program, 209–223.

Graciansky, P.-C. de, Poag, C. W., Cunningham, Jr., R., Loubere, P., Masson, D. G., Mazzullo, J. M., Montadert, L., Müller, C., Otsuka, K., Reynolds, L., Sigal, J., Snyder, S., Townsend, H. A., Vaos, S. P., and Waples, D. 1985.

Initial Reports of the Deep Sea Drilling Project, Volume 80: Washington, DC: U.S. Government Printing Office.

Grow, J. A. 1981. The Atlantic margin of the United States. Geology of Passive Continental Margins: History, Structure and Sedimentologic Record (With Special Emphasis on the Atlantic Margin). *American Association of Petroleum Geologists Education Course Note Series 19*, 3: 1–41.

Keen, C. E., Peddy, C., de Voogd, B., and Matthews, D. 1989. Conjugate margins of Canada and Europe: Results from deep reflection profiling. *Geology* 17: 173–176.

Kuznir, N. J., and Egan, N. N. 1989. Simple-shear and pure-shear models of extensional sedimentary basin formation: application to the Jeanne d'Arc Basin, Grand Banks of Newfoundland. *In* Extensional Tectonics and Stratigraphy of the North Atlantic Margins, A. J. Tankard and H. R. Balkwill, Eds., *American Association of Petroleum Geologists Memoir 46*, 305–322.

Lallemand, S., and Sibuet, J.-C. 1986. Tectonic implications of canyon directions over the Northeast Atlantic continental margin. *Tectonics* 5: 1125–1143.

Le Pichon, X., and Barbier, F. 1987. Passive margin formation by low-angle faulting within the upper crust: The northern Bay of Biscay margin. *Tectonics* 6: 133–150.

Le Pichon, X., and Sibuet, J.-C. 1981. Passive margins: a model of formation. *Journal of Geophysical Research* 86: 3708–3720.

Le Pichon, X., Angelier, J., and Sibuet, J.-C. 1983. Subsidence and stretching. *In* Studies in Continental Margin Geology, J. S. Watkins and C. L. Drake, Eds., *American Association of Petroleum Geologists Memoir 34*, 731–741.

Louden, K. E., Sibuet, J.-C. and Foucher, J.-P. 1991. Variations in heat flow across the Goban Spur and Galicia Bank continental margins. *Journal of Geophysical Research* 96: 16131–16150.

Mauffret, A., and Montadert, L. 1987. Rift tectonics on the passive continental margin off Galicia (Spain). *Marine and Petroleum Geology* 4: 49–70.

McKenzie, D. P. 1978. Some remarks on the development of sedimentary basins. *Earth and Planetary Science Letters* 40: 25–32.

Montadert, L., Roberts, D. G., Auffret, G. A., Bock, W. D., Dupeuble, P.-A. Hailwood, E. A., Harrison, W. E., Kagami, H., Lumsden, D. N., Müller, C. M., Schnitker, D., Thompson, R. W., Thompson, T. L., and Timofeev, P., 1979. Initial Reports of the Deep Sea Drilling Project, Volume 48: Washington DC: U.S. Government Printing Office.

Montadert, L., de Charpal, O., Roberts, D. G., Guennoc, P., and Sibuet, J.-C. 1979. Northeast Atlantic passive margins: Rifting and subsidence processes. *In* Deep Drilling Results in the Atlantic Ocean: Continental Margins and Paleoenvironments, M. Talwani, W. W. Hay, and W. B. F. Ryan, Eds., Maurice Ewing Series 3: Washington, DC: American Geophysical Union, 164–186.

Moretti, I., and Froidevaux, C. 1986. Thermomechanical models of active rifting. *Tectonics* 5: 501–511.

Mougenot, D. 1988. Géologie de la marge portugaise. Thesis, Université Pierre et Marie Curie, Paris.

Mougenot, D., Monteiro, J. H., Dupeuble, P. A., and Malod, J. A. 1979. La marge continentale sud-portugaise; Evolution structurale et sédimentaire, *Ciencas Terra* 5: 223–246.

Mougenot, D., Capdevila, R., Palain, C., Dupeuble, P.-A. and Mauffret, A. 1985. Nouvelles données sur les sédiments anté-rift et le socle de la marge continentale de Galice. *Comptes Rendus Académie des Sciences Paris* 301: 323–328.

Moullade, M., Brunet, M.-F., and Boillot, G. 1988. Subsidence and deepening of the Galicia margin: the paleoenvironmental control. *In* Proceedings of the Ocean Drilling Program, Scientific Results, Volume 103, G. Boillot, E. L. Winterer, et al.: College Station, TX: Ocean Drilling Program, 733–740.

Parga, J.-R. 1969. Sistemas de fracturas tardihercinias del macizo hesperico. *Trabalos Laboratorio Geologica Lage* 37: 1–15.

Radel, G., and Melosh, H. J. 1987. A mechanical basis for low-angle normal faulting in the Basin Range. *EOS Transactions, American Geophysical Union* 68 (44): 1449.

Recq, M., Whitmarsh, R. B., and Sibuet, J.-C. 1991. Anatomy of a lherzolite ridge, Galicia margin. European Union of Geosciences, Strasbourg, 24–28 March, 1991, *Terra Abstracts* 3 (1): 122.

Royden, L., and Keen, C. E. 1980. Rifting processes and thermal evolution of the continental margin of eastern Canada determined from subsidence curves. *Earth and Planetary Science Letters* 51: 343–361.

Savostin, L. A., Sibuet, J.-C., Zonenshain, L.-P., Le Pichon, X., and Roulet, M.-J. 1986. Kinematic evolution of Tethys belt from the Atlantic Ocean to Pamir since the Triassic. *Tectonophysics* 123: 1–35.

Sibuet, J.-C. 1988. Marges passives de l'Atlantique Nord-Est: estimation des paléocontraintes lors du rifting. *Bulletin de la Société Géologique de France* 4: 515–527.

Sibuet, J.-C. 1989. Paleoconstraints during rifting of the northeast Atlantic passive margins. *Journal of Geophysical Research* 94: 7265–7277.

Sibuet, J.-C., Ryan, W. B. F., Arthur, M. A., Lopatin, B. G., Moore, D. G., Maldonado, A., Rehault, J.-P., Iaccarino, S., Sigal, J., Morgan, G. E., Blechschmidt, G., Williams, C. A., Johnson, D., Barnes, R. O., and Habib, D. 1979. Initial Reports of the Deep Sea Drilling Project, Volume 47, Part 2: Washington, DC: U.S. Government Printing Office.

Sibuet, J.-C., Mazé, J.-P., Amortila, P., and Le Pichon, X. 1987. Physiography and structure of the western Iberian continental margin off Galicia from Sea-Beam and seismic data. *In* Proceedings of the Ocean Drilling Program, Initial Reports, Volume 103, G. Boillot, E. L. Winterer, et al.: College Station, TX: Ocean Drilling Program, 77–97.

Tankard, A. J., and Welsink, H. J. 1990. Mesozoic extension and styles of basin formation in Atlantic Canada. *In* Extensional Tectonics and Stratigraphy of the North Atlantic Margins, A. J. Tankard and H. R. Balkwill, Eds., *American Association of Petroleum Geologists Memoir 46*, 175–195.

Témine, D. 1984. Contribution à l'étude géologique de la marge Nord-Ouest de l'Espagne. Thèse 3ème cycle, Université Pierre et Marie Curie, Paris.

Thommeret, M., Boillot, G., and Sibuet, J.-C. 1988. Structural map of the Galicia margin. *In* Proceedings of the Ocean Drilling Program, Scientific Results, Volume 103, G. Boillot, E. L. Winterer, et al.: College Station, TX: Ocean Drilling Program, 31–36.

Watts, A. B., and Steckler, M. S. 1979. Subsidence and eustacy at the continental margin of eastern North America. *In* Deep Drilling Results in the Atlantic Ocean: Continental Margin and Paleoenvironment, M. Talwani, W. W. Hay, and W. B. F. Ryan, Eds., Maurice Ewing Series 3: Washington, DC: American Geophysical Union, 218–234.

Wernicke, B. 1985. Uniform-sense normal simple shear of the continental lithosphere. *Canadian Journal of Earth Sciences* 22: 108–125.

Wernicke, B., and Burchfield, B. C. 1982. Modes of extensional tectonics. *Journal of Structural Geology* 4: 105–115.

Whitmarsh, R. B., Avedik, F., and Saunders, M. R. 1986. The seismic structure of thinned continental crust in the northern Bay of Biscay. *Geophysical Journal of the Royal Astronomical Society* 86: 589–602.

Whitmarsh, R. B., Miles, P. K., and Mauffret, A. 1990. The ocean–continent boundary off the western continental margin of Iberia. I—Crustal Structure at 40°30′N. *Geophysical Journal International* 103: 509–531.

Withjack, M. O., and Jamison, W. R. 1986. Deformation produced by oblique rifting. *Tectonophysics* 126: 99–124.

Ziegler, P. A. 1982. Faulting and graben formation in western and central Europe. *In* The Evolution of Sedimentary Basins, P. Kent, M. H. P. Bott, D. P. McKenzie, and C. A. Williams, Eds., *Philosophical Transactions of the Royal Society of London* A305: 113–143.

2

Southwest African Plate Margin: Thermal History and Geodynamical Implications

Frédéric Walgenwitz, J. P. Richert and P. Charpentier

There has been substantial improvement in the appraisal of paleotemperatures in sedimentary basins during the last decade, using different methods that allow low temperature determination. Fluid-inclusion geothermometry and apatite fission-track chronothermometry provide new avenues for the investigation of geologic processes in the uppermost part of the earth's crust. Measuring low temperature events makes it possible to determine subsidence rates, uplifts, and heat flow. Previous works (Burrus, Cerone, and Harris, 1983; Meyer et al., 1989; Naeser, Naeser, and McCulloh 1990; Kamp and Green, 1990; Walgenwitz et al., 1990) have shown the potentialities of such methods.

We have selected the Gabon–Congo margin, which has been extensively explored for hydrocarbons for several decades, to appraise the contribution of paleotemperatures to the understanding of structural evolution. Deep drill holes, some of which were cored to a depth of 4 km, help determine the time versus temperature changes over a large stratigraphic interval. Based on a regional thermal survey, margin evolution during the Late Cretaceous and Tertiary can be described clearly. Especially important are inferences from the convergence of postrift cooling ages in onshore areas, and from sedimentary and tectonic features known in the offshore domain, where a more complete stratigraphic record exists.

GEOLOGIC SETTING

General Framework

The Gabon Basin consists of two subbasins: (1) the South Gabon–Congo subbasin, which opens into the West Congo fold belt; and (2) the North Gabon sub-

basin above the old Proterozoic Congo Craton (Fig. 2-1). Sedimentation on this margin is controlled by a permanent feature called the 200-m water-depth hinge line in the South Gabon–Congo subbasin. This deep-seated feature corresponds to a crustal change and possibly to a step of the Moho (Figs. 2-2, 2-3). The hinge line delineates two areas: (1) a continental domain, mainly characterized by shallow shelf deposits; and (2) an Atlantic domain, featuring deeper basinal deposits accumulated during the drift period. In the North Gabon subbasin, the 200-m hingeline is much less conspicuous, and is marked by several "en échelon" faults (Fig. 2-3). As this hinge zone is partly onshore in the North Gabon subbasin, the Atlantic domain has been drilled at numerous sites in this area, whereas the domain remains mostly unknown in the South Gabon–Congo subbasin due to its position offshore under great water depths. Therefore, our paleothermometric survey deals only with the continental domain.

Sedimentary Record

Two main lithostratigraphic units illustrate the evolution of the coastal basins from the rift stage (Neocomian–Aptian) to the drift stage (with development of the accretion slope as early as the Albian; Fig. 2-4). Earlier deposits of Permian to Jurassic age are known only in the Internal Basin in North Gabon. The rift and drift stages can be dated in relation to the Atlantic opening and global plate movements. Several phases can be defined and dated by magnetic anomalies (Fig. 2-4). The first phase, at 180 Ma, marks the onset of seafloor spreading in the South Atlantic, whereas rifting did not take place in the Gabon–Congo area until about 127 Ma.

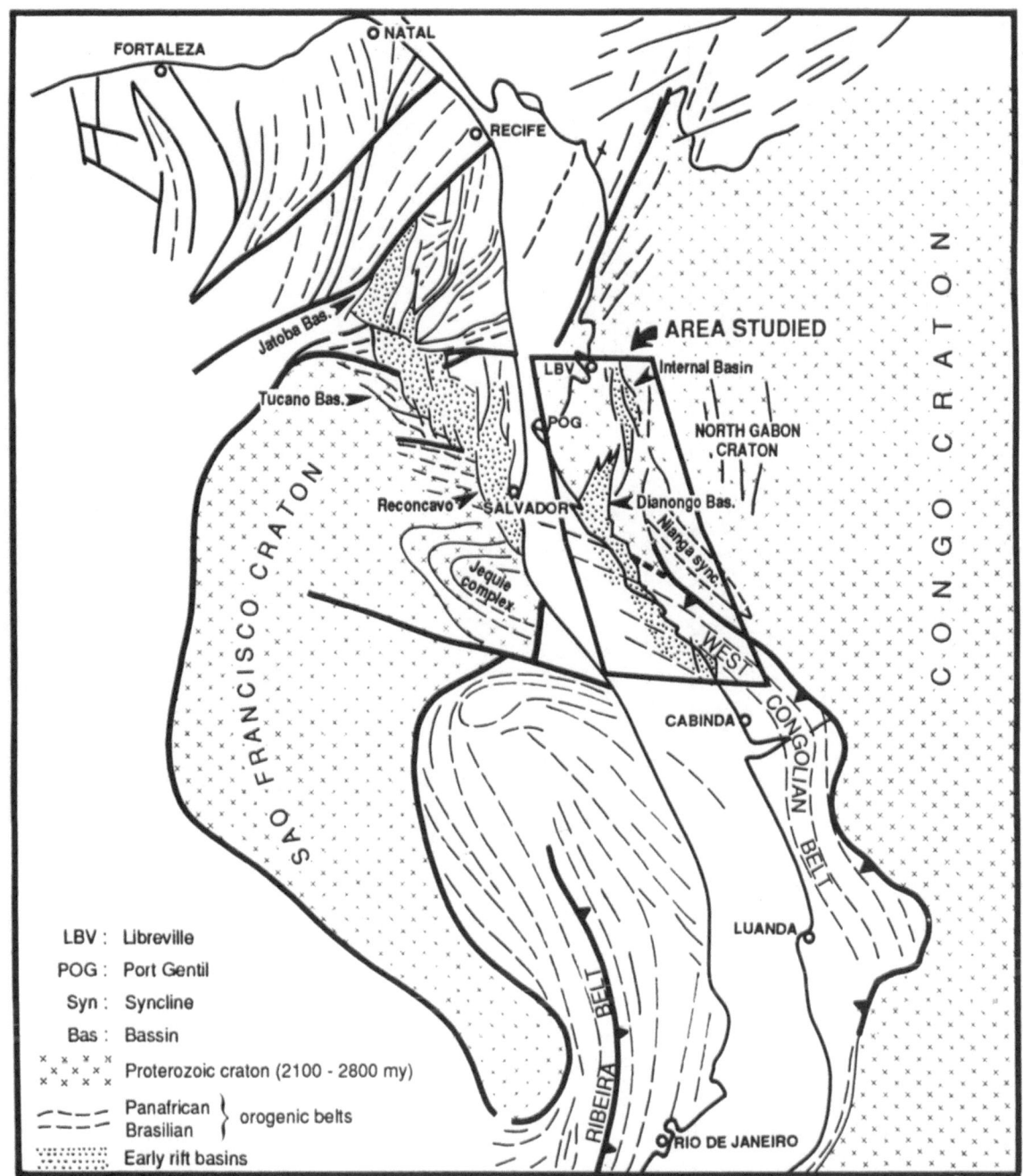

FIGURE 2-1.   Paleogeography and structural features of the Neocomian to Barremian rifting stage.

Rift Period

Two distinct stages of rifting can be defined (Fig. 2–4):

- Rift 1, from 127 to 115 Ma. The proto-Atlantic Ocean did not succeed in breaking through the Sao Francisco–Congo craton, and two rift branches (now aborted) originated from this locked zone. The Reconcavo–Jatoba branch extends to the west (Milani and Davidson, 1988), and the Dianongo–Internal Basin branch extends to the east (Fig. 2-1).
- Rift 2, from 115 to 110 Ma. The Atlantic opening succeeded in breaking through the craton at about

115 Ma. As a consequence of the abrupt tectonic change, the large Dentale basins are nearly independent of the preceding horsts and grabens. This rift period lasted until about 110 Ma, at which time widespread erosion occurred. First postdrift deposits of the Aptian salt sequence then sealed the normal fault systems.

The fluviolacustrine synrift deposits are composed of alternating sandstones and shales. Regionally the entire sequence exceeds 5,000 m in thickness. Facies distribution and subsidence rates are organized spa-

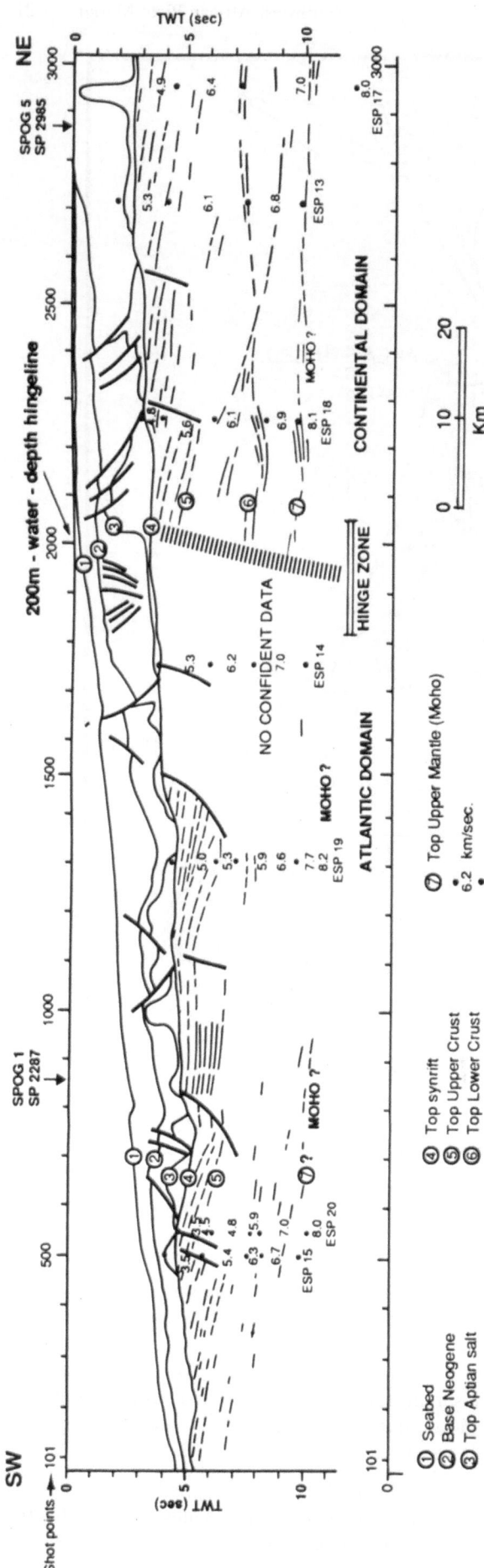

FIGURE 2-2. SPOG2 deep seismic profile (after J. C. Icart and J. Wanesson, Internal Elf Aquitaine–IFP report). Location of profile shown in Fig. 2-3.

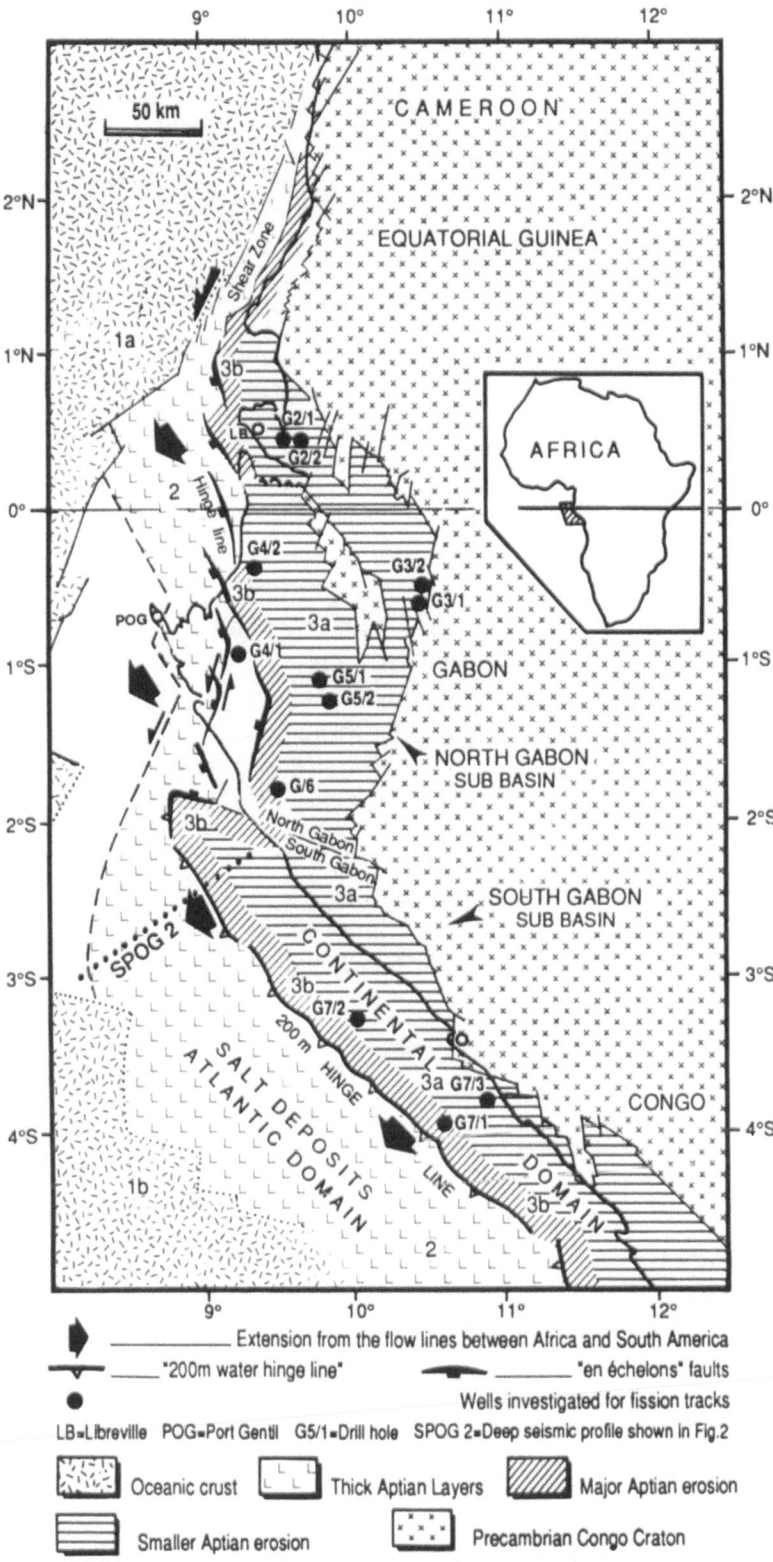

FIGURE 2-3. Tectonic framework during the Aptian: (1a) oceanic crust in North Gabon subbasin: sharp contact with continental crust; (1b) oceanic crust in South Gabon subbasin: gradual transition to continental crust; (2) thick Aptian deposits (early Aptian Coniquet clastics, late Aptian salt); (3a) sealed shoulder; weak Aptian uplift and erosion, quiescence during late Aptian, with low and regular subsidence; (3b) zone of maximum early Aptian uplift.

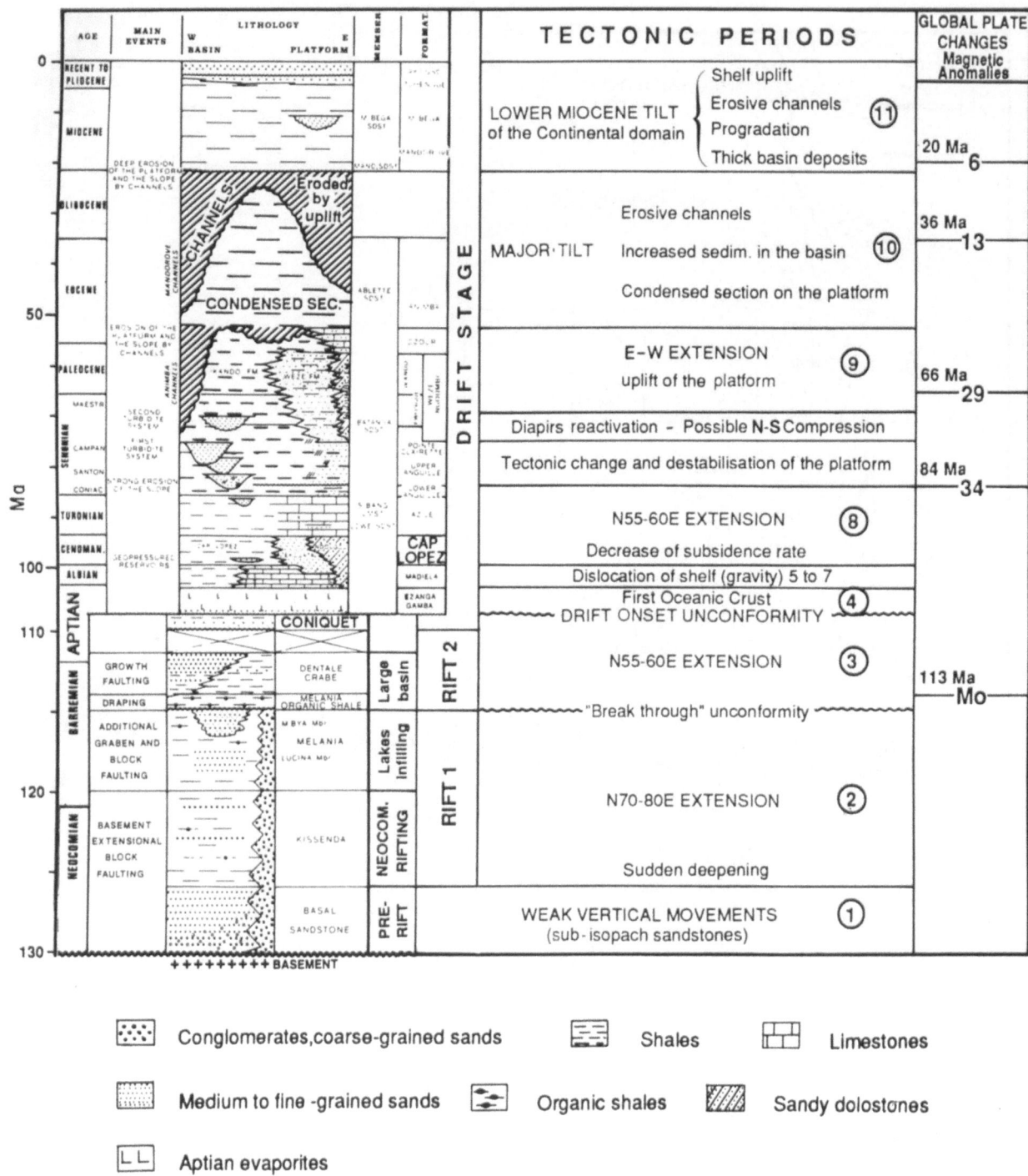

FIGURE 2-4. Stratigraphic and tectonic chart (after Teisserenc and Villemin, 1990). Numbers 1–11 refer to sedimentary features shown in Fig. 2-5.

tially and temporally according to the different stages of structural setting during the early and late Neocomian and the late Barremian. More than 2,000–3,000 m of clastic sediments were deposited in less than 2–3 m.y. during the late Barremian period of maximum crustal stretching (Dentale Basin; Figs. 2-4, 2-5). Rapid uplift and erosion at about 110 Ma (of greater magnitude near the 200-m hingeline; Figs. 2-3, 2-5), and the resulting lower Aptian stratigraphic hiatus, mark the

drift onset unconformity (= postrift unconformity of some authors; see Chapters 5, 6).

Drift (Postrift) Period

Thermal subsidence was rapid during the Albian, and was accompanied by deposition of 1,000–2,000 m of postrift sediments. Subsidence strongly decreased, however, in the Cenomanian and Turonian. Facies distribution at this time (clastic, lagoonal, platform

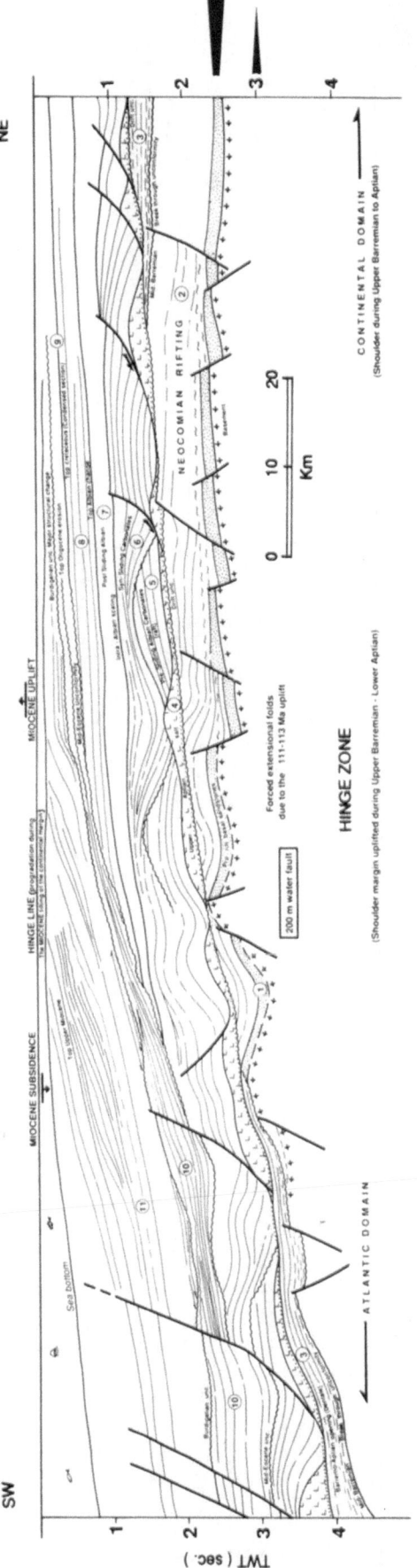

FIGURE 2-5. Geologic section across the "200-m-water-depth hingeline" in South Gabon–Congo subbasin. Numbers refer to tectonic stages on stratigraphic chart (Fig. 2-4).

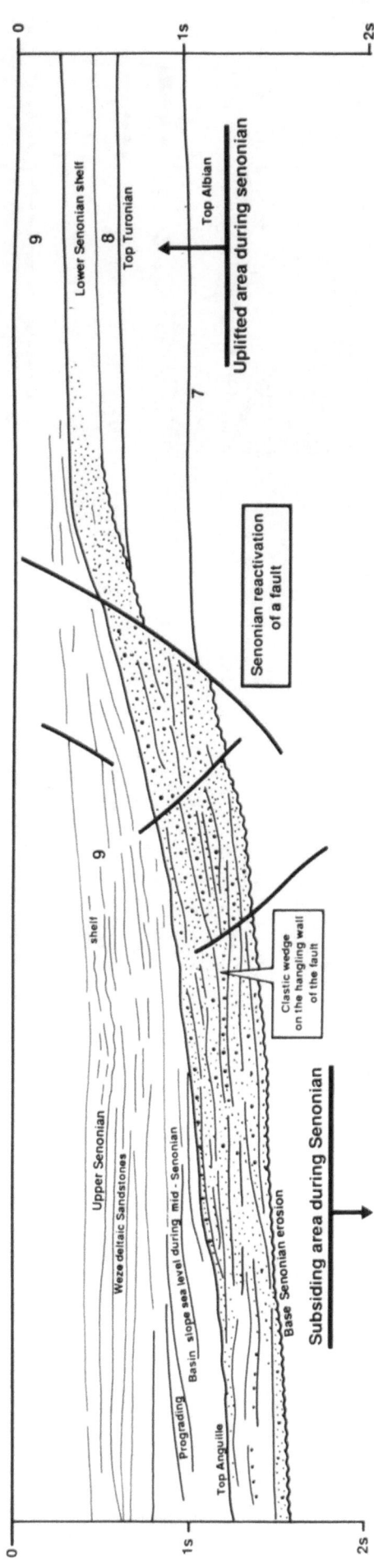

NORTH GABON
HINGE LINE DURING SENONIAN

FIGURE 2-6. Diagrammatic geological interpretation of seismic reflection profile showing the North Gabon hingeline during the Senonian (numbers refer to tectonic stages shown in Fig. 2-4). Total thickness of Anguille turbidites (Coniacian–Turonian), related to uplift of eastern source area, is 400–500 m. Late Senonian deposits reach maximum thickness of 2,000 m in Port Gentil area.

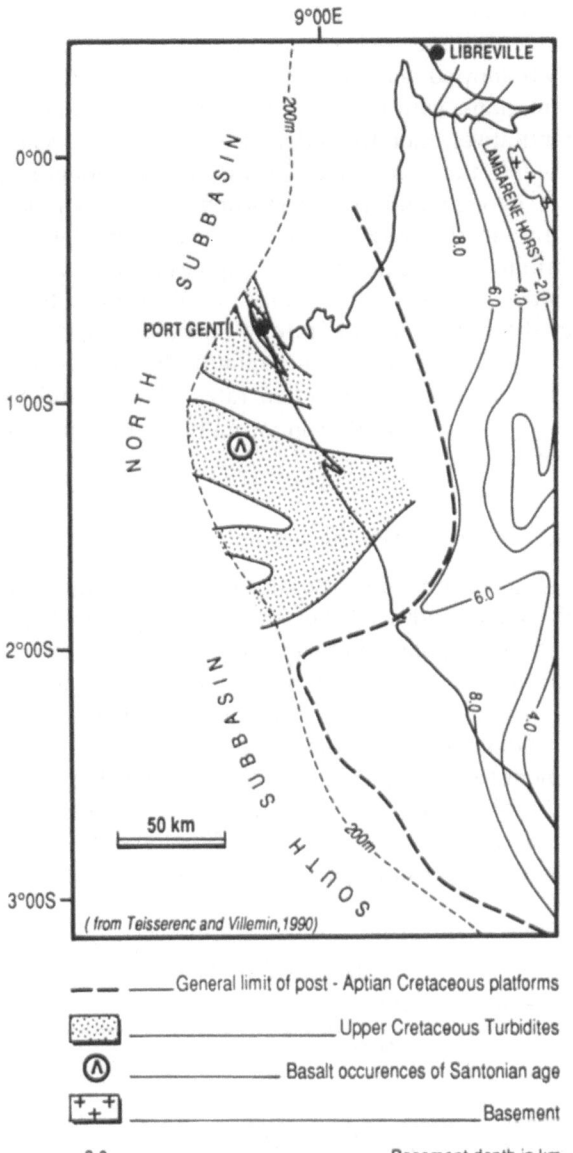

9°00E

LIBREVILLE

0°00

NORTH SUBBASIN

PORT GENTIL

1°00S

2°00S

SOUTH SUBBASIN

50 km

3°00S

(from Teisserenc and Villemin, 1990)

———— General limit of post - Aptian Cretaceous platforms

Upper Cretaceous Turbidites

Ⓐ ———— Basalt occurences of Santonian age

Basement

8.0 ———— Basement depth in km

FIGURE 2-7. Schematic distribution of Senonian turbidites, which suggests that sand currents were diverted by salt tectonics.

carbonates) reflects the first evidence of E–W polarity in sedimentation.

The Senonian interval is characterized by the late Santonian tectonic phase near 80 Ma (magnetic anomaly 34), which is related to a change in the plate movement vector from N60°E to N90°E. As a result of the reactivation of faults and regional uplift of areas east of the hingeline (Fig. 2-6), the platform was deeply eroded, and the slope was incised by channels. The distribution of turbidites in the basin (Fig. 2-7) suggests that sand currents could have been diverted and limited by salt tectonics (Teisserenc and Villemin, 1990; see also Chapter 5). Some Santonian basalt occurrences in distal offshore areas also reveal the instability of the margin during the Late Cretaceous.

Paleocene and Eocene deposits are mainly composed of shales, silts, and silexites. Little is known about tectonic activity during the Paleogene. A major tilt of the margin is thought to have occurred, evidenced by the Animba channels in the western part of the North Gabon subbasin (Fig. 2-4). This event may have been related to the worldwide middle/late Eocene geodynamic episodes at ~45 and ~36 Ma, which reflect a change in the organization of the plate tectonic pattern (Schwan, 1985).

The Paleocene deposits are overlain unconformably by Miocene deposits, without evidence of Oligocene sediments in the continental domain. This stratigraphic gap is related to the second major tilting stage of the margin at about 24 Ma. A striking change took place in the African plate during the early Miocene, when both the East African and Suez–Red Sea rifts were initiated (Omar et al., 1989).

The Gabon–Congo margin tilted along the 200-m hingeline (Fig. 2-5), with an uplifted domain to the east and a strongly subsident domain to the west. Miocene shales and silts, which seal the previous erosion surfaces and channels, exceed 2,000–3,000 m in thickness at the exit of the Ogooue and Congo deltas. Scarce basalt occurrences of middle Miocene age (16 ± 2 Ma), south of Libreville, are related to this Neogene tectonism.

METHODS IN PALEOTHERMOMETRY

Apatite Fission-Track Chronothermometry

Fission tracks are randomly distributed linear radiation damage zones resulting from the spontaneous fission of uranium-238, which is usually present as trace amounts in apatite. The number of tracks in a crystal depends on the uranium concentration and time. Tracks are made visible for optical microscopy by suitable etching of a polished grain surface. Track density (number per unit area) is used to calculate an age. The uranium concentration of crystals is measured from tracks formed by induced fission (via thermal neutron irradiation) in a uranium-free external detector placed in close contact with apatite samples.

Fission tracks are reversible damage zones, which are highly sensitive to temperature. The etchable length of tracks, constant and close to 16 μm when produced, becomes progressively shorter as thermal effect increases. Consequently, the probability for tracks to be revealed is reduced and the calculated age is lowered (Bigazzi, 1967; Storzer, 1970; Naeser, 1979; Gleadow, Duddy, and Lovering, 1983). Each track shrinks to a length that essentially depends on the maximum temperature it has experienced, and, to a lesser extent, on the duration of the thermal effect. Because the half-life of uranium is on the order of 10^{10} yr, the rate of track production is considered to be constant through time, and the final length distribution reflects the complete thermal history.

Fading characteristics of fission tracks (Fig. 2-8), established from laboratory annealing experiments and data from deep drill holes, show that total damage repair in fluorapatites occurs between 40 and 160°C for geological durations of 10^8 to 10^6 yr (Storzer and Selo, 1984; Meyer, 1990).

Track-length data can be derived from horizontal confined tracks fully contained in the crystal, etched by acid from cracks or from other tracks intersecting the polished surface. Laslett et al. (1982) have shown that length distribution of confined tracks may be strongly biased against shorter tracks because of lower probability of intersecting an acid-transmitting conduit. On the other hand, confined tracks cannot be used for measuring a statistically significant number of tracks in low-density or strongly annealed apatite samples.

Alternatively, the projected length of those tracks crossing the etched surface (surface tracks used for dating) can be measured under the strict conditions described by Meyer (1990). Because part of the total length is always missing (due to polishing and etching), the projected length of spontaneous tracks has no absolute meaning. Therefore, the length reduction is obtained by comparison with fresh induced tracks, which are measured under the same condition on an

apatite aliquot previously heated for 1 h at 500°C, for complete annealing of spontaneous tracks. Meyer (1990) showed that projected lengths allow two track populations, with different degrees of annealing, to be quantitatively determined in a single apatite sample, provided the differences between mean lengths are large enough and the proportion of one population relative to the other is not too small. Bimodal length distribution indicates that the apparent age measured includes two age components, which should be corrected for their specific rate of annealing. Age corrections are made using the relationship between the length and density reduction rates (Storzer and Selo, 1984; Meyer, 1990), thus allowing the dating of cooling events.

Fluid – Inclusion Geothermometry

Fluid inclusions in minerals are microcavities filled with liquid and(or) other phases (gas, solids), from which crystals have grown. If no chemical or physical alteration of the system has occurred subsequent to entrapment, the inclusions can provide valuable keys to the understanding of the thermal and chemical environment that prevailed at the time of inclusion closure.

Fluid-inclusion thermometry (FIT) is based upon the differential shrinkage of the host crystal and fluid upon cooling from the conditions of entrapment to

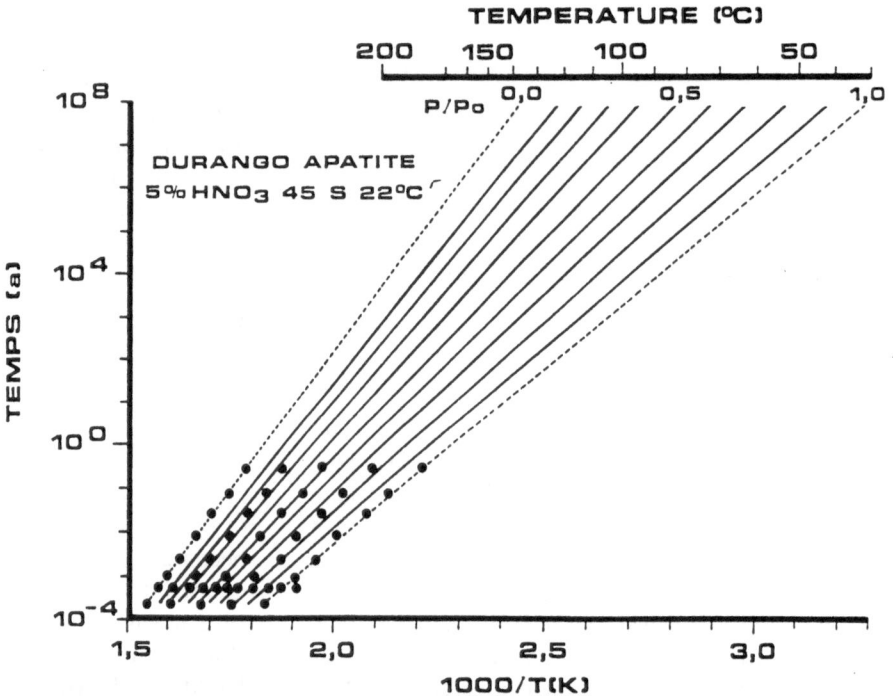

FIGURE 2-8. Fading characteristics of fission tracks in fluorapatite in form of Arrhenius plot (logarithm of time in years versus inverse absolute temperature). Data expressed according to track density reduction rate (from Meyer, 1990).

laboratory temperature. An initially homogeneous fluid (gas or liquid) will show a gas bubble and a liquid phase at laboratory temperature. The relative gas–liquid proportions illustrate the bulk density of the system. Upon heating, the process is reversed and the system evolves toward homogenization into a single fluid phase. Homogenization temperature (T_h) is rarely equivalent to the temperature of trapping (T_t), as most of the fluids are trapped at PT levels above the liquid–vapor equilibrium curve. Therefore, a temperature correction (ΔT) should be added to T_h as a function of pressure. Determination of ΔT is generally difficult due to fluid density and pressure uncertainties at the time of trapping. Furthermore, to determine ΔT requires assumptions about the burial depth and pressure gradient. A more favorable situation occurs when two immiscible fluids with different PVT properties are trapped simultaneously as separate inclusions (Narr and Burruss, 1984; Walgenwitz et al., 1990). In such a case, both pressure and temperature can be estimated, regardless of burial and pressure-gradient assumptions. Dating the fluid entrapment needs additional radiometric analysis of cogenetic K minerals or, when possible, of the host mineral itself (Walgenwitz et al., 1990). An extensive review of the applications derived from fluid inclusions was published by Roedder (1984).

ANALYTICAL PROCEDURES

We obtained fission-track chronothermometry data on apatites extracted from about 1 kg of core sample. Ages were calculated using the isochron technique, which is particularly suitable for dating detrital minerals included in sediments. About 50–100 apatite grains were mounted in epoxy resin, polished and etched in 5 wt% HNO_3 for 45 sec at 23°C, to reveal spontaneous tracks. Spontaneous tracks were counted at a total magnification of 1000× in oil immersion.

Grain mounts were then placed in close contact with a kapton polyamide external detector and irradiated. Induced tracks in the detector were etched using a boiling solution of 14% NaOCl + 12% NaCl for 8 min. Induced tracks were counted at total magnification of 1600× with a dry objective. Ages were calculated taking into account the different recording geometries of spontaneous tracks (4π solid angle) and induced tracks (2π solid angle). Irradiation characteristics were determined from induced tracks recorded in kapton external detectors in contact with NBS uranium dosimeter glasses (SRM 613) intercalated between apatite mounts. The neutron fluence was calibrated by comparison with the Dukovany III tektites (Storzer, Wagner, and King, 1973). Projected track lengths were measured at a total magnification of 3200× in oil

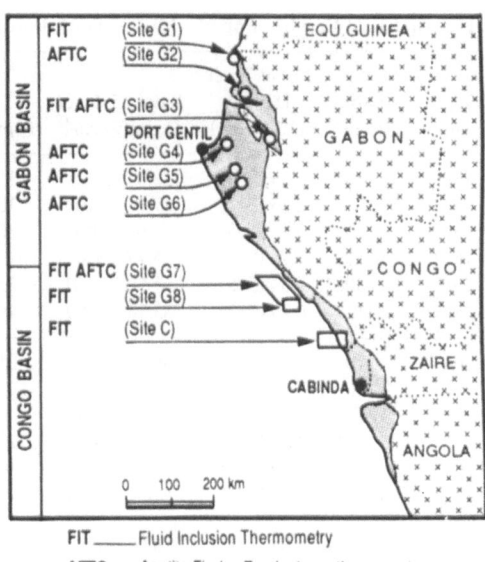

FIGURE 2-9.   Location map of geologic profiles investigated for paleotemperatures.

immersion, in strict compliance with the observation criteria set forth by Meyer (1990).

Fluid-inclusion analysis was performed on rock slides about 100 μm thick and polished on both sides. Microthermometric measurements (homogenization and ice-melting temperatures) were achieved on a Chaixmeca heating and freezing stage (Poty, Leroy, and Jachimowicz, 1976) set on an Ortholux Leitz optical microscope. Thermometric calibrations were made using substances with known melting points and fluid inclusions from synthetic and natural quartz crystals. Fluorescence under reflected ultraviolet light enabled us to easily distinguish oil inclusions from aqueous inclusions.

AREA OF INVESTIGATION AND SAMPLE DETAILS

The study area extends over nearly 800 km along the African margin, from the northernmost part of the Gabon Basin to the southern Congo Basin (Fig. 2-9). Our data are derived from eight sites located onshore and offshore. With the exception of outcrop samples from site G1, all analyses were performed on core materials from oil exploration wells. For each site, we collected samples either from a single well, which was cored through a large depth interval (G6), or, more frequently, from several adjacent wells, which were cored over restricted stratigraphic intervals at different depths.

Data collected over great vertical distances provide a better statistical mean for the interpretation of paleotemperatures. Exploration holes, well cored for stratigraphic purposes, enabled us to investigate over a

TABLE 2-1   Analytical Results of Apatite Fission-Track Dating.*

Wells	Depth (m)	Apatite Grains[†]	Fossil Track Density (10^5/cm²)	Induced Track Density/cm²[‡]	U (ppm)	Fission Track Ages (Ma)	Stratigraphic Ages (Ma)
G2/1	701	11/104	5.14 (565)	20.34 (211)	12	76.1 ± 6.1	~ 108–110 Aptian
	940	14/87	2.65 (371)	27.75 (340)	15	31.0 ± 2.3	~ 112–115 Barremian
	1315	15/104	3.83 (575)	51.60 (730)	30	22.4 ± 1.3	~ 112–115 Barremian
	1637	6/10	1.78 (99)	25.77 (199)	15	20.8 ± 2.6	~ 112–115 Barremian
	2012	31/306	1.64 (575)	61.50 (2770)	35	8.1 ± 0.4	~ 112–115 Barremian
G2/2	3679	33/117	0.29 (95)	85.14 (3506)	49	1.0 ± 0.1	~ 112–115 Barremian
G3/1	660	3/10	3.03 (91)	36.47 (104)	21	25.1 ± 3.6	~ 120–125 Neocomian
	1190	3/10	1.47 (44)	34.71 (99)	20	12.7 ± 2.3	~ 120–125 Neocomian
	1405	4/9	2.90 (116)	76.53 (291)	44	11.4 ± 1.2	140–120 Jurassic
G3/2	990	98/401	3.53 (4722)	47.16 (7986)	27	22.6 ± 0.4	~ 120–125 Neocomian
	1053	45/132	2.60 (182)	52.41 (269)	30	14.9 ± 1.4	~ 120–125 Neocomian
	1266	100/399	1.86 (354)	50.07 (1073)	29	11.2 ± 0.7	140–200 Jurassic
G4/1	211	1/52	1.80 (18)	6.88 (7)	4	79.0 ± 3.5	66–71 Maestrichtian
	475	11/89	7.19 (791)	34.15 (382)	20	63.4 ± 4.0	71–83 Campanian
	950	15/98	3.57 (535)	26.89 (519)	16	40.0 ± 2.5	71–83 Campanian
	1599	9/42	2.18 (196)	26.55 (243)	15	24.7 ± 2.4	83–89 Sant.–Conia.
	2422	25/185	3.42 (854)	68.64 (1745)	39	15.0 ± 0.6	95–105 Albian
	2501	16/48	1.63 (261)	34.77 (582)	21	13.7 ± 1.0	95–105 Albian
	3830	28/100	0.32 (89)	64.03 (1828)	39	1.4 ± 0.1	~ 112–115 Barremian
G4/2	3258	24/95	0.42 (101)	64.21 (1983)	37	1.9 ± 0.2	~ 112–115 Barremian
G5/1	1571	14/93	6.71 (940)	60.17 (1084)	35	33.6 ± 1.5	Basement (Proterozoic)
G5/2	2276	11/98	7.09 (780)	80.82 (1144)	46	26.4 ± 1.2	Basement (Proterozoic)
G6	428	12/93	19.02 (2282)	34.81 (515)	20	164.6 ± 8.0	91–95 Cenomanian
	1309	4/47	26.20 (917)	61.74 (634)	40	127.8 ± 6.7	91–105 Albian
	1567	5/98	9.44 (425)	23.10 (305)	15	123.2 ± 9.2	95–108 Albian
	2557	25/34	5.18 (845)	32.25 (865)	19	48.4 ± 2.3	~ 108–110 Aptian
	2588	68/196	4.03 (698)	17.99 (913)	12	67.6 ± 3.4	~ 112–115 Barremian
	2689	24/46	3.38 (1091)	19.20 (1820)	13	53.0 ± 2.0	~ 112–115 Barremian
	2798	24/92	3.34 (756)	28.61 (1897)	19	35.2 ± 1.5	~ 112–115 Barremian
	3524	57/104	1.07 (557)	28.04 (4278)	18	11.5 ± 0.5	~ 112–115 Barremian
	3612	33/106	1.83 (605)	53.4 (2175)	31	10.3 ± 0.5	~ 112–115 Barremian
G7/1	4020	20/93	0.91 (183)	75.02 (1850)	43	3.7 ± 0.3	~ 112–115 Barremian
G7/2	2245	73/289	3.04 (2234)	48.65 (4602)	28	18.8 ± 0.5	~ 108–110 Barremian
	2498	2/23	6.80 (136)	156.21 (297)	90	13.1 ± 1.4	~ 112–115 Barremian
	2911	21/98	2.49 (522)	66.24 (1790)	38	11.3 ± 0.6	~ 112–115 Barremian
	3030	3/38	2.10 (63)	68.02 (194)	39	9.3 ± 1.3	~ 112–115 Barremian
	3252	1/17	2.90 (29)	122.02 (116)	70	7.2 ± 1.5	~ 112–115 Barremian
G7/3	1930	70/140	4.53 (3913)	42.34 (10715)	27	32.3 ± 0.6	~ 115–120 Barremian

*Parentheses show number of tracks counted. Ages calculated with: $\lambda_f = 8.46 \times 10^{-17}$ yr^{-1}, $4\pi/2\pi$ geometry factor = 1.651, $^{235}U/^{238}U = 7.253 \times 10^{-3}$, and ^{235}U cross section = 580 barns.
[†]Number of grains analyzed with respect to the total number of grains.
[‡]Number of tracks normalized for a neutron fluence of 10^{11} neutrons/cm².

maximum depth range of about 3,800 m. Present-day temperature of the deepest samples is close to 140°C. When possible, we carried out both AFTC and FIT measurements on the same core samples to obtain better thermal constraints.

The stratigraphic intervals covered by sampling range from the Jurassic to Neocomian for the Gabon Internal Basin (site G3, Fig. 2-9), and from the Barremian to Maestrichtian for the coastal basins (Table 2-1).

APATITE FISSION-TRACK INVESTIGATION

Apparent Ages

Apparent ages rapidly decrease with depth (Fig. 2-10); even in shallow samples where significant annealing would not be expected (temperature less than 40–50°C). In most cases, fission-track ages are younger than stratigraphic ages (Fig. 2-10), except in the upper part of profile G6. Fission-track ages are always much younger than expected for apatites inherited from the

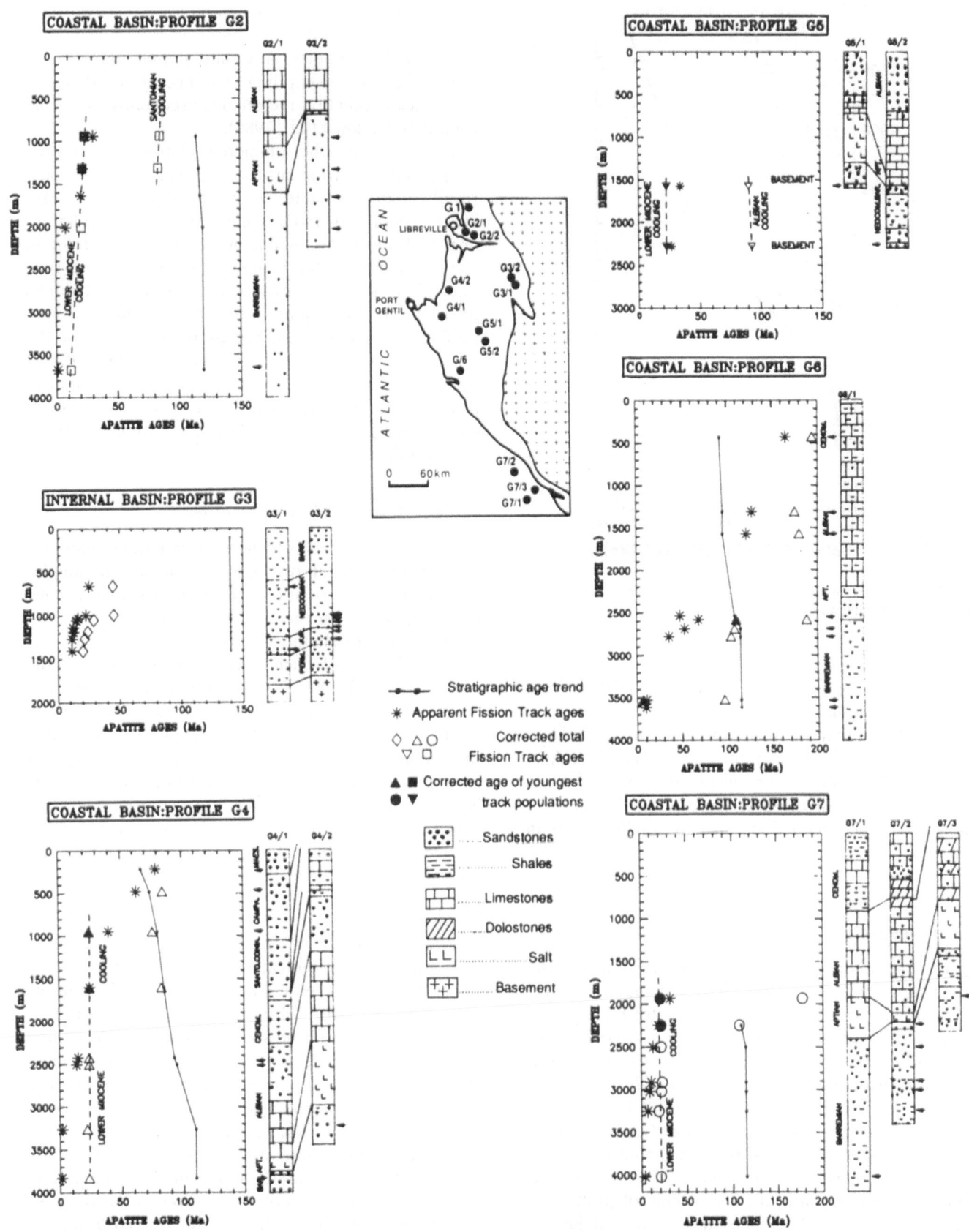

FIGURE 2-10.   Summary of lithostratigraphic data and fission-track ages of seven profiles from Gabon basins.

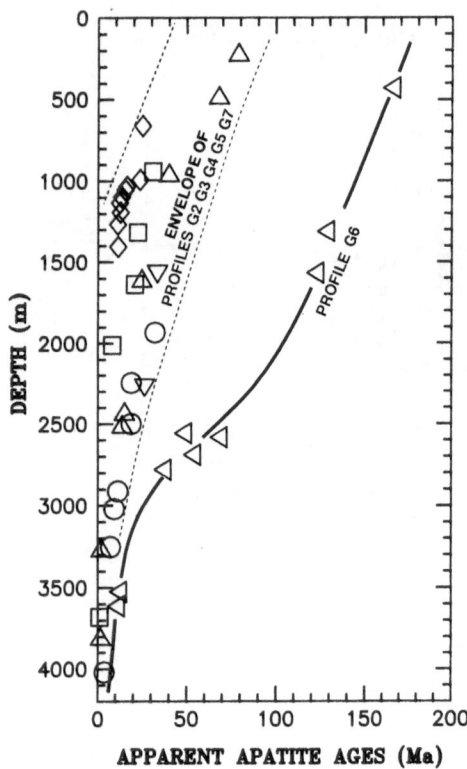

FIGURE 2-11.   Synthetic plot of apparent fission-track ages versus depth (symbols refer to profiles of Fig. 2-10).

Proterozoic Congo craton or from the Panafrican Congolian belt (Fig. 2 1).

Although apparent ages evolved asymptotically to zero, we did not observe complete track annealing, even at a depth of 4,000 m, where the temperature is close to 140°C. All the data reported on Figure 2-11 show lower age reduction rates in profile G6 than elsewhere.

Length Analysis

More constraints are imposed on thermal history by the measurement of projected track lengths. Spontaneous tracks are always shorter than freshly induced tracks, and length distributions show a typical evolution with depth, as illustrated by profile G4 (Fig. 2-12). Length distributions are either unimodal or bimodal; the latter occur mostly at the shallowest levels of the profiles, with two exceptions: profile G3 shows only unimodal distributions and profile G6 shows bimodal distributions in some deep samples (Fig. 2-13). The unimodal distribution in the top sample of profile G4 (Fig. 2-12) is actually composed of two populations with similar mean lengths; projected tracks are unable to discriminate between them.

Bimodal length distributions indicate that two track populations with different degrees of annealing coexist within apatite samples. Any short population refers to tracks that underwent higher temperatures than the long track population recorded subsequently to a rapid cooling (Gleadow et al., 1986).

Consequently, the unimodal distributions in the deeper samples from profiles G2, G4, and G7 would result from complete annealing (at temperatures greater than 140–160°C) of the shortest populations observed above. Unimodal distributions thus refer to tracks recorded after cooling, during the same time interval as longer track populations in the bimodal distributions. The temperature decrease through these profiles should be more recent than the age of the uppermost stratigraphic level analyzed (Fig. 2-10). A different situation exists in profile G6, where bimodal distributions in deep Barremian sandstones suggest cooling before deposition of Albian carbonates (Figs. 2-10, 2-13).

Corrected Ages and Temperatures

Length analysis has shown that apparent ages are actually the sum of two age components in samples with bimodal distributions. Each component is lowered, depending on the specific annealing of the track population, and should be corrected separately. The relative density of each population depends on the fraction of total time elapsed on both sides of the last stage of cooling, and on the temperature. Age components are determined by decomposition of the length histograms, assuming that distributions of the longest track populations are similar to those of induced tracks. The measured ages are split into two apparent ages, which are then corrected for the density reduction related to length shortening, as directly done for samples with unimodal distributions (Table 2-2).

Profile G2
The fact that corrected ages are always younger than stratigraphic ages indicates that temperatures were higher than 140 to 160°C throughout the interval investigated (Figs. 2-10, 2-14). First cooling occurred at about 84–85 Ma during the Santonian, but complete annealing lasted until the early Miocene in Barremian strata. This second cooling started at about 24 Ma, but slow deepening of the isotherms is evidenced by a corrected age of 12 Ma in the deepest level. For the 84–24 Ma period, temperatures ranged from 125–135°C at 940 m to more than 160°C in samples deeper than about 2,000 m. Paleotemperatures ranged from 70–80°C to 140–150°C between 700 and 3,679 m after cooling in the early Miocene. Present-day temperatures vary from 40 to about 140°C.

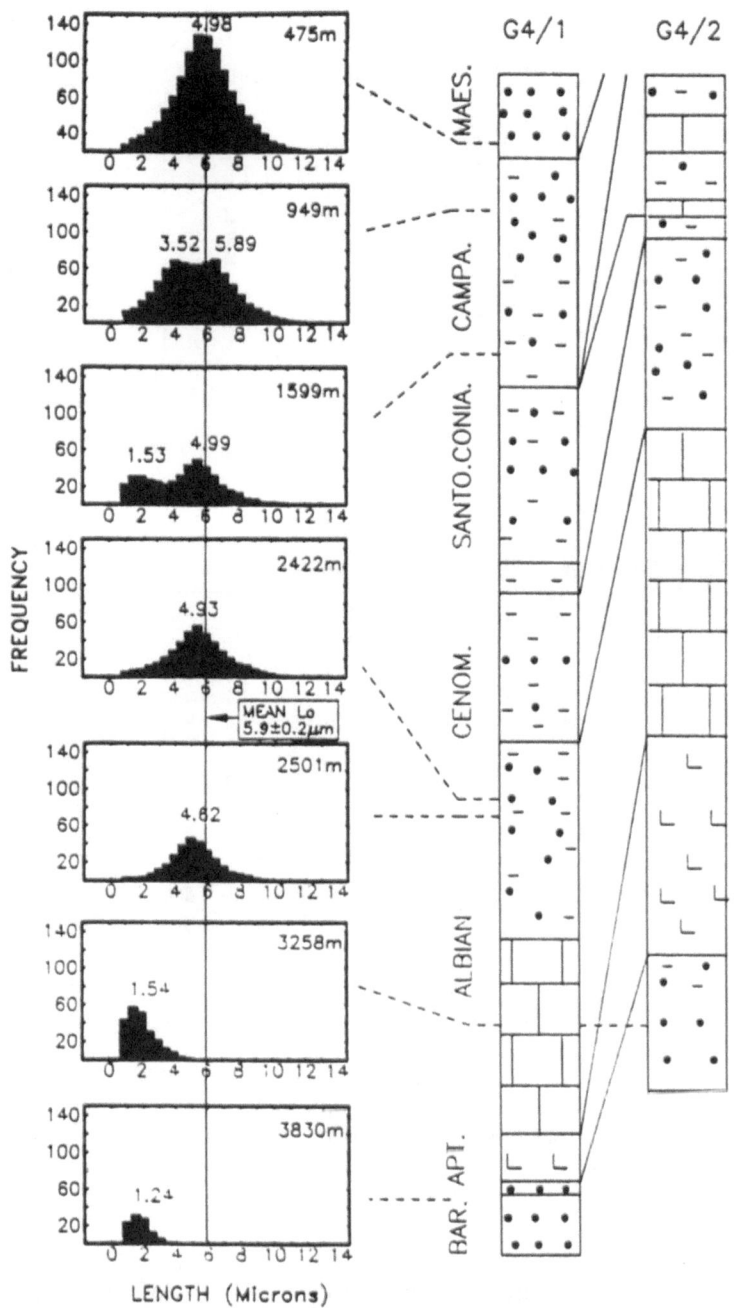

Shales   Limestones   Aptian Salt

BAR. : Barremian   APT. : Aptian   CENOM. : Cenomanian   SANTO. : Santonian

CONIA. : Coniacian   CAMPA. :Campanian   MAES. : Maestrichtian

FIGURE 2-12.   Track length distributions of samples from profile G4. Lo refers to mean length of fresh tracks induced via thermal neutron irradiation (see text for further details on analytical procedures). Apparent fission-track ages corrected for length distributions.

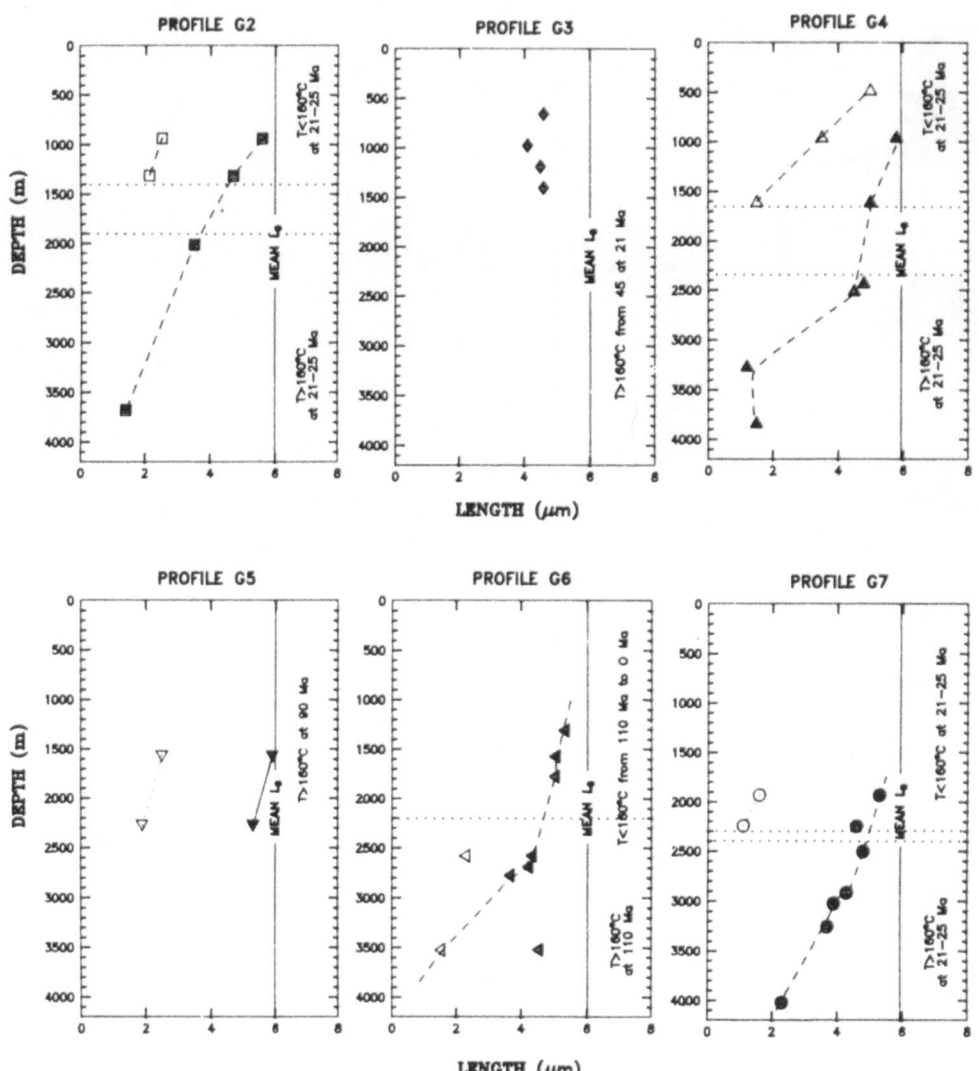

FIGURE 2-13.   Plot of mean track lengths versus depth. Open symbols = mean length of shortest track populations in bimodal distributions; solid symbols = mean length of longest track populations in unimodal distributions.

Profile G3

Temperatures higher than 40 to 160°C are also evidenced in the Internal Basin (Fig. 2-14). The fission-track record in apatites from Jurassic and Neocomian sediments began at about 45 Ma. The rapid decrease of corrected ages to values of 20 Ma over a short depth interval (Fig. 2-10) suggests high paleogeothermal gradient and very slow deepening of the isotherms, at a virtual rate of about 2.10^{-2} mm/yr. Fission-track temperatures after cooling are similar throughout the profile (close to 100°C, whereas they currently range from 40 to 60°C).

Profile G4

Total corrected ages are organized into two distinct groups according to depth (Fig. 2-10). They range from 76 to 84 Ma in the Maestrichtian–Santonian interval,

and decrease sharply to 21–25 Ma in Albian and Barremian sediments, where values much younger than stratigraphic ages date cooling at temperatures below 40–160°C. Early Miocene cooling ages are also derived from the Senonian deposits, and indicate that temperature decreased suddenly throughout the profile.

Corrected ages are virtually identical to stratigraphic ages in Senonian turbidites; this indicates that the fission-track record began close to the time of sediment deposition. Corrected ages thus refer to the thermal history of the source of clastic materials, indicating that the source region for apatite grains had undergone a complete track annealing by Santonian time.

Maximum temperatures experienced under deeper paleoburial conditions in the Late Cretaceous to early Miocene interval, ranged from 110 to 120°C at 950 m to > 160°C for samples deeper than 2,000–2,500 m

TABLE 2-2   Fission-Track Age Corrections and Paleotemperatures.

Wells	Depth (m)	Age Component* t_1 (Ma)	Age Component* t_2 (Ma)	L/L_0†	L_1/L_0‡	L_2/L_0‡	P/P_0†	P_1/P_0‡	P_2/P_0‡	t_{cor}§ (Ma)	t_{2cor}¶ (Ma)	T_1‖ (°C)	T_2** (°C)
G2/1	701			0.9			0.74			103 ± 11			70–80
	940	13.5	17.5		0.43	0.90		0.22	0.73	85 ± 15	24 ± 3	125–135	75–85
	1315	9.9	12.5		0.35	0.79		0.16	0.57	84 ± 13	22 ± 3	135–145	95–105
	2012			0.63			0.39			21 ± 2		≥ 160	110–120
G2/2	3679			0.65			0.09			12 ± 2		≥ 160	140–150
G3/1	660			0.78			0.56			45 ± 11		≥ 160	95–105
	1190			0.76			0.53			24 ± 7		≥ 160	105–110
	1405			0.78			0.56			20 ± 4		≥ 160	95–105
G3/2	990			0.73			0.50			45 ± 3		≥ 160	110–120
	1053			0.75			0.53			29 ± 3		≥ 160	105–110
	1266			0.75			0.53						105–110
G4/1	475			0.91			0.75			84 ± 8			70–80
	950	18.0	22.0		0.58	0.97		0.35	0.90	76 ± 12	24 ± 3	110–120	60–70
	1599	6.4	18.3		0.28	0.90		0.11	0.74	83 ± 16	25 ± 3	130–140	70–80
	2422			0.83			0.63			24 ± 2		≥ 160	90–100
	2501			0.79			0.57			24 ± 3		≥ 160	95–105
	3830			0.21			0.06			23 ± 4		≥ 160	145–155
G4/2	3258			0.25			0.09			21 ± 3		≥ 160	140–150
G5/1	1571	13.7	19.9		0.41	0.97		0.20	0.89	91 ± 13	21 ± 3	125–135	60–70
G5/2	2276	9.9	16.5		0.32	0.90		0.14	0.74	93 ± 14	22 ± 3	135–145	75–85
G6	428			0.95			0.85			194 ± 30			60–70
	1309			0.90			0.73			175 ± 14			70–80
	1567			0.87			0.68			180 ± 15			80–90
	2588				0.38	0.72		0.18	0.49	188 ± 15	110 ± 6	130–140	90–100
	2689			0.72			0.50			108 ± 7		≥ 160	95–105
	2798			0.58			0.35			104 ± 6		≥ 160	110–120
	3524	8.0	3.5		0.25	0.74		0.09	0.51	96 ± 9	7 ± 1	≥ 160	130–140
G7/1	4020			0.37			0.18			21 ± 3		≥ 160	135–145
G7/2	2245	4.5	14.3		0.20	0.86		0.05	0.67	108 ± 20	22 ± 2	145–155	90–100
	2498			0.8			0.59			22 ± 4		≥ 160	95–105
	2911			0.71			0.48			24 ± 3		≥ 160	105–110
	3030			0.66			0.43			22 ± 5		≥ 160	105–110
	3252			0.61			0.38			19 ± 6		≥ 160	110–120
G7/3	1930	15.9	16.9		0.27	0.93		0.10	0.79	180 ± 11	21 ± 1	140–150	75–80

*Fission-track age components in samples with bimodal length distributions.

†Length and density reduction rates in samples with unimodal length distributions.

‡Length and density reduction rates of track populations in samples with bimodal length distribution.

§Total corrected ages: $t_{cor} = t_{ap}/(P/P_0)$ or $t_{cor} = t_{1ap}/(P_1/P_0) + t_{2ap}/(P_2/P_0)$.

¶Corrected ages of youngest track populations: $t_{2cor} = t_{2ap}/(P_2/P_0)$.

‖Maximum temperature between stratigraphic age and cooling age t_{2cor}.

**Maximum temperature between age t_{2cor} and today. As for the preceding note (‖), the temperatures are estimated assuming apatite compositions close to Durango apatite.

(Fig. 2-14). Temperatures after cooling at 21–25 Ma were close to present-day temperatures, varying from 70–80°C to 140–150°C between 475 and 3,830 m.

Profile G5

Complete track annealing until 91–93 Ma (Fig. 2-10) occurred in the gneiss samples from the Congo craton. After a first cooling stage in the late Albian, high temperatures of about 130–140°C still prevailed until the early Miocene. An abrupt temperature decrease

occurred at 21–22 Ma. Annealing degrees of fission-track populations recorded since the early Miocene indicate temperatures of about 60–70°C to 75–85°C between 1,570 and 2,275 m. These values are very close to present-day temperatures.

Profile G6

In the post–salt-depositional sequence, total corrected ages of about 175 to 194 Ma (much greater than stratigraphic ages; Fig. 2-10) indicate temperatures

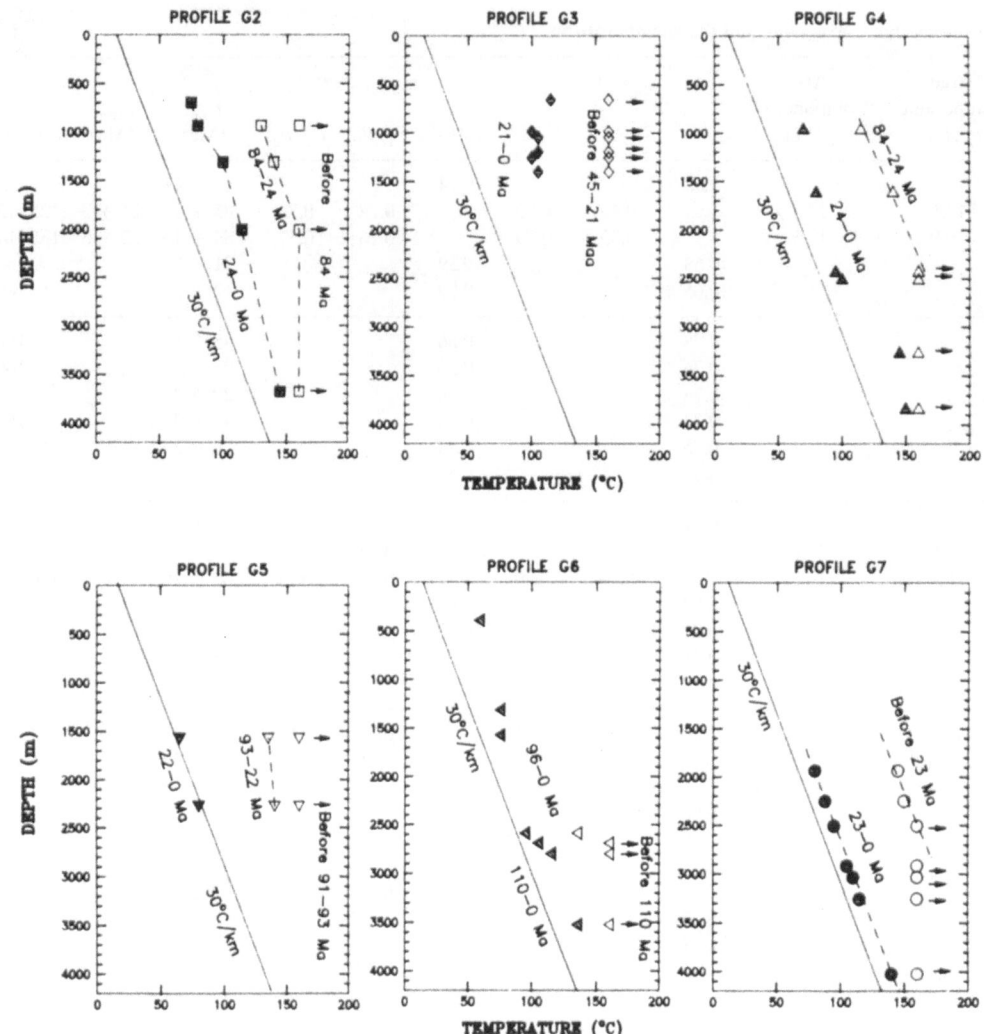

FIGURE 2-14. Summary of fission-track paleotemperatures. Solid and open symbols refer to those of track populations indicated in Fig. 2-13. Arrows indicate temperatures greater than 160°C.

always lower than 40 to 160°C since deposition. These ages refer to the thermal history of the provenance area of apatite grains during the Jurassic.

A similar age close to 188 Ma in the first Barremian sample at 2,588 m indicates a common origin for minerals throughout the Early Cretaceous, and temperature always lower than 40 to 160°C. Nevertheless, bimodal length distribution and annealing rates indicate maximum temperatures of 130–140°C between sediment deposition at about 112 Ma and sudden cooling at 110 Ma. The temperature decrease occurred at a time corresponding to the lower Aptian stratigraphic hiatus (drift onset unconformity) located at about 100 m above the sample analyzed. Consequently, cooling at 110 Ma illustrates a temperature variation from 130–140°C to near surface conditions.

Corrected ages are younger than stratigraphic ages in deeper Barremian samples and decrease with depth from 108 to 96 Ma. This trend, combined with the above-mentioned cooling age of 110 Ma, reflects the slow relative deepening of the isotherms in rift sediments from the early Aptian until the late Albian. Maximum temperatures during the drift stage (Fig. 2-14) ranged from 60–70°C to 130–140°C between 428 and 3,524 m, whereas present-day values vary from 30–35°C to 120–130°C.

Profile G7

A corrected age of 180 Ma in well G7/3 (much older than stratigraphic age; ~ 115 Ma; Fig. 2-10) indicates that grains were inherited from source regions with a thermal history similar to that of apatites from profile G6. In well G2/1, apatites from Aptian sands above the drift onset unconformity have a total corrected age of about 108 Ma, which is virtually identical with the stratigraphic age. Grains, therefore, were inherited from deeply buried Barremian sandstones ($T > 40$ to 160°C) prior to the early Aptian.

Deeper Barremian samples both from wells G7/2 and G7/1 show total corrected ages of 19 to 24 Ma, which indicates that complete track annealing took place after sediment deposition. A similar age for the beginning of the fission-track record over an interval of about 1,500 m indicates that cooling occurred suddenly during the early Miocene. A temperature decrease is also consistently observed at 21–22 Ma in the shallower samples (Fig. 2-10). Maximum paleotemperatures since sediment deposition range from 140–150°C at 1,930 m to more than 160°C in samples deeper than 2,300–2,800 m. Temperatures after cooling during the early Miocene are close to present values, varying from about 80°C to 140°C (Fig. 2-14).

INTERPRETATION

Prerift History

Provenance ages of about 176–194 Ma (mean value of 181 Ma) in apatites from profiles G6 and G7, indirectly reflect the Jurassic tectonic setting of the basement. No metamorphic or igneous event is known at that time in the Congo craton, and cooling ages could, therefore, indicate uplift and erosion in connection with the initiation of rifting in southern Africa. Sediments were deposited in the Internal Basin during the Jurassic, but the subsequent thermal evolution in this area prevents any approach to provenance ages on a larger scale. Though little is known of the prerift history, we can nevertheless conclude that the Jurassic was probably a period of major tectonic activity in the basement.

Cretaceous Tectonothermal History

Early Cretaceous Cooling

Cooling during the rift 2 stage occurred at about 110 Ma, which corresponds to the early Aptian stratigraphic hiatus (Fig. 2-4). Nevertheless, due to complete track annealing in most rift sediments by Tertiary time, the Cretaceous thermal history is essentially known only from data obtained in Dentale Basin samples from profile G6, which may be irrelevant to other areas. In this connection, we stress that the lower Barremian Melania sandstones (Fig. 2-4) in well G7/3 (located to the east of the "200-m-water-depth hingeline"; Fig. 2-3) do not show any evidence of a significant rise in temperature or of a subsequent cooling during the rift stage (Fig. 2-10).

The thermal history in the Dentale Basin (Figs. 2-4, 2-5), as indicated by well G6 data, is summarized as follows:

- Subsidence rates were very high after 115 Ma, during the late Barremian overstretching stage. Com-

plete track annealing, at a distance shorter than 100 m below the Aptian unconformity, shows that upper Barremian–lower Aptian deposits were at least 2–3 km thicker than now observed. Considering that high sedimentation rates would prevent high geothermal gradients in spite of a presumed heat-flow increase, this estimate appears to be a minimum.

Uplift and 2–3 km of erosion is consistent with the structural setting, and should have occurred during a maximum time interval of ~ 1 m.y. considering the duration of the early Aptian hiatus at ~ 112–111 Ma (Reyre, 1984). An average uplift rate equal or higher than 2–3 mm/yr is likely.

The 160°C isotherm slowly deepened with respect to the Aptian unconformity between 110 Ma and 96–91 Ma in profiles G6 and G5. Actually, within the same time interval, evaporites initially 500–800 m thick (de Ruiter, 1979) and nearly 2,000 m of Albian sediments were deposited (Fig. 2-10). As a result, the apparent geothermal gradient decreased from an extreme value during the early Aptian (≥ 100°C/km during the first appearance of oceanic crust) to a more common value of about 40–45°C/km during the Albian. The origin of such a gradient soon after erosion is still speculative. It could result partly from a transient thermal regime related to the high uplift rate, but an additional heat source of hydrothermal origin cannot be ruled out. Alternatively, the gradient could have originated from the slow decay of heat flow subsequent to crustal stretching (McKenzie, 1978; Jarvis and McKenzie, 1980). Geological data provide further details on the Aptian erosion at 110 Ma. Seismic lines show clearly the drift onset unconformity at the bottom of the salt sequence (Fig. 2-5); the Aptian erosion corresponds to a general uplift of the continental domain (contrast with Chapter 6), whereas subsidence may have continued in the Atlantic domain (Coniquet sandstones and litharenites, Fig. 2-4). Unlike later erosion episodes, especially during the Neogene, the area of maximum uplift during the Aptian was the hinge zone (domain 3b on Fig. 2-3).

Late Cretaceous Cooling

Major uplift and erosion occurred during the Santonian in the North Gabon subbasin (contrast with Chapter 6). The temperature decrease at ~ 84 Ma in profile G2, from values greater than 160°C to values of about 130–140°C at most (maximum temperatures between 84 Ma and the early Miocene), would reflect uplift and erosion of about 1,000 m, assuming a geothermal gradient close to 30°C/km. The stratigraphic age of sediments removed may range from Albian to Turonian.

Regional uplift during the Santonian (Fig. 2-6) is also evidenced by the corrected ages of apatites from Senonian turbidites in profile G4 (Fig. 2-10). Apatite ages are similar to stratigraphic ages, which indicates that grains are inherited from deeply buried, completely annealed source regions ($T \geq 160°C$). Uplift and erosion in such source regions should be rapid. The provenance of clastic materials deposited to the west of the hingeline should be the innermost domains of the Senonian margin, which were uplifted higher than the domains near the eastern border of the hingeline (Fig. 2-6). Although these areas cannot be delin-

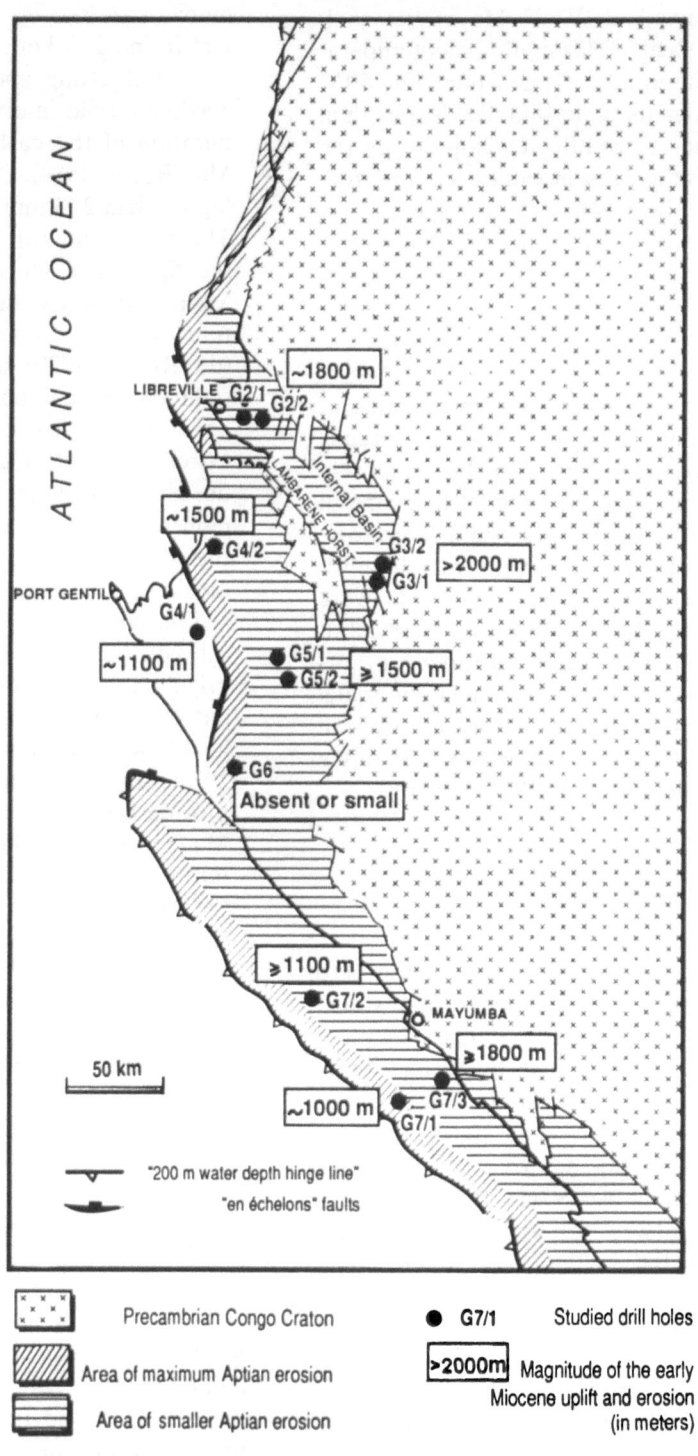

FIGURE 2-15.   Regional variation of magnitude of uplift and erosion during early Miocene (Burdigalian) in Gabon Basin, determined from fission-track chronothermometry. Total thickness of removed sediments decreases from craton border toward 200-m-water-depth hingeline, which approximates eastern limit of continuously subsiding domain during drift stage.

eated from available data, we conclude that the Santonian tectonism in North Gabon was probably a catastrophic event. It corresponds to a progressive change in the global plate motion initiated during magnetic anomaly 34 (84 Ma) and ended during the Maestrichtian (Fig. 2-4). Before anomaly 34, the extension direction in Gabon was N55°E, but it evolved into an E–W trend at the end of the Maestrichtian.

Tilting of the margin probably increased the uplift rate east of the hingeline. Scarce basalt occurrences in the offshore area near Port Gentil yield K-Ar ages of 78 ± 2 Ma (Walgenwitz and Pagel, 1990) and give further evidence of deep crustal fracturing during the Senonian.

Tertiary Tectonothermal History

In most of the profiles investigated, AFTC data show that cooling from high temperatures to values close to, or nearly identical to, present-day temperatures occurred during the Tertiary. Considering that the geothermal gradient decline cannot account for a sudden cooling, deep changes in the morphostructural characteristics of the margin must be the reason.

Paleogene
The middle Eocene cooling age (~ 45 Ma), which was obtained only in the uppermost samples from profile G3 in the Internal Basin (Fig. 2-10), is interpreted as the result of an uplift probably in connection with tilting of the margin. Eocene tectonism, illustrative of a worldwide geodynamic episode (Schwan, 1985; see also Chapter 6), is characterized west of the hingeline by the occurrence of the deep Animba erosion channels (Fig. 2-4). Nevertheless, very little is known of this tectonism. Temperatures were higher than 160°C until the Miocene in most samples from the coastal basin (Fig. 2-10), which does not allow precise determination of the regional distribution and magnitude of the Eocene cooling.

Neogene Cooling
The youngest basinwide stage of cooling occurred during the early Miocene. Ages range from 12 to 25 Ma (Table 2-2; Fig. 2-10), with a mean value of 22.5 ± 2 Ma (Burdigalian). Similar ages, both regionally and vertically, clearly indicate uplift and erosion. The lack of evidence for significant cooling during the early Miocene in profile G6 (Fig. 2-10) could be related to its location near the hingeline (Figs. 2-3, 2-5).

We estimated erosion by using a simple one-dimensional model, assuming a purely conductive system and vertical heat flow. Heat flow was assumed to be constant and similar to the present value ~ 70 mW m^{-2}). Other parameters considered were specific heat and thermal conductivity of the rock matrix and fluids, and

porosity variations due to burial. The lithological characteristics of eroded sediments were roughly extrapolated from offshore wells, where a more complete stratigraphic record exists. Temperatures were calculated in the different wells by adding or removing successive sedimentary strata, in order to obtain the best fit with AFTC data. Profile G3 was discarded due to the great uncertainty about the nature of the post-Barremian sedimentary cover (Fig. 2-10).

Results show that about 1–1.8 km of sediments were removed during the early Miocene uplift (Fig. 2-15). Probably more than 2 km was eroded in the Internal Basin, where sediments now located at a 600–700-m depth were subjected to a temperature higher than 160°C. The relative thickness of sediments removed is roughly dependent on distance from the hingeline and increases toward the east (Fig. 2-15). The stratigraphic position of missing strata remains speculative. Tertiary deposits may have existed in the area covered by profiles G2, G4, and G7/1, whereas thick deposits of indeterminate post-Aptian age should have been deposited in the Internal Basin. Deep erosional channels (Mandorove channels, Fig. 2-4) incised the slope beyond the hingeline that delineated the eastern boundary of the continuously subsiding Atlantic domain during the Neogene.

FLUID-INCLUSION INVESTIGATION

Among the profiles investigated for fission tracks, only sites G3 and G7 provided samples suitable for the microthermometric study of fluid inclusions. The lack, or at least the scarcity, of authigenic minerals apt to enclose observable inclusions, is the main reason for unavailing attempts at cross-checking the results obtained from different techniques. Additional data were, therefore, acquired from geological materials with more favorable diagenetic evolution: silicified carbonates with oil seepage and sphalerite occurrences cropping out north of Libreville (site G1), and lower Barremian arkosic arenites from profiles G8 in the South Gabon Basin and C1 in the Congo Basin (Fig. 2-9).

North Gabon

Sphalerite and quartz cements in silicified bioclastic carbonates of Early Cretaceous age (site G1, Fig. 2-16), contain two-phase aqueous and petroleum inclusions respectively. Both are primary, randomly distributed (Roedder, 1984), and homogenize within a similar temperature range (50–150°C). Decreasing frequencies on both sides of the maximum at 90–95°C, probably reflect the entrapment of inclusions throughout thermal evolution. No radiometric data are available to allow interpretation of these temperatures. Nevertheless, T_h values (which, on the other hand, are minimum esti-

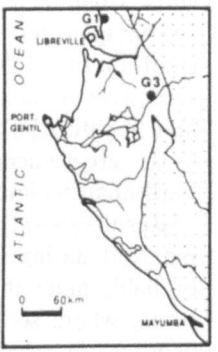

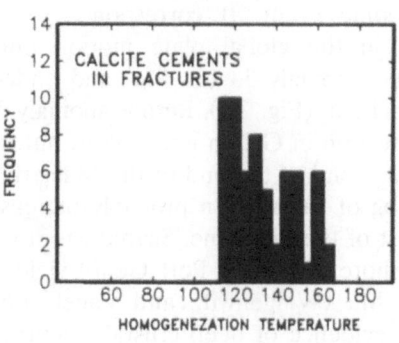

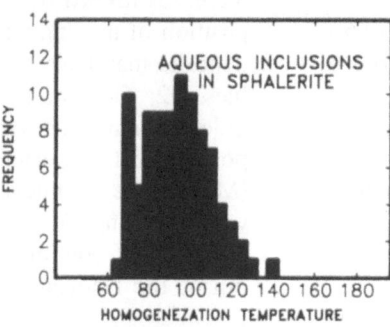

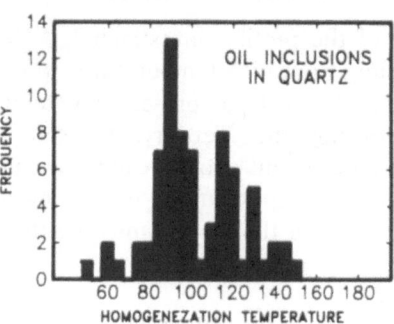

FIGURE 2-16.   Homogenization temperatures of aqueous and hydrocarbon fluid inclusions in authigenic minerals from site G1 and in northern Gabon subbasins.

mates of the fluid entrapment conditions) obviously indicate that outcropping host rocks were deeply buried at some time in the past. Uncertainties about the origin and characteristics of the process responsible for high fluid temperatures, restrict appraisal of the age and magnitude of uplift and erosion. Ice-melting temperatures of aqueous inclusions in sphalerite, ranging from −16 to −28°C, indicate high salt concentrations, which seem unlikely to have occurred before the Aptian evaporite sequence was deposited.

On the assumption that fluid salinities originated from salt leaching, the entrapment of inclusions would have thus taken place during subsidence in the drift stage. Cooling down to surface conditions should have occurred either during the Santonian, or more probably during the Burdigalian, as indicated by fission-track data from profile G2.

In the Internal Basin, calcite cements in narrow subvertical fractures at 1,050–1,053 m in well G2/2, enclose secondary two-phase aqueous inclusions distributed among healed microcracks. Homogenization takes place in the liquid phase between 110 and 155°C, with a mean value of about 130°C (Fig. 2-16). Deep burial until the Tertiary, evidenced from the fission-track investigation, suggests that homogenization temperatures are minimum estimates of conditions

prevailing at the time of trapping. Values as high as 155°C are consistent with complete track annealing before cooling at about 21 Ma.

South Gabon and Congo

The lower Barremian arkosic arenites are characterized by the widespread occurrence of authigenic feldspars and carbonates. Calcite and dolomite are of an early diagenetic origin (Walgenwitz and Pagel, 1990) and contain secondary fluid inclusions on healed microcracks. Authigenic feldspars occur either as adularia overgrowths or as recrystallized zones within the detrital K feldspars or plagioclases. Similar feldspar authigenesis has previously been observed in Barremian and Albian sediments from Angola (Walgenwitz et al., 1990).

Samples in seven wells, onshore and offshore, were taken from oil-, gas-, and water-bearing reservoirs between 1,463 and 3,694 m (Fig. 2-17). Three types of fluid inclusions have been identified in carbonates and feldspars:

1. Two-phase aqueous inclusions, widely distributed. The gas phase is estimated at about 5% of the bulk volume, but may amount to 20–25% in some inclu-

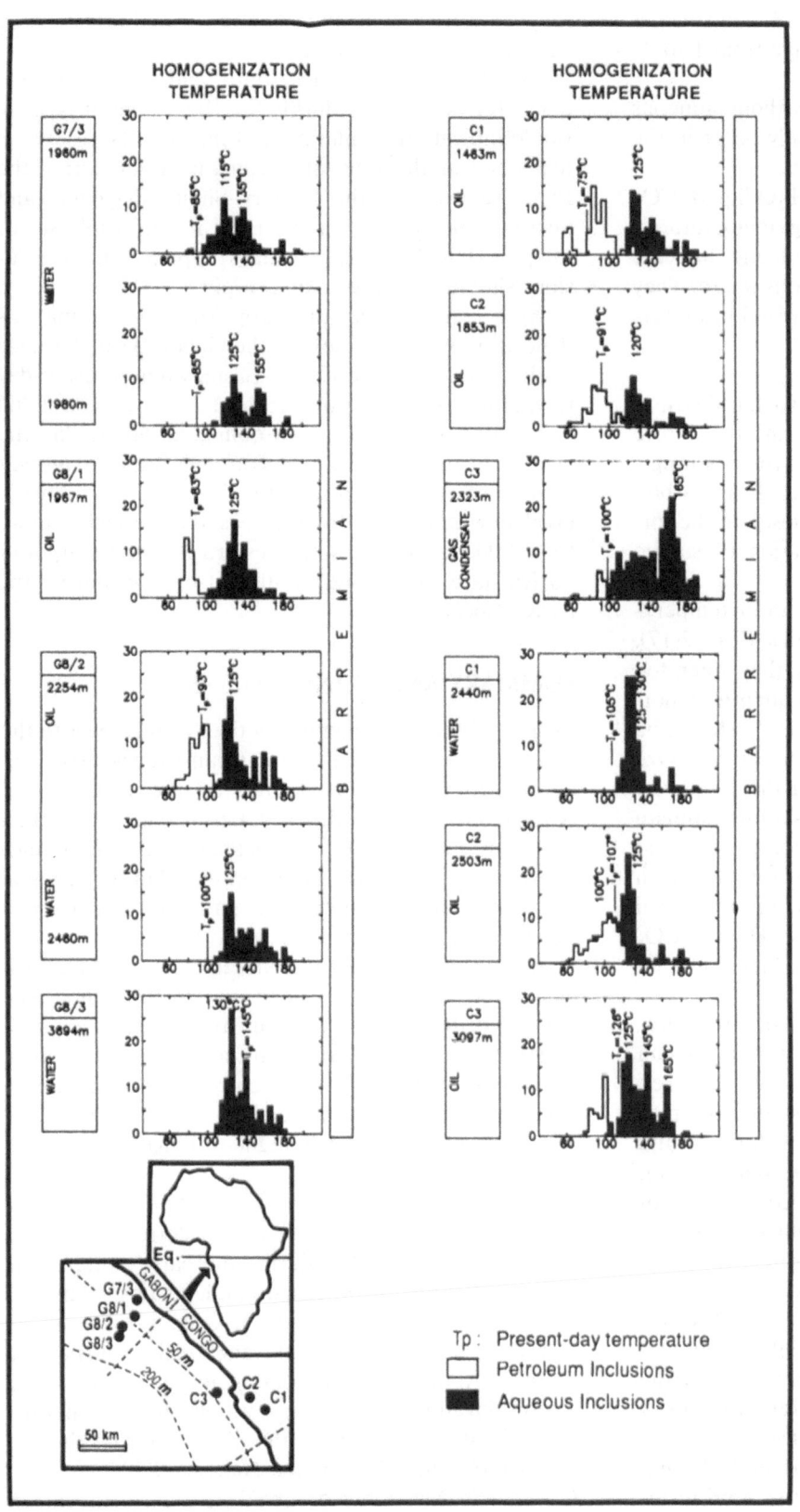

FIGURE 2-17.   Homogenization temperatures of aqueous and hydrocarbon fluid inclusions in Barremian reservoirs from southern Gabon and Congo.

sions from the gas reservoir of well C3. Salt concentrations are highly variable, and range from 1 to 23 wt.% eq. NaCl.

2. Two-phase petroleum inclusions, without aqueous component optically visible. They only occur in the oil-bearing reservoirs.

3. Three-phase inclusions containing H_2O liquid, CO_2 liquid, and CO_2 gas. Low melting temperatures, sometimes below the H_2O-NaCl eutectic, indicate high salt concentrations in the aqueous phase. They are strictly limited to the overpressurized gas reservoir (370 bars at 2,320 m).

The close relationship between the spatial distribution of the different types of inclusions and the nature of the continuous fluid phase in reservoirs, strongly suggests that entrapment is related to hydrocarbon migration. Fluid inclusions, therefore, postdate the formation of structural traps and maturation of source rocks in a late diagenetic stage.

Petroleum inclusions homogenize at lower temperatures than coexisting aqueous inclusions (Fig. 2-17). The system of two immiscible fluids that migrated simultaneously was, therefore, most frequently undersaturated with respect to a gas phase, and T_h should be corrected for pressure. Aqueous inclusions homogenize between 95 and 190°C, with a peak in the frequency distributions at 115–125°C, which yields a low estimate of the trapping temperatures. Secondary peaks at higher temperatures may occur in some samples, with the exception of well C3, where a maximum is observed at 160–165°C (Fig. 2-17). Three-phase $H_2O + CO_2$ inclusions failed to homogenize, due to decrepitation upon heating.

Greater frequency of high homogenization temperatures in a gas-bearing reservoir, such as in well C3, can be used as a key in interpreting the whole data set. Hanor (1980) has shown that CH_4 saturation may occur rapidly in highly saline aqueous solutions, and homogenization then takes place on the bubble-point line. Consequently, the temperature correction for pressure becomes negligible, or even unnecessary, and T_h yields a reliable estimate of the trapping temperature, which would be close to 160–165°C at 2,323 m in well C3. No direct detection of methane was possible, and whether or not the range of T_h in all the samples reflects a large spectrum of CH_4 concentrations or the entrapment of several populations at different stages of the thermal evolution, remains speculative.

Data from well C3 would support the first alternative. In this assumption, the entrapment of fluid inclusions would have occurred within a narrow range of temperatures, whose closest estimates would be given by greatest values of the secondary homogenization peaks. In well G7/3, temperatures of trapping were probably slightly greater than 155°C (Fig. 2-17) and thus are similar to temperatures required to account

for fission-track annealing before ~ 21 Ma in nearby samples (Table 2-2; Fig. 2-14).

The consistency of these data, and the lack of evidence for deep burial during the Barremian or Aptian, would indicate that migration of fluids in the reservoirs happened at the time of maximum burial during the Late Cretaceous or the Tertiary. Similar diagenetic and microthermometric characteristics in the whole set of samples (Fig. 2-17) suggest that trapping of fluid inclusions should have a common origin.

As a consequence, the early Miocene cooling evidenced from fission tracks in the Gabon Basin, has also likely occurred in the Congo Basin, where present-day temperatures range from 75 to 125°C between 1,460 and 3,100 m (Fig. 2-17). Cooling at about 25 Ma, evidenced from a fluid inclusion and ^{39}Ar-^{40}Ar investigation of Albian reservoirs in northern Angola (Walgenwitz et al., 1990), shows that probably more than 1,000 km of the southwestern Atlantic margin of Africa have undergone uplift and erosion during the early Miocene.

SUMMARY AND CONCLUSIONS

In spite of the small number of data with respect to the large area investigated, a great consistency exists between geological events known on the southwestern African margin and the key periods in its thermal evolution. Furthermore, it appears that chronothermometric data also allow a better appraisal of the regional distribution and magnitude of structural and depositional/erosional phenomena.

Among the principal tectonosedimentary stages evidenced on the Gabon–Congo margin throughout the rift and drift periods, only four are synchronized with drastic temperature changes in the continental domain of the basin (Fig. 2-18). They correspond to successive major events of uplift and erosion, spatially organized with respect to the so-called 200-m hingeline in South Gabon:

1. ~ 110 Ma. Regional erosion occurred along the hinge zone (Fig. 2-3) during the early Aptian. The unusually high geothermal gradients observed after uplift slowly decayed until the Albian. Easternmost areas do not show significant cooling during the early Aptian, thus indicating lower erosion rates and invalidating the assumption that geothermal gradients were high in this region during the rift stage (Giroir, Merino, and Nahon, 1989).

2. ~ 84–76 Ma. Only observed in the North Gabon Basin, the Senonian tectonism, in connection with the geodynamic change at 84 Ma (chron 34), is expressed as catastrophic uplift and erosion in some inner areas of the margin. Less erosion near the North Gabon hingeline suggests a first westward tilt. Upper Cretaceous turbidites and some volcanic

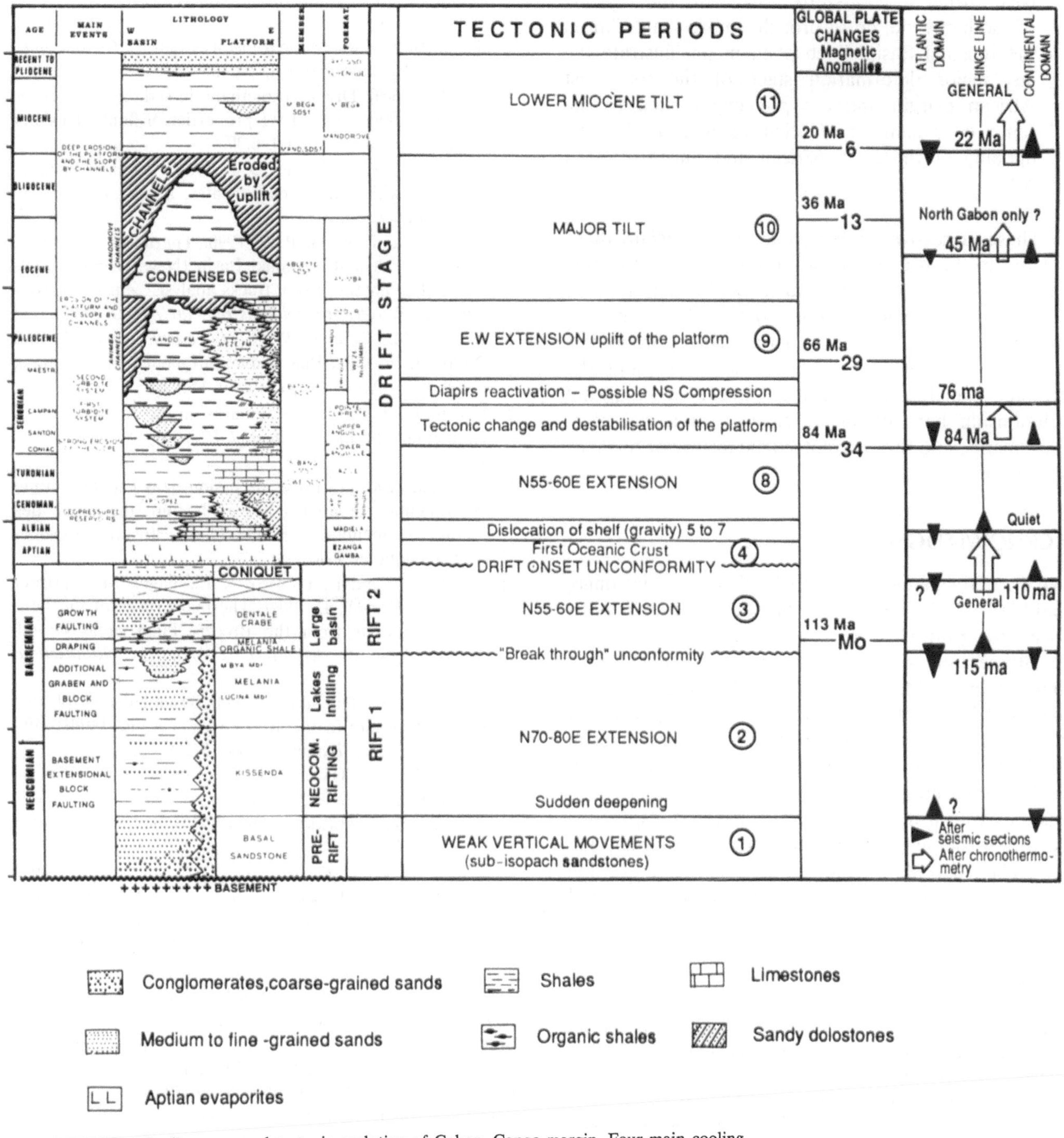

FIGURE 2-18. Cooling ages and tectonic evolution of Gabon–Congo margin. Four main cooling stages evidenced at 110 Ma, 84–76 Ma, 45 Ma, and 22 Ma on continental domain consistently agree with major tectonosedimentary features of margin. Aptian and early Miocene uplift and erosion are fairly well understood on regional scale; distribution and characteristics of Senonian and Eocene thermal events remain poorly documented. Numbers 1–11 refer to tectonosedimentary features shown in Fig. 2-5.

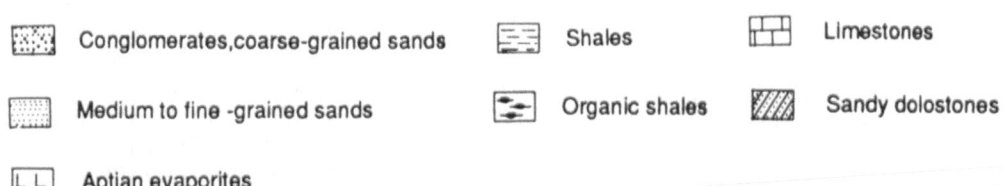

Conglomerates, coarse-grained sands   Shales   Limestones

Medium to fine-grained sands   Organic shales   Sandy dolostones

Aptian evaporites

events are related to the Santonian tectonism. This stage ends with a possible N–S compression, better known northward in Nigeria, where metamorphism and folding in the Abakaliki region is also of Santonian age (Benkhelil, 1986).

3. ~ 45 Ma. The middle Eocene tectonothermal event is still poorly known. Its occurrence in the Gabon Internal Basin is synchronous with incision of the Animba channels on the platform.

4. 25–21 Ma. A complete morphostructural modification giving the margin its present shape. Further westward tilt is shown from the marked temperature change within a short time in the continental domain. Uplift and erosion increase toward the

east, from a near-zero value in the vicinity of the hingeline, up to probably more than 2 km in the Internal Basin. Such phenomena illustrate the last major deformation stage of the southwest African margin and are probably related to the same geodynamic revolution responsible for the opening of the East African and Suez–Red Sea rifts.

These successive tilting periods, and especially those related to the drift stage, still do not explain the past sedimentary record in inner areas of the margin. Considerable improvement in the understanding of margin history should be expected from a further chronothermometric investigation of both the basins and the craton. Available data strongly indicate a significant relationship between major changes in global plate movements, and the thermotectonic history of this passive margin.

ACKNOWLEDGMENTS

We thank the management of ELF Gabon, ELF Congo, and SNEA(P) for supporting this study and for permission to publish. We gratefully acknowledge D. Storzer and M. Selo who performed fission-track analyses in the Museum d'Histoire Naturelle de Paris, and provided helpful discussions. E. George performed most of the the microthermometric measurements in the ELF Aquitaine laboratory. We are also grateful to G. Vasseur who performed numerical modelling. Discussions with a number of colleagues, and, especially, A. Meyer, helped to clarify the views on thermal history. Thanks to M. J. Lamanou, who helped in drawing some of the figures, and Mrs. Amespil who proofread the English version.

REFERENCES

Benkhelil, J. 1986 Structure et évolution géodynamique du bassin intracontinental de la Benoue (Nigeria). Thesis, Université de Nice.

Bigazzi, G. 1967. Length of fission tracks and age of muscovite samples. *Earth and Planetary Science Letters* 3:434–438.

Burruss, R. C., Cerone, K. R., and Harris, P. M. 1983. Fluid inclusions petrography and tectonic-burial history of the Al Ali no. 2 well: geohistory analysis and timing of oil migration, northern Oman Foredeep. *Geology* 11:567–570.

Giroir, G., Merino, E., and Nahon, D. 1989. Diagenesis of Cretaceous sandstone reservoirs of the South Gabon rift basin, West Africa: mineralogy, mass transfer and thermal evolution. *Journal of Sedimentary Petrology* 47:482–493.

Gleadow, A. J. W., Duddy, I. R., and Lovering, J. F. 1983. Fission track analysis: a new tool for the evaluation of thermal histories and hydrocarbon potential. *Australian Petroleum Engineering Association* 23:93–102.

Gleadow, A. J. W., Duddy, I. R., Green, P. F., and Lovering, J. F. 1986. Confined fission track lengths in apatite: a diagnostic tool for thermal history analysis. *Contributions to Mineralogy and Petrology* 94:405–415.

Hanor, J. S. 1980. Dissolved methane in sedimentary brines: potential effect on the PVT properties of fluid inclusions. *Economic Geology* 75:6.03–6.09.

Jarvis, G. T., and McKenzie, D. P. 1980. Sedimentary basin formation with finite extension rates. *Earth and Planetary Science Letters* 1:42–52.

Kamp, J. J., and Green, P. F. 1990. Thermal and tectonic history of selected Taranaki Basin (New Zealand) wells assessed by apatite fission tracks analysis. *American Association of Petroleum Geologists Bulletin* 74:1401–1419.

Laslett, G. M., Kendall, W. S., Gleadow, A. J. W., and Duddy, I. R. 1982. Bias in measurement of fission track distribution. *Nuclear Tracks* 6:79–85.

McKenzie, D. P. 1978. Some remarks on the development of sedimentary basins. *Earth and Planetary Science Letters* 40:25–32.

Meyer, A. J. 1990. Les traces de fission dans l'apatite: étude expérimentale et application a l'histoire thermique des bassins sédimentaires. Ph.D. Thesis, Institut National Polytechnique de Lorraine.

Meyer, A. J., Landais, P., Brosse, E., Pagel, M., Carisey, J. C., and Krewdl, D. 1989. Thermal history of the Permian formations from the Breccia Pipes area (Grand Canyon Region, Arizona). *Geologische Rundschau* 78(1):427–438.

Milani, E. J., and Davison, I. 1988. Basement control and transfer tectonics in the Reconcavo–Tucano–Jatoba rift, Northeast Brazil. *Tectonophysics* 154:41–70.

Naeser, C. W. 1979. Thermal history of sedimentary basins in fission track dating of subsurface rocks. *Society of Economic Paleontologists and Mineralogists Special Publication 26*, 109–112.

Naeser, N., Naeser, C. W., and McCulloh, T. H. 1990. Thermal history of rocks in southern San Joaquin Valley, California: evidence from fission track analysis. *American Association of Petroleum Geologists Bulletin* 74(1):13–29.

Narr, W., and Burruss, R. C. 1984. Origin of reservoir fractures in Little Knife Field, North Dakota. *American Association of Petroleum Geologists Bulletin* 62:1087–1100.

Omar, G. I., Steckler, M. S. Buck, W. R., and Kohn, B. P. 1989. Fission track analysis of basement apatites at the western margin of the Gulf of Suez rift, Egypt: evidence for synchroneity of uplift and subsidence. *Earth and Planetary Science Letters* 94:316–328.

Poty, B., Leroy, J., and Jachimowicz, L. 1976. A new device for measuring temperatures under microscope: the Chaizmeca microthermometry apparatus. *Bulletin Societe Francaise Mineralogie et Cristallographie* 99:173–178.

Reyre, D. 1984. Caracteres pétroliers et évolution géologique d'une marge passive. Le cas du bassin Bas Congo–Gabon. *Bulletin Centres Recherch Exploration Production ELF Aquitaine* 6:303–332.

Roedder, E. 1984. Fluid Inclusions. Mineralogical Society of America, P. H. Ribbe, Ed. *Reviews in Mineralogy* 12:644.

Ruiter P. A. C. de. 1979. The Gabon and Congo basins salt deposits. *Economic Geology* 74:419–431.

Schwan, W. 1985. The worldwide active Middle/Late Eocene geodynamic episode with peaks at ± 45 and ± 37 Ma, and implications and problems of orogeny and sea floor spreading. *Tectonophysics* 115:197–234.

Storzer, D. 1970. Fission track dating of volcanic glasses and the thermal history of rocks. *Earth and Planetary Science Letters* 8:55–60.

Storzer, D., and Selo, M. 1984. Towards a new tool in hydrocarbon resource evaluation: the potential of the apatite fission track chrono-thermometer. *In* Thermal Phenomena in Sedimentary Basins, B. Durand, Ed.: Paris: Technip Editions, 89–110.

Storzer, D., Wagner, G. A., and King, E. A. 1973. Fission track ages and stratigraphic occurrence of Georgia tektites. *Journal of Geophysical Research* 78:4915–4919.

Teisserenc, P., and Villemin, J. 1990. Sedimentary basin of Gabon. Geology and oil systems. *In* Divergent/Passive Margin Basins, J. D. Edwards and P. A. Santogroosi, Eds. *American Association of Petroleum Geologists Memoir 48*, 117–199.

Walgenwitz, F., and Pagel, M. 1990. Diagenesis of Cretaceous sandstone reservoirs of the South Gabon basin: Mineralogy, mass transfer, and thermal evolution. Discussion. *Journal of Sedimentary Petrology* 60:471–476.

Walgenwitz, F., Pagel, M., Meyer, A., Maluski, H., and Monie, P. 1990. Thermo-chronological approach to reservoir diagenesis in the offshore Angola basin: a fluid inclusion, ^{40}Ar-^{39}Ar and K-Ar investigation. *American Association of Petroleum Geologists Bulletin* 74:547–563.

II
Early Postrift Evolution

3

Antarctic Continental Margin: Geologic Image of the Bransfield Trough, An Incipient Oceanic Basin

Juan Acosta, P. Herranz, J. L. Sanz, and Elazar Uchupi

On the landward side of some Atlantic continental rises, two principal sedimentary sequences are separated by an unconformity (the breakup unconformity), which is associated with the onset of seafloor spreading. Below the unconformity is a synrift sequence deposited atop an extended continental crust that forms the foundation of the landward margin of the present continental rise. This attenuated basement is bounded on the seaward side by an oceanic crust, which was emplaced by seafloor spreading processes. The boundary between these crustal units displays various geometries, ranging from a gradational contact, to a landward-facing scarp, to a volcaniclastic wedge of seaward-dipping reflections, to a linear high (Uchupi, 1989; see also Chapter 1). Associated with the lower synrift unit is a series of tilted fault blocks and transverse strike-slip faults, which may represent transform structures (Montadert et al., 1979; Klitgord, Hutchinson, and Schouten, 1988; see also Chapter 11). Initially, rifting was highly diffused, but with time it became concentrated in marginal basins on what now is the lower continental slope–upper continental rise region, seaward of a hinge in the continental basement (Dunbar and Sawyer, 1989). For example, in Labrador, Greenland, and Newfoundland, the rift originally varied in width from 100 to 1,460 km; in the North Atlantic, it varied from 240 to 700 km. Continental attenuation was later constrained within a zone of 60 km between Greenland and Labrador, of more than 600 km between Newfoundland and Europe, of 225 km between North America and northwest Africa, and of more than 600 km between southeastern North America and Senegal, Africa (Dunbar and Sawyer, 1989). Such a narrowing of crustal attenuation is compatible with the

model described by Lin and Parmentier (1990). Prior to seafloor spreading, the rift topography had a relief of 1,500 to 2,000 m in the northern Bay of Biscay, 2,750 to 2,900 m off northwest Iberia, 2,000 m along the southern edge of the Galicia Plateau, and 3,000 m off the Grand Banks of Newfoundland (Montadert et al., 1979; Sibuet, and Ryan, 1979; Keen and Barrett, 1981; Moullade, Brunet, and Boillot, 1988; see also Chapter 1). According to Jansa (1986), the Early Jurassic rift trough in the North Atlantic was approximately 500 km wide, about 1,700 km long and probably less than 1,000 m deep. The Early Cretaceous rift trough in the equatorial Atlantic probably was even shallower, because sediments deposited during the transition from rifting to drifting are indicative of outer-shelf or shelf-break depths (The Shipboard Scientific Party, 1978).

Sedimentation in the highly unstable rift system reflects topographic relief of the rift, changes in configuration of the rift with time, paleoclimatic changes due to latitudinal migration of the underlying plates, changes in the CCD (calcite compensation depth), and oceanic circulation (Uchupi and Emery, 1991). The few data available indicate that rotation of fault blocks along the listric faults stopped at different times in different areas; rotation probably created voids to be filled by slumping or detachment faulting within the half-graben fill (Mann, 1989). Volcanism was pervasive throughout the rift system, and volcanic sequences are related to particular extension events (Uchupi, 1989). Sediment facies patterns in the topographic lows were quite variable and discontinuous as a result of the complex fabric of the rift system. As the rift subsided and was flooded by marine waters, continental sediments gave way to evaporites, which in turn were replaced by carbonates in some areas when circulation became less restricted. As the rift continued to sink, shallow-water sediment facies gave way to deep-water turbidites. Where conditions were ideal, as in the equatorial Atlantic, the topographic lows were sites of anoxic sedimentation. In the South Atlantic, where paleoclimatic conditions precluded the deposition of evaporites or carbonates, only terrigenous sediments accumulated. As rifting gave way to seafloor spreading, much of the Atlantic rift system was affected by a massive igneous event. The resultant flux of excessive heat led to uplift, which was followed by erosion of the flanks of the rift system, and the deeper parts of the system became sites of turbidite deposition. Where resedimented deposits were trapped in marginal topographic lows formed during the thermal uplift, the rifts became sites of nondeposition (Uchupi and Emery, 1991).

Above the breakup unconformity is the continental rise depositional sequence, which, except for deformation due to the plastic flow of evaporites from beneath the unconformity, is generally tectonically undisturbed. The rise sequence reflects depositional conditions maintained since the onset of seafloor spreading (see Chapters 4–7). Sampling by the Deep Sea Drilling Project and the Ocean Drilling Program, coupled with an extensive grid of seismic reflection and refraction measurements, has yielded much information concerning the transition from synrift to drift (postrift) deposition and emplacement of the continental-rise wedge, but numerous features and processes need clarification. Unfortunately, geologic events associated with early phases of seafloor spreading are obscured by more recent ones. Furthermore, in the older segments of the Atlantic, the geological transition is so deeply buried that it is either beyond the reach of present-day drilling technology or too expensive to sample. One way of understanding more fully these poorly documented events, is to look at a present-day analog. We

will briefly describe the geologic setting of the seismically active Bransfield Trough, which separates the South Shetland Islands from the Antarctic Peninsula, where such a transition can be documented (Fig. 3-1). In contrast to the Atlantic rift system, which was affected by paleoclimates ranging from arid to glacial, the whole paleoclimatic history of the Bransfield Trough was one of glacial or near-glacial conditions.

TECTONIC SETTING

The Antarctic Peninsula and the South Shetland Islands had similar geologic histories up to the Jurassic, a record of subductions of their Pacific margins extending back into the Paleozoic (Barker and Dalziel, 1983). The peninsula and islands were morphologically contiguous with the southern Andes Mountains in the Mesozoic, but were separated during formation of the Scotia Sea in the Teritary (Dalziel and Elliot, 1973). Opening of Drake Passage and most of the western Scotia Sea took place between 28 and 6 Ma, when the North and South Scotia ridges separated from southernmost South America. As a result of this opening, continental fragments making up the South Scotia Ridge were transported eastward from their original positions adjacent to the northeastern end of the Antarctic Peninsula. The South Shetland Trench, northwest of the islands (Fig. 3-2), is a surviving segment of a subduction zone that originally extended the length of the Antarctic Peninsula. Subduction in the trench ceased as segments of a mid-ocean ridge, Aluk Ridge, collided with the peninsula. Magnetic anomalies indicate that the time the spreading axis reached the trench varies from 50 Ma to the southwest to 4 Ma to the northeast (Herron and Tucholke, 1976; Barker, 1976). Continued oblique subduction to the northeast, at the same time that it ceased to the southwest, led to extensional formation of Boyd Strait (Fig. 3-2) along the boundary between contrasting tectonic regimes

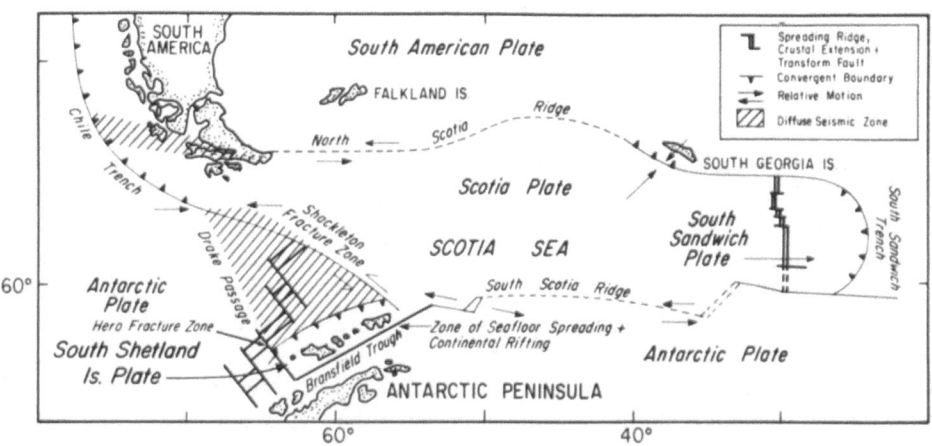

FIGURE 3-1. Tectonic setting of Bransfield Trough and vicinity. Modified from Pelayo and Wiens (1989, Fig. 11).

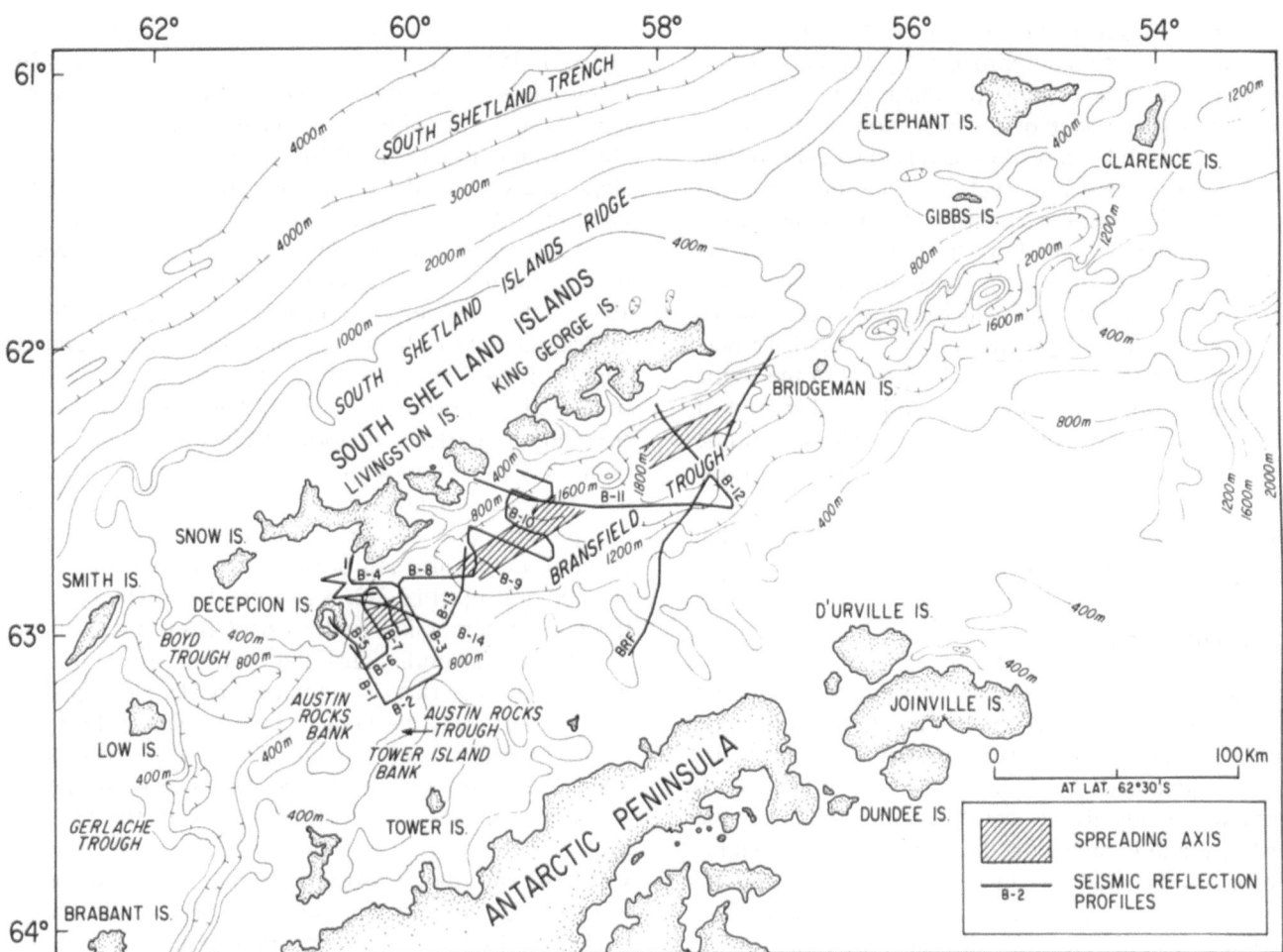

FIGURE 3-2.   Bathymetric chart of Bransfield Trough showing positions of incipient spreading axis and seismic reflection profiles recorded during the present investigation. Chart modified from Ashcroft (1972, Fig. 2).

(Jeffers, Anderson, and Lawler, 1991). Subduction, which was coupled to spreading in Drake Passage, is assumed to have stopped at 4 Ma (Barker, 1976). The northeast segment, where subduction ceased about 4 Ma displays a topographic trench; this segment is the site of late Tertiary volcanism on the South Shetland Islands and of the Bransfield Trough, a back-arc basin. The trough contains several Holocene volcanoes (Deception, Bridgman, and Penguin Islands, and probably the seamounts between Bridgeman and Deception Islands) and a prominent mid-trough magnetic anomaly associated with a line of submerged peaks (Weaver, Saunders, and Tarney, 1982; Barker and Daziel, 1983; Gonzáles-Ferrán, 1991). This large positive anomaly has been interpreted by Roach (1978) to be the result of narrow strips of magnetized material emplaced along the flank of the trough. Roach also stated that the high degree of disorder observed along the flanks of Bransfield Trough reflects the immaturity of the spreading system. That the trough is an extensional feature also is suggested by seismic refraction measurements

(Ashcroft, 1972; Guterch et al., 1991). Whereas the crust of the South Shetland Islands is 30–35 km thick and the coast of the Antarctic Peninsula crust is 38–45 km thick, the crust of Bransfield is anomalous; a seismic discontinuity of 7.0–7.2 km/sec is present at a depth of 10 km, and a second discontinuity of 7.6–7.7 km/sec is at present 20–25 km (Guterch et al., 1991). According to Barker and Dalziel (1983) and Jeffers, Anderson, and Lawler (1991), the trough supposedly began to open about 4 Ma in response to cooling and sinking of the subducted oceanic crust; the change from rifting to seafloor spreading took place during the last 1–2 m.y. (Roach, 1978). Since its inception in the Pliocene, the Bransfield Trough has opened between 5 and 15 km, at a rate of 0.25–0.75 cm/yr (González-Ferrán, 1991). Assuming that the depositional sequences identified in the Bransfield Trough are correlative with the five Pilocene and Pleistocene third-order eustatic cycles described by Haq, Hardenbol, and Vail (1988), Jeffers and Anderson (1990) proposed that subsidence preceded back-arc rifting, that subsidence began about 3 Ma, and that seafloor spreading was initiated approximately 1.6 Ma. If subsidence, rifting, and seafloor spreading in the Bransfield Trough are the result of the cessation of mid-ocean ridge-driven subduction, then extension in the trough should be short lived, as such a driving mechanism is incapable of sustaining extension far into the future (Jeffers, Anderson, and Lawver (1991). This scenario, however, is contradicted by focal mechanisms from earthquakes in the region, which indicate that subduction is continuing along the South Shetland Trench. Underthrusting is largely aseismic because of the young age of the plate being subducted and the slow rate of subduction (Pelayo and Wiens, 1989). The magnitude and depth of normal faulting in the southern Bransfield Trough also is indicative of diffuse extension (typical of transitional or continental crust) rather than of an organized spreading axis.

MORPHOLOGY

Bransfield Trough is asymmetrical in cross section, having a steep northwest slope and a gently sloping southeast margin (Fig. 3-2). The northwest side of the trough is dominated by the northeast-trending South Shetland Islands Ridge, whose crest is tilted to the northwest and is capped by the South Shetland Islands. Except for the Gibbs Islands, the South Shetland Islands are aligned along the southeast edge of the ridge; the Gibbs Islands are atop an east–west ridge, which Ashcroft (1972) described as a fault-bound block. Northwest of the South Shetland Islands is a broad, smooth, featureless, 40- to 60-km-wide shelf with a shelfbreak at 500 m; southeast of the islands, the continental shelf is narrow or not well developed (Fig. 3-2; Ashcroft, 1972). The continental slope northwest of the South Shetland Islands leads down to the South Shetland Trench, whose depth is more than 5,100 m; this slope displays only minor topographic irregularities. In contrast, the slope southeast of the South Shetland Islands is steep and rough and is disrupted by sharp, high-relief peaks and canyons, which form seaward extensions of insular bays (Jeffers and Anderson, 1990). One of these peaks (on the slope) rises above sea level to form Penguin Island, a recent, uneroded scoria cone south of King George Island (Barker and Daziel, 1983). This steep linear scarp is believed to be of fault origin. Normal faulting, but in a broader zone, also underlies the more gradual southeast slope (Ashcroft, 1972). The southeast side of Bransfield Trough is characterized by a broad shelf with a 250-m-deep shelfbreak. Rising above its surface are islands and numerous rocky shoals, which indicate that the shelf has a relatively thin cover of recent sediment (Ashcroft, 1972). The gentler continental slope on the southeast side of the trough consists of two segments

separated by a broad platform, which extends from a depth of 750 m to 900 m. The slope segment seaward of the platform, which descends to the floor of Bransfield Trough, is quite steep, having a declivity of 9° (Jeffers and Anderson, 1990). Incised on the shelf and slope are U-shaped canyons, which can be traced to near the 1,000-m isobath. Their shapes suggest that they were glacially carved, their orientation having been structurally controlled (Jeffers and Anderson, 1991; Jeffers, Anderson, and Lawver, 1991). The floor of Bransfield Trough, at the foot of the South Shetland Islands Ridge, consists of three narrow depressions separated by northwest-trending highs, one of which rises above sea level to form Bridgeman Island. The floor of the eastern depression, between Bridgeman and Clarence islands is irregular; several seamounts are present, one having a relief of 1,800 m. Three separate basins also are present, the deepest of which is 2,800 m deep (Fig. 3-2; Ashcroft, 1972). The floor of the central depression southwest of Bridgeman Island is relatively smooth. This flat bottom is disrupted, however, by a series of submarine peaks along a line between Bridgeman Island (a young volcanic feature; González-Ferrán and Katsui, 1970), and Deception Island (a composite stratovolcano), which displayed renewed activity in 1967–70 (Baker et al., 1975). The presence of magnetic anomalies over these topographic features, and their association with known volcanic islands, indicate that they also are of volcanic origin (Ashcroft, 1972; Acosta et al., 1989 Jeffers, Anderson, and Lawver, 1991). The western depression shoals abruptly from 1,300 to 800 m at its western limit, and branches southwestward into Gerlache Trough and northwestward into Boyd Trough (Fig. 3-2).

GEOLOGIC PROCESSES

The seismic reflection profiles (Figs. 3-3 to 3-8) displaying the results of geologic processes active in Bransfield Trough were obtained with a sparker having a capacity of 4,500 or 8,000 J. It was fired at 3- to 6-sec intervals and discharged via 99 electrodes to obtain the best resolution possible. Signals were received via a 24-element hydrophone single-channel array; they were amplified, filtered between 70 and 700 Hz, and recorded on a EPC 3200 dry paper recorder (Acosta et al., 1989). Navigation was achieved by GPS (global positioning system) and satellite.

All seismic reflection profiles crossing the southeastern flank of the South Shetland Islands Ridge display a patchy sedimentary apron resting unconformably on acoustic basement, which is broken by faults into a series of horsts and grabens; sediments are thickest in the structural lows (Profile B-11, Figs. 3-3, 3-6 and Profile B-7, Figs. 3-4, 3-7). Intrusions by dikes and plugs have added to the structural complexity of the

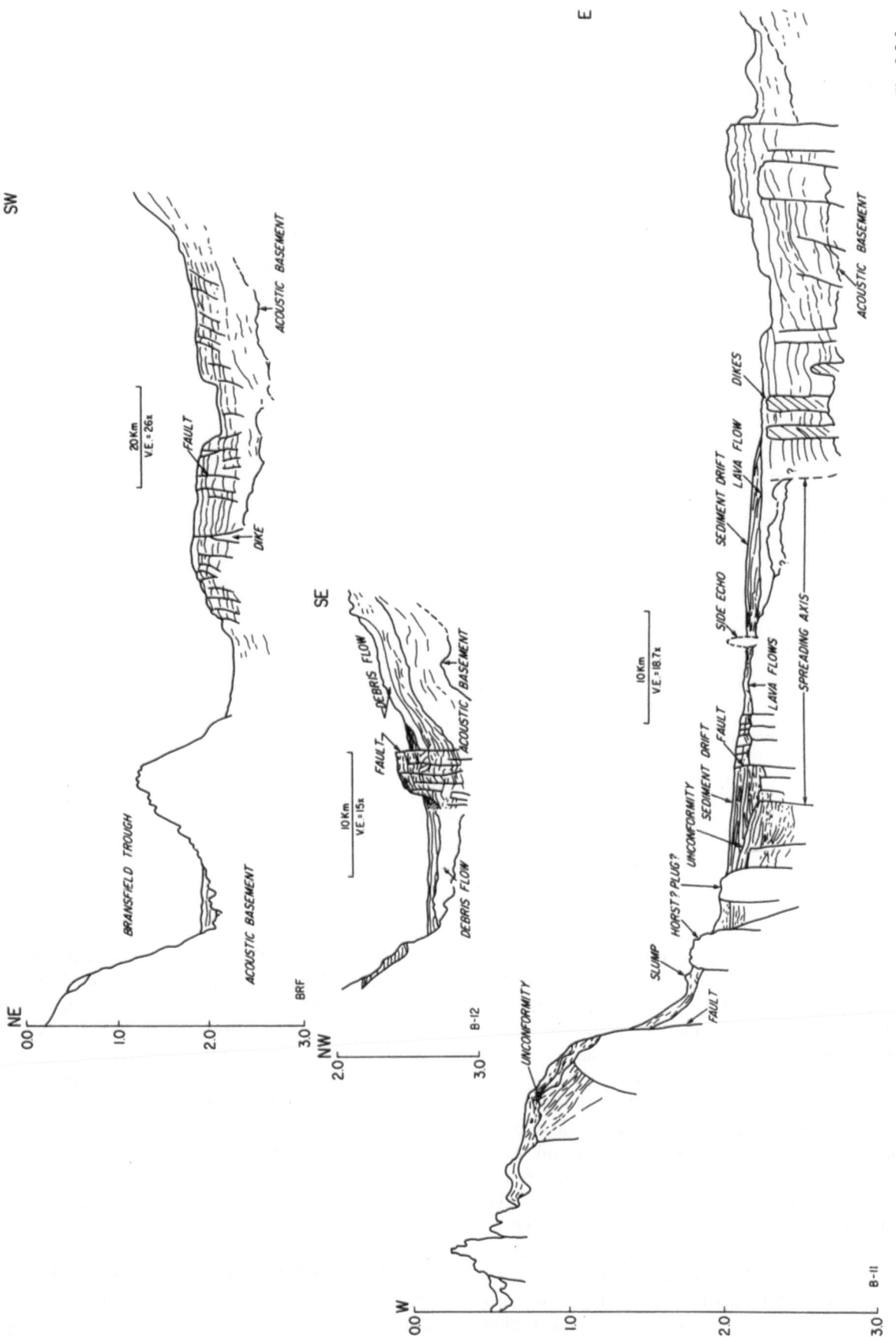

FIGURE 3-3. Tracings of single-channel seismic reflection profiles BRF, B-12, and B-11. Vertical exageration indicated is based on a water velocity of 750 m/sec. See Fig. 3-2 for locations of profiles. Compare with photographs in Fig. 3-6.

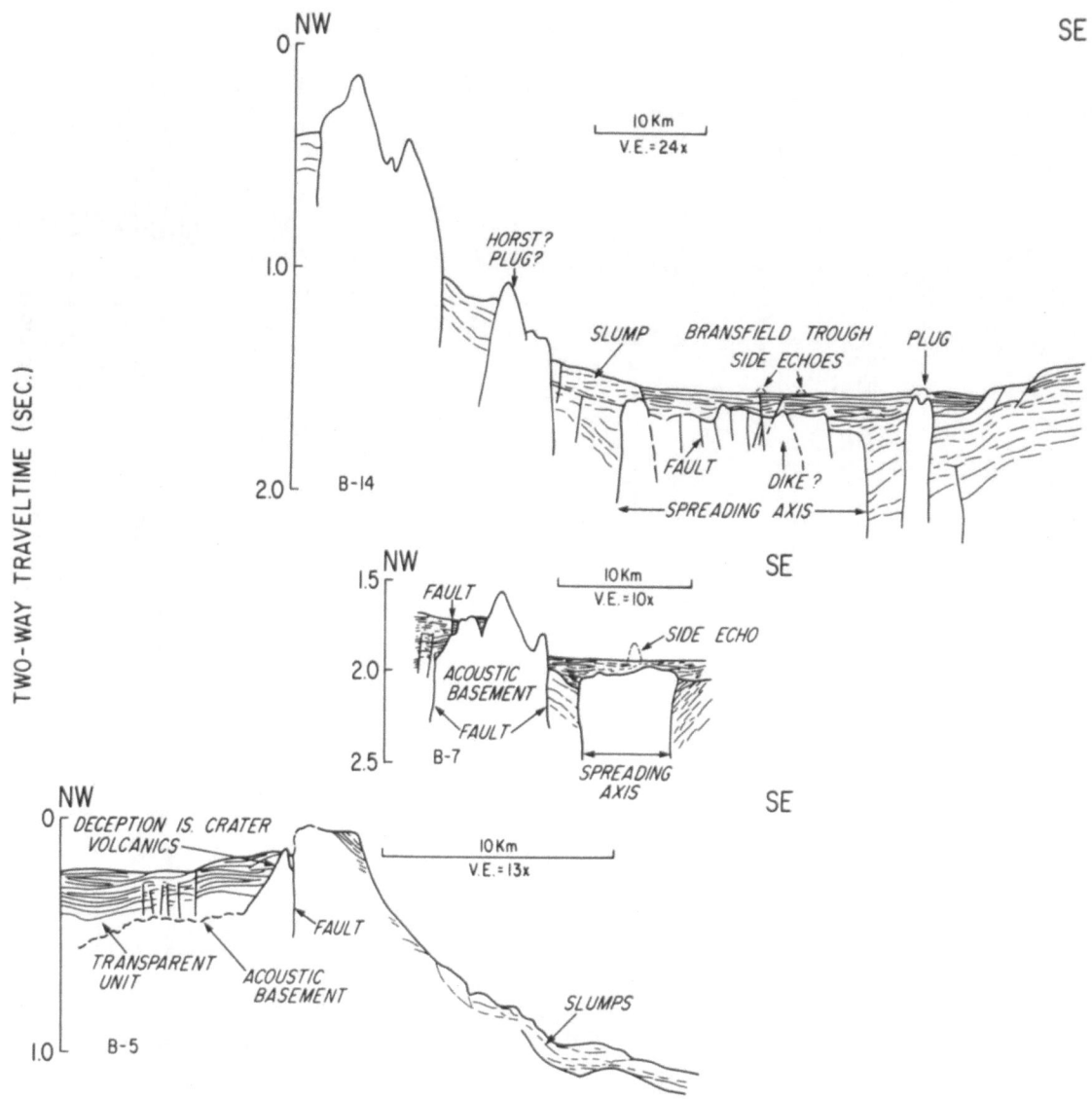

FIGURE 3-4.   Tracings of single-channel seismic reflection profiles B-14, B-7, and B-5 Vertical exaggeration indicated is based on a water velocity of 750 m/sec. See Fig. 3-2 for locations of profiles. Compare with photographs in Fig. 3-7.

scarp. The sharp transition from the ridge to the floor of the Bransfield Trough indicates that the linear southeastern edge of the ridge is fault-line scarp. Although basement was not sampled, we assume that it is composed of the same Mesozoic–Cenozoic rocks and Cenozoic volcanics exposed on the islands. Along Profile B-11 (Figs. 3-3, 3-6) the nearly 0.5-sec-thick sedimentary section on a horst is divided by an unconformity; along Profile B-7 (Figs. 3-4, 3-7) the fill in a structural low is disrupted by faulting. Gravitational processes have played a major role in depositing the unconsolidated sediment cover on the ridge. These processes are evidenced by debris flows and slumps recognized in many of the profiles, such as the strike Profile B-6 (Fig. 3-5). Debris-flow and slump facies also have been recovered in cores taken from the region (Jeffers and Anderson, 1991). Sediment fill in the

breached crater of Deception Island (Profile B-5, Figs. 3-4, 3-7) probably is composed of volcaniclastics; these sediments rest on an acoustic basement, which Ashcroft (1972) attributed to subsidence *en bloc*. Recent investigations show, however, that the crater has an orthogonal fracture system, more compatible with extensional rifting than with simple collapse of a pre-existing structure (Marti, Baraldo, and Rey, 1989).

The gentler southeastern flank of Bransfield Trough is formed by sediments displaying several depositional sequences separated by recognizable unconformities. Along profiles BRF and B-11 (Figs. 3-3, 3-6), the strata are disrupted by closely spaced faults, but along Profiles B-12 and B-14 (Figs. 3-3, 3-4, 3-6, 3-7) sediments are relatively undeformed. The seafloor, particularly along profiles BRF and B-11, is entrenched by flat-bottomed submarine canyons; along Profile B-12 a

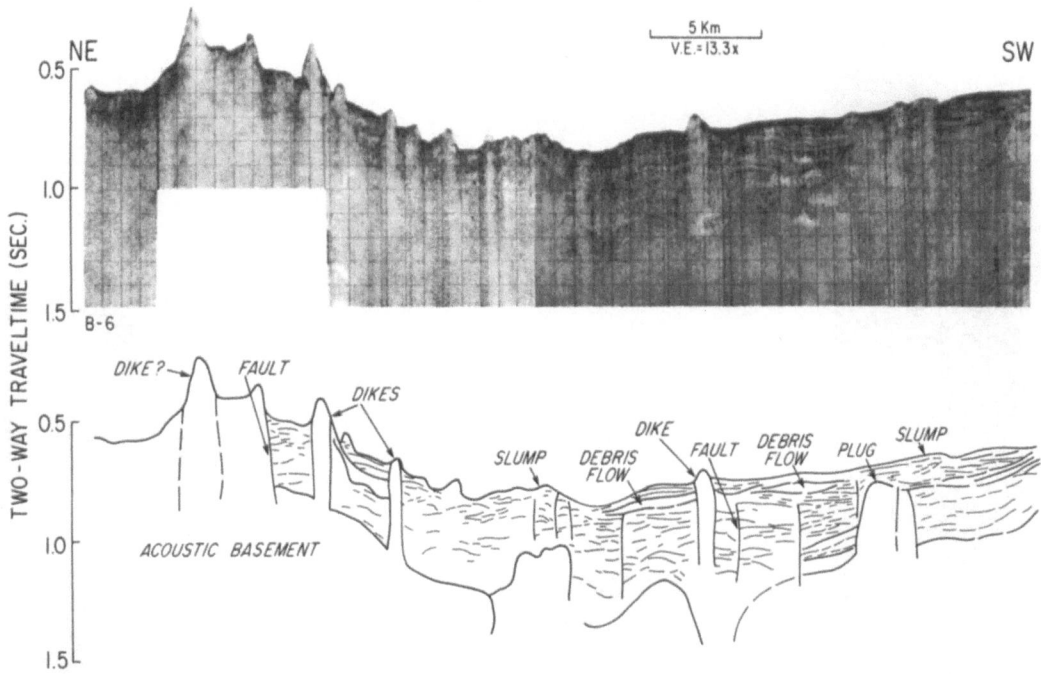

FIGURE 3-5. Photograph and tracing of single-channel seismic reflection profile B-6. Vertical exaggeration indicated is based on a water velocity of 750 m/sec. See Fig. 3-2 for location of profile.

thick debris flow mantles the southeastern flank of Bransfield Trough. This seismic facies can be traced to the floor of the trough, where it is overlain by horizontal strata; sampling indicates that these strata are of pelagic origin (Jeffers and Anderson, 1990). Jeffers and Anderson (1990) stated that glacial transport of sediment via the U-shaped troughs off Antarctica led to the formation of sediment wedges of onlapping prograding sequences at the mouths of the canyons. These wedges are fronted by 600-m-high slopes whose declinations are approximately 9°. The sequences in Bransfield Trough probably represent the distal facies of these wedges; presumably they are turbidites deposited as the canyon-mouth wedges oversteepened and failed, creating slumps, slides, and debris flows (Jeffers and Anderson, 1990). East of Bridgeman Island, Jeffers and Anderson identified four distinct acoustic sequences on the South Shetland Islands Ridge and three acoustic sequences off the Antarctic Peninsula. These sequences display basinward shifts of onlap, which they believe represent glacially induced sea-level regressions.

Strata in Bransfield Trough are disrupted by normal faults along its flanks and by a median ridge rising several hundred meters above its floor. This ridge, which is twin-peaked in places, and is associated with a prominent magnetic anomaly, has been interpreted as an incipient spreading axis. The tectonic style of the normal faults ranges from a single major fault, along which a large block has been dropped, to numerous

closely spaced faults, to low-angle faults on the shelf (Jeffers, Anderson, and Lawver, 1991). Transverse structures illustrated by the fault-bounded, glacially eroded, U-shaped canyons off the Antarctic Peninsula, divide the trough into three subbasins, which adds to the structural complexity of the region. Both our study and that of Jeffers, Anderson, and Lawver (1991) indicate that the intersection of these transverse structures with those parallel to the axial rift are sites of diffuse volcanic activity.

The present investigation has provided additional details on the tectonic style of the Bransfield Trough. Along Profile BRF, the trough's floor is dominated by a seamount, and on Profile B-12, a well stratified horst is prominent (Figs. 3-3, 3-6). In both profiles, acoustic basement is quite shallow. Southwest of Profile B-12, the northwest slope of the trough is underlain by two basement highs, which may represent basement horsts or volcanic intrusives (Profile B-11; Figs. 3-3, 3-6). Seaward of the outermost high is a structural low, whose sediment fill is divided by an unconformity. Eastward of this low, acoustic basement under the main part of Bransfield Trough is shallow and relatively smooth. A side echo, representing a low hill, was detected near the center of this shallow basement platform. The southeast side of the platform also appears to be fault-bounded, and the strata beyond it are terminated abruptly and injected by structures we interpret as dikes. A similar basement configuration is displayed by profiles B-14 and B-7. This basement high

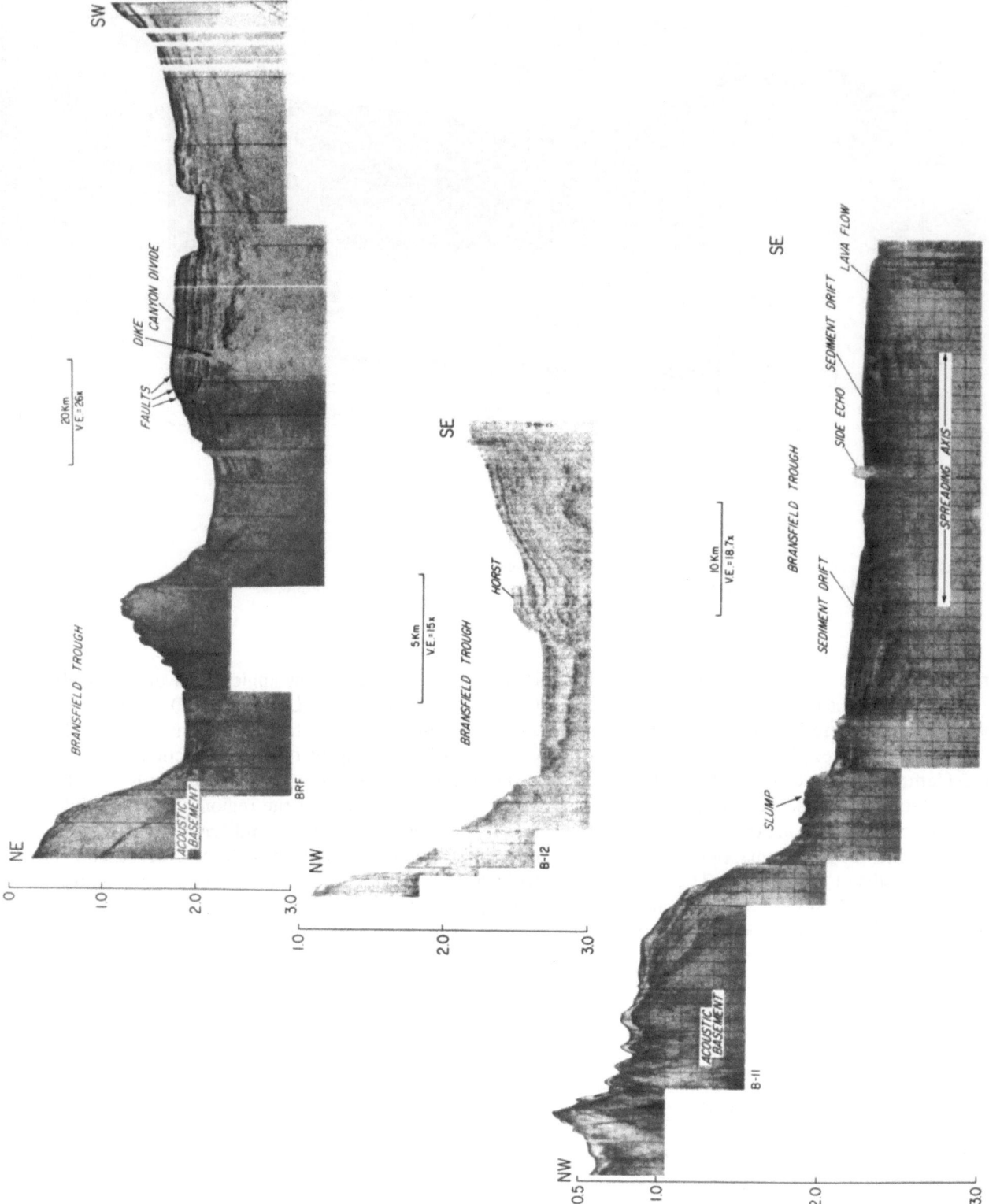

FIGURE 3-6. Photographs of seismic reflection profiles BRF, B-12, and B-11. Note sediment horst on Profile B-12; spreading axis and probable sediment drifts along B-11. The horst on Profile B-12 is on strike with the volcanic ridge on Profile B-11. The seamount on Profile BRF is along the southwest flank of the Bridgeman Island cross-structure. Compare with tracings in Fig. 3-3.

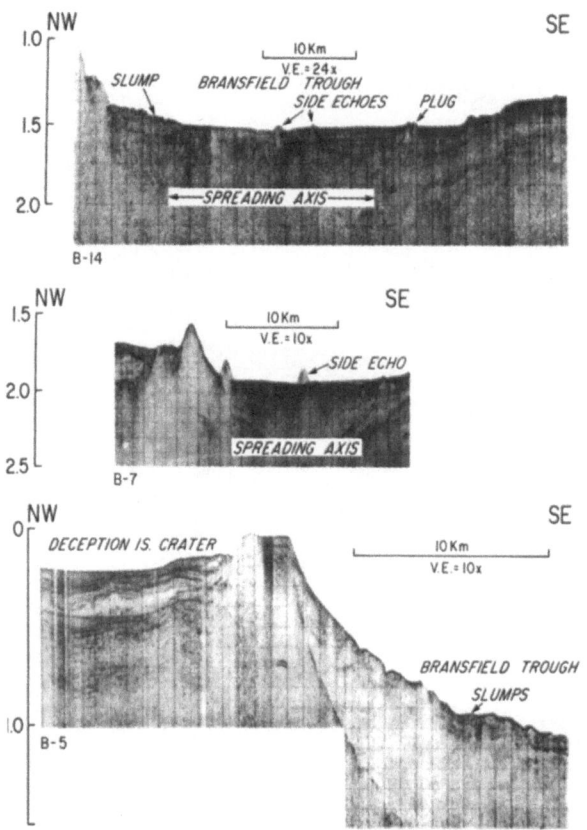

FIGURE 3-7. Photographs of seismic reflection profiles B-14, B-7, and B-5 displaying the spreading axis and showing the nature of the sediments in Deception Island crater. Note slump structures on the slope southeast of Deception Island. Compare with line interpretations in Fig. 3-4.

is the incipient spreading axis aligned along the axis of Bransfield Trough. Along Profile B-3, extending from South Shetland Islands Ridge to Tower Island Bank (Figs. 3-2, 3-8), the spreading axis is a twin-peaked basement high rising nearly 300 m above the trough's floor. No evidence of such an axis was encountered along Profile B-1, the most southwestern traverse recorded during the present investigation.

Seismic reflection profiles recorded during the present investigation and those described by Jeffers, Anderson, and Lawver (1991) suggest that the incipient spreading axis in Bransfield Trough is not continuous, but consists of discrete igneous centers separated by gaps and disrupted by structures aligned at right angles to the trough. The juncture of the transverse structures with those aligned parallel to the axis of the trough appear to be sites of diffuse magmatic activity, which formed Bridgeman Island and the seamounts along Profile B-5 and southeast off King Georges and Livingston Islands. To date, this incipient spreading axis has been traced to the southwest margin of the eastern depression (Jeffers, Anderson, and Lawver, 1991). The northeast extension of the axis and the nature of its

juncture with the southern boundary of the Scotia Plate (Fig. 3-1) have yet to be determined. To the southwest, the spreading axis extends to the middle of the western depression off Deception Island. The southwest end of Bransfield Trough is quite distinct from the rest of the structural low. Here (Profile B-1; Fig. 3-8) the trough contains a thick, slightly folded sedimentary section; there is no evidence of a seafloor spreading axis. It is possible that rifting still predominates in this part of the trough, a conclusion supported by the nature of seismicity in the area (Pelayo and Wiens, 1989). The two structural highs in this region also display different acoustic signatures. Whereas Austin Rocks Bank is stratified (Profile B-2; Fig. 3-8), Tower Island Bank is acoustically opaque, which is indicative of its volcanic origin (Choubert, Faure-Muret, and Chantaux, 1976).

As reported by Jeffers and Anderson (1990), the central part of the Bransfield Trough contains sediment accumulations over 0.75 sec thick. According to them, this fill consists of two acoustic units: (1) a draping unit of continuous reflections of pelagic origin, and (2) a unit of discontinuous reflections, which represent turbidites, onlapping and thinning toward the trough's margins. Deposition of the pelagic unit took place during interglacials; the turbidites accumulated during glacials. Sampling has revealed that the pelagic unit consists of ash-bearing diatomaceous ooze. The turbidite sequence has yet to be sampled. Seismic reflection profiles taken during the present investigation suggest that the stratigraphy of Bransfield Trough is more complex than that described by Jeffers and Anderson (1990), and that the spatial distribution of sediment is quite variable. Sediments along the axis of the Bransfield Trough are thick only in the lows flanking the central basement high and at the southwest end of the trough. East of the central seamount, on Profile B-5 (Figs. 3-3, 3-6), is a thin stratified unit of probable turbidity-current origin. Next to the seamount itself, the sediments are thin, and they rest on a high-impedance reflector, which may represent a lava flow. A similar reflector also is found east of the high. On Profile B-12 (Figs. 3-3, 3-6), the sediment fill northwest of the horst is made up of two units: a lower acoustically transparent unit, which resembles a debris-flow deposit, and a well-stratified upper unit, which sampling (Jeffers and Anderson, 1990) indicates is of pelagic origin.

The most interesting seismic facies were found along Profile B-11 (Figs. 3-3, 3-6). Along the southeast edge of the trough the thick sediment fill is divided by an unconformity. Below the unconformity is a sequence dipping toward the spreading axis, which fills a structural low between the southeast flank of the South Shetland Island Ridge and the spreading axis. A similar sequence also occurs southeast of the basement

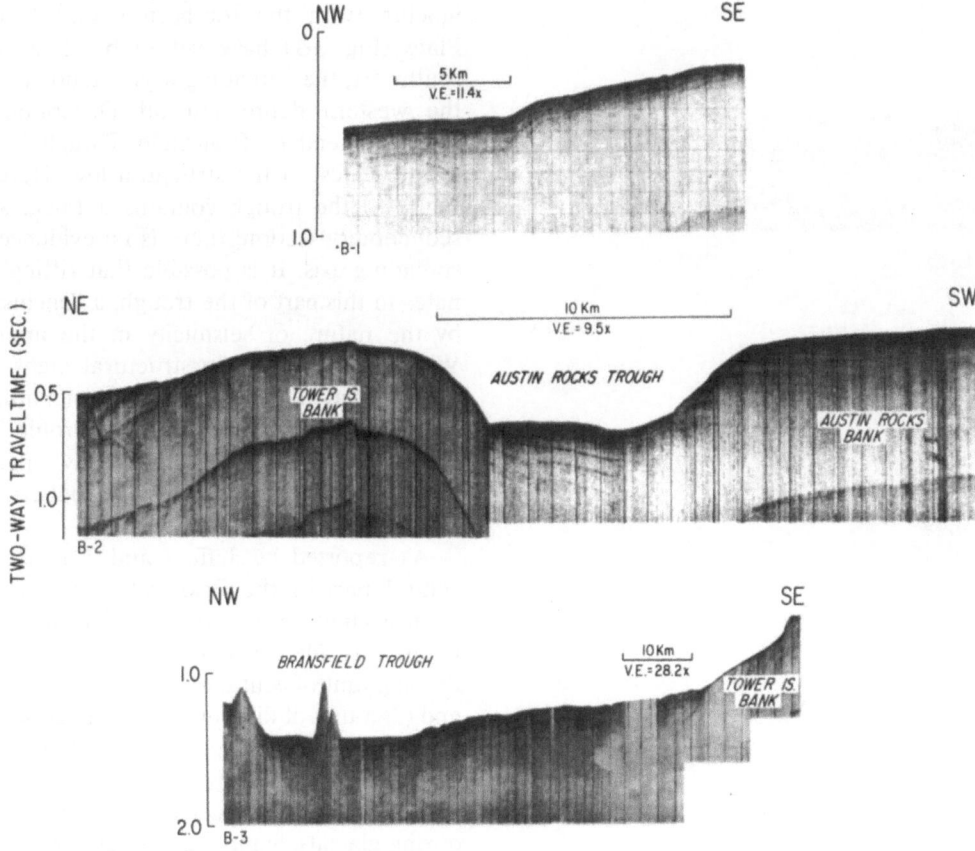

FIGURE 3-8.   Photographs of seismic reflection profiles B-1, B-2, and B-3. See Fig. 3-2 for locations of profiles. Note that Profile B-1, the most southwestern profile recorded in the Bransfield Trough, shows no evidence of a mid-ocean ridge, whereas Profile B-3, northeast of Profile B-1, displays a twin-peaked spreading axis. Strata along Profile B-1 also appear to be corrugated, which may indicate some compression in the region. Strata in Austin Rocks Trough (Profile B-2) appear to terminate against a fault off Austin Rocks Bank.

platform, which is continuous with the sediments forming the southeast slope of Bransfield Trough. We postulate that these beds are part of a synrift sequence emplaced prior to the initiation of seafloor spreading, and later divided by emplacement of the spreading axis. To date, no data are available to verify this postulate; the sequence could just as well have been deposited after seafloor spreading began. Above the unconformity, northwest of the spreading axis, a sediment wedge resembles a sediment drift plastered against the northwest side of the trough (Profile B-11; Figs. 3-3, 3-6). The geometry of this drift is similar to that displayed by Hatton and Fini drifts in the North Atlantic (McCave and Tucholke, 1986; see also Chapters 7, 13). Sediments above the spreading axis (in the middle of the trough) along this profile bulge upward; this geometry commonly indicates deposition by bottom currents. This detached configuration is comparable to that displayed by the Greater Antilles Outer Ridge in the North Atlantic (McCave and Tucholke, 1986). Profiles B-14 and B-7 display the same sedimentary setting; a lower unit of reflections dipping toward the volcanic ridge is truncated by an unconformity,

above which a horizontally bedded unit extends across the ridge (Figs. 3-4, 3-7). Along Profile B-14, the lower unit southeast of the ridge is intruded by a plug and is faulted. Southwest of Profile B-14, along Profile B-3 (Fig. 3-8), the trough's fill consists of a possible sediment drift northwest of the spreading axis, a twin-peaked high, and a horizontal fill, which rests on an irregular acoustic basement southeast of the axis. The nature of the contact between the basement and the sediment blanket above cannot be determined from our profile; we infer that it is either an unconformity or an intrusive boundary. On Profile B-1 (Fig. 3-8), the trough's sediment fill is much more uniform acoustically, and is continuous with the lower strata whose corrugations suggest compression. Strata in Austin Rocks Trough between Tower Island and Austin Rocks Banks (Profile B-2; Fig. 3-8) dip toward Austin Rocks Bank and terminate against a fault.

DISCUSSION AND CONCLUSION

The geologic relationships imaged by seismic reflection profiles recorded during the present investigation, along

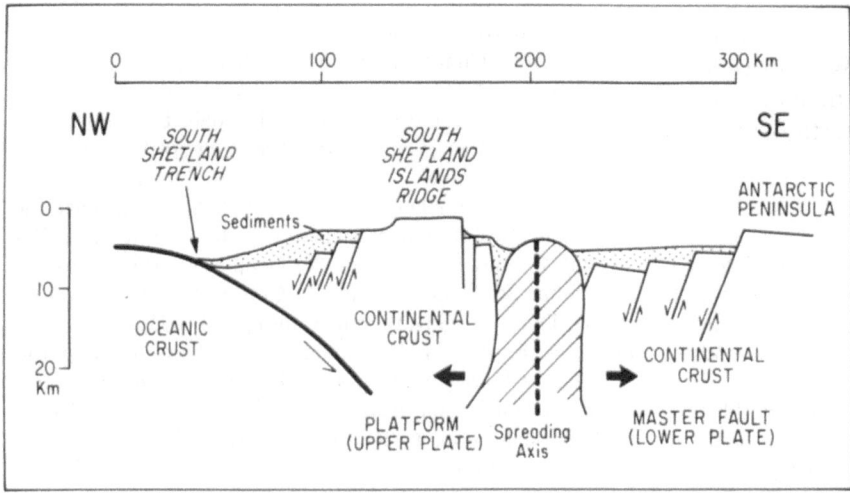

FIGURE 3-9.   Schematic cross section of Bransfield Trough. Modified from Ashfield (1972, Fig. 36) using data obtained during the present study.

with data obtained during previous studies, indicate that the Bransfield Trough is a continental rift structure in transition to becoming an oceanic one. The asymmetry of the trough suggests that it is a half-graben structure, whose master fault is on the Antarctic Peninsula side, away from the zone of subduction. The fault is opposed by a platform, the South Shetland Islands Ridge (Fig. 3-9). This tectonic model, however, needs verification. Crustal attenuation led to the displacement of a narrow continental fragment away from Antarctica to form the ridge. Sediments on the southeast side of the trough, off Antarctica, are involved in the faulting, which we take as evidence either of synsedimentary faulting, or of sagging and deposition that preceded rifting. The tectonic style displayed by Bransfield Trough is morphologically comparable to that of Arctic Eurasia, where the narrow continental Lomonosov Ridge was broken way from the Barents Shelf by the Nansen Ridge spreading axis. Unlike the Arctic counterpart, however, extension in the Bransfield Trough is taking place in a back-arc setting behind the South Shetland Trench. The cross-section geometry of the spreading axis probably is controlled by the dimensions of the Bransfield Trough, which is only 30 km wide. If extension is the result of the cooling and sinking of a subducted oceanic slab, then spreading should cease in the trough in the geologically near future, as such a mechanism is incapable of sustaining sea-floor spreading for very long. If on the other hand, spreading is the result of active subduction, then sea-floor spreading should persist longer and the spreading axis should become more organized with time. At present, the spreading axis in the Bransfield Trough consists of discrete igneous centers separated by gaps that may have allowed sediments from Antarctica to reach the northwest side of the trough. Continuity of the axis also is interrupted by structures aligned transversely to the trough axis. The intersection of these cross-structures with features parallel to the trough axis appear to be sites of pronounced magmatic activity. Sediment contribution from the South Shetland Islands Ridge to Bransfield Trough was probably less than that from Antarctica; some of the detritus was trapped in structural lows perched along the southeast flank of the ridge.

If the 30-km-wide Bransfield Trough is representative of conditions that exist when deposition changes from synrift to drift (postrift), then we can make some inferences regarding the analogous Jurassic/Cretaceous transitions in the North and South Atlantic. For example, the change from synrift to drift (postrift) deposition was not coeval along the strike of the trough, and the mid-ocean ridge (during its initial phases) may have consisted of discrete segments, gaps between which served as portals for sediment transport across the floor of the trough. Above all, our data show that the transition from rift to drift may be commonly marked by a complex distribution of synrift and postrift lithofacies.

ACKNOWLEDGMENTS

We wish to thank the captains and crews of the *R/V Pescapuerto IV* and *R/V Las Palmas* for their cooperation during the ANTÁRTIDA 8611 and EXANTARTE 88/89 expeditions, and to express our gratitude to fellow participants in the expedition, who helped collect the geologic data. Comments and suggestions by John B. Anderson and John D. Jeffers of Rice University, L. A. Lawver of the University of Texas, and C. W. Poag of the U.S. Geological Survey are acknowledged. John B. Anderson provided us with preprints of his articles on Bransfield Trough, which were useful in revising earlier drafts of the manuscript.

Uchupi's participation in this study was made possible by funds provided via the J. Seward Johnson Chair in Oceanography awarded to him by Woods Hole Oceanographic Institution. This is contribution No. 7647 of the Woods Hole Oceanographic Institution.

REFERENCES

Acosta, J., Canals, M., Herranz, P., and Sanz, J. L. 1989. Investigación geologica-geofísica y sedimentologica en el Arco de Scotia y Península Antarctica. Resultados de la Campaña "ANTARTIDA 8611": *Publicacion Especial del Instituto Español de Oceanografia 2*, 9–82.

Acosta, J., Catalan, M., Harranz, P., Sanz, L. 1989. Perfiles sismicos en las Shetlands del Sur y Estrecho de Bransfield. Estructura y Dinámica reciente. *Actas del IIIer Symposium Español de Estudios Antárticos*: Gredos, España: Comisión Interministerial de Ciencia y Techologica, 281–296.

Ashcroft., W. A. 1972. Crustal structure of the south Shetland Islands and Bransfield Strait. *British Antarctic Survey Scientific Report* 66:43.

Baker, P. E., McReath, I., Harvey, M. R., and Roobol, M. J. 1975. The geology of the South Shetland Islands: V. Volcanic evolution of Deception Island. *British Antarctic Survey Science Reports* 78:81.

Barker, P. F. 1976. The tectonic framework of the Cenozoic volcanism in the Scotia Sea region, a review. *In* Symposium on Andean and Antarctic Volcanology Problems, O. Gonzalez-Ferran, Ed.: Santiago, Chile: IAVCEI Special Series, 330–346.

Barker, P. F., and Dalziel, I. W. D. 1983. Progress in geodynamics in the Scotia Arc Region. *In* Geodynamics of the Eastern Pacific Region, Caribbean and Scotia Arcs, S. J. R. Cabre, Ed., *American Geophysical Union Geodynamics Series* 9:137–170.

Choubert, G., Faure-Muret, A., and Chanteux, P. 1976. *Geological Word Atlas*. Paris: UNESCO: Sheet 17, Scale 1:10,000,000.

Dalziel, I. W. D., and Elliot, D. H. 1973. The Scotia Arc and Antarctic Margin. *In* The Ocean Basins and Margins, Volume 1, The South Atlantic, A. E. M. Nairn and F. G. Stehli, Eds.: New York: Plenum Press, 171–246.

Dunbar, J. A., and Sawyer, D. A. 1989. Patterns of continental extension along conjugate margins of the central and North Atlantic oceans and Labrador Sea. *Tectonics* 8:1059–1077.

González-Ferrán, O., 1991. The Bransfield rift and its active volcanism. *In* Geological Evolution of Antarctica, M. R. A. Thomson, J. A. Came, and J. W. Thomson, Eds.: New York: Cambridge University Press, 505–509.

González-Ferrán, O., and Katsui, Y. 1970. Estudio integral del volcanismo Cenozoico superior de las Islas Shetland del Sur, Antarctica. *Instituto Antarctico Chile Series Cientificas* 1(2):123–174.

Guterch, A., Grad, M., Janik T. and Perchuć, E. 1991. Tectonophysical models of the crust between the Antarctic Peninsula and the South Shetland Trench. *In* Geological Evolution of Antarctica M. R. A. Thomson, J. A. Came, and J. W. Thomson, Eds.: New York: Cambridge University Press, 499–504.

Haq, B. U., Hardenbol, J., and Vail., P. R. 1988. Mesozoic and Cenozoic chronostratigraphy and eustatic cycles. *In* Sea-Level Changes: An Integrated Approach, C. K. Wilgus, B. S. Hastings, C. A. Ross, H. Posamentier, J. Van Wagoner, and C. G. St. C. Kendall, Eds.: *Society of Economic Paleontologists and Mineralogists Special Publication 42*, 71–108.

Herron, E. M., and Tucholke, B. E. 1976. Sea floor magnetic patterns and basement structure in the southeastern Pacific. *In* Initial Reports of the Deep Sea Drilling Project, Volume 35, C. D. Hollister, C. Craddock, et al.: Washington, DC: U.S. Government Printing Office, 263–287.

Jansa, L. F. 1986. Paleoceanography and evolution of the North Atlantic Ocean during the Jurassic. *In* The Geology of North American, Volume M, The Western North Atlantic Region, P. R. Vogt and B. E. Tucholke, Eds.: Boulder, CO: Geological Society of America, 603–616.

Jeffers, J. D., and Anderson, J. B. 1990. Sequence stratigraphy of the Bransfield Basin, Antarctica: implications for tectonic history and hydrocarbon potential: *In* Antarctica Exploration Frontier—Hydrocarbon Potential, Geology, and Hazards, Bill St. John, Ed.: *American Association of Petroleum Geologists Studies in Geology*, No. 31, 13–29.

Jeffers, J. D., Anderson, J. B., and Lawver, L. A. 1991. Evolution of the Bransfield Basin, Antarctic Peninsula. *In* Geological Evolution of Antarctica, M. R. A. Thomson, J. A. Crame, and J. W. Thomson, Eds.: New York: Cambridge University Press, 481–485.

Keen, C. E., and Barrett, D. L. 1981. Thermal subsided crust on the rifted margin of eastern Canada; crustal structure, thermal evolution, and subsidence history. *Royal Astronomical Society, Geophysical Journal* 65:443–465.

Klitgord, K. D., Hutchinson, D. R., and Schouten, H. 1988. U.S. Atlantic continental margin. The Geology of North America, Volume I-2, The Atlantic Continental Margin: Boulder CO: Geological Society of America, 19–55.

Lin, J., and Paramentier, E. M. 1990. A finite amplitude necking model of rifting in brittle lithosphere. *Journal of Geophysical Research* 95:4909–4923.

Mann, D. C. 1989. Thick-skin and thin-skin detachment faults in continental Sudanese rift basins. *Journal of African Earth Sciences* 8:307–322.

Marti, J., Baraldo, A. and Rey, J., 1989. Origen y estructura de la Isla Decepión (Islas Shetland del Sur). *Actas del IIIer Symposium Español de Estudios Antárticos*: Gredos, España: Comisión Interministerial de Ciencia y Tecnologica, 187–194.

McCave, I. N., and Tucholke, B. E. 1986. Deep-current controlled sedimentation in the western North Atlantic. The Geology of North America, Volume M, The Western North Atlantic Region: Boulder, CO: Geological Society of America, 451–468.

Montadert, L., Roberts, D. G., De Charpal, O., and Guennoc, P. 1979. Rifting and subsidence of the northern continental margin of the Bay of Biscay. *In* Initial Reports of the Deep Sea Drilling Project, Volume 48, L. Montadert, D. G. Roberts, et al.: Washington, DC: U.S. Government Printing Office, 1025–1060.

Moullade, M., Brunet, M.-F., and Boillot, G. 1988. Subsidence of the Galician margin: the paleoenvironmental control. *In* Proceedings Ocean Drilling Program, Scientific Results, Volume 103, G. Boillot, E. L. Winterer, et al.: College Station, TX: Ocean Drilling Program, 733–740.

Pelayo, A. M., and Wiens, D. A. 1989. Seismotectonics and relative plate motions in the Scotia Sea region. *Journal of Geophysical Research* 94:7293–7320.

Roach, P. J. 1978. The nature of back-arc extension in Bransfield Strait. *Royal Astronomical Society, Geophysical Journal (Abs.)*, 165.

Sibuet, J. -C., and Ryan, W. B. F., 1979. Site 398: Evolution of the west Iberian passive continental margin in the framework of the early evolution of the North Atlantic. *In* Initial Reports of the Deep Sea Drilling Project, Volume 47, Part 2, J. -C. Sibuet, W. B. F. Ryan, et al.: Washington, DC: U.S. Government Printing Office, 761–776.

The Shipboard Scientific Party. 1978. Angola continental margin—sites 364 and 365. *In* Initial Reports of the Deep Sea Drilling Project, Volume 40, H. M. Bolli, W. B. F. Ryan, et al.: Washington, DC: U.S. Government Printing Office, 357–390.

Uchupi, E. 1989. The tectonic style of the Atlantic rift system. *Journal of African Earth Sciences* 8:134–164.

Uchupi, E., and Emery, K. O. 1991. Pangaean divergent margins. *Marine Geology* special issue, 102:1–28.

Weaver, S. D., Saunders, A. D., and Tarney, J. 1982. Mesozoic–Cenozoic volcanism in the South Shetland Islands and Antarctic Peninsula: geochemical nature and plate tectonic significance. *In* Antarctic Geosciences, C. Craddock, Ed.: Madison, WI: University of Wisconsin Press, 263–274.

4

Southwestern Africa Continental Rise: Structural and Sedimentary Evolution

Richard V. Dingle and Simon H. Robson

On the eastern side of the Cape Basin, between Walvis Bay (lat. 22°S) and the southern tip of Africa (lat. 36°S) (a distance of ca. 1,500 km), the continental shelf is relatively wide (maximum 200 km) and deep (maximum 500 m), and the slopes and rises extend as far as 600 km offshore (Fig. 4-1).

This passive margin formed during the opening of the South Atlantic, an event which began not before magnetic anomaly M9 (126–121 Ma) south of the Orange River, and M4 (123–117 Ma) north of the Orange River (Austin and Uchupi, 1982). Major influences on the development of the outer continental margin (slopes and rises) were the location of the main postrift sediment depocenters (Fig. 4-2), variations in hinterland climate, and the establishment of deep-water ocean circulation. A further important factor has been alterations in the drainage pattern of the Orange River system, which is the only major modern river discharging into the Cape Basin. Local changes in relative sea level have had an influence on outer margin modification, but a full appreciation of this factor awaits publication of more detailed stratigraphies.

SEDIMENTARY HISTORY OF THE OUTER MARGIN

Early Phase: Late Aptian to Maestrichtian

The earliest recognizable shelf–slope–rise relationship occurs above a thin, relatively smooth seismic reflector that is widely developed in the southeastern Atlantic (Figs. 4-3, 4-4). This reflector (AII of Uchupi and Emery, 1972; Emery et al., 1975; = Horizon P of Gerrard and Smith, 1982) lies at the top of a series of

seaward-dipping reflections, and represents the drift-onset unconformity lying over irregular basement. At Deep Sea Drilling Project Site 361, on the rise southwest of Cape Town, Bolli et al. (1978) dated reflector AII as late Aptian, and showed that it marked the transition from a lower sequence of organic–carbon-rich quartz sands and pelagic shales, to overlying shaly turbidites. Under the present-day shelf (Kudu 9A boreholes; Fig. 4-1), the transition is from organic-rich, poorly oxygenated, deep-water shales with planktonic assemblages, to oxygenated deep-water shales with abundant benthic and planktonic assemblages (McMillan, 1990).

The post-AII succession consists of a thick sequence of discontinuous and variously deformed reflections that are terminated by a further regionally developed reflector (D of Uchupi and Emery, 1972; Emery et al., 1975; = Horizon L of Gerrard and Smith, 1982). Bolli et al. (1978) dated D as latest Maestrichtian. Within the AII to D sequence, the seaward advance of the base of slope (BOS) and base of rise (BOR) can be well documented for the Late Cretaceous interval in the southeastern Atlantic.

Subsidence of the continental margin during the early opening of the southeastern Atlantic was greatest in the sector between the Orange River and Cape Columbine, where more than 3.4 sec (ca. 5.1 km) of late Aptian–Maestrichtian sediment (i.e., between reflectors AII and D) accumulated in the Orange Basin (Figs. 4-2, 4-5). Subsidiary depocenters developed off Walvis Bay and Luderitz, whereas relatively buoyant basement underlay the continental margin south of Cape Columbine (Emery et al., 1975; Gerrard and Smith, 1982; Dingle, Siesser, and Newton, 1983). These

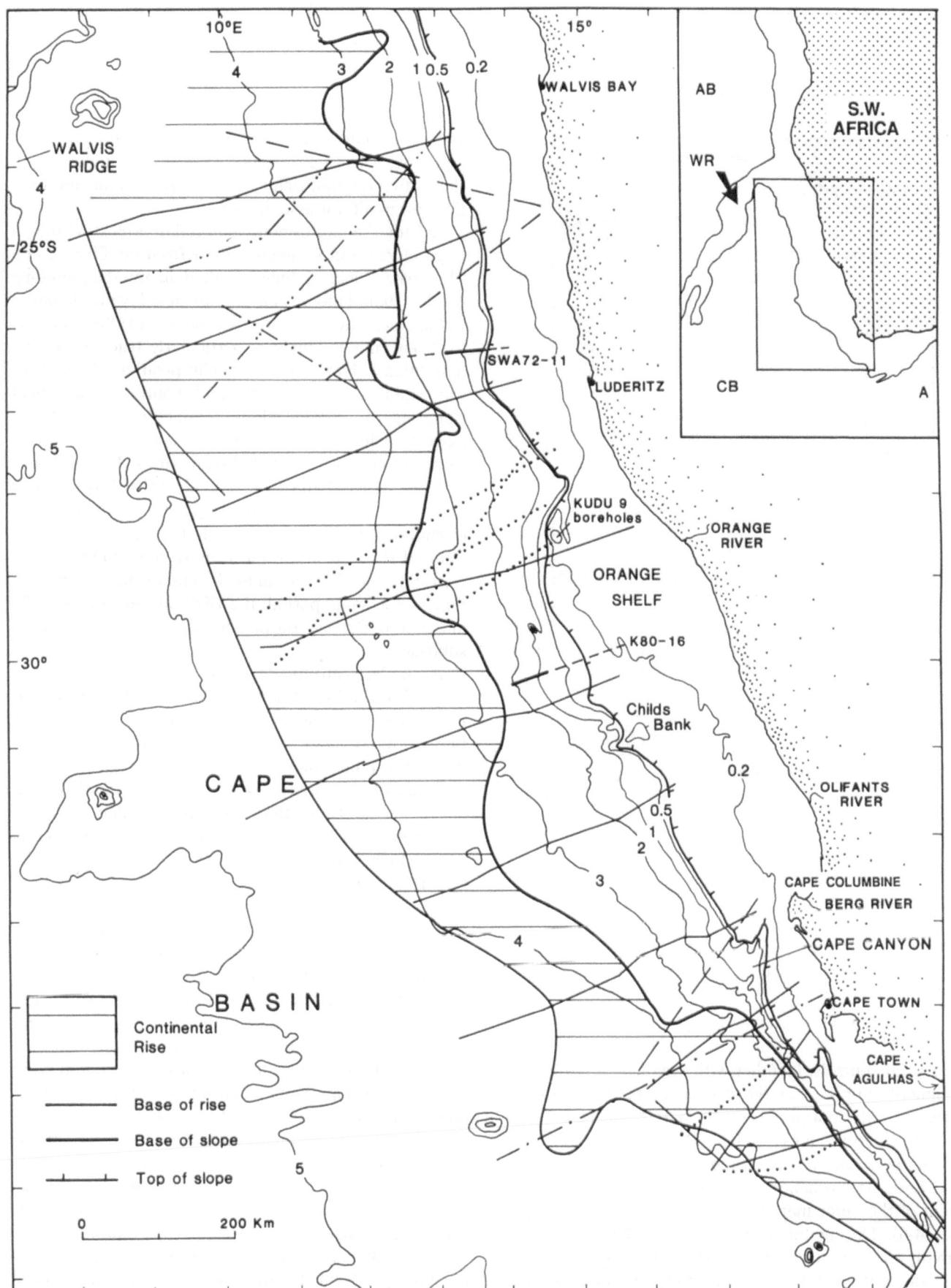

FIGURE 4-1.   Modern bathymetry and continental rise off southwestern Africa, SE Atlantic Ocean. Isobaths (km) after Dingle et al. (1987). Traverses are seismic lines used to construct Figs. 4-5, 4-9, 4-10: solid = Uchupi and Emery (1972), long dashed = Austin and Uchupi (1982), short dash SWA72-11 = Geological Survey of Namibia, short dash K80-16 = SOEKOR, dot-dash = Bolli et al. (1978), double dot-dash = Summerhayes, Bornhold, and Embley (1979), dots = Dingle (1980). Thickened sections of SWA72-11 and K80-16 are illustrated in Figs. 4-3 and 4-4, respectively. Abbreviations in inset: AB = Angola Basin, WR = Walvis Ridge, CB = Cape Basin, A = Agulhas Basin.

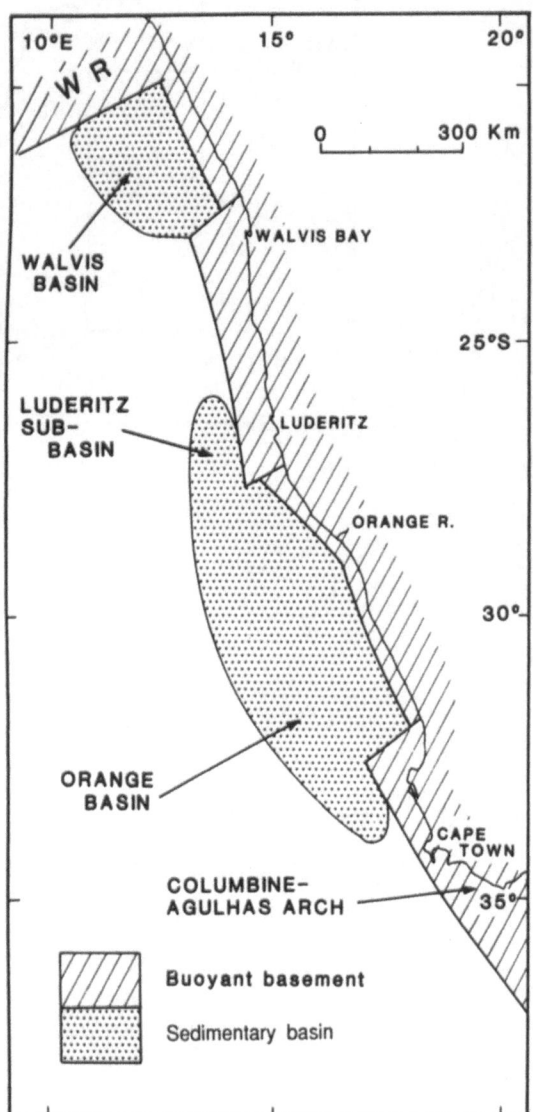

FIGURE 4-2.   Sedimentary basins and positive basement features on the continental margin off southwestern Africa (from Dingle, Siesser, and Newton, 1983).

structural controls, together with regional variations in sediment supply from the continent, produced three styles of outer-margin accretion during the middle to Late Cretaceous (Figs. 4-5 to 4-7).

The best known area is the central sector (Orange and Luderitz basins), where Late Cretaceous development of the outer margin occurred in two stages (Figs. 4-5, 4-6). During the first stage, the BOS advanced rapidly (> 2.0 km/m.y.; Fig. 4-8) from about an early Albian position in the central Orange Basin to a late Cenomanian position on the western side of the depocenter. At the same time, upbuilding of the shelf was relatively slow [44 m/m.y. under the middle shelf; 80 m/m.y. under the outer shelf (compacted values)]. During this period, the continental slope was ca. 0.4 sec high (1,200 m decompacted) and 10 to 12 km

wide, and there was no obvious continental rise physiography. The BOS advanced across low-angle abyssal shales that contain rich calcareous planktonic and benthic foraminiferal assemblages.

During the late Cenomanian to late Maestrichtian, a large increase in sediment supply from the Orange and "Luderitz" river systems resulted in rapid upbuilding of the shelf [140 m/m.y. (compacted value)], but a virtual cessation of top of slope (TOS) advance (0.09 km/m.y.). This was particularly true during the Santonian (McMillan, 1990). The position of the TOS was stabilized by massive and continual rotational slumping of unstable slope sediments, whose downslope transfer constructed and advanced a lower-slope wedge of complex, cascade-like features (Figs. 4-3, 4-4, 4-6; see also Chapters 9, 10, 12, 13). More mobile, distal components of this large-scale allochthonism (turbidites(?), debris flows(?), etc.) created a thick continental rise, which advanced more than 200 km westward into the Cape Basin by the end of the Cretaceous. During the same period, the BOS advanced a mere 25 km in the Orange Basin and 55 km in the Luderitz subbasin.

A similarly abrupt decrease in sediment supply at the end of the Maestrichtian/earliest Paleocene terminated this phase of margin construction. Seismic profiles show reflector D passing relatively unaffected across the fault zones.

South of the Orange Basin, the continental margin between Cape Columbine and Cape Agulhas is backed by a geomorphically complex and mountainous hinterland, which appears to have always kept the offshore area relatively starved of sediment (contrast with Chapters 2, 5, 6, 7). In addition, the margin in this area is underlain by granites, and is an extension of the buoyant Agulhas Arch (Fig. 4-2), on which pre-Mesozoic basement lies relatively close to the surface of the continental shelf and upper slope. These factors ensured that sediment accumulation on this part of the margin was confined largely to the lower slope and rise.

During Late Cretaceous time, a narrow slump zone on the steep slope maintained the position of the TOS and transferred material, which appears on seismic records to be discontinuous, jumbled structures, onto a narrow lower-slope zone. Distal components of this jumbled material prograded westward as a thick, prominent, continental-rise wedge. The chaotic reflections are interspersed with more continuous reflections, which may indicate alongslope transport from the Orange Basin. South of Cape Town, these lower slope–rise sediments are dammed behind large basement peaks associated with the Agulhas Fracture Zone (that fronts the margin off southeastern Africa) and(or) the seamount province at the southern tip of the continent (Figs. 4-7, 4-8). The complex nature of this succession makes it virtually impossible to confidently trace

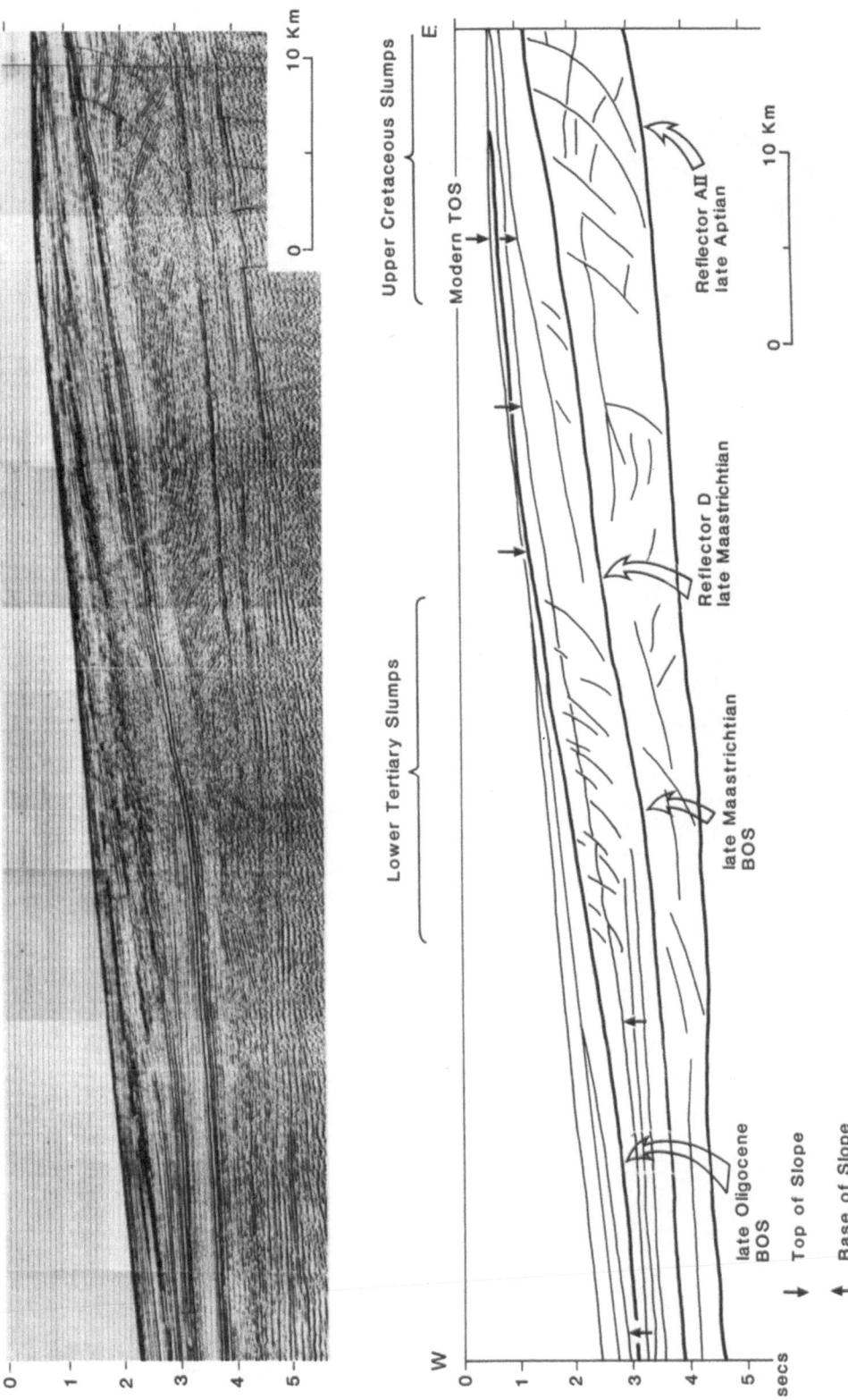

FIGURE 4-3. Seismic profile (above) and interpreted section (below) across the modern lower slope NW of Luderitz (SWA72-11: see Fig. 4-1 for location). Note the following features: (1) In Upper Cretaceous [between reflectors AII (late Aptian) and D (late Maestrichtian)]: outer part of upper slope zone of rotational faults (marked Upper Cretaceous Slumps), lower slope of chaotic slumped material, late Maestrichtian BOS, upper rise turbidites and debris flows. (2) In Paleogene (between D and overlying high-amplitude event), fault zone on lower slope (marked Lower Teritary Slumps), late Oligocene BOS, upper rise turbidites. (3) Neogene slope sediments. Small arrows show successive positions of BOS and TOS. Original seismic profile supplied by the Geological Survey of Namibia.

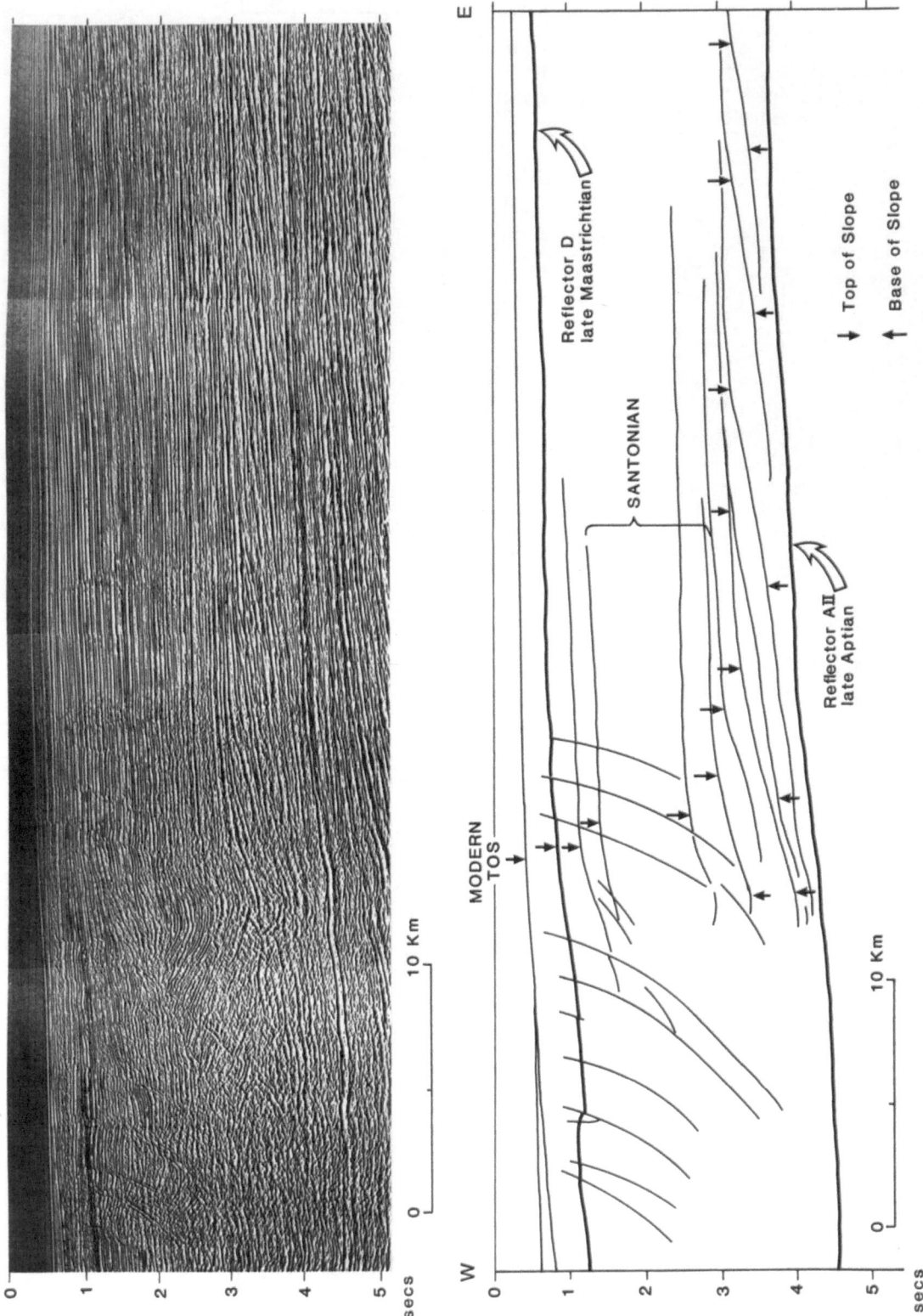

FIGURE 4-4. Seismic profile and interpreted section across the modern outer shelf and upper slope SW of the Orange River (K80-16: see Fig. 4-1 for location). Note the following features in the Upper Cretaceous (between reflectors AII (late Aptian) and D (late Maestrichtian)]: rapidly advancing Albian–late Cenomanian slope (with successive positions of TOS and BOS), rapidly upbuilding Turonian–Maestrichtian outer shelf, stabilized position of TOS and advancing BOS, upper slope zone of large rotational faults, some of which intersect the modern seafloor. Vertical exaggeration = ×4.2 (in water). Original seismic profile supplied by SOEKOR.

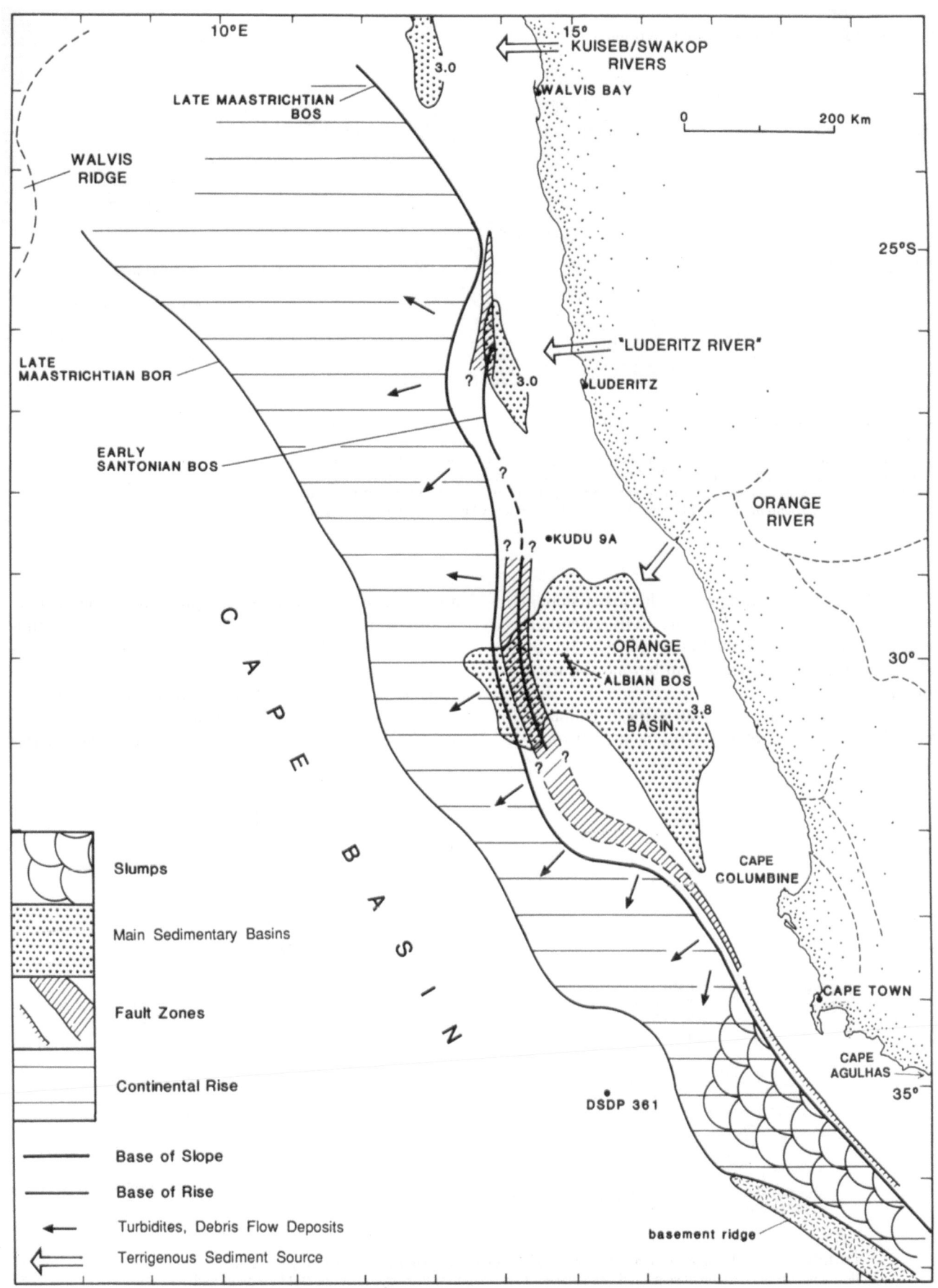

FIGURE 4-5. Upper Cretaceous continental rises, main depocenters, and sources of terrigenous sediment. Isopachs (km) from Gerrard and Smith (1982), converted to thicknesses using stratigraphy for KUDU 9A in McMillan (1990). BOR = base of rise; BOS = base of slope.

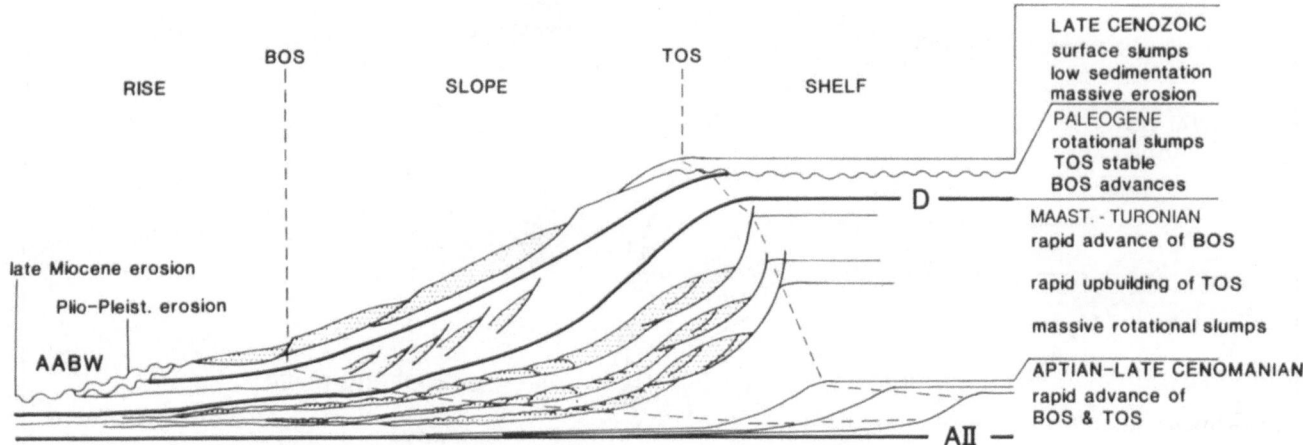

FIGURE 4-6. Schematic section across Orange Basin to illustrate post-Aptian development of continental slope and rise adjacent to area that underwent significant basement subsidence, and received large sediment input. Note relative positions through time of the top of slope (TOS) and bottom of slope (BOS). Shaded units are allochthonous sediments; wavy lines are erosion surfaces; thick lines are major seismic reflecting horizons; curved medium lines are glide planes of rotational faults; Maast. = Maestrichtian; Plio. = Pliocene; Pleist. = Pleistocene; AABW = Antarctic Bottom Water.

reflector AII inshore of the present-day lower slope. By the end of Maestrichtian time, the BOR extended as far as 160 km west of the shelf edge, whereas the BOS had advanced no more than 20 km beyond the narrow slump zone.

West of Walvis Bay, the Walvis Basin contains at least 3.0 km of Late Cretaceous sediment, but there is

no evidence of large-scale slumping on the outer margin (Uchupi and Emery, 1972; Austin and Uchupi, 1982). The Late Cretaceous continental slope advanced as a relatively low-angle feature, and smoothly passed into the rise, whose continuous, very-low-angle reflections extend nearly 400 km southwestward. We have insufficient borehole data to understand details of the

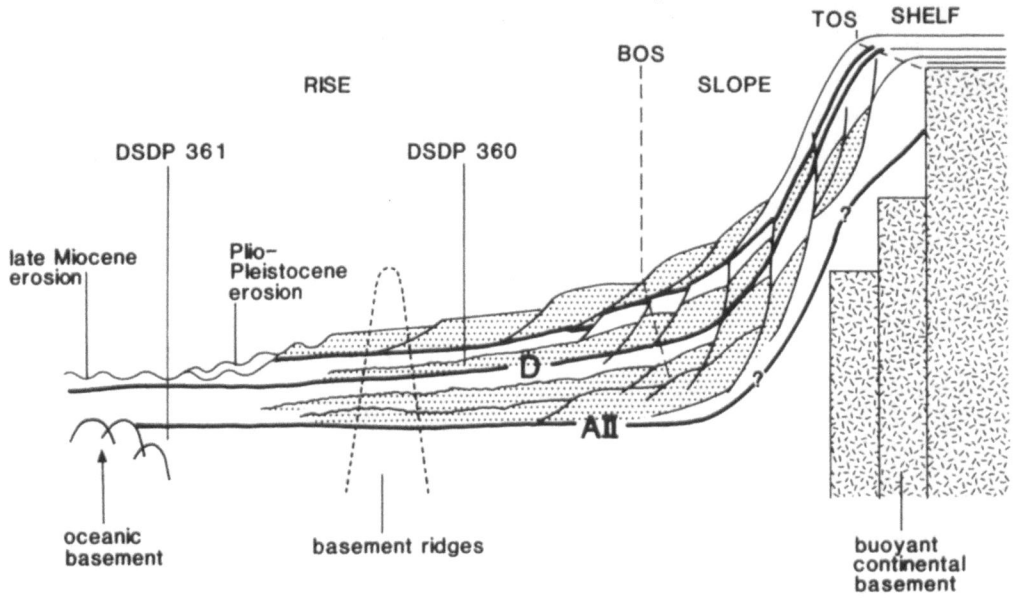

FIGURE 4-7. Schematic section across margin SW of Cape Agulhas to illustrate post-Aptian development of continental slope and rise adjacent to a margin with buoyant continental basement and relatively low sediment input. Note relative positions through time of the top of slope (TOS) and bottom of slope (BOS). DSDP Sites 360 and 361 penetrated to Eocene and Aptian, respectively (Bolli et al., 1978). Key as in Fig. 4-6. The dotted outline of basement ridges signifies their relative position in areas farther to the southeast.

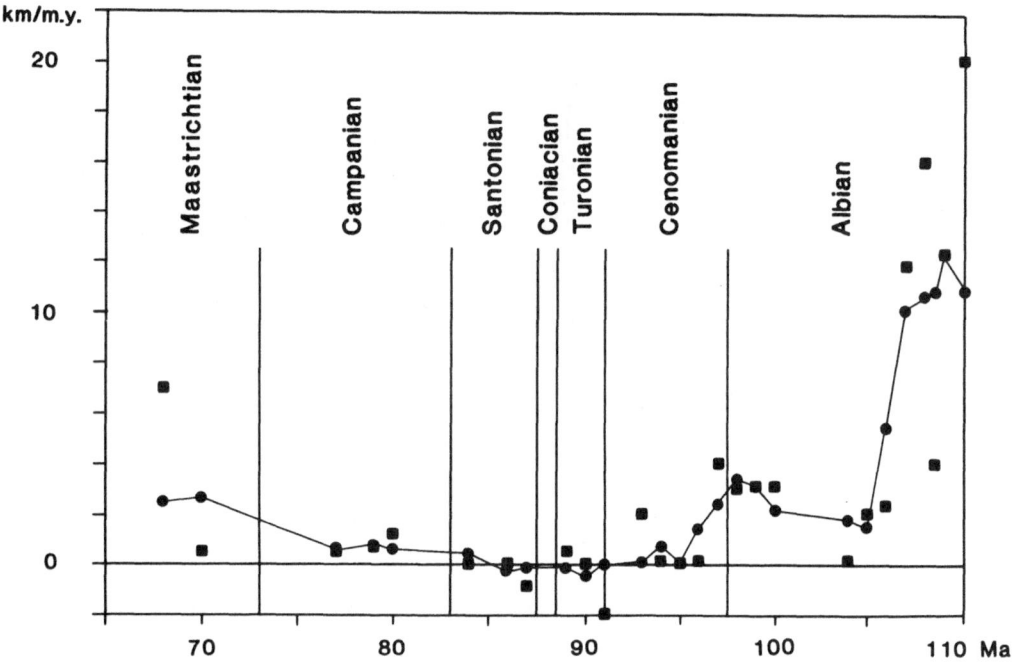

FIGURE 4-8.   Upper Cretaceous top-of-slope advancement rates in kilometers per million years in the Orange Basin. Values are averages from SOEKOR seismic profiles. Squares = original data points, smoothed curve is for three-point means. Stage boundaries dated according to Harland et al. (1982).

early history of this margin north of Walvis Bay, where it presumably was strongly affected by the development of the Walvis Ridge abutment (e.g., Goslin et al., 1974).

Middle Phase: Paleogene

Paleogene outer margin sedimentation was affected by end-of-Cretaceous alteration in the drainage pattern of the Orange River system (Dingle and Hendey, 1984). The upper and middle sections of the Orange River were diverted to the modern Olifants River mouth (Upper Orange/Olifants River), whereas the modern Orange exit point continued to receive runoff from the lower and Namibian areas only (Lower Orange River). As a result, the thickest sediment accumulations (ca. 3 km) lie at the southern end of the Orange Basin and in the Luderitz subbasin, with a somewhat thinner deposit (ca. 2 km) west of the Lower Orange River (Fig. 4-9). In addition, as aridity increased, early Tertiary sediment yields from the subcontinent were significantly lower than those for the Late Cretaceous (probably less than 40%: Dingle and Hendey, 1984), which resulted in lower accumulation rates (compare with Chapters 6, 12).

The main Paleogene depocenters formed seaward-prograding lenses on the lower Maestrichtian slope (defined by reflector D), and zones of rotational faults developed above or west of the Late Cretaceous BOS (Figs. 4-3, 4-6, 4-9). Slumped material from these zones

advanced the Paleogene BOS 5 to 10 km in the Orange–Luderitz area, and > 40 km west of Cape Columbine, opposite the Orange/Olifants sediment source. In both areas, the continental rise is relatively thick (ca. 0.5 sec = 1 km on the upper rise off Cape Columbine) and displays medium-amplitude, continuous reflections, presumably composed of interbedded, distal, slump debris and hemipelagic material. The precise location of the late Paleogene BOR west of the Lower Orange River has been obscured by later faulting and slumping.

West and south of Cape Town, adjacent to the margin underlain by buoyant basement, relatively thick Paleogene sediments lie at the foot of the Late Cretaceous slope. Here, faulting was a continual feature of early Tertiary sedimentation, transferring material from the upper slope to the lower slope and rise, and maintaining a very steep, rugged, outer margin. As in the Late Cretaceous, this allochthonous sediment was dammed behind basement ridges in the extreme south, and produced a narrow, convex continental rise composed of irregular, slumped blocks, which maintained a high degree of internal cohesion.

Sea level dropped significantly in middle Oligocene (?) time, and the Upper Orange/Olifants river incised a large canyon obliquely across the shelf and slope west of Cape Town, removing a portion of the Paleogene sediment lens. A large sedimentary cone developed at the distal end of the canyon and extended the continental rise an unknown distance toward the southwest.

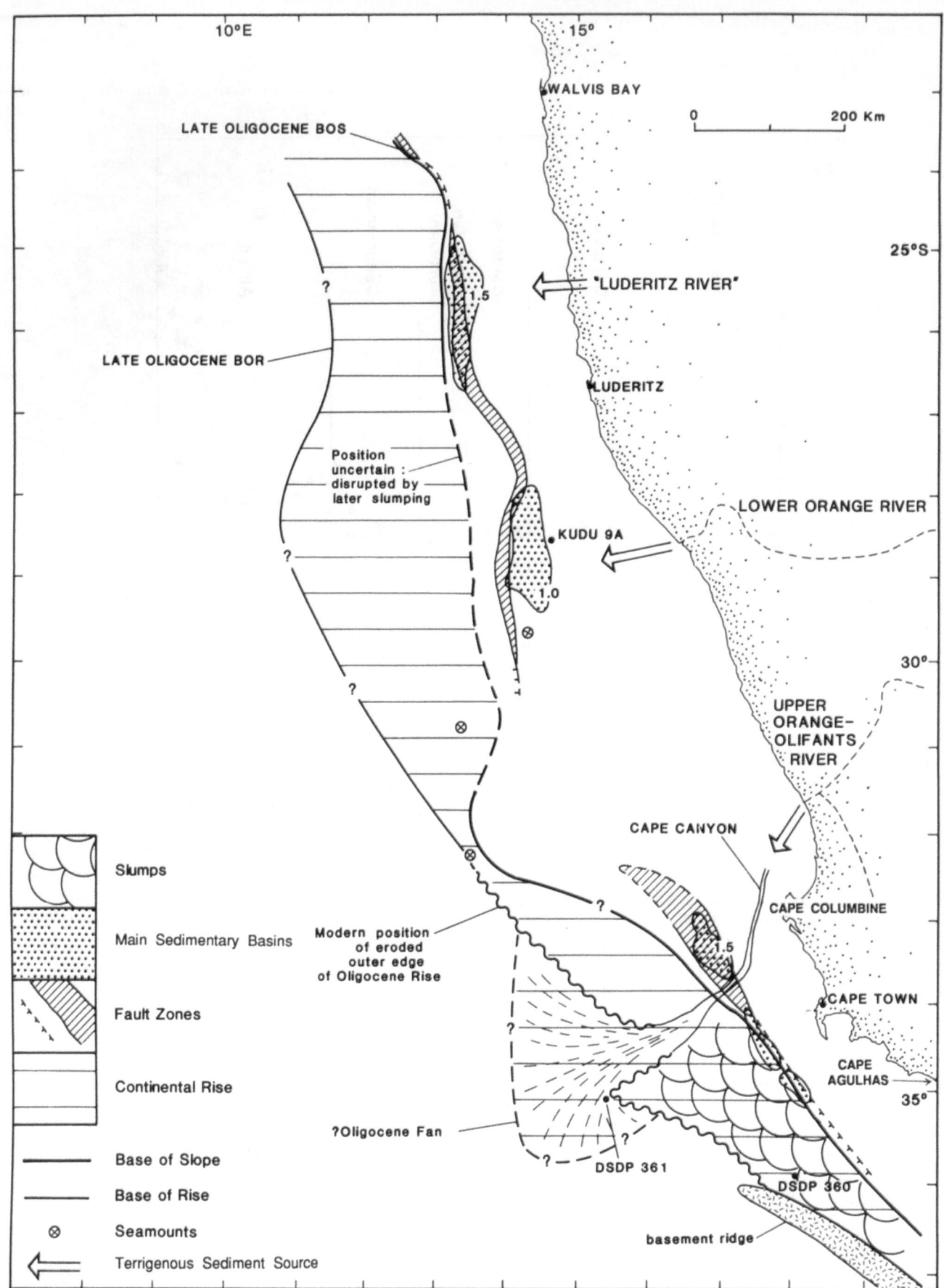

FIGURE 4-9. Paleogene continental rises in relation to main depocenters and terrigenous sediment sources. Note that during this period, the Upper Orange–Olifants River was the main sediment input, and that the Cape Canyon was cut by it during the Oligocene lowstand. Neogene–Quaternary erosion has modified distal outline of parts of continental rise. Isopachs (km) from Gerrard and Smith (1982), converted thicknesses using velocity of 2 km/sec. BOR = base of rise; BOS = base of slope.

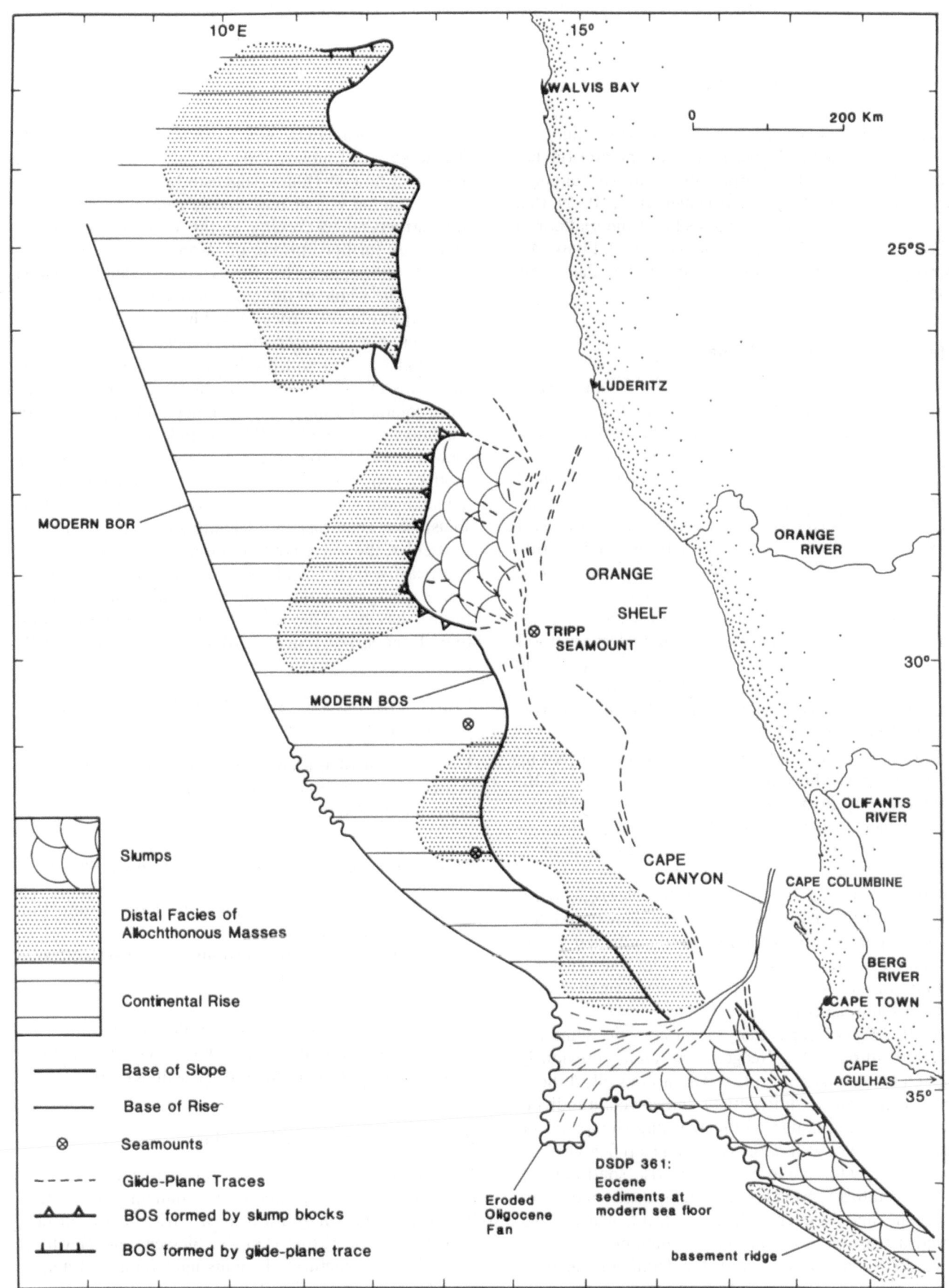

FIGURE 4-10. Neogene–Quaternary continental rises in relation to Pliocene–Pleistocene allochthonous masses (positions from Summerhayes, Bornhold, and Embley 1979; Dingle 1980). Upper surfaces and distal edges of rises have been extensively scoured by Antarctic Bottom Water flow from a northwesterly direction. Throughout this period, the whole of Orange River catchment debouched through modern estuary, but because paleoclimate of hinterland north of Olifants River was arid or semi-arid, sediment yield was very low. Areas shown as "Slumps" consist of faulted blocks; "Distal Facies" are debris-flow deposits and turbidites. BOR = base of rise; BOS = base of slope.

71

The original westward extent of the late Paleogene BOR is not known, because subsequent erosion has removed so much of the upper surface and distal edges (Tucholke and Embley, 1984; Dingle et al., 1987), that overcompacted Eocene sediments lie at the seafloor at DSDP Site 361 (Bolli et al., 1978) (Fig. 4-7). Figure 4-9 shows the present position of the eroded Paleogene rises.

Late Phase: Neogene and Quaternary

Since the end of the Paleogene, the Orange River drainage system has been composed of its modern sectors, although in the central subcontinent, their courses progressively shifted during Tertiary time (Dingle and Hendey, 1984). However, intensification of aridity in the hinterland since the late Miocene, occasioned by large-scale oceanic upwelling along the west coast (e.g., Siesser, 1978; Diester-Haass and Rothe, 1987), has resulted in a relatively low sediment yield from this source (< 5% of Late Cretaceous rates; Dingle and Hendey, 1984). As a consequence, post-Paleogene sediments on the continental margin off southwestern Africa are relatively thin (probably < 250 m; Dingle and Hendey, 1984), and with only limited borehole data available, it is difficult to correlate shelf and deep-sea deposits.

Practically all the outer margin has been affected by extensive post-Pliocene allochthonism (Fig. 4-10; Summerhayes, Bornhold, and Embley, 1979; Dingle, 1980), which locally has modified the morphology of the slope and upper rise (Fig. 4-6; see also Chapters 3, 7, 9, 10, 12, 13). The distal parts of these slumps extended as far as 250 km across the Neogene rises, and much of the sediment involved probably originated on the upper sectors of the rise. In the affected areas, the seafloor frequently shows microrelief that reflects a tectonic origin: for example, hummocks, small faults, and tensional depressions.

South of the Cape Canyon, large-scale slumping occurred during the Neogene in areas previously affected by Paleogene and Late Cretaceous allochthonism. It is difficult to differentiate and correlate individual structures, although relatively young slumps occur along the eastern wall of the canyon. Deep Sea Drilling Project Site 360, on the continental rise southwest of Cape Agulhas, provides stratigraphical data on Oligocene and Miocene–Pliocene strata, which have moved downslope over Oligocene sediments (see Bolli et al., 1978, and Dingle, 1980, for details of the microtectonism and stratigraphy).

Late Miocene scour by Antarctic Bottom Water (AABW) removed about 500 m of post-upper Eocene sediment from the vicinity of DSDP Site 361 (Bolli et al., 1978), and widely affected the sediments of the Paleogene continental rises farther afield off south-western Africa (Figs. 4-9, 4-10; Emery et al., 1975; Tucholke and Embley, 1984; see also Chapters 5, 11). Dingle et al. (1987) have suggested that the late Cenozoic slumping was at least partly caused by the resultant structural weakening at the base of continental slopes. A further episode of strong AABW scour occurred subsequent to the youngest allochthonous events. The whole of the modern outer continental rise area off southwest Africa probably lies under a zone of vigorous bottom-water flow, which has maintained omission surfaces and produced ferromanganese nodule fields over large areas of the eastern Cape Basin since middle Tertiary time (Embley and Morley, 1980; Tucholke and Embley, 1984; Dingle et al., 1987). In view of this bottom-water activity, the base of the modern continental rise (Fig. 4-10) is probably an erosional feature formed of sediments no younger than Pleistocene (contrast with Chapters 8, 9, 10, 12, 13). In particular, the lower continental margin in the extreme southeast, which lies at the entrance to narrow deep-water passages connecting the Cape and Agulhas basins, has been the site of especially intense scour (e.g., Rogers, 1987; Siesser, Rogers, and Winter, 1988).

PRINCIPAL FACTORS IN CONTINENTAL-RISE EVOLUTION

Development of the continental rise off southwestern Africa has been controlled by three principal factors since middle Cretaceous time:

1. Relative buoyancy of the basement along the continental margin.
2. Rate of dispersal of continental detritus onto the outer continental margin.
3. Energy of deep-water currents in the eastern Cape Basin (see Chapter 6).

Each of these factors has been dominant from time to time in determining the location of the base of slope and the nature of continental-rise sedimentation.

Buoyancy of Continental Margins

Regional tectonic trends in the pre-Mesozoic basement determined the disposition of sediment basins and basement highs off southwestern Africa at the inception of Jurassic rifting (Dingle, Siesser, and Newton, 1983). These structural elements have remained dominant features of continental margin development ever since (Fig. 4-2), and the width and structure of the outer margins off southwestern Africa reflect these fundamental differences.

Sedimentary processes along the outer margin south of Cape Columbine have operated in a regime that predetermined a thin sediment sequence on the shelf and a steep continental slope. This slope was maintained by continual faulting of the unstable sediment pile, probably compounded by relatively high seismicity in the southwestern part of the continent. This region is crossed by numerous long fault zones, trending subparallel to the Cape Fold Belt orogenic trend, some parts of which are still active (Dingle, Siesser, and Newton, 1983). Rise accretion adjacent to the Columbine–Agulhas Arch has been caused primarily by slumping, with probably some injection along the lower rise of distal turbidites from the Orange Basin. To a large extent, therefore, development of the outer margin here has been independent of environmental factors, and the rise remained narrow and rugged from Late Cretaceous to Neogene time. As a result, the main depocenter in this region lies under the lower slope and upper rise.

In contrast, the sedimentary profile that developed on the outer margin of the Orange Basin was effectively a succession of massive prograding strata, which accumulated on rapidly subsiding continental crust. These features were susceptible to the types of paleoenvironmental change that determined the route and quantity of sediment input (e.g., those affecting the paleoclimate and geomorphology of the hinterland). Consequently, development of the continental rise off the Orange Basin reflects similar major environmental changes through aspects such as speed of advancement of base of slope, thickness of sedimentary column, and width of rise (compare with Chapter 6).

Dispersal of Continental Detritus

During Cretaceous time, several drainage systems (the largest of which was the Orange River) delivered large quantities of detritus to the southwest African continental margin (compare with Chapters 2, 5, 6). Temporal variations in the quantity of sediment delivered produced dramatic changes in the nature of the outer margin. In the Albian–late Cenomanian interval, the continental slope advanced rapidly across thin foresets that hardly warrant the term continental rise. However, a large increase in sediment supply during the early Turonian rapidly established a large bathymetric discontinuity on the outer margin. Along the face of the sediment wedge, faulting stabilized the position of the shelf edge (Figs. 4-6, 4-8) and injected allochthonous material into deep water. This created a continental rise that prograded far into the Cape Basin. The reason for the increase in sedimentation rate in post-Cenomanian (and particularly Santonian) time is not clear. Rust and Summerfield (1990) suggested that the paleo-Orange River was not established prior to Coniacian time, but Dingle, Siesser, and Newton (1983, p. 232) identified the Coniacian–Santonian interval

as one of high sediment yields around the whole subcontinent. So the increase may have been related to a regional paleoclimatic change. A similarly dated period of increased sediment supply on the middle Atlantic continental margin of the United States has been ascribed to Appalachian tectonism (e.g., Poag and Sevon, 1989; see also Chapter 6), but an element of global paleoclimatic change could have influenced both margins.

Post-Cretaceous sediment supply to the southwest African continental margin declined stepwise as aridity increased along the western seaboard and the hinterland. Reorganization of the Orange River drainage patterns at the end of the Cretaceous, and again in the Oligocene, resulted in a temporary locus of sediment accumulation at the southern end of the Orange Basin. But the style of outer margin sedimentation and advance was regionally similar, albeit at somewhat lower rates and in less spectacular fashion, to the late Cretaceous situation. It is apparent that throughout the Albian to Oligocene interval (while there was an adequate supply of continental detritus), the continental rise developed by accretion of the distal facies of continental-slope–generated slumps, and not by contour-parallel accretion through deep-sea currents.

Correlation of alterations in relative sea level with changes in deep-water sedimentation requires a detailed stratigraphic and sedimentologic analysis of cored sequences (e.g., Shanmugam and Moiola, 1982; Cotillon, 1987), and this is not available for the pre-Quaternary succession off southwestern Africa. Consequently, only major events, loosely tied in with sea-level fluctuations can be assessed. In this category, the incision of the Cape Canyon during an Oligocene low has already been cited. In addition, the end-of-Cretaceous to early Tertiary lowstand may correlate with the sedimentary event associated with reflector D, and the Late Cretaceous highstand may correlate with the smooth, post-slump reflections immediately below reflector D on the lower slope (Fig. 4-3; Siesser and Dingle, 1981).

The effect of Quaternary glacial (and by definition, eustatic) events on upwelling-related shelf sediments off Namibia has been discussed by Diester-Haass and Rothe (1987), and possible outer shelf erosion by the Benguela Current has been described by Van Andel and Calvert (1971; see also Chapter 5). In addition, both Summerhayes, Bornhold, and Embley (1979) and Dingle (1980) speculated that some of the Pleistocene (?) outer margin slumping may be related to eustatic fluctuations. However, changes in the relatively complex paleo-oceanographic (and paleoclimatic) environment off southwestern Africa are poorly understood. The efficacy of Agulhas water penetration into the southeast Atlantic and the location and intensity of Benguela system upwelling have not been satisfactorily established for glacial episodes, and no overall Quaternary sedimentary model for the southwest African margin has been presented.

Circulation of Deep-Water Masses

The abundant sediment supplies necessary to maintain high-angle continental slopes through slumping of rapidly deposited sediment, and the consequent debris-flow– and turbidite-generated rises, ceased by early Neogene time. Siesser (1978) detected evidence of upwelling along the shelf in the Oligocene, and intense upwelling in the late Miocene (see also Chapters 5, 12). From this time onward, the dominant factor in the development and modification of the lower slopes and rises off southwestern Africa became deep-water current activity in the Cape Basin.

There is some uncertainty as to when the modern deep-water circulation patterns were initiated, but suggestions of the Eocene/Oligocene boundary for North Atlantic Deep Water (NADW), and early Oligocene for Antarctic Bottom Water (AABW) (Johnson, 1982) accord well with phenomena ascribed to their activity around southern Africa (Tucholke and Embley, 1984; Dingle et al., 1987).

Because the Walvis Ridge forms an almost complete barrier between the Cape and Angola Basins, AABW circulation off southwest Africa follows a clockwise route around the Cape Basin, flowing southeastward off southwest Africa (Embley and Morley, 1980; Tucholke and Embley, 1984; McCave, 1986). North Atlantic Deep Water also moves southeastward, with the result that, at present, the whole water column below about 1,500 m off southwestern Africa is moving in the same direction. Because this situation is primarily controlled by basin geometry, we suggest that it has prevailed probably since the inception of these two deep-water masses.

Despite these strong currents, there is no evidence of large current-generated bedforms in the southeast Cape Basin (contrast with Chapters 6, 8, 9, 12). On the contrary, their effect, and in particular that of the AABW, has been to cause massive erosion of the continental rise and lower slope over large areas, at least as far west as the Walvis Ridge (Bornhold and Summerhayes, 1977; Tucholke and Embley, 1984). Rogers (1987) has reported erosion on the continental rise off the southern tip of Africa in the form of channels as deep as 35 m, spaced 2 km apart in level terrain. Some channels form 100-m-deep and 20-km-wide moats around basement features.

Dingle et al. (1987) considered a corollary of this erosion to be the late Neogene slumping that affected the youngest sediments on the slopes and rises off southwest Africa. It should be noted that although this allochthonism is considered to have been facilitated by

lower slope undercutting, south of Luderitz the head regions of these young slumps frequently coincide with still-active extensions of both the Paleogene and Late Cretaceous rotational faults [e.g., north of the Tripp Seamount (Fig. 4-10), on the outer edge of Childs Bank (Fig. 4-1), and at the shelfbreak immediately west of the Kudu 9 boreholes (Figs. 4-1, 4-10); see also Chapters 9, 10, 12, 13].

A further factor affecting Neogene outer margin sedimentation is the sediment starvation to which this area has been subjected, with the possible exceptions of Pleistocene glacial episodes. Even in the latter case, it is probable that Orange River muds, which are presently carried southward by nearshore currents (Rogers 1977; Bremner, Rogers, and Willis, 1990; Shillington, Brundrit, and Lutjeharms, 1990), would have been funnelled off the shelf via the Cape Canyon, only to be entrained in mid-water currents (NADW) and carried southeastward out of the region, in the manner suggested by Gardner (1989).

The large amounts of sediment removed from the continental rises and lower slopes off southwest Africa by deep-sea erosion since late Miocene time, appear to have been swept out of the Cape Basin through the Agulhas Passage (off the southern tip of Africa), and deposited in the southwest Indian Ocean as a series of massive longitudinal bedforms on the continental rises off southeast Africa (Dingle and Camden-Smith, 1979; Westall, 1984; Dingle and Robson, 1985; Dingle et al., 1987). There is evidence of limited Pleistocene deposition on the continental rise in the southern part of the area in the form of coarse detritus (sand to boulder size) deposited from icebergs (Needham, 1962; Rogers, 1987).

Finally, although intensified AABW scour took place in early Pleistocene time (with sediment erosion and ferromanganese nodule formation), mud drapes over nodule pavements are evidence that Holocene energy levels on the continental rise in the southern part of the study area (Fig. 4-10) are somewhat lower (Rogers 1987; compare with Chapters 8, 9, 12, 13).

SUMMARY

Development of the outer continental margin off southwest Africa can be traced from Albian time. Contrasts in the degree of continental basement subsidence to the north and south of Cape Columbine, gave rise to two styles of rise accretion.

In the Orange–Luderitz basin, rapid basement subsidence and large sediment supply resulted in wide continental rises constructed of distal, slope-derived allochthonous material during Turonian to Oligocene time. Successive migration of the Orange River exit point switched the main centers of sediment accumulation from north to south, and back, and resulted in

Oligocene incision of the Cape Canyon and creation of a deep-water fan. The onset of severe coastal aridity in late Miocene time approximately coincided with the establishment of vigorous scour by Antarctic Bottom Water in the eastern Cape Basin, with the result that the development of Neogene–Quaternary continental rises has been primarily an interval of intense erosion of their upper surfaces and distal margins. Young (Pleistocene ?) slumps have further modified the slope and rise profiles.

Farther south, adjacent to the buoyant Columbine–Agulhas continental margin, rise and lower slope accretion during the Late Cretaceous–Oligocene period was by superposition of large slump blocks (in the south dammed behind basement ridges). In common with the rest of the margin, there was intense Neogene–Quaternary erosion by AABW of the outer margin off the southwest tip of Africa.

Material removed from the continental rises and lower slopes in the Cape Basin by late Cenozoic AABW flow has been transferred to the southwestern Indian Ocean, where large longitudinal bedforms were constructed on the continental rises off south and southeastern Africa.

ACKNOWLEDGMENTS

Our own original observations at sea, which we undertook while at the University of Cape Town, were funded by the South African National Committee for Oceanographic Research and the South African Geological Survey. RVD acknowledges funding from the Foundation for Research Development and South African Museum for compilation. We are grateful to George Smith of the Southern Oil Exploration Corporation (SOEKOR) and to Dr. Roy Miller, Director of the Geological Survey of Namibia, for providing unpublished seismic data for our study, and to SOEKOR for allowing SHR to participate in the project. We thank Judy Woodford for drafting the figures.

REFERENCES

Austin, J. A., and Uchupi, E. 1982. Continental–oceanic crust transition off southwest Africa. *American Association of Petroleum Geologists Bulletin* 66:1328–1347.

Bolli, H. M., Ryan, W. B. F., Foresman, J. B., Hottman, W. E., Kagami, H., Longoria, J. F., McKnight, B. K., Melguen, M., Natland, J., Proto-Decima, F., and Siesser, W. G. 1978. Cape Basin continental rise—sites 360 and 361. *In* Initial Reports of the Deep Sea Drilling Project, *Volume 40*, H. M. Bolli, W. B. F. Ryan, et al.: Washington DC: U.S. Government Printing Office, 29–182.

Bornhold, B. D., and Summerhayes, C. P. 1977. Scour and deposition at the foot of the Walvis Ridge in the northernmost Cape Basin, South Atlantic, *Deep-Sea Research* 24:743–752.

Bremner, J. M., Rogers, J., and Willis, J. P. 1990. Sedimentological aspects of the 1988 Orange River floods. *Transactions of the Royal Society of South Africa* 47:247–305.

Cotillon, P. 1987. Bed-scale cyclicity of pelagic Cretaceous successions as a result of world-wide control. *Marine Geology* 78:109–123.

Diester-Haass, L., and Rothe, P. 1987. Plio-Pleistocene sedimentation on the Walvis Ridge, southeast Atlantic (DSDP Leg 75, Site 532)—influence of surface currents, carbonate dissolution and climate. *Marine Geology* 77:53–85.

Dingle, R. V. 1980. Large allochthonous sediment masses and their role in the construction of the continental slope and rise off southwestern Africa. *Marine Geology* 37:333–354.

Dingle, R. V., and Camden-Smith, F. 1979. Acoustic stratigraphy and current-generated bedforms in deep ocean basins off southeastern Africa. *Marine Geology* 33:239–260.

Dingle, R. V., and Hendey, Q. H. 1984. Late Mesozoic and Tertiary sediment supply to the eastern Cape Basin (SE Atlantic) and palaeo-drainage systems in southwestern Africa. *Marine Geology* 56:13–26.

Dingle, R. V., and Robson, S. H. 1985. Slumps, canyons and related features on the continental margin off East London, SE Africa (SW Indian Ocean). *Marine Geology* 67:37–54.

Dingle, R. V., Siesser, W. G., and Newton, A. R. 1983. *Mesozoic and Tertiary Geology of Southern Africa*. Rotterdam: Balkema.

Dingle, R. V., Birch, G. F., Bremner, J. M., de Decker, R. H., du Plessis, A., Engelbrecht, J. C., Fincham, M. J., Fitton, T., Flemming, B. W., Gentle, R. I., Goodlad, S. W., Martin, A. K., Mills, E. G., Moir, G. J., Parker, R. J., Robson, S. H., Rogers, J., Salmon, D. A., Siesser, W. G., Simpson, E. S. W., Summerhayes, C. P., Westall, F., Winter, A., and Woodborne, M. W. 1987. Deep-sea sedimentary environments around southern Africa (South-East Atlantic and South-West Indian Oceans). *Annals of the South African Museum* 98:1–27.

Embley, R. W., and Morley, J. J. 1980. Quaternary sedimentation and palaeoenvironmental studies off Namibia (south-west Africa). *Marine Geology* 36:183–204.

Emery, K. O., Uchupi, E., Bowin, C. O., Phillips, J., and Simpson, E. S. W. 1975. Continental margin off western Africa: Cape St. Francis (South Africa) to Walvis Ridge (South-West Africa). *American Association of Petroleum Geologists Bulletin* 59:3–59.

Gardner, W. D. 1989. Baltimore Canyon as a modern conduit of sediment to the deep sea. *Deep-Sea Research* 36:323–358.

Gerrard, I., and Smith, G. C. 1982. Post-Paleozoic succession and structures of the southwestern African continental margin. *In*: Studies in Continental Margin Geology, J. S. Watkins and C. L. Drake, Eds.: *American Association of Petroleum Geologists Memoir 34* 49–74.

Goslin, J., Mascle, J., Sibuet, J. C., and Hoskins, H. 1974. Geophysical study of the easternmost Walvis Ridge, South Atlantic: morphology and shallow structure. *Geological Society of American Bulletin* 85:619–632.

Harland, W. B., Cox, A. V., Llewellyn, P. G., Pickton, C. A. G., Smith, A. G., and Walters, R. 1982. *A Geologic time scale*. Cambridge: Cambridge University Press.

Johnson, D. A. 1982. Abyssal teleconnections II. Initiation of Antarctic Bottom Water flow in the southwestern Atlantic. *In*: South Atlantic Paleoceanography, K. J. Hsu and H. J. Weissert, Eds.: Cambridge: Cambridge University Press, 243–281.

McCave, I. N. 1986. Local and global aspects of the bottom nepheloid layers in the World Ocean. *Netherlands Journal of Sea Research* 20:167–181.

McMillan, I. K. 1990. Foraminiferal biostratigraphy of the Barremian to Miocene rocks of the KUDU 9A-1, 9A-2 and 9A-3 boreholes. *Communications of the Geological Survey of Namibia* 6:23–29.

Needham, H. D. 1962. Ice-rafted rocks from the Atlantic Ocean off the coast of the Cape of Good Hope. *Deep-Sea Research* 9:475–486.

Poag, C. W., and Sevon, W. D. 1989. A record of Appalachian denudation in postrift Mesozoic and Cenozoic sedimentary deposits of the U.S. middle Atlantic continental margin. *Geomorphology* 2:119–157.

Rogers, J. 1977. Sedimentation on the continental margin off the Orange River and the Namib Desert. *Bulletin Joint Geological Survey/University of Cape Town Marine Geoscience Unit* 7:1–212.

Rogers, J. 1987. Seismic, bathymetric and photographic evidence of widespread erosion and a manganese-nodule pavement along the continental rise of the southeast Cape Basin. *Marine Geology* 78:57–76.

Rust, D. J., and Summerfield, M. A. 1990. Isopach and borehole data as indicators of rifted margin evolution in southwestern Africa. *Marine and Petroleum Geology* 7:277–287.

Shanmugam, G., and Moiola, R. J. 1982. Eustatic control of turbidites and winnowed turbidites. *Geology* 10:231–235.

Shillington, F. A., Brundrit, G. B., and Lutjeharms, J. R. E. 1990. The coastal current circulation during the Orange River flood 1988. *Transactions of the Royal Society of South Africa* 47:307–330.

Siesser, W. G. 1978. Aridification of the Namib Desert: evidence from oceanic cores. *In* Antarctic Glacial History and World Palaeo-environments, E. M. Van Zinderen-Bakker, Ed.: Rotterdam: Balkema, 105–113.

Siesser, W. G., and Dingle, R. V. 1981. Tertiary sea level movements around Southern Africa. *Journal of Geology* 89:83–96.

Siesser, W. G., Rogers, J., and Winter, A. 1988. Late Neogene erosion of the Agulhas Moat and the Oligocene position of Subantarctic surface water. *Marine Geology* 80:119–129.

Summerhayes, C. P., Bornhold, B. D., and Embley, R. W. 1979. Surficial slides and slumps on the continental slope and rise of southwest Africa: a reconnaissance study. *Marine Geology* 31:265–277.

Tucholke, B. E., and Embley, R. W. 1984. Cenozoic regional erosion of the abyssal sea floor off South Africa. *American Association of Petroleum Geologists Memoir 36* 145–164.

Uchupi, E., and Emery, K. O. 1972. Seismic reflection, magnetic, and gravity profiles of the eastern Atlantic continental margin and adjacent deep-sea floor. I Cape Francis (South Africa), to Congo Canyon (Republic of Zaire). *Woods Hole Oceanographic Institution, Reference No. 72-95*: 8 sheets.

Van Andel, Tj. H., and Calvert, S. E. 1971. Evolution of the sediment wedge, Walvis Shelf, southwest Africa. *Journal of Geology* 27:585–602.

Westall, F. 1984. Current-controlled sedimentation in the Agulhas Passage, SW Indian Ocean. *Bulletin Joint Geological Survey/University of Cape Town Marine Geoscience Unit* 12:1–276.

5

Angola Basin: Geohistory and Construction of the Continental Rise

Elazar Uchupi

The Angola Basin off western equatorial Africa extends from the Guinea Ridge/Cameroon Volcanic Ridge, near the Equator, to the Walvis Ridge at lat. 18°S and forms a geologic link between the North and South Atlantic (Fig. 5-1). This basin is the deepest depression in the eastern South Atlantic, and is nearly isolated from the other basins in the South Atlantic; its bottom-water mass is significantly warmer than that of the other basins and the CCD (carbonate compensation depth) is much deeper (Bolli et al., 1978). Congo Canyon (Fig. 5-2), the only active submarine canyon in the Atlantic, serves as a pathway for turbidity currents to reach the distal parts of the basin. Antarctic Bottom Water (AABW) reaches the region from west of the Mid-Atlantic Ridge via the west–east trending Romanche Fracture Zone at lat. 1°S (Fig. 5-2). This water mass enters the Guninea Basin (Fig. 5-1) east of the ridge, and thence flows southward across a 4,800-m-deep sill on the Guinea Ridge (Figs. 5-1, 5-2; Bolli et al., 1978). Once in the Angola Basin, the bottom water circulates clockwise within a central gyre. Some AABW also reaches the southwest corner of the Angola Basin from the Cape Basin (Fig. 5-1) via a 3,900-m-deep gap at lat. 36°S, long. 7°W, west of the Walvis Ridge. A barrier or barriers located north of lat. 32°S prevent much of this minor flow from spreading into the main part of the basin north of lat. 20°S (Connary, 1972). The bottom-water flow that does reach the basin is very weak, as evidenced by the Miocene and younger sediments at Deep Sea Drilling Project Site 364, which are indicative of partially anoxic bottom conditions (The Shipboard Scientific Party, 1978b). Deep and intermediate water masses reach the Angola Basin from the south via sills at lat. 36.5°S, long. 5°W and lat.

30°S, long. 1°E, and a deeply eroded saddle (3,200 m deep) across the Walvis Ridge (Figs. 5-1, 5-2; Bolli, et al., 1978). The cold north-flowing Benguela Current (a surface current; Fig. 5-2) is deflected westward away from the Angola Basin by the Walvis Ridge. These conditions, weak bottom-current activity, isolation from cold north-flowing surface currents, and the presence of the Congo Canyon, make the Angola Basin particularly significant to our understanding of the depositional history of the Atlantic. Unfortunately, data that can be used to reconstruct its geology are limited. Because of their petroleum reserves the Cuanza, Congo–Cabinda, and Gabon coastal basins (Fig. 5-3) and adjacent inner shelves are relatively well known (Belmonte, Hirtz, and Wenger, 1965; Brognon and Verrier, 1966; Reyre, 1966; Hedberg, 1968; Franks and Nairn, 1973; Brink, 1974; Brice et al., 1982; Dailly, 1982; Sieglie and Baker, 1982; see also Chapter 2). The other nonpetroliferous coastal basins (Douala and Moçâmedes; Fig. 5-3), however, are not well documented. Knowledge seaward of the inner continental shelf is based mainly on single and multichannel seismic reflection profiles and oblique reflection and refraction measurements obtained through the use of sonobuoys (Baumgartner and van Andel, 1971; Leyden, Bryan, and Ewing, 1972; Uchupi and Emery, 1972, 1974; Von Herzen, Hoskins and van Andel, 1972; Pautot et al., 1973; Beck and Lehner, 1974; Emery et al., 1975b; Leyden et al., 1976; Lehner and de Ruiter, 1977). The Deep Sea Drilling Project has occupied three sites in Angola Basin, two near the Angola Escarpment (Sites 364 and 365; Bolli, Ryan, et al., 1978; The Shipboard Scientific Party, 1978b) and one (Site 530; Hay, Sibuet, et al., 1984) at the southern end

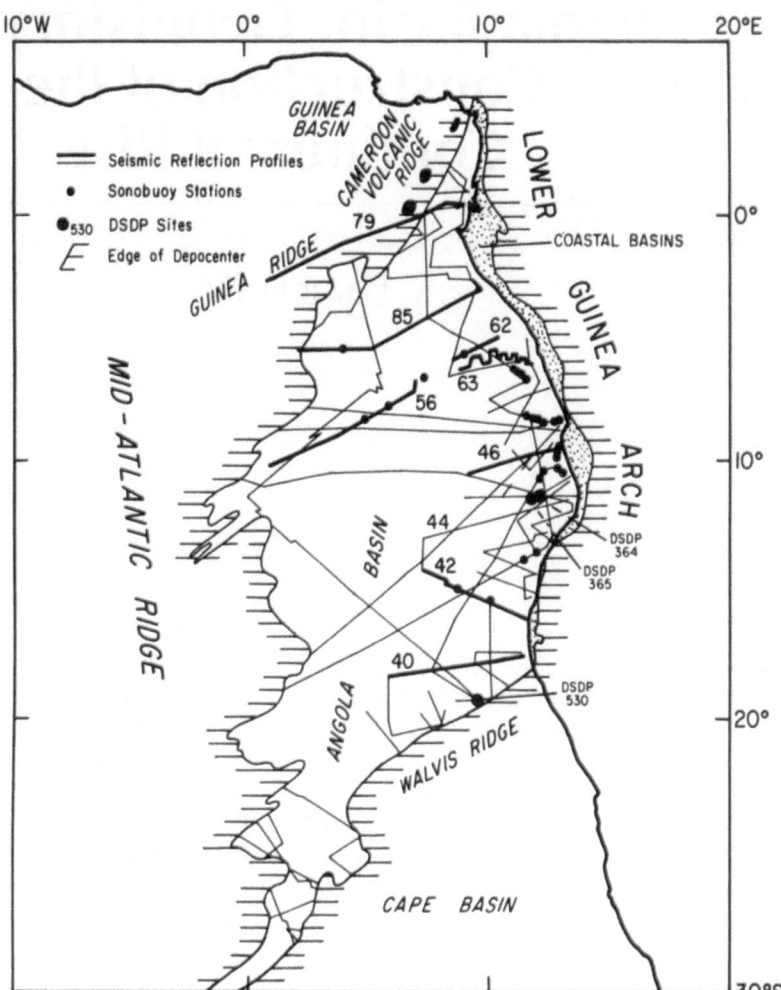

FIGURE 5-1.   Distribution of single and multichannel seismic reflection profiles, sonobuoy stations, and Deep Sea Drilling Project sites in the Angola Basin. Compiled from Leyden, Bryan, and Ewing (1972), Uchupi and Emery (1972, 1974), The Shipboard Scientific Party (1978b), and Hay, Sibuet, et al. (1984). Numbered heavier lines show positions of seismic reflection profiles in Figs. 5-8 and 5-11. Also indicated is the location of profile 44, a segment of which is shown in Fig. 5-7.

of the Angola Basin, immediately north of the Walvis Ridge (Fig. 5-1).

MORPHOLOGY

The coastal upland of equatorial Africa north of lat. 17°S is dominated by the north-trending Lower Guinea Arch, which reaches an elevation greater than 2,000 m (Figs. 5-2, 5-3). South of lat. 17°S, the land rises eastward in wide pediments to elevations of 1,000 to 1,200 m at the base of the Great Escarpment (Fig. 5-3), 140 km from the coast (Martin, 1973). From the crest of the scarp at elevations as great as 2,300 m, the region descends gradually to the interior Kalahari and Etosha basins (Fig. 5-3). Extending their courses across the north–south-trending Great Escarpment are four major river systems: Congo, Cuanza, Cunene, and Orange (Fig. 5-2). The coastal plain is narrow, being widest

(150 km) in Angola (Fig. 5-2; Franks and Nairn, 1973). The north–south continuity of the coastal plain is disrupted by basement spurs, which reach the coast; the largest is Luanda High, which separates Moçâmedes Basin from Cuanza Basin (Fig. 5-3). The continental shelf is narrower than 100 km, and the average depth of the shelfbreak is about 100 m (Fig. 5-2; Emery et al., 1975b). The continental slope seaward of the shelf consists of two provinces, a very hilly upper part, which becomes somewhat subdued seaward, and a steep lower part, the Angola Escarpment (Fig. 5-3). Although there are no well data to indicate the nature of the features on the upper slope, the presence of Aptian salt diapirs in the coastal basins, the high salinity of interstitial waters encountered at Deep Sea Drilling Project sites 364 and 365 (Fig. 5-1; The Shipboard Scientific Party, 1978b), and high heat-flow measurements (Von Herzen, Hoskins, and van Andel, 1972), are evidence that these

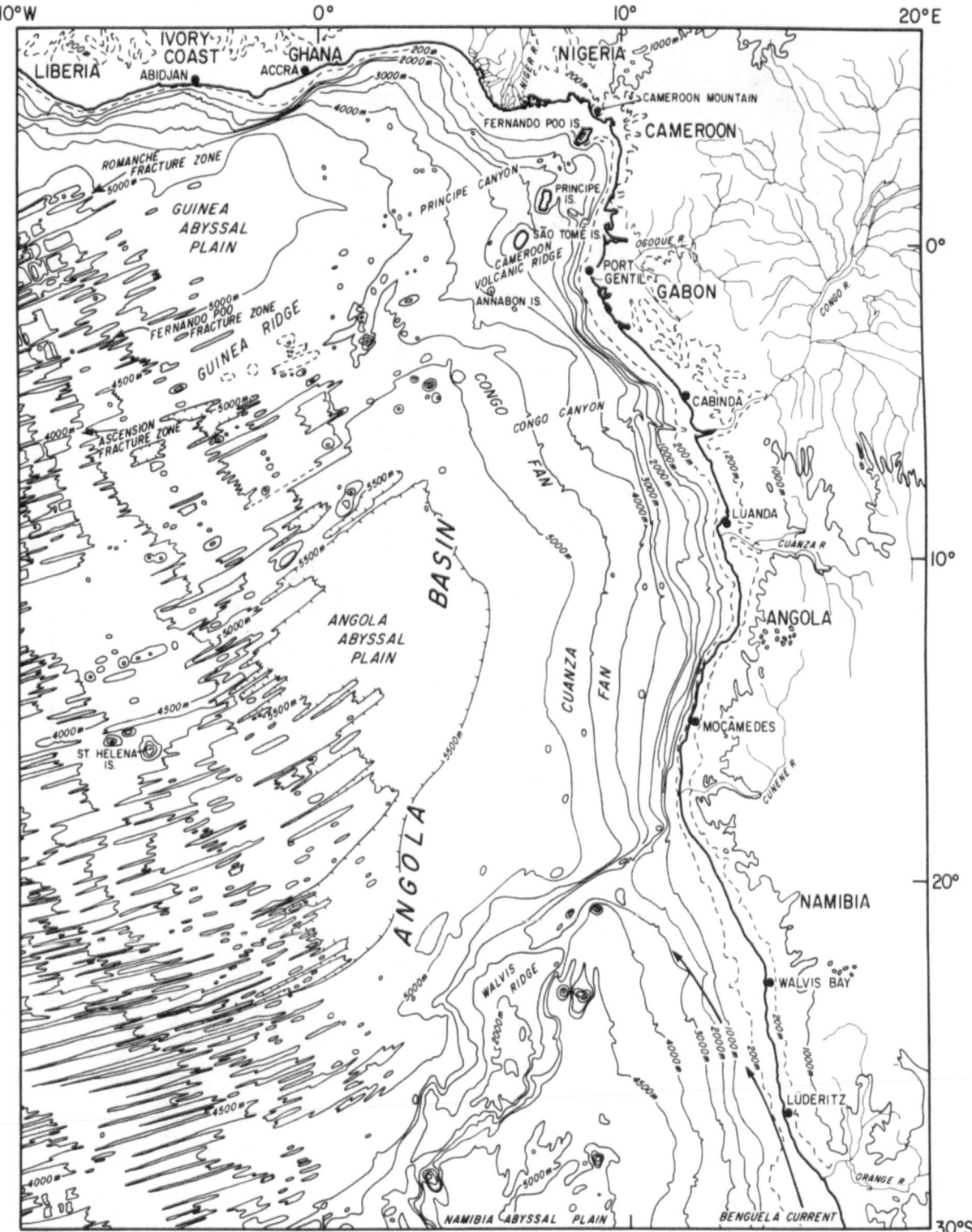

FIGURE 5-2.   Bathymetry of the Angola Basin and vicinity. Adopted from Heezen et al. (1978). Contours in meters.

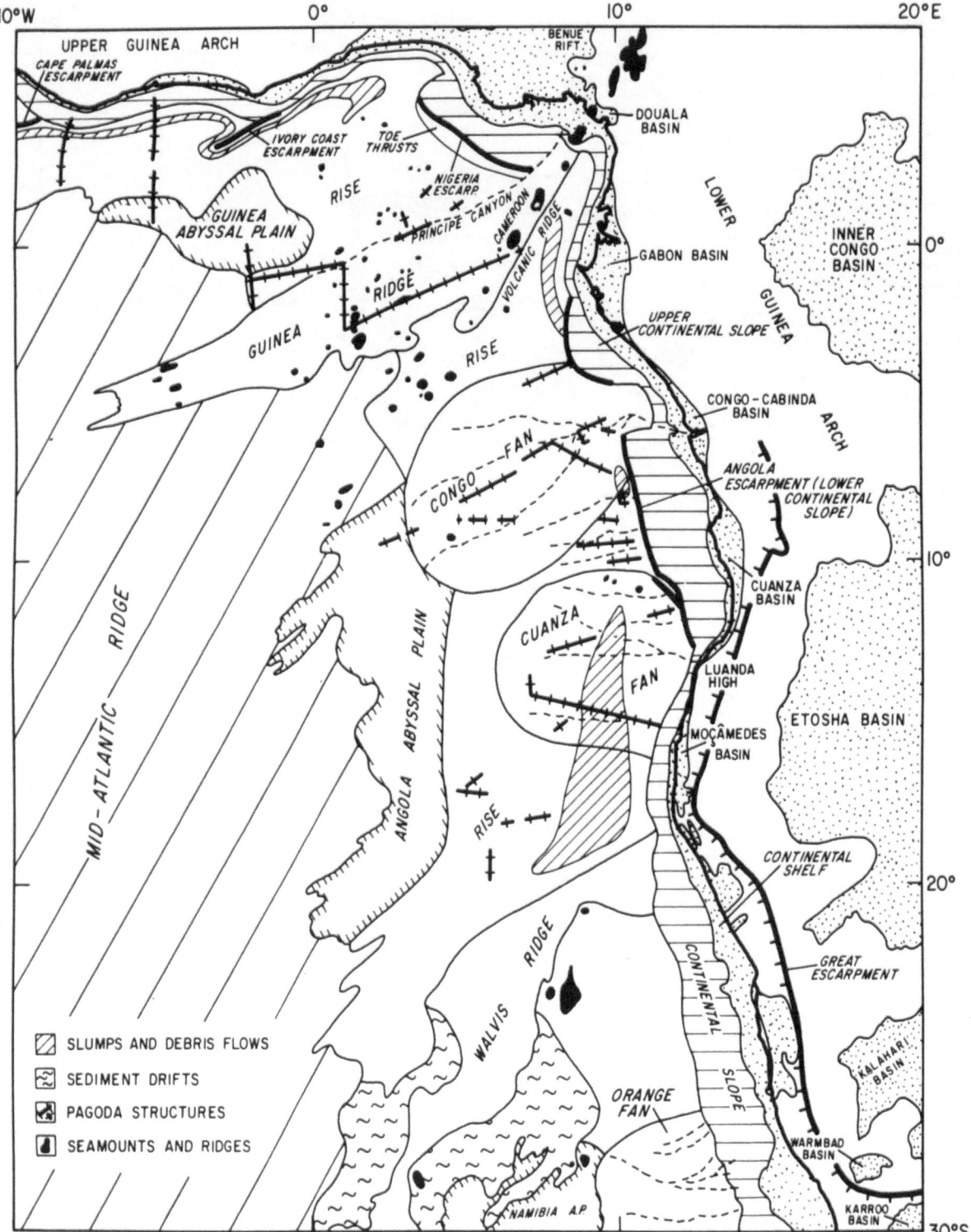

FIGURE 5-3. Physiographic provinces of the Angola Basin and vicinity and distribution of pagoda structures, sediment drifts, slumps, and slides. Compiled from Fig. 5-2, Emery et al. (1975b), and Emery and Uchupi (1984, Chart IIB).

features are salt-piercement structures. The smoother lower part of the upper slope represents a salt massif, and the Angola Escarpment is a salt front. The southern limit of the offshore diapiric structures and the Angola Escarpment is at about lat. 13°S, a limit related to a western extension of the Luanda High (Fig. 5-3). The Angola Escarpment has a gradient of 3–4° and extends between lats. 13° and 3°S. South of lat. 10°S, the scarp is flanked on the seaward side by an oceanic basement ridge (Fig. 5-3); north of Congo Canyon, the scarp gradually decreases in height until it disappears. A gap in the scarp off the Congo River is filled by the apex of the Congo Fan, into which the Congo Canyon has been incised. The Congo, the largest submarine canyon in the South Atlantic, and the only active one today in the Atlantic, has been the main conduit for sediments to the continental rise since the end of the Pleistocene. Recent turbidity-current activity along this canyon has been responsible for numerous submarine cable breaks in the entrance to the Congo estuary (Heezen et al., 1964).

The continental rise beyond the Angola Escarpment grades westward into the Angola Abyssal Plain, which, in turn, terminates along the foothills (abyssal hills) of the Mid-Atlantic Ridge (Fig. 5-3). The 100- to 880-km-wide continental rise is dominated by the Congo and Cuanza submarine fans. The largest of these two depocenters, the Congo Fan, is incised by two nearly parallel channels and several leveed distributaries (Heezen et al., 1964; Shepard and Emery, 1973). Cores taken from this section of the fan contained turbidite sand layers; two layers from a water depth of 4,000 m

contained leaves and twigs of land plants (Heezen et al., 1964). South of the Congo Fan is another sediment buildup, for which I propose the name Cuanza Fan. This poorly surveyed constructional feature is made of sediments transported to the site via submarine canyons associated with the Cunene River system (Fig. 5-2). The few echo-sounding profiles available from this region indicate that the surface of the fan also is marked by a complex distributary system. In the middle of this fan is a massive debris-flow deposit, which extends southward almost to the northern flank of the Walvis Ridge (Fig. 5-3). Similar deposits also occur on the upper part of the Congo Fan and on the upper continental rise off Gabon.

At the sill between the Cape and Angola basins, an extensive sediment drift field has been generated by AABW that flows into the southern Angola Basin. In addition to debris-flow deposits and sediment drifts, much of the continental rise surface is marked by bumpy topography; areas having poor internal acoustic reflectivity are separated by ones of high internal reflectivity in 3.5-kHz echo-sounding recordings (Figs. 5-3, 5-4). At least three origins have been proposed for these features. Emery (1974), who first described them and named them "pagoda structures," believes that they resulted from cementation by methane clathrates. Flood (1983) suggested that the structures may represent furrows eroded by bottom currents, or that they possibly represent debris-flow deposits. I am doubtful of Emery's explanation, because none of the 3.5-kHz recordings in the region has revealed a bottom simulating reflector (BSR) or acoustic wipeouts, which are

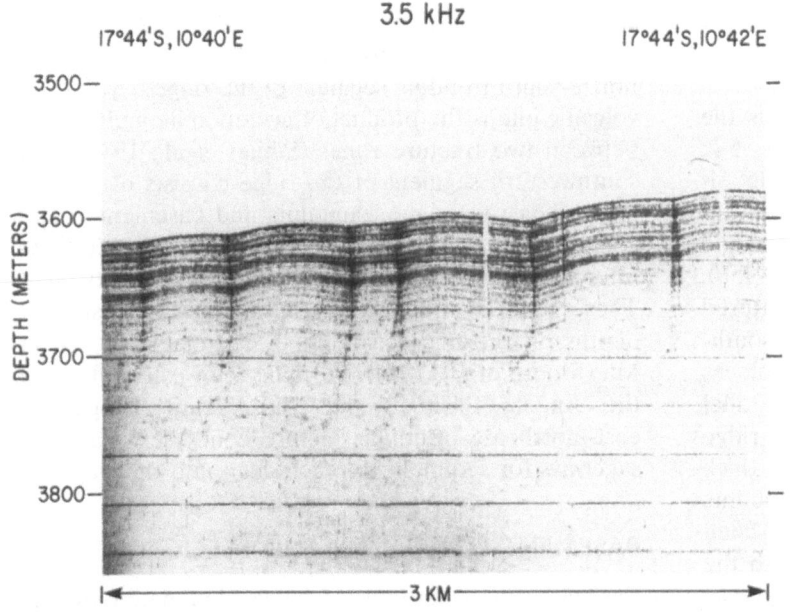

FIGURE 5-4.   A 3.5-kHz echo-sounding profile of pagoda structures in the Angola Basin. Vertical exaggeration 6 ×.

characteristic of areas known to be rich in clathrates. The structures on the Guinea Ridge could have been produced by AABW cascading down the topographic high, but the wide distribution of the features, from the upper rise to the distal edges of the Angola Abyssal Plain, and the sluggish nature of the bottom currents in the Angola Basin, indicate that most of the pagodas are not the result of erosion. An origin from debris-flow erosion is not unreasonable for some of the pagodas, particularly those associated with identifiable debris-flow deposits. Hyperbolic echoes associated with debris-flow deposits tend to overlap, however, and they display varying vertex elevations above the seafloor. Subbottom reflectors, where present, are conformable, an acoustic feature not characteristic of pagodas. I propose a fourth possible origin for these features, as inferred from their distribution. The 3.5-kHz echo-sounding recordings indicate that the pagodas tend to be concentrated between channels. Profile spacing in the Angola Basin is, unfortunately, too wide to determine the orientation of pagodas, but studies of the continental rise off the eastern United States suggest that they occur in lobes whose axes trend at right angles to the continental slope. Off the eastern United States, one field of undulating topography on the lower rise, trends parallel to the isobaths (the Lower Continental Rise Hills); another field comprises a series of lobes that trend at right angles to the isobaths, and are associated with deep-sea channels (Vassalo, Jacobi, and Shor, 1984; Laine, Damuth, and Jacobi, 1986). Numerous studies have demonstrated that the undulating topography on the lower rise is related to the southwesterly flowing Western Boundary Undercurrent. I suspect that the bedforms in lobes trending at right angles to the isobaths (both off the eastern United States and in the Angola Basin) were formed not by bottom currents, but by turbidity currents that overflowed their banks.

The northern terminus of the Angola Basin is the Guinea Ridge–Cameroon Volcanic Ridge (Figs. 5-2, 5-3). The east–northeast trending Guinea Ridge is flanked by the Ascension Fracture Zone on the southeast and by the Fernando Poo Fracture Zone on the northwest (Fig. 5-2). According to Emery et al. (1975b), at least 15 seamounts and abyssal hills are scattered along the crest of the Guinea Ridge. Along the southern side of the ridge, at least three basement blocks support three seamounts, which are elongate parallel to the trend of the ridge. Roughness of the ridge topography has been smoothed by a relatively thick sedimentary veneer. The presence of many seamounts is evidence that the Guinea Ridge may have been formed by a mantle plume, but its alignment with the fracture zones is more indicative of leakage along these zones of structural weakness. Cutting obliquely across the Guinea Ridge at its eastern end is the northeast-trending Cameroon Volcanic Ridge (Fig. 5-3). Rising above the crest of the ridge are the oceanic islands of Fernando Poo, Principe, São Tomé, and Annabon (Fig. 5-2). Between the islands are oceanic basement highs and several seamounts, which can be traced to the intersection of the Cameroon Volcanic Ridge and the Guinea Ridge. Annabon, at the southwest end of the island chain, consists of Miocene or younger basaltic rocks (Hedberg, 1968). São Tomé may be an oceanic basement block uplifted after having been covered by Cretaceous and younger continental-rise sediments (Grunau et al., 1975; Gorini, 1977). The basaltic Fernando Poo was formed by volcanic activity during the Pleistocene. A basement high northeast of Fernando Poo links the oceanic part of the Cameroon Volcanic Ridge with its continental part (a northeast-trending chain of alkalic volcanoes). This basement high defines the northwest boundary of the Douala Basin (Fig. 5-3).

The 2,000-km-long Walvis Ridge, at the southern end of the Angola Basin, extends from the continental shelf to the eastern flank of the Mid-Atlantic Ridge, and is made of two northeast-trending segments separated by a north–south trending one (Figs. 5-2, 5-3). The eastern end of the northeast-oriented segment near the coast consists of continental crust rifted during the Late Jurassic (Oxfordian–Kimmeridgian); the rifted low was filled with a volcaniclastic sequence of seaward-dipping reflectors. According to Sibuet et al. (1984), the axis of this rift basin trends 350°, and links the Angola and Namibia margins. The Walvis Ridge segment, beyond this rift basin, is an oceanic basement high, having been emplaced either near sea level (as suggested by the presence of late Aptian shallow-water calcareous algae), or above sea level (as suggested by the presence of erosional surfaces). Sibuet et al. (1984) believed that part of this crust was emplaced before any oceanic crust was formed in the Angola Basin. The north–south trending segment of the ridge beyond this volcanic pile is the product of accretion along its edges between two fracture zones (Sibuet et al., 1984). The southwestern segment of the ridge consists of a chain of northeast-trending seamounts and basement blocks composed of pillow basalts and massive volcanic flows intercalated with nannofossil chalks and limestones. These pillows and flows were emplaced at shallow depths on the mid-ocean ridge, approximately 69–71 Ma (Moore et al., 1984). Forming the western flank of the Angola Basin is the Mid-Atlantic Ridge. Its east–northeast trending fracture zones serve as passageways for sediment to reach deep into the ridge.

BASEMENT LITHOLOGY AND STRUCTURE

Basement in equatorial Africa is dominated by the Congo craton, a product of many orogenies that took place during the Precambrian. The youngest tectonic belt around the periphery of the craton, the Pan-African belt, was formed during the opening and subsequent closing of a proto-South Atlantic in Late Precambrian time, which involved the collision of South America and Africa and the closure of the intracontinental Damara aulacogen in the Early Cambrian (Porada, 1979; Martin, 1983). These terranes have undergone rifting to form the foundation of the coastal sedimentary basins. Earlier, I proposed (Uchupi, 1989) that the fundamental structure of this rift is a chain of half-grabens (with the master fault dipping westward), which are opposed by platforms on the Brazilian conjugate margin. Contemporaneous with this extensional phase was development of the Cretaceous West and Central African rifts, which are associated with wrench faults that diverge from the Gulf of Guinea into central Africa at the northern end of the equatorial Atlantic rift (Fairhead and Greene, 1989). Basement highs separating the coastal rift basins may represent traces of fracture zones that were synchronous with rifting of the equatorial Atlantic (Le Pichon and Hayes, 1971). This extension, combined with subsequent subsidence due to thermal decay and to sediment loading in the coastal basins, has depressed the basement along the continental/oceanic crustal boundary to depths in excess of 9 km (Emery and Uchupi, 1984, Charts VA, VB).

The nature of the crust on which the evaporites were deposited is a point of contention. Some authors believe that the offshore evaporites were deposited on oceanic basement; they place the continental/oceanic crustal boundary along or near the landward edge of the offshore massive diapiric field. Cande and Rabinowitz (1978), for example, believed that they could distinguish seafloor spreading magnetic anomalies beneath the evaporites; they proposed that spreading in the southern part of the Brazil/Angola basins began before the formation of either magnetic anomaly MO or M3. They supposed that seafloor spreading continued until the end of evaporite deposition in late Aptian time, when the mid-ocean ridge changed its position. According to Rabinowitz and LaBrecque (1979), differences in the widths of evaporite basins on either side of the equatorial Atlantic are due to a series of complex jumps by the Mesozoic mid-ocean ridge. Emery and Uchupi (1984), on the other hand, believe that the evaporites were deposited on continental crust, and that subsequent updip sediment loading caused the evaporites to flow seaward over the oceanic crust. South of lat. 3°S, I place the crustal boundary within the diapiric field; I believe that differences in the amount of seaward flow account for differences in the widths of the evaporite basins off Africa and South America. From lat. 3°S to the equator, the crustal boundary migrates landward to the shelf edge; in the re-entrant between the Gabon margin and the Cameroon Volcanic Ridge, the boundary also is along

the shelf edge (Emery and Uchupi, 1984, Charts X1A, X1B). I also postulate that breaching of the Torres/Walvis Ridge barrier, which blocked circulation from the south, and thereby enhanced evaporite deposition to the north, was due to the initiation of seafloor spreading and thermal subsidence in the Angola Basin. Rabinowitz and LaBrecque (1979), in contrast, proposed that such breaching, which occurred after seafloor spreading began in the Angola Basin, was the result of thermal subsidence or some other process, which took place shortly after a change in the pole of rotation.

Oceanic basement seaward of the Angola Basin rises gradually westward from a depth of over 9 km at the continental/oceanic crustal boundary, to 4 km at the base of the Mid-Atlantic Ridge, and from there to less than 2 km at the crest of the ridge. The basement surface is broken into east–northeast-trending ridges and troughs associated with fracture zones. Off Luanda, Angola, this trend is disrupted by north–south ridges and swales that flank the offshore diapiric field (Rabinowitz, 1972; Emery et al., 1975b). Seamounts also disrupt the basement configuration in the region, and one of these, St. Helena, rises above sea level. Much of the surface of oceanic basement in the region is quite rough, but off the Congo, the surface is relatively smooth (Emery et al., 1975b). Oceanic basement velocities range from 4.0 to 7.0 km/sec; highest values occur on the uppermost continental rise off the Congo (Emery et al., 1975b). Deep reflections within basement have been reported from the southern Angola Basin. They appear to originate from the layer 2B/layer 2C boundary and from within layer 3 (Musgrove and Austin, 1983).

STRATIGRAPHY

Coastal Basins and Shelf

Infrarift and Synrift Sequences

The > 8-km-thick stratigraphic section (Fig. 5-5) on the coastal basins and inner shelf consists of three sedimentary sequences separated in places by unconformities. At the base of the section is an infrarift sequence, deposited in non-faulted sags. Above the infrarift strata is a synrift sequence, deposited in synsedimentary fault basins. The third (youngest) sequence, the drift (postrift) sequence, accumulated during a seafloor spreading regime, when tectonics on the margin were dominated by thermal subsidence. The infrarift sequence has been described only from the Congo–Cabinda and Gabon coastal basins (see Chapter 2). In the Congo–Cabinda Basin, the infrarift sequence rests unconformably on basement and consists of Late Jurassic(?) fluviolacustrine sediments (Brice et al., 1982). A layer of volcanic rocks, radiomet-

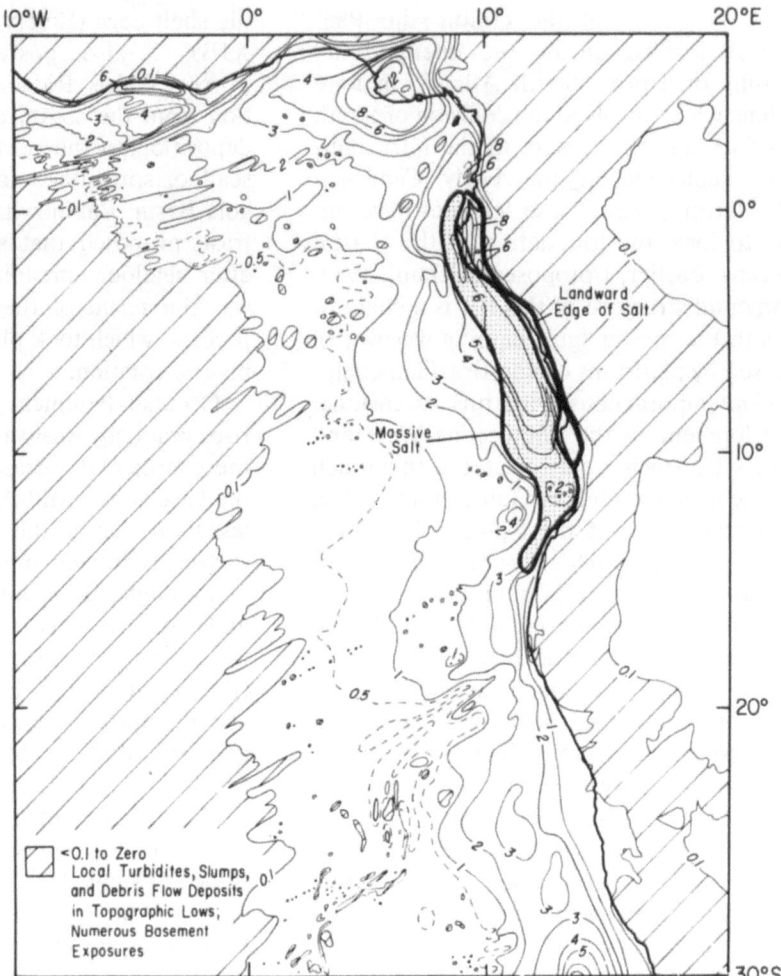

FIGURE 5-5.   Isopach map of Mesozoic–Cenozoic sediments in the Angola Basin and vicinity. Compiled from Emery and Uchupi (1984, Chart VIB) and Divins and Rabinowitz (in press). Contours in kilometers.

rically dated at 140 ± 5 Ma, is often present at the top of the sequence. In the Gabon Basin, the infrarift sequence is represented by Late Jurassic(?) continental siliciclastics.

The synrift sequence has been identified in all the coastal basins. Along the eastern side of Moçâmedes Basin the sequence consists of Late Jurassic(?) or Early Cretaceous(?) volcanic and intrusive rocks, and Aptian or Albian fanglomerates, which grade upward into sandstones, marls, clays, and sandy limestones with dense limestone interbeds (Franks and Nairn, 1973). At the northern end of the basin, red beds and a sequence of Barremian (and older) sandstone and limestone is capped by a basalt flow. Along the landward edge of the Cuanza Basin, north of the Luanda High, are synrift conglomerates, sandstones with asphaltic horizons, lagoonal gypsum, calcareous and argillaceous beds, and late Aptian evaporites and dolomites capped by gypsiferous clays that grade upward into oolitic and dolomitic limestone (Brognon and

Verrier, 1966). These beds grade basinward into pre-Aptian conglomeratic sandstone (with ash layers) immediately above basement, continental-lagoonal sandstones, dolomitic limestones, and Aptian evaporites ("massive" salt), the Quianga sequence (calcarenites, limestones, dolomite, and anhydrite), and the Binga sequence (evaporites, carbonates, and euxinic black and argillaceous limestone). Above these are the Albian saline and anhydritic Tuenza sequence, and the dolomitic Tuenza, Catumbela, and Quissonde sequences. Termination of evaporite deposition in the earliest Albian marks the initiation of seafloor spreading. During the Aptian Quianga cycle (synrift) and the Albian Tuenza and Catumbela (drift) cycles, reefs developed along topographic highs along the western and northern edges of Cuanza Basin.

The synrift unit in the Congo–Cabinda Basin is associated with major tectonic events. During the Neocomian, a graben formed, underwent rapid subsidence, and was filled with lacustrine turbidites (Brice et al.,

1982). These strata grade upward and laterally into organic-rich lacustrine, dolomitic shale, which was deposited in an anoxic lake that extended from Angola to northern Gabon. As the lake shoaled, it became the site of shale deposition, whereas algal limestones accumulated on neighboring topographic highs. A tectonic uplift and erosional phase terminated this depositional cycle. Barremian–Aptian sediments associated with a second period of subsidence consist of lacustrine carbonates and sands and of continental clastics. These lithofacies grade upward into evaporites that mark the onset of a marine transgression from the south. At least four desiccation cycles can be recognized in the Congo–Cabinda Basin (Belmonte, Hirtz and Wenger, 1965; Brice et al., 1982). Termination of evaporitic deposition is marked by a latest Aptian dolomitic layer, which is 50 m thick. The second synrift episode also ended with uplift and erosion. This erosion, followed by subsequent subsidence in the earliest Albian, marks the onset of seafloor spreading.

In the Gabon Basin, which is split into a series of subbasins by northwest-trending basement highs, the lower part of the synrift sequence consists of Valanginian–Barremian continental shales and early to middle Aptian continental sandstones and organic-rich shales (see also Chapter 2). Above these strata is an Aptian marine transgressive unit of continental siliciclastics, which grades upward into lagoonal clastics and evaporites. In the northernmost coastal basin, the Douala Basin, which is separated from the Benue Trough by the Cameroon Volcanic Ridge, pre-Aptian and Aptian continental siliciclastics rest on Precambrian basement. Offshore, the synrift sequence may include Aptian(?) evaporites (Grunau et al., 1975; Lehner and de Ruiter, 1977).

Drift (Postrift) Sequence

The drift sequence in the Moçâmedes Basin consists of Upper Cretaceous to Eocene clastics and carbonates. Eocene strata extend inland to an elevation as high as 250 m above sea level; this inland occurrence documents the maximum marine transgression (Franks and Nairn, 1973). The lower part of the drift unit in the Cuanza Basin is made up of three sequences:

1. A Cenomanian–early Senonian interval of silty shales and banded limestones.
2. A Turonian deltaic sequence (deposited during a regression), plus Turonian limestones, Coniacian silty shales, and Santonian shales.
3. A sequence containing upper Senonian and Paleocene limestones, lower Eocene shales and siltstones with chalk interbeds, and middle and upper Eocene shales and siltstones with chert, along the eastern edge of the basin, grading basinward into shales (Brognon and Verrier, 1966).

The Oligocene was a time of an extensive regression, which formed a regional unconformity. The drift sequence above this unconformity consists of Aquitanian gypsiferous shales grading upward into black and brown shales, and Burdigalian shales that grade eastward to sandstones near a flexure, which is covered by shallow-water carbonates and algae. Miocene deposition was terminated by another regression, which was followed by paralic and continental deposition.

In the southern part of the Congo–Cabinda Basin (the Cabinda Basin), the drift sequence consists of a transgressive unit that began with red beds, followed by a nearshore carbonate–siliciclastic sequence grading upward into deep-water shales and marls (Brice et al., 1982). The transgression slowed at the end of the Campanian, and was followed by a regional regression through the Paleogene, during which a sequence of nearshore siliciclastics and carbonates was deposited. At the end of Paleogene, the seaward edge of the shelf foundered, which produced a strong westward tilt. An Oligocene to Holocene regressive unit was unconformably deposited across the older shallow-water strata. This rapidly emplaced Oligocene–Holocene unit is marked by gravitational tectonics, differential loading, and growth faults associated with high-pressure shale movement. As a result of changes in sea level, the Oligocene–Holocene unit was subjected to channeling, which produced cut and fill structures.

In the northern part of the Congo–Cabinda Basin (Congo Basin), the drift sequence consists of Albian clastics, marls, and limestones, and Cenomanian–Turonian carbonates, sandy marls, and calcareous sandstones. The western margin of this Cretaceous depocenter is a basement ridge (Landana Ridge) near the present coast (Franks and Nairn, 1973). West of this topographic high are more than 1,000 m of clastics. The uppermost Cretaceous deposits are marked by a transgressive unit composed of a basal conglomerate, above which are phosphatic and calcareous strata. Resting unconformably on the Cretaceous strata are Paleocene calcareous sediments with phosphatic interbeds. The Eocene beds are separated from Paleocene beds by an unconformity of Ypresian age. Eocene beds consist of carbonates, siliceous limestones, and muddy sandstones. Oligocene deposits are missing, but the Miocene is represented by a regressive unit of carbonates and siliciclastics. Valleys eroded into Miocene strata are filled with conglomerates. Elsewhere along the coast, a Pliocene–Pleistocene unit is made up of siliciclastics. Along the coast also are two, or possibly three, terraces; the youngest terrace is covered at a high tide and contains a fossil taxon whose successors today live farther north (Franks and Nairn, 1973).

In the Gabon Basin, the Albian drift sequence contains mainly carbonate rocks (see Chapter 2). Near the present coast, these carbonates grade upward into deeper-water shales and marls (Brink, 1974). The Cenomanian section, separated from the Albian section by an unconformity, consists of red beds grading westward into marine siliciclastics. Above these red beds is a Turonian carbonate facies that grades westward into siliciclastics. Resting unconformably on the Turonian section is a Senonian–Eocene siliciclastic transgressive unit. This transgression was followed by a continuous Oligocene regression, which terminated with erosion. Beneath the present inner shelf, the Senonian–Eocene sequence is made of several depositional cycles, which are bounded by unconformities (Brink, 1974). Each cycle begins with estuarine sediments followed by deeper-water shales and terminates with regression and erosion. The Miocene section, which is indicative of a basal transgression followed by a regression, rests on the Oligocene unconformity and consists of a thick sequence of shales that grade upward into alternations of shales and sandstones; the section is capped by sandstones. The present depocenter is restricted to the outer shelf. On the inner shelf, the basinward slope of an Albian–Paleocene carbonate platform is dissected by westward-draining submarine canyons, which served as passageways for terrigenous sediments to reach the upper continental rise during the Cenomanian, Turonian, Maestrichtian, and Paleogene. According to Seiglie and Baker (1982), the sea-level curve defined by these regressions is more comparable to that of the eastern Arabian peninsula and of southern England than that of the Exxon model described by Haq, Hardenbol, and Vail (1988).

The Douala Basin, southeast of the Cameroon Volcanic Ridge, parallels this northeast-trending high. Its geohistory was largely influenced by vertical oscillations along the Cameroon trend, particularly during early Santonian–Campanian time, and during the Burdigalian (Reyre, 1966). The drift sequence in the Douala Basin consists of Albian(?) to Cenomanian continental siliciclastics overlain unconformably by Turonian–Senonian fluviomarine siliciclastics, and Senonian black shales. Campanian deposition was followed by erosion, the resultant unconformity is capped by Maestrichtian marine siliciclastics. The late part of Maestrichtian time is not represented by sediments. The unconformity is mantled by Paleogene marine siliciclastics. Resting unconformably on the older Paleogene deposits are Miocene strata, which, in turn, underwent erosion during the Burdigalian time.

Coastal Basin Development

As summarized by Brice et al. (1982) for Cabinda, coastal basin development in equatorial Africa consists of four distinct depositional episodes, one infrarift, two rift, and one drift. During a Late Jurassic infrarift episode, basement subsided slowly and the resultant

low was filled with fluviolacustrine siliciclastics. In the first rift episode, during the latest Jurassic(?) to Neocomian, a deep anoxic lake extended from Angola to northern Gabon and was filled with turbidites, organic-rich carbonates, green shales, and carbonates. This episode closed with basement uplift and erosion. In a second rifting episode, extending from the Barremian to the Aptian, secondary rift basins were superimposed on the older structures and were filled with lacustrine siliciclastics and carbonates, which grade eastward into continental siliciclastics. With continued subsidence, the basins were flooded by marine waters from the south during the Aptian, and an episode of evaporitic deposition was initiated. Termination of evaporite deposition is marked by the emplacement of a 50-m-thick dolomitic unit. In the Cuanza Basin, this dolomite layer is interlayered with evaporites, bioclastic calcarenites, and partially euxinic black and argillaceous siliciclastics. Evaporitic conditions in the Angola rift system were the result of restricted water circulation. A shallow topographic barrier, Torres/Walvis Ridge, lay to the south, and Africa and South America were juxtaposed in the equatorial region to the north.

The second rift episode was followed by uplift and erosion, which, in turn, was followed by subsidence. An increase in the rate of subsidence in the Albian is concomitant with the beginning of seafloor spreading. During a transgression of marine waters from the southwest, which peaked during the Campanian, a sequence of nonmarine siliciclastics, nearshore sands, shallow-water platform carbonates, and deep-water marls and shales (seaward of the carbonate platforms) was deposited in the coastal lows. Associated with this depositional cycle are salt movements, which produced growth faults and minor domes. When subsidence terminated in the Campanian, a regressive cycle of nearshore carbonates and sands was deposited in the coastal basins, and they prograded over the offshore carbonate platforms. During the Neogene, a regressive shale sequence was rapidly deposited, and the region tilted westward. As a result of this rapid deposition, shales became overpressurized, and the Aptian salt was reactivated. This updip sediment loading and salt reactivation formed the massive offshore diapiric field in the manner described by West (1989) for diapiric fields in the northwest Gulf of Mexico.

Continental Slope

Synrift and drift sediments on the upper continental slope are over 6 km thick (Fig. 5-5). Sediments above the diapiric structures have been greatly deformed by discontinuous movement of the diapirs; their deformation increases with burial depth. Some of the diapirs display collapse structures that resulted from solution or faulting. Two sites have been cored by the Deep Sea

Drilling Project on the continental slope: Site 364 on a plateau 10 km east of the Angola Escarpment; and Site 365 along the side of a partially buried canyon, 10 km southeast of Site 364 (The Shipboard Scientific Party, 1978b). The section cored at Site 364 consists of upper Aptian–lower Albian dolomitic limestones (deposited on an outer shelf or at the shelfbreak), and marly limestones interbedded with black shales (coeval to the dolomitic unit above the evaporites in the Cuanza Basin?), lower to middle Albian massive limestone and shale, upper Albian to Coniacian–Santonian black shales interbedded with burrowed limestones, Santonian to middle Eocene nannofossil chalk, middle Eocene to middle Miocene pelagic clay and red mud (deposited below the CCD), middle Miocene to lower Pliocene nannofossil ooze and mud, and upper Pliocene to Pleistocene calcareous mud. The dark color, odor of H_2S, and presence of pyrite in the youngest two units indicate that reducing conditions have prevailed at the site since the middle Miocene. The sedimentary section is disrupted by an unconformity at the Cretaceous/Tertiary boundary, and at the middle Eocene–middle Oligocene contact. If the Cretaceous/Tertiary erosional event is the result of a global rise in the CCD, as described by Worsley (1974), then by the end of the Cretaceous, the Angola Basin was interchanging waters with the rest of the oceanic basin system (The Shipboard Scientific Party, 1978b). The sharp drop in sedimentation rate during the Eocene–Miocene interval may have been due to erosion, nondeposition, or dissolution of planktonic carbonates (The Shipboard Scientific Party, 1978b). I infer erosional origin from an unconformity seen on seismic profiles from the area, and from the presence of the submarine canyon at Site 365. Major dissolution cycles within upper Eocene, middle to lower Oligocene, and upper Miocene sediments on Walvis Ridge are evidence that the slow deposition also resulted from dissolution. Other factors that may have reduced the influx of terrigenous sediments include isolation of the site by the vertical flow of Aptian evaporites and erosion of submarine canyons. Plastic flow of the salt probably was the result of massive updip siliciclastic sedimentation during the Miocene, concurrent with uplift and erosion of the coastal plain and central Africa (formation of the Victoria Falls erosional surface, the largest erosional surface in Africa; Holmes, 1965, pages 605–609). This uplift was instrumental in establishing the present Congo River drainage system, and initiating the marked influx of terrigenous sediments to Site 364 following the Miocene. Sediments recovered at Site 365 consist of Neogene and Paleogene siliciclastic canyon-fill, Coniacian–Santonian nannofossil ooze, and Cenomanian–upper Albian allochthonous blocks of sapropelic mudstone displaced from the canyon walls (The Shipboard Scientific Party, 1978b). The cored

Miocene and Oligocene sediments are indicative of deposition in a deep, well-oxygenated environment below the CCD.

The occurrence of upper Aptian–lowermost Albian euxinic sediments both in the Cuanza Basin and on the continental slope favors the proposal by The Shipboard Scientific Party (1978b), that these sapropels reflect a transition from restricted evaporite conditions to less restricted ones characterized by bottom stagnation and oxygen starvation. This change was the result of a gradual widening and subsiding of the rift system, just prior to initiation of seafloor spreading and breaching of the southern topographic barrier. Whereas the first period of black shale deposition was characterized by gradual ventilation, euxinic deposition during the Albian and Turonian–Santonian (as described in succeeding text) was the result of intermittent salinity stratification due to excessive evaporation (The Scientific Shipboard Party, 1978b; Natland, 1978). Episodic spillover of this dense saline water across the remnants of the southern topographic barrier (Walvis Ridge) caused erosion, not only of the barrier, but also along the eastern edge of the Cape Basin south of the barrier.

Continental Rise – Abyssal Plain

The continental rise in the Angola Basin, which is the focus of the present synthesis, is the least known of the geomorphic units. Knowledge of this 100- to nearly 800-km wide sedimentary wedge is based on widely spaced single-channel seismic profiles and stratigraphic data from Deep Sea Drilling Project Site 530, at the southern end of the Angola Basin. Sediments immediately north of Walvis Bay, in the vicinity of DSDP Site 530, are made up of four acoustic units (Fig. 5-6; Musgrove and Austin, 1984). This acoustic stratigraphy is recognizable throughout much of the Angola Basin south of the Congo Fan (Fig. 5-6). Unit I, at the base, consists of low-to-moderate amplitude reflections onlapping, downlapping, and terminating abruptly against the oceanic basement. Unit II consists of a series of uniform, continuous, high-amplitude, flat to gently undulating reflections. The continuous reflections of this unit become less distinct north of Walvis Ridge, and are more widely spaced in the direction of the African margin. Unit III, above the high-amplitude reflections of Unit II, is acoustically transparent, and is subdivided by a discontinuous, low-amplitude seismic unconformity. Unit IV, at the top, is distinguished by moderate-amplitude, discontinuous, subparallel, or hummocky to lenticular reflections, which onlap, downlap, or are parallel to those of Unit III. The lower part of Unit I consists of upper Albian hemipelagic clays, and upper Albian–Santonian red, green, and black mudstones, marlstones, and thin-bedded siltstones (Stow, 1984a).

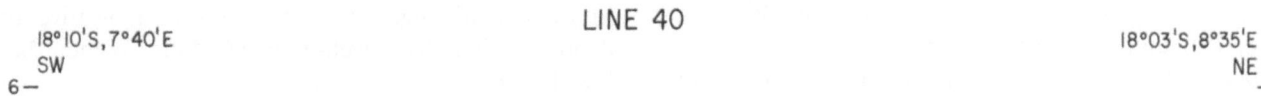

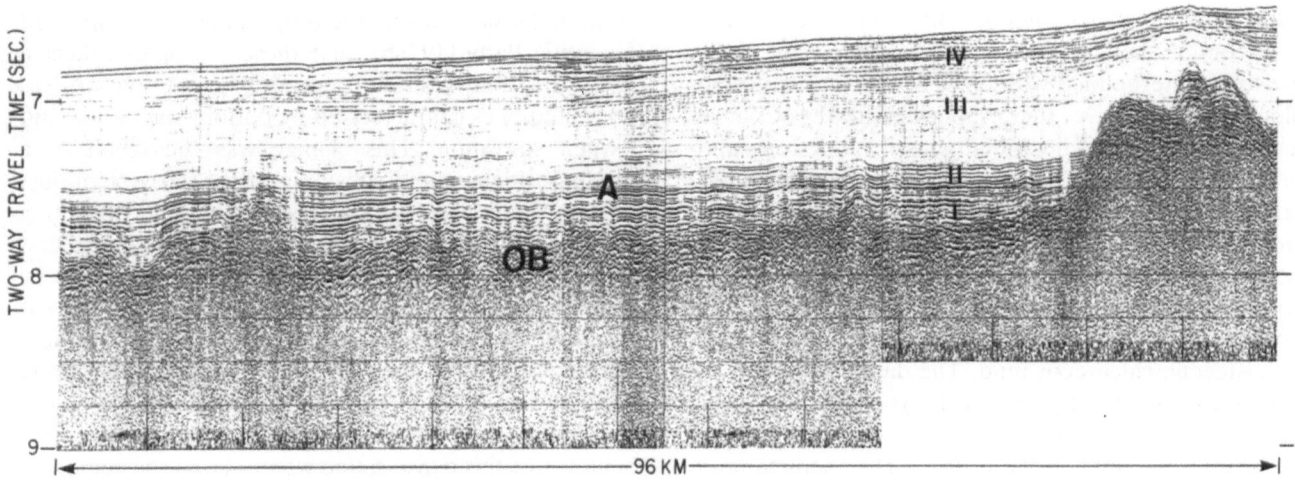

FIGURE 5-6. Segment of single-channel seismic reflection profile 40. This and profile 44 in Fig. 5-7 were recorded using a 300-in³ (40) and a 10-in³ (44) air gun as a sound source. The guns were fired at 10-sec intervals. Signals were received by two 30-m arrays (each containing 200 sensing elements), summed, amplified, filtered at 15–100 Hz, and recorded on a dry paper printer. This record and the one in Fig. 5-7 illustrate the nature of the four seismic sequences (I–IV) and of Horizon A (A) in the southern Angola Basin. OB = oceanic basement. Vertical exaggeration of the seafloor on this and other photographic copies (Figs. 5-7, 5-9, 5-10, 5-12, and 5-13) is 15×, using a velocity of 750 m/sec.

The black sediments at this site are similar to those at Site 364; maximum black sediment developed nearly synchronously at both sites (Stow and Dean, 1984). At Site 363 on Walvis Ridge, black sediments also were cored: five thin black beds occur within a sequence of marls and muds (The Shipboard Scientific Party, 1978a). The black sediments of Site 350 show evidence of both turbiditic and pelagic processes. Seismic reflections associated with the lower Cenomanian–middle Coniacian black and green shales onlap oceanic basement to the west, a geometry that led Musgrove and Austin (1984) to suggest that these deposits may be distal turbidites derived from the African margin. The organic-rich sediments may have been derived from restricted shelf basins or coastal lagoons that were temporary sinks for organic-rich sediments. As the black sediments are associated with red ones, anoxic or near anoxic conditions presumably occurred only periodically; when oxygenated conditions prevailed, the sediments were oxidized to produce the red deposits. When the anoxic unit was deposited, the Angola Basin was only 300–400 km wide, had a maximum depth of 3,500 m, and was restricted to the south by the Torres/Walvis Ridge and to the north by the juxtaposition of South America against Africa (Stow and Dean, 1984). The climate at that time was warm, surface and bottom waters were warm (30 and 15°C, respectively; Barron, Salzmanb, and Price, 1984) and saline (Brass,

Southman, and Peterson, 1982), and sea level was high (Haq, Hardenbol, and Vail, 1988).

The top part of seismic Unit I and the lower part of Unit II (Fig. 5-6) correlate with a Santonian–Campanian mudstone–siltstone–sandstone fan sequence composed dominantly of volcanic debris (Green Fan; Stow, 1984a). Stow believed that the source of this fan was the Walvis Ridge. Musgrove and Austin (1984), on the basis of seismic reflection geometry, concluded that the prominent sediment source of the fan was not the Walvis Ridge to the south, but the African continental margin to the east. The distribution of sediments shown by an isopach map of Unit I (Musgrove and Austin's fig. 11), however, is more indicative of a Walvis Ridge source than an African margin one. The top part of Unit II corresponds to another submarine fan sequence of mudstone, marlstone, and thin bedded limestone of Campanian–Eocene age (White Fan, Stow, 1984a). This unit is continuous with the underlying Green Fan, but is disrupted at the top by a middle Eocene to middle Oligocene unconformity or condensed section (Dean, Hay, and Sibuet, 1984). The calcarenites are composed of shallow-water carbonate debris, and the calcilutites and mudstones of calcareous microfossils, illite, and some smectite. Stow (1984a) believed that reefs and volcanic islands on Walvis Ridge supplied the shallow-water carbonate debris and minor volcanics of the White Fan; he concluded that the illite component was derived from the African continental margin. At the Cretaceous–Tertiary boundary there is a significant increase in the concentration of iridium, and immediately above the iridium anomaly is a marked decrease in calcium carbonate. This reduction of shallow-water carbonates was ascribed by Dean, Hay, and Sibuet (1984) to the same process that produced the iridium anomaly, which they believed was an asteroid impact. As in the Cape Basin to the south (Tucholke and Embley, 1984), the middle Eocene to middle Oligocene unconformity (or condensed section) in the southern Angola Basin is probably due to a combination of events; reduced deposition caused by a high sea-level stand and shoaling of the CCD in the early Eocene (Horizon D; Emery et al., 1975a) and the onset of abyssal circulation in the early to middle Oligocene [Horizon LO of Tucholke and Embley (1984), which correlates with Horizon E of Connary (1972)]. Sediment drifts in the region of the sill between the Cape and Angola basins were formed as thermohaline circulation decayed in the Miocene. The upper Oligocene to upper Miocene sediments, deposited beneath a shallow CCD, consist mainly of pelagic clays with an occasional silty turbidite, a few rare foraminifera–nannofossil ooze layers, and several middle Miocene volcanic ash layers (Musgrove and Austin, 1984). At the top of the unit are calcareous biogenic sediments of early late middle Miocene age. Reflections at the top of the unit are truncated, and sediments at several locations along the base of the Walvis Ridge are deeply incised by erosion. This erosional event may have been coincident with establishment of the Benguela Current and the associated upwelling system during the Miocene (Siesser, 1980; Musgrove and Austin, 1984). Seismic Unit IV, above this pelagic sequence, consists of another submarine fan (Brown Fan) whose development began in the late Miocene and probably continues today (Stow, 1984a). Beginning in the late Miocene, sediments eroded from the Walvis Ridge by the northward flowing Benguela Current were displaced down the northern scarp of the ridge in the form of turbidity currents and debris flows (Stow, 1984b). A rise in the CCD near the end of the Pliocene led to formation of an unconformity within seismic Unit IV (Fig. 5-6). This high CCD was accompanied by maximum upwelling and deposition of diatom-rich pelagic sediments and clay–diatom turbidites in the southern Angola Basin during the Pliocene–early Pleistocene. As the CCD deepened through the middle Pleistocene to its present depth of 4,750 m (van Andel et al., 1977), a mixture of nannofossils, diatoms, and clay accumulated at Site 530. Thus deposition in the southeastern Angola Basin

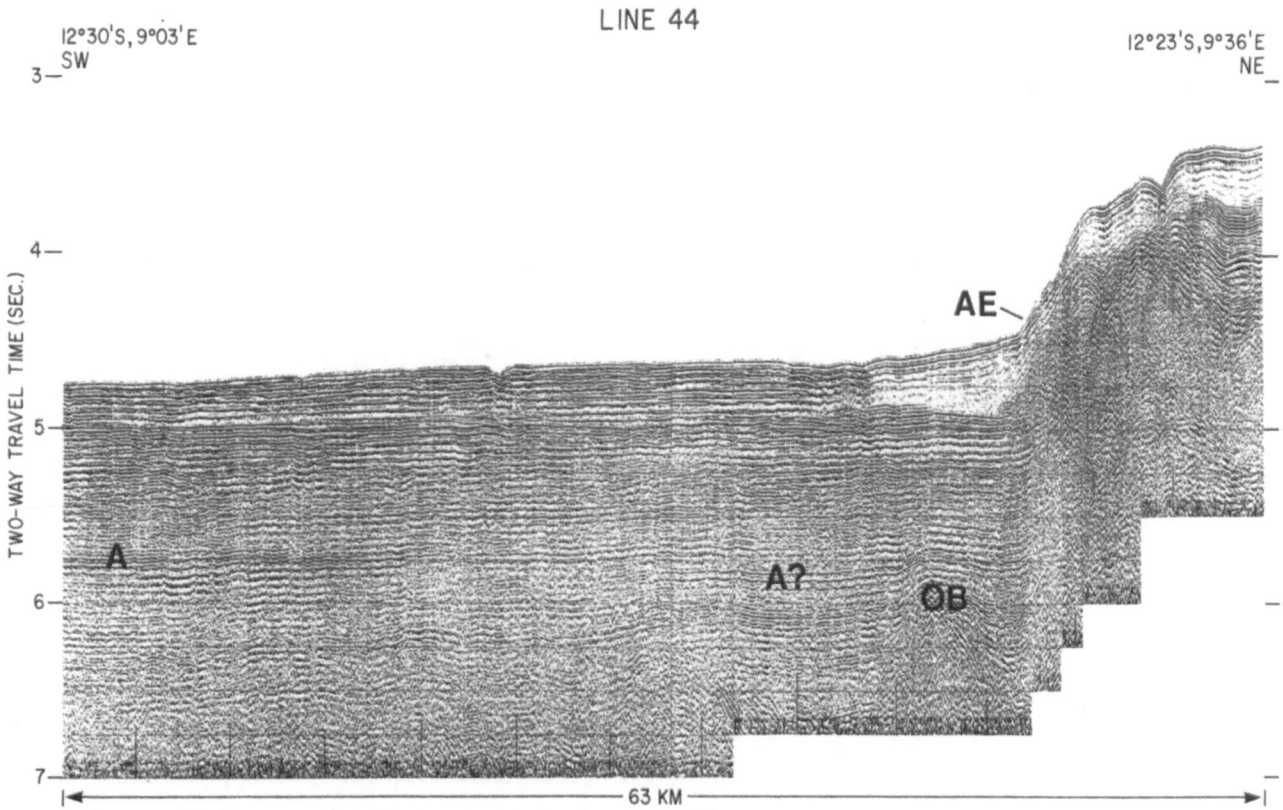

FIGURE 5-7.   Segment of single-channel seismic reflection profile 44. See Fig. 5-6 for additional details. OB = oceanic basement; AE = Angola Escarpment; A = Horizon A.

since the Cretaceous has been the result of sediment gravity flows, whose dominant source was the Walvis Ridge to the south. Sediments of the older submarine fan (Green Fan) were derived from submarine volcanoes and subaerial volcanic islands on the ridge (Stow, 1984b). As the ridge migrated northward into warm tropical waters, reef detritus and volcanic debris provided sediments to the White Fan. With the establishment of the Benguela Current and enhancement of biological productivity, a new source of sediments was provided for construction of the Brown Fan.

The sedimentary section of the continental rise east of Site 530 (lines 40, 42, 44, 46; Figs. 5-7, 5-8) thickens eastward and the units onlap westward onto oceanic basement. These relationships indicate that the source

of siliciclastic sediment was the African margin (Figs. 5-6, 5-7, 5-8). A prominent horizon near the base of the sedimentary sequence (Horizon A of Emery et al., 1975b; Figs. 5-6, 5-7, 5-8) may be correlative with the Eocene/Oligocene unconformity at Site 530. Sediments below the unconformity have velocities of 3.0–3.6 km/sec, whereas those above the unconformity have velocities of less than 2.5 km/sec. If my correlation of the reflector along lines 40, 42, 44, and 46 is correct, then most of the strata of this part of the Angola Basin continental rise, including the Cuanza Fan, are younger than Eocene. Sediments below this unconformity are 1.0 sec or less thick, and their depocenter is east of the oceanic basement highs on the upper rise. Strata on the slope and uppermost rise on line 42 (Figs. 5-8, 5-9)

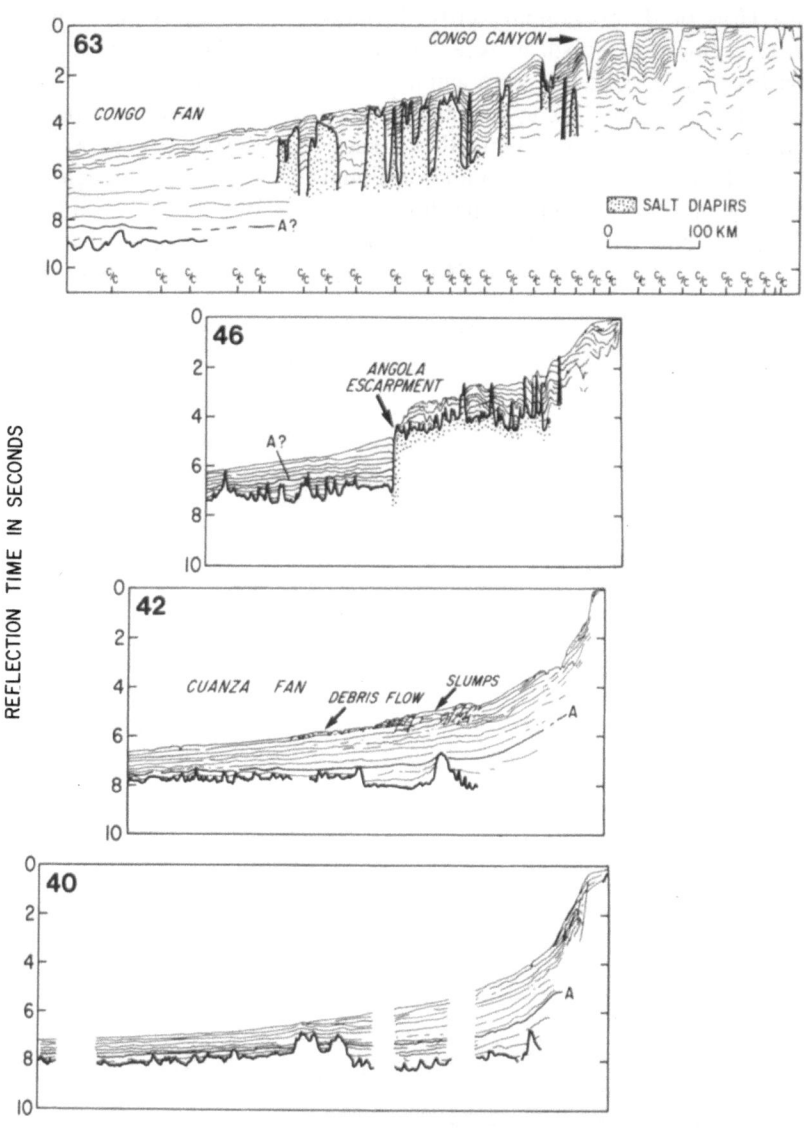

FIGURE 5-8.   Line interpretations of single-channel seismic reflection profiles of the southern and central Angola Basin. See Fig. 5-1 for locations of profiles. Vertical exaggeration of seafloor in profiles is approximately 33 ×, using a velocity of 750 m/sec. (Compiled and modified from Emery et al., 1975b).

LINE 42

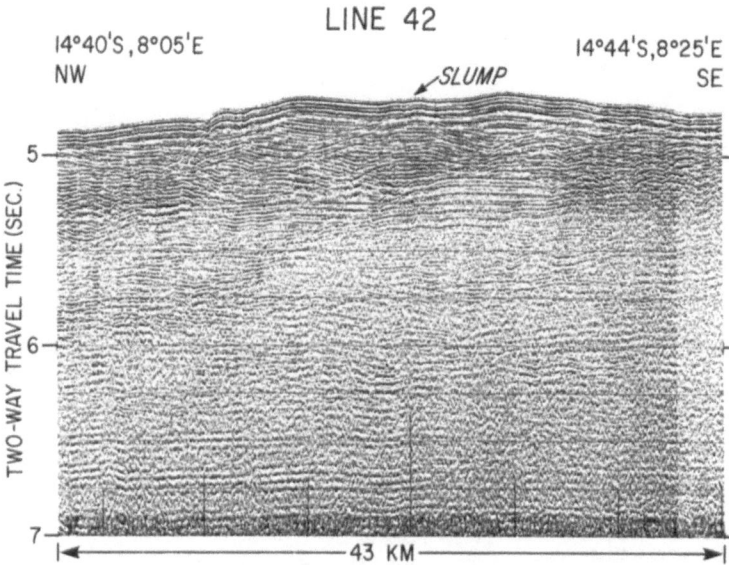

14°40′S, 8°05′E
NW

14°44′S, 8°25′E
SE

SLUMP

TWO-WAY TRAVEL TIME (SEC.)

5

6

7

← 43 KM →

FIGURE 5-9. Segment of single-channel reflection profile 42 displaying slumps on the middle part of the continental rise. Profile was recorded using a 10-in³ air gun. See Fig. 5-8 for additional interpretation.

LINE 63

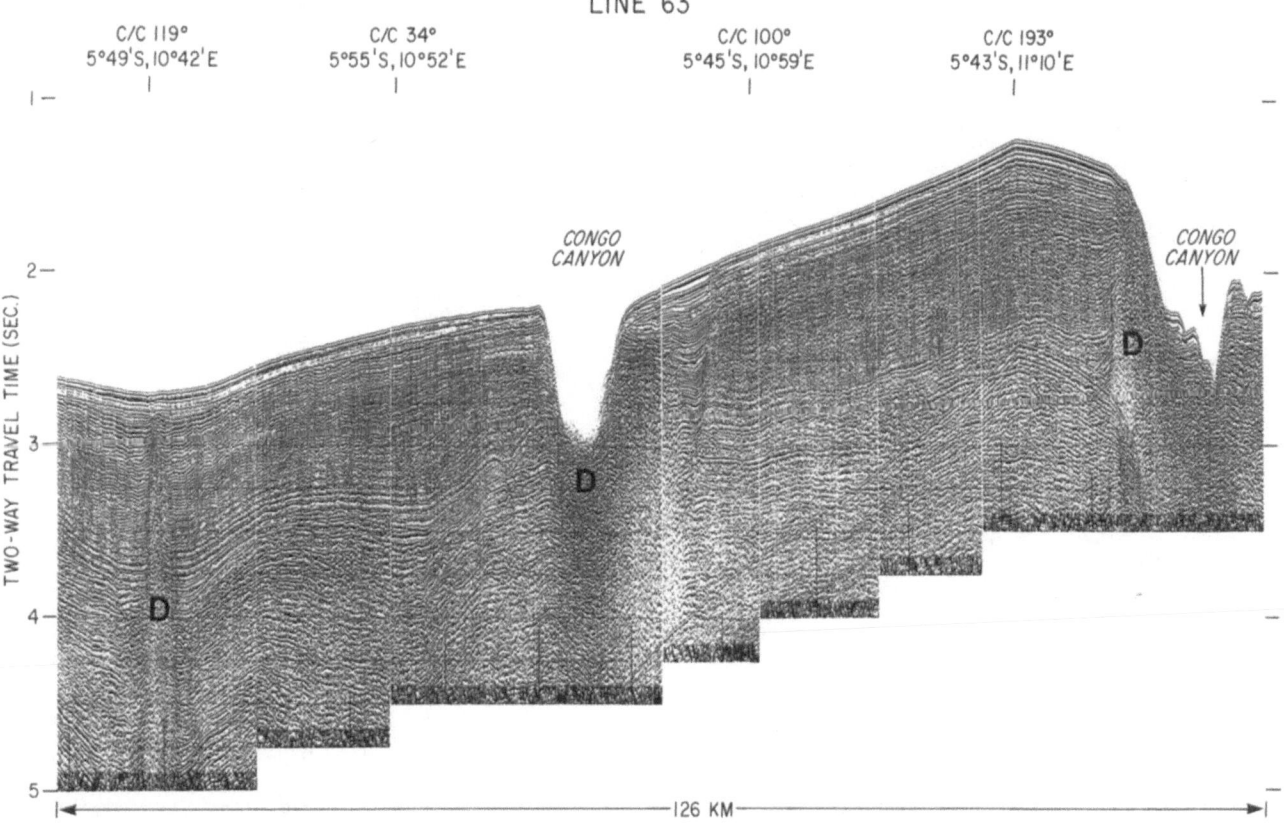

C/C 119°
5°49′S, 10°42′E

C/C 34°
5°55′S, 10°52′E

C/C 100°
5°45′S, 10°59′E

C/C 193°
5°43′S, 11°10′E

1 —

CONGO
CANYON

CONGO
CANYON

2 —

TWO-WAY TRAVEL TIME (SEC.)

3

D

D

D

4

D

5

← 126 KM →

FIGURE 5-10. Segment of seismic-reflection profile 63 recorded in Congo Canyon region. Note that on one crossing the canyon incises a diapiric structure. See Fig. 5-8 for additional interpretation. D = diapir.

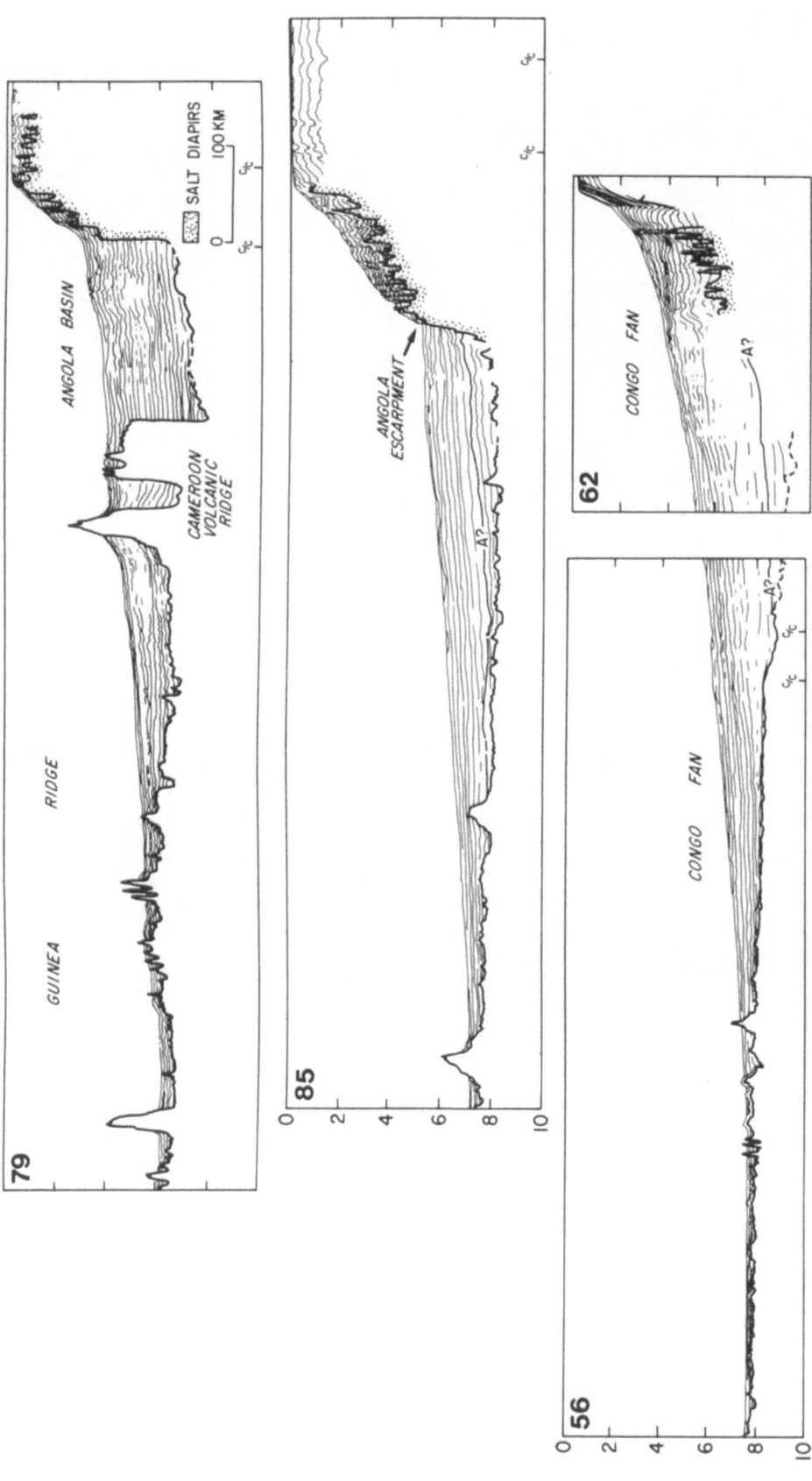

FIGURE 5-11. Line interpretations of seismic reflection profiles from the northern Angola Basin. See Fig. 5-1 for locations of profiles. Vertical exaggeration of seafloor is 33 × using a velocity of 750 m/sec. (Compiled and modified from Emery et al., 1975b).

REFLECTION TIME IN SECONDS

LINE 56

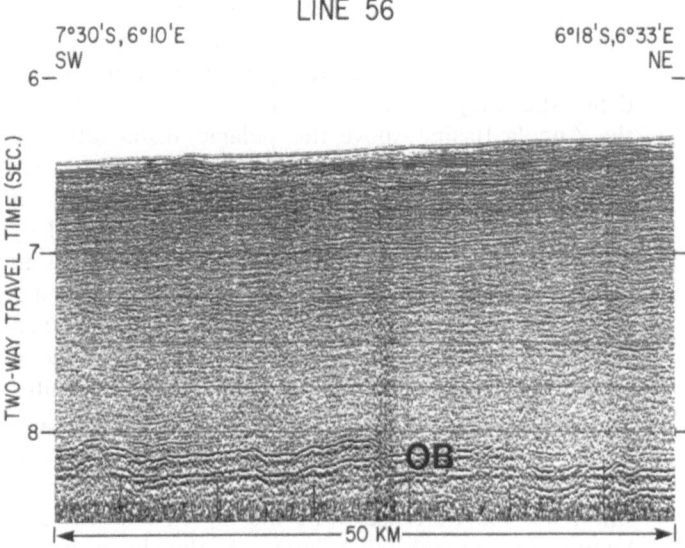

FIGURE 5-12.   Segment of single-channel seismic reflection profile 56 displaying nature of the seismic stratigraphy of the Congo Fan. A characteristic feature of the fan is its poor stratification. Profiles recorded using a 300-in^3 air gun. See Fig. 5-11 for additional interpretation. OB = oceanic basement.

display evidence of slumping (Fig. 5-9). Seaward of the slump structures is an extensive debris field whose surface (Fig. 5-8) is crenulated by overlapping hyperbolic echoes.

The Congo Canyon, the main conduit for the Congo Fan, cuts irregularly across the offshore diapiric field on the upper continental slope (Profile 63, Figs. 5-8, 5-10). The canyon's course through the field is largely controlled by the distribution of diapirs, as the canyon appears to zig-zag around the highs. At one location, the canyon appears to cut into one of the diapirs (Figs. 5-8, 5-10). This incision is evidence either that vertical

motion of the diapirs had ceased at the time the canyon was cut, or that erosion equalled or exceeded the vertical flow of the Aptian evaporites at the time of canyon cutting. Associated with the canyon is the Congo Fan, the major depocenter of the Angola Basin. Its sediments are acoustically poorly stratified (Profiles 56, 62; Figs. 5-11, 5-12). A horizon near the base of the fan, which separates sediments with velocities as high as 3.5 km/sec (below) from strata with velocities of less than 2.2 km/sec (above), may be coeval with Horizon A farther south (Fig. 5-8). If this correlation is correct, then much of the Congo Fan was deposited since the

LINE 85

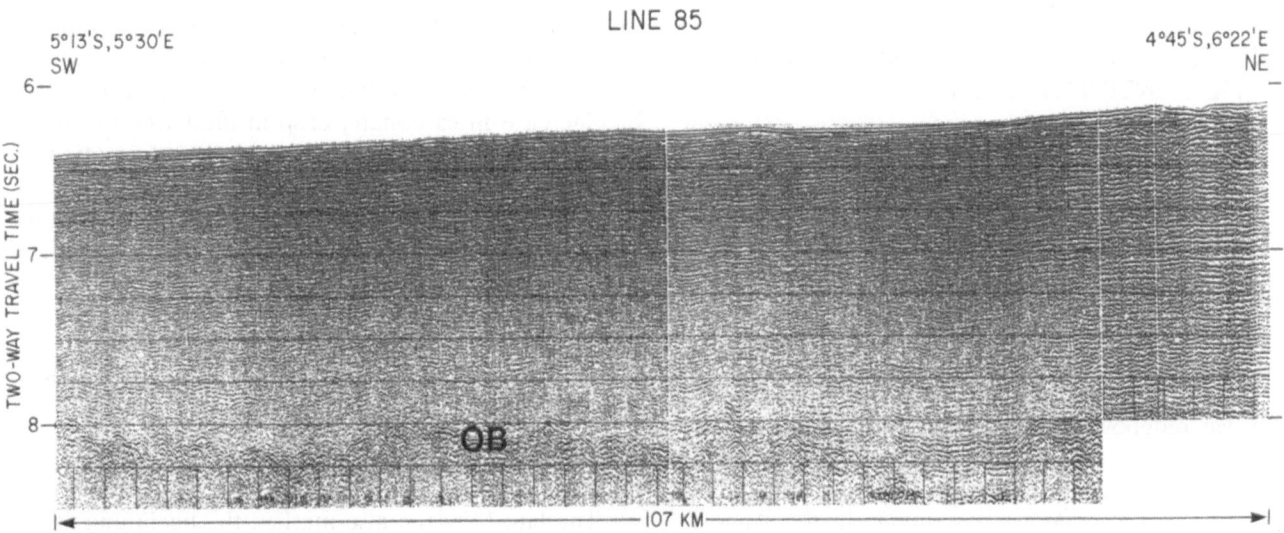

FIGURE 5-13.   Segment of seismic reflection profile 85 displaying acoustic stratigraphy in the northern Angola Basin using a 120-in^3 air gun. OB = oceanic basement.

Eocene, probably as a consequence of the uplift of the East African Rift System and establishment of the present Congo River drainage during the Miocene. A Neogene age for both the Congo and Cuanza fans also is indicated by the distribution of sediments at DSDP Site 364; terrigenous sediments became important there only in the Neogene. The Congo Canyon incision at the apex of the fan probably reflects glacially induced regressions during the Pleistocene. The acoustic stratigraphy displayed by profile 85 (Figs. 5-11, 5-13) shows westward-dipping reflections onlapping oceanic basement, which indicates that the sediments were derived from the African margin. A prominent reflector near the base of the section, which terminates against a basement high on the outer rise, separates sediments with velocities ranging from 3.1 to 2.4 km/sec (below) from sediments having velocities of less than 2.1 km/sec (above). I have tentatively correlated this reflector with Horizon A. Such a correlation implies that the bulk of the continental rise seaward of the Angola Escarpment in this region also has been constructed since the Paleogene. At present, little is known about the narrow segment of the continental rise seaward of the Douala Basin (Profile 79; Fig. 5-11). Like the continental rise north of Walvis Ridge, sedimentation on the rise east of the Cameroon Volcanic Ridge probably was influenced by tectonism along this volcanic crustal lineament. Deformation in the Benue Trough (northwest of the ridge) and in the Douala Basin was most pronounced during Santonian–Campanian time. Another compressional phase took place during the Burdigalian, when much of the Douala Basin was uplifted and eroded. Magmatic activity that accompanied these compressional phases increased from northeast to southwest. In the Cameroon Volcanic Ridge, this magmatic activity is as recent as 1.1 Ma (Hedberg, 1968; Grunau et al., 1975).

BASIN DEVELOPMENT AND CONSTRUCTION OF THE CONTINENTAL RISE

Construction of the continental rise in the Angola Basin was controlled by sediment derived both from ridge barriers to the south and north, and from the African margin to the east. Throughout much of its depositional history, the continental rise in the southern Angola Basin has received large volumes of redeposited sediments in the form of turbidites and debris-flow deposits. Most of this material was derived from the Walvis Ridge (to the south); additional sediment was redeposited from the African margin (to the east). The basal unit (Albian to Santonian) immediately above oceanic basement consists of black, green, and red claystones. Anoxic conditions in the Angola Basin appear to have been intermittent, and most of the time, bottom waters and interstitial waters were

oxic enough to allow the accumulation of red sediment. The extent of these anoxic sediments is yet to be resolved, but the lack of topographic barriers is evidence that they blanket most, if not all, of the floor of the Angola Basin. Above this pelagic/distal turbidite sequence in the southern Angola Basin are three distinct submarine fans. The oldest of the fans is of Santonain to Campanian age, and is characterized by volcanic sediments. The middle fan, of Maestrichtian to Eocene age, overlaps the oldest fan, and is dominated by redeposited shallow-water carbonates mixed with volcaniclastics. The youngest fan, of late Miocene to Holocene age, consists of muddy pelagic turbidites rich in siliceous biogenic debris and debris-flow deposits. Some of the sediments in this fan sequence were derived from the African margin. Turbidite deposition was most intense while the region was within a humid to subhumid paleoclimate. As the site migrated into an arid subtropical zone, the sedimentation regime changed to a predominantly basinal one (Dean, Hay, and Sibuet, 1984). Sediment deposition of the continental rise, both in the southern Angola Basin (Site 530) and in the central basin (Site 364), is marked by a drastic decrease in carbonate content and by either an unconformity or a condensed middle Eocene to middle Oligocene section. The carbonate decrease in the early middle Eocene probably was due to shoaling of the CCD and a rise in sea level; the middle Eocene to late Oligocene unconformity or condensed section correlates with initiation of thermohaline circulation. The decreased intensity of bottom currents entering the southern Angola Basin in the Neogene led to construction of sediment drifts in the sill area between the Cape and Angola basins. Other changes in the sedimentary regime of the southern part (and probably other parts) of the Angola continental rise occurred in:

1. The middle Oligocene (worldwide glacially induced regression; Haq, Hardenbol, and Vail, 1988).
2. The middle to the late Miocene (unconformity and increase in carbonate, drop in the CCD, beginning of a major ice cap in Antarctica, formation of pre-AABW, and sea–level drop).
3. The early Pliocene to late Miocene (unconformity, drop in sea level, development of AABW, and development of a thick Antarctic ice sheet.
4. The late Pliocene (unconformity, development of northern hemisphere glaciation).
5. The early Pleistocene (decrease in calcium carbonate, increase in diatom deposition: shoaling of the CCD) (Dean, Hay, and Sibuet, 1984).

The late Miocene also marked the beginning of the Benguela Current and associated upwelling (based on the relative abundance of diatoms). Upwelling reached its maximum at Site 530 and on the crest of Walvis Ridge during the late Pliocene and Pleistocene.

North–northeast trending linear ridges and troughs also were formed along the northern edge of the Angola Basin as Africa and South America slid past one another. Since rifting began in the equatorial Atlantic, these troughs have served as passageways for turbidity currents to reach the deep sea. Compressional phases (i.e., Santonian and Miocene) associated with changes in plate geometry, coupled with changes in sea level, must have influenced the sedimentary regime of the narrow continental rise between the continental slope off Gabon and the Cameroon Volcanic Ridge. Construction of the Cameroon Volcanic Ridge during the Neogene and Pleistocene also must have affected sedimentation on the rise south of the high, but no information is available at present to determine the significance of this contribution.

Although the timing was different, construction of the continental rise in the central part of the Angola Basin resembles that off eastern North America (see Chapter 6). Soon after seafloor spreading began in the earliest Albian, a gentle regional subsidence was initiated, and a carbonate platform developed on the seaward side of the coastal basins (Brink, 1974; Brice et al., 1982; Dailly, 1982). The front of this carbonate platform was located along the inner part of the present shelf. According to Brice et al., loading due to the construction of the carbonate platform triggered flow of the Aptian evaporites, creating growth faults ("carbonate growth faults"), which, in turn, influenced local depositional processes. With continued subsidence, the carbonate platform drowned and was covered by deep-water shales and marls. Landward of the carbonate platform, restricted lagoonal sediments graded to nonmarine siliciclastics on the eastern edge of the coastal basins. Deposition on the carbonate platform and in the landward lagoon was disrupted by regressions, which created unconformities or disconformities of regional extent. These unconformities were best developed during the middle Cenomanian, late Turonian, late Santonian, middle Campanian to late Maestrichtian, and Paleocene. Landward of the carbonate platforms, the regression/transgression cycle is marked by an erosional surface, above which are brackish-water estuarine deposits, which may contain channel sandstones and deeper-water marine sediments (Brink, 1974). Seaward of the carbonate platforms, the regressive cycles are represented by submarine fans, whose sediments reached the base of the platform via canyons carved into the platform front (Dailly, 1982). Turbidity-current deposition on the Cretaceous continental rise was concentrated at the base of the scarp fronting the carbonate platform; sediments at DSDP Sites 364 and 365 (on the seaward side of the offshore diapiric field) are mainly pelagic carbonates, whereas

at Site 530 (immediately north of Walvis Ridge), contributions from the African margin are represented by distal turbidites. Like the rise segment north of the Walvis Ridge, deposition on the rise seaward of the carbonate platform also was influenced by the Cretaceous anoxic events.

Sediment input to the Angola continental rise increased considerably after subsidence of the coastal basins slowed at the end of the Campanian. During the regional regression which followed, Maestrichtian and Paleogene nearshore to marginal marine clastics and littoral carbonates prograded seaward over the carbonate platform. As a result of this progradation, the landward margin of the continental rise migrated to a position near the outer edge of the present shelf (Fig. 5-14). Construction of this regressive wedge appears to have been disrupted by the Eocene–Oligocene sediment break north of the Walvis Ridge. An elevated sea level in the Eocene, when the continental rise became sediment starved, is documented by the Eocene silicified shales and porcellanites beneath the present shelf. The initiation of thermohaline circulation in the early to middle Oligocene is evidenced by a basinwide unconformity (Horizon A). This part of the margin also appears to have been affected by the middle Oligocene drop in sea level postulated by Haq, Hardenbol, and Vail (1988). The top of the Paleogene regressive unit in the coastal basins and on the continental shelf is marked by a high-relief surface (Brice et al., 1982). A complex canyon system was carved on the slope by turbidity currents during this regression (Dailly, 1982, fig. 12), and they served as passageways for turbidity currents to reach the upper rise, where they deposited their loads on Horizon A.

The combination of a strong westward tilt of the outer coastal basins and an uplift of the East African Rift System, which established the present drainage regime, led to deposition of an Oligocene to Miocene regressive siliciclastic regime, following formation of the middle Oligocene unconformity. Building of this clastic wedge led to the present morphology of the continental margin (Fig. 5-14). On the outer shelf, sedimentation during this regional subsidence (Brice et al., 1982) produced a regressive, shaley siliciclastic sequence in which deep-water muds and turbidites grade upward to paralic sediments, which are capped by continental deposits. The bulk of the Congo and Cuanza fans also was deposited during this regressive phase. A massive influx of siliciclastics to the outer shelf led to remobilization and seaward flow of the evaporites over oceanic crust to produce the huge offshore diapiric field. Off Luanda this seaward flow terminated against a north-trending oceanic basement ridge. Plastic flow of the evaporites elevated the upper rise strata (including parts of the Congo and Cuanza fans) to form the upper continental slope and the

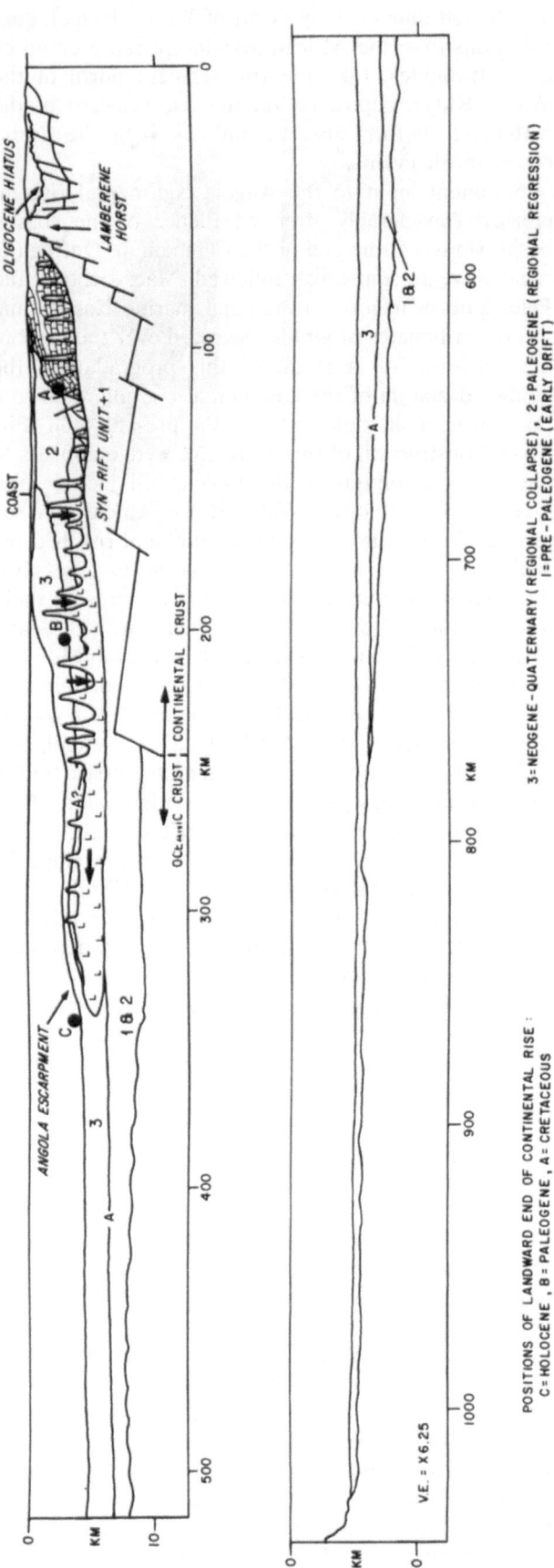

POSITIONS OF LANDWARD END OF CONTINENTAL RISE :
C=HOLOCENE , B=PALEOGENE , A=CRETACEOUS

3=NEOGENE-QUATERNARY (REGIONAL COLLAPSE), 2=PALEOGENE (REGIONAL REGRESSION)
I=PRE-PALEOGENE (EARLY DRIFT)

FIGURE 5-14. Schematic cross section of equatorial African continental margin. Compiled from Dailly (1982) and seismic reflection profiles described in this report. As a result of seaward flow of evaporites (due to updip sediment-loading), the landward edge of the continental rise migrated westward to its present position. The former upper rise was uplifted and deformed to form the present upper continental slope (offshore diapiric field) and the Angola Escarpment. Heavy arrows indicate evaporite flow directions.

Angola Escarpment (lower continental slope), which is a salt front. Sediments dispersed by the Congo River were sufficient to keep pace with this uplift in the immediate vicinity of the river. The Congo Canyon, incised on the apex of the Congo Fan, formed in response to glacially induced lowerings of sea level in the Pleistocene. Sediments eroded during this canyon-cutting phase were transported beyond the Angola Escarpment and deposited on the present upper continental rise as the middle part of the Congo Fan. Turbidity current deposition during the Pleistocene also contributed to the buildup of the Cuanza Fan. Whereas deposition still continues on the Congo Fan today, the Cuanza depocenter is probably inactive.

CONCLUSION

The Albian–Holocene depositional history of the continental rise in the Angola Basin is distinct from other continental rises in the Atlantic, in that sediment was derived not only from the continental margin (to the east), but also from oceanic basement ridges north and south of the rise (compare with Chapter 4). The southern part of the Angola Basin received large volumes of turbidites and debris-flow deposits from the Walvis Ridge to the south. Mixed with this redeposited material was some sediment derived from the African continental margin to the east. The Coniacian–Campanian turbidites were derived from submarine to subaerial volcanic peaks atop Wavis Ridge; downslope sediment transport was controlled by tectonic activity on the ridge flank. The Campanian–Eocene sediments redeposited from Walvis Ridge are a mixture of reef talus and volcanic debris, whereas the upper Miocene–Holocene turbidites consist of organic-rich calcareous and siliceous oozes deposited on the ridge in response to worldwide cooling and development of the Benguela Current and upwelling system off Namibia. To date, little is known regarding any contribution made by the northern linear highs to the construction of the northern continental rise, but a history of tectonic instability and recent volcanism indicate that these highs must have played a significant role in the formation of this part of the rise. Rise sediment derived from the African margin reached its maximum volume during the Neogene and Quaternary as a result of uplift of the East African Rift System and establishment of the present drainage system. This massive influx of terrigenous sediment remobilized the Aptian evaporites to form the offshore diapiric field and the salt-supported Angola Escarpment (Fig. 5-14). Only off the Congo estuary was terrigenous deposition rapid enough to exceed the vertical flow of salt and form a continuous sedimentary apron from the outer shelf to the deep sea; this apron is the Congo Fan. Elsewhere, vertical flow of the evaporites has elevated the upper rise to form an irregular upper continental slope off equatorial Africa. In addition to this unusual contribution from three distinct provinces, construction of the rise in the Angola Basin also was influenced by episodic intervals of basinwide anoxia throughout the middle Cretaceous. Additional depositional modifications arose from three short regressions at the end of the middle Cenomanian, late Turonian, and late Santonian; from a general regression during the middle Campanian to late Maestrichtian; from a rise in sea level during the Eocene; from initiation of thermohaline circulation in the Oligocene; from a glacially induced regression in the middle Oligocene; from decay of the thermohaline circulation in the Miocene; and from repetitive Pleistocene transgressions and regressions. During the Pleistocene regressions, the Congo Canyon was incised on the Congo Fan, and the present seafloor morphology was imprinted on the continental rise.

ACKNOWLEDGMENTS

I wish to thank W. P. Dillon and C. W. Poag of the U.S. Geological Survey, Atlantic Marine Geology Branch, Woods Hole, MA, for suggestions during the preparation of this chapter. Funds provided by the J. Seward Johnson Chair in Oceanography, awarded to me by Woods Hole Oceanographic Institution, made possible the writing of this chapter. This is contribution No. 7632 of the Woods Hole Oceanographic Institution.

REFERENCES

Barron, E. J., Salzman, E., and Price, D. A. 1984. Occurrence of *Inoceramus* in the South Atlantic and oxygen isotopic paleotemperatures in Hole 530A. *In* Initial Reports of the Deep Sea Drilling Project, Volume 75, Part 2, W. W. Hay, J. C. Sibuet, et al.: Washington, DC: U.S. Government Printing Office, 893–904.

Baumgartner, T. R., and van Andel, Tj. H. 1971. Diapirs of the continental margin of Angola, Africa. *Geological Society of America Bulletin* 82:793–802.

Beck, R. H., and Lehner, P. 1974. Ocean, new frontier in exploration. *American Association of Petroleum Geologists Bulletin* 58:376–395.

Belmonte, Y., Hirtz, P., and Wenger, R. 1965. The salt basins of the Gabon and the Congo. Salt Basins Around Africa: London: Institute of Petroleum, 55–74.

Bolli, H. M., Ryan, W. B. F., McKight, B. K., Kagami, H., Melgnen, M., Siesser, W. G., Natland, J. H., Longoria, J. F., Proto Decima, F., Foresman, J. B., and Hottman, W. E. 1978. Initial Reports of the Deep Sea Drilling Project, Volume 40: Washington, DC: U.S. Government Printing Office.

Brass, G. W., Southam, J. R., and Peterson, W. H. 1982. Warm saline bottom water in the ancient ocean. *Nature* 296:620–623.

Brice, S. E., Cochran, M. D., Pardo, G., and Edwards, A. D. 1982. Tectonics and sedimentation in the South Atlantic rift sequence: Cabinda, Angola. *In* Studies in Continental Margin Geology, J. S. Watkins and C. L. Drake, Eds., *American Association of Petroleum Geologist Memoir 34*, 5–18.

Brink, A. H. 1974. Petroleum geology of Gabon Basin. *American Association of Petroleum Geologists Bulletin* 58:216–235.

Brognon, G., and Verrier, G. 1966. Oil and geology in Cuanza Basin of Angola. *American Association of Petroleum Geologists Bulletin* 50:108–158.

Cande, S., and Rabinowitz, P. D. 1978. Mesozoic seafloor spreading bordering conjugate continental margins of Angola and Brazil. *Proceeding of Offshore Technology Conference, Report 3268*, 1869–1876.

Connary, S. D. 1972. Investigations of the Walvis Ridge and environs. Ph.D. Thesis, Columbia University.

Dailly, C. G. 1982. Slope readjustment during sedimentation on continental margins. *In* Studies in Continental Margin Geology, J. S. Watkins, and L. Drake, Eds., *American Association of Petroleum Geologists Memoir 34*, 593–608.

Dean, W. E., Hay, W. W., and Sibuet, J.-C. 1984. Geologic evolution and paleoenvironments of the Angola Basin and adjacent Walvis Ridge. Synthesis of results of Deep Sea Drilling Project Leg 75. *In* Initial Reports of the Deep Sea Drilling Project, Volume 75, Part 1, W. W. Hay, J.-C. Sibuet, et al.: Washington, DC: U.S. Government Printing Office, 509–542.

Divins, D. L., and Rabinowitz, P. D. in press. Thickness of sedimentary cover. Sheet 3. South Atlantic. Geological–Geophysical Atlas of the Atlantic Ocean.

Emery, K. O. 1974. Pagoda structures in marine sediments. *In* Natural Gases in Marine Sediments, I. R. Kaplan, Ed.: New York: Plenum Press, 309–317.

Emery, K. O., and Uchupi, E. 1984. *The Geology of the Atlantic Ocean*: New York: Springer-Verlag.

Emery, K. O., Uchupi, E., Bowin, C. O., Phillips, J., and Simpson, E. S. W. 1975a. Continental margin off western Africa: Cape St. Francis (South Africa) to Walvis Ridge (South-West Africa). *American Association of Petroleum Geologists Bulletin* 59:3–59.

Emery, K. O., Uchupi, E., Phillips. J., Bowin, C., and Mascle, J. 1975b. Continental margin off western Africa: Angola to Sierra Leone. *American Association of Petroleum Geologists Bulletin* 59:2209–2265.

Fairhead, J. D., and Green, C. H. 1989. Controls on rifting in Africa and the regional tectonic model for Nigeria and East Niger rift basins. *Journal of African Earth Sciences* 8:231–250.

Flood, R. D. 1983. Classification of sedimentary furrows and a model for furrow initiation and evolution. *Geological Society of America Bulletin* 94:630–639.

Franks, S., and Nairn, A. E. M. 1973. The equatorial marginal basins of west Africa. *In* The Ocean Basins and Margins, Volume 1, The South Atlantic, A. E. M. Nairn and F. G. Stehli, Eds.: New York: Plenum Press, 301–350.

Gorini, M. A. 1977. The tectonic fabric of the equatorial Atlantic and adjoining margins: Gulf of Guinea and northeast Brazil. Ph.D. Thesis, Columbia University.

Grunau, H. R., Lehner, P., Cleintaur, M. R., Allenbach, P., and Baskker, G. 1975. Radiometric ages and seismic data from Fuerteventura (Canary Islands) and São Tome (Gulf of Guinea). *In* Progress in Geodynamics: Royal Netherlands Academy of Arts and Sciences, 90–119.

Haq, B. U., Hardenbol, J., and Vail, P. R. 1988. Mesozoic and Cenozoic chronostratigraphy and cycles of sea level. *In* Sea-Level Changes: An Integrated Approach, C. K. Wilgus, B. S. Hastings, C. A. Ross, H. Posamentier, and C. G. St. C. Kendall, Eds., *Society of Economic Paleontologists and Mineralogists Special Publication No. 42*, 71–108.

Hay, W. W., Sibuet, J. C., Barron, E. J., Boyce, R. E., Biassell, S., Dean, W. E., Huc, A. Y., Keating, B. H., McNulty, C. L., Meyers, P. A., Nohara, M., Schallrenter, R. E., Steinmetz, J. C., Stow, D., and Stradner, H. 1984. Initial Reports of the Deep Sea Drilling Project, Volume 75, Part 2: Washington, DC: U.S. Government Printing Office.

Hedberg, J. D. 1968. A geological analysis of the Cameroon trend. Ph.D. thesis, Princeton University.

Heezen, B. C., Menzies, R. J., Schneider, E. D., Ewing, W. M., and Granelli, N. C. L. 1964. Congo submarine canyon. *American Association of Petroleum Geologists Bulletin* 48:1126–1149.

Heezen, B. C., Tharp, M., Jicha, H., and McClellan, M. 1978. General Bathymetric Chart of the Oceans (GEBCO), Chart 5.12.: Ottawa: Geomarine Mapping Section, Canadian Hydrographic Service.

Holmes, A. 1965. *Principles of Physical Geology* (rev. ed.): New York: George Ronald Pub.

Laine, E. P., Damuth, J. E., and Jacobi, R. 1986. Surficial sedimentary processes revealed by echo-character mapping in the western North Atlantic. *In* The Geology of North America, Volume M, The Western North Atlantic, P. R. Vogt and B. E. Tucholke, Eds.: Boulder, CO: Geological Society of America, 427–436.

Lehner, P., and de Ruiter, P. A. C. 1977. Structural history of Atlantic margin of Africa. *American Association of Petroleum Geologists Bulletin* 61:961–981.

Le Pichon, X., and Hayes, D. E. 1971. Marginal offsets, fracture zones, and the early opening of the South Atlantic. *Journal of Geophysical Research* 76:6283–6293.

Leyden, R., Asmus, H., Zembrushi, S., and Bryan, G. 1976. South Atlantic diapiric structures. *American Association of Petroleum Geologists Bulletin* 60:682–693.

Leyden, R., Bryan, G., and Ewing, M. 1972. Geophysical reconnaissance on the African shelf: 2. Margin sediments from the Gulf of Guninea to Walvis Ridge. *American Association of Petroleum Geologists Bulletin* 56:319–343.

Martin, H. 1973. The Atlantic margin of southern Africa between Latitude 17° South and the Cape of Good Hope. *In* The Ocean Basin and Margins, Volume 1, The South Atlantic, A. E. M. Nairn and F. G. Stehli, Eds.: New York: Plenum Press, 277–300.

Martin, H. 1983. Alternative geodynamic models for the Damara Orogeny. A critical discussion. *In* Intracontinental Foldbelts, H. Martin and F. W. J. Elder, Eds.: New York: Springer-Verlag, 913–945.

Moore, T. C., Rabinowitz, P. D., Borella, P. E., Shackleton, N. J., and Boersma, A. 1984. History of the Walvis Ridge. *In* Initial Reports of the Deep Sea Drilling Project, Volume 74, T. C. Moore, Jr., P. D. Rabinowitz, et al.: Washington, DC: U.S. Government Printing Office, 874–894.

Musgrove, L. E., and Austin, J. A., Jr. 1983. Intrabasement structure in the southern Angola Basin. *Geology* 11:169–173.

Musgrove, L. A., and Austin, J. A., Jr. 1984. Multichannel seismic reflection survey of the southeastern Angola Basin. *In* Initial Reports of the Deep Sea Drilling Project, Volume 75, Part 2, W. W. Hay, J.-C. Sibuet, et al.: Washington, DC: U.S. Government Printing Office, 1191–1210.

Natland, J. H. 1978. Composition, provenance, and diagenesis of Cretaceous clastic sediments drilled on the Atlantic continental rise off southern Africa, DSDP Site 361—implications for early circulation of the South Atlantic. *In* Initial Reports of the Deep Sea Drilling Project, Volume 40, H. M. Bolli, W. B. F. Ryan, et al.: Washington, DC: U.S. Government Printing Office, 487–524.

Pautot, G., Renard, V., Daniel, J., and Dupont, J. 1973. Morphology, limits, origin, and age of salt layer along the South Atlantic African margin. *American Association of Petroleum Geologists Bulletin* 57:1658–1671.

Porada, H. 1979. The Damara–Rebeira orogen of the Pan-African–Braziliano cycle in Namibia (southwest Africa) and Brazil as interpreted in terms of continental collision. *Tectonophysics* 57:237–265.

Rabinowitz, P. D. 1972. Gravity anomalies on the continental margin of Angola, Africa. *Journal of Geophysical Research* 77:6327–6347.

Rabinowitz, P. D., and LaBrecque, J. 1979. The Mesozoic South Atlantic Ocean and evolution of its continental margins. *Journal of Geophysical Research* 84:5973–6003.

Reyre, D. 1966. Historie geologique du bassin de Douala (Cameroun). *In* Sedimentary Basins of African Coasts, Part 1, Atlantic Coast, D. Reyre, Ed.: Paris: Association of African Geological Surveys, 143–161.

Seiglie, G. A., and Baker, M. B. 1982. Foraminiferal zonation of the Cretaceous off Zaire and Cabinda, west Africa and its geological significance. *In* Studies in Continental Margin Geology, J. S. Watkins and C. L. Drake, Eds., *American Association of Petroleum Geologists Memoir 34*, 651–658.

Shepard, F. P., and Emery, K. O. 1973. Congo submarine canyon and fan valley. *American Association of Petroleum Geologists Bulletin* 57:1679–1691.

Sibuet, J.-C., Hay, W. W., Prunier, A., Montadert, L., Hinz, K., and Fritsch, J. 1984. The eastern Walvis Ridge and adjacent basins (South Atlantic): morphology, stratigraphy, and structural evolution in the light of the results of Legs 40 and 75. *In* Initial Reports of the Deep Sea Drilling Project, Volume 74, Part 2, W. W. Hay, J.-C. Sibuet, et al.: Washington, DC: U.S. Government Printing Office, 483–508.

Siesser, W. G. 1980. Late Miocene origin of the Benguela Current (sic) system of northern Namibia. *Science* 208:283–285.

Stow, D. A. V. 1984a. Turbidite facies, associations, and sequences in the southeastern Angola Basin. *In* Initial Reports of the Deep Sea Drilling Project, Volume 75,

Part 2, W. W. Hay, J.-C. Sibuet, et al.: Washington, DC: U.S. Government Printing Office, 785–800.

Stow, D. A. V. 1984b. Anatomy of debris-flow deposits. *In* Initial Reports of the Deep Sea Drilling Project, Volume 75, Part 2, W. W. Hay, J.-C. Sibuet, et al.: Washington, DC: U.S. Government Printing Office, 801–808.

Stow, D. A. V., and Dean, W. E. 1984. Middle Cretaceous black shales at Site 530 in the southeastern Angola Basin. *In* Initial Reports of the Deep Sea Drilling Project, Volume 75, Part 2, W. W. Hay, J.-C. Sibuet, et al.: Washington, DC: U.S. Government Printing Office, 809–818.

The Shipboard Scientific Party. 1978a. Walvis Ridge—sites 362 and 363. *In* Initial Reports of the Deep Sea Drilling Project, Volume 40, H. M. Bolli, W. B. F. Ryan, et al.: Washington, DC: U.S. Government Printing Office, 183–356.

The Shipboard Scientific Party. 1978b. Angola continental margin—sites 364 and 365. *In* Initial Reports of the Deep Sea Drilling Project, Volume 40, H. M. Bolli, W. B. F. Ryan, et al.: Washington, DC: U.S. Government Printing Office, 357–390.

Tucholke, B. E., and Embley, R. W. 1984. Cenozoic regional erosion of the abyssal floor off South Africa. *In* Interregional Unconformities and Hydrocarbon Accumulation, J. S. Schlee, Ed., *American Association of Petroleum Geologists Memoir 36*, 145–164.

Uchupi, E. 1989. The tectonic style of the Atlantic rift system. *Journal of African Earth Sciences* 8:143–164.

Uchupi, E., and Emery, K. O. 1972. Seismic reflection, magnetic, and gravity profiles of the eastern Atlantic continental margin and adjacent deep-sea floor. I. Cape Francis (South Africa) to Congo Canyon (Republic of Zaire). Woods Hole Oceanographic Institution Reference No. 72-95, 8 sheets.

Uchupi, E., and Emery, K. O. 1974. Seismic reflection, magnetic, and gravity profiles of the eastern Atlantic continental margin and adjacent deep-sea. II. Congo Canyon (Republic of Zaire) to Lisbon (Portugal). Woods Hole Oceanographic Institution Reference No. 74-19, 14 sheets.

van Andel, Tj., Thiede, H., Sclater, J. G., and Hay, W. W. 1977. Depositional history of the South Atlantic during the last 125 million years: *Journal of Geology* 85:651–698.

Vassalo, K., Jacobi, R. D., and Shor, A. N. 1984. Echo character, microphysiography, and geologic character. *In* Ocean Drilling Program, Regional Atlas Series 4, J. I. Ewing and P. D. Rabinowitz, Eds.: Woods Hole, MA: Marine Science International, 31.

Von Herzen, R. P., Hoskins, H., and van Andel, Tj. H. 1972. Geophysical studies in the Angola diapiric field. *Geological Society of America Bulletin* 83:1910–1910.

West, D. B. 1989. Model for salt deformation on deep margins of central Gulf of Mexico. *American Association of Petroleum Geologists Bulletin* 73:1472–1482.

Worlsey, T. 1974. The Cretaceous–Tertiary boundary event in the Ocean. *In* Studies in Paleo-Oceanography, W. W. Hay, Ed., *Society of Economic Paleontologists and Mineralogists Special Publication No. 20*, 94–125.

6

U.S. Middle Atlantic Continental Rise: Provenance, Dispersal, and Deposition of Jurassic to Quaternary Sediments

C. Wylie Poag

During the past 20 years, drilling and geophysical profiling have added immensely to our knowledge of the stratigraphic framework and subsurface depositional patterns of the middle segment of the U.S. Atlantic continental margin, from the coastal plain to the continental rise (e.g., Hollister, Ewing, et al., 1972; Benson, Sheridan, et al., 1978; Grow, Mattick, and Schlee, 1979; Hathaway et al., 1979; Tucholke and Mountain, 1979, 1986; Tucholke and Vogt, 1979; Schlee, 1981; Tucholke and Laine, 1982; Uchupi and Shor, 1984; Ward and Krafft, 1984; Cleary, Pilkey, and Nelson, 1985; Mountain and Tucholke, 1985; Poag, 1985, 1987, 1991; Jansa, 1986; Vogt and Tuckolke, 1986; Poag et al., 1987; Schlee and Hinz, 1987; Van Hinte, Wise et al., 1987; Grow, Klitgord, and Schlee, 1988; Sheridan and Grow, 1988; Locker, 1989; McMaster, Locker, and Laine, 1989; Meyer, 1989; Poag and Sevon, 1989; Pratson and Laine, 1989; Danforth and Schwab, 1990; Poag et al., 1990; Prather, 1991; Locker and Laine, 1991). Poag and Sevon (1989) recently described the postrift depositional history of this ~ 500,000-km² region and highlighted the main features of its sedimentary evolution. Poag (1991) provided additional details regarding the Jurassic and Early Cretaceous development of a large, outer shelf carbonate platform and its deep-sea depositional equivalents in this region.

For this chapter, I draw on all these studies, particularly those of Poag and Sevon (1989) and Poag (1991), but I focus more directly on the sedimentary development of the continental rise in the northern part of the Hatteras Basin. This continental rise is by no means an isolated depositional province, as several authors in this volume (as well as many others) clearly demonstrate; its sediments have been derived from various landward source terrains, dominated by the central Appalachian, Adirondack, and New England Appalachian highlands, but also generally including a well-defined coastal plain, a broad continental shelf, and (much of the time) a relatively steep continental slope. The siliciclastic sediments initially were dispersed through coastal river systems and eventually reached the continental rise by means of various sediment gravity flows (debris flows, turbidity currents) and mass movements (slumps, slides). Once on the rise, some of these sediments have been redistributed by bottom currents or mixed and interlayered with pelagic biocarbonates or biosilicates.

This chapter begins with a description of the general geologic setting of the study area and the methods and problems of analysis. Then I shall address the relations among all the sediment sources and dispersal systems, define their primary regulating agents (tectonism, eustacy, paleoclimate), and describe each resultant depositional unit composing the continental rise. These unit descriptions are followed by a comprehensive summary of eight proposed stages of depositional history and a discussion of primary conclusions.

GEOLOGIC SETTING

The study area covers approximately 500,000 km² of the U.S. eastern seaboard from 90 km north of Cape Hatteras, North Carolina, to Cape Cod, Massachusetts (lat. 36°00′–42°00′N; long. 69°30′–78°00′W; Figs. 6-1, 6-2). The boundaries of this area have been selected by the U.S. Geological Survey to delineate the middle Atlantic sheet of its Continental Margin Map (CONMAP) series. Three contiguous postrift sedimentary

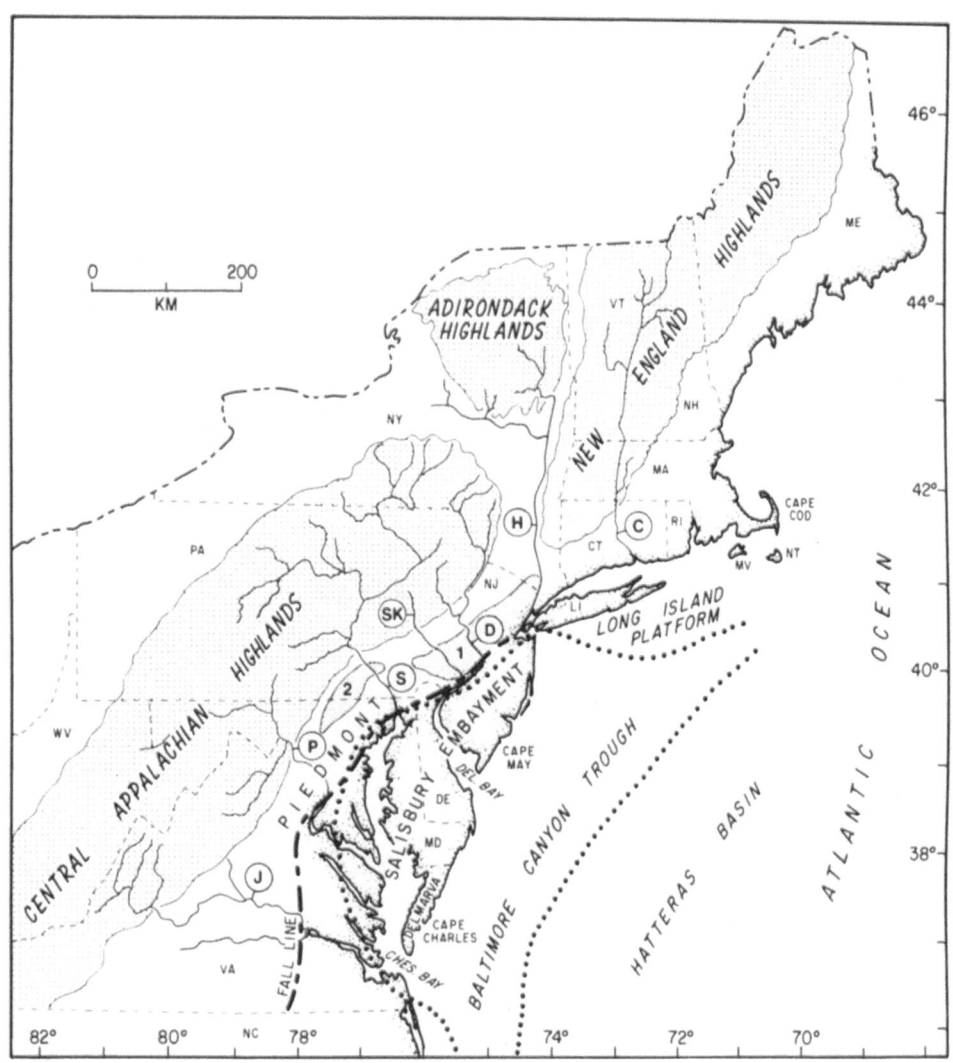

FIGURE 6-1. Physiographic, geologic, and oceanographic features of study area. MV = Martha's Vineyard; NT = Nantucket; LI = Long Island; Del. Bay = Delaware Bay; Ches. Bay = Chesapeake Bay; C = Connecticut River; H = Hudson River; D = Delaware River; SK = Schuylkill River; S = Susquehanna River; P = Potomac River; J = James River; 1 = Newark Basin; 2 = Gettysburg Basin; Fall Line = inner edge of coastal plain. (From Poag and Sevon, 1989).

basins underlie this region:

1. The Salisbury Embayment underlies the coastal plain.
2. The Baltimore Canyon Trough underlies the continental shelf and the upper part of the continental slope.
3. The northern part of the Hatteras Basin underlies the lower part of the continental slope and the continental rise.

The sedimentary deposits of this basin complex can be divided into 23 allostratigraphic units, which are mappable stratiform bodies of sedimentary rock defined and identified on the basis of their bounding unconformities (North American Commission on Stratigraphic Nomenclature, 1983; Poag and Sevon, 1989). The 23 units range in age from early Middle Jurassic [Aalenian(?)] to Quaternary (Figs. 6-3, 6-4). These allostratigraphic units correspond in a general way to second-order sequences of the Exxon model of sequence stratigraphy (Vail, Mitchum, and Thompson, 1977; Haq, Hardenbol, and Vail, 1987, 1988; Van Wagoner et al., 1988); they also correspond to first-order depositional complexes of Mutti and Normark (1987). Correlation with sequences of the Exxon model is accurate for younger allostratigraphic units, because the requisite planktonic biozonation is available, but for older units [Aalenian(?) to Oxfordian], correlation is less precise.

The Oxfordian unit is the oldest drilled in the study area, but all units have been drilled in the adjacent Georges Bank Basin (Poag, 1991). Seismostratigraphic

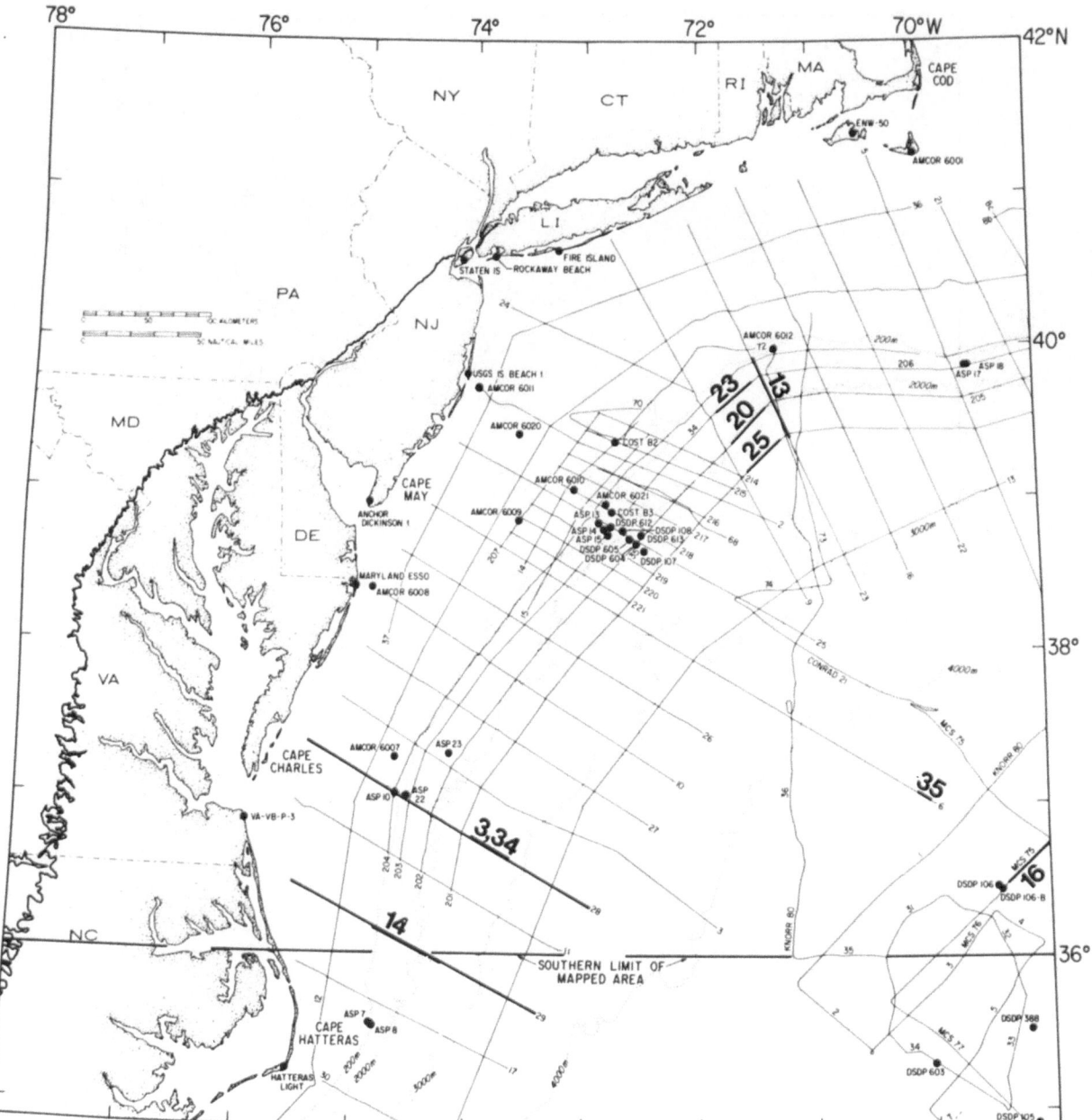

FIGURE 6-2.   Grid of multichannel seismic-reflection profiles and locations of key boreholes used for analyses presented herein. Numbered heavy black segments of seismic lines indicate geographic locations and figure numbers of profile segments illustrated herein. AMCOR = Atlantic Margin Coring Project; ASP = Atlantic Slope Project; COST = Continental Offshore Stratigraphic Test; DSDP = Deep Sea Drilling Project; see Poag (1985) for further discussion of borehole lithology and biostratigraphy.

correlations between the Baltimore Canyon Trough and Georges Bank Basin are generally straightforward, with only a few structural complications in the oldest units. All units except the Aalenian(?) have been dated by microfossils in one or both basins [see Poag and Sevon (1989) and Poag (1991) for further discussion of the allostratigraphic framework].

The postrift sedimentary column is thickest (> 15 km) in the Baltimore Canyon Trough, where it overlies

a thick (> 5-km) synrift (mainly Triassic) succession of presumed sedimentary and volcaniclastic rocks (Manspeizer and Cousminer, 1988). The zone of greatest thermotectonic subsidence lies between Paleozoic continental crust (to the northwest) and Mesozoic oceanic crust (to the southeast) and approximates the position of the East Coast Magnetic Anomaly (ECMA), which marks the continent–ocean boundary (Fig. 6-3; Klitgord, Hutchinson, and Schouten, 1988).

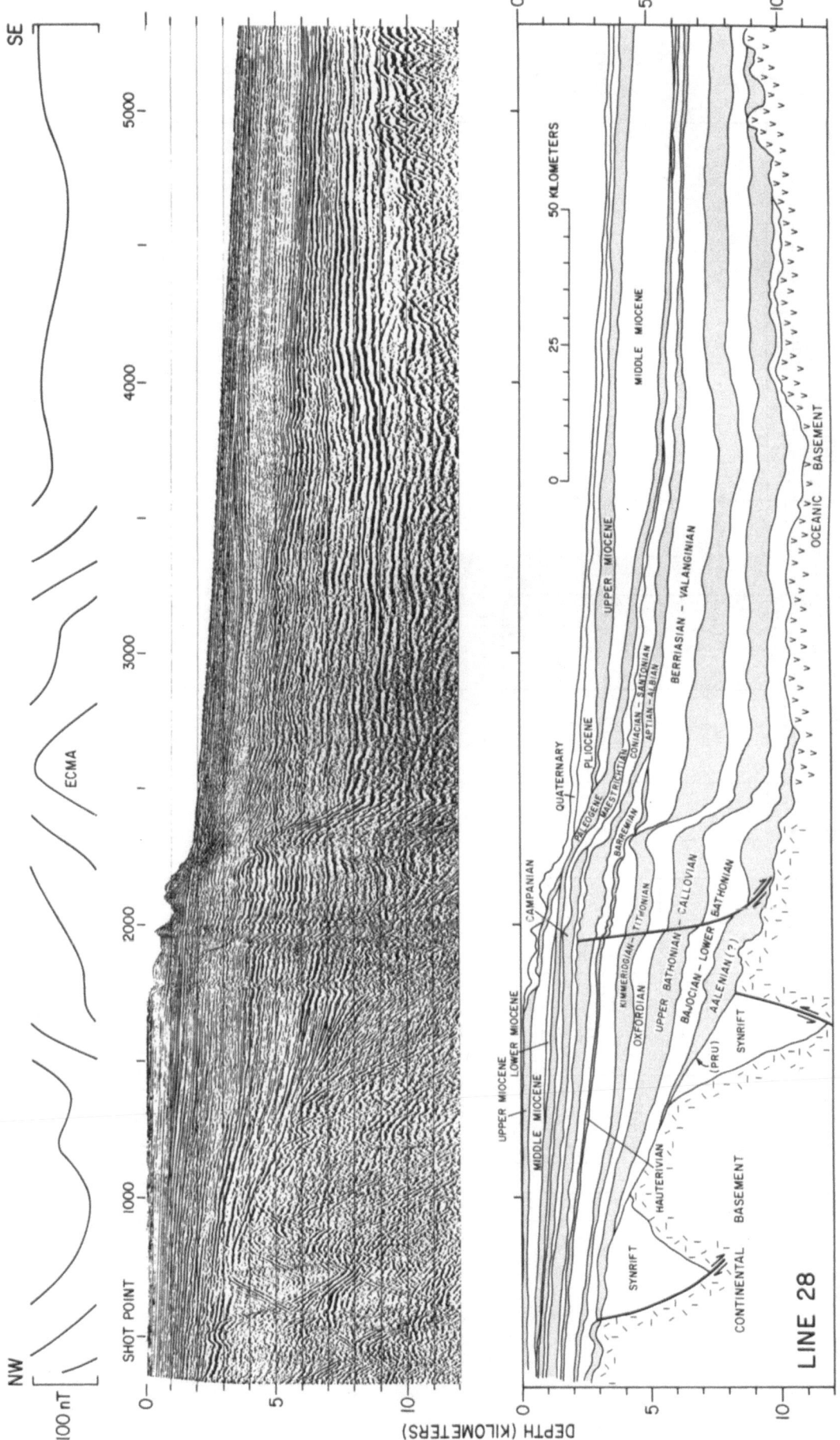

FIGURE 6-3. Depth section along multichannel seismic-reflection profile 28 crossing continental shelf, slope, and rise parallel to depositional dip, seaward of Cape Charles, Virginia (see Fig. 6-2 for location). Above is uninterpreted profile. Below is interpreted tracing of profile; only thicker allostratigraphic units labelled; thinner units are difficult to discern at this scale. Alternate units shaded for visual aid: ECMA = East Coast Magnetic Anomaly; PRU = postrift unconformity.

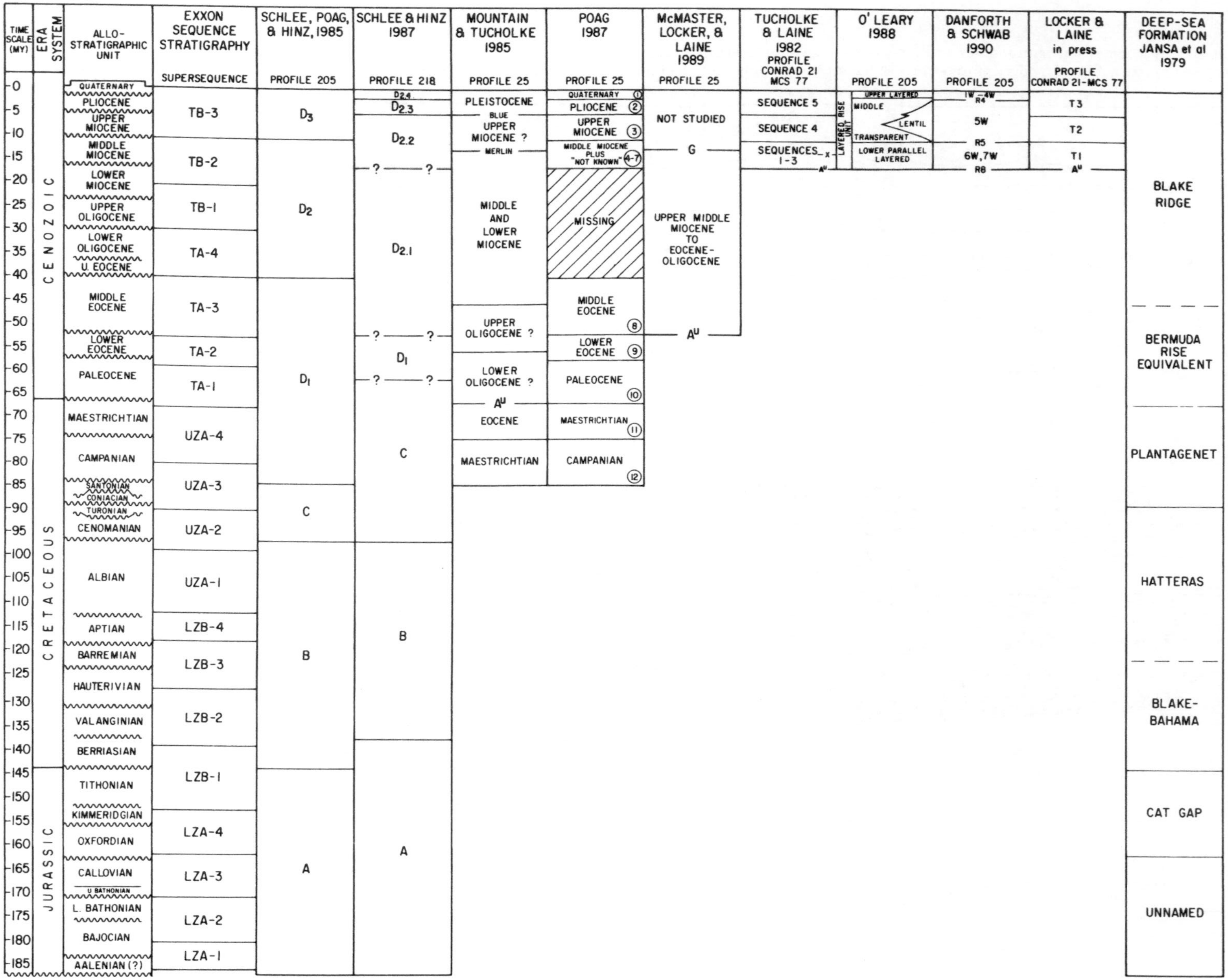

FIGURE 6-4. Stratigraphic chart showing the 23 allostratigraphic units mapped herein and their relation to the Exxon model of sequence stratigraphy (Haq, Hardenbol, and Vail, 1987) and to stratigraphic frameworks constructed by previous authors. A^u, X, G, Merlin, and Blue are designations for widespread deep-sea seismic boundaries described from the study area; R8, R5, and R4 are local seismic unconformities designated by Danforth and Schwab (1990). Circled numerals shown in column for Poag (1987) indicate depositional sequence numbers. Profile numbers on column headings (e.g., Profile 205) indicate profiles on which I directly compared positions of cited authors' seismic boundaries with boundaries I defined for allostratigraphic units. Ages of units based on biostratigraphy extrapolated from boreholes (e.g., Poag, 1987, 1991; Poag and Sevon, 1989).

The largest highland source terrains (central and New England Appalachian highlands; Denny, 1982; Hack, 1982) are parts of the Appalachian Orogen, which was faulted and uplifted during the Paleozoic suturing of North American and African tectonic plates. The third, smaller, source terrain is composed of Precambrian igneous rocks now exposed in the Adirondack dome (Isachsen, 1975). Present elevations in these highlands reach 1–2 km above sea level. Each source terrain is drained today by one or more major river system (Fig. 6-1), several of which appear to have been established early in the Middle Jurassic (Poag and Sevon, 1989).

METHODS AND PROBLEMS OF ANALYSIS

The chief basis for my interpretations is a series of 25 isochron maps (scale 1:1,000,000; contours represent equal two-way traveltime) presented in this paper. The maps are drawn from a network of ~ 10,000 line-km of multichannel seismic-reflection profiles, which are calibrated stratigraphically with 88 key boreholes and outcrops (coastal-plain and seafloor exposures; Fig. 6-2; Poag, 1985, 1987, 1991; Poag and Sevon, 1989). Areal distribution of seismic profiles and boreholes is most comprehensive on the continental shelf, slope, and upper rise. Only a few seismic profiles extend seaward of the 3,500-m bathymetric contour, however,

and only three key Deep Sea Drilling Project (DSDP) core sites (Sites 604, 605: Van Hinte et al., 1987; Site 613: Poag, Watts, et al., 1987) are on the part of the upper continental rise that I have mapped. Two coreholes on the slower continental rise (Fig. 6-2; DSDP Sites 105 and 603; Hollister, Ewing, et al., 1972; Van Hinte, Wise, et al., 1987) provide crucial stratigraphic and paleoenvironmental data, which are incorporated into this study.

Isochron maps are used to identify primary (greatest volume) and secondary depocenters, to delineate sediment-dispersal routes and distribution patterns, to estimate offshore sediment volumes (and thus to derive sediment-accumulation rates; Table 6-1; Fig. 6-5), and to determine the dominant sources of terrigenous detritus. Features as small as 0.5–1.0 km can be distinguished on the original seismic profiles and isochron maps, but most features smaller than 5–10 km are barely discernible in page-size illustrations. I, therefore, discuss mainly features larger than 10 km. Thicknesses of allostratigraphic units cited in the text are extrapolated from published seismic-reflection profiles (e.g., Fig. 6-3) in which two-way traveltime has been converted to depth (e.g., Grow, Mattick, and Schlee, 1979; Schlee, Dillon, and Grow, 1979; Ewing, 1984; Grow et al., 1988; McMaster, Locker, and Laine, 1989).

Approximate sediment volumes (Table 6-1) were calculated by treating each mapped unit as a series of

TABLE 6-1   Net Volumes of Compacted Siliciclastic Sediment Derived from Depth-Converted Isochron Maps. Values in Cubic Kilometers.

Allostratigraphic Unit	Duration (m.y.)	Rate (10^3 km^3/m.y.)	Total Volume	Volume on Rise	Percent on Rise
Quaternary	1.6	17	26,600	23,340	95
Pliocene	3.7	15	56,749	51,149	90
Upper Miocene	5.9	9	54,061	40,600	75
Middle Miocene	5.4	26	140,770	93,100	66
Lower Miocene	7.1	3	22,000	4,000	18
Upper Oligocene	6.3	1	6,000	0	0
Upper Eocene–Lower Oligocene	10.0	0.7	7,000	200	3
Middle Eocene	12.0	2	29,590	10,000	34
Lower Eocene	5.8	4	24,530	16,000	65
Paleocene	8.6	3	26,015	14,750	57
Maestrichtian	8.1	9	73,951	62,500	85
Campanian	9.5	9	89,190	28,700	32
Coniacian–Santonian	4.5	19	83,848	31,000	37
Cenomanian–Turonian	9.0	2	14,415	2,000	14
Aptian–Albian	21.5	4	82,648	31,000	38
Barremian	5.0	11	53,030	11,800	22
Hauterivian	7.0	1	7,500	1,500	20
Berriasian–Valanginian	13.0	6	72,000	32,800	46
Kimmeridgian–Tithonian	12.0	6	71,184	40,360	57
Oxfordian	7.0	9	63,255	43,000	68
Upper Bathonian–Callovian	9.5	13	122,675	102,500	84
Bajocian–Lower Bathonian	10.5	13	139,650	97,750	70
Aalenian(?)	4.0	18	70,700	40,520	57
Total	187.0		1,337,361	780,569	58

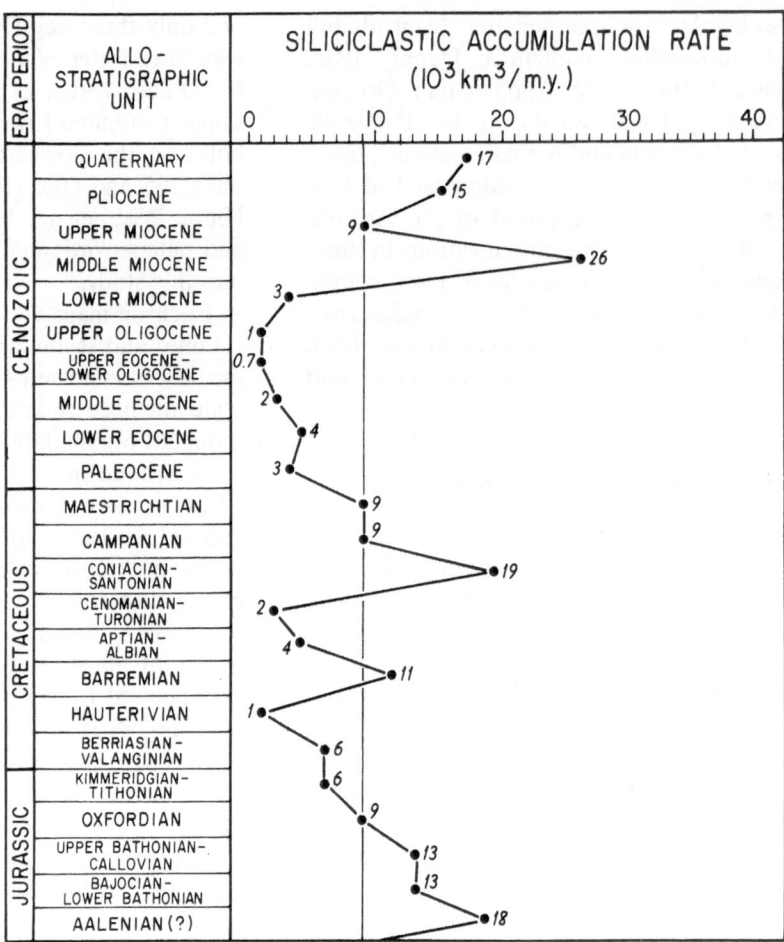

FIGURE 6-5.   Net siliciclastic sediment-accumulation rates for 23 postrift allostratigraphic units of study area. Values are volumetric rates given in thousands of cubic kilometers per million years. Most values have been recalculated from updated isochron maps (contrast with original values of Poag and Sevon, 1989); raw values for middle Miocene through Quaternary units have been reduced by 30% to make them compatible with volumes of older, more deeply buried (more compacted) units (see Table 6-1). Vertical line represents mean value of 9,000 km³/m.y.

stacked tabular slices. The volume of each 100-m-thick slice was calculated individually, and the results were summed to produce the total volume of the mapped unit. From those units that contain significant volumes of carbonate deposits, I subtracted the carbonate values to obtain the volume of siliciclastics alone. I also made a volume adjustment (30% reduction) for middle Miocene and younger units, which are undercompacted by roughly 30% (30% more porosity) compared to older, more deeply buried deposits. Resultant volume estimates are imprecise but sufficiently accurate for the analyses and interpretations applied herein.

Values obtained for accumulation rates are directly dependent upon the time scale used. I have selected the scale adopted by the Geological Society of America for the Decade of North America (DNAG) series of publications (Palmer, 1983). Accuracy of rate calculations also depends on whether or not a unit represents continuous deposition during its entire geological time

span. Because the mapped units are allostratigraphic units, the upper and lower boundary of each is a widespread unconformity; additional unconformities exist *within* each unit. As a result, each mapped unit represents an incomplete portion of its original sedimentary prism. We cannot know exactly how much time is represented by an unconformity except at outcrops and boreholes; furthermore, the duration of hiatuses varies from place to place. Thus, the calculated volume for each unit is a *net* volume.

It should be noted that I have recalculated volumes based on revisions of several isochron maps; thus, the volumes and accumulation rates given herein are somewhat different from those originally published by Poag and Sevon (1989).

The resolution of multichannel seismic profiles is generally insufficient to distinguish detailed morphological and facies relations ascribed to modern gravity-flow systems. What is more, each mapped unit results from the interaction of numerous depositional and erosional processes. Even though some of the dominant sedimentary features may have formed in a geological instant, comparison with short-lived analogs on the modern seafloor is risky. Much more drilling is required before ancient features can be quantitatively measured by the stringent sedimentological and morphometric criteria applied to modern counterparts. Nevertheless, the gross features (e.g., geometry, seismic facies) of each allostratigraphic unit resemble modern features (fans, aprons, plains, canyons, deltas) and are presumed to be roughly equivalent.

Terminology used to describe depositional and erosional, features of modern continental rises, such as "fan," "apron," "ramp," "canyon," "channel," "lobe,"

is by no means uniformly defined or applied by experts in this discipline (e.g., Bouma, Normark, and Barnes, 1985). Thus, when applied to ancient analogs, these terms are even more fraught with potential ambiguity or misinterpretation (Shanmugam and Moila, 1985; Mutti and Normark, 1987; Bouma, 1990). I have relied on properties such as size, volume, cross-sectional and three-dimensional geometry, isochron patterns, sediment type (where known), seismic expression, and relative shelf-to-basin location to distinguish the major features I describe. Selected representative examples are illustrated from multichannel seismic reflection profiles taken from a variety of depositional settings in the study area. For purposes of clarity, I define below these principal depositional and erosional features.

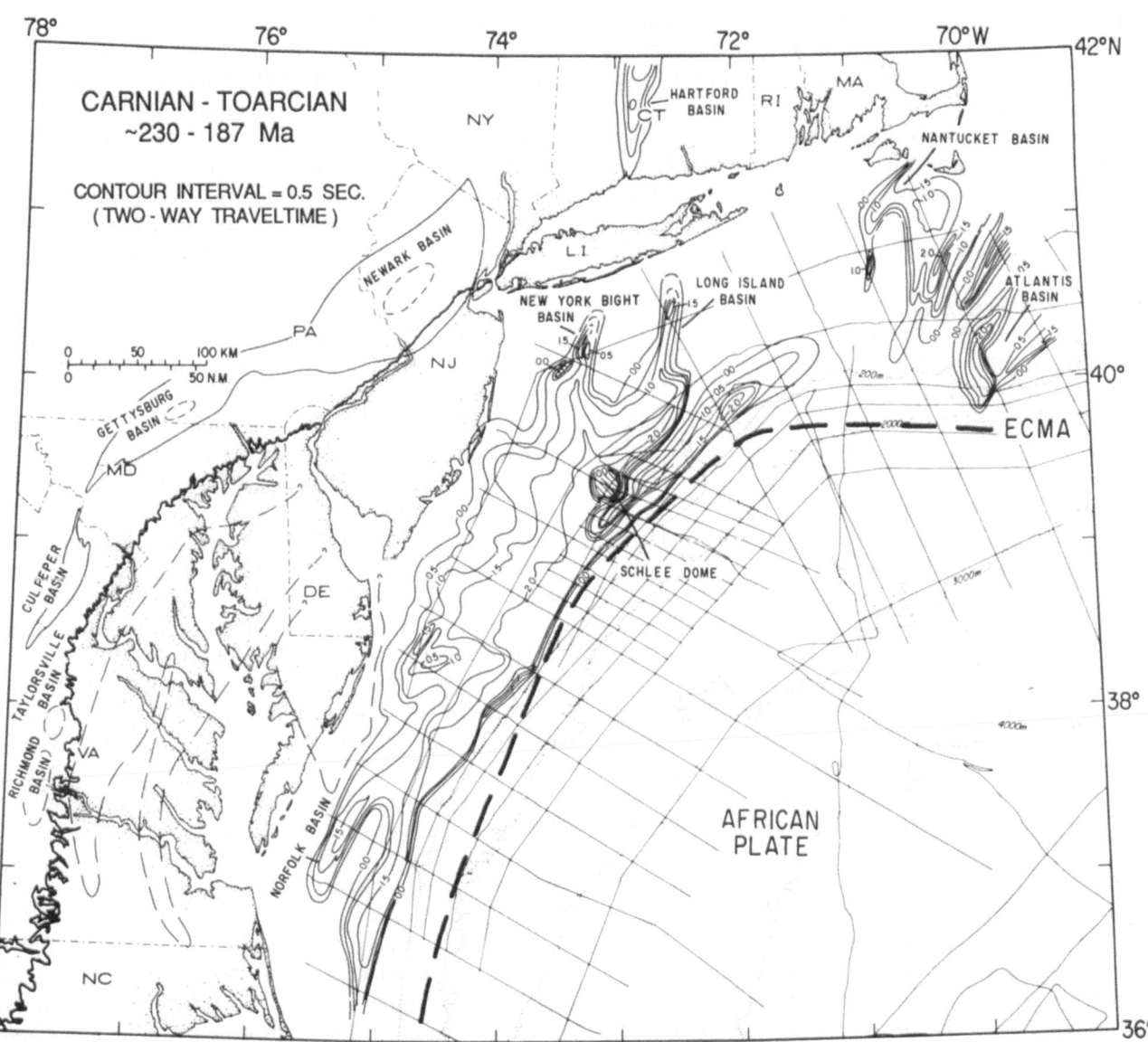

FIGURE 6-6.   Isochron map (offshore contours in 0.5 sec of two-way traveltime) of synrift deposits in study area. Thickness of sediments in several onshore basins not known. Boundaries of synrift basins beneath coastal plain (long dashes) poorly known. ECMA = East Coast Magnetic Anomaly.

These are working definitions, not intended for application outside the context of this paper.

BASIN PLAIN: Widespread, relatively thin sedimentary body, generally deposited on lower or middle continental rise, distal to fans.

CANYON: V-shaped (cross section) sediment conduit usually incising shelf edge, continental slope, and upper rise.

CHANNEL: U-shaped (cross section) sediment conduit usually incising continental slope and upper rise.

DEBRIS APRON: Lenticular, alongslope–trending sedimentary body, generally deposited on continental slope and(or) upper rise; derived from multiple sources.

DELTA: Thick, mounded, prograded, sedimentary body deposited on continental shelf, derived from fluvial drainage system (equivalent to delta complex of some authors); confluent delta = lateral merger of two or more deltas.

FAN: Mounded, downslope–trending sedimentary body, generally elongate or fan-shaped, commonly distally lobed, generally deposited on shelf edge, slope, or upper rise from single or multiple sources (equivalent to fan complexes of Mutti and Normark, 1987):

small fan = 50–200 km long
medium fan = 200–300 km long
large fan = 300–400 km long
megafan = > 400 km long
confluent fan = lateral merging of two or more fans on middle or lower continental rise.

INTERFAN PLAIN: Relatively narrow, thin, sedimentary body deposited between fans.

SEDIMENT POND: Depression on continental rise filled with stratified sediments.

SYNRIFT DEPOSITS

Most synrift deposits of the study area occupy northeast-trending grabens or half-grabens, formed as the North American and African tectonic plates diverged during the initial breakup of Pangaea (Fig. 6-6; Manspeizer, 1981, 1985, 1988; Klitgord and Hutchinson, 1985; Benson and Doyle, 1988; Hansen, 1988; Klitgord, Hutchinson, and Schouten, 1988; Manspeizer and Cousminer, 1988; several papers in the 1988 book edited by Manspeizer). Offshore rift basins occupy a zone 100–150 km wide landward of the East Coast Magnetic Anomaly and seaward of their counterparts exposed along the inner margin of the coastal plain. The largest (50–100 km × 500 km) and deepest (~ 5 km) graben system lies beneath the central Baltimore Canyon Trough off New Jersey. These grabens are believed to be filled mainly with coarse conglomerates,

arkosic sandstones, lacustrine siltstones and shales, and basaltic lavas and volcaniclastic rocks, which accumulated prior to oceanization of this region (Poag, 1985; Manspeizer and Cousminer, 1988); they, therefore, contain no continental rise deposits. A sedimentary volume of ~ 265,000 km³ was deposited in these grabens during the ~ 43 m.y. of their development (siliciclastic accumulation rate = ~ 6,000 km³/m.y.; compare with Fig. 6-5; see also Chapters 1, 2).

POSTRIFT DEPOSITS

Aalenian(?) Allostratigraphic Unit

The oldest identifiable postrift deposits in the study area were laid down on an extensive "postrift unconformity" (PRU; Fig. 6-3), which formed probably during late Toarcian(?) and(or) early Aalenian(?) time. The transition from rifting to seafloor spreading took place in the early part of the Middle Jurassic [~ 187 Ma(?); early Aalenian(?); Poag and Sevon, 1989]. Initial postrift marine deposits have not been sampled in situ in this region; on seismic profiles these deposits are represented by an areally restricted series of very high-amplitude reflections separated by relatively thick, acoustically transparent layers, which together are interpreted as bedded halite and anhydrite (Swift et al., 1990; Poag, 1991; see also Chapters 2, 5). Halite has been drilled in stratigraphically equivalent sections in the Goerges Bank and Scotian basins (Poag, 1982a, b; Holser et al., 1988).

In the Baltimore Canyon Trough, bedded evaporites occupy an elongate narrow tract (~ 50 km wide × 300 km long × 2–3 km thick), which was the center of the nascent trough (Fig. 6-7). Three small evaporite depocenters (0.5–0.7 sec thick) are present along the eastern margin of the main evaporite basin, and several even smaller (but thicker) ones surround Schlee Dome, an igneous pluton near the northwestern edge of the evaporite basin (Grow, 1980; Lippert, 1983; Poag, 1987; Grow et al., 1988; Jansa and Pe-Piper, 1988). One exploratory well (Houston Oil and Minerals 676 No. 1) drilled halite on the flank of Schlee Dome (Grow, 1980).

Most of the evaporites lie parallel to and just west of the ECMA (approximate continent–ocean boundary), and thickest sections generally are on the basin's eastern margin. Maximum thickness of bedded evaporites is ~ 3,000 m, but diapirs protrude at least 8–10 km into overlying strata (e.g., Klitgord Dome; Poag, 1987). In at least two places, mobile evaporitic strata appear to have subsequently flowed downslope from the main evaporite basin and now protrude as isolated diapir fields within the continental rise (Fig. 6-7; see also Chapter 5). At the time these evaporitic strata formed, however, the African plate was the eastern margin of

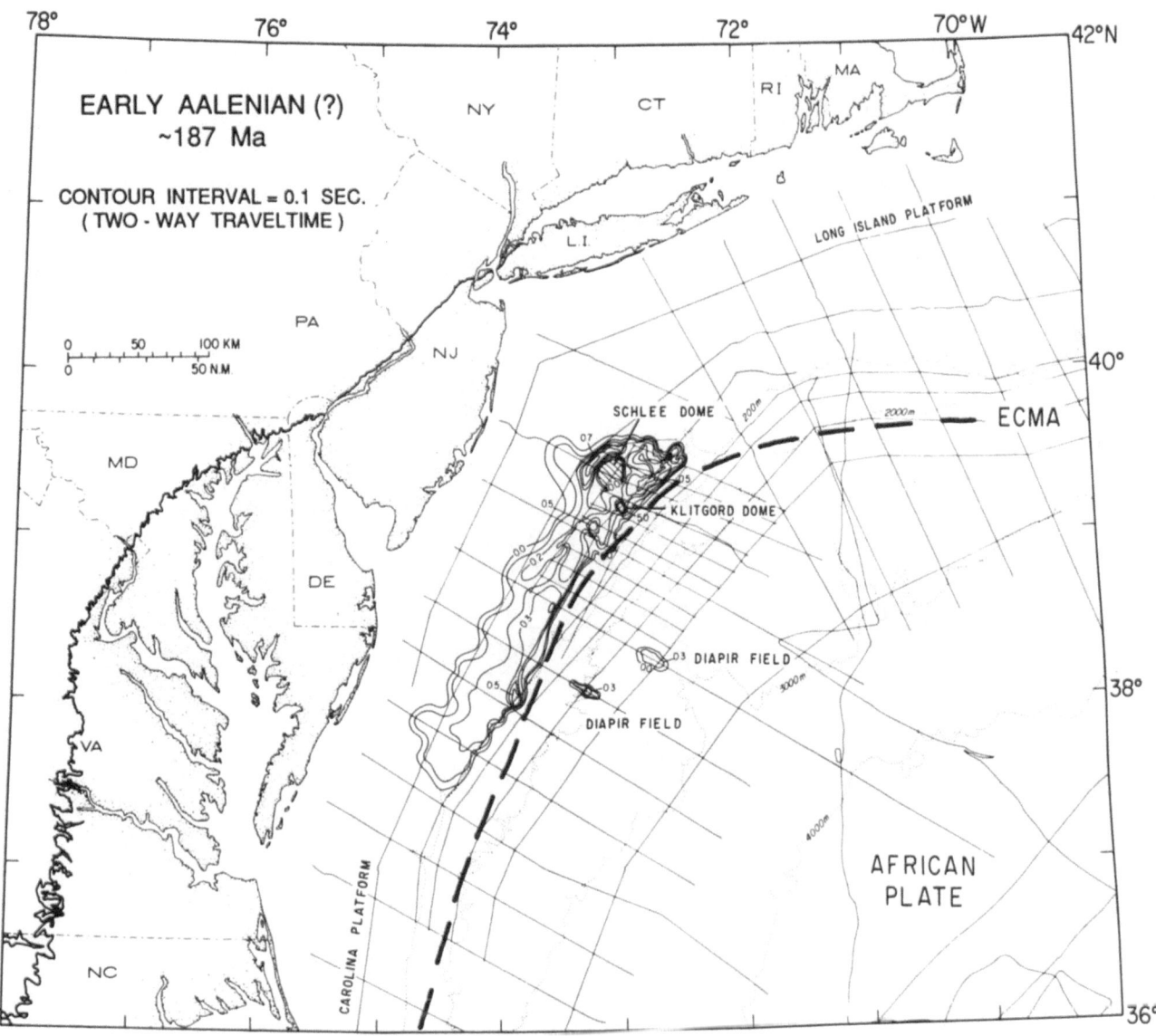

FIGURE 6-7.   Isochron map (contours in 0.1 sec of two-way traveltime) of evaporite beds that form initial postrift deposits in study area. ECMA = East Coast Magnetic Anomaly.

the narrow basin, and there was no physiographically identifiable continental rise.

By the end of the first major postrift depositional episode [Aalenian(?); 187–183 Ma] in the study area, the mid-Atlantic spreading center lay ~ 100–150 km east of the ECMA. Landward of the spreading center, rapidly subsiding continental crust had formed a broad, ~ 45,000-km² (100 km × 450 km) depression (three times the size of the preceding evaporite basin), which collected sediments derived mainly from highland source terrains to the west (Fig. 6-8). According to the Exxon model of sequence stratigraphy (e.g., Haq, Hardenbol, and Vail, 1987), sea levels were generally low, but rising. Paleoclimate was changing from arid [early Aalenian(?) evaporites] to seasonally wet–dry (Poag and Sevon, 1989; Cecil, 1990). The siliciclastic

accumulation rate was higher (~ 18,000 km³/m.y.) during this interval than at any other time prior to the Late Cretaceous (Fig. 6-5; Poag and Sevon, 1989).

In map view (Fig. 6-8), the depositional area is elongate and wedge-shaped; it is narrowest to the southwest, where it abuts the structurally high Carolina Platform, and broadest to the northeast, where it abuts the Long Island Platform. The continental rise, which made up about half this area, consisted of a trough (500 km × 100 km) that was part of a narrow proto-Atlantic seaway (Jansa, 1986; Tucholke and McCoy, 1986; Poag, 1991).

Terrigenous detritus built a series of five outer shelf deltas, three of which merged into a confluent delta (~ 2.5–6.0 km thick; 1.2–2.2 sec) in the northern half of the Baltimore Canyon Trough. Isochron patterns

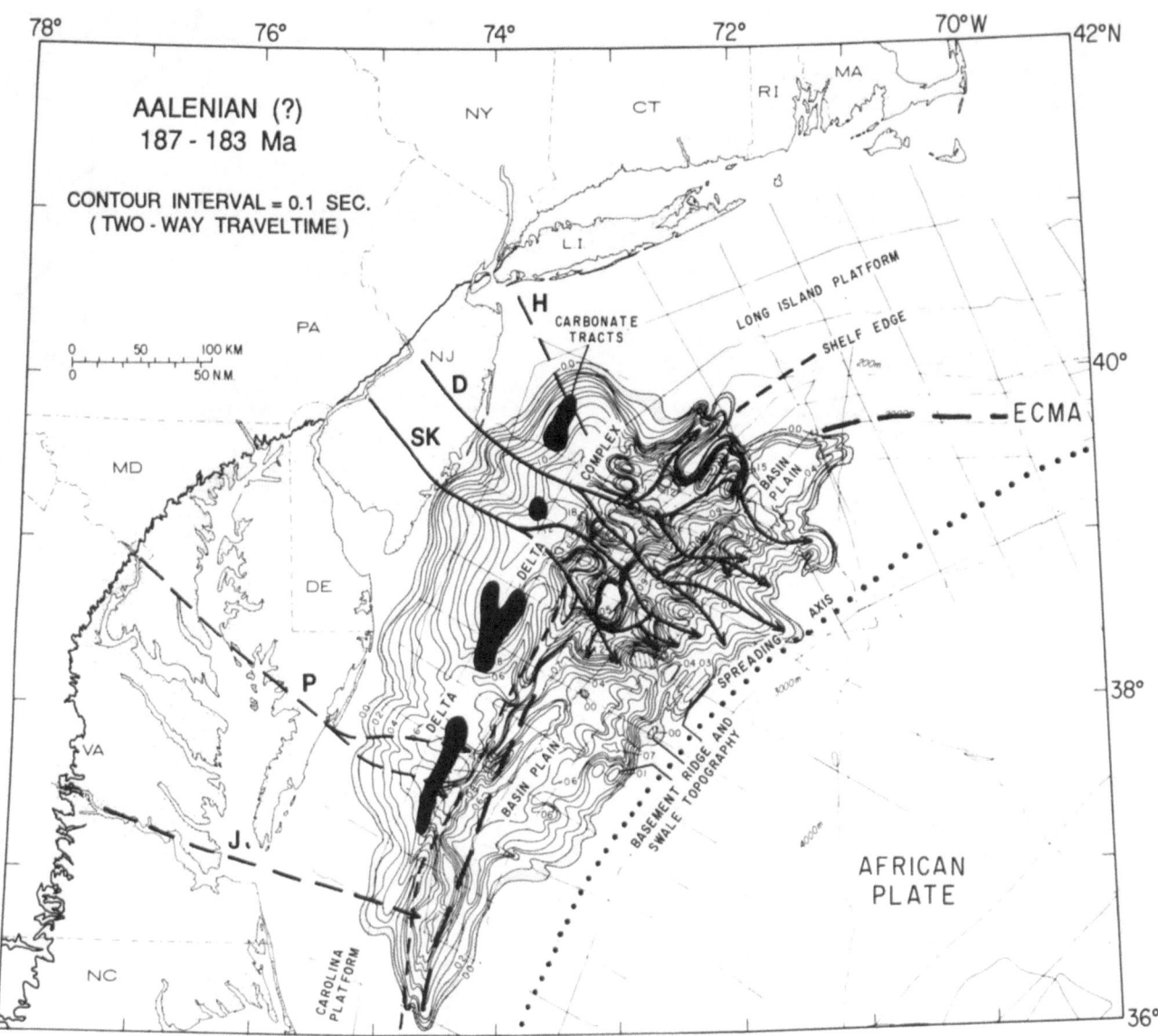

FIGURE 6-8.   Isochron map (contours in 0.1 sec of two-way traveltime) of Aalenian(?) allostratigraphic unit. Solid arrows indicate inferred primary routes of sediment dispersal; dashed arrows
indicate secondary routes. Shaded contour interval (0.0–0.1 sec) indicates thinnest parts of unit.
ECMA = East Coast Magnetic Anomaly; J = ancient James River; P = ancient Potomac River;
SK = ancient Schuylkill River; D = ancient Delaware River; H = ancient Hudson River.

indicate that this terrigenous sediment was dispersed
mainly through the ancient Schuylkill and Delaware
rivers. Sediment from the confluent delta formed thick
debris aprons (~ 2.5–4.5 km thick; 1.2–1.6 sec) along
the base of a prominent continental shelfbreak and
gave rise to five small submarine fans, which extended
100 km down slope to the southeast. Distal sediments
from these fans coalesced laterally along a 180-km
front to construct a confluent fan on the lower continental rise. Terminal sediments from the confluent fan
lapped onto the elevated western flank of the mid-
Atlantic spreading center and filled narrow swales in
the basement surface. Alternate sedimentary thickening and thinning across the basement ridges and swales

forms a characteristic ribbed pattern on the isochron
map (Fig. 6-8).

The continental rise was narrower (20–60 km wide)
in the southwestern half of the basin (Fig. 6-8), seaward of two small shelf deltas (~ 2.5 km thick; 1.0 sec),
formed by the ancient James and Potomac rivers. The
Long Island and Carolina platforms received little or
no sediment belonging to this allostratigraphic unit.

Though siliciclastic sediments presumably dominated this depositional unit, I interpret high-amplitude
reflections in the middle of the unit (along some profiles) to represent a linear trend of four low, narrow
(~ 20 km × 70–80 km) carbonate tracts (oolite
shoals?), which occupied the middle and outer parts of

the 100-km-wide continental shelf (see also Lawrence, Doyle, and Aigner, 1990; Poag, 1991). These carbonate buildups appear not to have reached the shelf edge, however, and probably contributed little or no sediment to the continental rise.

Bajocian–Lower Bathonian Allostratigraphic Unit

By the end of the next major depositional episode (Bajocian–early Bathonian; 183–172.5 Ma), deposition in the study area extended 50–130 km farther to the southeast and northeast; the northern Hatteras Basin

now covered 500 km × 175 km, as it continued to expand via seafloor spreading (Figs. 6-3, 6-9). The rate of siliciclastic accumulation declined, however, to 13,000 km³/m.y. (Fig. 6-5). I take this decline as evidence of reduced tectonic uplift in the central Appalachian highlands accompanied by reduced rates of thermotectonic subsidence of the trough (Watts and Steckler, 1979; Poag and Sevon, 1989). Major siliciclastic depocenters (now represented by sediments ~ 1.7–2.0 km thick; 0.7–0.8 sec) developed on the middle part of the continental shelf seaward of the ancient Schuylkill, Delaware, and Hudson rivers.

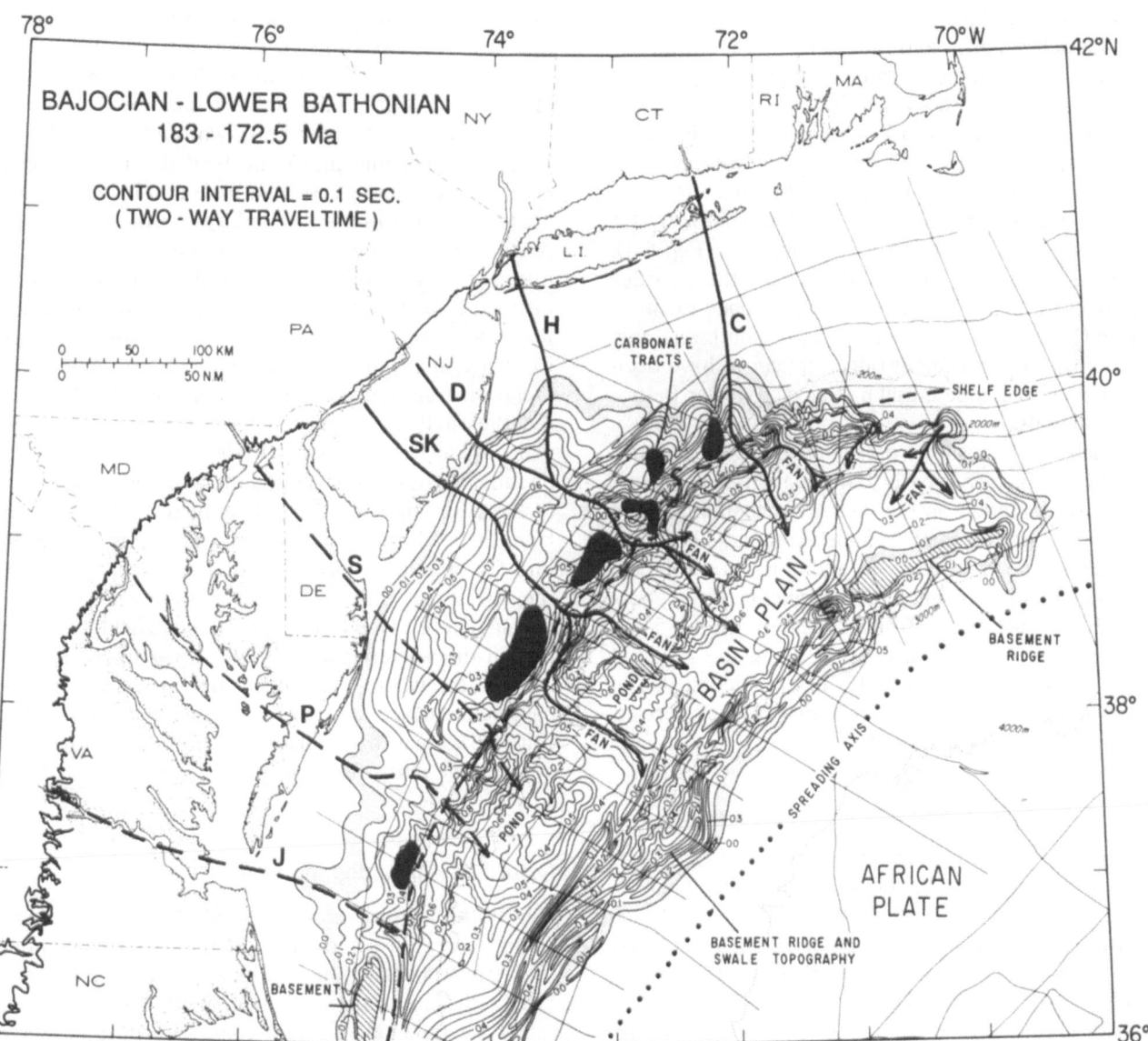

FIGURE 6-9.   Isochron map (contours in 0.1 sec of two-way traveltime) of Bajocian–lower Bathonian allostratigraphic unit. Solid arrows indicate inferred primary routes of sediment dispersal; dashed arrows indicate secondary routes. Diagonal hachures indicate allostratigraphic unit absent or too thin to resolve on seismic profiles. Shaded contour interval (0.0–0.1 sec) indicates thinnest parts of unit. J = ancient James River; P = ancient Potomac River; S = ancient Susquehanna River; SK = ancient Schuylkill River; D = ancient Delaware River; H = ancient Hudson River; C = ancient Connecticut River.

Thickest siliciclastic deposits (2.5–4.0 km; 1.0–1.2 sec), however, formed as debris aprons along the continental slope, though the Exxon model postulates that sea level was rising and that no major lowstands occurred.

The ancient James, Potomac, and Susquehanna rivers formed secondary dispersal systems that fed siliciclastic sediments to smaller debris aprons. To the north, the ancient Connecticut River supplied terrigenous detritus directly to a large debris apron, which onlapped the Long Island Platform.

Sediments on the upper rise generally are thinner (0.9–1.2 km; 0.2–0.4 sec) than the debris aprons, but the section thickens again to ~ 1.7 km (0.6 sec) in several sediment ponds along the middle part of the continental rise. Five small submarine fans extended from the debris apron and skirted the sediment ponds, and their distal sediments formed an elongate, 1.7-km-thick basin plain on the lower rise.

Along the southeastern margin of the Hatteras Basin, continental-rise and basin-plain sediments onlapped the western flank of the mid-Atlantic spreading center and, as in the Aalenian(?), filled narrow swales whose long axes parallel the spreading axis (Fig. 6-3). Alternate thickening and thinning across these swales and ridges form an even more extensive ribbed isochron pattern (Fig. 6-9). To the northeast and southwest, two particularly large basement ridges appear to have been high enough to escape burial by Bajocian–lower Bathonian sediments.

Siliciclastic deposition presumably still was dominant during Bajocian–early Bathonian time, but carbonate accumulation on the shelf was expanding. Poag (1991) noted several small, rounded carbonate buildups landward of the shelf edge. In addition, several larger shelf-edge buildups formed early segments of the enormous Bahama–Grand Banks Gigaplatform (> 5,000 km long) that had begun to grow along the Atlantic margin of North America at this time. Thickest debris aprons abut carbonate tracts; presumably, they derived detritus from the gigaplatform.

Upper Bathonian – Callovian Allostratigraphic Unit

During late Bathonian–Callovian time (172.5–163 Ma), the depositional regime of the Baltimore Canyon Trough and northern Hatteras Basin widened 200 km toward the southeast, and extended 150 km to the northeast toward the Georges Bank Basin (Fig. 6-10; the Hatteras Basin was 600 km long and 350 km wide). The siliciclastic accumulation rate remained constant at 13,000 km^3/m.y. (Fig. 6-5; Poag and Sevon, 1989), even though a major lowstand is postulated for the late Bathonian by the Exxon model. Carbonate deposition on the outer shelf gigaplatform accelerated, however, and the shelf edge prograded seaward during late Bathonian–Callovian time. Numerous mounded seismic reflections along the seaward margin of the gigaplatform indicate that reefs constructed an elevated rim there during the later part of this interval (Fig. 6-3), under a high and rising sea level (Haq, Hardenbol, and Vail, 1987). The maximum thickness of upper Bathonian–Callovian strata (~ 5 km; 1.8 sec) accumulated on the outer margin of the gigaplatform. This outer margin has not yet been drilled, but some of the deeper exploratory wells in the Baltimore Canyon Trough have documented biomicritic limestones along its shoreward margin (e.g., Tenneco 642-2 well; Bielak, ed., 1986).

Siliciclastic detritus from the northern part of the central Appalachian highlands (supplied mainly by the ancient Delaware, Schuylkill, and Susquehanna rivers) and from the Adirondack highlands (via the ancient Hudson River) collected in a thick confluent delta (1.7–2.5 km thick; 0.8–1.0 sec) landward of the gigaplatform along the northern two-thirds of the trough (Fig. 6-10). The ancient Connecticut River and rivers in eastern Massachusetts (which have no obvious modern counterparts) formed secondary northern dispersal routes, and the ancient Potomac dispersed sediments into secondary depocenters to the southwest. Four distinct submarine fans are associated with the ancient Potomac, Susquehanna, Delaware–Hudson, and Connecticut drainage systems. The Susquenhanna, Delaware, and Hudson systems appear to have transitted the shelf edge through large submarine canyons, which dissected the seaward edge of the gigaplatform. Sediments from the smaller Potomac and Connecticut systems, on the other hand, appear to have passed from the shelf into debris aprons, and from the aprons into fan systems, which extended as far as 200 km to the lower rise (Fig. 6-10). Distal sediments from the Susquehanna, Delaware, and Connecticut fans onlapped the western flanks of the mid-Atlantic spreading center, where they filled narrow swales in the basement surface and buried all but the highest ridges (Fig. 6-3). Small basin plains developed on either side of the ridge and swale domain.

Oxfordian Allostratigraphic Unit

During the Oxfordian (163–156 Ma), siliciclastic deposition decreased somewhat and the net accumulation rate was 9,000 km^3/m.y. (Fig. 6-5; Poag and Sevon, 1989); siliciclastic deposits encroached less rapidly on carbonate tracts at the shelf edge. This decline was in part a consequence of decreased thermotectonic subsidence of the basin, as sediment loading began to dominate subsidence on this passive margin (Watts and Steckler, 1979; Poag and Sevon, 1989). The Oxfordian segment of the Bahama–Grand Banks Gigaplatform was 20–25 km wide, ~ 1.5 km thick (0.8–0.9 sec), and

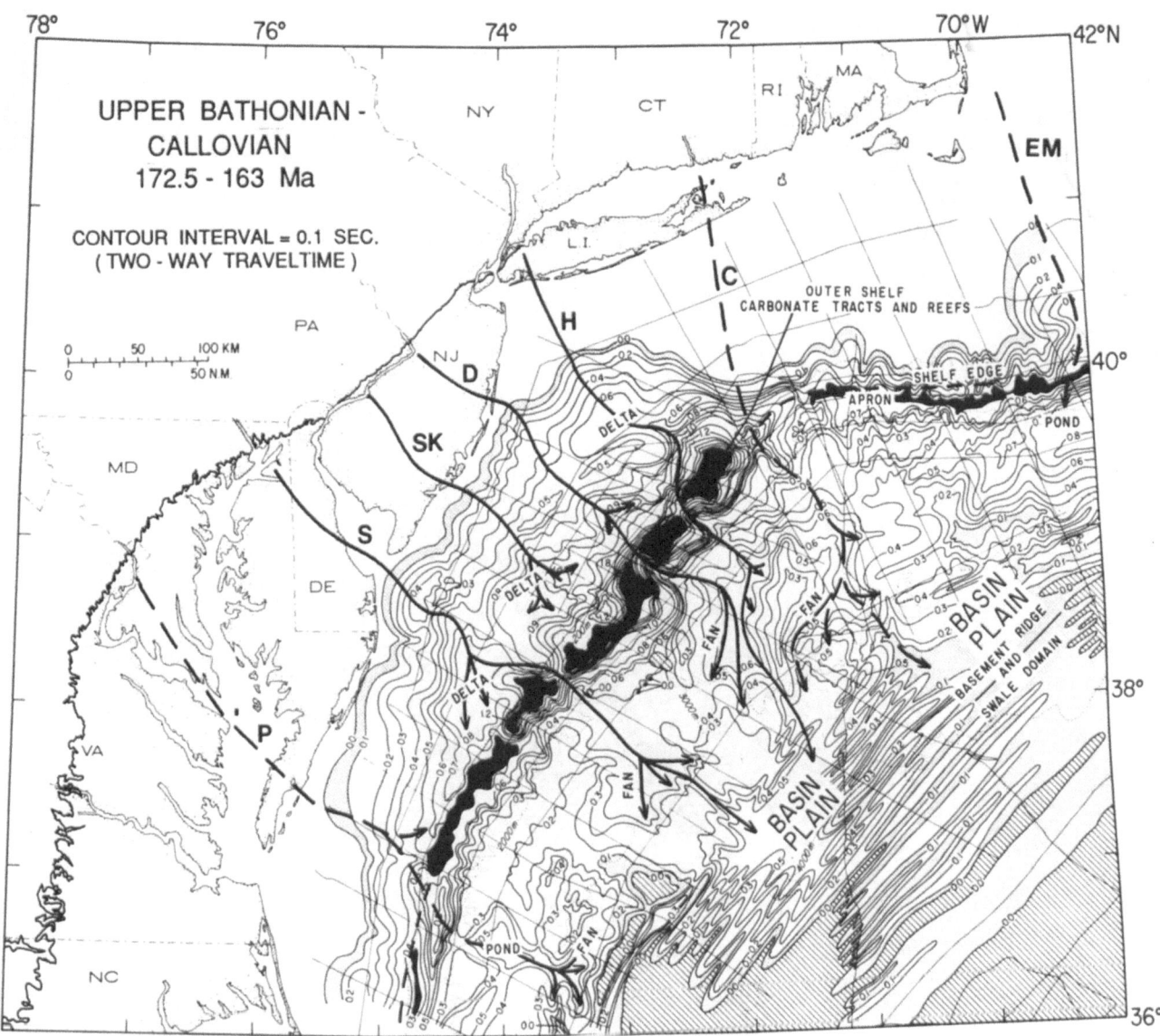

FIGURE 6-10.   Isochron map (contours in 0.1 sec of two-way traveltime) of upper Bathonian–Callovian allostratigraphic unit. Solid arrows indicate inferred primary routes of sediment dispersal; dashed arrows indicate secondary routes. Diagonal hachures indicate unit absent or too thin to resolve on seismic profiles. Shaded contour interval (0.0–0.1 sec) indicates thinnest parts of unit. P = ancient Potomac River; S = ancient Susquehanna River; SK = ancient Schuylkill River; D = ancient Delaware River; H = ancient Hudson River; C = ancient Connecticut River; EM = ancient river(s) in eastern Massachusetts.

was nearly continuous along the outer shelf (Figs. 6-3, 6-11). Platform expansion may have been enhanced by a long-term eustatic rise, which the Exxon model postulates began in the Callovian and peaked in the Tithonian. In the central part of the Baltimore Canyon Trough, the shelf edge had prograded as far as ~ 50 km eastward of the ECMA.

Landward of the gigaplatform, off what now is New Jersey, a prominent, confluent, mid-shelf delta was built by the ancient Hudson, Delaware, Schuylkill, and Susquehanna rivers (Fig. 6-11). Terrigenous detritus moved over the shelf edge through two submarine

canyons associated with the ancient Delaware and Hudson rivers, and this detritus eventually filled the canyons. Thick (~ 1.0 km; 0.7 sec) debris aprons were prominent on the slope and upper rise, and one medium and seven small submarine fans spread siliciclastic sediment into the northern Hatteras Basin. A ninth fan (375 km long × 100 km wide × 0.4–1.0 km thick), constructed seaward of the ancient Delaware and Hudson submarine canyons, is the oldest large fan yet documented in the Hatteras Basin. Debris aprons and upper rise parts of this fan undoubtedly include carbonate debris derived from the gigaplatform.

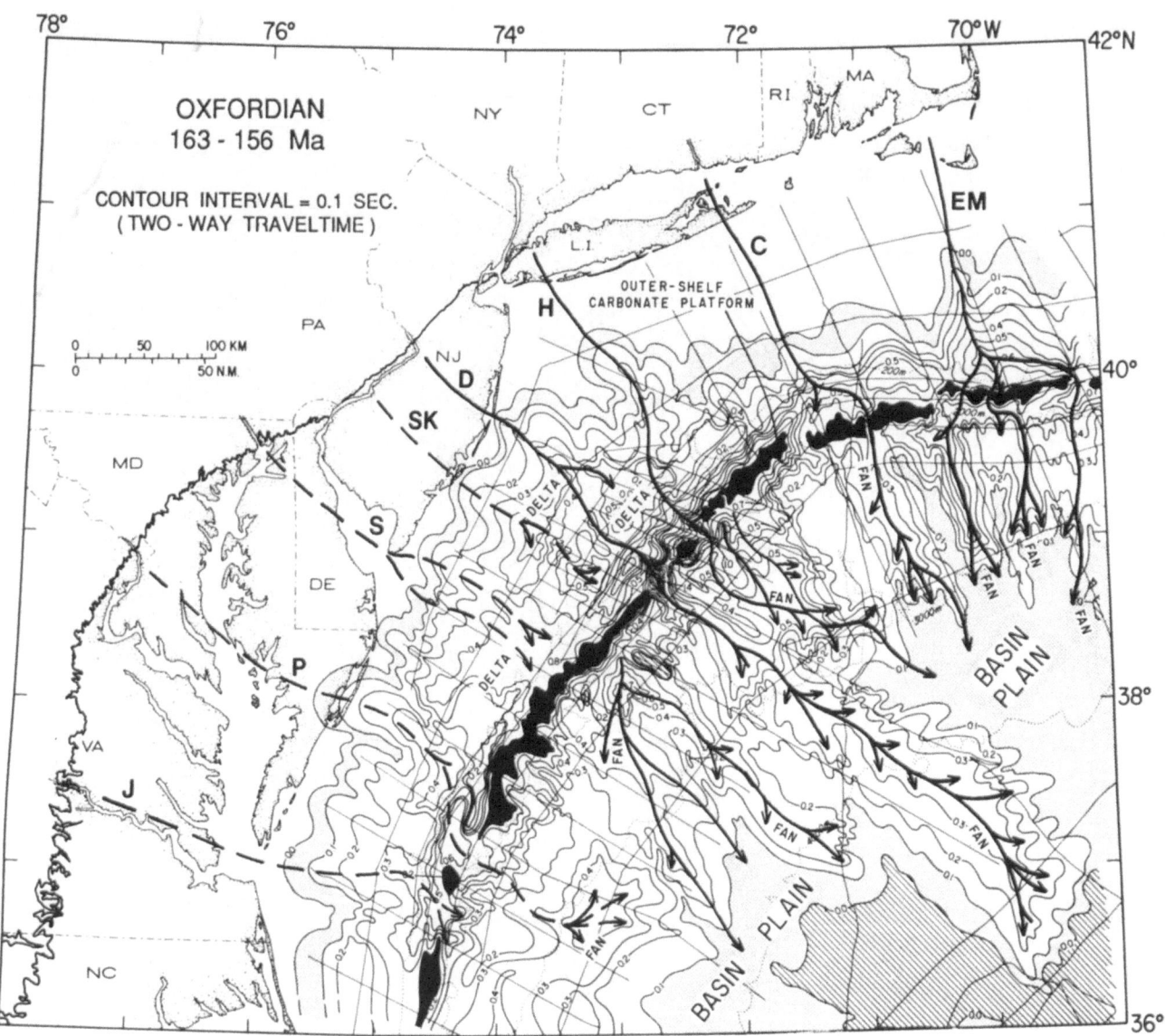

FIGURE 6-11.   Isochron map (contours in 0.1 sec of two-way traveltime) of Oxfordian allostratigraphic unit. Solid arrows indicate inferred primary routes of sediment dispersal; dashed arrows indicate secondary routes. Diagonal hachures indicate unit absent or too thin to resolve on seismic profiles. Shaded contour interval (0.0–0.1 sec) indicates thinnest parts of unit. J = ancient James River; P = ancient Potomac River; S = ancient Susquehanna River; SK = ancient Schuylkill River; D = ancient Delaware River; H = ancient Hudson River; C = ancient Connecticut River; EM = ancient river(s) in eastern Massachusetts.

Gravity-flow deposits of the Oxfordian allostratigraphic unit have been sampled at DSDP Site 105 (125 km south of the southeast corner of the study area; Fig. 6-2). There, Hollister, Ewing, et al. (1972) cored a mixture of variegated calcareous claystones, marls, and clayey limestones of the Cat Gap Formation, which contain clay minerals (chiefly montmorillonite), quartz, heavy minerals, and terrestrial plant debris.

Kimmeridgian – Tithonian Allostratigraphic Unit

The acme of gigaplatform construction and reef growth in the Baltimore Canyon Trough, as well as on the rest of the North American Atlantic margin, spanned the peak of sea level postulated by the Exxon model for Kimmeridgian–Tithonian time (156–144 Ma; Erlich et al., 1988; Meyer, 1989; Lawrence, Doyle, and Aigner, 1990; Poag, 1991). The carbonate rim had by this time prograded to its maximum eastward position (55 km onto oceanic basement off New Jersey; Fig. 6-12), and sediments of the adjacent Hatteras Basin nearly covered the entire mapped area seaward of the shelf edge (~ 600 km × 400 km). At the end of the Tithonian, aggradation had produced a 10- to 30-km-wide carbonate platform, whose elevated seaward rim was topped by a discontinuous, pinnacled, barrier-reef system

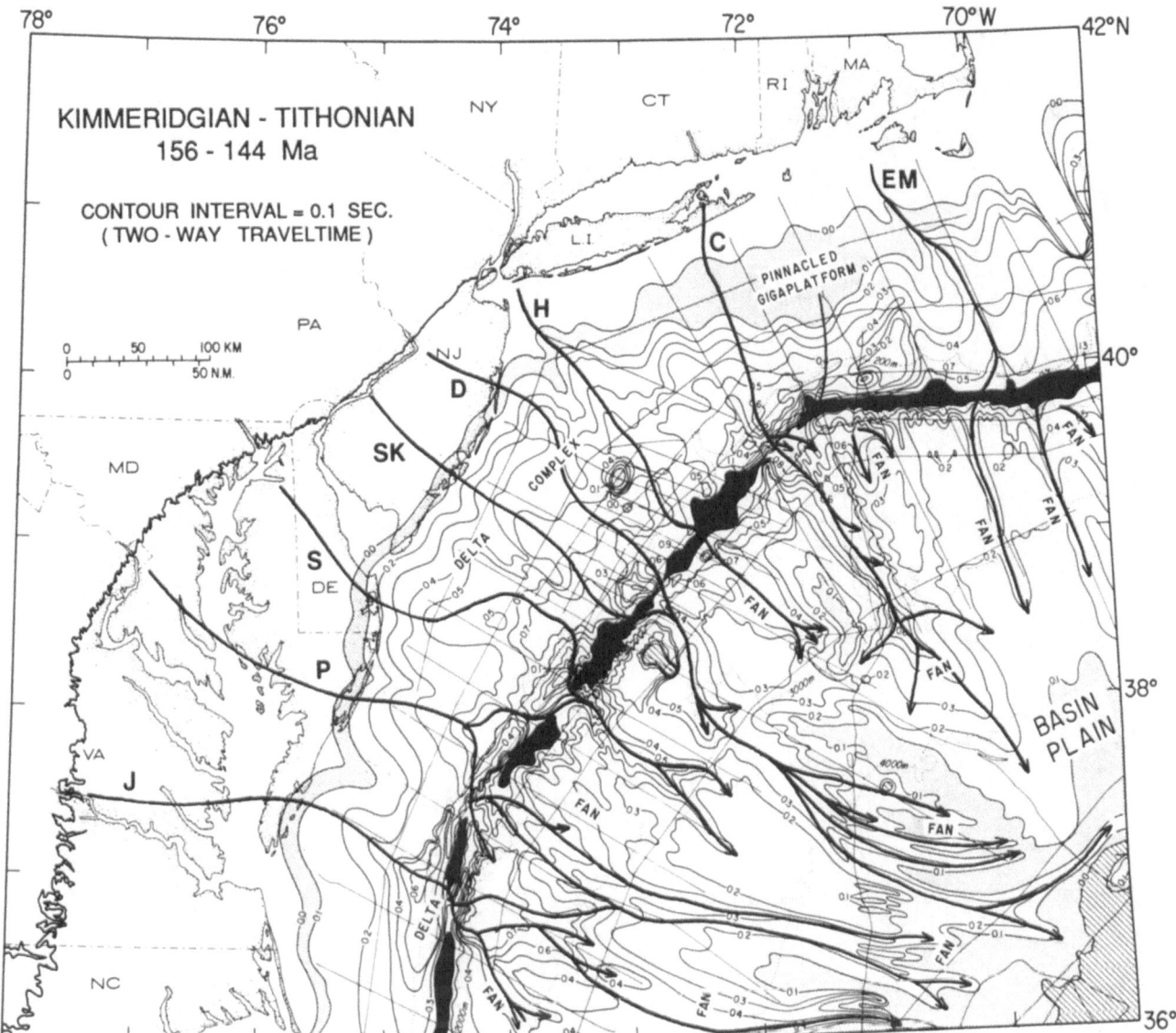

FIGURE 6-12.   Isochron map (contours in 0.1 sec of two-way traveltime) of Kimmeridgian–Tithonian allostratigraphic unit. Arrows indicate inferred primary routes of sediment dispersal. Diagonal hachures indicate unit absent or too thin to resolve on seismic profiles. Shaded contour interval (0.0–0.1 sec) indicates thinnest parts of unit. J = ancient James River; P = ancient Potomac River; S = ancient Susquehanna River; SK = ancient Schuylkill River; D = ancient Delaware River; H = ancient Hudson River; C = ancient Connecticut River; EM = ancient river(s) in eastern Massachusetts.

(Meyer, 1989; Prather, 1991), shown on Figure 6-12 as a narrow linear trend of closely spaced isochrons (Poag, 1991). The reef crest rose as high as 3 km above the floor of the Hatteras Basin and 150 m above the adjacent backreef shelf (Figs. 6-3, 6-13, 6-14).

Landward of the gigaplatform were a series of siliciclastic deltas. One large confluent delta extended 400 km from Long Island to Maryland and was more than 100 km wide. Separate deltas in this complex were fed by the ancient Hudson, Delaware, Schuylkill, Susquehanna, and Potomac rivers, carrying detritus from the Adirondack and central Appalachian highlands (Poag and Sevon, 1989). The siliciclastic accumulation rate

had dropped from 9,000 km³/m.y. to 6,000 km³/m.y. by this time, however (Fig. 6-5; Poag and Sevon, 1989), which slowed encroachment toward the shelf-edge carbonate tracts and allowed the gigaplatform to aggrade as sea level peaked.

Along the Kimmeridgian–Tithonian continental slope, debris aprons collected detritus from the reef tract, and several small (presumably carbonate-enriched) submarine fans formed on the upper continental rise (Poag, 1991). Four larger siliciclastic submarine fans derived their constituents mainly from shelf deltas associated with the ancient Connecticut, Delaware–Schuylkill–Susquehanna, Potomac, and

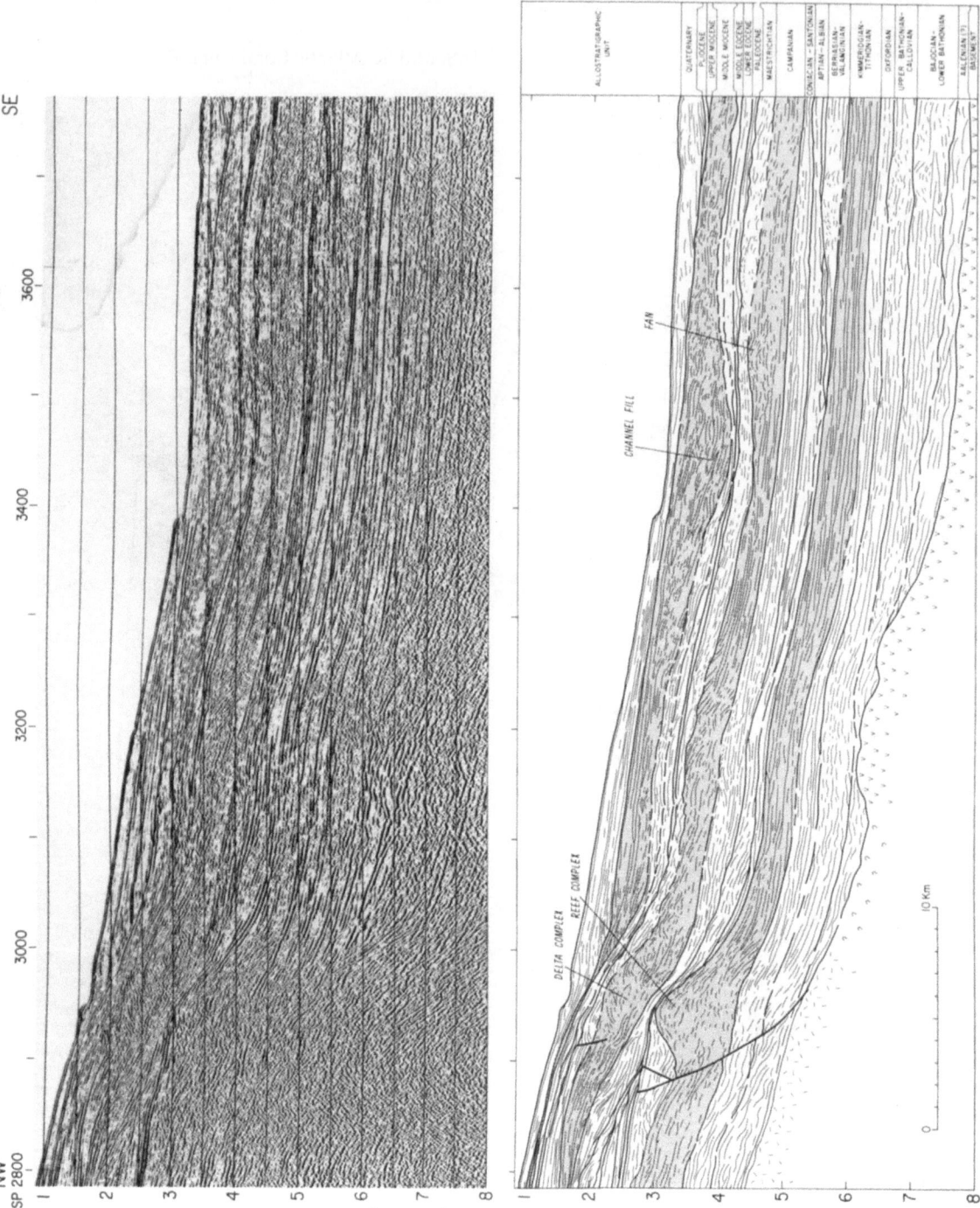

FIGURE 6-13. Stratigraphic section parallel to depositional dip along seismic profile 23 crossing the shelf edge, continental slope, and upper rise seaward from northern end of Long Island (see Fig. 6-2 for location). Above is uninterpreted seismic profile; below is interpreted tracing of seismic profile. Three allostratigraphic units shaded to facilitate comparison with same units on strike sections 206 (Fig. 6-23), 205 (Fig. 6-20), and 201 (Fig. 6-25), and to emphasize contrasts in geometry, seismic facies, and depositional style. Kimmeridgian–Tithonian unit comprises mounded chaotic facies within shelf-edge reef complex and in forereef talus; parallel, discontinuous reflections in slope and rise facies are interpreted to represent sediment gravity flows (mainly turbidites). Campanian unit shows prograded clinoform reflections in shelf-edge delta complex changing to chaotic, mounded reflections on slope and upper rise (slumps, debris-flow deposits) and finally becoming parallel, nearly continuous reflections in turbidite facies basinward of intersection with profile 205. Near shot-point 3600, section cuts diagonally across a mound interpreted to represent submarine fan. Note truncated reflections along irregular upper surface of Campanian allostratigraphic unit (interpreted to result from marine erosion). Middle Miocene allostratigraphic unit is very thin on continental slope (interpreted to have been eroded), but thickens rapidly at base of slope. Thins again near intersection with profile 205, but thickens to maximum in channel near shot-point 3450. Chaotic seismic facies in channel-fill is interpreted to represent slumps and debris-flow deposits.

116

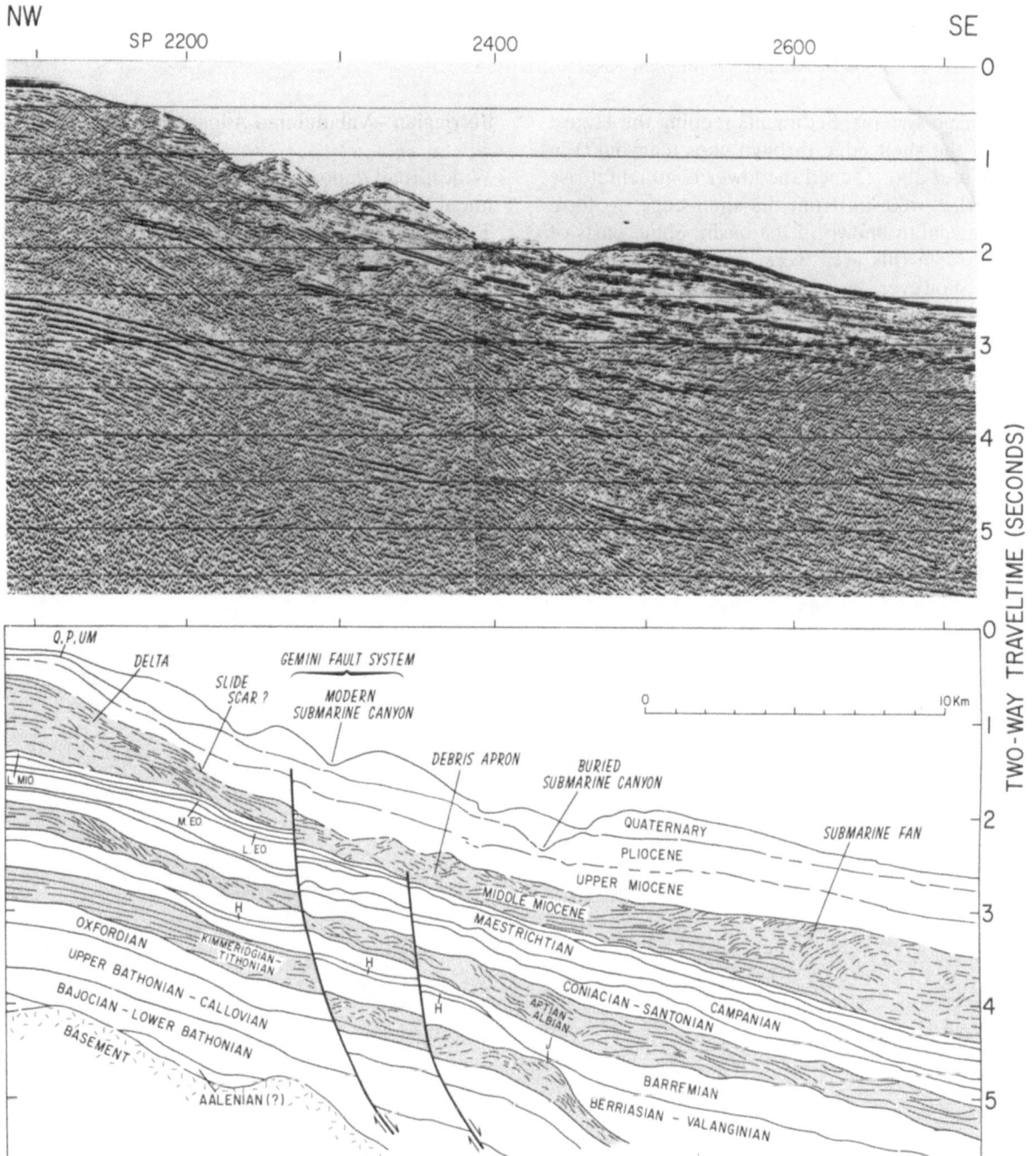

FIGURE 6-14. Stratigraphic section parallel to depositional dip along seismic profile 29 crossing the shelf edge, continental slope, and upper rise seaward of the North Carolina–Virginia border (see Fig. 6-2 for location). Above is uninterpreted seismic profile; below is interpreted tracing of seismic profile. Three allostratigraphic units are shaded and reflections are traced to emphasize contrasts and similarities in geometry, seismic facies, and depositional style. Kimmeridgian–Tithonian unit between shot-points 2100 and 2350 shows high-amplitude, parallel, continuous reflections from Bahama–Grand Banks Gigaplatform. Reflections become mounded and chaotic near Late Jurassic shelf edge (J). Aptian–Albian unit displays prograding clinoform reflections of delta complex near shelf edge at shot-point 2150. Reflections become more chaotic (slumps, debris-flow deposits) in slope-rise apron between shot-points 2300 and 2500, then become subparallel and wavy (turbidites) on upper rise. Middle Miocene unit displays several sets of prograding clinoform reflections in thick shelf-edge delta near shot-point 2100. Reflections become chaotic (slumps, debris-flow depisits) in debris apron along slope and upper rise. Between shot-point 2500 and 2600, middle Miocene unit thickens rapidly into mounded submarine fan whose base displays continuous reflections typical of turbidites; turbidites are overlain by chaotic zone of slumps and debris-flow deposits. Note truncated reflections along irregular (eroded) upper surface of unit. Quaternary allostratigraphic unit is incised by several modern submarine canyons and, near shot-point 2425, fills canyon excavated into Pliocene unit. Note migration of shelf edge from "J" at end of Jurassic to "Q" at end of latest Quaternary deposition. H = Hauterivian; L. Mio. = Lower Miocene; M. Eo. = Middle Eocene; L. Eo. = Lower Eocene; UM = Upper Miocene; P = Pliocene.

117

James drainage systems. Sediments feeding the largest fan crossed the shelf edge through gaps (canyons?) in the barrier reef and reached the lower continental rise (basin plain) > 400 km from the shelf edge, to form the oldest megafan known in the basin. Some parts of the lower rise in this area were accumulating pelagic carbonates, however, as shown by the presence of fine-grained pelagic limestones of the Cat Gap Formation in the Kimmeridgian–Tithonian section cored at DSDP Site 105 (Hollister, Ewing, et al., 1972). But even at Site 105, parts of the Kimmeridgian–Tithonian section contain terrigenous clays.

Berriasian–Valanginian Allostratigraphic Unit

Widespread deposition of shallow-water carbonate sediment along the outer margin of the Baltimore Canyon Trough and adjacent basins essentially ended when the Bahama–Grand Banks Gigaplatform drowned at the end of the Tithonian (Meyer, 1989; Poag, 1991). In isolated patches, shallow-water carbonate production continued into the Berriasian (Ryan et al., 1978; Erlich et al., 1988; Meyer, 1989; Prather, 1991; Poag, 1991), but most of the earliest Cretaceous shelf carbonate appears to have been of relatively deep-water pelagic

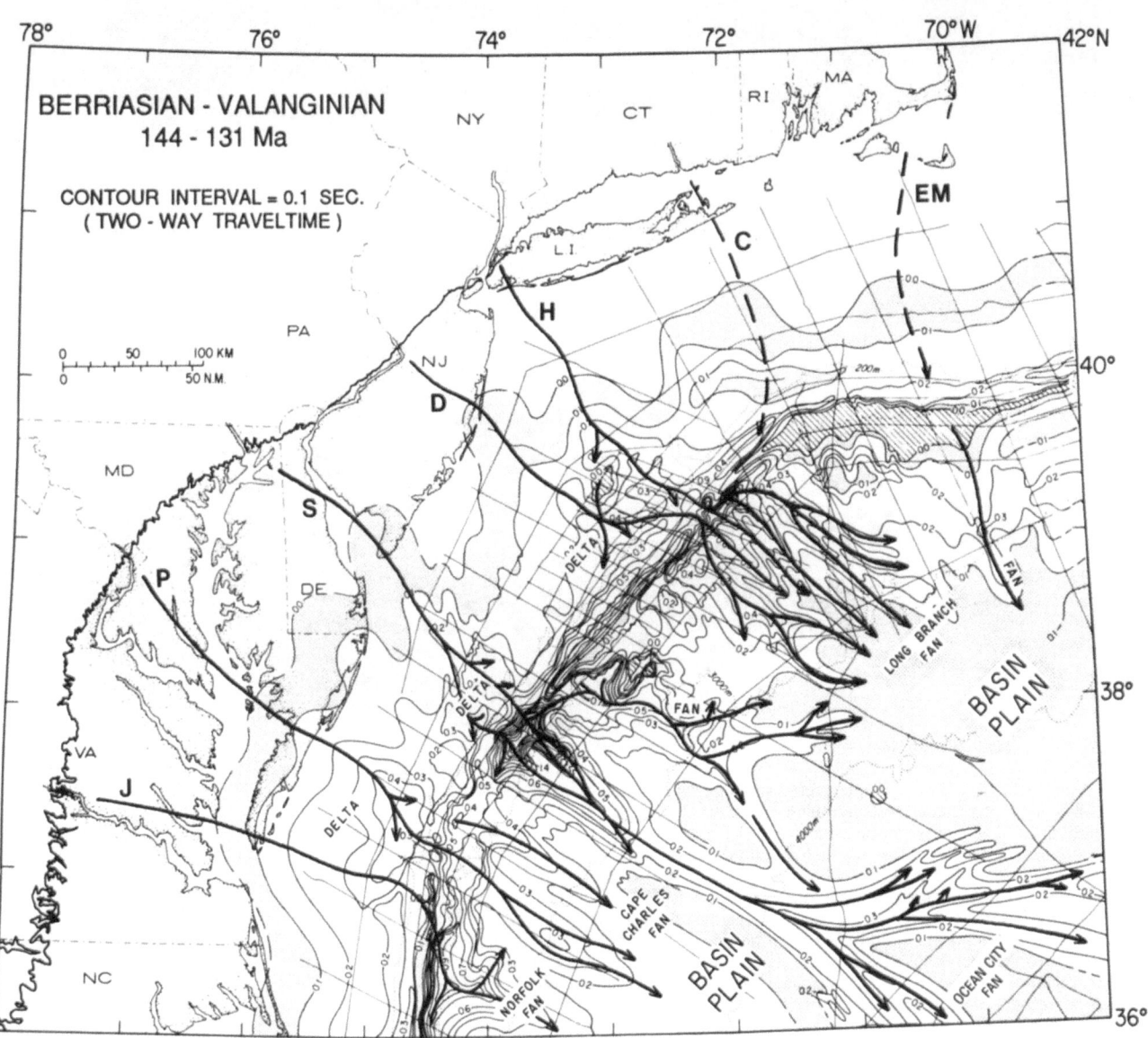

FIGURE 6-15.   Isochron map (contours in 0.1 sec of two-way traveltime) of Berriasian–Valanginian allostratigraphic unit. Solid arrows indicate inferred primary routes of sediment dispersal; dashed arrows indicate secondary routes. Diagonal hachures indicate unit absent or too thin to resolve on seismic profiles. Shaded contour interval (0.0–0.1 sec) indicates thinnest parts of unit. J = ancient James River; P = ancient Potomac River; S = ancient Susquehanna River; D = ancient Delaware River; H = ancient Hudson River; C = ancient Connecticut River; EM = ancient river(s) in eastern Massachusetts.

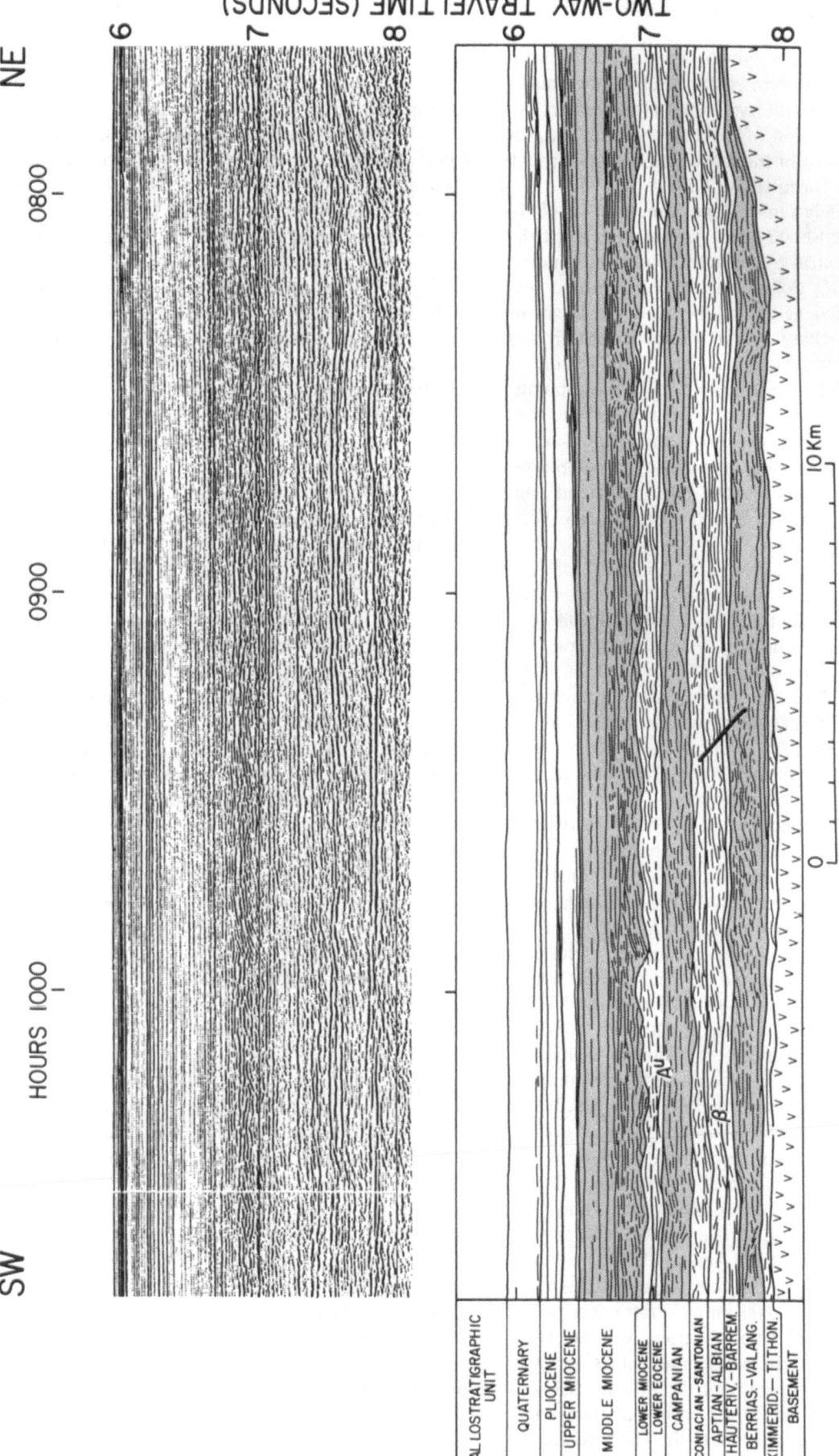

FIGURE 6-16. Stratigraphic section parallel to depositional strike along seismic profile Conrad 21-MCS 75, crossing the lower part of continental rise in southeast corner of study area, ~ 10 km northeast of intersection with DSDP Site 106 (see Fig. 6-2 for location). Berriasian–Valanginian, Aptian–Albian, Coniacian–Santonian, Campanian, lower Miocene, and lower part of middle Miocene allostratigraphic units display broadly mounded, discontinuous, somewhat chaotic seismic facies interpreted to represent mixtures of debris-flow deposits and turbidites of submarine fan complexes (Berriasian–Valanginian unit represents principal distal lobe of Ocean City fan at this locality; see Fig. 6-15). Middle Miocene unit displays mounded fan facies only in its lower half; upper half of middle Miocene (as well as all younger units) displays continuous, parallel reflections representative of basin-plain turbidite deposits coeval with contourites deposited near DSDP Site 603. A^u and β are well-known deep-sea seismic boundaries of the western North Atlantic (see Mountain and Tucholke, 1985; Tucholke and Mountain, 1986). Hauteriv.-Barrem. = Hauterivian–Barremian; Berrias.-Valang. = Berriasian–Valanginian; Kimmerid.-Tithon. = Kimmeridgian–Tithonian.

origin (Meyer, 1989). Though the Exxon model postulates a significant long-term eustatic fall and two major lowstands during this interval, they alone would not account for the cessation of carbonate production, for it continued unabated in the adjacent Carolina Trough and Blake Plateau Basin (Poag, 1991).

During the Berriasian–Valanginian interval (144–131 Ma), large shelf-edge delta systems built out from sources in the central Appalachian and Adirondack highlands (Fig. 6-15; Poag et al., 1990). Elevated reef structures prevented access to the Hatteras Basin by these shelf-edge deltas, except where deposition was rapid enough to bury them, or where gaps existed in the reef system (Figs. 6-3, 6-13, 6-14). At these points, large volumes of siliciclastic sediment reached the continental slope and rise (Fig. 6-16) in the form of extensive, thick debris aprons and submarine fan complexes. The siliciclastic accumulation rate remained unchanged, however, from the Late Jurassic rate (Fig. 6-5; Poag and Sevon, 1989).

The largest delta–apron–fan complex of the Berriasian–Valanginian allostratigraphic unit is the Ocean City complex (Poag et al., 1990). The bilobate Ocean City delta occupied ~ 3,000 km² on the Berriasian–Valanginian outer shelf east of Ocean City, Maryland. In this delta, maximum sediment thickness is > 900 m (> 0.3 sec), but it increases markedly to > 2,000 m (> 1.4 sec) in the heart of the contiguous debris apron. The Ocean City submarine fan, a megafan, is > 1,000 m thick as far as 100 km seaward of the shelf edge; its long axis trends N140° E, directly toward DSDP Site 603 (Fig. 6-2), where terrigenous turbidites of the Blake–Bahama Formation constitute distal elements of the complex (Holmes, Breza, and Wise, 1987; Sarti and von Rad, 1987). Alignment of the ancient Susquehanna River, the Ocean City delta, the Ocean City submarine fan, and the Site 603 turbidites supports the contention of Holmes, Breza, and Wise (1987) that heavy-mineral suites in the Site 603 turbidites came from the central Appalachian highlands. The debris apron of the Ocean City delta–fan complex was penetrated near its thickest part by the Shell 93-1 well (Prather, 1991), which revealed ~ 2 km of gray to brown, shallow-marine to nonmarine, calcareous to noncalcareous siltstones interbedded with thin calcareous shales, poorly consolidated sandstones, and thin chalky limestones. Organic kerogens in these strata are nearly 100% terrestrial types (Amato, 1987).

The > 900-m-thick (> 0.4 sec) Cape Charles confluent delta covered ~ 8,000 km² on the outer continental shelf seaward of Cape Charles, Virginia (Fig. 6-15; Poag et al., 1990). Its contiguous debris apron is > 1,000 m thick (> 0.9 sec), and the 600–m (0.3-sec) isochron of the associated Cape Charles submarine fan extends at least 130 km farther into the Hatteras Basin. Isochron patterns indicate that most of the terrigenous sediment for this delta–apron–fan complex probably

was delivered by the ancient Potomac and James rivers. No boreholes penetrate any element of this delta–fan complex.

The Long Branch confluent delta occupied ~ 5,000 km² on the outer shelf seaward of Long Branch, New Jersey (Fig. 6-15; Poag et al., 1990), where maximum sediment thickness is > 900 m (> 0.4 sec). The elongate debris apron associated with this delta reaches a maximum thickness of > 1,500 m (> 0.9 sec). The trilobate Long Branch submarine fan complex is > 1,000 m thick (> 0.6 sec) as far as 50 km from the shelf edge. Isochrons extend landward from the Long Branch delta toward the mouth of the ancient Hudson and Delaware rivers, evidence that these rivers were the main conduits for terrigenous detritus in this region. Lovegreen (1974) and Libby-French (1984) reached similar conclusions.

The Long Branch delta has been penetrated by 15 deep wells on the continental shelf, including the Continental Offshore Stratigraphic Test (COST) B-2 well (Scholle, 1977). Lower Cretaceous sediments in this well are chiefly shallow-marine to nonmarine sandstone, siltstone, and shale, commonly interbedded with lignite and coal. Organic carbon is abundant, comprising dominantly terrestrial types of kerogen. Smith (1980) found that Lower Cretaceous heavy-mineral suites of the COST B-2 well include constituents characteristic of the New England Appalachian highlands and the Long Island Platform, which lends further support to the hypothesis that these strata came mainly from northern source terrains.

A fourth prominent delta–apron–fan complex is associated with the ancient James River system. Its delta lies ~ 100 km east of Norfolk, Virginia (Fig. 6-15), from which I herein derive its name. The Norfolk delta is coalescent with the Cape Charles delta, but the small Norfolk submarine fan is a distinctly separate, much thicker feature (~ 900 m thick; 0.6 sec) than the Cape Charles fan.

Hauterivian Allostratigraphic Unit

In the Hauterivian (131–124 Ma), the net siliciclastic accumulation rate decreased sixfold to 1,000 km³/m.y. (Fig. 6-5; Poag and Sevon, 1989). The Hauterivian deposits formed a thin blanket, which can be discerned mainly on the continental shelf and slope (Figs. 6-3, 6-13, 6-14, 6-17). It is absent from most of the continental rise, or it is too thin to be resolved on multichannel seimic profiles. Small shelf depocenters (~ 100 m thick; 0.1 sec) developed in front of the ancient Hudson, Susquehanna, and Potomac drainage systems, but, elsewhere, the Hauterivian sequence is generally thinner than 100 m.

A ~ 140-m-thick section of Hauterivian sediments was cored at DSDP Site 603, however, where it consists of four turbiditic units of the Blake–Bahama Forma-

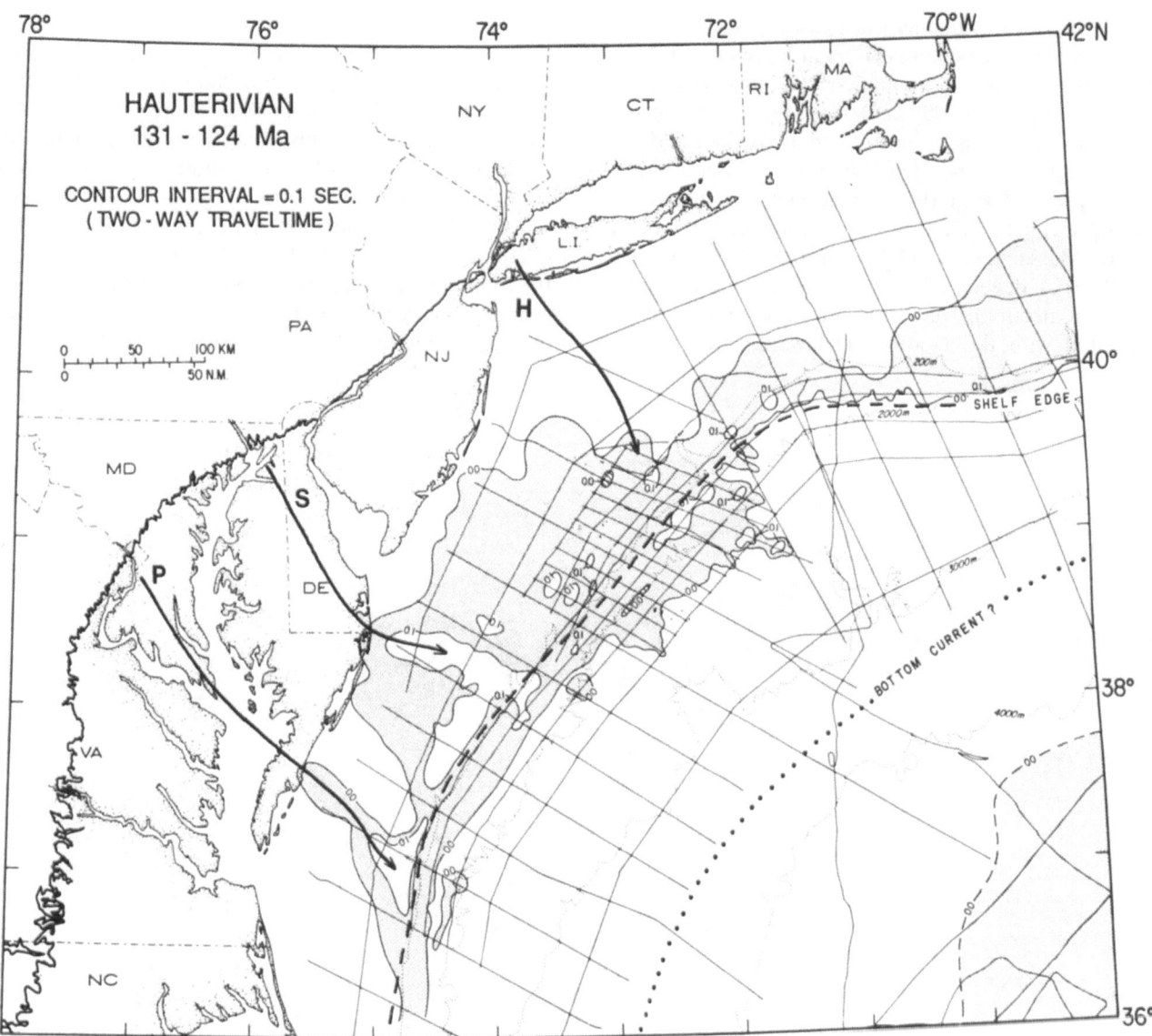

FIGURE 6-17.   Isochron map (contours in 0.1 sec of two-way traveltime) of Hauterivian allostratigraphic unit. Arrows indicate inferred primary routes of sediment dispersal. Shaded contour interval (0.0–0.1 sec) indicates thinnest parts of unit. Dotted line represents possible axis of eroding bottom current. P = ancient Potomac River; S = ancient Susquehanna River; H = ancient Hudson River.

tion. Each turbidite unit contains pelagic limestone and marl layers alternating with various combinations and amounts of mudstone, sandstone, slurried beds, and sandy debris-flow deposits. Starti and von Rad (1987) interpreted these lithic successions as part of an elongate channel-levee submarine fan, which was emplaced during high sea level. At DSDP Site 105, Hauterivian sediments comprise ~ 35 m of gray or white limestone interbedded with laminated, dark gray, clayey limestone, which contains terrestrial organic matter (Hollister, Ewing, et al., 1972). The lower rise Hauterivian sediments thin rapidly a few hundred kilometers to the north and northwest of these sites and cannot be traced farther on the seismic records.

I interpret the Hauterivian depositional pattern as evidence that this unit accumulated mainly during a highstand, which, when combined with the low siliciclastic accumulation rate, prevented a significant contribution to continental rise deposition, except on a local scale. In addition, the broad swath of sparse or no Hauterivian sediment across the middle continental rise might be evidence that abyssal currents were actively scouring the seafloor.

Barremian Allostratigraphic Unit

During the Barremian (124–119 Ma), a broad confluent delta (250 km wide × > 3 km thick) built out on the middle and outer continental shelf seaward of what now is North Carolina to Delaware (Fig. 6-18). A delayed response to renewed uplift in the southern part of the central Appalachian highlands channeled siliciclastic detritus mainly through the ancient James, Potomac, and Susquehanna rivers at a greatly accelerated rate; the net accumulation rate increased from 1,000 km^3/m.y. to 11,000 km^3/m.y. (Fig. 6-5; Poag and Sevon, 1989) during a postulated peak in sea level (Haq, Hardenbol, and Vail, 1987). Lesser depocenters were associated with the ancient Delaware, Hudson, and Connecticut rivers, and undesignated rivers in eastern Massachusetts. Debris aprons, 3–5 km thick (0.3–0.5 sec), developed seaward of the confluent delta and to the north off New England. From debris aprons associated with the Hudson, Delaware, and Susque-

hanna deltas, two large submarine fans extended across the upper part of the continental rise and coalesced farther downslope to form a broad confluent megafan, which covered the lower rise. There, the megafan formed several small distributaries, one or more of which supplied ~120 m of muddy and sandy turbidites to DSDP Site 603 (Fig. 6-2; part of the Blake–Bahama Formation; Sarti and von Rad, 1987). Barremian strata at DSDP Site 105 (Fig. 6-2) comprise 10–15 m of black, highly carbonaceous (terrestrial plant debris), zeolitic clay interbedded with soft, laminated, zeolitic limestone (Hollister, Ewing, et al., 1972).

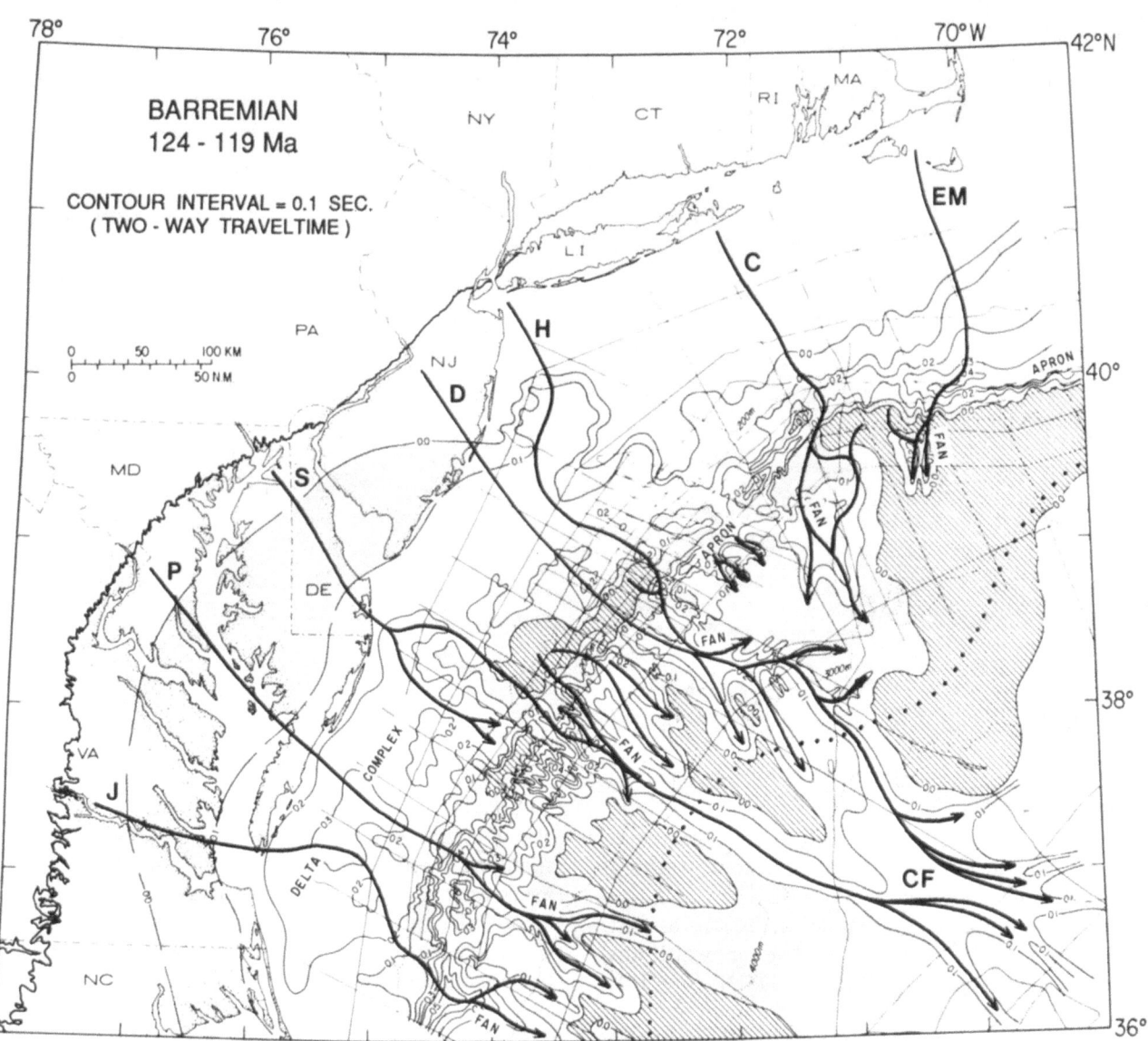

FIGURE 6-18.   Isochron map (contours in 0.1 sec of two-way traveltime) of Barremian allostratigraphic unit. Arrows indicate inferred primary routes of sediment dispersal. Shaded contour interval (0.0–0.1 sec) indicates thinnest parts of unit. Diagonal hachures indicate unit absent or too thin to resolve on seismic profiles. Heavy dotted line indicates possible axis of eroding bottom current. J = ancient James River; P = ancient Potomac River; S = ancient Susquehanna River; D = ancient Delaware River; H = ancient Hudson River; C = ancient Connecticut River; EM = ancient river(s) in eastern Massachusetts; CF = confluent fan.

Holmes, Breza, and Wise (1987) noted a distinct change in heavy-mineral suites and an acme in turbidite deposition in the middle of the Barremian succession at Site 603. These events presumably reflect renewed uplift of the central Appalachian highlands and, perhaps, the increased importance of Adirondack detritus supplied by the ancient Hudson drainage system.

At several places on the upper and middle continental rise, however, large patches of the seafloor appear to be devoid of Barremian sediments (or they are too thin to resolve on the seismic profiles). This condition may indicate that pelagic deposition was minimal in this region or that strong bottom currents subsequently removed or prevented accumulation of pelagic sediments.

Aptian–Albian Allostratigraphic Unit

By the end of Aptian–Albian deposition (119–97.5 Ma), the confluent delta seaward of North Carolina to southern New Jersey stretched for > 300 km along the continental shelf, was 150–100 km wide, and was ~ 500 m (0.5 sec) thick (Fig. 6-19; see Chapters 2, 5). The

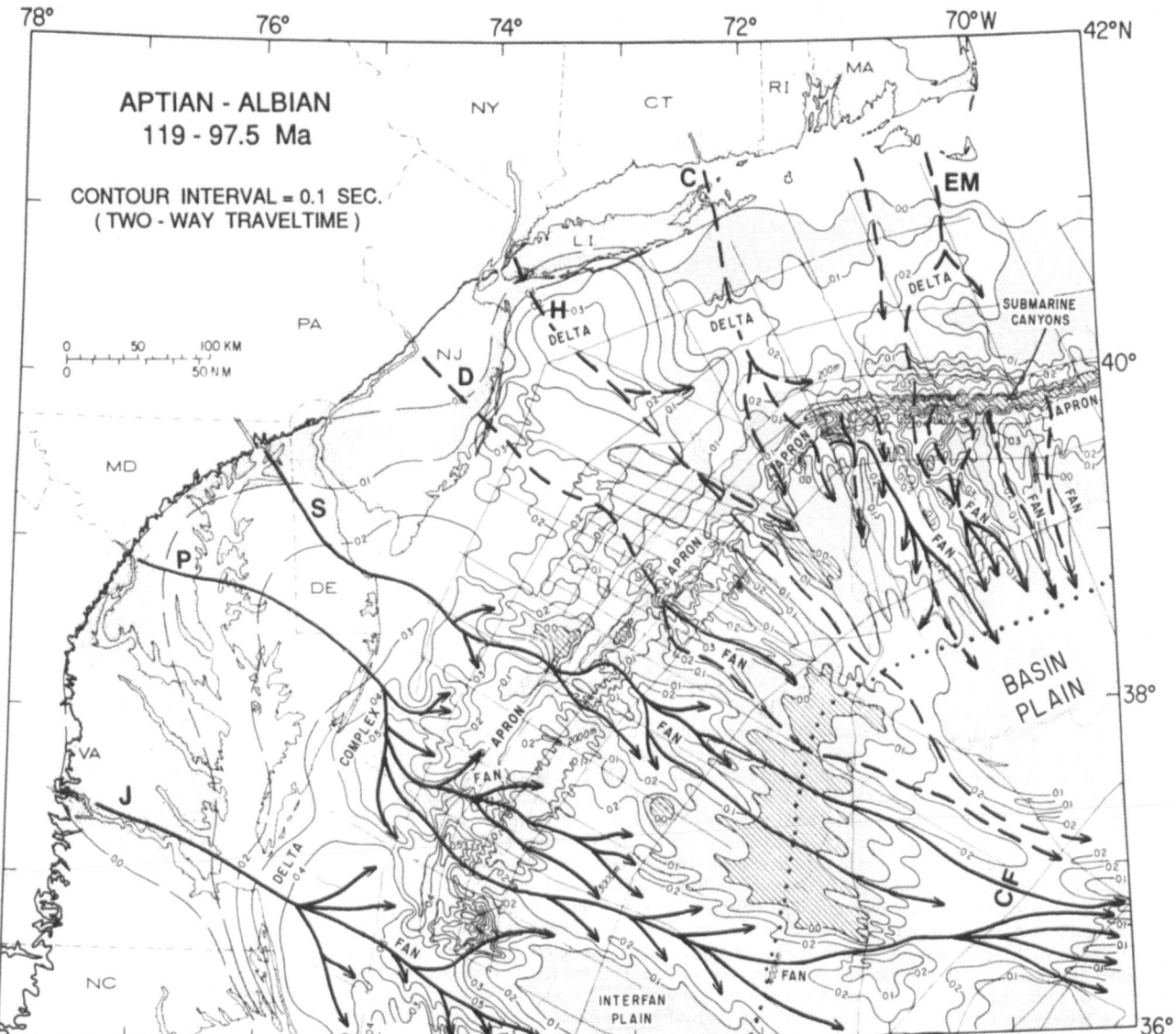

FIGURE 6-19.   Isochron map (contours in 0.1 sec of two-way traveltime) of Aptian–Albian allostratigraphic unit. Solid arrows indicate inferred primary routes of sediment dispersal; dashed arrows indicate secondary routes. Diagonal hachures indicate unit absent or too thin to resolve on seismic profiles. Shaded contour interval (0.0–0.1 sec) indicates thinnest parts of unit. Heavy dotted line indicates possible axis of eroding bottom current. J = ancient James River; P = ancient Potomac River; S = ancient Susquehanna River; D = ancient Delaware River; H = ancient Hudson River; C = ancient Connecticut River; EM = ancient river(s) in eastern Massachusetts; CF = confluent fan.

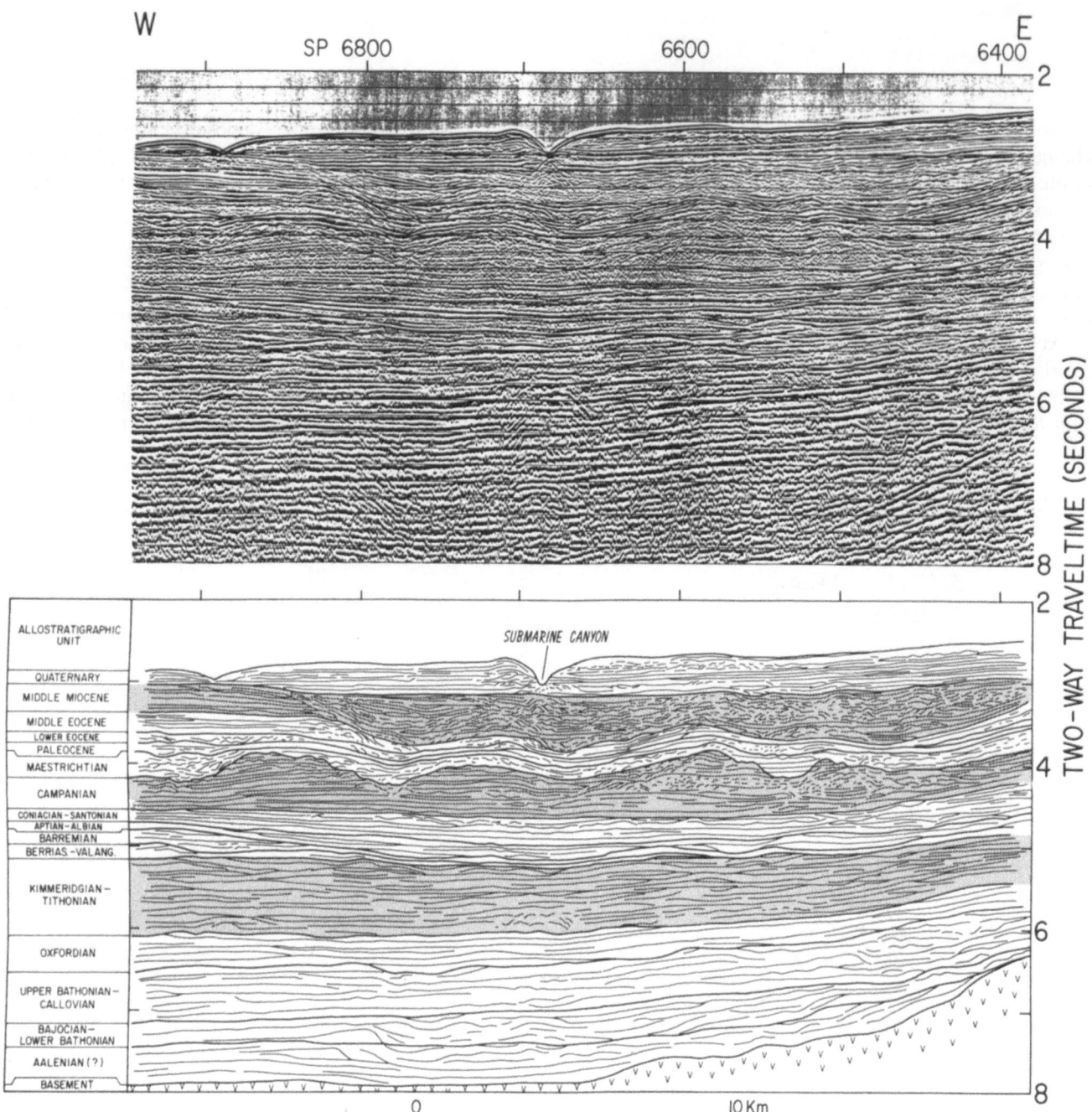

FIGURE 6-20. Stratigraphic section parallel to depositional strike along seismic profile 205 crossing the upper part of New England continental rise between seismic profiles 9 and 23 (see Fig. 6-2 for location; compare with Fig. 6-13). Above is uninterpreted seismic profile; below is interpreted tracing of seismic profile showing stacked allostratigraphic units, unconformable unit boundaries, seismic facies, and large submarine channels. Three units shaded to facilitate comparison with profiles 23 (Fig. 6-13), 206 (Fig. 6-23), and 201 (Fig. 6-25). Aalenian(?) through Kimmeridgian–Tithonian units are characteristically thick (but older units thin eastward onto basement high). Nearly continuous, high-amplitude reflections represent mainly turbidite aprons. Some units display subtle channeling of upper surfaces and within units. Berriasian–Valanginian through Coniacian–Santonian units are significantly thinner than older units, but are quite variable across section; several units display subtle channeling. Campanian allostratigraphic unit is unusually thick; it displays mainly continuous-parallel or clinoform reflections (stratified turbidites); upper surface of unit is deeply channeled and reflections are truncated along walls of U-shaped channels; truncation was caused by submarine erosion. Between shot-points 6600 and 6400, chaotic reflections in middle of the unit probably represent slumps and(or) debris-flow deposits. Maestrichtian through middle Eocene units are thin; they partly fill channels with turbidites (parallel and clinoform reflections) and slumps and debris-flow deposits (chaotic reflections). Thick middle Miocene allostratigraphic unit displays two distinct seismic facies. At left, parallel, continuous reflections are typical of turbidites. Turbidites truncated near shot-point 6800; the rest of middle Miocene unit to east is mainly chaotic channel-fill [slumps and(or) debris-flow deposits]. Quaternary unit contains mainly continuous, parallel reflections of turbidite facies, with occasional chaotic facies typical of slumps and debris-flow deposits. The seafloor (Quaternary unit) is incised near shot-point 6700 by deep V-shaped submarine canyon; smaller V-shaped canyon is present near shot-point 6900. Berrias.-Valang. = Berriasian–Valanginian.

ancient James, Potomac, and Susquehanna rivers remained the primary dispersal routes for detritus eroded mainly from the southern part of the central Appalachian highlands. The rate of sediment delivery had slowed by approximately two-thirds, to 4,000 km³/m.y. (Fig. 6-5; Poag and Sevon, 1989), however, even though long- and short-term sea-level lows are postulated for this interval by the Exxon model. Smaller deltas were present in front of the ancient Delaware, Hudson, Connecticut, and eastern Massachusetts rivers. The James and eastern Massachusetts delta systems also built significant debris aprons (500–600 m thick; 0.4–0.5 sec) along the continental slope and upper rise (Fig. 6-13), from which small, elongate, submarine fan complexes (400–500 m thick; 0.4–0.5 sec) extended across the upper rise. Off New England, the Aptian–Albian allostratigraphic unit has been eroded from much of the shelf edge and upper slope. A series of broad submarine channels characterizes this erosional tract (Fig. 6-20) and creates a scalloped isochron pattern along the continental slope.

The most extensive submarine fan system emanated from the Potomac delta system and extended in a broad swath (90 km wide) to the lower continental rise. There it merged with the Susquehanna, Delaware, and Hudson dispersal systems to form a confluent megafan. Distal sediments from this megafan were cored at DSDP Site 603, where a 170-m succession begins with

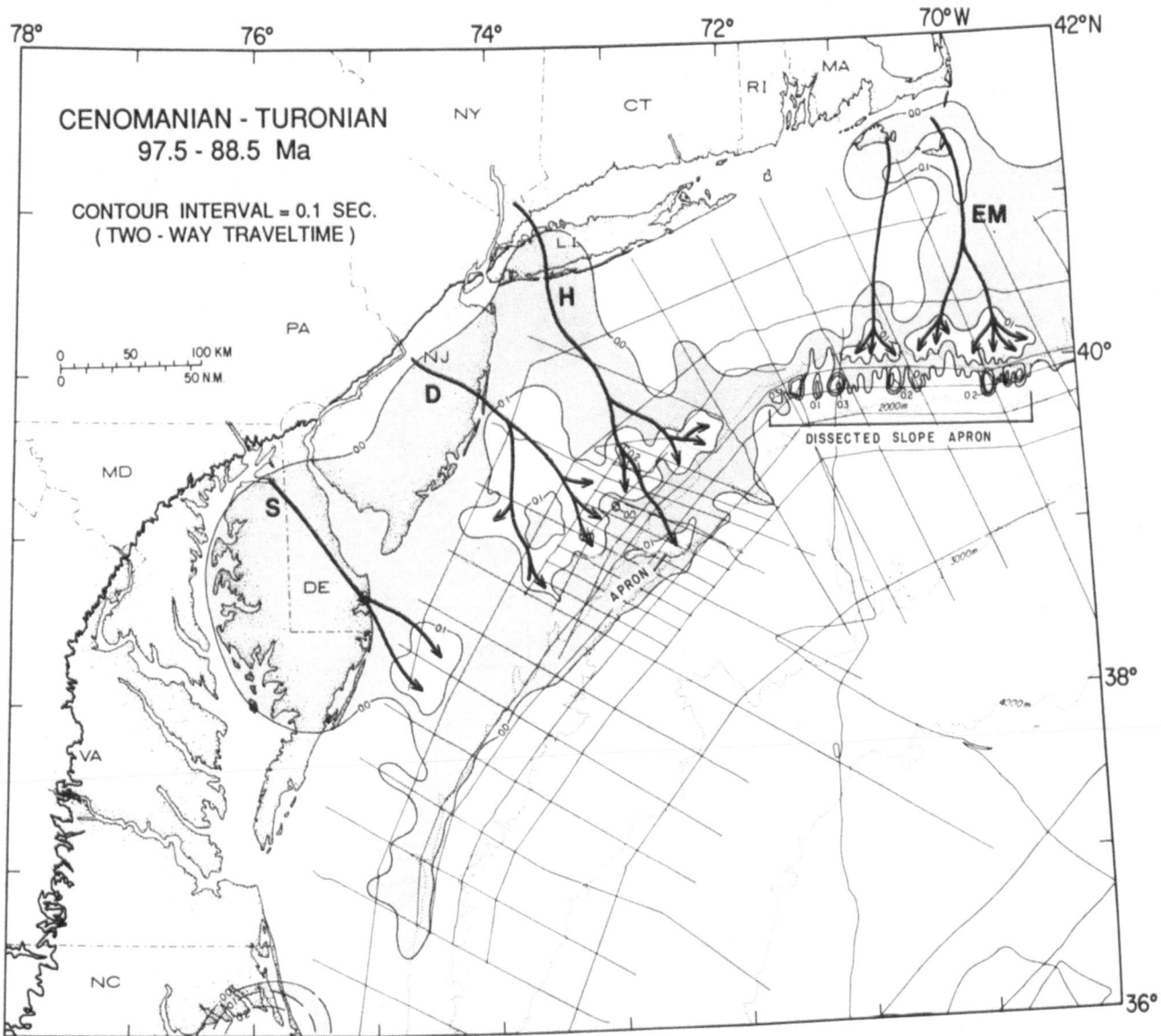

FIGURE 6-21.   Isochron map (contours in 0.1 sec of two-way traveltime) of Cenomanian–Turonian allostratigraphic unit. Arrows indicate inferred primary routes of sediment dispersal. Shaded contour interval (0.0–0.1 sec) indicates thinnest parts of unit. S = ancient Susquehanna River; D = ancient Delaware River; H = ancient Hudson River; EM = ancient river(s) in eastern Massachusetts.

coarse, unlithified sand, changes upward into thin turbidites, and culminates in red, greenish-gray, and black
claystones (lower part of the Hatteras Formation; Sarti
and von Rad, 1987). At DSDP Site 105, Aptian–Albian
sediments comprise ~ 70 m of black, highly carbonaceous (terrestrial plant debris), zeolitic clay (Hollister,
Ewing, et al., 1972; compare with Chapter 5).

A large, irregular tract of little or no Aptian–Albian
sediment occupies the middle continental rise seaward
of Virginia to New Jersey. The long axis of this tract is
perpendicular to the general downslope trend of gravity-flow deposition, an orientation which I take as
evidence that abyssal currents were actively eroding
here during Aptian–Albian time.

Cenomanian – Turonian Allostratigraphic Unit

Cenomanian–Turonian deposition (97.5–88.5 Ma)
slowed even more than Aptian–Albian deposition; the
net accumulation rate decreased from 4,000 km³/m.y.
to 2,000 km³/m.y. (Fig. 6-5; Poag and Sevon, 1989).
The deposits were largely restricted to a thin blanket
on the continental shelf and slope (Fig. 6-21; contrast
with Chapter 4). Four modest depocenters thicker than
100 m (0.1 sec) developed on the shelf: three small
individual ones (50–100 km long × 10–40 km wide) in
front of the ancient Susquehanna and eastern Massachusetts rivers and a confluent one (175 km long ×
50–100 km wide) associated with the ancient Delaware

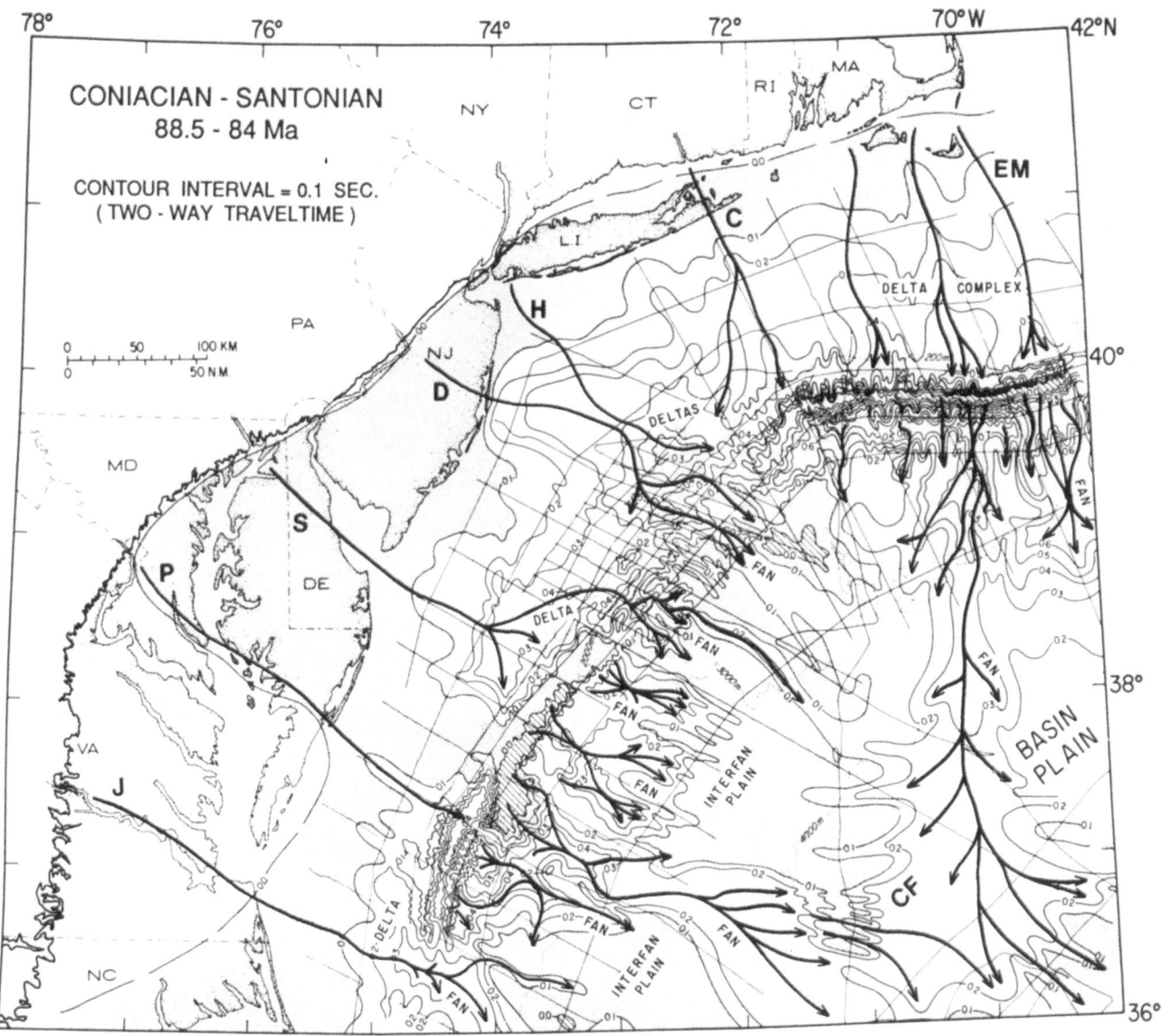

FIGURE 6-22. Isochron map (contours in 0.1 sec of two-way traveltime) of Coniacian–Santonian
allostratigraphic unit. Arrows indicate inferred primary routes of sediment dispersal. Shaded contour
interval (0.0–0.1 sec) indicates thinnest parts of unit. Diagonal hachures indicate unit absent or too thin
to resolve on seismic profiles. J = ancient James River; P = ancient Potomac River; S = ancient
Susquehanna River; D = ancient Delaware River; H = ancient Hudson River; C = ancient Connecticut River; EM = ancient rivers in eastern Massachusetts; CF = confluent fan.

and Hudson drainage systems. A narrow debris apron, locally as thick as 300 m (0.3 sec), appears to have developed along the New England slope, but, subsequently, it has been dissected by submarine canyons.

Cenomanian–Turonian sediments are present locally in the Hatteras Basin but generally are too thin to resolve on multichannel seismic profiles (Figs. 6-3, 6-13, 6-14, 6-16). At DSDP Site 105, Cenomanian–Turonian deposition is represented by at least 20 m of silty, zeolitic, carbonaceous (terrestrial plant debris), black clay of the Hatteras Formation (Hollister, Ewing, et al., 1972). At DSDP Site 603, coeval deposits comprise > 50 m of similar black clays, interbedded with variegated clays and silty and sandy turbidites (Haggerty et al., 1987).

I interpret this distribution pattern (similar to the Hauterivian pattern) to be the result of high sea level (as postulated by the Exxon model) and low rates of sediment delivery [as postulated by Cecil's (1990) model, when high values are used for Cretaceous warmth and wetness (Barron, 1989)]. There is no evidence that debris aprons or fans formed during a postulated late Turonian lowstand (Haq, Hardenbol, and Vail, 1987).

Much of the lower continental rise at this time was deeper than the carbonate compensation depth (CCD; Tucholke and Vogt, 1979); thus, carbonate sediments did not accumulate.

Coniacian – Santonian Allostratigraphic Unit

Siliciclastic deposition accelerated nearly 10-fold in the Coniacian–Santonian (88.5–84 Ma; see also Chapters 4, 5, 11); the net accumulation rate increased from 2,000 km³/m.y. to 19,000 km³/m.y. (Fig. 6-5; Poag and Sevon, 1989), as the New England Appalachian highlands experienced renewed uplift. This tectonic pulse caused primary depocenters of the continental shelf (now supplied by the Connecticut and several eastern Massachusetts river systems) to migrate from the North Carolina–southern New Jersey region to northern New Jersey and New England (Fig. 6-22). Several large deltas developed on the continental shelf, but thickest depocenters (> 700 m thick; > 0.7 sec) were debris aprons on the continental slope and upper rise (Figs. 6-13, 6-20, 6-23), though no major lowstands are postulated for this interval by the Exxon model. An elongate submarine megafan emanated from New England debris aprons and extended > 400 km to the lower rise.

A second debris apron and a large submarine fan complex developed on the upper and middle continental rise seaward of Virginia (Figs. 6-3, 6-22), but these deposits were not associated with a prominent shelf delta. Sediments from the eastern Massachusetts megafan and the large Virginia fan merged to form a 200-km-wide confluent megafan on the lower rise. This confluent fan was cored at DSDP Site 603; the Coniacian–Santonian unit consists of 35 m of variegated and black, quartz- and mica-rich clays containing a few sandy silts (lower part of the Plantagenet Formation; Haggerty et al., 1987).

Erosion, presumably during a late Santonian lowstand, removed strata from a narrow swath along the upper to middle continental slope seaward of Virginia. I suspect that much of this eroded material is incorporated in the two small upper rise fans located downslope from the erosional tract.

On the New England slope, a scalloped, downslope-oriented pattern of isochrons represents, in part, turbidite mounds, fans, and debris-flow deposits (Figs. 6-13, 6-20, 6-22, 6-23). West of the thickest upper rise apron (700 m thick; 0.7 sec), however, many of the elongate sediment lenses represent filled submarine channels within the Coniacian–Santonian unit. In addition, a few small channels incised the upper surface of the unit.

Campanian Allostratigraphic Unit

Siliciclastic deposition continued to dominate during the Campanian (84–74.5 Ma). Principal depocenters of the continental shelf, slope, and upper rise remained off New England (Fig. 6-24), though the supply of siliciclastic detritus from the New England Appalachian highlands had dwindled approximately by half; the net accumulation rate decreased from 19,000 km³/m.y. to 9,000 km³/m.y. (Fig. 6-5; Poag and Sevon, 1989). The thickest depocenters (1–1.2 km; 0.9–1.1 sec) were debris aprons and several small to large submarine fans on the upper and middle continental rise. Many of the downslope-oriented, scalloped, isochron patterns along the New England continental slope and upper rise represent filled submarine channels or turbidite and debris-flow mounds within the Campanian unit; others represent subsequent dissection of the upper surface of the unit, perhaps during a postulated mid-Campanian lowstand (Haq, Hardenbol, and Vail, 1987; Figs. 6-13, 6-20, 6-23, 6-24, 6-25).

The middle continental rise was characterized by a large confluent fan off New England and an elongate digitate fan off Virginia. Sedimentary sections of the interfan plains and lower rise basin plain are quite thin (< 50 m; < 0.1 sec), however, and probably comprise mainly pelagic and hemipelagic sediments. The volume of siliciclastic detritus from the central Appalachian and Adirondack highlands was further reduced (compared to the Coniacian–Santonian volume), although the ancient Delaware and Hudson rivers still were important dispersal routes. A wide swath along the southwestern shelf edge, similar to that of the Coniacian–Santonian, appears to have been subsequently

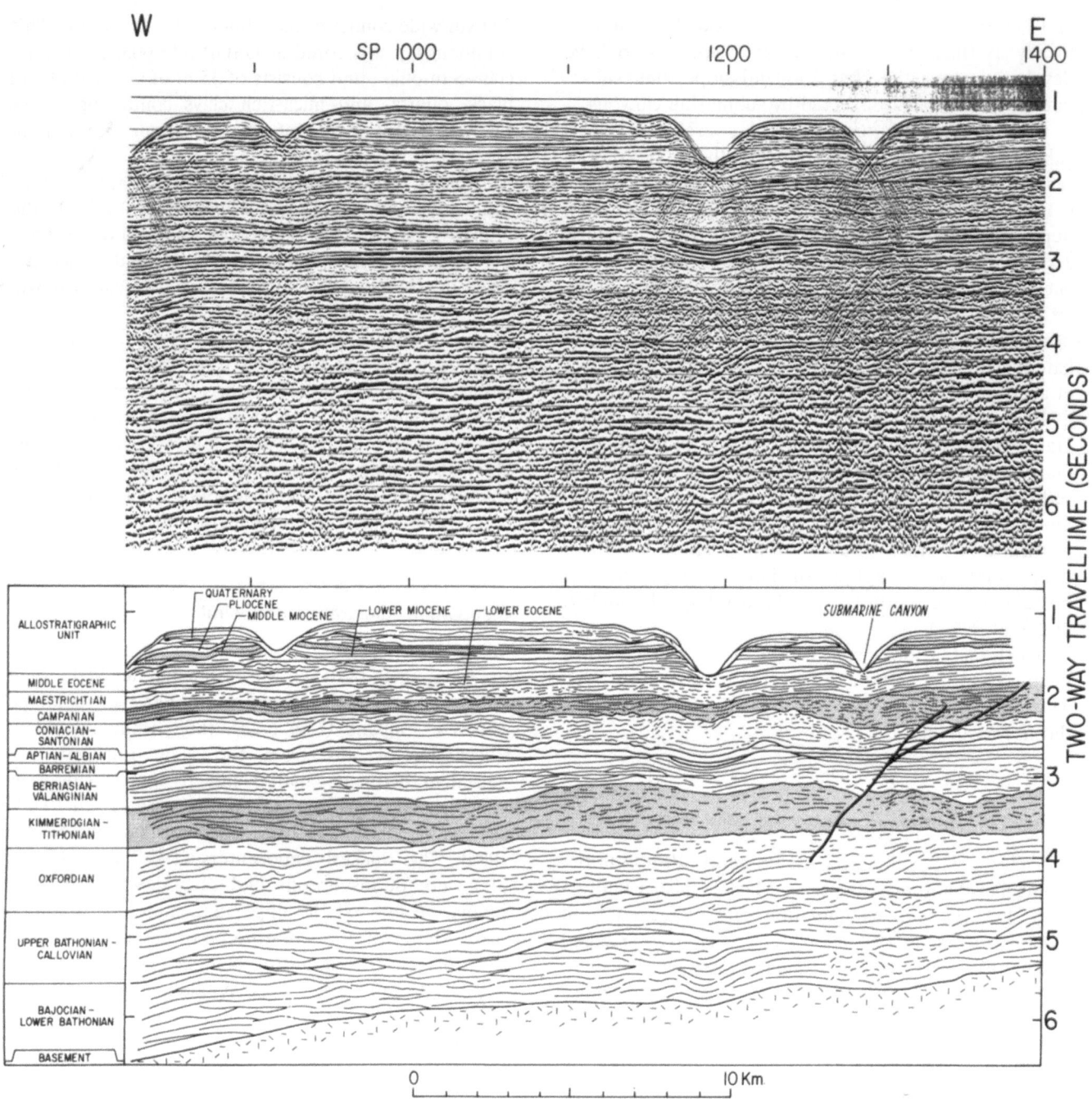

FIGURE 6-23.   Stratigraphic section parallel to depositional strike along seismic profile 206 crossing the New England shelf edge between seismic profiles 9 and 23, 20-km updip from profile 205 (see Fig. 6-2 for location; compare with Fig. 6-13). Above is uninterpreted seismic profile; below is interpreted tracing of seismic profile. Three units shaded to facilitate comparison with profiles 23 (Fig. 6-13), 205 (Fig. 6-20), and 201 (Fig. 6-25), and to emphasize contrasting geometry, seismic facies, and depositional styles. Kimmeridgian–Tithonian unit (upper part of Bahama–Grand Banks Gigaplatform) is crossed in back-reef position. At left, gently convergent, high-amplitude reflections are typical of mixed carbonate–siliciclastic back-reef facies; poorly organized reflections east of shot-point 1100 are typical of mounded carbonate buildups (patch reefs?). Campanian allostratigraphic unit is thin; it contains parallel, continuous reflections in western half of section, but thickens eastward where it fills small depressions and large channels eroded into the underlying Coniacian–Santonian surface. Campanian unit is part of prograding shelf-edge delta complex east of shot-point 1200. Middle Miocene unit is a very thin sheet whose inner reflections cannot be discerned at the scale of this figure. Four unfilled modern submarine canyons incise the shelf edge and excavate as deeply as middle and lower Eocene units.

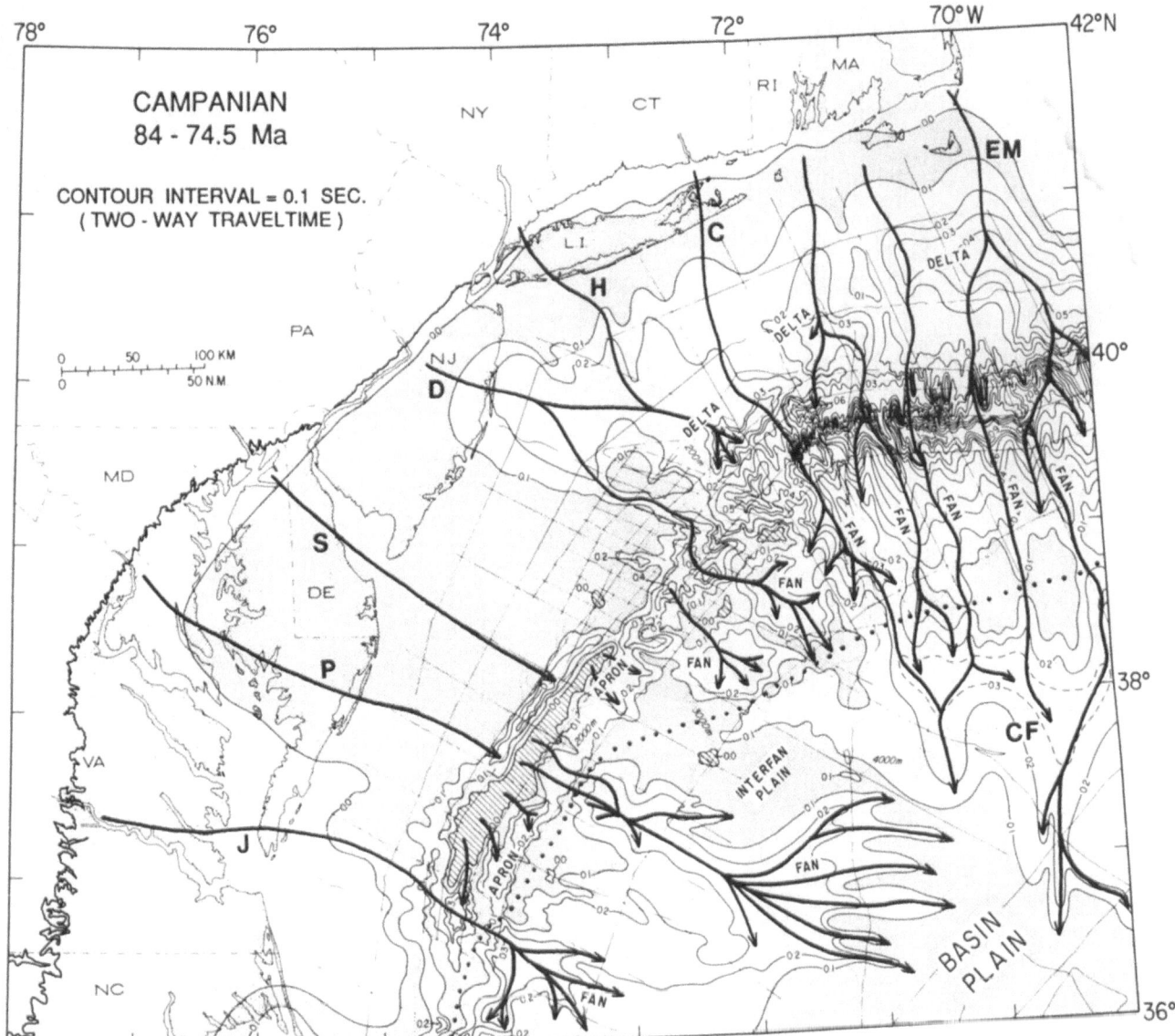

FIGURE 6-24. Isochron map (contours in 0.1 sec of two-way traveltime) of Campanian allostrati-graphic unit. Arrows indicate inferred primary routes of sediment dispersal. Shaded contour interval (0.0–0.1 sec) indicates thinnest parts of unit. Diagonal hachures indicate unit absent or too thin to resolve on seismic profiles. Heavy dotted line indicates possible axis of eroding bottom current. J = ancient James River; P = ancient Potomac River; S = ancient Susquehanna River; D = ancient Delaware River; H = ancient Hudson River; C = ancient Connecticut River; EM = ancient river(s) in eastern Massachusetts; CF = confluent fan.

stripped of Campanian strata, which I presume was added to the associated mid-rise fan.

Principal Campanian sediments at DSDP Site 603 are variegated and black clays interbedded with sandy and silty turbidites (middle part of the Plantagenet Formation; Haggerty et al., 1987). No coeval sediments have been firmly identified at Site 105, though an 18-m section of undated, brown, silty clay may include Campanian beds (Hollister, Ewing, et al., 1972).

A lobate swath of thin sediment parallels the shelf edge along the middle part of the continental rise, cutting directly across the New England confluent fan.

This may be evidence that contour-following bottom currents eroded this area during the Campanian.

Maestrichtian Allostratigraphic Unit

Sedimentary deposits are thin over most of the Maestrichtian (74.5–66.4 Ma) continental shelf. This sparsity may be related in part to a late Maestrichtian lowstand (postulated by the Exxon model) and(or) to extensive disruption of the shelf depositional regime by impact-wave phenomena caused by an end-of-Creta-ceous bolide impact, as envisioned by some authors

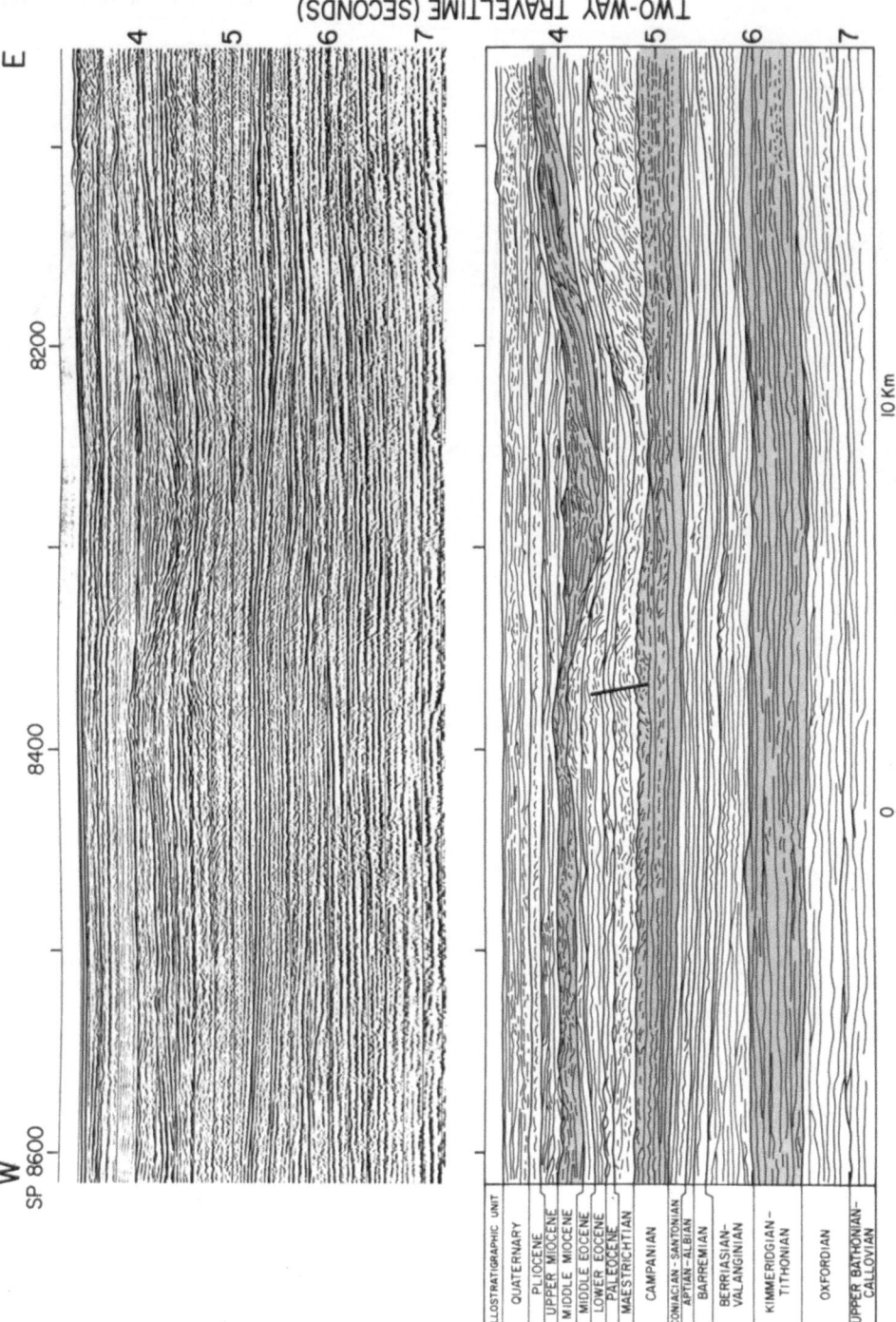

FIGURE 6-25. Stratigraphic section parallel to depositional strike along seismic profile 201 crossing upper part of the New England continental rise between seismic profiles 9 and 23, 28-km downdip from profile 205 (see Fig. 6-2 for location; compare with Fig. 6-13). Above is uninterpreted seismic profile; below is interpreted tracing of seismic profile showing stacked allostratigraphic units. Three allostratigraphic units are shaded to facilitate comparison with profiles 23 (Fig. 6-13), 206 (Fig. 6-23), and 205 (Fig. 6-20), and to emphasize contrasting geometry, seismic facies, and depositional styles. Older units display gently convergent, high-amplitude reflections typical of turbidites, with occasional chaotic intervals representing slumps and(or) debris-flow deposits and lenticular mounds representing submarine fans. Campanian unit shows mainly continuous, parallel reflections of turbidite facies near its base, but reflections become more chaotic (debris-flow deposits) near the middle and in the upper part, especially west of shot-point 8100. Maestrichtian unit displays a striking mound of chaotic and clinoform reflections (slumps, debris-flow deposits) interpreted to form a large fan east of 8200; mounded reflections of a smaller submarine fan are seen near shot-point 8350. Middle Miocene unit contains thick channel fill [mixed convergent and chaotic reflections indicate turbidites and(or) debris-flow deposits] near shot-point 8300. Quaternary unit displays gently convergent reflections of turbidite facies west of shot-point 8250, but reflections become chaotic (slumps, debris-flow deposits) to the east.

(e.g., Van Hinte, Wise, et al., 1985). A moderately large middle-shelf delta system developed in front of the ancient Connecticut River, and a broad sedimentary prism formed along the shelf edge off the ancient James and Potomac rivers (Fig. 6-26). The net rate of siliciclastic sediment accumulation remained unchanged at 9,000 km^3/m.y. (Fig. 6-5; Poag and Sevon, 1989). The largest depocenters (> 600 m thick; > 0.6 sec) continued to occupy the New England continental slope and upper rise (see also Chapters 4, 5). The corehole at DSDP Site 605 penetrated the upper 76 m of a 200-m-thick debris apron off New Jersey and found dark-gray, clayey, pelagic limestones containing silt-size quartz and rare glauconite in the upper beds. The cut-and-fill pattern of the isochron contours on the New England slope is caused mainly by the erosion and filling of submarine channels (Figs. 6-13, 6-20, 6-23, 6-25).

A conspicuous debris apron (~ 300 m thick; 0.3 sec) also is present along the margin from Delaware to New Jersey. A large submarine fan system is associated with this apron and probably contains large amounts of detritus eroded from the adjacent continental shelf and slope. Another large submarine fan complex is seaward of the Virginia margin and extends into the basin plain to the southeast. Two large upper rise fans along the

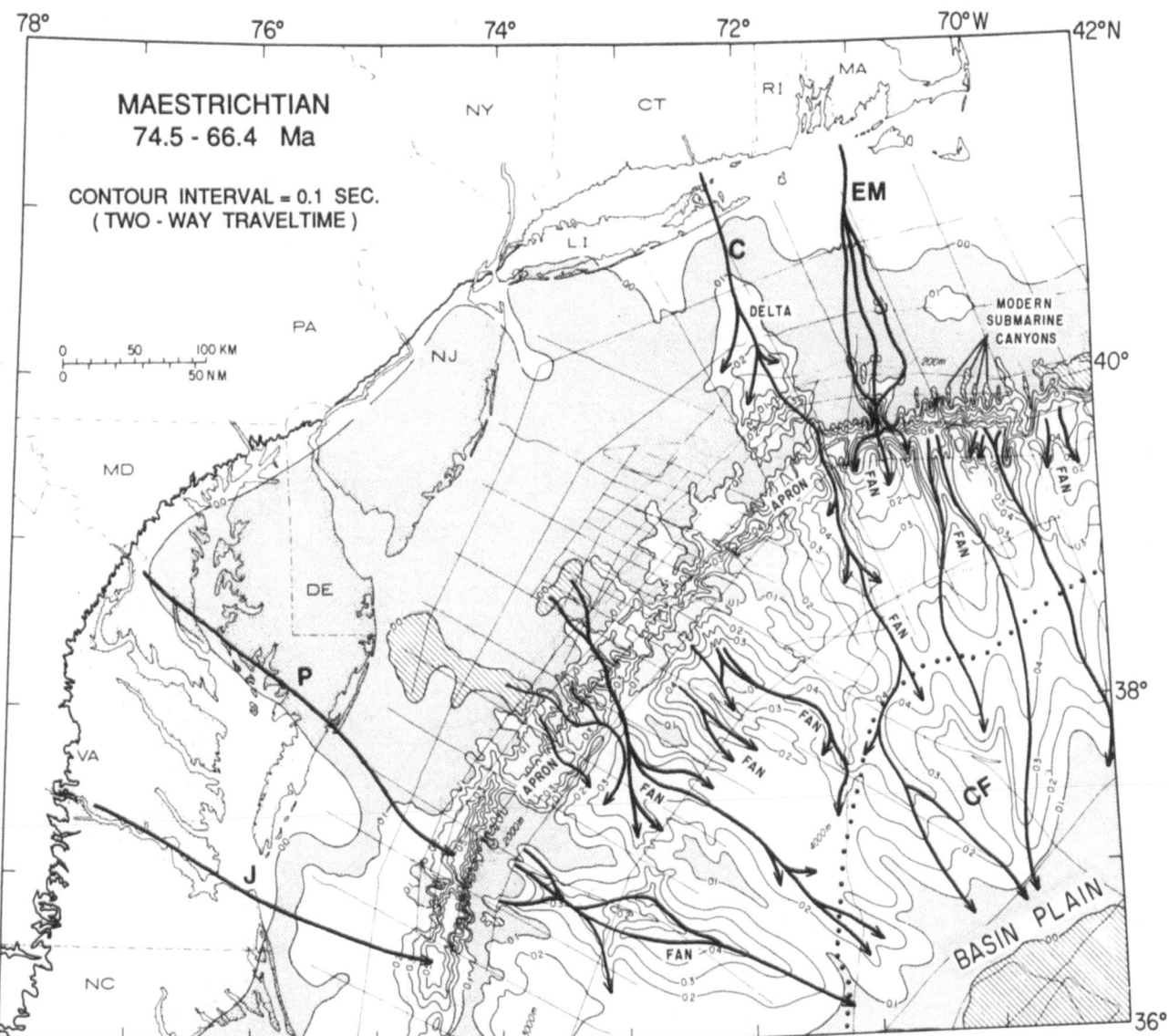

FIGURE 6-26.   Isochron map (contours in 0.1 sec of two-way traveltime) of Maestrichtian allostratigraphic unit. Arrows indicate inferred primary routes of sediment dispersal. Diagonal hachures indicate unit absent or too thin to resolve on seismic profiles. Shaded contour interval indicates thinnest parts of unit. Heavy dotted line indicates possible axis of depositing contour current. J = ancient James River; P = ancient Potomac River; C = ancient Connecticut River; EM = ancient river(s) in eastern Massachusetts; CF = confluent fan.

New England margin coalesced on the middle and lower rise to form a large confluent fan. On the middle rise part of this confluent fan is an elongage, deeply lobed, sedimentary prism (~ 400 m thick; 0.4 sec), whose long axis parallels the New England slope. This alongslope orientation may be evidence that bottom currents actively redistributed sediments into a contourite drift along this track in the Maestrichtian.

The southeastern corner of the mapped area appears to be devoid of Maestrichtian sediment, partly because a high CCD precluded carbonate accumulation (Haggerty et al., 1987), but perhaps also because the unit may be too thin to resolve on the seismic

profiles. No Maestrichtian strata have been firmly dated at DSDP Sites 105 and 603 (Hollister, Ewing, et al., 1972; Haggerty et al., 1987). The undifferentiated latest Cretaceous beds at those sites consist mainly of noncalcareous, variegated clays and black clays, but sandy and silty turbidites are present at the top of the Upper Cretaceous section at Site 603.

Paleocene Allostratigraphic Unit

The siliciclastic accumulation rate dropped by two-thirds (from 9,000 to 3,000 km^3/m.y.; Fig. 6-5; Poag and Sevon, 1989) during the Paleocene (66.4–57.8 Ma)

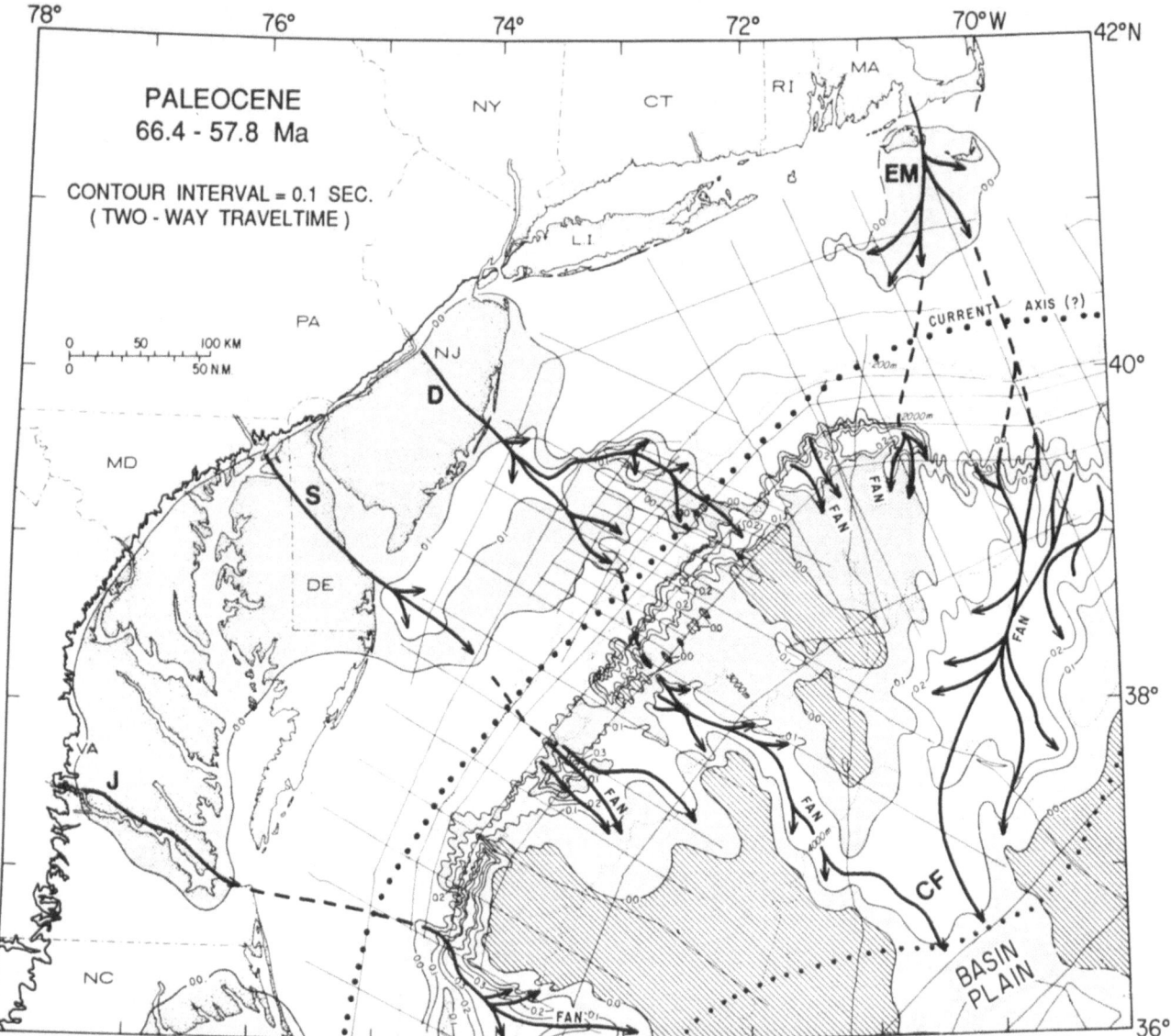

FIGURE 6-27.   Isochron map (contours in 0.1 sec of two-way traveltime) of Paleocene allostratigraphic unit. Arrows indicate inferred primary routes of sediment dispersal. Diagonal hachures indicate unit absent or too thin to resolve on seismic profiles. Shaded contour interval (0.0–0.1 sec) indicates thinnest parts of unit. Heavy dotted lines indicate possible axes of eroding bottom currents. J = ancient James River; S = ancient Susquehanna River; D = ancient Delaware River; EM = ancient river(s) in eastern Massachusetts; CF = confluent fan.

and accumulation rates remained quite low for the next ~ 42 m.y. (into the early Miocene; contrast with chapter 4). All highland source terrains appear to have been topographically low and tectonically quiescent during most of this interval (Poag and Sevon, 1989). As a result of this stability and a high sea level (Poag, 1987), carbonate sediments were more abundant in the study area during the Paleogene than at any time since the Jurassic.

Accompanying these changes were significant shifts in latitudinal and bathymetric positions of main depocenters (Fig. 6-27). The thickest Paleocene depocenter (> 300 m thick; > 0.3 sec), for example, developed as a debris apron and upper rise submarine fan complex seaward of the ancient James River, reflecting a shift of the primary siliciclastic source terrain from New England to the southern part of the central Appalachian highlands. A secondary depocenter (small debris apron and fan > 200 m thick; > 0.2 sec) still occupied the New England slope and rise (Figs. 6-13, 6-20, 6-25). Debris aprons, which also developed along other segments of the margin, may contain, in part, mixed carbonate and siliciclastic detritus eroded from the Paleocene shelf. Evidence that some terrigenous sediments still reached the upper rise is found in silty, micaceous, quartzose, glauconitic Paleocene strata cored in a 196-m pelagic limestone succession at DSDP Site 605 (Van Hinte, Wise, et al., 1987).

At present, most of the Paleocene shelf sediments are on the Delaware to New Jersey margin, where an elongate prism (> 100 m thick; > 0.1 sec), supplied by the ancient Susquehanna and Delaware rivers, developed just seaward of what now is the shoreline. The present sparse distribution of Paleocene deposits on the shelf, however, appears to be the result of extensive erosion, perhaps by strong outer shelf currents (Olsson and Wise, 1987). Extensive tracts on the continental rise also appear to be devoid of Paleocene sediments, but some of these may contain strata too thin to resolve on the seismic profiles.

The main contribution of Paleocene sediments to the lower rise and the abyssal basin plain appears to have come from a medium submarine fan complex, which derived sediments from the New England margin (Fig. 6-27). This fan joined another, which extended from the New Jersey margin, to form a large confluent fan in the southeast corner of the study area.

Deep Sea Drilling Project Site 603 was still below the CCD, and Paleocene strata there (dated by radiolarians; Nishimura, 1987) consist primarily of thin (~ 3 m) variegated, biosiliceous claystones containing small amounts of quartz, mica, and heavy minerals in thin sand and silt turbidites (local equivalents of the Bermuda Rise Formation; Haggerty et al., 1987). Possible equivalents (dated by ichthyoliths; Kaneps, Doyle, and Riedel, 1981) at DSDP Site 105 comprise ~ 3 m of similar variegated, silty clay with occasional turbiditic layers of quartz sand, mica, and heavy minerals (Hollister, Ewing, et al., 1972).

Lower Eocene Allostratigraphic Unit

Shelf deposition continued to be relatively sparse, and low siliciclastic accumulation rates were maintained (4,000 km^3/m.y.; Fig. 6-5; Poag and Sevon, 1989) during the postulated long-term sea-level peak of the early Eocene (57.8–52 Ma). As in the Paleocene, an elongate depocenter [300 km long × 30–40 km wide × > 100 m thick (> 0.1 sec)] occupied the middle shelf from Delaware to New Jersey (Fig. 6-28). Seaward of this primary shelf depocenter, a carbonate-dominated apron–fan complex (> 200 m thick; > 0.2 sec) developed along the outer shelf, the slope, and the upper rise. The coreholes at DSDP Sites 605 and 613 penetrated the upper rise fan off New Jersey and revealed 150–218 m of light-gray pelagic limestone and chalk.

Two additional submarine fan systems formed seaward of New England (Figs. 6-13, 6-20, 6-23, 6-25), one of which extended ~ 300 km to the basin plain in the southeast corner of the study area. As in the Paleocene, several areas on the outer shelf are devoid of lower Eocene deposits and may indicate the presence of erosive currents. Even larger areas on the continental slope and upper to middle rise lack lower Eocene strata (or the strata are too thin to resolve on seismic profiles).

Lower Eocene strata at DSDP Site 603 (dated with radiolarians; Nishimura, 1987) comprise ~ 47 m of variegated, biosiliceous clays containing thin sandy and silty turbidites (Haggerty et al., 1987). Equivalent strata at DSDP Site 106 are biosiliceous, hemipelagic muds dominated by clay minerals (Hollister, Ewing, et al., 1972). Similar sediments at DSDP Site 105 appear to be coeval, but have not yet been firmly dated (Hollister, Ewing, et al., 1972).

Middle Eocene Allostratigraphic Unit

During the middle Eocene (52–40 Ma), the siliciclastic accumulation rate continued its decline, dropping from 4,000 to 2,000 km^3/m.y. (Fig. 6-5; Poag and Sevon, 1989), in spite of a significant pulse of tectonic uplift in the central Appalachians (as indicated by magmatic and topographic evidence; De Boer et al., 1988; Roden and Miller, 1989; Vogt, 1991; compare with Chapters 7, 11). The lack of accelerated siliciclastic deposition may be explained by Cecil's (1990) model of sedimentological response to climate. Cecil indicated that heavy vegetation, characteristic of ever-wet tropical climates, would minimize availability and dispersal of siliciclastic detritus. The coastal part of the study area was occu-

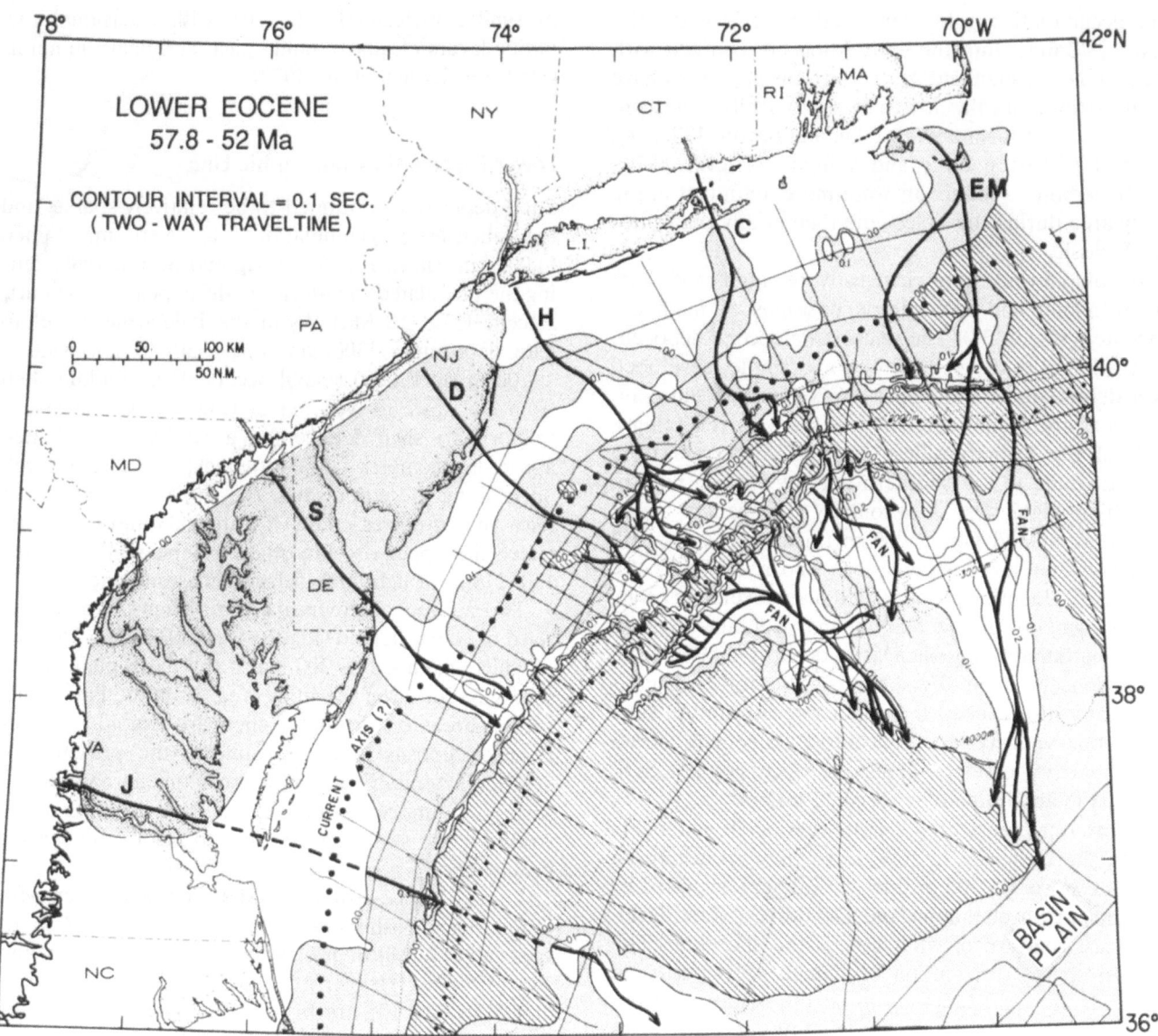

FIGURE 6-28.   Isochron map (contours in 0.1 sec of two-way traveltime) of lower Eocene allostrati-graphic unit. Arrows indicate inferred primary routes of sediment dispersal. Diagonal hachures indicate unit absent or too thin to resolve on seismic lines. Shaded contour interval (0.0–0.1 sec) indicates thinnest parts of unit. Heavy dotted line indicates possible axis of eroding, outer shelf bottom current; light dotted line indicates possible axis of eroding bottom current along upper continental rise. J = ancient James River; S = ancient Susquehanna River; D = ancient Delaware River; H = ancient Hudson River; C = ancient Connecticut River; EM = ancient river(s) in eastern Massachusetts.

pied by a tropical rain forest during the middle Eocene tectonic event (Wolfe, 1978).

Despite the reduced supply of siliciclastic sediment, shelf deposition was rejuvenated somewhat during the middle Eocene. A relatively large confluent delta [250 km long × 100 km wide × > 400 m thick (> 0.4 sec)] built out seaward of Virginia to southern New Jersey, and a smaller delta occupied the shelf from northern New Jersey to Long Island (Fig. 6-29). Together, these deltas formed a distinct, active, "hinter" shelf edge 100 km landward of the subdued, inactive, "fore" shelf

edge (Poag, 1987). Construction of this hinter shelf marked the beginning of a progressive seaward progradation of continental-shelf depocenters, which lasted into the Quaternary.

Extensive downslope channeling characterizes the fore slope (mainly carbonate-dominated substrates), particularly off New Jersey (Mountain, 1987; Poag and Mountain, 1987), and carbonate-dominated fore-slope aprons 200–300 m thick are common. Coreholes at DSDP Sites 605 and 613 penetrated one apron, finding 145 and 173 m, respectively, of light–greenish-gray,

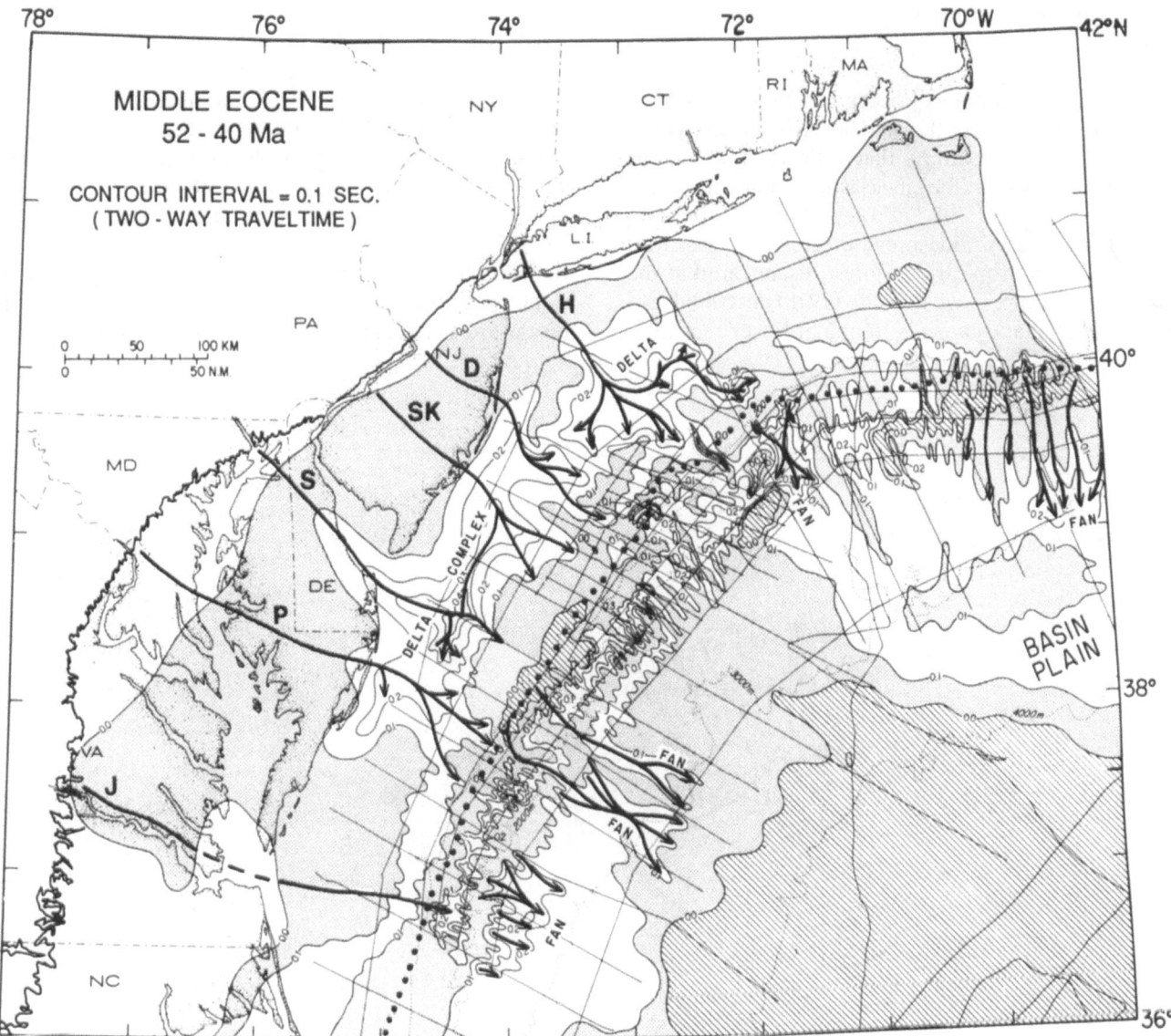

FIGURE 6-29. Isochron map (contours in 0.1 sec of two-way traveltime) of middle Eocene allostratigraphic unit. Arrows indicate inferred primary routes of sediment dispersal. Diagonal hachures indicate unit absent or too thin to resolve on seismic profiles. Shaded contour interval (0.0–0.1 sec) indicates thinnest parts of unit. Heavy dotted line indicates possible axis of eroding bottom current along outer shelf. J = ancient James River; P = ancient Potomac River; S = ancient Susquehanna River; SK = ancient Schuylkill River; D = ancient Delaware River; H = ancient Hudson River.

biosiliceous, nannofossil chalk (Van Hinte, Wise, et al., 1987; Poag, Watts, et al., 1987). The outer fore shelf in front of the main middle Eocene delta system has been stripped of middle Eocene strata, probably by the same current system that eroded the older Paleogene shelf deposits. Erosion also has stripped middle Eocene strata from the fore slope off Long Island and New England.

Two small, multilobed (presumably carbonate) fans derived sediment from fore-slope debris aprons off Virginia and New England (Figs. 6-13, 6-29). Two other elongate fans appear to be related to the conflu-

ent delta fed by the ancient Susquehanna and Potomac rivers and may, therefore, contain siliciclastic sediment.

The carbonate-rich middle Eocene section thins rapidly on the middle continental rise, and is missing or too thin to resolve over most of the lower rise. Drilling at DSDP Sites 105 and 603 indicates that much of the lower rise was still deeper than the CCD during middle Eocene time, though no middle Eocene sediments have yet been confirmed there. As a result, middle Eocene deposits, if present, presumably are represented by thin, noncalcareous, radiolarian-rich claystones.

Upper Eocene – Lower Oligocene Allostratigraphic Unit

The distribution of upper Eocene and lower Oligocene strata (40–30 Ma) on the middle Atlantic margin is quite restricted, a pattern that reflects slow, mainly carbonate deposition (the siliciclastic accumulation rate dropped to a low of 700 $km^3/m.y.$; Fig. 6-5), periodically interrupted by widespread erosion (contrast with Chapter 7). The tropical rain forest disappeared at the end of the Eocene, in what Wolfe (1978) described as a major paleoclimatic change ("terminal Eocene event").

No siliciclastic acceleration accompanied this "event," but lower Oligocene strata on much of the shelf contain significantly less carbonate than Eocene and Paleocene strata (Poag, 1987). As in the middle Eocene, the primary depocenter [> 200 m (> 0.2 sec) of mixed carbonate and siliciclastic sediment] was on the shelf, where a long (250 km), narrow (30–40 km) confluent delta developed (Fig. 6-30). Scattered occurrences of this allostratigraphic unit have been documented on the present coastal plain, outer shelf, and continental slope by drilling and by seafloor sampling. Remnants of a seaward-thickening sedimentary wedge) (dominantly

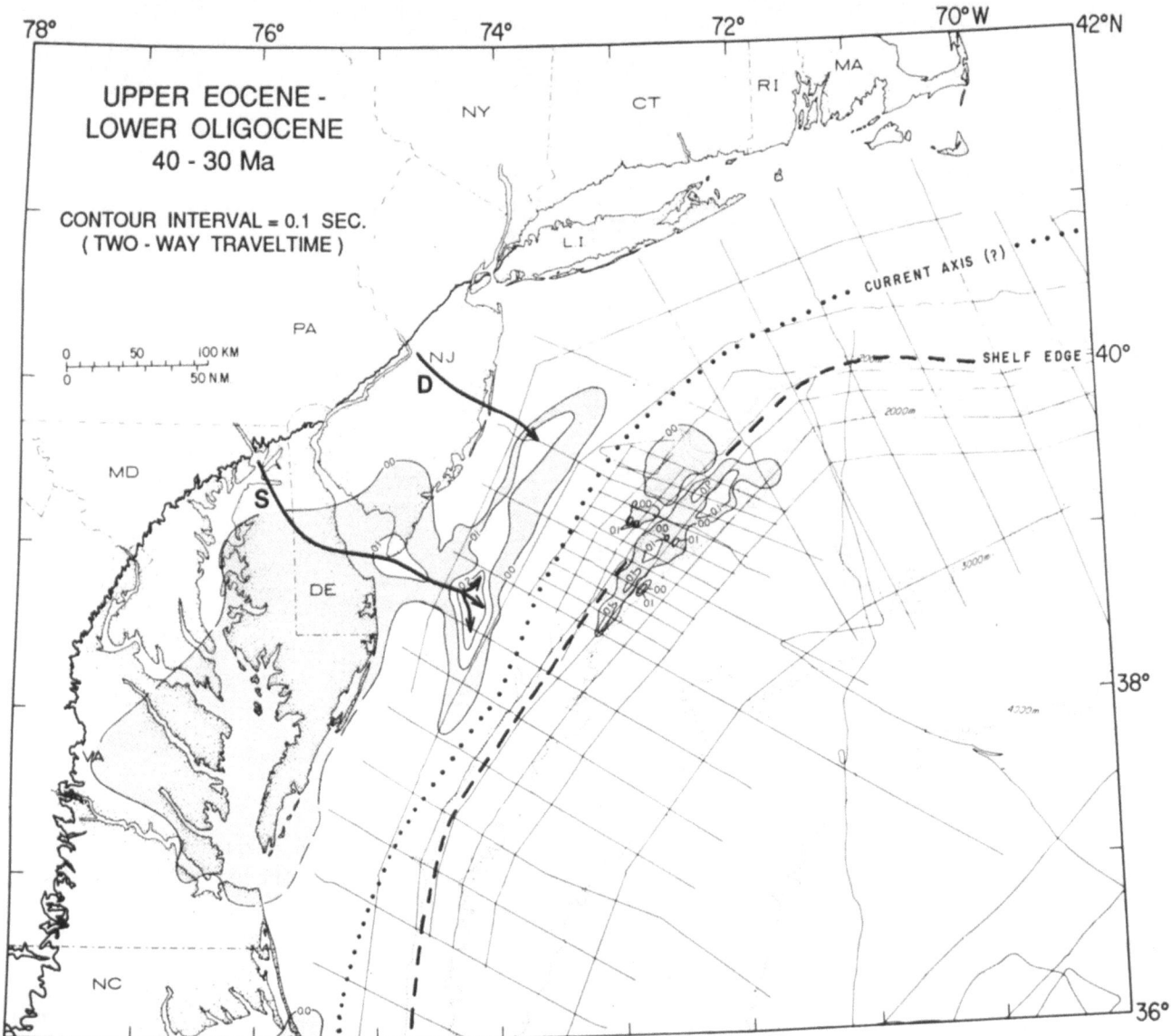

FIGURE 6-30.   Isochron map (contours in 0.1 sec of two-way traveltime) of upper Eocene–lower Oligocene allostratigraphic unit. Arrows indicate inferred primary routes of sediment dispersal. Shaded contour interval (0.0–0.1 sec) indicates thinnest parts of unit. Heavy dotted line indicates possible axis of eroding bottom current on outer shelf. S = ancient Susquehanna River; D = ancient Delaware River.

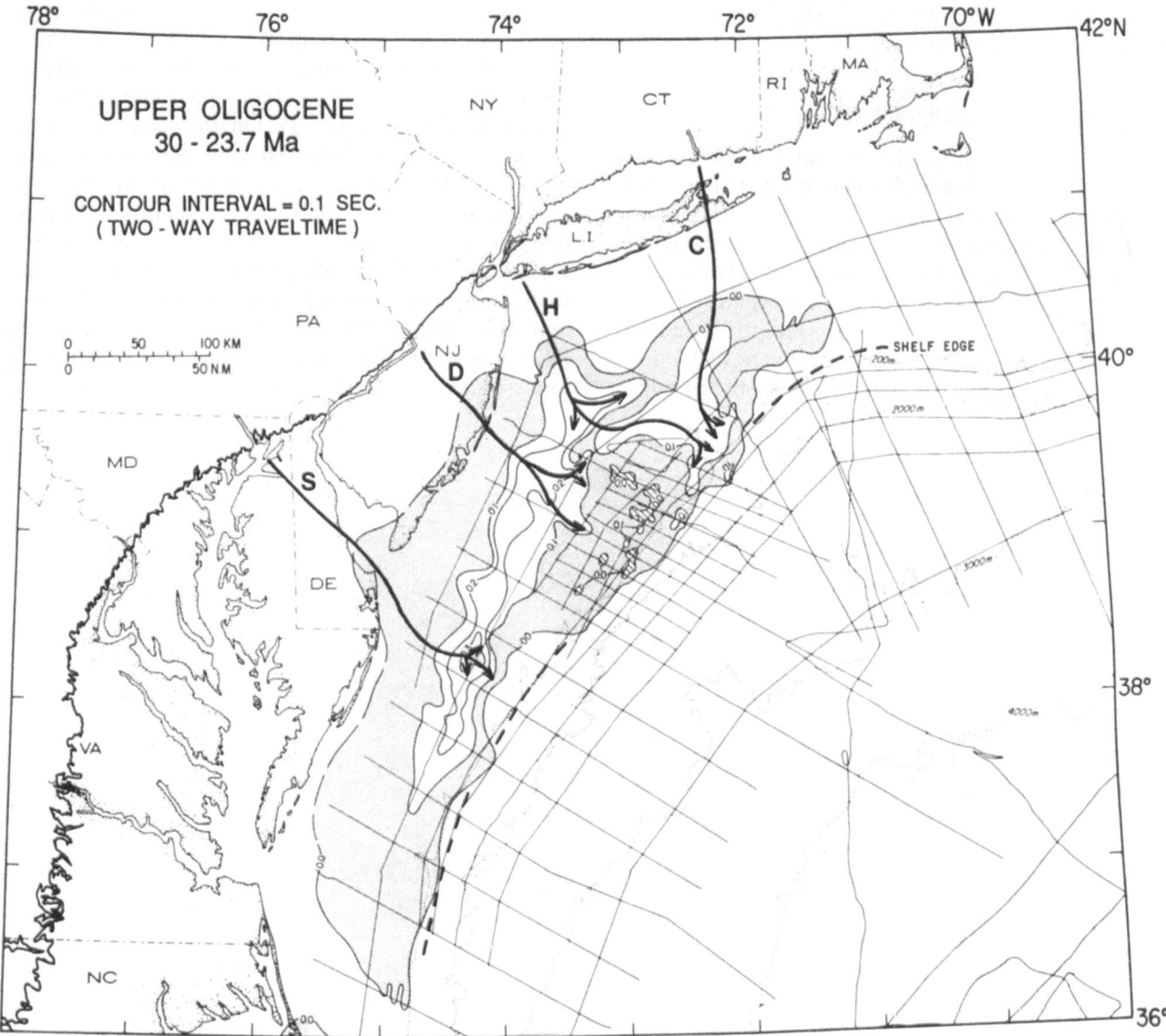

FIGURE 6-31.   Isochron map (contours in 0.1 sec of two-way traveltime) of upper Oligocene allostratigraphic unit. Arrows indicate inferred primary routes of sediment dispersal. Diagonal hachures indicate unit absent or too thin to resolve on seismic profiles. Shaded contour interval (0.0–0.1 sec) indicates thinnest parts of unit. S = ancient Susquehanna River; D = ancient Delaware River; H = ancient Hudson River; C = ancient Connecticut River.

pelagic carbonates) are present at the shelf edge seaward of New Jersey. This unit cannot be recognized at present over most of the mapped area; it has contributed little to construction of the continental rise.

Upper Oligocene Allostratigraphic Unit

According to the Exxon sequence-stratigraphy model, a dramatic lowstand at the beginning of the late Oligocene (30–23.7 Ma) initiated a significant long-term eustatic fall. This did not, however, trigger a depositional pulse in the study area; the siliciclastic accumu-

lation rate remained essentially unchanged at 1,000 km^3/m.y. (Fig. 6-5). An elongate delta system developed seaward of Virginia to Long Island, and continued the seaward progradation of the Paleogene hinter shelf (Fig. 6-31). The main depocenter was 400 km long and reached maximum thickness [~ 300 m; 0.3 sec) seaward of Delaware. Upper Oligocene strata pinch out along the outer shelf and middle continental slope and cannot be traced farther into the Hatteras Basin, nor have they been firmly identified at any DSDP Site in the study area. However, an ichthyolith assemblage recovered at DSDP Site 105 within an interval of variegated, zeolitic, silty clay (part of the Blake Ridge

Formation) may prove to be of late Oligocene age (Kaneps, Doyle, and Riedel, 1981).

Lower Miocene Allostratigraphic Unit

Continued seaward progradation and a moderate increase in the siliciclastic accumulation rate (from 1,000 to 3,000 km^3/m.y.; Fig. 6-5) led to a thicker (500 m; 0.5 sec) continental shelf depocenter during the early Miocene (23.7–16.6 Ma) than during any of the Paleogene intervals (Fig. 6-32). The main depocenter shifted ~ 100 km northward to a position seaward of middle New Jersey, but the associated confluent delta extended as far south as Virginia. This allostratigraphic

unit thins to an erosional termination along the continental slope, ~ 2-km undip from DSDP Site 612 (Poag, 1987). The unit is absent, or too thin to resolve an seismic profiles, over most of the upper and middle rise (compare with Chapter 11). On the lower rise, however, the margin of a lower Miocene depocenter (> 200 m thick; > 0.2 sec) was cored at DSDP Site 603 (also see Fig. 6-16). The section there comprises 14 m of gray, brown, and yellow, sideritic, gassy claystones (assigned to the Blake Ridge Formation), whose only fossils are early to middle Miocene ichthyoliths (Hart and Mountain, 1987) and early Miocene radiolarians (Nishimura, 1987; site was still below the CCD). No lower Miocene sediments were sampled at DSDP Site

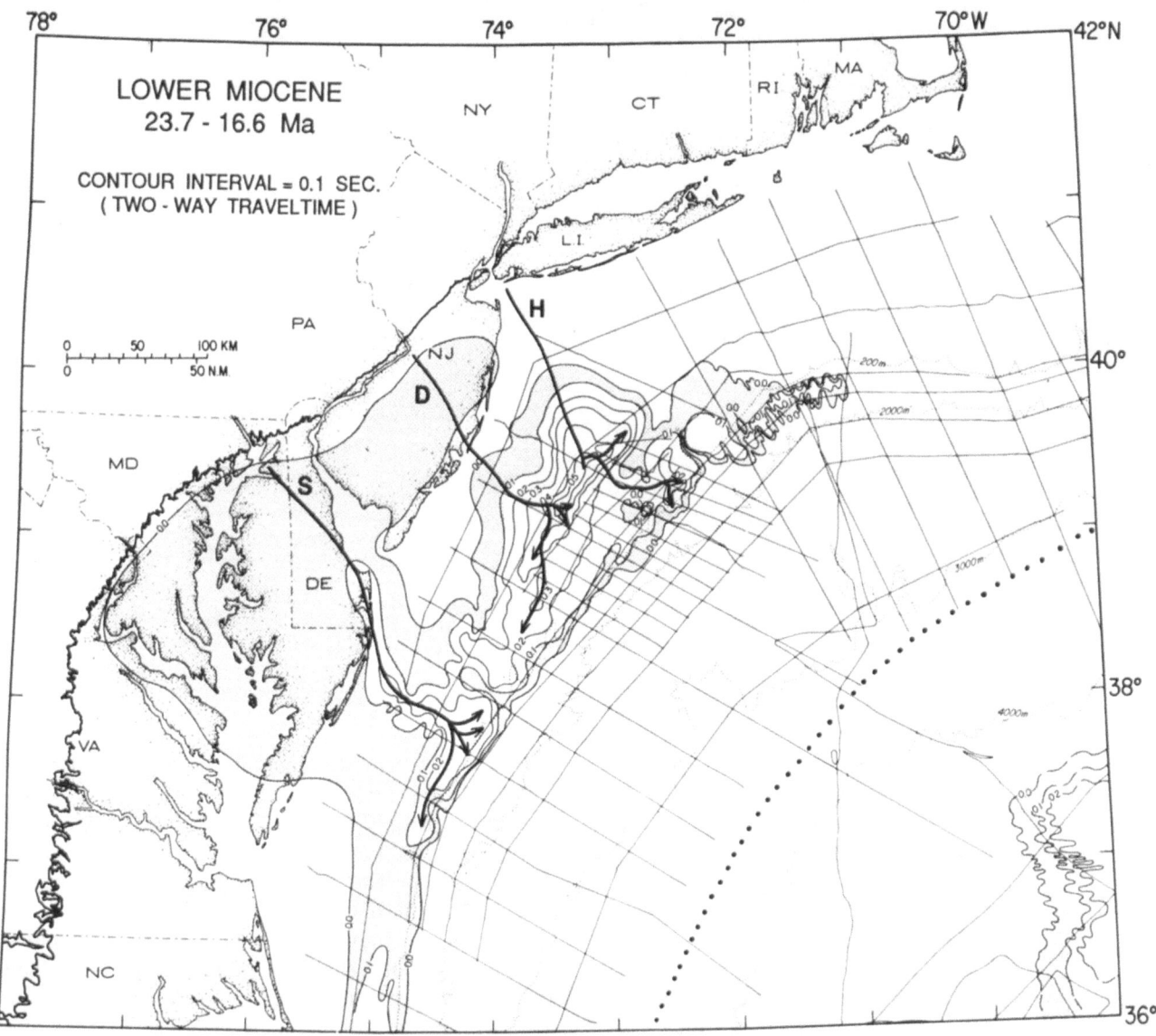

FIGURE 6-32.   Isochron map (contours in 0.1 sec of two-way traveltime) of lower Miocene allostratigraphic unit. Arrows indicate inferred primary routes of sediment dispersal. Shaded contour interval (0.0–0.1 sec) indicates thinnest parts of unit. Heavy dotted line indicates possible axis of eroding bottom current. S = ancient Susquehanna River; D = ancient Delaware River; H = ancient Hudson River.

105, because the appropriate interval was drilled instead of cored.

Middle Miocene Allostratigraphic Unit

A major pulse of tectonic uplift in all highland source terrains (but especially evident in the northern part of the central Appalachians), coupled with increasingly more temperate paleoclimates and the continued fall of long-term sea level, brought a massive infusion of siliciclastic detritus into the Baltimore Canyon Trough and Hatteras Basin during the middle Miocene (16.6–11.2 Ma; Poag and Sevon, 1989). The siliciclastic accumulation rate soared from 3,000 to 26,000 km^3/m.y. (Fig. 6-5; see "Methods and Problems of Analysis" for discussion of rate adjustments to compensate for undercompaction of middle Miocene to Quaternary sediments). An unusually large outer shelf depocenter (1.0–1.3 km thick; 1.0–1.3 sec) developed seaward of Maryland, Delaware, and southern New Jersey, and brought an end to the double–shelf-edge physiography that characterized most of the earlier Cenozoic (Fig. 6-33). This confluent delta dumped huge volumes of siliciclastic sediment onto the continental

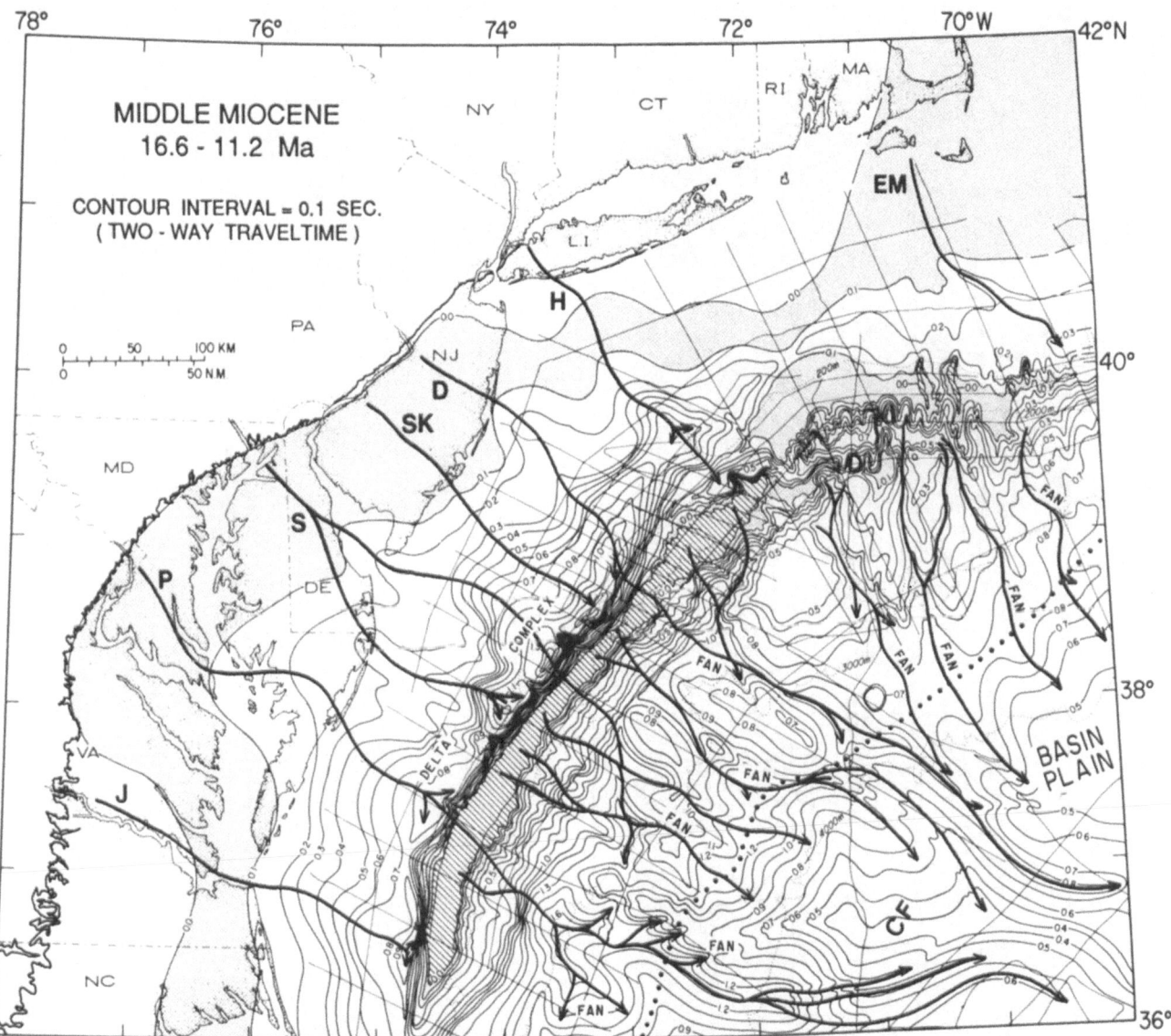

FIGURE 6-33.   Isochron map (contours in 0.1 sec of two-way traveltime) of middle Miocene allostratigraphic unit. Arrows indicate inferred primary routes of sediment dispersal. Diagonal hachures indicate unit absent or too thin to resolve on seismic profiles. Shaded contour interval (0.0–0.1 sec) indicates thinnest parts of unit. Heavy dotted line indicates approximate axis of depositing contour current. J = ancient James River; P = ancient Potomac River; S = ancient Susquehanna River; SK = ancient Schuylkill River; D = ancient Delaware River; H = ancient Hudson River; EM = ancient river(s) in eastern Massachusetts; CF = confluent fan.

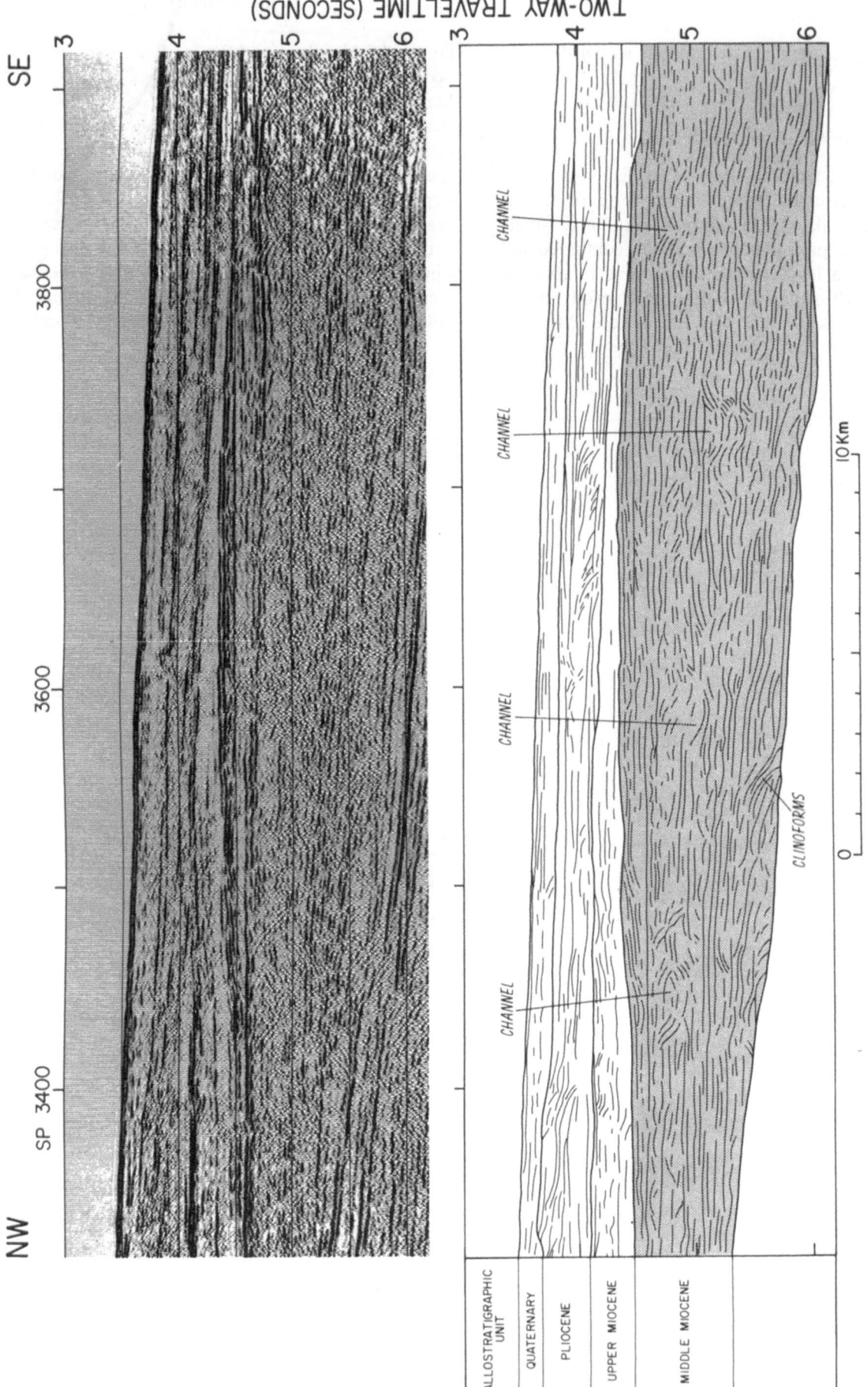

FIGURE 6-34. Stratigraphic section parallel to depositional dip along segment of seismic profile 28 crossing upper part of the continental rise seaward of Cape Charles, Virginia (see Fig. 6-2 for location; compare with Fig. 6-3). Above is uninterpreted seismic profile; below is interpreted tracing of part of seismic profile. Thick, rapidly deposited, shallowly buried middle Miocene sediments display relatively undisturbed depositional features better than most other continental-rise units. For example, note clinoform reflections near base of unit (= edge of submarine fan system) and four depressions bordered by mounded reflections (= leveed channels of other submarine fan systems).

slope and rise (Figs. 6-3, 6-14; Locker and Laine, in press; see also Chapter 10). In fact, some upper rise depocenters were thicker than 1.6 km (1.6 sec; e.g., Delmarva depocenter of McMaster, Locker, and Laine, 1989), and several small, medium, and large submarine fan complexes spread thick turbidite successions across the middle and lower rise. Five fans converged to form a confluent megafan in the southeast corner of the study area. Much of this sediment appears to be of early middle Miocene age (planktonic foraminiferal zones N8 and N9; Olsson, Melillo, and Schreiber, 1987; Poag, unpublished data); thus, the seaward shift of depocenters may have been promoted by two major lowstands postulated for the early middle Miocene by the Exxon model (Haq, Hardenbol, and Vail, 1987).

Rapid deposition and shallow burial of the middle Miocene unit have preserved many depositional and erosional features that were obscured in thinner or more deeply buried units (see also Locker and Laine, in press). For example, prograding clinoforms and leveed channel systems typical of modern turbidite fans are visible within middle Miocene strata on several seismic profiles (Figs. 6-14, 6-34, 6-35).

Distal extremities of the confluent megafan were sampled at DSDP Sites 603, 106, and 388, where they constitute part of the Blake Ridge Formation. At Site 603, Haggerty et al. (1987) found numerous silty and sandy turbidites in 325 m of silty, micaceous claystones, which they interpreted to have accumulated in the absence of significant bottom currents. At Site 106, Hollister, Ewing, et al. (1972) estimated that the middle Miocene section was ~ 300 m thick. Sparse coring there (only 14 m recovered) revealed faintly bedded hemipelagic mud, which lacked obvious turbidites. At Site 388, coring penetrated 29 m into the top of the middle Miocene section and recovered greenish-gray clay and silty clay containing quartz and heavy minerals.

Tucholke and Mountain (1979, 1986), Tucholke and Laine (1982), Mountain and Tucholke (1985), McMaster, Locker, and Laine (1989), and Locker and Laine (in press) have described distinctive seismic-reflection features (hummocky, shingled, and sediment-wave seismic facies, plus widespread erosional truncation of reflections) related to bottom-current activity. They conclude that during the late part of the middle Miocene (see Fig. 6-16), geostrophic bottom currents began to redistribute gravity-flow sediments of the middle and lower continental rise into huge contourite drifts (e.g., Hatteras Outer Ridge, whose main buildup is just beyond the southeast corner of the study area). These drifts formed linear seafloor elevations whose long axes are normal to the downslope direction of gravity flows. The pervasive downslope-oriented isochron pattern of the middle Miocene allostratigraphic unit, however, is evidence that sediment gravity

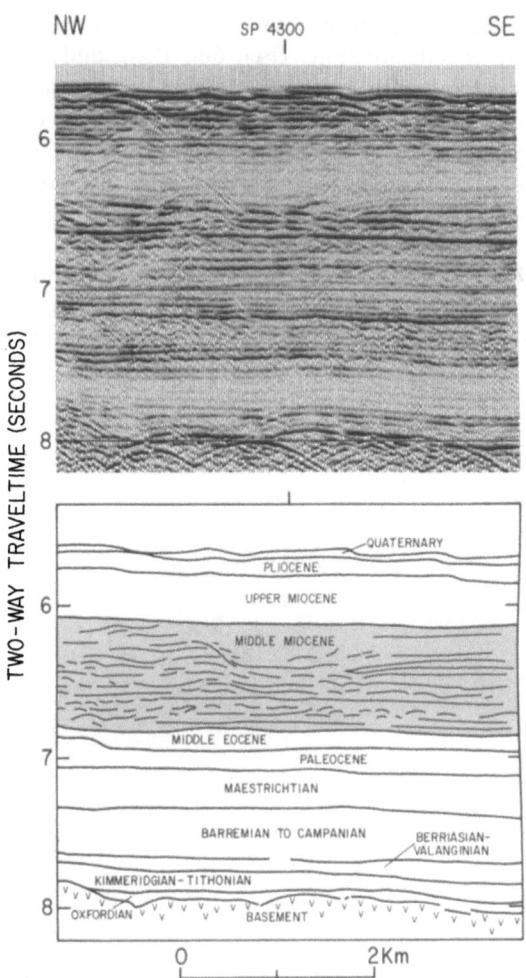

FIGURE 6-35. Stratigraphic section parallel to depositional dip along seismic profile 6 crossing lower part of the continental rise seaward of Cape May, New Jersey (see Fig. 6-2 for location). Above is uninterpreted seismic section; below is interpreted tracing of seismic profile showing allostratigraphic units overlying oceanic basement. Internal reflections of middle Miocene unit shown (shaded) to illustrate well-preserved leveed channel within submarine fan complex in upper part of unit.

flows still dominated deposition over most of the study area. There is, however, a possible indication of current-controlled deposition along the middle part of the continental rise (see also Chapters 5, 9). There, two lobate depocenters [> 800 and > 1,200 m thick (> 0.8 and > 1.2 sec); approximately equivalent to the southern New England and Wilmington depocenters, respectively, of McMaster, Locker, and Laine, 1989] are located within the fan complexes. Long axes of both depocenters are oriented subparallel to the shelf edge, as if aligned with a western boundary undercurrent.

Upper Miocene Allostratigraphic Unit

During the late Miocene (11.2–5.3 Ma), despite a postulated major lowstand and renewed long-term eustatic decline (Haq, Hardenbol, and Vail, 1987), the siliciclastic accumulation rate decreased by nearly two-thirds to 9,000 km³/m.y. (Fig. 6-5; Poag and Sevon, 1989; contrast with Chapters 5, 7). Continental-shelf depocenters were restricted to a narrow, seaward-thickening wedge at the shelf edge (Fig. 6-36), which has been subsequently subdivided into discrete segments by incision of submarine canyons. A broad swath (~ 20–25 km wide) on the continental slope, which widens (to ~ 50 km) to the north, has been stripped of upper Miocene sediments (Figs. 6-13, 6-20, 6-23). Thick debris aprons and submarine fan complexes (500–

> 800 m thick) developed along the margin from Virginia to New Jersey (Figs. 6-3, 6-14, 6-34, 6-35). The upper and middle continental rise off New Jersey are crossed by numerous downslope-trending channels filled with chaotic upper Miocene seismic facies (Poag, 1987). At DSDP Site 604, the upper part of such a channel-fill yielded coarse conglomeratic sands containing large quartz pebbles, igneous and metamorphic lithoclasts, and white chunks of reworked Eocene chalk (Van Hinte, Wise, et al., 1987). Site 613, located nearby on an interchannel ridge, cored fine, glauconitic,

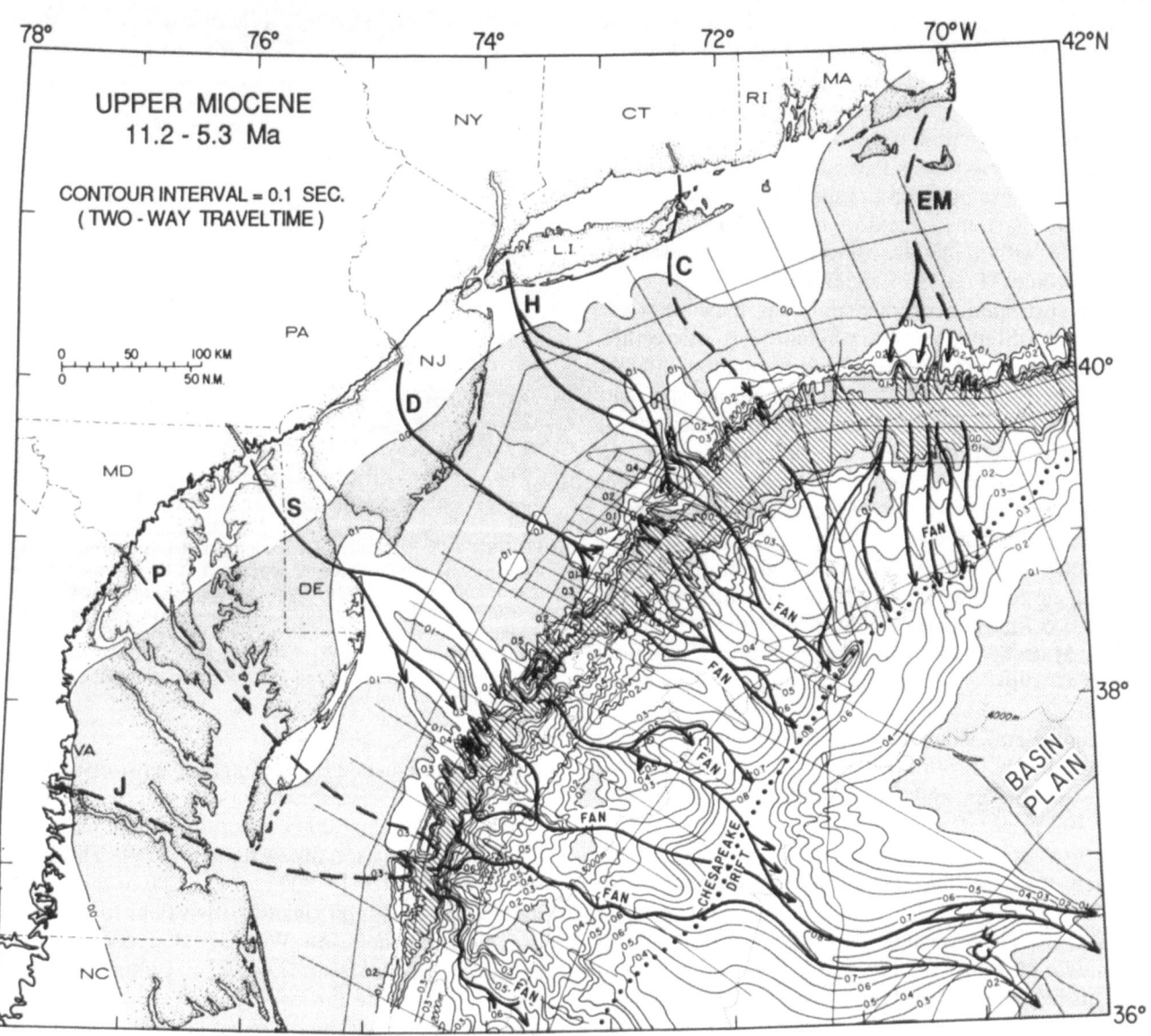

FIGURE 6-36.   Isochron map (contours in 0.1 sec of two-way traveltime) of upper Miocene allostratigraphic unit. Solid arrows indicate inferred primary routes of sediment dispersal; dashed arrows indicate secondary routes. Diagonal hachures indicate unit absent or too thin to resolve on seismic profiles. Heavy dotted line indicates inferred axis of depositing contour current. Shaded contour interval (0.0–0.1 sec) indicates thinnest parts of unit. J = ancient James River; P = ancient Potomac River; S = ancient Susquehanna River; D = ancient Delaware River; H = ancient Hudson River; C = ancient Connecticut River; EM = ancient river(s) in eastern Massachusetts; CF = confluent fan.

quartzose sand and conglomeratic sand (Poag, Watts, et al., 1987).

One small fan and two megafans merged on the middle continental rise to form a confluent megafan, which extends into the southeast corner of the study area (Fig. 6-36). At Site 603, distal sediments from this confluent megafan comprise 311 m of dark–greenish-gray, quartz-bearing, micaceous, sideritic, silty clay-stones of the Blake Ridge Formation (Haggerty et al., 1987). Sporadic preservation of nannofossils in this interval indicates a fluctuating CCD, which reduced the carbonate content of the upper Miocene section. At Site 105, an 80–100-m section of presumably similar sediments (only one 9-m core was recovered) may include upper Miocene beds. At Site 106, upper Miocene beds may be 160 m thick, but only one 9-m core was taken there, also; it contained dark–greenish-gray hemipelagic mud (Hollister, Ewing, et al., 1972). Four cores taken in the ~ 100-m upper Miocene section at nearby Site 388 recovered similar sediment.

During the late Miocene, a second large contourite drift, known as the Chesapeake Drift (Mountain and Tucholke, 1985; Tucholke and Mountain, 1986; McMaster, Locker, and Laine, 1989, in press; see also Chapters 5, 9), was a prominent feature of the middle continental rise. One axis of its crescentic depocenter (> 800 m thick; > 0.8 sec) parallels the shelf edge and indicates the presumed principal flow direction of its associated bottom current. The current flow did not, however, obscure the dominant pattern of downslope gravity-flow deposition.

Pliocene Allostratigraphic Unit

The siliciclastic accumulation rate in the study area nearly doubled in the Pliocene (5.3–1.6 Ma) to reach 15,000 km^3/m.y. (Fig. 6.5; Poag and Sevon, 1989), perhaps in response to intensification of northern continental glaciation (Miller, Fairbanks, and Mountain, 1987) and the continued fall of long-term sea level (Haq, Hardenbol, and Vail, 1987). Pliocene shelf deposits are thin (< 100 m; < 0.1 sec), however, throughout the region, except in a narrow shelf-edge prism seaward of Long Island and northern New Jersey (Fig. 6-37). Like its late Miocene counterpart, this prism has been subsequently dissected by numerous submarine canyons (e.g., Fig. 6-23), and Pliocene sediments have been stripped from a broad swath of the continental slope (Fig. 6-13).

Thickest depocenters (700–> 900 m thick; 0.7–0.9 sec) developed as debris aprons seaward of Virginia (Figs. 6-3, 6-14). Small, elongate submarine fans extended from these aprons across the upper and middle rise to form a confluent fan (Norfolk–Washington Fan

of Locker and Laine, in press). Gravity-flow deposits from the New England margin also contributed significantly to development of the Pliocene continental rise (Fig. 6-37) by means of three medium fan complexes, which stretched toward the southeastern corner of the study area (Fig. 6-16; see also Chapter 9).

Distal sediments from the Virginia and New England fan complexes debouched into a broad basin plain and formed a large sediment pond (~ 300 m thick; 0.3 sec) along the northwestern flank of the Hatteras Outer Ridge (southeast corner of Fig. 6-37; see also Tucholke and Laine, 1982; Mountain and Tucholke, 1985; Locker and Laine, in press). At DSDP Site 603, at least 200 m of Pliocene dark–grayish-green, nannofossil-bearing clay and claystone was cored from the upper part of the Blake Ridge Formation. At Site 106, ~ 120 m of inferred Pliocene section was sampled by two 9-m cores, whose constituents are similar to those at Site 603. At Site 105, similar sediments also were present in a 9-m core taken at the base of an estimated 60-m-thick Pliocene section. This unit was not cored at Site 388.

Though the Chesapeake Drift continued to build into the Pliocene (Mountain and Tucholke, 1985; Locker and Laine, in press), there is no clear expression of it on the isochron map. In fact, the Pliocene sediments thin over the late Miocene crest of the drift, whereas a thick turbidite section fills the topographic low between the drift and the elevated Hatteras Outer Ridge. However, the northeast–southwest orientation of a > 100-m-thick (> 0.1-sec) sediment lens, crossing the basin plain in the southeast corner of the study area, may be evidence of bottom-current activity. Additional evidence is the southwestward diversion of fan complexes associated with the Hudson and Delaware dispersal systems along the middle continental rise, though such diversion may be controlled in part by antecedent topography (Locker and Laine, in press).

Quaternary Allostratigraphic Unit

During the Quaternary (1.6 Ma to present), siliciclastic accumulation remained high (17,000 km^3/m.y.; Fig. 6-5; Poag and Sevon, 1989; see also Chapters 3–5, 7, 9, 10, 12, 13). Much of the sediment was undoubtedly derived from continental ice sheets, which scoured the source terrains. A narrow, seaward-thickening, sedimentary wedge (500–600 m thick; 0.5–0.6 sec) developed at the shelf edge and was areally more extensive than its Pliocene predecessor (Fig. 6-38). This wedge too, has been notably dissected by submarine canyons and a host of related downslope channel systems (Figs. 6-3, 6-14, 6-20, 6-23; see Robb, 1980; Robb et al., 1981; Kirby, Robb, and Hampson, 1982; McGregor et al., 1982; Twichell and Roberts, 1982; Hampson and Robb, 1984; Farre and Ryan, 1987; O'Leary, 1988; Danforth

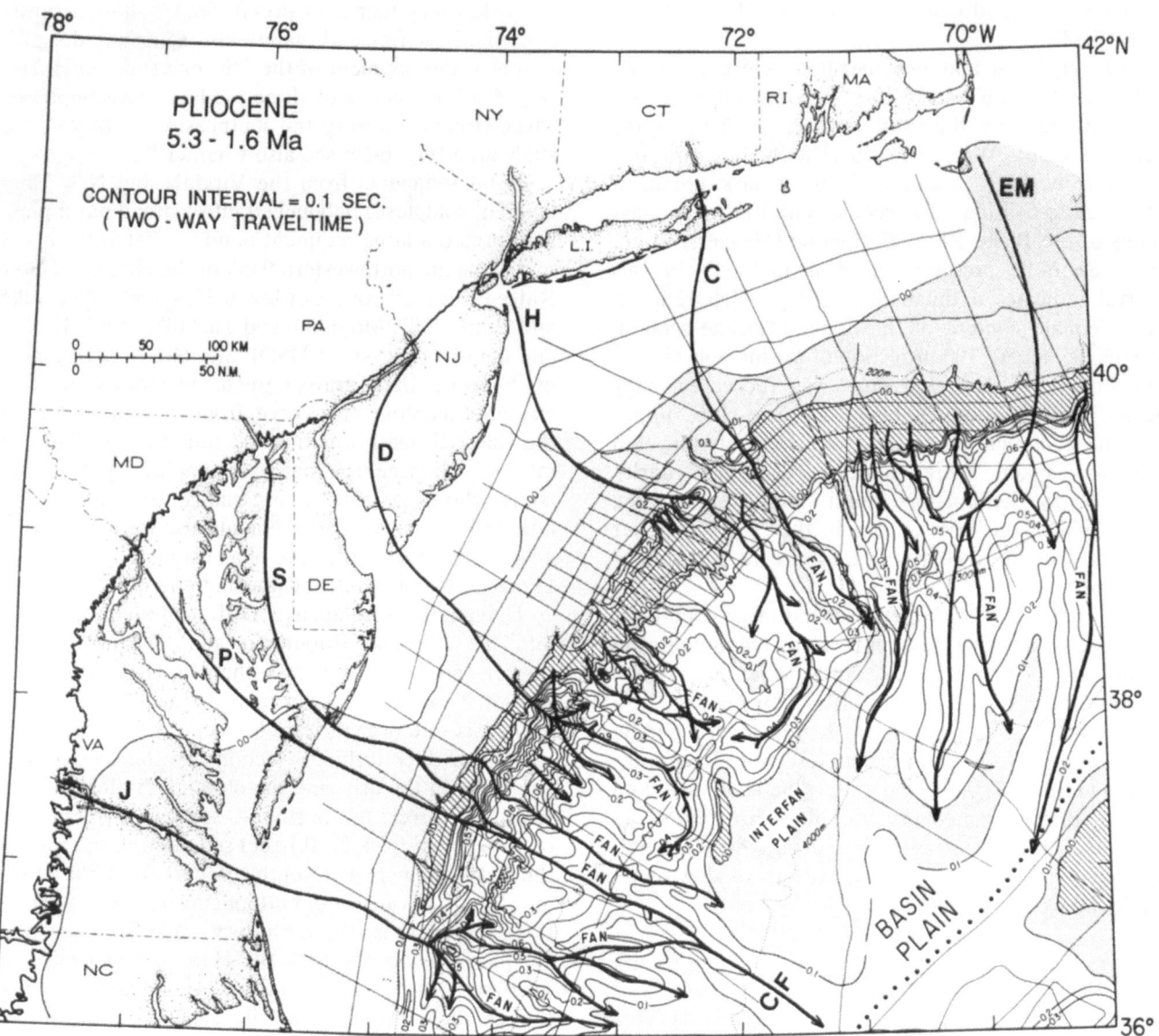

FIGURE 6-37.   Isochron map (contours in 0.1 sec of two-way traveltime) of Pliocene allostratigraphic unit. Arrows indicate inferred primary routes of sediment dispersal. Diagonal hachures indicate unit absent or too thin to resolve on seismic profiles. Heavy dotted line indicates inferred axis of depositing contour current. Shaded contour interval (0.0–0.1 sec) indicates thinnest parts of unit. J = ancient James River; P = ancient Potomac River; S = ancient Susquehanna River; D = ancient Delaware River; H = ancient Hudson River; C = ancient Connecticut River; EM = ancient river(s) in eastern Massachusetts; CF = confluent fan.

and Schwab, 1990; see also Chapter 9). Some canyons and channels have been filled and buried with as much as 700 m (0.7 sec) of Quaternary siliciclastic sediment.

The Quaternary sequence is missing (or is quite thin; < 100 m thick; < 0.1 sec) along much of the lower continental slope from southern New Jersey to New England, where white Eocene chalks are exposed at the seafloor (Robb et al., 1983; Farre and Ryan, 1985; see also Chapter 9). Slope and upper-rise debris aprons as thick as 600–800 m (0.6–0.8 sec) developed between Virginia and Delaware and seaward of New England, but coeval debris aprons seaward of New Jersey and Long Island are notably thinner (300–400 m

thick; 0.3–0.4 sec). At DSDP Sites 604, 605, and 613 (drilled on the upper rise seaward of New Jersey), dark, greenish-gray, homogeneous, gassy, organic-rich, commonly diatomaceous mud, interbeds of quartzose, glauconitic sand, occasional conglomeratic beds, chunks of white Eocene chalk, and displaced shelf biota are characteristic of Quaternary deposits.

Several prominent submarine fan complexes developed on the upper and middle continental rise (Figs. 6-3, 6-13, 6-14, 6-16, 6-20, 6-23, 6-25), and the most prominent derived sediment from the Virginia to Long Island margin; several of these coalesced as they crossed the middle rise. Two megafans related to the

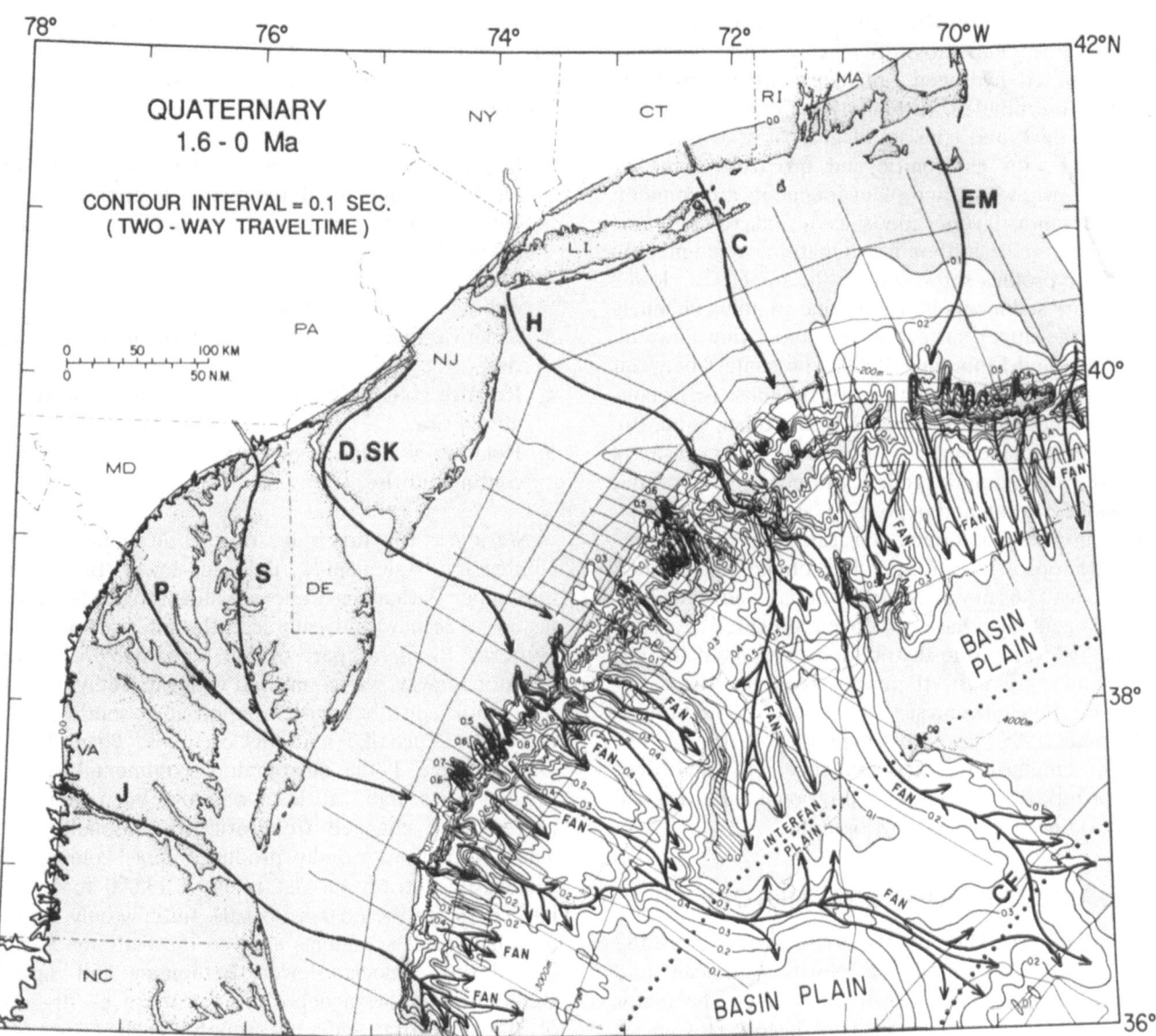

FIGURE 6-38. Isochron map (contours in 0.1 sec of two-way traveltime) of Quaternary allostratigraphic unit. Arrows indicate inferred primary routes of sediment dispersal. Diagonal hachures indicate unit absent or too thin to resolve on seismic profiles. Heavy dotted line indicates inferred axis of depositing contour current. Shaded contour interval (0.0–0.1 sec) indicates thinnest parts of unit. J = ancient James River; P = ancient Potomac River; S = ancient Susquehanna River; D = ancient Delaware River; SK = ancient Schuylkill River; H = ancient Hudson River; C = ancient Connecticut River; EM = ancient river(s) in eastern Massachusetts; CF = confluent fan.

Hudson–Connecticut and Delaware–Schuylkill–Susquehanna–Potomac dispersal systems converged to form a huge confluent megafan in the southeast corner of the study area. Distal distributaries of this confluent megafan continued to pond turbidites behind the Hatteras Outer Ridge and to bury the ridge crest. Quaternary sediments, like those of the Pliocene, are thin over the crest of the buried Chesapeake Drift. Possible evidence of continued bottom-current activity comes from the northeast–southwest axial trend of a 300-m-thick (0.3 sec) sediment lens on the lower rise, and from a similar orientation of the interfan plain. In addition, the Connecticut, Hudson, and Delaware–

Schuylkill fan systems appear to have been diverted toward the southwest by bottom currents and(or) antecedent drift topography (Locker and Laine, in press; see also Chapter 9).

At DSDP Site 388, ~ 54 m of turbiditic Quaternary sand, silty clay, and clay of the Blake Ridge Formation contain rounded (shallow-water origin) quartz grains, heavy-mineral suites, and reworked Cretaceous, Eocene, and Miocene nannofossils. At Site 105, the Quaternary section is ~ 90 m thick (stratigraphy revised herein on the basis of seismic correlations with Site 603) and comprises similar sediments. At Site 603, ~ 31 m of lower Quaternary strata are present seaward of the Hatteras Outer Ridge. The sediments are mainly greenish-gray, nannofossil-rich clay and claystone, which emitted hydrogen sulfide gas when cored. At Site 106, the ponded, turbiditic Quaternary section is ~ 360 m thick and consists of gray to brown terrigenous mud with glauconitic and quartzose sand interbeds. Mica, wood, and plant fragments are common in some sandy layers, and siliceous microfossils are especially notable in the lower Quaternary sediments.

Seismic profiles show that on the upper rise, lower Quaternary sediments fill downslope-trending channels cut into the upper surface of the underlying Pliocene unit (Poag and Mountain, 1987). The contact between the lower and upper Quaternary subunits, is, in contrast, a rather smooth surface. On the other hand, the upper Quaternary surface (modern seafloor) displays marked physiographic relief, caused by differential downslope erosion and deposition (see also Chapter 9).

Because the Quaternary allostratigraphic unit represents the top of the stratigraphic column, and formed in the last 1.6 m.y., its depositional and erosional features generally have not been obscured by later events. The composite distribution patterns of the entire Quaternary unit, therefore, resemble those observed on the Holocene seafloor (Locker, 1989; Pratson and Laine, 1989; see also Chapters 8, 9, 10, 12, 13), thereby emphasizing the influence of antecedent seafloor topography (buried depositional and erosional features) on the present physiography.

SUMMARY OF DEPOSITIONAL HISTORY

Deposition in the Hatteras Basin commenced at ~ 187 Ma [Aalenian(?)] when the North American and African tectonic plates began to diverge. The major events of subsequent depositional history (Fig. 6-39;

Table 6-1), as described herein, can be used to divide the continental rise evolution into eight distinct stages. The main sedimentary aspects used to recognize stage boundaries are:

1. Variation of siliciclastic accumulation rate (rate varied in response to source-terrain tectonism, paleoclimate change, and eustatic change; Poag and Sevon, 1989).
2. Presence or absence of large carbonate platform on the outer continental shelf.
3. Relative thickness of depocenters on continental rise.
4. Relative volume of siliciclastics reaching the continental rise.
5. Relative abundance of pelagic carbonate on the continental rise.

Stage 1. The first stage of rise development [encompassing Aalenian(?), Bajocian–lower Bathonian, and upper Bathonian–Callovian allostratigraphic units] lasted ~ 24 m.y. and embraced the interval when the Hatteras Basin, as part of the long, narrow, proto-Atlantic seaway, was connected with the Tethys Ocean to the northeast, but was closed off at its southwestern extremity (Tucholke and McCoy, 1986; Poag, 1991). During stage 1 the surrounding continental source terrains were high, sea level was low, but rising, and paleoclimate changed from arid to seasonally wet. These conditions rapidly produced huge volumes of siliciclastic detritus (at net rates of 13,000 to 18,000 km^3/m.y.) for the offshore basins. Initially only a little over half these sediments entered the Hatteras Basin, but during Bajocian–lower Bathonian and upper Bathonian–Callovian deposition, as much as 70–84% of the terrestrial sediment supply reached the rise

FIGURE 6-39. Summary chart of allostratigraphic units of U.S. middle Atlantic continental rise (northern part of Hatteras Basin) and their characteristic sedimentary features. Time scale from Palmer (1983). Column 1 = stages in sedimentary evolution of the continental rise. Column 2 = approximate length and width of area covered by rise deposits, as calculated from isochron maps. Shaded envelope emphasizes temporal changes; "patchy" indicates areas of no (or very thin) sediment. Column 3 = maximum thickness of allostratigraphic units on continental rise. Shaded intervals indicate thickness exceeds 1,000 m. Column 4 = principal source terrains for siliciclastic sediments as inferred from isochron maps. Solid dots indicate principal highland source terrains; solid line traces migration of northernmost principal highland source terrain through time. C = central Appalachian highlands; A = Adirondack highlands; N = New England Appalachian highlands; Crosses indicate secondary lowland source terrains within the basin complex (SH = continental shelf; SL = continental slope). Column 5 = primary dispersal routes for siliciclastic sediments as inferred from isochron maps. J = ancient James River; P = ancient Potomac River; S = ancient Susquehanna River; SK = ancient Schuylkill River; D = ancient Delaware River; H = ancient Hudson River; C = ancient Connecticut River; M = ancient river(s) in eastern Massachusetts. Shaded envelope emphasizes temporal migration of southernmost (left sawtooth line) and northernmost (right sawtooth line) dispersal routes through time. Column 6 = approximate volumetric rates of siliciclastic sediment accumulation as calculated from isochron maps. Shaded intervals indicate rate exceeds mean postrift value of 9,000 km^3/m.y. Column 7 = percent (by volume) of siliciclastic sediments that reached continental rise. Shaded intervals indicate values exceed 50%. Column 8 = number and type of deltas on continental shelf. S = single delta; C = confluent delta; SE = shelf-edge delta; number in parentheses indicates how many individual deltas compose confluent delta. Column 9 = relative abundance of submarine canyons (V-shaped cross section) and submarine channels (U-shaped cross section) on continental slope. A = abundant; C = common; F = few; ? = difficult to ascertain presence or absence. Column 10 = number and types (as defined herein) of submarine fan complexes. S = small; M = medium; L = large; ME = megafan; C = confluent fan; SI = siliciclastic sediments; CA = carbonate sediments. Column 11 = evidence of and type of bottom currents as inferred from isochron patterns or published records. E = dominantly eroding; C = dominantly redistributing sediments to form contourite drifts. Column 12 = dominant lithology of lower rise sediments as derived from DSDP coreholes. LS = limestone; CLS = claystone; CLY = clay; ? = not sampled or age uncertain; blk = black; biosilica = mainly radiolarians in planktonic assemblage. Column 13 = name of deep-sea formation (from Jansa et al., 1979; Van Hinte, Wise, et al., 1987). Column 14 = elevation of carbonate compensation depth (CCD) relative to seafloor. A = CCD above seafloor (carbonate dissolved); B = CCD below seafloor (carbonate preserved). Column 15 = temporal changes in sea level (from Haq, Hardenbol, and Vail, 1987). High-frequency curve = short-term sea level; smoother curve = long-term sea level; datum is present sea level.

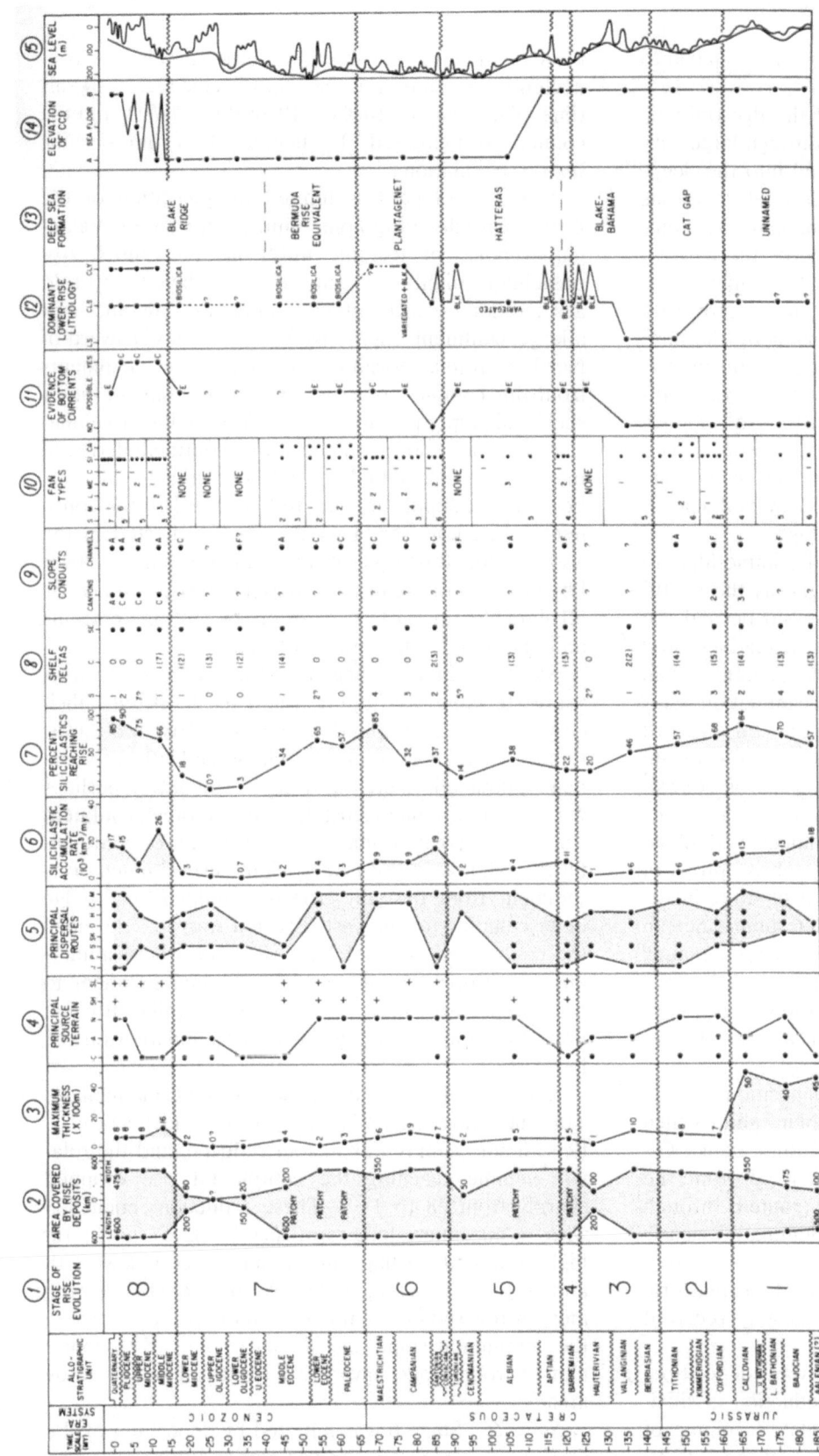

(Table 6-1). Rise depocenters were as thick as 4,000 to 5,000 m.

As these sediments crossed the shelf edge (initially through shelf-edge deltas, and later through large submarine canyons), they were distributed into the deepest parts of the Hatteras Basin to form debris aprons and submarine fan complexes. Dominant fan complexes were 100–200 km long (limited by the narrowness of the Hatteras Basin), and four to six major fans are notable in each allostratigraphic unit. Carbonate sediments formed small debris aprons along the base of the Bahama–Grand Banks Gigaplatform during the Bajocian, Bathonian, and Callovian, but most rise sediment is inferred to have been siliciclastic. No evidence of significant redistribution by bottom currents is seen in the isochron patterns of stage 1, nor would one expect strong bottom currents in the confined seaway that characterized this evolutionary stage.

Stage 2. Stage 2 (Oxfordian and Kimmeridgian–Tithonian allostratigraphic units) began as the southeast margin of the Hatteras Basin moved beyond the study area (due to continued seafloor spreading), and submarine fan complexes reached lengths greater than 400 km (= megafans). Siliciclastic accumulation rates declined substantially, and the maximum thickness of primary depocenters decreased fivefold. Carbonate production on the gigaplatform increased as source-terrain tectonism diminished and sea level reached its Jurassic peak. As a result, greater volumes of shallow-water carbonate detritus collected in debris aprons of the slope and upper rise and also probably were incorporated into proximal parts of smaller submarine fan complexes. Siliciclastic sediment, still derived mainly from the central Appalachian highlands, probably was dominant in larger submarine fans, however. Approximately 57–68% of available siliciclastic sediment reached the continental rise during stage 2.

Stage 3. Stage 3 began with deposition of the Berriasian–Valanginian allostratigraphic unit, which followed drowning of the Bahama–Grand Banks Gigaplatform. Siliciclastic sediments, derived from the central Appalachian highlands and routed through large confluent shelf-edge deltas, dominated nearly everywhere (46% reached the rise). Exceptions were a few debris aprons at the base of the dead gigaplatform and the broad basin plain, which probably received hemipelagic sediments.

Stage 3 terminated with Hauterivian deposition, when the basin underwent a sharp reduction in siliciclastic sediment supply and in the relative volume of siliciclastics reaching the rise. Only 20% of the terrigenous sediment reached the Hatteras Basin, where it resided mainly in debris aprons of the slope and upper rise. These reductions accompanied a long-term increase in relative sea level.

Vigorous bottom currents may have contributed to the absence (or thinness) of Hauterivian sediments on most of the rise. The proto-Atlantic seaway was connected to the Gulf of Mexico and Pacific Ocean by this time (Tucholke and McCoy, 1986; Poag, 1991), and the connection increased the likelihood of interoceanic bottom circulation.

Stage 4. Stage 4 is limited to deposition of the Barremian allostratigraphic unit. A Berriasian–Valanginian pulse of tectonic uplift in the central Appalachian highlands caused a delayed, 10-fold acceleration of siliciclastic accumulation rate, and built a large, confluent, shelf-edge delta complex seaward of North Carolina, Virginia, and Delaware. Sediments from this complex, plus debris eroded from the outer shelf and slope, produced the oldest confluent submarine megafan known on the lower continental rise of the Hatteras Basin. Only ~ 22% of available siliciclastic sediment supply appears to have reached the continental rise, however. This sparsity is reflected in the absence (or thinness) of Barremian sediments over large patches of the middle continental rise and may be attributable, in part, to erosion by bottom currents and(or) to high long-term sea level.

Stage 5. Stage 5 began in the Aptian–Albian as source terrains were uplifted and sea level fell. Siliciclastic sediment supply declined by almost two-thirds, though the relative volume reaching the Hatteras Basin nearly doubled (increase from 22 to 38%), perhaps because of two major eustatic falls. Another confluent submarine megafan built out onto the lower continental rise, and, as in the Hauterivian, derived much of its sediment from the still growing confluent shelf-edge delta complex to the west. Erosion was notable along the continental slope of New England, where numerous downslope channel systems supplied sediment to small upper rise fans. The CCD rose significantly and prevented appreciable carbonate accumulation on the lower rise.

Stage 5 ended with deposition of the Cenomanian–Turonian allostratigraphic unit, during which siliciclastic sediment supply was further reduced, and the relative volume reaching the continental rise dropped sharply from 38 to 14%. These reductions coincided with a maximum long-term sea level, a short-term highstand, and, perhaps, maximum Cretaceous warmth and wetness (Barron, 1989). As in the Hauterivian, most of this sediment ended up in debris aprons along the continental slope; no submarine fans of Cenomanian–Turonian age have been identified in the study area.

Stage 6. Stage 6 was initiated by deposition of the Coniacian–Santonian allostratigraphic unit, when the siliciclastic accumulation rate increased 10-fold, in delayed response to source-terrain uplift. A notable shift of major shelf deltas to the northeast followed renewed tectonic uplift in the New England Appalachian highlands. Shelf-edge deltas and numerous submarine channels supplied a particularly large, north–south oriented submarine megafan, which joined a large fan coming from the Virginia margin to form a broad ($\sim$300-km-wide) confluent megafan on the lower continental rise. This megafan, along with several smaller ones, nearly tripled the relative sediment volume reaching the rise (increase from 14 to 37%).

Though siliciclastic sediment supply was reduced approximately by half during the Campanian and Maestrichtian (from 19,000 to 9,000 km^3/m.y.), the New England Appalachian highlands remained the dominant source terrain. The relative volume of sediments retained on the shelf also diminished; during the Maestrichtian, 85% of the siliciclastic sediment reached the continental rise. Large confluent submarine fans, supplied from the heavily incised New England margin, dominated sediment distribution on the continental rise throughout stage 6. The dearth of Maestrichtian sediments on the shelf and their absence from large patches seaward of Virginia and Delaware might be an indication that shelf deposition was significantly interrupted by impact-wave phenomena, believed by some to have accompanied an end-of-Cretaceous bolide impact (e.g., Van Hinte et al., 1985).

A large scalloped lens of sediment on the middle continental rise, whose long axis is normal to the dominant downslope gravity-flow direction, may indicate that the oldest contourite drift in the Hatteras Basin developed during the Maestrichtian.

Stage 7. Stage 7 (Paleocene to early Miocene) was characterized by unusually low rates of siliciclastic accumulation (700–4,000 km^3/m.y.) and thin depocenters (maximum thickness = 100–400 m), conditions which persisted for $\sim$49 m.y. During the latter half of stage 7, only 3–18% of the siliciclastic volume reached the continental rise. Initially, long-term sea level was high, source terrains appear to have been quiescent and topographically low, and a trophical rain forest occupied the coastal region. Consequently, major depocenters shifted to the continental shelf.

Though a magmatic event in the middle of stage 7 (middle Eocene) signals an uplift in the central Appalachians, the dense vegetation of the tropical rain forest appears to have severely limited transport of siliciclastic sediment to the offshore basins. The result of these developments was construction of an actively depositing "hinter" shelf, shoreward of an inactive, morphologically subdued "fore" shelf. A generally progressive decrease in the relative volume of siliciclastic sediment reaching the continental rise and a high CCD (which caused pelagic carbonate sediments to dissolve) contributed to patchy sediment distribution in the Hatteras Basin. The number and size of submarine fans also progressively decreased during stage 7; from the late Eocene to early Miocene, for example, no fans appear to have developed.

Terrigenous sediments of possible early Miocene age are present on the lower continental rise, but are separated from the shelf edge by a 300-km-wide swath barren of appreciable early Miocene deposits. The presence of this swath may be evidence that bottom currents were active in the Hatteras Basin near the end of stage 7.

Stage 8. Stage 8 was initiated dramatically in the middle Miocene when tectonic uplift of the central Appalachian highlands, accompanied by paleoclimatic change (Miller, Fairbanks, and Mountain, 1987; Poag and Sevon, 1989), increased the rate of siliciclastic sediment supply nearly 10-fold. All the major ancient river systems joined to built a huge 500-km-wide confluent delta at the shelf edge. In addition, the continental slope was heavily eroded, and as a result, 66% of the available siliciclastic sediment collected in huge depocenters (maximum thickness = > 1,600 m) on the rise. A confluent submarine megafan, supplied by the shelf-edge deltas and fed through numerous submarine canyons and channels, formed on the lower rise.

After a significant decline in deposition during the late Miocene, siliciclastic accumulation rates again reached high values (15,000 to 17,000 km^3/m.y.) in the Pliocene and Quaternary. Shelf sedimentation was reduced progressively throughout stage 8, until in the Quaternary, 95% of siliciclastic sediment was dumped onto the continental rise. Confluent submarine fans continued to form in the Quaternary, supplied mainly by submarine canyons, which derived sediment from the incised outer shelf and slope. The large Quaternary sediment volume was supplied by continental ice sheets that scoured the northern and northwestern margins of the study area and by subaerial erosion of the continental shelf and upper slope, which were repeatedly exposed during glacial advances.

Stage 8 also is characterized by a notable vertical fluctuation in the CCD and an increase in the activity of bottom currents. Bottom currents redistributed continental rise sediments into margin-parallel contourite drifts, particularly during the late middle Miocene, late Miocene, and Pliocene, and created widespread abyssal unconformities, such as the much discussed Horizon A^u (Tucholke and Mountain, 1986).

CONCLUSIONS

Importance of Seafloor Physiography

Many authors (e.g., Bouma, 1990; Locker and Laine, in press; see also Chapters 9, 12, 13) have acknowledged that antecedent basin physiography has greatly influenced the distribution patterns, depositional processes, and lithofacies of continental rises. The gross effects of seafloor physiography in the Hatteras Basin are particularly evident during the earliest and latest stages of its evolution. During stage 1, its importance is strongly reflected in the ridge and swale patterns that characterize the western flank of the spreading center. The basaltic oceanic basement had a relatively high relief at this time, which strongly controlled deposition patterns, especially along the southeastern margin of the basin. Also, the narrowness of the Hatteras Basin at this time severely limited the length of submarine fans that could develop there.

During stage 8, elevated contourite drifts exerted significant control on the subsequent distribution of abyssal sediments. Isochron patterns clearly show that successive allostratigraphic units formed thick sediment ponds in topographic lows around the drifts, whereas, in some areas, sedimentary sections over the drift crests were thin in comparison (see also Locker and Laine, in press, and Chapter 9).

Other examples could be cited, and no doubt numerous smaller features exist, but do not show on the isochron maps because of their relatively coarse spatial and temporal scales. The maps also tend to smooth topography by compositing successive features over intervals of several million years.

Relative Importance of Tectonism, Eustacy, and Paleoclimate

Arguments abound over whether tectonism is more influential than eustacy in the development of continental rise depositional systems (e.g., Stow, Howell, and Nelson, 1985, versus Shanmugam, Moiola, and Damuth, 1985). There appears to be no generally applicable rule of thumb, however; each basin must be evaluated on its own merits. In the Hatteras Basin, even though it evolved on a passive margin, the prime control on regional depositional patterns appears to have been relative uplift of source terrains. Initially, relative uplift determined rates and volumes of siliciclastic detritus reaching the basins. The relative sediment supply from one or more source terrains determined, in turn, where the main delta–apron–fan systems would develop (i.e., the latitudinal position of primary depocenters along the margin) and what their ultimate size (length, width, thickness) and principal constituents (e.g., heavy-mineral suites) would be. At times, such as during the Early Cretaceous and middle Miocene, tectonically accelerated deposition helped push confluent deltas to the shelf edge, where voluminous sediments could be dumped directly into the Hatteras Basin rather than bypassing the shelf edge through submarine canyons.

Effects of sea-level changes (be they eustatic or tectonic) on the formation of continental rises also are important and have been quite diverse and complex. In the Hatteras Basin, the effects of postulated long-term eustatic changes appear to have been subordinate to effects of source-terrain tectonism, except when source terrains were quiescent and(or) topographically low (e.g., stages 2, 3, 5, 7 of basin evolution). At those

times, however, sea level appears to have directly determined the bathymetric position of primary depocenters and the relative volume of siliciclastic sediment reaching the continental rise. During the Paleogene, for example, high sea levels, coupled with a sparsity of siliciclastics, kept major depocenters on the continental shelf, and a ~ 43-m.y. condensed section [minimal deposition and(or) widespread erosion and carbonate dissolution] characterized the continental rise. As a result, at the coarse scale of my analysis, I detected no significant submarine fans of Paleogene age in the Hatteras Basin.

Short-term sea-level changes (e.g., third-order cycles of the Exxon sequence-stratigraphy model) appear to have been most important in regulating the *temporal* development of sedimentary features on the continental rise. That is, as most authors agree, the largest submarine fans and debris aprons probably developed mainly during lowstands (they constitute part of the lowstand systems tract of the Exxon model). Clearly, however, postulated major lowstands did not always produce large debris aprons and(or) submarine fans in the study area (e.g., Paleogene units). Because most of the allostratigraphic units (first-order features of Mutti and Normark, 1987) I have mapped are composites of several depositional sequences (second-order and finer scale features of Mutti and Normark, 1987), and thus encompass more than one lowstand, I cannot generally relate the mapped fan systems directly to any particular highstand or lowstand. Effects of two short-term highstands can be seen, however, in the lack (or sparsity) of rise deposition during the Hauterivian and Cenomanian–Turonian.

Though many authors pay little attention to paleoclimate when considering regional depositional patterns, this agent was a significant regulator of sedimentation in the study area. Paleoclimate was particularly effective in its extremes (e.g., unusual aridity early in stage 1; tropical rain forests in stage 7; extreme continental glaciation in stage 8). In fact, if Cecil's (1990) model is correct, the effects of a tropical paleoclimate (rain forest development) were sufficient to prevent appreciable volumes of siliciclastic sediment from reaching offshore basins during the central Appalachian uplift of stage 7.

Types of Gravity-Flow Deposits

Lacking extensive sampling of the allostratigraphic units, I relied mainly on seismic facies and isochron maps to interpret the depositional implications of ancient continental rise features. I found it difficult to consistently recognize boundaries of individual fans, or to always separate individual fans from composite or confluent ones. Nevertheless, most mapped units display sediment thickenings on the scale of 10 km or more in diameter whose general morphology and geometry resemble those of either elongate submarine fans or debris aprons that characterize many modern sediment gravity flows and mass movements (turbidites, debris-flow deposits, slumps, slides; Stow, Howell, and Nelson, 1985; see also Chapters 9, 12, 13). What is more, most large fans and megafans (as I have defined them) appear to be easily separable into upper, middle, and lower subdivisions, as commonly applied to modern fan systems.

Most of these features are interpreted to consist dominantly of siliciclastic sediments. Carbonate-rich deposits were periodically important on the rise mainly as debris aprons or as hemipelagic blankets on interfan plains and basin plains.

Importance of Submarine Canyons

In the study area, large channels with V-shaped cross sections, which incise the continental slope and shelf edge (typical of modern submarine canyons), are generally restricted to middle Miocene and younger allostratigraphic units. This apparent restriction may be partly due to the lack of sufficient cross-channel seismic profiles. On the other hand, large channels with U-shaped cross sections are common along the shelf edge, slope, and upper rise of many of the mapped units. Individual channels commonly appear to have been an integral part of updip sediment dispersal systems of the Hatteras Basin fans, but in many cases, several channels appear to have formed multipoint sources for large debris aprons. Foremost examples of the latter are the Berriasian–Valanginian, Barremian, Santonian–Coniacian, Campanian, and middle Miocene allostratigraphic units, in which large confluent deltas prograded to the shelf edge and debouched large sediment loads directly onto the slope and upper rise in the manner described by Field and Gardner (1990), Heller and Dickinson (1985), and Suter and Berryhill (1985) (see also discussion by Collinson, 1986).

Modification by Oceanographic Processes

Downslope-oriented isochron patterns predominate for most of the mapped units, an indication that gravity-flow mechanisms dominated continental-rise deposition as long as sediment supply was sufficient to allow mass movement. However, two oceanographic processes, carbonate dissolution and geostrophic bottom-current flow, periodically exerted considerable control over depositional patterns and lithofacies of the Hatteras Basin. During the first ~ 74 m.y. [Aalenian(?) to Aptian] of basin evolution, abyssal water masses appear to have been saturated with respect to calcium

carbonate, and continental-rise sediments were able to incorporate abundant pelagic carbonate, as particularly recorded in the Upper Jurassic and Lower Cretaceous allostratigraphic units. For the next ~ 100 m.y. (Albian to middle Miocene), however, rise sediments (sampled mainly on the lower rise) were depleted in carbonate, due presumably to a raised CCD. This elevation contributed to the lack of (or thinness of) abyssal strata in many of the allostratigraphic units. The CCD began to fluctuate in the late Miocene, and for the last ~ 10 m.y., it has remained depressed enough for carbonate-rich abyssal sediments to accumulate once again.

Effects of bottom currents in the study area (such as the Western Boundary Undercurrent and deep elements of the Gulf Stream) have been well documented for Miocene and younger strata. Alongslope-oriented erosional swaths and(or) patches of thin sediment, which may have resulted from bottom-current activity, are present as early as the Hauterivian, however, and with few exceptions, appear to have persisted for ~ 80 m.y. (until the early Eocene). The fact that unconformable allostratigraphic boundaries are so widely recognizable throughout the Hatteras Basin is additional evidence of episodic erosion and redistribution of abyssal sediments.

The oldest evidence of sedimentary mounds possibly constructed by contour currents appears as an alongslope-oriented thickening of the Maestrichtian allostratigraphic unit. Following the Maestrichtian, no drift-like contour patterns appear until the middle Miocene, late Miocene, and Pliocene. Unusual southwestward axial trends of several submarine fans in the Pliocene and Quaternary allostratigraphic units, however, may be evidence that bottom currents were partly responsible for deflecting gravity-flow deposition.

The Quaternary as an Anomalous Depositional Unit

Several authors have recently urged caution when using analyses of modern continental-rise deposits to interpret ancient counterparts. Evidence derived from the present study shows that, indeed, the entire Quaternary allostratigraphic unit in the Hatteras Basin constitutes a geologically anomalous sedimentary deposit. For example, the siliciclastic accumulation rate is unusually high, due in large part, not to uplift of source terrains, but to the anomalous condition of widespread North American continental glaciation. In addition, the shelf received minimal Quaternary deposition, because of anomalously frequent exposure of the shelf by anomalously low sea levels. So an anomalously high percentage (95%) of available sediment reached the continental rise. Also, submarine canyons are anomalously large, V shaped, and abundant. Thus, it would be hazardous to use fine-scale models of Quaternary deposition in the Hatteras Basin to interpret its ancient sedimentary deposits. Models based on Holocene sedimentary systems are even more inappropriate.

ACKNOWLEDGMENTS

I am indebted to Dennis O'Leary, Ann Swift, Peter Popenoe, John Schlee, James Robb, Dave Twichell, and William Dillon for discussions and technical advice that led to improvement of this paper. Brian Tucholke kindly provided several seismic reflection profiles. Special thanks go to Robert McMaster, Stanley Locker, Michael Field, and Pierre Charles de Graciansky for thorough reviews of the manuscript.

REFERENCES

Amato, R. V., Ed. 1987. Shell Baltimore Rise 93-1 well, geological and operational summary. *U.S. Minerals Management Service, OCS Report MMS 86-0117.*

Barron, E. J. 1989. Climate variations and the Appalachians from the late Paleozoic to the Present: Results from model simulations. *Geomorphology* 2:99–118.

Benson, R. N., and Doyle, R. G. 1988. Early Mesozoic rift basins and the development of the United States middle Atlantic continental margin. *In* Triassic–Jurassic Rifting, W. Manspeizer, Ed.: Amsterdam: Elsevier, 99–127.

Benson, W. E., Sheridan, R. E., Pastouret, L., Enos, P., Freeman, T., Murdmaa, I. O., Worstell, P., Gradstein, F. M., Schmidt, R. R., Weaver, F. M., and Stuermer, D. H. 1978. Initial Reports of the Deep Sea Drilling Project, Volume 44: Washington, DC: U.S. Government Printing Office.

Bielak, L. E., Ed. 1986. Tenneco Hudson Canyon 642-2 well, geological and operational summary. *U.S. Minerals Management Service, OCS Report MMS 86-0077.*

Bouma, A. H. 1990. Clastic depositional styles and reservoir potential of Mediterranean basins. *American Association of Petroleum Geologists Bulletin* 74:532–546.

Bouma, A. H., Normark, W. R., and Barnes, N. E. 1985. COMFAN: Needs and results. *In* Submarine Fans and Related Turbidite Systems, A. H. Bouma, W. R. Normark, and N. E. Barnes, Eds.: New York: Springer-Verlag, 7–11.

Cecil, C. B. 1990. Paleoclimate controls on stratigraphic repetition of chemical and siliciclastic rocks: *Geology* 18:533–536.

Cleary, W. J., Pilkey, O. H., and Nelson, J. C. 1985. Wilmington Fan, Atlantic Ocean. *In* Submarine Fans and Related Turbidite Systems, A. H. Bouma, W. R. Normark, and N. E. Barnes, Eds.: New York: Springer-Verlag, 157–164.

Collinson, J. D. 1986. Submarine ramp facies model for delta-fed, sand-rich turbidite systems: Discussion. *American Association of Petroleum Geologists Bulletin* 70:1742–1743.

Danforth, W. W., and Schwab, W. C. 1990. High-resolution seismic stratigraphy of the upper continental rise seaward of Georges Bank. *U.S. Geological Survey Miscellaneous Field Studies Map MF-2111.*

De Boer, J. Z., McHone, J. G., Puffer, J. H., and Ragland, P. C. 1988. Mesozoic and Cenozoic magmatism. *In* The Geology of North America, Volume I-2, The Atlantic Continental Margin, U.S., R. E. Sheridan and J. A. Grow, Eds.: Boulder, CO: Geological Society of America, 217–241.

Denny, C. S. 1982. The geomorphology of New England. *U.S. Geological Survey Professional Paper 1208.*

Erlich, R. N., Maher, K. P., Hummel, G. A., Benson, D. G., Kastritis, G. J., Linder, H. D., Hodr, R. S., and Neely, D. H. 1988. Baltimore Canyon Trough, Mid-Atlantic OCS: Seismic stratigraphy of Shell/Amoco/Sun wells. *In* Atlas of Seismic Stratigraphy, A. W. Bally, Ed.: *American Association of Petroleum Geologists Studies in Geology No. 27* 2:51–65.

Ewing, J. I. 1984. Sections along U.S.G.S. Line 27, Baltimore Canyon Trough. *In* Eastern North American Continental Margin and Adjacent Ocean Floor, 34° to 41°N and 68° to 78°W. Ocean Margin Drilling Program Regional Atlas Series, J. I. Ewing and P. D. Rabinowitz, Eds.: Woods Hole, MA: Marine Science International, 30.

Farre, J. A., and Ryan, W. B. F. 1985. 3-D view of erosional scars on U.S. mid-Atlantic continental margin. *American Association of Petroleum Geologists Bulletin* 69:923–932.

Farre, J. A., and Ryan, W. B. F. 1987. Surficial geology of the continental margin offshore New Jersey in the vicinity of Deep Sea Drilling Project Sites 612 and 613. *In* Initial Reports of the Deep Sea Drilling Project, Volume 95, C. W. Poag, A. B. Watts, et al.: Washington, DC: U.S. Government Printing Office, 725–759.

Field, M. E., and Gardner, J. V. 1990. Pliocene–Pleistocene growth of the Rio Ebro margin, northeast Spain: A prograding-slope model. *Geological Society of America Bulletin* 102:721–733.

Grow, J. A. 1980. Deep structure and evolution of the Baltimore Canyon Trough in the vicinity of the COST No. B-3 well. *In* Geological Studies of the COST No. B-3 Well, United States Mid-Atlantic Continental Slope Area, P. A. Scholle, Ed.: *U.S. Geological Survey Circular 833,* 117–126.

Grow, J. A., Klitgord, K. D., and Schlee, J. S. 1988. Structure and evolution of Baltimore Canyon Trough. *In* The Geology of North America, Volume I-2, The Atlantic Continental Margin, U.S., R. E. Sheridan and J. A. Grow, Eds.: Boulder, CO: Geological Society of America, 269–290.

Grow, J. A., Klitgord, K. D., Schlee, J. S., and Dillon, W. P. 1988. Representative seismic profiles of U.S. Atlantic continental margin. *In* The Geology of North America, Volume I-2, The Atlantic Continental Margin, U.S., R. E. Sheridan and J. A. Grow, Eds.: Boulder, CO: Geological Society of America, Plate 4.

Grow, J. A., Mattick, R. E., and Schlee, J. S. 1979. Multichannel seismic depth sections and interval velocities over outer continental shelf and upper continental slope between Cape Hatteras and Cape Cod. *In* Geological and Geophysical Investigations of Continental Margins, J. S. Watkins, L. Montadert, and P. W. Dickerson, Eds.: *American Association of Petroleum Geologists Memoir 29,* 65–84.

Hack, J. T. 1982. Physiographic divisions and differential uplift in the Piedmont and Blue Ridge. *U.S. Geological Survey Professional Paper 1265.*

Haggerty, J., Sarti, M., von Rad, U., Ogg, J. G., and Dunn, D. A. 1987. Late Aptian to Recent sedimentological history of the lower continental rise off New Jersey, Deep Sea Drilling Project Site 603. *In* Initial Reports of the Deep Sea Drilling Project, Volume 93, J. E. Van Hinte, S. W. Wise, Jr., et al.: Washington, DC: U.S. Government Printing Office, 1285–1304.

Hampson, J. C., Jr., and Robb, J. M. 1984. A geologic map of the continental slope off New Jersey: Lindenkohl Canyon to Toms Canyon. *U.S. Geological Survey Miscellaneous Investigations Map Series I-1608*, scale 1:50,000.

Hansen, H. 1988. Buried rift basin underlying coastal plain sediments, central Delmarva Peninsula, Maryland. *Geology* 16:779–782.

Haq, B. U., Hardenbol, J., and Vail, P. R. 1987. The chronology of fluctuating sea level since the Triassic. *Science* 235:1156–1167.

Haq, B. U., Hardenbol, J., and Vail, P. R. 1988. Mesozoic and Cenozoic chronostratigraphy and cycles of sea-level change. *In* Sea-Level Changes—An Integrated Approach, C. K. Wilgus, B. S. Hastings, C. G. St. C. Kendall, H. W. Posamentier, C. A. Ross, and J. C. Van Wagoner, Eds.: *Society of Economic Paleontologists and Mineralogists Special Publication No. 42*, 71–108.

Hart, M. B., and Mountain, G. S. 1987. Ichthyolith evidence for the age of reflector A^u, Deep Sea Drilling Project Site 603. *In* Initial Reports of the Deep Sea Drilling Project, Volume 93, J. E. Van Hinte, S. W. Wise, Jr., et al.: Washington, DC: U.S. Government Printing Office, 739–750.

Hathaway, J. C., Poag, C. W., Valentine, P. C., Miller, R. E., Schultz, D. M., Manheim, F. T., Kohout, F. A., Bothner, M. H., and Sangrey, D. A. 1979. U.S. Geological Survey core drilling on the Atlantic shelf. *Science* 206:515–527.

Heller, P. L., and Dickinson, W. R. 1985. Submarine ramp facies model for delta-fed, sand-rich turbidite systems. *American Association of Petroleum Geologists Bulletin* 69:960–976.

Hollister, C. D., Ewing, J. I., Habib, D., Hathaway, J. C., Lancelot, Y., Luterbacher, H., Paulus, F. J., Poag, C. W., Wilcoxon, J. A., and Worstell, P. 1972. Initial Reports of the Deep Sea Drilling Project, Volume 11: Washington, DC: U.S. Government Printing Office.

Holmes, M. A., Breza, J. R., and Wise, S. W., Jr. 1987. Provenance and deposition of Lower Cretaceous turbidite sands at Deep Sea Drilling Project Site 603, lower continental rise off North Carolina. *In* Initial Reports of the Deep Sea Drilling Project, Volume 93, J. E. Van Hinte, S. W. Wise, Jr., et al.: Washington, DC: U.S. Government Printing Office, 941–959.

Holser, W. T., Clement, G. P., Jansa, L. F., and Wade, J. A. 1988. Evaporite deposits of the North Atlantic rift. *In* Triassic–Jurassic Rifting, W. Manspeizer, Ed.: Amsterdam: Elsevier, 525–556.

Isachsen, Y. W. 1975. Possible evidence for contemporary doming of the Adirondack Mountains, New York, and suggested implications for regional tectonics and seismicity. *Tectonophysics* 29:161–181.

Jansa, L. F. 1986. Paleoceanography and evolution of the North Atlantic Ocean basin during the Jurassic. *In* The Geology of North America, Volume M, The Western North Atlantic Region, P. R. Vogt and B. E. Tucholke,

Eds.: Boulder, CO: Geological Society of America, 603–616.

Jansa, L. F., and Pe-Piper, G. 1988. Middle Jurassic and Early Cretaceous igneous rocks along eastern North American continental margin. *American Association of Petroleum Geologists Bulletin* 72:347–366.

Jansa, L. F., Enos, P., Tucholke, B. E., Gradstein, F. M., and Sheridan, R. E. 1979. Mesozoic–Cenozoic sedimentary formations of the North American Basin, western North Atlantic. *In* Deep Drilling Results in the Atlantic Ocean: Continental Margins and Paleoenvironments, M. Talwani, W. W. Hay, and W. B. F. Ryan, Eds., Maurice Ewing Series 3: Washington, DC: American Geophysical Union, 1–58.

Kaneps, A. G., Doyle, P. S., and Riedel, W. R. 1981. Further ichthyolith age determinations of otherwise unfossiliferous Deep Sea Drilling cores. *Micropaleontology* 27:317–331.

Kirby, J. R., Robb, J. M., and Hampson, J.C., Jr. 1982. Detailed bathymetric map of the United States continental slope between Lindenkohl Canyon and Toms Canyon, offshore New Jersey. *U.S. Geological Survey Miscellaneous Field Studies Map MF-1443*, scale 1:50,000, 1 sheet.

Klitgord, K. D., and Hutchinson, D. R. 1985. Distribution and geophysical signatures of early Mesozoic rift basins beneath the U.S. Atlantic continental margin. *U.S. Geological Survey Circular 946*, 45–61.

Klitgord, K. D., Hutchinson, D. R., and Schouten, H. 1988. U.S. Atlantic continental margin; structural and tectonic framework. *In* The Geology of North America, Volume I-2, The Atlantic Continental Margin, U.S., R. E. Sheridan and J. A. Grow, Eds.: Boulder, CO: Geological Society of America, 19–55.

Lawrence, D. T., Doyle, M., and Aigner, T. 1990. Stratigraphic simulation of sedimentary basins: Concepts and calibrations. *American Association of Petroleum Geologists Bulletin* 74:273–295.

Libby-French, J. 1984. Stratigraphic framework and petroleum potential of northeastern Baltimore Canyon Trough, mid-Atlantic outer continental shelf. *American Association of Petroleum Geologists Bulletin* 68:50–73.

Lippert, R. H. 1983. The "Great Stone Dome"—A compaction structure. *In* Seismic Expression of Structural Styles—A Picture and Word Atlas, A. W. Bally, Ed.: *American Association of Petroleum Geologists Studies in Geology No. 15*, 1.3-1 to 1.3-4.

Locker, S. D. 1989. Cenozoic depositional history of the middle U.S. Atlantic continental rise. Ph.D. Thesis, University of Rhode Island.

Locker, S. D., and Laine, E. P. (in press). Paleogene–Neogene depositional history of the middle U.S. Atlantic continental rise: Mixed turbidite and contourite depositional systems. *Marine Geology*.

Lovegreen, J. R. 1974. Paleodrainage history of the Hudson estuary. Master's Thesis, Columbia University.

Manspeizer, W. 1981, Early Mesozoic basins of the central Atlantic passive margins. *In* Geology of Passive Continental Margins: History, Structure, and Sedimentologic Record, A. W. Bally, Ed.: *American Association of Petroleum Geologists Education Course Notes Series No. 19*, 4-1 to 4-60.

Manspeizer, W. 1985. Early Mesozoic history of the Atlantic passive margins, *In* Geologic Evolution of the United States Atlantic Margin, C. W. Poag, Ed.: New York: Van Nostrand Reinhold, 1–23.

Manspeizer, W. 1988. Triassic–Jurassic rifting and opening of the Atlantic: An overview. *In* Triassic–Jurassic Rifting, W. Manspeizer, Ed.: Amsterdam: Elsevier, 41–80.

Manspeizer, W., Ed. 1988. *Triassic–Jurassic Rifting*: Amsterdam: Elsevier.

Manspeizer, W., and Cousminer, H. L. 1988. Late Triassic–Jurassic synrift basins of the U.S. Atlantic margin. *In* The Geology of North America, Volume I-2, The Atlantic Continental Margin, U.S., R. E. Sheridan and J. A. Grow, Eds.: Boulder, CO: Geological Society of America, 179–216.

McGregor, B. A., Stubblefield, W. C., Ryan, W. B. F., and Twichell, D. C. 1982. Wilmington submarine canyon: A fluvial-like system. *Geology* 10:27–30.

McMaster, R. L., Locker, S. D., and Laine, E. P. 1989. The early Neogene continental rise off the eastern United States. *Marine Geology* 87:137–163.

Meyer, F. O. 1989. Siliciclastic influence on Mesozoic platform development: Baltimore Canyon Trough, western Atlantic. *In* Controls on Carbonate Platform and Basin Development, P. D. Crevello, J. L. Wilson, J. F. Sarg, and J. F. Read, Eds.: *Society of Economic Paleontologists and Mineralogists Special Publication No. 44*, 213–232.

Miller, K. G., Fairbanks, R. G., and Mountain, G. S. 1987. Tertiary oxygen isotope synthesis, sea-level history, and continental margin erosion. *Paleoceanography* 2:1–19.

Mountain, G. S. 1987. Cenozoic margin construction and destruction offshore New Jersey. *In* Timing and Depositional History of Eustatic Sequences: Constraints on Seismic Stratigraphy, C. A. Ross and D. Haman, Eds.: *Cushman Foundation for Foraminiferal Research Special Publication No. 24*, 57–84.

Mountain, G. S., and Tucholke, B. E. 1985. Mesozoic and Cenozoic geology of the U.S. Atlantic continental slope and rise. *In* Geologic Evolution of the United States Atlantic Margin, C. W. Poag, Ed.: New York: Van Nostrand Reinhold, 293–341.

Mutti, E., and Normark, W. R. 1987. Comparing examples of modern and ancient turbidite systems: Problems and concepts. *In* Marine Clastic Sedimentology, J. K. Leggett and G. G. Zuffa, Eds.: London: Graham and Trotman, 1–38.

Nishimura, A. 1987. Cenozoic radiolaria in the western North Atlantic, Site 603, Leg 93 of the Deep Sea Drilling Project. *In* Initial Reports of the Deep Sea Drilling Project, Volume 93, J. E. Van Hinte, S. W. Wise, Jr., et al.: Washington, DC: U.S. Government Printing Office, 713–737.

North American Commission on Stratigraphic Nomenclature. 1983. North American stratigraphic code. *American Association of Petroleum Geologists Bulletin* 67:841–875.

O'Leary, D. W. 1988. Shallow stratigraphy of the New England continental margin. *U.S. Geological Survey Bulletin 1767*, 40.

Olsson, R. K., and Wise, S. W., Jr. 1987. Upper Maestrichtian to middle Eocene stratigraphy of the New Jersey slope and coastal plain. *In* Initial Reports of the Deep Sea Drilling Project, Volume 93, J. E. Van Hinte, S. W. Wise,

Jr., et al.: Washington, DC: U.S. Government Printing Office, 1343–1366.

Olsson, R. K., Melillo, A. J., and Schreiber, B. L. 1987. Miocene sea-level events in the Maryland Coastal Plain and the offshore Baltimore Canyon Trough. *In* Timing and Depositional History of Eustatic Sequences: Constraints on Seismic Stratigraphy, C. A. Ross and D. Haman, Eds.: *Cushman Foundation for Foraminiferal Research Special Publication No. 24*, 85–98.

Palmer, A. R., Compiler. 1983. The Decade of North American Geology 1983 geologic time scale. *Geology* 11:503–504.

Poag, C. W. 1982a. Foraminiferal and seismic stratigraphy, paleoenvironments, and depositional cycles in the Georges Bank Basin. *In* Geological Studies of the COST Nos. G-1 and G-2 Wells, United States North Atlantic Outer Continental Shelf, P. A. Scholle and C. R. Wenkam, Eds.: *U.S. Geological Survey Circular 861*, 43–91.

Poag, C. W. 1982b. Stratigraphic reference section for Georges Bank Basin—Depositional model for New England passive margin. *American Association of Petroleum Geologists Bulletin* 66:1021–1041.

Poag, C. W. 1985. Depositional history and stratigraphic reference section for central Baltimore Canyon Trough. *In* Geologic Evolution of the United States Atlantic Margin, C. W. Poag, Ed.: New York: Van Nostrand Reinhold, 217–263.

Poag, C. W. 1987. The New Jersey Transect: Stratigraphic framework and depositional history of a sediment-rich passive margin. *In* Initial Reports of the Deep Sea Drilling Project, Volume 95, C. W. Poag, A. B. Watts, et al.: Washington, DC: U.S. Government Printing Office, 763–817.

Poag, C. W. 1991. Rise and demise of the Bahama-Grand Banks Gigaplatform, northern margin of the Jurassic proto-Atlantic seaway. *Marine Geology* special issue, 102:63–130.

Poag, C. W., and Mountain, G. S. 1987. Upper Cretaceous and Cenozoic evolution of the New Jersey continental slope and rise: An integration of borehole data with seismic reflection profiles. *In* Initial Reports of the Deep Sea Drilling Project, Volume 95, C. W. Poag, A. B. Watts, et al.: Washington, DC: U.S. Government Printing Office, 673–724.

Poag, C. W., and Sevon, W. D. 1989. A record of Appalachian denudation in postrift Mesozoic and Cenozoic sedimentary deposits of the U.S. middle Atlantic Continental margin. *Geomorphology* 2:119–157.

Poag, C. W., Swift, B. A., Schlee, J. S., Ball, M. M., and Sheetz, L. L. 1990. Early Cretaceous shelf-edge deltas of the Baltimore Canyon Trough: Principal sources for sediment gravity deposits of the northern Hatteras Basin. *Geology* 18:149–152.

Poag, C. W., Watts, A. B., Cousin, M., Goldberg, D., Hart, M. B., Miller, K. G., Mountain, G. S., Nakamura, Y., Palmer, A., Schiffelbein, P. A., Schreiber, B. C., Tarafa, M., Thein, J. E., Valentine, P. C., and Wilkens, R. H. 1987. Initial Reports of the Deep Sea Drilling Project, Volume 95: Washington, DC: U.S. Government Printing Office.

Prather, B. E. 1991. Petroleum geology of the Upper Jurassic and Lower Cretaceous, Baltimore Canyon Trough, Western North Atlantic. *American Association of Petroleum Geologists Bulletin* 75:258–277.

Pratson, L. F., and Laine, E. P. 1989. The relative importance of gravity-induced versus current-controlled sedimentation during the Quaternary along the middle U.S. continental margin revealed by 3.5 kHz echo character. *Marine Geology* 89:87–126.

Robb, J. M. 1980. High-resolution seismic-reflection profiles collected by R/V *James M. Gillis*, Cruise GS 7903-4, in the Baltimore Canyon outer continental shelf area, offshore New Jersey. *U.S. Geological Survey Open-File Report 80-934*, 3.

Robb, J. M., Hampson, J. C., Jr., Kirby, J. R., and Twichell, D. C. 1981. Geology and potential hazards of the continental slope between Lindenkohl and South Toms canyons, offshore mid-Atlantic United States. *U.S. Geological Survey Open-File Report 81-600*.

Robb, J. M., Kirby, J. R., Hampson, J. C., Jr., Gibson, P. R., and Hecker, B. 1983. Furrowed outcrops of Eocene chalk on the lower continental slope offshore New Jersey. *Geology* 11:182–186.

Roden, M. K., and Miller, D. S. 1989. Apatite fission-track thermochronology of the Pennsylvania Appalachian Basin. *Geomorphology* 2:39–52.

Ryan, W. B. F., Miller, E. L., Hanselmann, D., Nesteroff, W. D., Hecker, B., and Nibbelink, M. 1978. Bedrock geology in New England submarine canyons. *Oceanologica Acta* 1:233–254.

Sarti, M., and von Rad, U. 1987. Early Cretaceous turbidite sedimentation at Deep Sea Drilling Site 603, off Cape Hatteras (Leg 93). *In* Initial Reports of the Deep Sea Drilling Project, Volume 93, J. E. Van Hinte, S. W. Wise, Jr., et al.: Washington, DC: U.S. Government Printing Office, 891–940.

Schlee, J. S. 1981. Seismic stratigraphy of Baltimore Canyon Trough. *American Association of Petroleum Geologists Bulletin* 65:26–53.

Schlee, J. S., and Hinz, K. 1987. Seismic stratigraphy and facies of continental slope and rise seaward of Baltimore Canyon Trough. *American Association of Petroleum Geologists Bulletin* 71:1046–1067.

Schlee, J. S., Dillon, W. P., and Grow, J. A. 1979. Structure of the continental slope off eastern United States. *In* Geology of the Continental Slope, L. J. Doyle and O. H. Pilkey, Eds.: *Society of Economic Paleontologists and Mineralogists Special Publication No. 27*, 95–117.

Schlee, J. S., Poag, C. W., and Hinz, K. 1985. Seismic stratigraphy of the continental slope and rise seaward of Georges Bank. *In* Geologic Evolution of the United States Atlantic Margin, C. W. Poag, Ed.: New York: Van Nostrand Reinhold, 265–292.

Scholle, P. A., Ed. 1977. Geological studies on the COST No. B-2 well, U.S. mid-Atlantic outer continental shelf area. *U.S. Geological Survey Circular 750*.

Shanmugam, G., and Moiola, R. J. 1985. Submarine fan models: Problems and solutions. *In* Submarine Fans and Related Turbidite Systems, A. H. Bouma, W. R. Normark, and N. E. Barnes, Eds.: New York: Springer-Verlag, 29–34.

Shanmugam, G., Moiola, R. J., and Damuth, J. E. 1985. Eustatic control of submarine fan development. *In* Sub-

marine Fans and Related Turbidite Systems, A. H. Bouma, W. R. Normark, and N. E. Barnes, Eds.: New York: Springer-Verlag, 23–28.

Sheridan, R. E., and Grow, J. A., Eds. 1988. The Geology of North America, Volume I-2, The Atlantic Continental Margin, U.S.: Boulder, CO: Geological Society of America.

Smith, R. V. 1980. Provenance of mid-Atlantic continental margin sediments from the COST B-2 test well. Master's Thesis, University of Delaware.

Stow, D. A. V., Howell, D. G., and Nelson, C. H. 1985. Sedimentary, tectonic, and sea-level controls. *In* Submarine Fans and Related Turbidite Systems, A. H. Bouma, W. R. Normark, and N. E. Barnes, Eds.: New York: Springer-Verlag, 15–22.

Suter, J. R., and Berryhill, H. L., Jr. 1985. Late Quaternary shelf-margin deltas, northwest Gulf of Mexico. *American Association of Petroleum Geologists Bulletin* 69:77–91.

Swift, B. A., Lee, M. W., Poag, C. W., and Agena, W. F. 1990. Seismic properties and modeling of an early Aalenian salt(?) layer in Baltimore Canyon Trough, offshore New Jersey. *American Association of Petroleum Geologists Bulletin* 74:774.

Tucholke, B. E., and Laine, E. P. 1982. Neogene and Quaternary development of the lower continental rise off the central U.S. East Coast. *In* Studies in Continental Margin Geology, J. S. Watkins and C. L. Drake, Eds.: *American Association of Petroleum Geologists Memoir 34*, 295–305.

Tucholke, B. E., and McCoy, F. W. 1986. Paleogeographic and paleobathymetric evolution of the North Atlantic Ocean. *In* The Geology of North America, Volume M, The Western North Atlantic Region, P. R. Vogt and B. E. Tucholke, Eds.: Boulder, CO: Geological Society of America, 589–602.

Tucholke, B. E., and Mountain, G. S. 1979. Seismic stratigraphy, lithostratigraphy, and paleosedimentation patterns in the North Atlantic Basin. *In* Deep Drilling Results in the Atlantic Ocean: Continental Margins and Paleoenvironments, M. Talwani, W. W. Hay, and W. B. F. Ryan, Eds.: Maurice Ewing Series 3: Washington, DC: American Geophysical Union, 58–86.

Tucholke, B. E., and Mountain, G. S. 1986. Tertiary paleoceanography of the western North Atlantic Ocean. *In* The Geology of North America, Volume M, The Western North Atlantic Region, P. R. Vogt and B. E. Tucholke, Eds.: Boulder, CO: Geological Society of America, 631–650.

Tucholke, B. E., and Vogt, P. R. 1979. Western North Atlantic: Sedimentary evolution and aspects of tectonic history. *In* Initial Reports of the Deep Sea Drilling Project, Volume 43, B. E. Tucholke, P. R. Vogt, et al.: Washington, DC: U.S. Government Printing Office, 791–825.

Twichell, D. C., and Roberts, D. G. 1982. Morphology, distribution, and development of submarine canyons on the United States Atlantic continental margin between Hudson and Baltimore canyons. *Geology* 10:408–412.

Uchupi, E., and Shor, A. N., Eds. 1984. Eastern North American Continental Margin and Adjacent Ocean Floor, 39° to 46°N and 64° to 74°W. Ocean Margin Drilling

Program Regional Atlas Series: Woods Hole, MA: Marine Science International, Atlas 3:38 sheets.

Vail, P. R., Mitchum, R. M., and Thompson, S. III. 1977. Seismic stratigraphy and global changes of sea level. Part four: Global cycles of relative changes of sea level. *In* Seismic Stratigraphy—Applications to Hydrocarbon Exploration, C. E. Payton, Ed.: *American Association of Petroleum Geologists Memoir 36*, 83–98.

Van Hinte, J. E., Wise, S. W., Jr., Biart, B. N. M., Covington, J. M., Dunn, D. A., et al. 1985. Deep drilling on the upper continental rise off New Jersey: DSDP Sites 604 and 605. *Geology* 13:397–400.

Van Hinte, J. E., Wise, S. W., Jr., Biart, B. N. M., Covington, J. M., Dunn, D. A., Haggerty, J. A., Johns, M. W., Meyers, P. A., Moullade, M. R., Muza, J. P., Ogg, J. G., Okamura, M., Sarti, M., and von Rad, U. 1987. Initial Reports of the Deep Sea Drilling Project, Volume 93: Washington, DC: U.S. Government Printing Office.

Van Wagoner, J. C., Posamentier, H. W., Mitchum, R. M., Vail, P. R., Sarg, J. F., Loutit, T. S., and Hardenbol, J. 1988. An overview of the fundamentals of sequence stratigraphy and key definitions. *In* Sea-Level Changes— An Integrated Approach, C. K. Wilgus, B. S. Hastings, C. G. St. C. Kendall, H. W. Posamentier, C. A. Ross, and J. C. Van Wagoner, Eds.: *Society of Economic Paleontologists and Mineralogists Special Publication No. 42*, 39–45.

Vogt, P. R. 1991. Bermuda and Appalachian–Labrador rises: Common non-hotspot processes? *Geology* 19:41–44.

Vogt, P. R., and Tucholke, B. E., Eds. 1986. The Geology of North America, Volume M. The Western North Atlantic Region. Boulder, CO: Geological Society of America.

Ward, L. W., and Krafft, K., Eds. 1984. Stratigraphy and paleontology of the outcropping Tertiary beds in the Pamunkey River region, central Virginia Coastal Plain— Guidebook for Atlantic Coastal Plain Geological Association 1984 Field Trip: Atlantic Coastal Plain Geological Association.

Watts, A. B., and Steckler, M. S. 1979. Subsidence and eustasy at the continental margin of eastern North America. *In* Deep Drilling Results in the Atlantic Ocean: Continental Margins and Paleoenvironments, M. Talwani, W. W. Hay, and W. B. F. Ryan, Eds., Maurice Ewing Series 3: Washington, DC: American Geophysical Union, 273–310.

Wolfe, J. A. 1978. A paleobotanical interpretation of Tertiary climates in the northern hemisphere. *American Scientist* 66:694–703.

7

Norway – Svalbard Continental Margin: Structural and Stratigraphical Styles

Annik M. Myhre, Olav Eldholm, J. I. Faleide, J. Skogseid,
S. T. Gudlaugsson, S. Planke, L. M. Stuevold,
and E. Vågnes

The Norway–Svalbard passive continental margin, off the Norwegian mainland, the western Barents Sea, and Svalbard, forms the eastern perimeter of the Norwegian–Greenland Sea (Fig. 7-1). Development of the margin was initiated by continental breakup and accretion of oceanic crust between Eurasia and Greenland at the Paleocene–Eocene transition.

Although a variety of morphological, depositional, and structural styles characterize the margin, a clear geological division occurs at about lat. 70°N. The southern region, which predominantly developed as a rifted margin, separates the Cenozoic ocean basin from the Fennoscandian continental landmass, which consists almost exclusively of Precambrian and Caledonian basement rocks, covered by thin Quarternary deposits. Just offshore, however crystalline basement plunges seaward below thick late Paleozoic and younger sediments on the continental shelf and slope. The northern region is characterized by early Tertiary large-scale shear movements within the late Paleozoic and Mesozoic sedimentary basins and platforms of the western Barents Sea and the Svalbard archipelago.

The Norway–Svalbard margin has been extensively investigated both for commercial and scientific purposes. At present there is an excellent geophysical data base, including a large number of multichannel seismic profiles (MCS). In general, MCS coverage is most complete on the continental shelf south of lat. 74°N, where the exploration for hydrocarbons has been concentrated. Seaward of the shelf edge and north of lat. 74°N, where most profiles have been acquired by scientific institutions, the coverage is more regional. A large number of sonobuoys and two-ship expanded spread profiles (ESP) help to establish good velocity control in most areas.

Since 1980, drilling has been carried out in areas opened for commercial exploration on the Vøring shelf and in the southwestern Barents Sea, and some shallow stratigraphic holes have recently been drilled on the shelf. Though only limited results of the commercial drilling are publicly available, the published data allow a first-order correlation to seismic lines extending across the margin. In addition, several scientific holes at the outer margin and in the adjacent ocean basins have been drilled during Deep Sea Drilling Project (DSDP) Leg 38 and Ocean Drilling Program (ODP) Leg 104. These holes sampled the Cenozoic sedimentary sequence and the underlying basement rocks. The Vøring Plateau, in particular, has been the focus of scientific drilling.

In this study, we describe the Norway–Svalbard continental margin, focussing on the Cenozoic sediment distribution and depositional style. However, there is a strong link between the prerift geological configuration, the tectonism and magmatism during the early Tertiary continental breakup, and the present segmentation and depositional style of the margin. Therefore, we first discuss the physiography, prerift geological setting, and the magmatic—tectonic framework. Then, we address the syn- and postrift sedimentation and vertical movements, and illustrate them by a seismic type section across each of five margin provinces. The southern region is divided into the Møre, Vøring, and Lofoten–Vesterålen margins, and the northern region into the Barents Sea and Svalbard margins (Fig. 7-1).

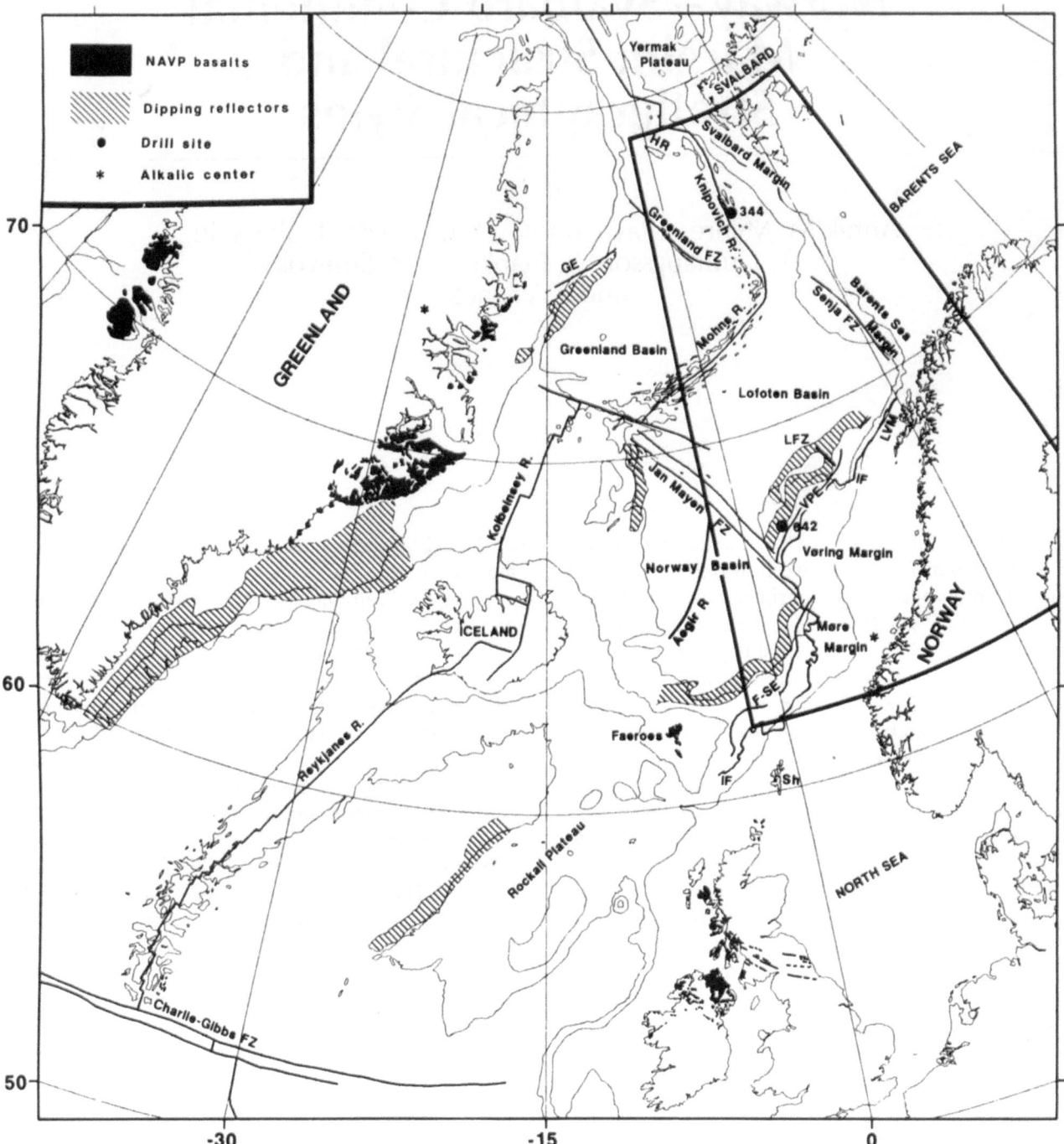

FIGURE 7-1. Northern North Atlantic continental margins and ocean basins as outlined by the 500- and 2,000-m contours. Note occurrences of early Tertiary igneous rocks of North Atlantic Volcanic Province (NAVP) and seaward-dipping reflector sequences along margins; extrusive constructions cover areas much larger than dipping wedges. Study area of this chapter shown within box. F-SE = Faeroe–Shetland Escarpment; GE = Greenland Escarpment; HR = Hovgaard Ridge; IF = inner flows; LFZ = Lofoten Fracture Zone; LVM-Lofoten–Vesterålen margin; Sh-Shetland Islands; VPE-Vøring Plateau Escarpment. Based on Eldholm (1991).

PHYSIOGRAPHY

The width of the continental shelf and the width and steepness of the continental slope vary considerably along strike (Figs. 7-1, 7-2). Both the base of the slope and the shelf edge are clearly recognized by a change in bathymetric gradient, except in local areas of large Neogene and Quaternary sediment supply. The young age and plate-tectonic segmentation of the Norwegian–Greenland Sea have produced only small, poorly developed abyssal plains in the Lofoten and Norway basins. Furthermore, the lower continental slope exhibits a variable, relatively steep declination. Therefore, some authors suggest that a typical continental rise has not yet developed (Johnson and Heezen, 1967; Eldholm and Windish, 1974), though others have shown a rise in this region (e.g., Kennett, 1982; Emery and Uchupi, 1984).

The continental shelf is typical for high-latitude regions, being relatively deep (100–400 m), with numerous banks and depressions, which reflect glacial influence. The shelf edge is often deeper than 300 m. Off the coasts of Norway and Svalbard, a characteristic system of elongate channels crosses the shelf, both parallel and transverse to the coastline (Holtedahl, 1960). The coast-parallel channels are often located in the vicinity of the pre-Quaternary sedimentary pinchout on the shelf, and separate an irregular seafloor morphology near the coast from the smooth, sediment-covered shelf.

Margin physiography reflects the combined effects of tectonism and magmatism during breakup and the post-Paleocene history of vertical movement, varying sediment supply, and erosion. Off Norway, these processes have segmented the margin into three physiographic provinces (Figs. 7-1, 7-2). The shelf is relatively narrow at the Møre and Lofoten–Vesterålen margins, with a minimum width of about 20 km near lat. 69°N, but widening significantly to about 220 km at the central Vøring margin. The continental slope is steepest off the northern part of the Lofoten–Vesterålen margin, and becomes gentler farther south. The slope is broken by the Vøring Plateau, a marginal plateau at a depth of 1,100–1,500 m, which covers an area of 35,000 km^2. The detailed seafloor morphology reveals large-scale mass movements at the northern Møre margin (Bugge et al., 1987). There is also evidence of small-scale sliding on the flanks of the Vøring Plateau, and local diapirs rise above the plateau surface (Talwani and Eldholm, 1972). The morphology of the narrow northernmost slope is particularly irregular, exhibiting canyon-like incisions from the shelf edge to the Lofoten Basin.

The Barents Sea and Svalbard shelves are transected by wide depressions extending to the 250 to 400-m-deep shelf edge (Fig. 7-2). The two most prominent transverse depressions, Bjørnøyrenna and Storfjordrenna, attain depths of 500–600 and 350 m, respectively. The continental slope bulges westward at the mouth of the depressions, evidence that they have acted as main drainage systems. The transverse channels west of Svalbard appear to be extensions of the fjord systems. The detailed slope morphology off the Barents Sea is irregular, reflecting both small- and large-scale sliding, debris-flow lobes, channels, and gullies as much as 150 m deep and 1 km wide (Damuth, 1978; Vorren et al., 1989). In particular, SeaMARC studies show that the giant fan off the Bjørnøyrenna is decorated by lobate sediment-flow deposits, 3–10 km wide and as much as 100 km long (Sundvor, Vogt, and Crane, 1990).

The Svalbard margin narrows to the north, and the shelf is only about 30 km wide at lat. 79°N. There is only a small basin between the slope and the present day plate boundary (the Knipovich Ridge) southwest of Spitsbergen. The Knipovich Ridge progressively approaches the continental margin northward, and the ridge axis intersects the lower slope west of Spitsbergen (Figs. 7-1, 7-2). Thus, we have the unique setting in which a spreading axis with active volcanism (Sundvor, Vogt, and Crane, 1990) is part of a passive continental margin in an area easily accessible for surveying and drilling. The prominent Knipovich Ridge province, from Mohns Ridge to lat. 78°N, comprises a 3- to 10-km-wide and 3,200-m-deep rift valley surrounded by large asymmetric axial mountains, which rise higher on the western flank and are partly buried by sediments on the eastern flank. The Knipovich Ridge is often mapped as a segmented feature (Vogt, 1986; Max and Ohta, 1988), but recent SeaMARC surveys have confirmed a largely continuous rift valley (Crane et al., 1990). The eastern Norwegian–Greenland Sea margin terminates against the southwestern flank of the Yermak Plateau (Fig. 7-1), a marginal plateau at 600 to 900 m depth.

MAGMATIC – TECTONIC FRAMEWORK

Pre-Tertiary History

Prior to the Paleocene–Eocene opening of the Norwegian–Greenland Sea, the continental part of the present margin was part of a shallow epicontinental sea between Eurasia and Greenland, which extended into the North Sea and the Barents Sea (Fig. 7-3). The epicontinental sea was structured into regional basins and platforms; its geological history in a regional perspective has been discussed in a series of recent papers based on numerous MCS profiles and commercial drilling data in the North Atlantic realm (Doré and Gage, 1987; Larsen, 1987; Ziegler, 1988, 1989; Doré, 1991; Faleide, Vågnes, and Gudlaugsson, in press).

After the Caledonian orogeny, extensive erosion removed much of the relief, and sedimentation started on an epeirogenic, subsiding, post-Caledonian surface.

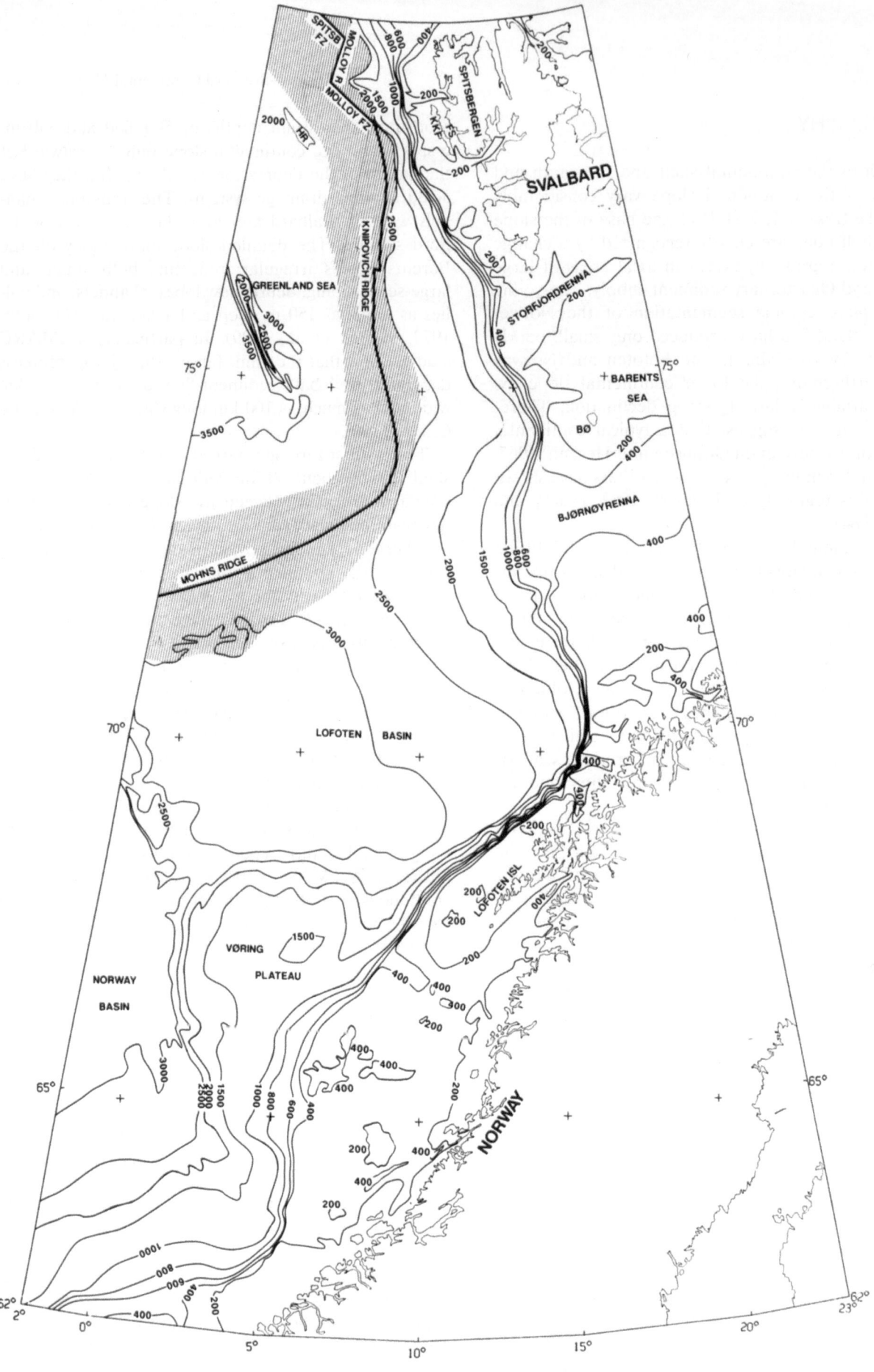

FIGURE 7-2. Bathymetry of Norway–Svalbard continental margin (modified from Perry et al., 1980). Contour interval: 200 m between 200 and 1,000 m; 500 m between 1,000 and 3,500 m. Mohns and Knipovich axial province is hachured. BØ = Bjørnøya; FS = Forlandsundet; HR = Hovgaard Ridge; KKF = Kong Karls Forland.

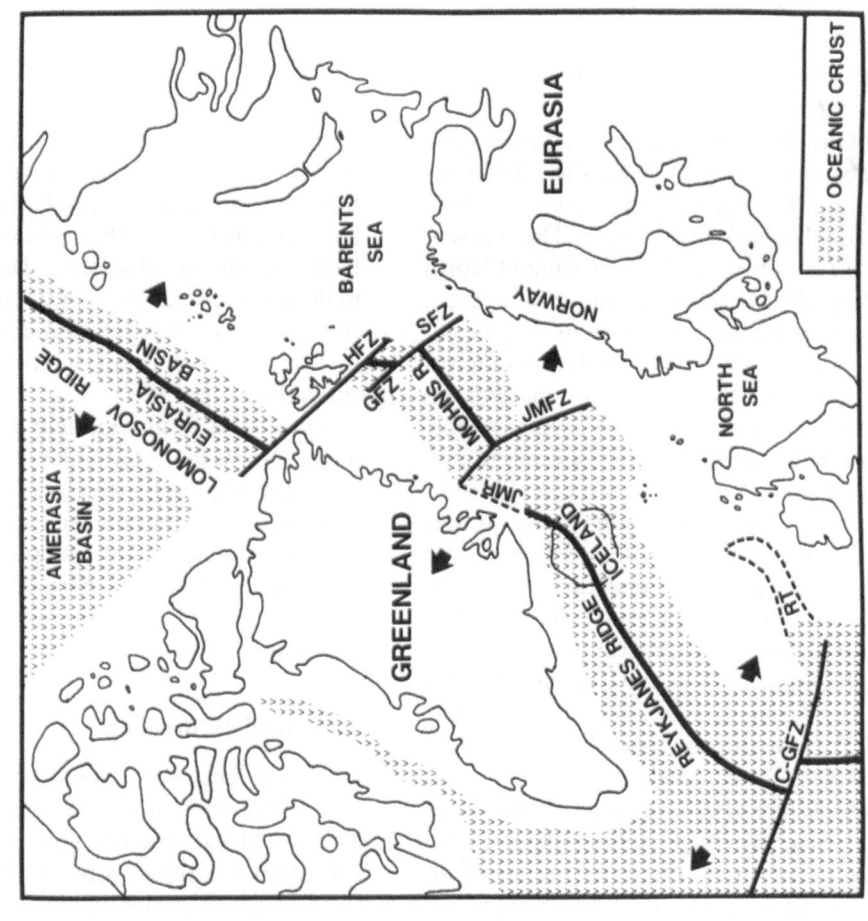

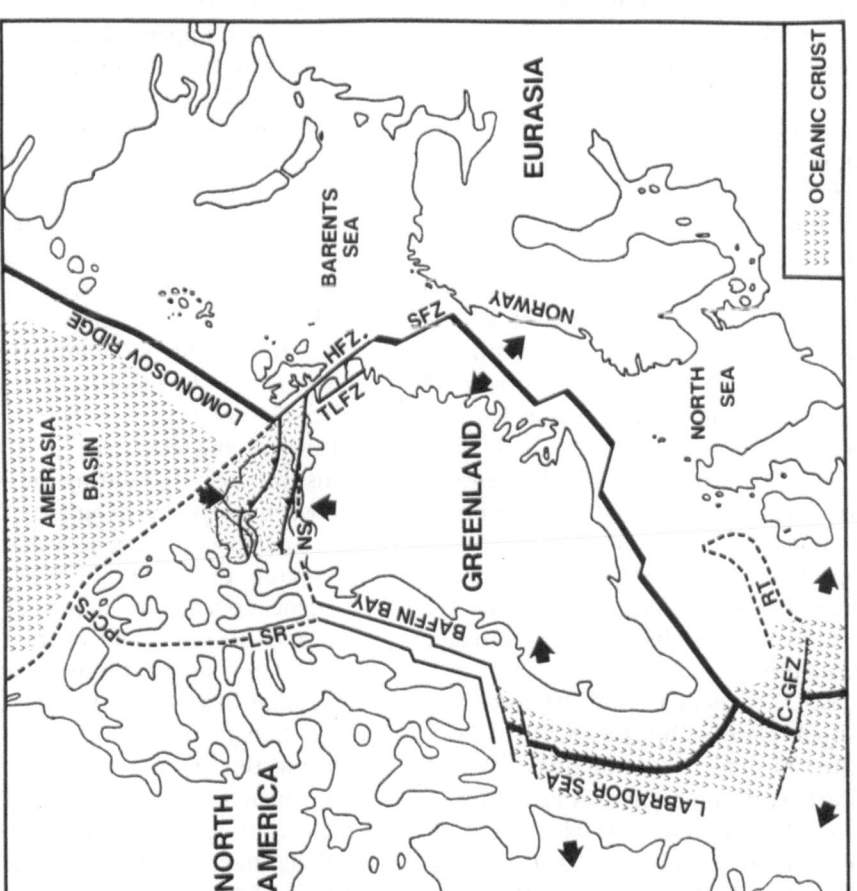

FIGURE 7-3. Plate-tectonic setting of North Atlantic–Artic region at Paleocene–Eocene (left) and Eocene–Oligocene (right) transitions, showing predominantly rifted margin setting south of 70°N and sheared setting farther north. C-GFZ = Charlie-Gibbs Fracture Zone; GFZ = Greenland Fracture Zone; HFZ = Hornsund Fault Zone; JMFZ = Jan Mayen Fracture Zone; JMR = Jan Mayen Ridge; LSR = Lancaster Sound Rift; NS = Nares Strait; PCFZ = Parry Channel Fracture Zone; RT = Rockall Trough; SFZ = Senja Fracture Zone; TLFZ = Trolle Land Fault Zone. Western boundary of De Geer Zone follows Senja Fracture Zone and Trolle Land Fault Zone. Arrows indicate relative plate motion. Based on Faleide, Vågnes, and Gudlaugsson (in press).

Seismic data and drilling results suggest that the region has been a depositional area at least since the Carboniferous (Rønnevik, Eggen, and Vollset, 1983; Larsen, 1987; Brekke and Riis, 1987). The present structure (Fig. 7-4) largely reflects the prominent North Atlantic Late Jurassic–Early Cretaceous rift episode, defined by the regional base-of-Cretaceous seismic reflector. In most places, the seismic resolution deteriorates below this reflector. Along the entire margin, the crystalline continental basement is recognized only locally in MCS profiles, but high-quality seismic refraction studies reveal considerable pre-Cretaceous sediment thicknesses in the main sedimentary basins (Eldholm and Mutter, 1986; Jackson, Faleide, and Eldholm, 1990; Planke, Skogseid, and Eldholm, 1991; Olafsson et al., 1992). The pre-Cretaceous crust has been subjected to several tectonic events of predominantly extensional character. Rift episodes in the Early Carboniferous to Permian and the Triassic/Early Jurassic to Middle/Late Jurassic have been described off Norway (Bukovics et al., 1984; Brekke and Riis, 1987). These episodes are also recognized in the southwestern Barents Sea, where Faleide, Vågnes, and Gudlaugsson (in press) show separate Middle/Late Jurassic, Early Cretaceous, and Late Cretaceous rift phases. These phases document an important temporal relationship between the post-Middle Jurassic tectonic history off Norway and Svalbard and plate tectonic events in the North Atlantic–Arctic realm.

The Late Jurassic–Early Cretaceous rifting and subsequent differential thermal subsidence divided the Norwegian margin into regional basins and platforms (Fig. 7-4). The Lofoten–Vesterålen shelf and the Trøndelag platform stayed relatively elevated with respect to the Vøring and Møre basins, in which rapid subsidence and contemporaneous sediment-fill created a 2 to 8-km-thick Cretaceous sequence. The southwestern Barents Sea developed similarly; rifting was followed by rapid subsidence and filling of the Harstad, Tromsø, and Bjørnøya basins by 5–6 km of shale, which covered most of the structural relief by Cenomanian time (Faleide, Vågnes, and Gudlaugsson, in press). It is important to note that the Barents Sea east of the Ringvassøy–Loppa Fault Complex (Fig. 7-4) was only moderately affected by the Mesozoic extensional tectonics, and has acted as a stable platform probably since late Paleozoic time.

In general, post-Caledonian crustal extension and basin formation took place within the North Atlantic region affected by Caledonian deformation. If we include the early Tertiary opening, there is an important spatial evolution of the rift zones. Off Norway, it appears that subsequent rift pulses and basin development migrated westward through time (Skogseid, Pedersen, and Larsen, in press). This is also the case farther north, where late Paleozoic rifting extending

northeast into the Barents Sea proper, whereas Mesozoic rifting gradually moved west to a position between Svalbard and Greenland (Faleide, Vågnes, and Gudlaugsson, in press). The presence of deep sedimentary basins in the southwestern Barents Sea implies that underlying crystalline continental crust is extremely thin (Jackson, Faleide, and Eldholm, 1990), evidence that crustal extension almost reached the seafloor spreading stage.

Thus, successful continental breakup and formation of the Norwegian–Greenland Sea took place in a region in which large areas had undergone late Paleozoic and Mesozoic thinning and extension. The crust was further attenuated by the prominent Late Jurassic–Early Cretaceous rift episode prior to the onset of early Tertiary seafloor spreading (compare with Chapters 1–6).

Early Tertiary Breakup

The incipient plate boundary between Eurasia and Greenland constitutes two structural megalineaments. The North Atlantic rift zone, or the Reykjanes–Mohn Line, between the Charlie-Gibbs and Greenland–Senja fracture zones, and a regional shear zone, the De Geer Zone (Harland, 1969), between Svalbard and Greenland, which apparently continues into the Arctic along the northern Greenland and Canadian continental margins. The two lineaments intersect near the oldest part of the Senja Fracture Zone (Figs. 7-1, 7-3). Hence, the eastern Norwegian–Greenland Sea margin is structurally composed of predominantly northeast-trending rifted segments in the southern region, and approximately north-northwest–trending sheared margin segments in the northern region.

Both the segmentation of the ocean basins (Fig. 7-1) and the margin structural framework (Fig. 7-4) are governed by the plate-tectonic evolution of the Norwegian–Greenland Sea, which comprises two main stages (Talwani and Eldholm, 1977). The earliest oceanic crust was accreted during the negative polarity interval between magnetic anomalies 25 and 24B, about 57.5 Ma (Paleocene–Eocene transition), according to the Berggren et al. (1985) time scale. Throughout the Eocene epoch, Greenland moved northwest with respect to Eurasia, and opened the Norway, Lofoten-Greenland, and southern Greenland Sea basins along a spreading axis offset by the Jan Mayen and Greenland–Senja Fracture Zone systems (Figs. 7-1, 7-3). During this period, the northern Greenland Sea had not opened, and plate motion was achieved by continent–continent translation along a regional transform linking the Atlantic with the Eurasia Basin in the Arctic Ocean. This plate translation, however, caused deformation along the plate boundary, which created a pull-apart setting in the south, whereas transpression

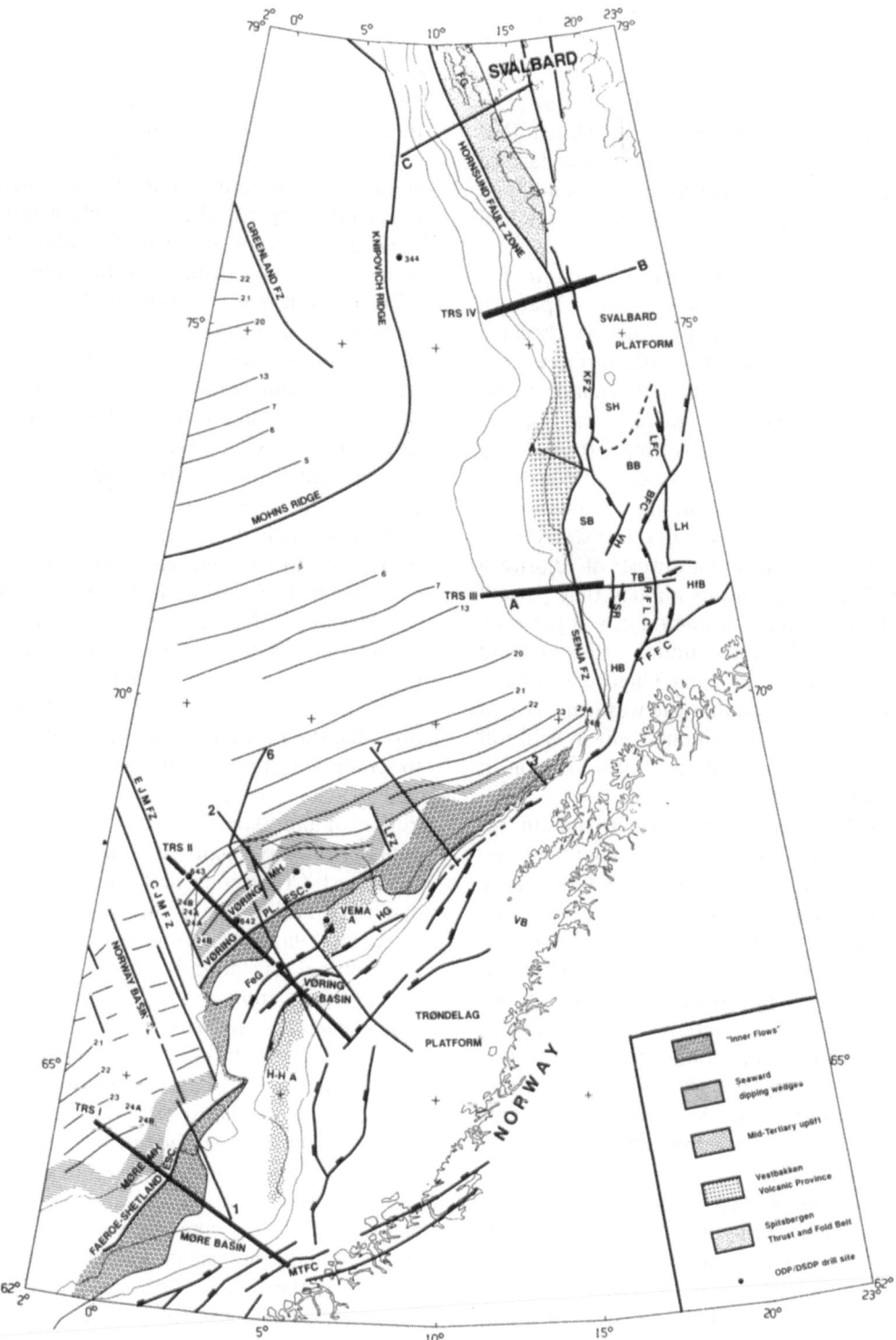

FIGURE 7-4. Major structural elements between 62 and 79°N. Principal margin transects I–IV shown by thick lines; other seismic sections indicated by thinner lines. BB = Bjørnøya Basin; BFC = Bjørnøyrenna Fault Complex; FG = Forlandsundet Graben; FeG = Fenris Graben; HB = Harstad Basin; HfB = Hammerfest Basin; HG = Hel Graben; HHA = Helland–Hansen Arch; KFZ = Knølegga Fault Zone; LFC = Leirdjupet Fault Complex; LFZ = Lofoten Fracture Zone; LH = Loppa High; MTFC = Møre–Trøndelag Fault Complex; RLFC = Ringvassøy–Loppa Fault Complex; SB = Sørvestsnaget Basin; SH = Stappen High; SR = Senja Ridge; TB = Tromsø Basin; TFFC = Troms–Finnmark Fault Complex; VB = Vestfjorden Basin; VH = Veslemøy High. Eastern boundary of De Geer Zone defined by Troms–Finnmark, Ringvassøy–Loppa, Bjørnøyrenna, and Leirdjupet Fault Complexes.

initiated the Spitsbergen Orogeny in the north (Eldholm, Faleide, and Myhre, 1987). The latter is manifested by the Spitsbergen Fold and Thrust Belt along the western coast of Svalbard. Only the eastern part of the deformed region is exposed on land (Fig. 7-4); the remainder is buried beneath the continental margin.

When Labrador Sea spreading terminated in the early Oligocene, the lithospheric stress pattern changed, and the relative plate motion became west–northwest, opening also the northern Greenland Sea. Nonetheless, complete continental separation was not achieved before the middle or late Miocene (Lawver et al., 1990; Eldholm, 1990). This setting left two large shear-margin segments, the Senja Fracture Zone and the Hornsund Fault Zone, extending to almost lat. 80°N (Sundvor and Austegard, 1990), offset by a north-east-trending rifted segment southwest of Bjørnøya (Figs. 7-3, 7-4). This relationship implies that part of the Hornsund Fault Zone, which acted as a shear zone during the Eocene, later came under extension and became a rifted margin in the early Oligocene. Myhre and Eldholm (1988) proposed that the western block of the Hovgaard Ridge is a continental sliver split off from the eastern continental margin during the onset of Oligocene crustal extension.

Early investigations of the Norwegian margin revealed buried basement highs in the vicinity of the rifted parts of the continent–ocean boundary. These marginal highs are elevated with respect to the adjacent oceanic crust, exhibit a smooth and strongly opaque acoustic basement surface, and are bounded landward by escarpment-like structures (Talwani and Eldholm, 1972). The Faeroe–Shetland, Vøring Plateau, and Greenland escarpments bound features of this kind. Another poorly developed high (Myhre and Eldholm, 1988) within the Vestbakken Volcanic Province offsets the two shear margins north of lat. 70°N (Fig. 7-4).

Multichannel seismic surveying has documented a smooth basement underlain by a distinctive, but laterally varying, seismic subbasement signature, which locally includes prominent wedges of seaward-dipping reflections (Hinz and Weber, 1976; Mutter, Talwani, and Stoffa, 1982; Talwani, Mutter, and Hinz, 1983; Mutter, Talwani, and Stoffa, 1984; Hinz et al., 1987; Skogseid and Eldholm, 1987). Deep Sea Drilling Project Leg 38 drilled at the Vøring marginal high and recovered basalts at the level of acoustic basement (Talwani, Udintsev, et al., 1976). However, a much improved understanding of the nature of the Vøring marginal high was obtained during ODP Leg 104, when Hole 642 E (Fig. 7-4) penetrated the innermost part of the main seaward-dipping wedge and recovered tholeitic flow basalts and interbedded sediments. Beneath the dipping wedge, Hole 642 E terminated in another interbedded series of flow basalts with dacitic

composition (Eldholm, Thiede, Taylor, et al., 1987, 1989). The entire interbedded sequence was emplaced in a terrestrial environment.

Subbasement reflection sequences near the the transition between continental and oceanic crust also are observed along rifted margins elsewhere in the North Atlantic north of the Charlie-Gibbs Fracture Zone (Fig. 7-1). Thus, drilling at the Vøring Plateau and West Rockall margin (Roberts, Schnitker, et al., 1984) appears to confirm the existence of large extrusive constructions along these margins. The extrusives, which are compositionally and temporally similar to the plateau basalts of the onshore North Atlantic Volcanic Province (Fig. 7-1; Upton, 1988), were emplaced during continental breakup by a transient magmatic pulse lasting about 3 m.y.

Igneous activity during breakup has led to a classification of passive margins into volcanic and nonvolcanic types; the Vøring margin represents the volcanic end-member (Eldholm, 1991). Although, there is much controversy about the processes causing the transient pulse (Mutter, Buck, and Zehnder, 1988; White and McKenzie, 1989), the well-studied Vøring margin has provided data for a breakup model that may be applied to other volcanic margins as well (Eldholm, Thiede, and Taylor, 1989; Skogseid and Eldholm, 1989; Pedersen and Skogseid, 1989). The rifted volcanic margins in the Norwegian–Greenland Sea evolved through the latest Cretaceous and/or Paleocene lithospheric extension and rifting within previously thinned crust. This was followed by significant decompressional partial melting, crustal underplating, and intrusions in the lower crust, enhanced uplift of the rift zone, and large-scale erosion. With time, the dikes and sills shallowed and merged toward the incipient line of breakup, emplacing dacitic lavas by shallow melting of crustal rocks during the latest rift stage. The asthenospheric plume reached the surface in the earliest Eocene, and formed Icelandic-type crust. The subaerial emplacement of the tholeiitic flows and high rates of magma production led to construction of the marginal high and created the subbasement reflection patterns. Moreover, the basaltic lavas overflowed large areas of the neighboring continental crust. The landward extent of these inner flows (Talwani, Mutter, and Hinz, 1983) is shown in Figures 7-1 and 7-4. The volcanism, which was most intense during breakup, gradually waned, and about 3 m.y. after breakup, the injection center had subsided to oceanic depth and was producing normal oceanic crust.

Because the marginal highs were elevated above the growing oceanic crust and the sedimentary basins on the landward side, they greatly influenced the Paleogene margin development; they exerted particular control on depositional conditions on the continental slope (Skogseid and Eldholm, 1989) and on the oldest oceanic crust (Johansen and Eldholm, 1989). The postopening history is generally characterized by sedimentation on a thermally subsiding margin without large-scale structural deformation. The margin was affected by local Tertiary deformation, however, and appreciable thermal subsidence, which varied along strike. On the other hand, there is evidence of regional late Tertiary continental uplift on the order of at least 1.5–2.0 km in the North Atlantic region from about lat. 55 to 80°N (Stuevold, 1989). This uplift, and northern hemisphere glaciation, dominated the Neogene and Quaternary depositional environment. In the following sections, we address the Cenozoic evolution of this region by describing five individual margin provinces.

MØRE MARGIN

The Møre margin includes the large Møre Basin and the Møre marginal high (Figs. 7-1, 7-4). The relatively unstructured Møre Basin is located between the Norwegian mainland, the Trøndelag Platform, and the Faeroe–Shetland Escarpment, and continues southward toward the West-Shetland Basin and the Viking Graben in the North Sea, and northward toward the Vøring Basin. There is no clear structural transition between these basins.

Seismic transect MVM-1 [Fig. 7-5 (foldout)] illustrates the margin geology and, changing sediment thickness between the oceanic Norway Basin and the Norwegian coast. The Møre Basin sediments can be divided into pre- and postopening sequences separated by an earliest Eocene ash marker (Jøgensen and Navrestad, 1981; Bøen, Eggen, and Vollset, 1984; Bukovics et al., 1984; Hamar and Hjelle, 1984; Brekke and Riis, 1987). This ash marker [reflector TP; Fig. 7-5 (foldout)] is correlated with a distinctive, well-dated tephra horizon in North Sea wells (Knox and Morton, 1988; Morton and Knox, 1990), and extends northward along the Norwegian margin. Most authors recognize two regional Mesozoic reflectors, mid-Cretaceous and base-Cretaceous. However, the great thickness of post-Jurassic sediment allows mapping of the pre-Cretaceous sediments only locally. A series of seismic horizons has been identified in the postopening sequence, of which the base-Pliocene and early Miocene reflectors [O′ and A; Fig. 7-5 (foldout)] are regionally most consistent. The eastern basin flank (adjacent to the coastline) and the Trøndelag Platform reveal tilted basement blocks and horst-like features, which locally trap late Paleozoic and early Mesozoic sediments; this geometry reflects a structural fragmentation during the Late Carboniferous (Rønnevik and Navrestad, 1977; Gravdal, 1985) or Triassic–Jurassic rift phases (Hagen, 1987).

The Late Jurassic–Early Cretaceous rift episode dominated the structural configuration, and created several intrabasinal highs, which are expressed at the base-Cretaceous level (Hamar and Hjelle, 1984; Blystad

et al., in press). Crustal extension led to rapid basin subsidence and deposition of marine shales with intercalated sandstones near the highs and along the eastern basin flank, to uplift of the adjacent rift flanks, and to Early Cretaceous wrenching along the Møre–Trøndelag Fault Zone (Hagevang and Rønnevik, 1986). Structural deformation did not reach the Upper Cretaceous sequence; a gentle basin relief developed by the end of the Cretaceous and has been maintained until present. Except for middle Tertiary doming along the Helland–Hansen Arch, the post-Paleocene Møre Basin is not structurally deformed.

West of the escarpment, the Møre marginal high is underlain by a series of irregular, flat-lying, subbasement reflections, which change to seaward dipping wedges of varying sizes and geometries before reaching the normal oceanic crust in the Norway Basin. The extrusive complex west of the Faeroe–Shetland Escarpment received much less sediment than the adjacent Møre Basin as it was progressively covered by sediment from south to north. The sediment column thins from 1.5 sec in the south to less than 0.5 sec in the northern region. It appears that the oldest sediments over the southern part of the marginal high were laid down during the late Oligocene to early Miocene, whereas the northern part is covered with mainly Pliocene–Pleistocene deposits [possibly floored by a thin veneer of Miocene sediments; Fig. 7-5 (foldout)]. Although the difference in sediment distribution may be related to the initial relief and subsidence history of the marginal high, we believe that there also was a considerable contribution from southern sediment sources, probably from platform areas around the Shetland Islands. A strongly reflecting sedimentary horizon in the eastern Norway Basin tends to mask the underlying section both in single-channel (Eldholm and Windisch, 1974) and in large-energy-source MCS profiles [Fig. 7-5 (foldout)] Consequently, a stratigraphy for the eastern Norway Basin, including most of the marginal high, has not yet been established.

There are many local, smooth, high-amplitude reflectors in the Upper Cretaceous and Paleocene sequences in the western part of the Møre Basin [Fig. 7-5 (foldout)]. These features, which parallel the stratification and shallow toward the inner flows east of the escarpment, we interpret as sills. Landward of the inner flows, there also is evidence of strong reflectors, probably dikes, cutting across Upper Cretaceous and Paleocene strata. The intrusions in the outer Møre Basin have been described by Gatliff et al. (1984), Hitchen and Ritchie (1987), and Gibb, Kanaris-Sotiriou, and Neves (1986).

The inner flows, defined by a band of irregular high-amplitude reflectors, are interpreted as tholeiitic lavas, which flowed onto the adjacent continental crust during breakup. This interpretation is consistent with the results of commercial drilling in the region (Gibb,

Kanaris-Sotiriou, and Neves, 1986; Gibb and Kanaris-Sotiriou, 1988). The fact that the earliest Eocene ash marker locally onlaps the inner flows, is evidence that the ashes are slightly younger than the flows. The Faeroe–Shetland Escarpment is more irregular and much less prominent than its counterpart at the Vøring margin, and local volcanic centers have been identified (Smythe et al., 1983; Gatliff et al., 1984). Although, the detailed igneous and sedimentary stratigraphy remains to be worked out, we infer that construction of the Faeroe–Shetland extrusive complex was preceded by a significant phase of intrusive activity. The actual location of the escarpment may have been governed by a pre-existing structure, but its morphology has been interpreted as the feather-edge of the extrusive flows (Smythe, 1983). Similarly, the landward edge of the inner flows may represent another extrusive feather-edge (Eldholm, Thiede, and Taylor, 1989).

Below the late Pliocene to base-Quaternary unconformity, the dip of the strata increases toward the coast on the inner shelf, and the Neogene sediment distribution documents a greatly increased sediment supply from a relatively elevated continent and inner shelf region. Although there is evidence of sedimentary progradation throughout the Tertiary, a markedly rapid seaward migration of the shelf edge started in the Miocene, similar to that observed along the U.S. Atlantic margin (see Chapter 6). Progradation culminated in the Pliocene, when about 2 km of shale, with increasing sand content toward the continent, was deposited. Gravdal (1985) documented a regional uplift of the eastern flank of the Møre Basin from middle Oligocene through Miocene time, and inferred that the uplift and subsequent erosion supplied the large volume of prograding sediments.

In the northern Møre Basin, rapid outbuilding of the shelf may have contributed to the huge downslope mass movements, which caused the recent erosional deformation shown in Figure 7-6. The 290-km-long submarine slide scarp is as deep as 450 m, and the slide volume is about 5,600 km^3, according to detailed investigations by Bugge (1983) and Bugge et al. (1987). These authors proposed that the scarp was excavated by three slide stages, in the period 50–6 ka, triggered by earthquakes in combination with the decomposition of gas hydrates. Redeposition of lower Tertiary to Quaternary sediments took place over most of the northern Norway Basin, including the extinct Aegir Ridge province (Fig. 7-1). We believe, in fact, that the poor seismic penetration in this region may be caused by redeposited, coarse, chaotic sequences of detritus derived from the outer shelf and upper slope.

VØRING MARGIN

The Vøring margin covers roughly the area between the Jan Mayen and Lofoten fracture zones and their prolongation toward the continent, and comprises three main geological provinces; the Vøring Marginal High, the Vøring Basin, and the Trøndelag Platform (Fig. 7-4). The transition to the north is largely unstructured, and consists of the thinning and shallowing of beds toward the Lofoten shelf and an extension into the Early Cretaceous Vestfjord Basin (Brekke and Riis, 1987). The structural and depositional style of the margin are shown in Figures 7-7 (foldout) and 7-8.

The Trøndelag Platform, defined at the base-Cretaceous level, underwent active rifting and major basin development prior to the Cretaceous. The deepest reflector is a probable middle Permian major erosional unconformity; another rift phase is documented in the Middle to Late Triassic (Brekke and Riis, 1987). The base-Cretaceous is the deepest regional horizon in the Vøring Basin, but Skogseid and Eldholm (1989) have inferred a pre-Cretaceous sequence of about the same thickness as on the Trøndelag Platform. They also

FIGURE 7-6.   Seismic profile NOR-JM-17 showing large-scale recent mass movements at northern Møre margin. Line 1 in Figure 7-4.

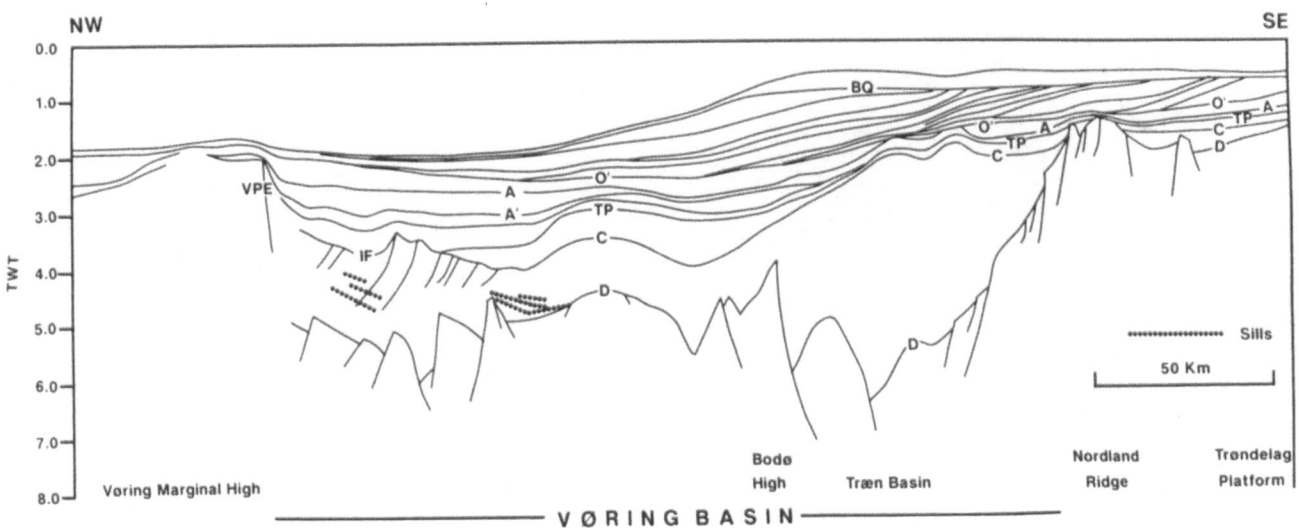

FIGURE 7-8. Interpretive line drawing of seismic profiles C164 and B18-83 crossing Vøring Basin and southern tip of Vema Arch. Line 2 in Figure 7-4. Note outbuilding of Pliocene sedimentary wedges and Pliocene depocenter beneath shelf edge. BQ = base-Quarternary; VPE = Vøring Plateau Escarpment; other abbreviations given in Figure 7-5 (foldout). (Modified from Stuevold, 1989.)

noted that seismic correlations are uncertain across the down-faulted transition zone between the platform and the Vøring Basin. The Late Jurassic–Early Cretaceous tectonism formed a number of intrabasinal structures (Fig. 7-4). Postrift regional subsidence and Early Cretaceous sedimentation smoothed the relief, but local highs stayed emergent until middle Cretaceous time. Subsequently, regional subsidence continued and allowed deposition of a 3 to 5-km-thick Upper Cretaceous sequence. The seismic character of this sequence is compatible with that of a basin dominated by homogeneous marine shale interspersed with local lenses of coarser sediment adjacent to the Trøndelag Platform (Hasting, 1986) and to the intrabasinal highs.

The Cenozoic history of the Vøring Basin reflects igneous activity associated with opening of the Norwegian–Greenland Sea. In the latest Cretaceous and(or) early Paleocene, the outer part of the basin, particularly the Fenris and Hel grabens and the region farther west (Fig. 7-4), experienced extension, which caused listric faulting, uplift, and erosion, followed by post-Paleocene regional subsidence; this allowed deposition of 1–2 km of Paleogene sediments. Post-opening subsidence was largely restricted to a 100 to 150-km-wide region east of the Vøring Plateau Escarpment (Skogseid and Eldholm, 1989). Thus, the outer part of the Vøring Basin was characterized by Paleogene differential extension and subsidence, whereas the inner part was stable. The rift-induced uplift is documented by: (1) the base-Tertiary reflector, which forms an erosional unconformity in the outer basin; and (2) the thinning of Paleocene and lower Eocene sequences to the east and onlapping to the west [Figs. 7-7 (foldout) to 7-9]. The

lowermost Eocene ash marker [reflector TP; Figs. 7-7 (foldout), 7-8] locally onlaps the inner flows, as in the Møre Basin. Here it is often broken by narrow conical depressions, which have been interpreted as volcanic vents (Skogseid and Eldholm, 1989). The vents extend southward into the northern Møre Basin. These vents are observed, however, only in the parts of the basins that experienced Paleogene subsidence. Here a series of discontinuous, high-amplitude reflectors within the Mesozoic and Paleocene sediments are interpreted as sills and low-angle dikes. These reflectors shallow toward the eastern boundary of the inner flows, and are considered to be an intrabasinal intrusive succession similar to that in the Møre Basin.

The relative subsidence of the margin is documented by the nature of the Paleogene and lower Miocene sequences. Actually, the inner Vøring Basin and the Trøndelag Platform exhibit a much more condensed pre-Pliocene section than that of the outer Vøring Basin [Figs. 7-7 (foldout) to 7-9]. On the other hand, the upper Neogene sequence is greatly expanded on the eastern margin. The base-Pliocene unconformity, in particular, floors a huge, up to 1.5-km-thick, prograding wedge of Pliocene–Pleistocene sediments, centered near the present shelf edge (compare with Chapter 6). These sediments prograde westward and downlap the unconformity, which gradually becomes less distinct and becomes a conformable horizon in the western basin region (Fig. 7-8). The Pliocene and older units are truncated by an unconformity formed by late Pliocene and Quaternary glacial erosion.

West of the Vøring Plateau Escarpment, the lower Miocene unconformity [Fig. 7-7 (foldout)] separates

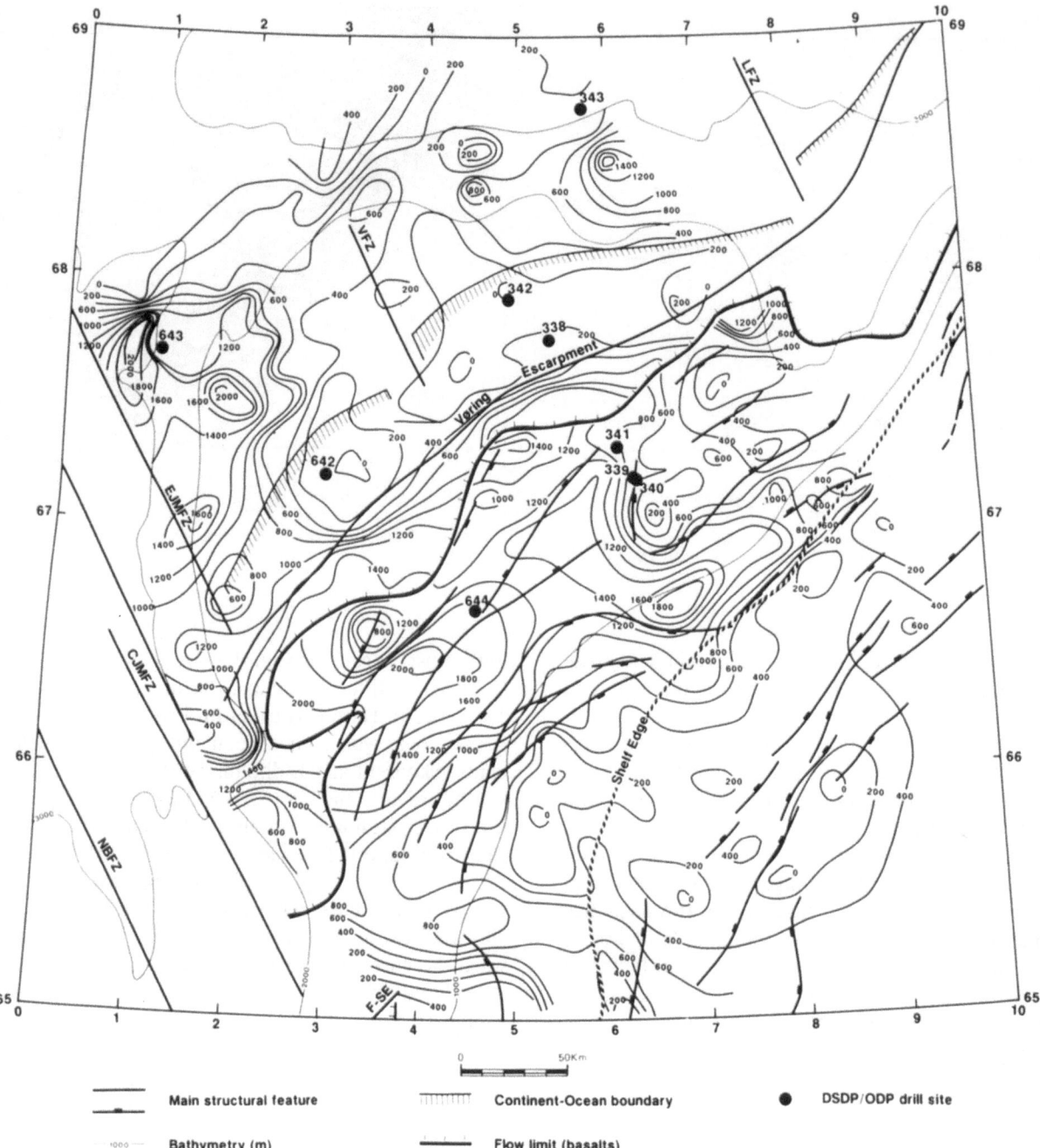

FIGURE 7-9.   Thickness of Paleogene and lowermost Miocene sediments on outer Vøring margin. Contour interval 0.2 sec two-way traveltime. Reflectors C, EE, and A defined in Figure 7-7 (foldout). Based on Skogseid and Eldholm (1989).

sediments above the basalts into a lower sequence of relatively uniform depositional units, and an upper irregular, locally disturbed sequence. The unconformity is also the youngest horizon documenting erosion of large areas of the marginal high. By correlation with DSDP and ODP drill sites, Skogseid and Eldholm (1989) proposed a division into seven stratigraphic sequences [1-7; Fig. 7-7 (foldout)]. The lower Eocene sequence is composed of erosional products from the

young lavas and volcaniclastic sediments trapped in a basement depression, whereas the overlying middle Eocene to middle Oligocene zeolitic mudstones were laid down during a regional transgression, which eroded adjacent basement highs. The middle Oligocene to lower Miocene sequence reflects a more rapid subsidence of the outer plateau, but the exposed summit region was still the main source area. Since early Miocene time, regional subsidence has caused a transition from shallow to deep marine environments over the entire plateau, but there is evidence of both local and regional erosion and redeposition, which probably reflects changing current patterns and sea-level fluctuations. The strongly disturbed lower to middle Miocene diatomaceous muds and oozes were laid down during a period of rapid progradation followed by two middle Miocene to upper Pliocene depositional sequences of primarily biogenic siliceous muds and oozes (compare with U.S. Atlantic margin; Chapter 6). The uppermost sequence consists of upper Pliocene and Pleistocene glacial and interglacial deposits, and there is evidence of erosion at the seafloor, both at the summit and along the outer flank of the plateau. The Paleogene damming effect of the marginal high is reflected by regionally thinner deposits off the Vøring Plateau than on oceanic crust of the same age farther north (Eldholm and Windisch, 1974).

The nature and origin of the spectacular linear Vøring Plateau Escarpment has been much discussed since Talwani and Eldholm (1972) related it to the continent–ocean boundary. That idea was supported by Talwani, Mutter, and Hinz (1983) and by Mutter, Talwani, and Stoffa (1984), whereas Smythe (1983) considered it simply to be a flow front, which demarcates a paleoshoreline. Relative movements along the escarpment throughout the Tertiary were proposed by Caston (1976); Theilen, Uenzelmann, and Gimpel (1984) have inferred recent motion. These interpretations contrast with that of Mutter (1984), who argued against significant movements since the end of the early Eocene. Skogseid and Eldholm (1987, 1989) suggested that the location was predetermined by a Late Jurassic–Early Cretaceous fault, which was reactivated during the early Tertiary rifting stage; movement gradually diminished through the Paleogene. Accordingly, it may be considered a structurally controlled flow boundary. Furthermore, Skogseid and Eldholm (1987) placed the continent–ocean boundary seaward of the escarpment at the termination of a base reflector for the innermost part of the seaward dipping wedge [reflector K; Fig. 7-7 (foldout)]. They interpreted the area between the landward limit of the inner flows and the continent–ocean boundary as strongly intruded or transitional crust. After the onset of seafloor spreading, the eastern shelf and the continent progressively became the dominant sediment sources for the continental slope. Nonetheless, terrigenous sediments first reached the escarpment in the middle Eocene, and gradually overflowed the marginal high, which except for local peaks, became sediment-covered from middle Oligocene to early Miocene (Skogseid and Eldholm, 1989).

There is evidence that intrabasinal deformation was superimposed on the postopening marginal subsidence. This resulted in relative uplift of parts of the margin.

These uplift events include intrabasinal arching, regional continental "Fennoscandian" uplift, and glacial rebound.

Early single-channel seismic surveying identified a local diapir province, the Vema Arch, on the northern Vøring Plateau, 40–50 km landward of the escarpment (Talwani and Eldholm, 1972). The prominent diapirs rise more than 70 m above the seafloor. Piston coring (Bjørklund and Kellogg, 1972) and DSDP Leg 38 drilling (Caston, 1976) recovered Eocene shales, which pierce from broad, elongate, buried anticlines within the early Tertiary sediments (Skogseid and Eldholm, 1989; Fig. 7-9). The diapir province forms the intersection of two elongate arch systems at the continental part of the margin (Figs. 7-4, 7-9), defined by a regional dome-like elevation of Tertiary reflectors. One arch is located in the outer Vøring Basin parallel to the escarpment; it reflects early Tertiary rift-induced movements. This arch subsided during the Eocene, but maintained a positive relief, covered only by thin hemipelagic sediments until the Oligocene. Presently, the base-Tertiary reflector is elevated 1–2 km with respect to the adjacent basin.

The Helland–Hansen Arch and local subparallel domes trend north–northwest in the central Vøring Basin, and continue south into the Møre Basin. Doming along the arch has been related to the mobilization of evaporites, to igneous intrusions, to deep crustal movements, and to tectonic events in the Norwegian–Greenland Sea (Mutter, 1984; Hinz, Dostman, and Hanisch, 1984; Rønnevik, Jørgensen, and Motland, 1979). Although Mutter (1984) indicated motion throughout the Cretaceous, and ending in the Oligocene, most studies favor Oligocene and Miocene movement (Bøen, Eggen, and Vollset, 1984; Bukovics et al., 1984; Hinz, Dostman, and Hanisch, 1984). On the other hand, Skogseid and Eldholm (1989) proposed a two-stage deformation model considering the position of the arch with respect to the areas affected by postopening thermal subsidence and Neogene sedimentary loading.

The Helland–Hansen Arch is a broad anticline with a relief of almost 2 km, located east of the Paleogene depocenter and under the distal part of the large Pliocene prograding sequence [Figs. 7-4, 7-7 (foldout)]. The central part of the arch is most elevated; from there it plunges toward the north and south following the trend of deep-seated Late Jurassic–Early Cretaceous faults. Paleocene–Eocene reverse faults are present along the western flank of the arch, whereas local Neogene normal faults are found on the eastern side. The base-Pliocene unconformity at the top of the arch indicates that relief was accentuated in Miocene–Pliocene time. Subsequently, the prograding Pliocene wedges downlapped onto the young arch. Differential loading of the underlying thick Cretaceous marine shales caused mass movements and further growth of the arch. The Pliocene prograding sequences did not reach the apex of the arch until the late Pliocene. We postulate that the growth of the arch kept pace with sediment loading and subsidence until this time (Stuevold, Skogseid, and Eldholm, 1990).

Early geomorphological studies revealed remnants of a pre-glacial Fennoscandian peneplane at an elevation about 1.0–1.5 km above sea level. These studies, summarized by Gjessing (1967), are interpreted in terms of a Tertiary continental uplift, but neither the timing, magnitude, nor mechanism are well understood. Because Tertiary continental uplift also has taken place along other margins in the North Atlantic, both Torske (1972) and Talwani and Eldholm (1972) suggested that crustal flexuring took place in response to the opening of the Norwegian–Greenland Sea, and that the continental margin subsequently subsided around a hingeline near the present shelf edge. Although some flank uplift may have occurred in response to the early Tertiary opening, the margin stratigraphy rules out a large-scale event. Stuevold (1989) showed, however, that the sequence stratigraphy at the Vøring margin is compatible with initiation of large-scale, flexural, continental uplift in the late Oligocene, continuing into the Pliocene. The uplifted continent and the exposed parts of the shelf provided the source for large late Neogene sediment wedges illustrated in Figure 7-8.

Neogene uplift and margin subsidence took place on either side of a hingeline near the Late Jurassic–Early Cretaceous fault systems separating the Trøndelag Platform from the Vøring Basin (Fig. 7-4). Farther south, the hingeline appears to follow the Møre–Trøndelag Fault Complex, whereas the hingeline is near the shelf edge at the Lofoten–Vesterålen margin (Stuevold, 1989). This led to the inference that the first-order pre-Tertiary structural framework may have governed the extent of Neogene uplift and subsidence. The onset of northern hemisphere glaciation at the Vøring margin about 2.6 Ma (Thiede, Eldholm, and Taylor, 1989) again caused increased erosion due to both glacial activity and crustal rebound. These events created the late Pliocene unconformity and enlarged the late Neogene sedimentary wedges.

LOFOTEN – VESTERÅLEN MARGIN

The southern shelf is characterized by relatively thin, block-faulted, pre-Cenozoic sediments, structured by the Late Jurassic–Early Cretaceous rift episode, and possibly reactivated in the Late Cretaceous (Brekke and Riis, 1987; Blystad and Sand, 1988). The basement-involved faulting has exposed Precambrian rocks on the Lofoten Islands; basement also crops out on a structural high just landward of the shelf edge

(Eldholm, Sundvor, and Myhre, 1979; Jørgensen and Navrestad, 1989; Mokhtari, Pegrum, and Sellevoll, 1989). The basement ridges also are clearly recognized in the gravity field. In most places less than 100 m of Quaternary sediments covers the seafloor (Lien, 1976). The Tertiary section is thin or absent, and a down-faulted basin filled with Lower Cretaceous sediments separates the islands from the outer shelf high (Blystad and Sand, 1988). A narrow Tertiary wedge pinches out below the Quaternary cover just east of the shelf edge. This platform-like shelf setting appears to be representative, however, only for the margin south of about lat. 69°N. At this latitude, the Mesozoic relief and the outer high disappear, or plunge beneath a thick Cenozoic wedge covering the main part of the shelf. Blystad and Sand (1988) attributed this change to Late Cretaceous–Early Tertiary down-faulting along large listric fault planes. Eldholm, Thiede, and Taylor (1989) indicated that an early transfer-fault might be responsible for the change in margin style.

The continental slope is underlain by a prominent acoustic basement reflector (Fig. 7-10), which continues into the oceanic crust in the Lofoten Basin, and is correlated with the lowermost Eocene basalts [reflector EE; Fig. 7-7 (foldout)] at the Vøring Plateau. The basement surface changes from an irregular oceanic crustal relief in the Lofoten Basin to a smooth basement of pre-anomaly 23 age in a 50-km-wide zone west of the foot of the continental slope. Distinct seaward-dipping reflections are observed in this region, where the basement elevation shallows gently toward the foot of the slope, but does not form a prominent marginal high similar to those farther south. In some places, this zone is terminated by basement relief resembling a flow front or a structural high (Johnson, Freitag, and Pew, 1971; Eldholm and Windisch, 1974; Eldholm, Sundvor, and Myhre, 1979; Hinz et al., 1987; Mokhtari, Sellevoll, and Olafsson, 1987). Many investigators have previously interpreted these features as an extension of the Vøring Plateau Escarpment. Because of the quite different geological and geophysical signature on either side of the Lofoten Fracture Zone, however, we restrict the use of the term Vøring Plateau Escarpment to the Vøring margin.

The acoustic basement reflector beneath the continental slope is broken by extensive block faulting north of about lat. 69°N. The fault displacement increases northward, but does not involve the overlying sediments. Fault scarps with throws of more than 700 m have been reported by Eldholm, Sundvor, and Myhre (1979; Fig. 7-10). The MCS profiles do not penetrate the strongly reflecting basalts, but results from ocean-bottom seismometer profiling (Goldschmidt-Rokita et al., 1988) and gravity modelling (Mokhtari, Pegrum, and Sellevoll, 1989) are compatible with the interpretation that thick sediments of probable Paleocene and

LINE L-II

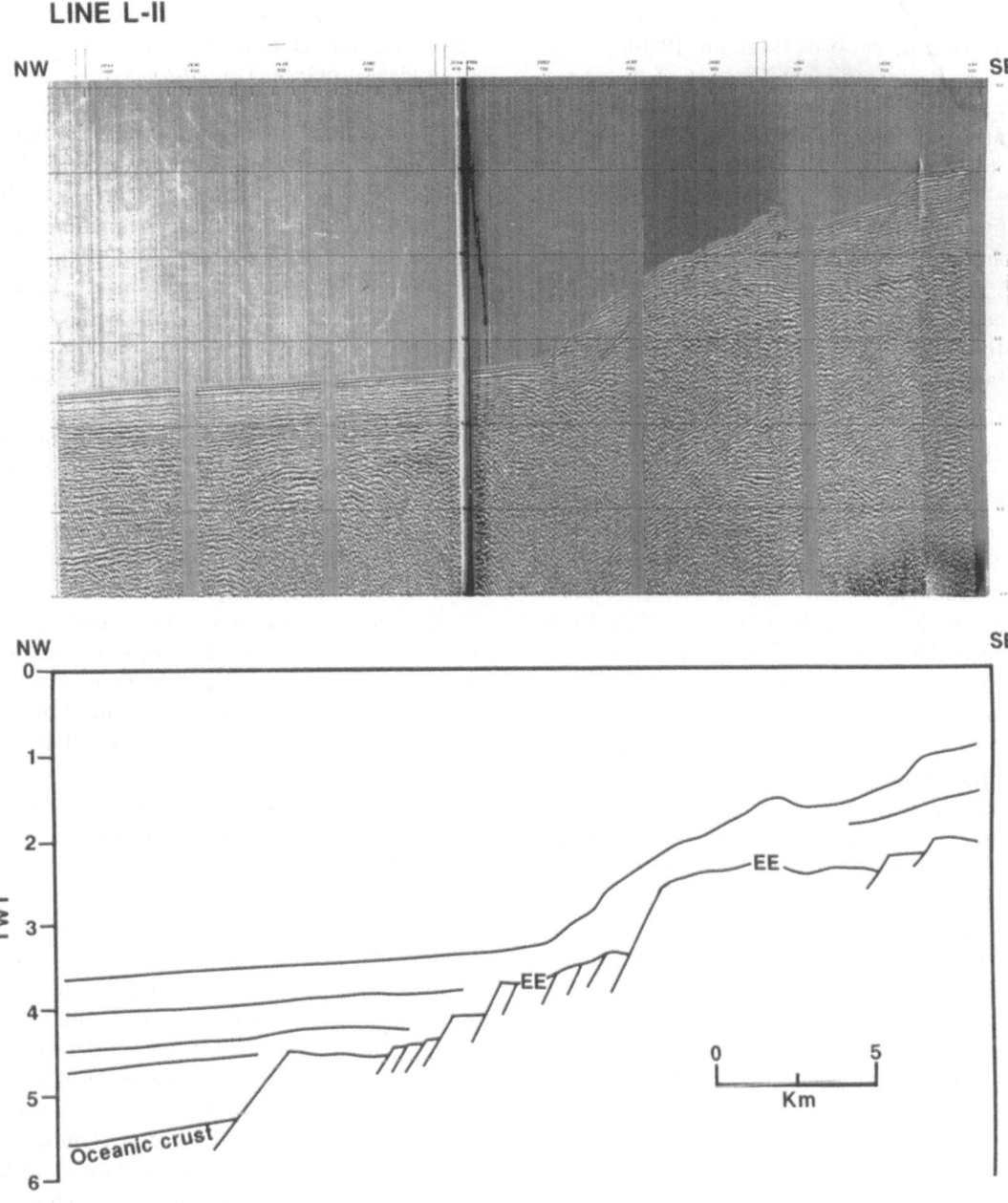

FIGURE 7-10.   Seismic profile L-II (above) and interpretive line drawing (below) across Lofoten–Vesterålen margin. Note that lower Eocene flow basalts (reflector EE) almost reach shelf edge. Line 3 in Figure 7-4. (Modified from Eldholm, Sundvor, and Myhre, 1979.)

Mesozoic age lie below the basalts. Mjelde et al. (in press) estimate that as much as 4 km of sediment is present below the smooth basement surface, which we believe represents flow basalts extruded and faulted during the initiation of seafloor spreading in the Norwegian–Greenland Sea. Hence, the basement reflector here is equivalent to the diachronous reflector EE on top of the Vøring Marginal High.

The character of the Cenozoic slope sediments above the basalts changes along strike (Fig. 7-10). To the

south, the lower sequence is generally conformable with basement, whereas the upper sequence shows upbuilding and progradation on the upper slope and evidence of erosion and some slumping on the lower slope. Farther north, the sediments show a complex depositional history of erosion and redeposition, including significant sliding and slumping (Sundvor, Vogt, and Crane, 1990), changing to a prograded setting near the shelf edge. The irregular slope bathymetry suggests that erosion is active also at present, and has exposed

Upper Cretaceous mudstones, which have been dredged in submarine canyons (Manum, 1966).

The Lofoten Basin sediments consist of an upper, partly transparent, homogeneous sequence and a lower, thicker, stratified sequence, in which beds thin and pinch out toward the gently rising basalts near the base of the slope. The sequences are separated by a regionally distinctive reflector (Johnson, Freitag, and Pew, 1971; Eldholm and Windwich, 1974; Eldholm, Sundvor, and Myhre, 1979). The lack of satisfactory seismic continuity, however, has precluded a direct tie to available drill sites. Thus, a basinwide stratigraphy has not yet been established here.

BARENTS SEA MARGIN

The geology of the western Barents Sea is relatively well known, and has been summarized by Riis, Vollset, and Sand (1986), Gabrielsen et al. (1990), and Faleide, Vågnes, and Gudlaugsson (in press). The post-Caledonian history is dominated by Late Devonian–Early Carboniferous, Middle Jurassic–Early Cretaceous, and early Tertiary rift episodes, each of which included several tectonic pulses. Since the late Paleozoic, when the entire Barents Sea was dominated by crustal extension, structural deformation has migrated westward with time, and has established a basin province in the southwest and a region dominated by wrenching farther north. Apart from epeirogenic movements causing the present elevation differences, the Svalbard Platform and the Barents Sea east of the basin province have been stable regions since the late Paleozoic. The late Mesozoic wrench regime, the De Geer Zone, is bounded by fault complexes on either side. To the east, these include the Troms–Finnmark, Ringvassøy–Loppa, Bjørnøyrenna, and Leirdjupet fault complexes, continuing north along the Knølegga Fault Zone and an inferred paleo-Hornsund Fault west of Spitsbergen (Fig. 7-4). The western boundary consists of the Senja Fracture Zone and the Trolle Land Fault Zone in northeast Greenland (Fig. 7-3; Faleide, Vågnes, and Gudlaugsson, in press).

The western Barents Sea is underlain by great thicknesses of upper Paleozoic to Cenozoic sediments, which locally are more than 15 km thick. Seismic profiles rarely allow definition of a crystalline basement, except west of the continent–ocean boundary, where oceanic basement is recognized. In a regional sense, the area is characterized by thick, relatively undeformed sediment sequences of great lateral extent, overlain by a huge postrift wedge, which extends over both continental and oceanic crust. Figure 7-11 (foldout) shows that this thick wedge has left 5–7 km of middle and upper Cenozoic deposits over the oldest oceanic crust in the Lofoten Basin.

In terms of sediment fill, tectonic setting, and crustal structure, the Barents Sea margin constitutes three geological provinces (Fig. 7-4):

1. The Svalbard Platform, covered by flat-lying upper Paleozoic and Mesozoic (mostly Triassic) sediments.
2. A Mesozoic basin province between the Svalbard Platform and Norway, comprising several highs (Senja Ridge, Veslemøy, and Stappen highs) and subbasins (Harstad, Tromsø, Bjørnøya, and Sørvestsnaget basins) filled with Jurassic, Cretaceous, and, in some places, Paleocene sediments.
3. The oceanic crust, bounded by the sheared margin segments along the Senja Fracture Zone and the southern Hornsund Fault Zone, and the volcanic rifted margin within the Vestbakken Volcanic Province.

The crustal transition is structurally defined by boundary faults along the Senja Fracture Zone and the listric fault complex bounding the Vestbakken Volcanic Province to the southeast. The continent–ocean transition along the Senja Fracture Zone is indeed narrow and distinct (Eldholm, Faleide, and Myhre, 1987), and the oceanic crust can be traced to within 5–10 km of a continental boundary fault along the fracture zone, which marks the onset of a steeply eastward-dipping base of the crust (Faleide et al., 1991).

The Harstad, Sørvestsnaget, Tromsø, and Bjørnøya basins (Fig. 7-4) have experienced large-scale Cretaceous subsidence, which kept pace with deposition of dominantly Lower Cretaceous shales and claystones and Upper Cretaceous claystones. The magnitude of subsidence is illustrated in the Tromsø Basin, where the Middle Jurassic reflector, at a depth of 12–14 km, overlies a considerable section of Triassic and Jurassic siliciclastics and Permo-Carboniferous mixed carbonates, evaporites, and siliciclastics (Gudlaugsson at al., 1987; Jackson, Faleide, and Eldholm, 1990; Faleide, Vågnes, and Gudlaugsson, in press). The configuration of the Sörvestsnaget and Harstad basins is also controlled by pre-Cenozoic structures, but there is a strong imprint of Paleogene deformation in the vicinity of the early Tertiary line of opening (Faleide, Vågnes, and Gudlaugsson, in press). For example, transform motion along the Senja Fracture Zone caused Eocene uplift of the neighboring continental crust, and created a narrow, elongate, outer high just landward of the continent–ocean boundary, which was eroded until Neogene time, when the margin became tectonically quiet.

The evolution of the intrabasinal highs is still poorly known, but they appear to have been rejuvenated as positive features as a result of Late Cretaceous and early Tertiary faulting and differential subsidence. Riis,

Vollset, and Sand (1986) and Brekke and Riis (1987) have suggested that the Senja Ridge developed by two phases of transpression, terminating in Late Cretaceous and post-Paleocene time, whereas Faleide, Vågnes, and Gudlaugsson (in press) postulate Late Cretaceous–early Tertiary normal faulting and salt mobilization. Furthermore, they believe that the Stappen High, protruding south from the Svalbard Platform (Fig. 7-4), was initiated in response to early Eocene rifting and volcanism, rather than to an earliest Oligocene change in stress pattern preferred by Wood, Edrich, and Hutchinson (1989). If Faleide, Vågnes, and Gudlaugsson (in press) are correct, the Stappen High became a source area for the thick Eocene sequences in the Vestbakken Volcanic Province and the Sørvestsnaget Basin, whereas subsequent downfaulting of the volcanic province and renewed volcanism coincided with the early Oligocene event.

The Vestbakken Volcanic Province is defined by a smooth acoustic basement surface elevated above the adjacent oceanic crust. Below the high are irregular subbasement reflections, particularly at the inner part (Faleide, Myhre, and Eldholm, 1988; Myhre and Eldholm, 1988). By analogy to the Vøring and Faeroe–Shetland marginal highs, it is assumed to comprise early Tertiary extrusives emplaced during breakup. Faleide, Myhre, and Eldholm (1988) concluded that the high is underlain by an outer oceanic part and an inner part of extended continental crust that is partly covered by flow basalts and interbedded sediments (Fig. 7-12). Moreover, they interpreted two buried basement peaks penetrating the flows as evidence of a renewed early Oligocene phase of volcanism, apparently related to structural adjustment caused by a change in relative plate motion. Flows within the listric fault complex, which terminates the high landward, may be evidence of large-scale Eocene–Oligocene movements.

Early Cretaceous oblique extension within the De Geer Zone reflects a structural link between the grow-

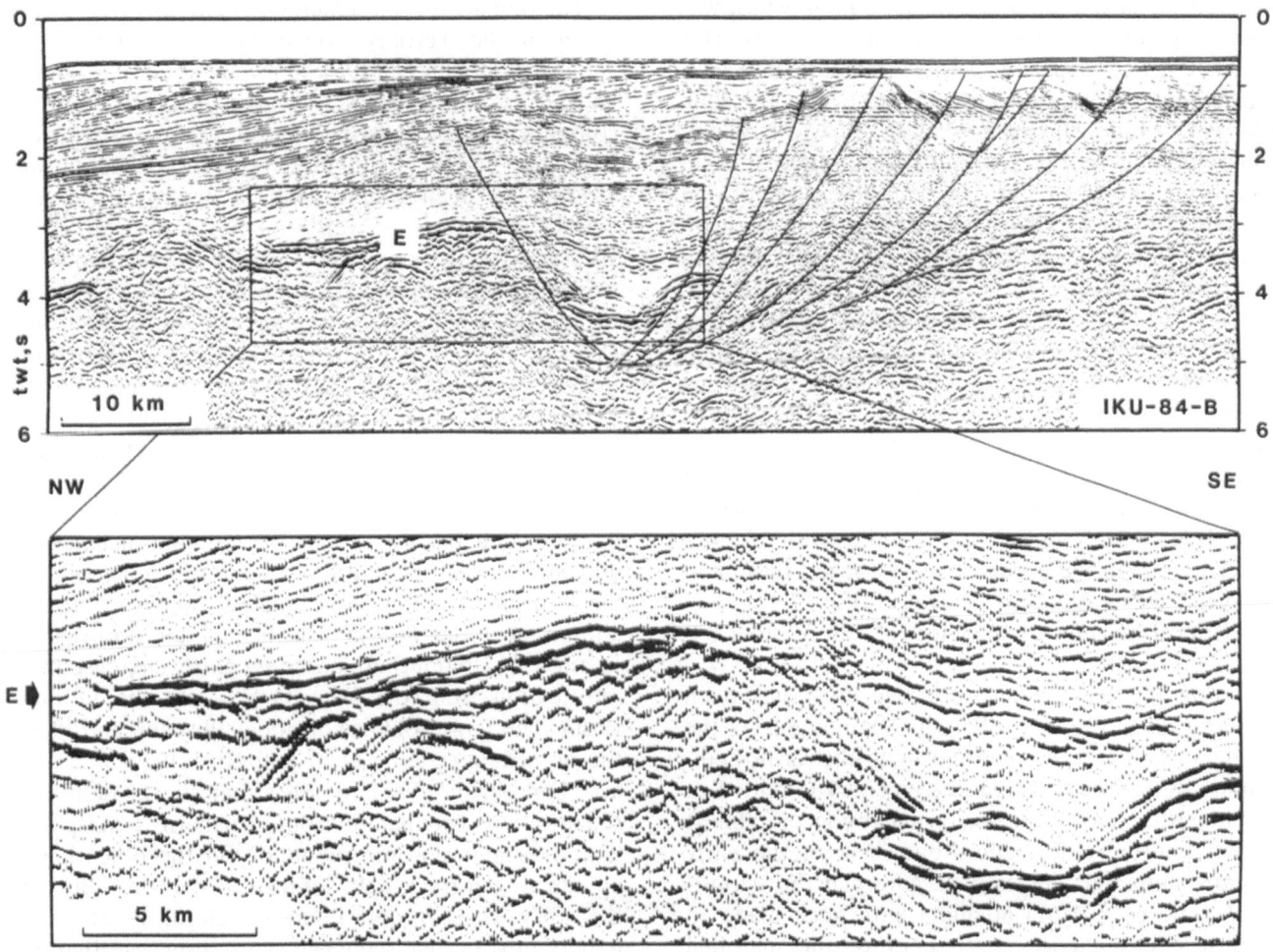

FIGURE 7-12.   Seismic profile IKU-84-B across Vestbakken Volcanic Province. Line 4 in Figure 7-4. E = early Eocene volcanics. (Modified from Faleide, Myhre, and Eldholm, 1988.)

ing North Atlantic and Arctic rifts. Moreover, opening of the Labrador Sea at 92 Ma (Roest and Srivastava, 1989) marks the transition to dextral strike-slip motion in the De Geer Zone. This motion caused continued subsidence of the Sørvestsnaget and Harstad basins, whereas the basins farther east became relatively stable (Faleide, Vågnes, and Gudlaugsson, in press). The Late Cretaceous–Paleocene extension is also documented by normal faults along the Veslemøy High and Senja Ridge. Following a hiatus at the Cretaceous–Tertiary transition, a uniform, widespread, upper Paleocene sequence was deposited.

The north–northwest early Tertiary line of breakup along the Senja Fracture Zone and the Hornsund Fault Zone follows the De Geer Zone (Figs. 7-3, 7-4). The general depositional pattern was dominated by regional westward progradation (Spencer, Home, and Berglund, 1984). However, early Eocene shear along the Senja Fracture Zone caused uplift and erosion of the southern Sørvestsnaget Basin, which shed sediments westward into the growing ocean basin and eastward into the Tromsø Basin. This local uplift was followed by subsidence and slow deposition, which was repeatedly interrupted by minor erosional events. This led to a Paleogene sequence, which thins and, in places, pinches out toward the high along the fracture zone [Fig. 7-11 (foldout)]. Farther north, however, extensional faulting in the northern Sørvestsnaget Basin created a pull-apart setting, which allowed deposition of up to 2–3 km of Paleogene sediments. (contrast with Chapter 6). The late Paleocene clinoforms indicate transport from the north (Rønnevik and Jacobsen, 1984), related to the breakup-induced uplift of the Stappen High. In the south, the Paleogene succession in the Harstad Basin thins to the south, indicating variable early Eocene uplift along the southern part of the margin. Prior to the early Oligocene change in relative plate motion, the Mohns Ridge axis had not moved the entire length of the present Senja Fracture Zone (Fig. 7-3). This induced faulting with a more westerly trend in the northern Sørvestsnaget Basin and the Vestbakken Volcanic Province (Fig. 7-12). Renewed local volcanism characterized the Vestbakken Volcanic Province.

During the last 30 m.y., oceanic crust has been produced along the entire Barents Sea margin. As the margin subsided, it was covered by a siliciclastic sedimentary wedge derived from the emerged Barents Sea region farther east. Regionally, the wedge is almost without internal deformation, but some of the early Tertiary faults and domes have governed the depositional patterns. Several attempts have been made to establish the sequence stratigraphy of the marginal wedge (Spencer, Home, and Berglund, 1984; Eidvin and Riis, 1989; Vorren et al., 1990, 1991; Richardsen, Henriksen, and Vorren, 1991; Faleide, Vågnes, and

Gudlaugssen, in press). These studies show that at least half the sequence is very young, probably deposited since 5.5 Ma [Fig. 7-11 (foldout)]. A regional unconformity, similar to the late Pliocene–Quaternary unconformity off Norway, but dated at 0.8 Ma, and caused by glacial erosion of the shelf, separates the clastic wedge from a blanket of glacigenic sediments. The glacigenic section consists of several seismostratigraphic units, which prograde seaward (Solheim and Kristoffersen, 1984; Vorren, Lebesbye, and Larsen, 1990). The sequence reaches 0.3 sec thickness on the main shelf, and increases to a maximum of 0.9–1.0 sec at the shelf edge and on the upper slope. Farther downslope, the glacigenic sediments become subparallel with underlying beds. There is seismic and bathymetric evidence of small- and large-scale mass movements on the upper slope and redeposition on the lower slope (Vorren et al., 1989; compare with Chapter 9). A particularly large, 300-m-deep, slide scar at about lat. 72°N, extends from the shelf edge to more than 1,500 m depth; it is estimated to have moved 1,000 km^3 of sediment (Bugge, 1983).

The impact of an emergent Barents Sea during much of the Tertiary, which was inferred by early geological (Freebold, 1951) and geophysical (Eldholm and Ewing, 1971) studies, was largely neglected until its consequences for hydrocarbon potential were realized. Recently, however, the large amount of uplift and erosion has been confirmed by a number of studies based on commercial well data (Nyland et al., 1989; Riis and Fjeldskaar, 1989), and by seismic-sequence geometries and mass-balance estimates, which document a regionally increasing uplift toward the north (Vorren et al., 1991). The magnitude of uplift is estimated at about 1 km in the southern Barents Sea (Vorren, Lebesbye, and Henriksen, 1988; Nøttvedt et al., 1988) and 2–3 km in Svalbard (Manum and Throndsen, 1978; Eiken and Austegard, 1987).

The fact remains, however, that available data have not provided precise estimates of timing, magnitude, and mechanisms of uplift. Present drill holes do not allow determination of detailed Neogene wedge stratigraphy; the age of seismic marker horizons in Figure 7-11 (foldout) are based on partly circumstantial information from the Barents Sea and the Lofoten Basin. The large volumes of upper Pliocene and Pleistocene sediment led Eidvin and Riis (1989) and Riis and Fjeldskaar (1989) to infer that the erosion and redeposition originated from northern hemisphere glaciation and isostatic rebound. On the other hand, apatite fission-track data indicate two main uplifts in the early Tertiary and Pliocene (Dorè, 1991). Most importantly, seismic-sequence analysis suggests that the increased late Cenozoic sedimentation started in latest Miocene time, 5.5 Ma (Richardson, Henriksen, and Vorren, 1991), and calculations by Vågnes, Faleide, and Gudlaugsson (1990) show that a tectonic uplift component must be considered. The latter authors also point out that the large postrift depocenters in the Vestbakken Volcanic Province and on oceanic crust in the Lofoten Basin require emergent eastern source areas since the Eocene. Nonetheless, the late Pliocene–Pleistocene glaciations appear to have contributed significantly to the construction of the Neogene and Quaternary depositional wedge by amplifying both the extent and intensity of erosion.

SVALBARD MARGIN

The Svalbard margin segment lies in front of Spitsbergen, which contains an upper Paleozoic to lower Tertiary sedimentary succession. The Tertiary sediments occur in the Central Basin and locally along the west coast of Spitsbergen (Fig. 7-4). The Central Basin contains more than 2 km of Paleocene–Eocene sediments (Steel et al., 1985; Manum and Throndsen, 1986) unconformably overlying Aptian–Albian rocks in the south, and Lower Triassic and Permian rocks in the north (Steel and Worsley, 1984). The early Tertiary breakup and subsequent continent–continent translation was accompanied by dextral transpression, which deformed most of the sediments in the Central Basin and the rocks below (Harland, 1969; Lowell, 1972; Birkenmajer, 1972; Kellogg, 1975).

Structurally, the entire margin north of lat. 75°N has been divided into two parts by the Hornsund Fault Zone below the central and outer continental shelf [Figs. 7-4, 7-13 (foldout)]. Below a veneer of glacial sediments, the inner shelf is composed of high-velocity sedimentary and basement rocks without distinct seismic structure and stratification (Schlüter and Hinz, 1978; Eiken and Austegard, 1987; Myhre and Eldholm, 1988). There are indications, however, of extensional features following the trend of the Hornsund Fault Zone (Myhre and Eldholm, 1988). For example, the island of Kong Karls Forland (Fig. 7-2) is a horst, separated from Spitsbergen by the Forlandsundet Graben (Fig. 7-4), which has been mapped south to about lat. 77°N (Eiken and Austegard, 1987; Austegard and Sundvor, 1991).

The Hornsund Fault Zone was first defined by a rapid lateral change in velocity-depth distribution (Sundvor and Eldholm, 1976, 1979), and later recognized as a continuous feature consisting of several, subparallel, down-faulted blocks having seismic characteristics similar to those of the inner shelf complex (Myhre, Eldholm, and Sundvor, 1982). Myhre and Eldholm (1988) placed the continent–ocean boundary just seaward of the outermost fault scarp and showed that the distance between it and the identified oceanic crust is indeed narrow along the margin. The distance is less than 10 km between lat. 75 and 76°N and is

10–30 km farther north. This interpretation of the continent–ocean transition was recently confirmed by seismic experiments (Austegard et al., 1988; Austegard and Sundvor, 1991). By analogy with the Senja Fracture Zone, we believe that the change from continental to oceanic basement occurs over such a narrow zone that the continent–ocean boundary is, in fact, a distinct geological feature at these sheared margins.

The Hornsund Fault Zone marks the eastern boundary of a thick sedimentary wedge (7 km thick), which extends into the Greenland Sea, and reaches the axial mountains of the Knipovich Ridge [Fig. 7-13 (foldout)]. At the northern margin, the sediments cover the entire eastern axial province; there is evidence of overflow into the rift valley, and locally onto the axial mountains west of the axis. Although a series of regional reflectors and depositional units has been recognized, the margin stratigraphy is not resolved due to absence of adequate tie-lines to the Barents Sea margin and because of an enigmatic correlation to DSDP Site 344 (Fig. 7-4), which recovered poorly dated glacial marine sediments (Talwani, Udintsev, et al., 1976). Schlüter and Hinz (1978) identified three depositional sequences separated by unconformities. These include an upper Pliocene–Pleistocene glacial sequence, a Pliocene sequence characterized by downslope mass movements, and a pre-middle Oligocene basal sequence. By studying the termination of the unconformities on oceanic basement, Myhre and Eldholm (1988) inferred that the lowermost unconformity was not older than 5.5 m.y. and that the basal sequence is composed of Miocene and Oligocene sediments. Regardless of the details, the great thickness of the two upper sequences shows that as much as 4 km of sediment has been deposited since the late Miocene (compare with Chapters 3, 6).

The evolution of this margin segment comprises the two stages shown in Figure 7-3. The southern part of the Hornsund Fault Zone developed as an Eocene sheared margin linked with the Eurasia Basin by a continental megatransform. A slightly delayed opening of the southern Greenland Sea (with respect to the area farther south), at 54–55 Ma, was postulated by Eldholm, Faleide, and Myhre (1987), but the latest Paleocene onset of the Spitsbergen Orogeny is compatible with a simultaneous opening of the North Atlantic basins at about 57.5 Ma. The maximum compressional deformation in Spitsbergen, related to transpression along the plate boundary, occurred in the early Eocene and decreased toward the late Eocene, leaving the Spitsbergen Fold and Thrust Belt (Fig. 7-4). Although Townsend and Mann (1989) preferred a late Paleozoic or Triassic age for the block faulting on the inner shelf, we believe that it was mainly postorogenic. This faulting reflects the change to an extensional structural setting in the earliest Oligocene, when the Hornsund

Fault Zone became the locus of rifting, and eventually allowed accretion of oceanic crust in the northern Greenland Sea. It is important to note that continent–continent translation inhibited a deep-water connection to the Arctic Ocean until middle to late Miocene time (Lawver et al., 1990; Eldholm, 1990). The opening of a deep-water passage has probably contributed to the complex Neogene depositional regime at the northern margin, where lateral thickness variations and sediment drifts document bottom-current activity, in addition to considerable downslope mass movement by submarine slides (Kristoffersen, 1990; Eiken and Hinz, 1990; compare with Chapters 6, 9).

This leaves a margin constructed by a wedge of postrift sediments over a down-faulted, predominantly pre-Cretaceous, substratum and Cenozoic oceanic crust. The wedge includes Paleogene sediments in the south, but changes to predominantly Neogene and Quaternary sediments farther north. Despite the lack of a regional margin stratigraphy, it is evident that the main part of the postrift wedge was deposited in late Neogene and Quaternary time, associated with uplift and subsequent erosion of the Svalbard archipelago. The sediment distribution suggests that the east–west trending fjords in Spitsbergen, particularly Storfjordrenna (Fig. 7-2), acted as main drainage systems.

CRUSTAL STRUCTURE

A large number of seismic refraction profiles, mostly unreversed sonobuoys, are available for this margin and the adjacent ocean basins. These profiles, summarized by Myhre and Eldholm (1981) and Myhre (1984), are adequate for mapping the uppermost crust, but rarely provide information on the velocity structure in the deep sediment basins and crystalline crust. Recently, however, digitally recorded sonobuoy and two-ship ESP profiles (Mutter, Talwani, and Stoffa, 1984; Eldholm and Mutter, 1986; Hinz et al., 1987; Mutter and Zehnder, 1988; Jackson, Faleide, and Eldholm, 1990; Planke, Skogseid, and Eldholm, 1991; Olafsson et al., in press) have yielded improved seismic resolution. To show the first-order crustal features and their relationships to the margin evolution and deep basin configuration, we have constructed simplified crustal transects (Figs. 7-14, 7-15).

The data off Norway show:

1. That the escarpments mark prominent boundaries with respect to the lateral velocity distribution in the upper crust.
2. The seaward extent of the large sediment basins.
3. Change in depth to the Moho. The extrusive complexes are characterized by a velocity of around 4 km/sec at the top of the basalts, which increases

to about 6 km/sec near the base. Moreover, crustal thickness increases to 20–22 km below the extrusives. The expanded crustal thickness is obviously related to the compensation of the extrusive pile, but the fact that a 7.2- to 7.4-km/sec velocity is found in the lower crust has been interpreted in terms of a high-density underplated magmatic body emplaced during the late Paleocene breakup (White et al., 1987). The underplated body also appears to extend landward of the continent–ocean boundary beneath the region of inner flows and strongly intruded crust [Figs. 7-4, 7-5 (foldout), 7-7 (foldout), 7-14]. West of the marginal highs, the crust in the Norway and Lofoten basins has a clear oceanic velocity signature, although the thickness is somewhat greater than the average of the world ocean.

The total sediment thickness in the outer Møre and Vøring basins is estimated to be as much as 10–14 km, and the depth to Moho decreases to 19–20 km at the outer margin, indicating a strongly attenuated crust. There is a distinct increase, however, to a 25–27-km-thick crust at the inner margin before it thickens to more than 30 km east of the coastline. Thus, there is significant structuring of the continental part of the margin in terms of crustal thickness. The change occurs near the shelf edge in the Møre Basin and along the inner flank of the Vøring Basin (Fig. 7-14). We believe that the outer thin crust corresponds to that part of the margin involved in early Tertiary crustal extension, whereas the less extended crust below the inner Møre and Vøring basins and Trøndelag Platform is the remnant of previous rift episodes.

Onshore refraction experiments have defined a 25–26-km-thick crust beneath the Lofoten–Vesterålen Islands. The crust shallows slightly toward the large, 120-mGal gravity anomaly over the southern islands (Sellevoll, 1983), where Proterozoic and Archean granulite facies are exposed (Griffin et al., 1978; Chroston and Brooks, 1989). The same Moho level is maintained on the shelf, but it shallows rapidly beneath the slope, and reaches a depth of about 15 km at the foot of the slope (Goldschmidt-Rokita et al., 1988; Mjelde et al., in press). Thus, the crustal thickness of the Lofoten–Vesterålen Islands and the adjacent shelf is similar to that of the Trøndelag Platform and the inner Møre Basin. A 7.4-km/sec velocity at the base of the crust has been measured at the outer margin, but there is no typical crustal thickening below the oldest oceanic crust covered by extrusives (Fig. 7-14). Sellevoll and Mokhtari (1988) made the important observation that a continuous intracrustal reflector is present at the top of the 7.4-km/sec layer, which they associated with oceanic crustal layer 3B. We believe that this layer is the equivalent of the underplated bodies reported elsewhere in the North Atlantic, which are characteristic of volcanic rifted margins; normal oceanic crust is present seaward of the underplated segment. Moreover, the shallower Moho (Fig. 7-14) appears to correspond to the absence of a prominent marginal high north of the Lofoten Fracture Zone.

The gravity field along the margin is dominated by elongate, positive, high-amplitude anomaly belts just landward of the shelf edge. The belts are offset right-laterally by the prolongation of the Jan Mayen Fracture Zone, and continue onshore into Scotland and the Lofoten Islands. Talwani and Eldholm (1972) interpreted the anomalies to originate from large intrabasement density contrasts combined with less important structural relief. A pre-Permian, probably Precambrian, age was inferred for the bodies. An ESP profile on the Møre shelf (Fig. 7-14) shows a basal crustal

pillow of 8.5–km/sec velocity, which may cause the gravity anomalies (Olafsson et al., in press). This question is not yet resolved, however. A particular problem is that early Tertiary alkaline mantle rocks dredged on the inner shelf (Bugge, Prestvik, and Rokoengen, 1980) may indicate a contribution from intrusions along a zone of weakness, which predetermined the location of the Jan Mayen Fracture Zone (Fig. 7-1; Eldholm, Thiede, and Taylor, 1989; Torske and Prestvik, 1991). Figure 7-14 shows some possible, local, high-velocity bodies in the lower crust and(or) upper mantle (Planke, Skogseid, and Eldholm, 1991; Olafsson et al., in press). The margin lies within the Caledonian Orogenic Zone, and we believe that the intrabasement velocity anomalies reflect Caledonian thrust sheets of varying physical properties.

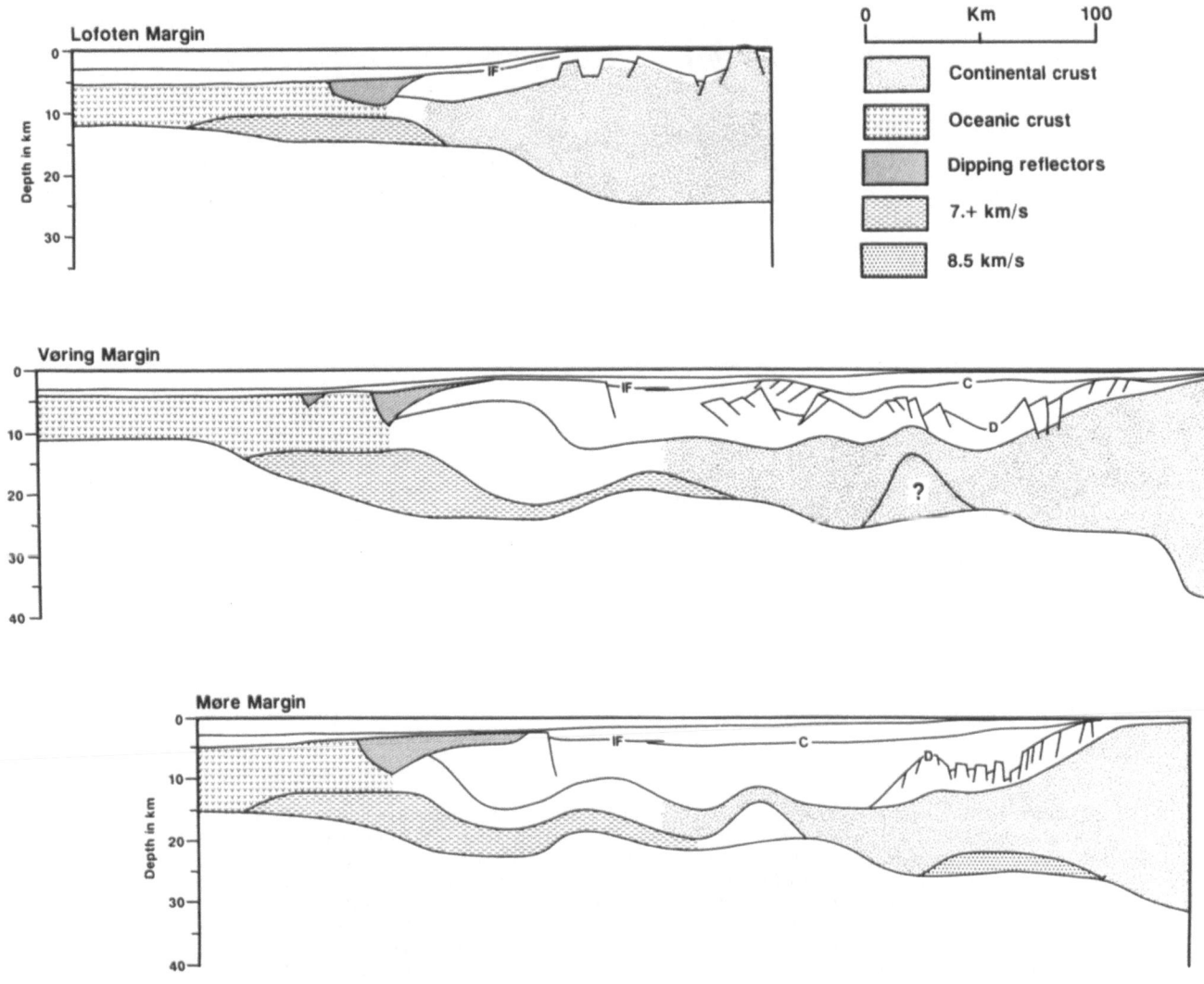

FIGURE 7-14.   Depth sections showing main crustal units across margin provinces south of 70°N. Location in Figure 7-4. Crustal section across Vøring margin shown as line 6 on Figure 7-4. Crustal section across Lofoten margin shown as line 7 on Figure 7-4. Note that Møre margin section follows seismic transect (MVM-1, Fig. 7-5 (foldout). Crystalline basement level in deep sediment basins is placed at 6.0 km/sec isovelocity. C = base-Tertiary; D = base-Cretaceous; IF = inner flows. Based on sources cited in text.

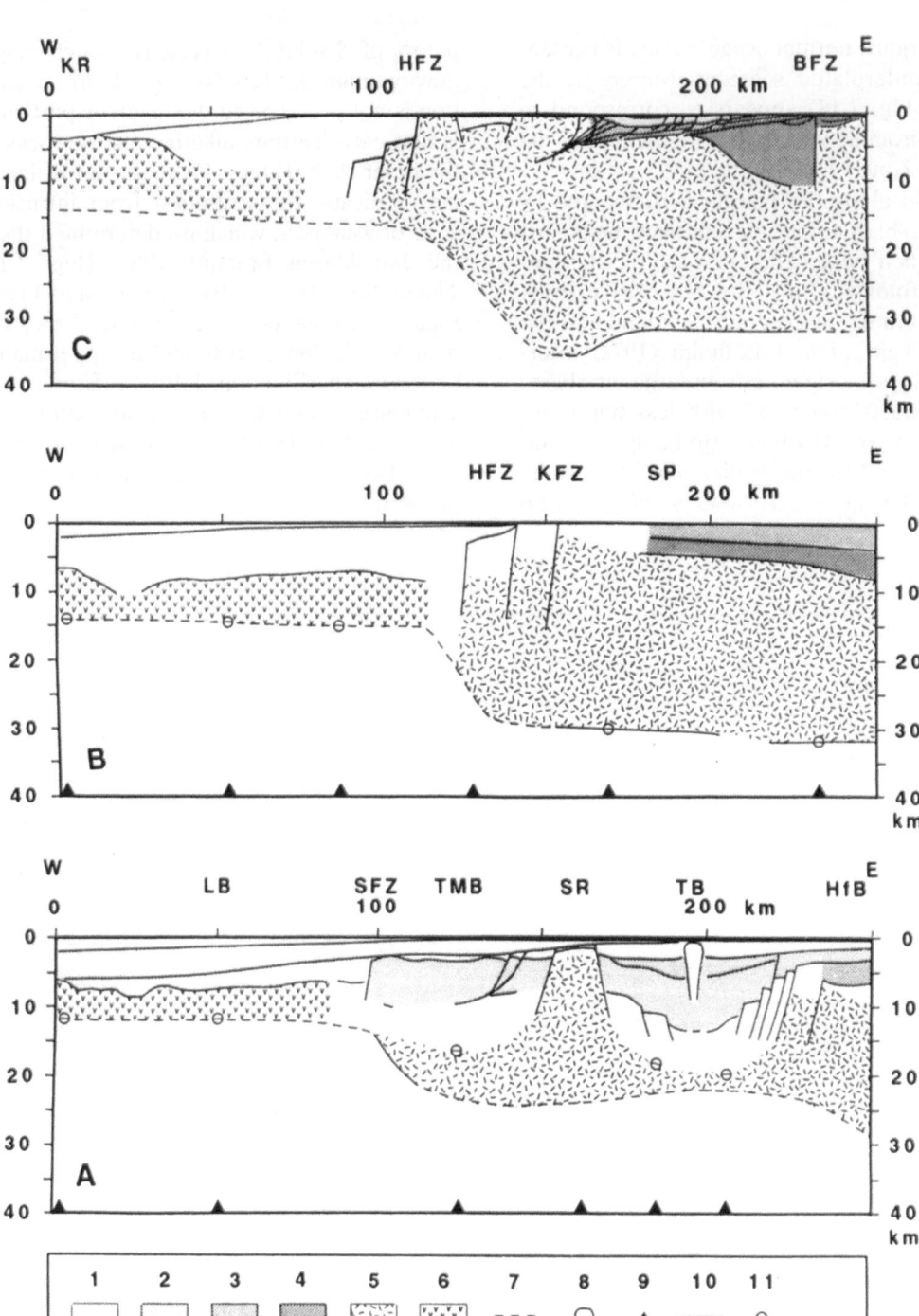

FIGURE 7-15.   Depth sections showing main crustal units across margin provinces north of 70°N. Location in Figure 7-4. 1 = Cenozoic; 2 = Cretaceous; 3 = Jurassic–Triassic; 4 = upper Paleozoic; 5 = continental (crystalline) crust; 6 = oceanic (crystalline) crust; 8 = salt; 9 = projected location of adjacent expanding spread profiles (ESP). Depth control based on: 10 = multichannel seismic data; 11 = ESP; 7 = gravity modelling. BFZ = Billefjorden Fault Zone; HfB = Hammerfest Basin; HFZ = Hornsund Fault Zone; KR = Knipovich Ridge; KFZ = Knølegga Fault Zone; LB = Lofoten Basin; SFZ = Senja Fracture Zone; SP = Svalbard Platform; SR = Senja Ridge; TB = Tromsø Basin; TMB = Tertiary marginal basin (Sørvestsnaget Basin). Based on Jackson, Faleide, and Eldholm (1990) and Faleide et al. (1991).

Analysis of ESP data in the western Barents Sea shows that the geological provinces we have described also constitute distinct crustal units, and that the well-defined change from oceanic to continental basement along the sheared margin segments also marks the onset of a rapid increase in crustal thickness (Fig. 7-15; compare with Chapter 1). The continental crust is divided into three units (Jackson, Faleide, and Eldholm, 1990; Faleide et al., 1991). The Svalbard Platform and the region east of the Cretaceous Tromsø and Bjørnøya basins are underlain by 30–35-km-thick crust. The Cretaceous basins are located on intermediate crust of 26-km average thickness, although the Moho relief is accentuated, reflecting the structural segmentation of the basin province. Finally, there is a western region where Moho shallows toward the continent–ocean boundary. The average crustal thickness is about 20 km, and the thinning is most pronounced beneath the boundary faults just east of the continent–ocean boundary. This crustal unit corresponds to the region that was affected by the early Tertiary deformation and subsequent thermal subsidence of the margin. Although there is poor velocity control in the Vestbakken Volcanic Province, a depth to Moho of 22 km was indicated by Faleide et al. (1991). The crustal setting off Spitsbergen (Austegard et al, 1988; Austegard and Sundvor, 1991) shows a slight crustal root below the area of main early Tertiary thrusting and rapid thinning in response to the post-early Oligocene extension west of the Hornsund Fault Zone (Fig. 7-15).

SUMMARY AND CONCLUSIONS

The Norwegian continental marginal lies within the region affected by the Caledonian Orogeny. Subsequently, the continental crust between Norway, Svalbard, and Greenland was denuded and structured by several tectonic episodes prior to the opening of the Norwegian–Greenland Sea at the Paleocene–Eocene transition. Although the present margin is a Cenozoic feature, we have shown that the post-opening depositional history of the margin to a large extent was predetermined by its pre-opening structural history. In particular, crustal extension and subsequent subsidence caused by the Late Jurassic–Early Cretaceous tectonic episode played a major role by creating the large Cretaceous sedimentary basins off Norway and in the southwestern Barents Sea. Moreover, the margin contains a clear structural and depositional imprint of late Mesozoic plate-tectonic events in the North Atlantic–Arctic realm.

Renewed lithospheric extension in the Late Cretaceous and(or) early Paleocene was the precursor of the subsequent plate boundary between the Atlantic and Arctic oceans through the Norwegian–Greenland Sea.

In a regional sense, the boundary consists of two lineaments, the Reykjanes–Mohns Line between the the Charlie-Gibbs and Greenland–Senja fracture zones (characterized by extension), and the De Geer Zone farther north (with predominantly dextral shear; Figs. 7-1, 7-3, 7-4). This setting was maintained during the actual breakup in the earliest Eocene (57–58 Ma) and led to a series of rifted margin segments south of lat. 70°N, and two major sheared segments offset by a small rifted margin off the Barents Sea and Svalbard. The plate geometry, however, caused continent–continent translation between northeast Greenland and Svalbard during the Eocene, and oceanic crust was first produced when the relative plate motion changed to west–northwest in the early Oligocene.

The structural and depositional style of the margin is the combined effect of the regional change from rift- to shear-dominated margins, the different lithospheric response to rifting in areas of varying crustal strength, and the intense early Tertiary magmatism at the rifted segments. In particular, the magmatic activity prior to, during, and after the crustal breakup, has led to a classification of the North Atlantic margins as volcanic rifted margins, as opposed to the nonvolcanic type, which lacks appreciable volcanism and uplift during breakup (Eldholm, Thiede, and Taylor, 1989; White and McKenzie, 1989). Prior to breakup, the incipient volcanic margins off Norway experienced lithospheric extension, listric faulting and uplift, and magmatism manifested by intrusions in the Cretaceous basins and underplating. Continental breakup and the initial formation of oceanic crust occurred during a transient pulse of intense volcanic activity, which constructed large extrusive edifices and covered both oceanic and continental crust. In addition, basaltic lavas overflowed large areas of the adjacent sedimentary basins. When volcanism abated, about 3 m.y. after breakup, the injection centers subsided to oceanic depths, leaving the extrusive constructions elevated above the growing oceanic crust.

The depositional history of the outer margin is dominated by the tectonomagmatic fragmentation of the Norwegian–Greenland Sea, which created barriers to the downslope transport of sediments and governed watermass circulation (Eldholm, 1991a). Although stratigraphic control is still considered poor, seismic facies show complex patterns of unconformities, sedimentary prisms, and drifts indicative of periods with strong bottom currents and sediment instability. As the ocean widened and deepened, the tectonomagmatic barriers gradually became less important, but the imprint of currents, slides, and slumps was maintained until present. We attribute the absence of well-developed abyssal plains and the vague distinction between the lower continental slope and continental rise to the youth of the Norwegian–Greenland Sea Basin, but we expect these incipient physiographic provinces to become more distinct as this continental margin matures.

The Paleogene depositional history along the rifted margins was dominated by rift-induced uplift, which allowed erosion of the highs and redeposition in the outer Møre and Vøring basins. As the marginal highs subsided during the Eocene and Oligocene, pelagic sediment sources gradually became dominant. The highs, however, continued to influence sedimentation on the outer margin and in the ocean basins until they were buried in the late Oligocene. Sedimentation along the sheared margins north of lat. 70°N has been dominated by nonrigid lithospheric behavior along the transforms. South of Svalbard, the main sediment source was in the east; this was supplemented by erosion of highs along the plate boundary and of local uplifted areas within a transtensional structural regime. Farther north, sedimentation was largely governed by the transpressional regime, which created the Spitsbergen Orogeny, and by the subsequent erosion and redeposition during rifting and subsidence of the developing margin.

Neogene and Quaternary sedimentation has been characterized by outbuilding on a subsiding margin. The sediment supply increased greatly during the late Miocene and Pliocene, when huge siliciclastic wedges were constructed along the present shelf edge. These wedges are capped by a sequence of glacial deposits. The most attractive sediment source for the widespread late Neogene wedges is erosion of an uplifted continent and inner shelf. Such a source is consistent with the geomorphology of Fennoscandia. Because increased sedimentation started prior to the onset of main northern hemisphere glaciation in the late Pliocene, we postulate that the Fennoscandian uplift predated the glaciation, but that glacial erosion and crustal rebound amplified the magnitude of the young sedimentary wedges.

Shale diapirism at the northern Vøring Plateau, and elongate domes in the Vøring and Møre basins, are indicative of local instability that caused deformation in the Cenozoic sediments. The intrabasinal arching appears to have originated from middle-Tertiary rejuvenation of older compressional structures by the loading effect of the late Neogene wedges. Furthermore, recent large-scale downslope mass movements observed at the Møre and western Barents Sea margins may be related to instabilities caused by the rapid shelf outbuilding.

Thermal subsidence and structural deformation, which followed the early Tertiary breakup, did not include the entire margin; they are restricted to a zone of varying width landward of the continent–ocean boundary, which divides the margin into an inner stable part and an outer subsiding part. This post-Paleocene marginal basin is narrow at the Lofoten–

Vesterålen margin and north of lat. 75°N, where the breakup took place in thick crust, but is much wider in the Cretaceous basins. Furthermore, the marginal basin corresponds to a region of crustal thinning. The late Neogene prograding wedges are superimposed on both the stable and subsiding parts of the margin, where they exert an excessive sedimentary load.

Early Tertiary reconstructions of the Norwegian–Greenland Sea show a clear asymmetry in the distribution of Mesozoic basins and platforms, as well as of early Tertiary extrusives, on either side of the plate boundary. It is therefore possible that the present margin segmentation south of lat. 70°N is caused by transfer faults related to preexisting crustal properties (variable along strike), which induced asymmetric rifting and magmatism.

ACKNOWLEDGMENTS

This study has benefitted from interaction with numerous colleagues in Norway and abroad. We also have relied strongly on ongoing research projects within the Marine and Applied Geophysics Research Group, Department of Geology, University of Oslo, and particularly acknowledge results of unpublished candidata scientiarum theses by Anne Fiedler, Nina Gravdal, Kjersti Grue, Merete Hagen, and Bente Vigen. This project has, in part, been supported by the Norwegian Research Council for Natural Science and the Humanities, VISTA, BP Norway, and Statoil. K. O. Emery and E. Uchupi kindly reviewed an early version of the manuscript.

REFERENCES

Austegard, A., and Sundvor, E. 1991. The Svalbard continental margin: Crustal structure from analysis of deep seismic profiles and gravity. Seismo-Series 53, Seismological Observatory, University of Bergen, 31.

Austegard, A., Eiken, O., Stordal, T., and Evertsen, E. C. 1988. Deep-seismic sounding and crustal structure in the western part of Svalbard. Extended abstract. *Norsk Polarinstitutt Rapportserie* 46:89–90.

Berggren, W. A., Kent, D. V., Flynn, J. J., and Van Couvering, J. A. 1985. Cenozoic geochronology. *Geological Society of America Bulletin* 96:1407–1418.

Birkenmajer, K. 1972. Tertiary history of Spitsbergen and continental drift. *Acta Geologica Polonica* 11:193–218.

Bjørklund, K. R., and Kellogg, D. 1972. Five new Eocene radiolarian species from the Norwegian Sea. *Micropaleontology* 18:386–396.

Blystad, P., and Sand, M. 1988. Strukturell utvikling av kontineantalsokkelen vest for Lofoten og Vesterålen. Abstract, 18. Nordic Geological Winter Meeting 1988, Copenhagen, Denmark.

Blystad, P., Færseth, R. B., Larsen, B. T., Skogseid, J., and Tørudbakken, B. in press. Structural elements of the

Norwegian continental margin between 62° and 69°N. *Norwegian Petroleum Directorate Bulletin*.

Bøen, F., Eggen, S., and Vollset, J. 1984. Structures and basins of the margin from 62°–69°N and their development. *In* Petroleum Geology of the North European Margin, A. M. Spencer, et al., Eds.: London: Graham and Trotman, 253–270.

Brekke, H., and Riis, F. 1987. Tectonics and basin evolution of the Norwegian shelf between 62°N and 72°N. *Norsk Geologisk Tidsskrift* 67:295–322.

Bugge, T. 1983. Submarine slides on the Norwegian continental margin, with special emphasis on the Storegga area. *Continental Shelf Institute Publication 110*, 152.

Bugge, T., Befring, S., Belderson, R. H., Eidvin, T., Jansen, E., Kenyon, N. H., Holtedal, H., and Sejrup, H. P. 1987. A giant three-stage submarine slide off Norway. *Geo-Marine Letters* 7:191–198.

Bugge, T., Prestvik, T., and Rokoengen, K. 1980. Lower Tertiary volcanic rocks off Kristiansund—mid Norway. *Marine Geology* 35:277–286.

Bukovics, C., Shaw, N. D., Cartier, E. G., and Ziegler, P. A. 1984. Structure and development of the mid-Norway continental margin. *In* Petroleum Geology of the North European Margin, A. M. Spencer, et al., Eds.: London: Graham and Trotman, 407–423.

Caston, V. N. D. 1976. Tertiary sediments of the Vøring Plateau, Norwegian Sea, recovered by Leg 38 of the Deep Sea Drilling Project. *In* Initial Reports of the Deep Sea Drilling Project, Volume 38: M. Talwani, G. Udintsev, et al.: Washington, DC: U.S. Government Printing Office, 761–782.

Chroston, P. N., and Brooks, S. G. 1989. Lower crustal seismic velocities from Lofoten-Vesterålen, north Norway. *Tectonophysics* 157:251–269.

Crane, K., Vogt, P., Sundvor, E., deMoustier, C., Doss, H., and Nishimura, C. 1990. SeaMARC II and associated geophysical investigation of the Knipovich Ridge, Molloy Ridge/Fracture Zone and Barents/Spitsbergen continental margin: Part II Volcanic/tectonic structure. Abstract, American Geophysical Union.

Damuth, J. E. 1978. Echo character of the Norwegian-Greenland Sea: relationship to Quarternary sedimentation. *Marine Geology* 28:1–36.

Doré, A. G. 1991. The structural foundation and evolution of Mesozoic seaways between Europe and the Arctic. *Palaeogeography, Palaeoclimatology, Palaeoecology* 87:441–492.

Doré, A. G., and Gage, M. S. 1987. Crustal alignments and sedimentary domains in the evolution of the North Sea, North-east Atlantic Margin and Barents Shelf. *In* Petroleum Geology of North West Europe, J. Brooks and K. Glennie, Eds.: London: Graham and Trotman, 33–46.

Eidvin, T., and Riis, F. 1989. Nye dateringer av de tre vestligste borehullene i Barentshavet. Resultater og konsekvenser for den tertiære hevningen. *Norwegian Petroleum Directorate Contribution* 27:44.

Eiken, O., and Austegard, A. 1987. The Tertiary orogenic belt of West-Spitsbergen: Seismic expressions of the offshore sedimentary basins. *Norsk Geologisk Tidsskrift* 67:383–394.

Eiken, O., and Hinz, K. 1990. Contourites in the Fram Strait. *In* Aspects of Reflection Seismics, O. Eiken, Dr. Scient. thesis, University of Bergen, Norway, 93–121.

Eldholm, O. 1990. Paleogene North Atlantic magmatic-tectonic events: environmental implications. *Memoire della Societa Geologica Italiana* 44:13–28.

Eldholm, O. 1991. Magmatic–tectonic evolution of a volcanic rifted margin. *In* Evolution of Mesozoic and Cenozoic Continental Margins, A. W. Meyer, T. A. Davies, and S. W. Wise, Jr., Eds.: *Marine Geology* special issue, 102:43–61.

Eldholm, O., and Ewing, J. 1971. Marine geophysical survey in the southwestern Barents Sea. *Journal of Geophysical Research* 76:3832–3841.

Eldholm, O., and Mutter, J. C. 1986. Basin structure of the Norwegian margin from analysis of digitally recorded sonobuoys. *Journal of Geophysical Research* 91:3763–3783.

Eldholm, O., and Windish, C. C. 1974. Sediment distribution in the Norwegian–Greenland Sea. *Geological Society of America Bulletin* 85:1661–1676.

Eldholm, O., Sundvor, E., and Myhre, A. M. 1979. Continental margin off Lofoten–Vesterålen, Northern Norway. *Marine Geophysical Research* 4:3–35.

Eldholm, O., Faleide, J. I., and Myhre, A. M. 1987. Continent-ocean transition at the western Barents Sea–Svalbard margin. *Geology* 15:1118–1122.

Eldholm, O., Thiede, J., and Taylor, E. 1989. Evolution of the Vøring Volcanic margin. *In* O. Eldholm, J. Thiede, E. Taylor, et al., Proceedings of the Ocean Drilling Program, Scientific Results, Volume 104: College Station, TX: Ocean Drilling Program, 1033–1065.

Eldholm, O., Thiede, J., Taylor, E., Barton, C., Bjørklund, K., Bleil, U., Ciesielski, P., Desprairies, A., Donnally, D., Froget, C., Goll, R., Henrich, R., Jansen, E., Krissek, L. Kvenvolden, K., LeHuray, A., Love, D., Lysne, P., McDonald, T., Mudie, P., Osterman, L., Parson, L., Phillips, J. D., Pittenger, A., Qvale, G., Schönharting, G., and Viereck, J. 1987. Proceedings of the Ocean Drilling Program, Initial Reports, Volume 104: College Station, TX: Ocean Drilling Program.

Eldholm, O., Thiede, J., Taylor, E., Barton, C., Bjørklund, K., Bleil, U., Ciesielski, P., Desprairies, A., Donnally, D., Froget, C., Goll, R., Henrich, R., Jansen, E., Krissek, L., Kvenvolden, K., LeHuray, A., Love, D., Lysne, P., McDonald, T., Mudie, P. Osterman, L., Parson, L., Phillips, J. D., Pittenger, A., Qvale, G., Schönharting, G., and Viereck, J. 1989. Proceedings of the Ocean Drilling Program, Scientific Results, Volume 104: College Station, TX: Ocean Drilling Program.

Emery, K. O., and Uchupi, E. 1984. *The Geology of the Atlantic Ocean*: New York: Springer-Verlag.

Faleide, J. I., Myhre, A. M., and Eldholm, O. 1988. Early Tertiary volcanism at the western Barents Sea margin. *In* Early Tertiary Volcanism and the Opening of the NE Atlantic, A. C. Morton and L. M. Parson Eds.: *Geological Society of London Special Publication 39*, 135–146.

Faleide, J. I., Vågnes, E., and Gudlaugsson, S. T. in press. Evolution of the southwestern Barents Sea in a regional rift-shear tectonic setting. *Marine and Petroleum Geology*.

Faleide, J. I., Gudlaugsson, S. T., Eldholm, O., Myhre, A. M., and Jackson, H. R. 1991. Deep seismic transects across the sheared western Barents Sea–Svalbard continental margin. *Tectonophysics* 189:73–89.

Freebold, H. 1951. Geologie des Barentsschelfes. *Abhandlung der deutschen Akadamie der Wissenschaften zu Berlin* 5:1–150.

Gabrielsen, R. H., Færseth, R. B., Jensen, L. N., Kalheim, J. E., and Riis, F. 1990. Structural elements of the Norwegian continental shelf—Part I: The Barents Sea Region. *Norwegian Petroleum Directorate Bulletin* 6:33.

Gatliff, R. W., Hitchen, K., Ritchie, J. D., and Smythe, R. K. 1984. Internal structure of the Erlend Tertiary volcanic complex north of Shetland, revealed by seismic reflection. *Journal of the Geological Society of London* 141:555–562.

Gibb, F. G. F., and Kanaris-Sotiriou, R. 1988. The geochemistry and origin of the Faeroes–Shetland sill complex. *In* Early Tertiary Volcanism and the Opening of the NE Atlantic, A. C. Morton and L. M. Parson, Eds.: *Geological Society of London Special Publication 39*, 241–252.

Gibb, F. G. F., Kanaris-Sotiriou, R., and Neves, R. 1986. A new Tertiary sill complex of mid-ocean ridge basalt type NNE of the Shetland Isles: a preliminary report. *Transactions of the Royal Society of Edinburgh: Earth Sciences* 77:223–230.

Gjessing, J. 1967. Norway's paleic surface. *Norsk Geografisk Tidsskrift* 21:69–132.

Goldschmidt-Rokita, A., Sellevoll, M. A., Hirschleber, H. B., and Avedik, F. 1988. Results of two seismic refraction profiles off Lofoten, Northern Norway. *Norges Geologisk Undersøkelse Special Publication* 3:49–57.

Gravdal, N. 1985. The Møre Basin. Candidata scientiarum thesis, University of Oslo, Norway, 111.

Griffin, W. L., Taylor, P. N., Hakkinen, J. W., Heier, K. S., Iden, I. K., Krogh, E. J., Malm, O., Olsen, K. I., Ormaasen, D. E., and Tveten, E. 1978. Archaean and Proterozoic crustal evolution in Lofoten–Vesterålen, N. Norway. *Journal of the Geological Society of London* 135:629–647.

Gudlaugsson, S. T., Faleide, J. I., Fanavoll, S., and Johansen, B. 1987. Deep seismic reflection profiles across the western Barents Sea margin. *Geophysical Journal of the Royal Astronomical Society* 89:273–278.

Hagen, M. 1987. Maringeofysiske undersøkelser i det østlige Mørebassenget. Candidata scientiarum thesis, University of Oslo, Norway, 145.

Hagevang, T., and Rønnevik, H. C. 1986. Basin development and hydrocarbon occurrence offshore mid-Norway. *American Association of Petroleum Geologists Memoir 40*, 599–613.

Hamar, G. P., and Hjelle, K. 1984. Tectonic framework of the Møre Basin and northern North Sea. *In*: Petroleum Geology of the North European margin, A. M. Spencer, et al., Eds.: London: Graham and Trotman, 349–358.

Harland, W. B. 1969. Contribution of Spitsbergen to understanding of tectonic evolution of North Atlantic region. *In* North Atlantic: Geology and Continental Drift, M. Kay, Ed.: *American Association of Petroleum Geologists Memoir 12*, 817–851.

Hastings, D. S. 1986. Cretaceous stratigraphy and reservoir potential, mid-Norway continental shelf. *In* Habitat of Hydrocarbons on the Norwegian Continental Shelf, A. M.

Spencer, et al., Eds.: London: Graham and Trotman, 287–298.

Hinz, K., and Weber, J. 1976. Zum geologischen Aufbau des Norwegischen Kontinentalrandes und der Barents-See nach reflexionsseismischen Messungen. *Erdoel und Kohle, Erdgas, Petrochemie*, 3–29.

Hinz, K., Dostman, H. J., and Hanisch, J. 1984. Structural elements of the Norwegian continental margin, *Geologisches Jahrbuch A* 75:193–221.

Hinz, K., Mutter, J. C., Zehnder, C. M., and the NAT Study Group. 1987. Symmetric conjugation of continent–ocean boundary structures along the Norwegian and East Greenland margins. *Marine and Petroleum Geology* 4:167–187.

Hitchen, K., and Ritchie, J. D. 1987. Geological review of the West Shetland area. *In* Petroleum Geology of North West Europe, J. Brooks and K. Glennie, Eds.: London: Graham and Trotman, 737–749.

Holtedahl, O. 1960. On supposed marginal faults and the oblique uplift of the landmass in Cenozoic time. *In* Geology of Norway, O. Holtedahl, Ed.: *Norges Geologiske Undersøkelse* 208:351–357.

Jackson, H. R., Faleide, J. I., and Eldholm, O. 1990. Crustal structure of the sheared southwestern Barents Sea continental margin. *Marine Geology* 93:119–146.

Johansen, B., and Eldholm, O. 1989. Sediment distribution and regional subsidence in the Norwegian–Greenland Sea. Unpublished manuscript, 73.

Johnson, G. L., and Heezen, B. C. 1967. Morphology and evolution of the Norwegian–Greenland Sea. *Deep-Sea Research* 14:755–771.

Johnson, G. L., Freitag, J. S., and Pew, J. A. 1971. Structure of the Norwegian Basin. *Norsk Polarinstitutt Årbok 1969*, 7–16.

Jørgensen, F., and Navrestad, T. 1981. The geology of the Norwegian shelf between 62°N and the Lofoten Islands. *In*: Petroleum Geology of the Continental Shelf of NW Europe, L. V. Illing and G. D. Hobson, Eds.: London: The Institute of Petroleum, 407–413.

Kellogg, H. E. 1975. Tertiary stratigraphy and tectonism in Svalbard and continental drift. *American Association of Petroleum Geologists Bulletin 59*, 465–485.

Kennett, J. P. 1982. *Marine Geology*: Englewood Cliffs, NJ: Prentice-Hall.

Knox, R. W. O'B., and Morton, A. C. 1988. The record of early Tertiary N Atlantic volcanism in sediments of the North Sea Basin. *In* Early Tertiary Volcanism and the Opening of the NE Atlantic, A. C. Morton and L. M. Parson, Eds.: *Geological Society of London, Special Publication 39*, 407–419.

Kristoffersen, Y. 1990. On the tectonic evolution and paleoceanographic significance of the Fram Strait gateway. *In* Geological History of the Polar Oceans: Arctic Versus Antarctic, U. Bleil and J. Thiede, Eds: Dordrecth: Kluwer Academic Publishers, 63–76.

Larsen, V. B. 1987. A synthesis of tectonically-related stratigraphy in the North Atlantic–Arctic region from Aalenian to Cenomanian time. *Norsk Geologisk Tidsskrift* 67:281–293.

Lawver, L. A., Muller, R. D., Srivastava, S. P., and Roest, W. 1990. The opening of the Arctic Ocean. *In* Geological

History of the Polar Oceans. Arctic Versus Antarctic, U. Bleil and J. Thiede, Eds.: Dordrecht: Kluwer Academic Publishers, 29–62.

Lien, R. 1976. Ingeniørgeologisk kartlegging på kontinentalsokkelen utenfor Lofoten-Versterålen. *Continental Shelf Institute Publication* 78:36.

Lowell, J. D. 1972. Spitsbergen Tertiary orogenic belt and the Spitsbergen Fracture Zone. *Geological Society of America Bulletin* 83:3091–3102.

Manum, S. B. 1966. Deposits of probable Upper Cretaceous age offshore from Andøya, Northern Norway. *Norsk Geologisk Tidsskrift* 46:246–247.

Manum, S. B., and Throndsen, T. 1978. Rank of coal and dispersed organic matter and its geological bearing in the Spitsbergen Tertiary. *Norsk Polarinstitutt Årbok 1977*, 159–177.

Max, M. D., and Ohta, Y. 1988. Did major fractures in continental crust control orientation of the Knipovich Ridge–Lena Trough segment of the plate margin? *Polar Research* 6:85–93.

Mjelde, R., Sellevoll, M. A., Shimamura, H., Iwasaki, T., and Kanazawa, T. in press. A crustal study off Lofoten, N. Norway, by use of 3-component ocean bottom seismographs: *Tectonophysics*.

Mokhtari, M., Pegrum, R. M., and Sellevoll, M. A. 1989. A geophysical study of the Norwegian continental margin between 67°N and 69°N. Seismo-Series 28, Seismological Observatory, University of Bergen, 22.

Mokhtari, M., Sellevoll, M. A., and Olafsson, I. 1987. Seismic study of Lofoten continental margin N. Norway. Seismo-Series 10, Seismological Observatory, University of Bergen, 25.

Morton, A. C., and Knox, R. W. O'B. 1990. Geochemistry of late Palaeocene and early Eocene tephras from the North Sea Basin. *Journal of the Geological Society of London* 147:425–437.

Mutter, J. C. 1984. Cenozoic and late Mesozoic stratigraphy and subsidence history of the Norwegian margin. *Geological Society of America Bulletin* 95:1135–1149.

Mutter, J. C., and Zehnder, C. M. 1988. Deep crustal structure and magmatic processes: the inception of seafloor spreading in the Norwegian–Greenland Sea. *In* Early Tertiary Volcanism and the Opening of the NE Atlantic, A. C. Morton and L. M. Parson, Eds.: *Geological Society of London Special Publication 39*, 407–419.

Mutter, J. C., Buck, W. R., and Zehnder, C. M. 1988. Convective partial melting. 1. A model for the formation of thick basaltic sequences during the initiation of spreading. *Journal of Geophysical Research* 93:1031–1048.

Mutter, J. C., Talwani, M., and Stoffa, P. L. 1982. Origin of seaward dipping reflectors in oceanic crust off the Norwegian margin by "subaerial sea-floor spreading". *Geology* 10:353–357.

Mutter, J. C., Talwani, M., and Stoffa, P. L. 1984. Evidence for a thick oceanic crust adjacent to the Norwegian Margin. *Journal of Geophysical Research* 89:483–502.

Myhre, A. M. 1984. Compilation of seismic velocity measurements along the margins of the Norwegian–Greenland Sea. *Norsk Polarinstitutt Skrifter* 180:41–61.

Myhre, A. M., and Eldholm, O. 1981. Sedimentary and crustal velocities in the Norwegian–Greenland Sea. *Journal of Geophysical Research* 89:5012–5022.

Myhre, A. M., and Eldholm, O. 1988. The Western Svalbard margin (74–80°N). *Marine and Petroleum Geology* 5:134–156.

Myhre, A. M., Eldholm, O., and Sundvor, E. 1982. The margins between Senja and Spitsbergen fracture zones: implications from plate tectonics. *Tectonophysics* 89:33–50.

Nøttvedt, A., Berglund, L. T., Rasmussen, E., and Steel, R. J. 1988. Some aspects of Tertiary tectonics and sedimentation along the western Barents Shelf. *In* Early Tertiary Volcanism and the Opening of the NE Atlantic, A. C. Morton and L. M. Parson, Eds.: *Geological Society of London Special Publication 39*, 421–425.

Nyland, B., Jensen, L. N., Skagen, J. I., Skarpnes, O., and Vorren, T. O. 1989. Tertiary uplift and erosion in the Barents Sea; magnitude, timing and consequences. Abstract. Structural and Tectonic Modelling and its Application to Petroleum Geology, 18–20 October 1989, Stavanger, Norway.

Olafsson, I., Sundvor, E., Eldholm, O., and Grue, K. in press. Møre Margin: Crustal structure from analysis of expanded spread profiles. *Marine Geophysical Research*.

Pedersen, T., and Skogseid, J. 1989. Vøring Plateau volcanic margin: Extension, melting and uplift. Seismic interpretation, stratigraphy and vertical movements. *In* Proceedings of the Ocean Drilling Program, Scientific Results, Volume 104: O. Eldholm, J. Thiede, E. Taylor, et al.: College Station, TX: Ocean Drilling Program, 985–991.

Perry, R. K., Fleming, H. S., Cherkis, N. Z., Feden, R. H., and Vogt, P. R. 1980. Bathymetry of the Norwegian–Greenland and western Barents seas. Map, U.S. Naval Research Laboratory, Washington, DC.

Planke, S., Skogseid, J., and Eldholm, O. 1991. Crustal structure off Norway, 62–70°N. *Tectonophysics* 189:91–107.

Richardsen, G., Henriksen, E., and Vorren, T. O. 1991. Evolution of the Cenozoic sedimentary wedge during rifting and sea-floor spreading west of the Stappen High, Western Barents Sea. *Marine Geology* 101:11–30.

Riis, F., and Fjeldskaar, W. 1989. The importance of erosion and mantle phase changes for the late Tertiary uplift of Scandinavia and the Barents Sea. Abstract. Structural and Tectonic Modelling and its Application to Petroleum Geology, 18–20 October, 1989, Stavanger, Norway.

Riis, F., Vollset, J., and Sand, M. 1986. Tectonic development of the western margin of the Barents Sea and adjacent areas. *In*: Future Petroleum Provinces of the World, M. T. Halbouty, Ed.: *American Association of Petroleum Geologists Memoir 40*, 661–676.

Roberts, D. G., Schnitker, D., Backman, J., Baldauf, J. G., Desprairies, A., Homrighausen, R., Huddlestun, P., Kattenback, A. J., Keene, J. B., Krumsiek, K. A. O., Morton, A. C., Murray, J. W., Westburg-Smith, J., and Zimmerman, H. B. 1984. Initial Reports of the Deep Sea Drilling Project, Volume 81: Washington, DC: U.S. Government Printing Office.

Roest, W. R., and Srivastava, S. P. 1989. Sea-floor spreading in the Labrador Sea: A new reconstruction. *Geology* 17:1000–1003.

Rønnevik, H. C., and Jacobsen, H. P. 1984. Structures and basins in the western Barents Sea. *In* Petroleum Geology of the North European Margin, A. M. Spencer et al., Eds.: London: Graham and Trotman, 9–32.

Rønnevik, H. C., and Navrestad, T. 1977. Geology of the Norwegian shelf between 62°N and 69°N. *GeoJournal* 1:33–46.

Rønnevik, H. C., Eggen, S., and Vollset, J. 1983. Exploration of the Norwegian Shelf. *In* Petroleum Geochemistry and Exploration of Europe, J. Brooks, Ed.: Oxford: Blackwell, 71–94.

Rønnevik, H. C., Jørgensen, F., and Motland, K. 1979. The geology of the northern part of the Vøring Plateau, Report NSS/12, Norwegian Petroleum Society, Oslo.

Schlüter, H.-U., and Hinz, K. 1978. The continental margin of west Spitsbergen. *Polarforschung* 48:151–169.

Sellevoll, M. A. 1983. A study of the earth's crust in the island area of Lofoten–Vesterålen, northern Norway. *Norges Geologisk Undersøkelse* 380:235–243.

Sellevoll, M. A., and Mokhtari, M. 1988. An intra-oceanic crustal seismic reflecting zone below the dipping reflectors on Lofoten margin. *Geology* 16:666–668.

Skogseid, J., and Eldholm, O. 1987. Early Cenozoic crust at the Norwegian continental margin and the conjugate Jan Mayen Ridge. *Journal of Geophysical Research* 92:11471–11491.

Skogseid, J., and Eldholm, O. 1989. Vøring Plateau continental margin: Seismic interpretation, stratigraphy and vertical movements. *In* O. Eldholm, J. Thiede, E. Taylor, et al., Proceedings of the Ocean Drilling Program, Scientific Results, Volume 104: College Station, TX: Ocean Drilling Program, 993–1030.

Skogseid, J., Pedersen, T., and Larsen, V. in press. Vøring Basin: Subsidence and tectonic evolution. *In* Structural and Tectonic Modeling and its Application to Petroleum Geology.

Smythe, D. K. 1983. Faeroe–Shetland Escarpment and continental margin north of the Faeroes. *In* Structure and Development of the Greenland–Scotland Ridge, H. M. P. Bott, S. Saxov, M. Talwani, and J. Thiede, Eds.: New York: Plenum Publishers, 109–120.

Smythe, D. K., Chalmers, J. A., Skuce, A. G., Dobinson, A., and Mould, A. S. 1983. Early opening history of the North Atlantic, I, Structure and origin of the Faeroe–Shetland Escarpments. *Geophysical Journal of the Royal Astronomical Society* 72:373–398.

Solheim, A., and Kristoffersen, Y. 1984. The physical environment, western Barents Sea, 1:1 500 000, sheet B; Sediments above the upper regional unconformity: thickness, seismic stratigraphy and outline of the glacial history. *Norsk Polarinstitutt Skrifter* 179B.

Spencer, A. M., Home, P. C., and Berglund, L. T. 1984. Tertiary structural development of the western Barents Shelf. *In*: Petroleum Geology of the North European Margin, A. M. Spencer, et al. Eds.: London: Graham and Trotman, 199–209.

Steel, R. J., and Worsley, D. 1984. Svalbard's post-Caledonian strata—an atlas of sedimentational patterns and paleogeographic evolution. *In* Petroleum Geology of the North European Margin, A. M. Spencer, et al., Eds.: London: Graham and Trotman, 109–135.

Steel, R. J., Gjelberg, J., Nøttvedt, A., Helland-Hansen, W., Kleinsphen, K., and Rye-Larsen, M. 1985. The Teritary strike-slip basins and orogenic belt of Spitsbergen. *Society of Economic Paleontologists and Mineralogists Special Publication* 37, 339–359.

Stuevold, L. M. 1989. Den tertiære fennoskandiske landhevning i lys av vertikalbevegelser på midnorsk kontinentalmargin. Candidata scientiarum thesis, University of Oslo, 175.

Stuevold, L. M., Skogseid, J., and Eldholm, O. 1990. Post-Cretaceous uplift events at the Vøring Margin. Abstract. Post-Cretaceous uplift and sedimentation along the Western Fennoscandian Shield. TSGS 7th annual meeting Stavanger, Norway, 3–5 October 1990.

Sundvor, E., and Austegard, A. 1990. The evolution of the Svalbard margins: Synthesis and new results. *In* Geological History of the Polar Oceans: Arctic Versus Antarctic, U. Bleil and J. Thiede, Eds.: Dordrecht: Kluwer Academic Publishers, 77–94.

Sundvor, E., and Eldholm, O. 1976. Marine geophysical survey on the continental margin from Bear Island to Hornsund, Spitsbergen. Science Report 3, Seismological Observatory, University of Bergen.

Sundvor, E., and Eldholm, O. 1979. The western and northern margin off Svalbard. *Tectonophysics* 59:239–250.

Sundvor, E., Vogt, P., and Crane, K. 1990. Preliminary Results from SeaMARC II investigations in the Norwegian-Greenland Sea. Seismo-Series 48, Seismological Observatory, University of Bergen.

Talwani, M., and Eldholm, O. 1972. Continental margin off Norway: A geophysical study. *Geological Society of America Bulletin* 83:3575–3606.

Talwani, M., and Eldholm, O. 1977. Evolution of the Norwegian–Greenland Sea. *Geological Society of America Bulletin* 88:969–999.

Talwani, M., Mutter, J. C., and Hinz, K. 1983. Ocean continent boundary under the Norwegian continental margin. *In* Structure and Development of the Greenland–Scotland Ridge—New Methods and Concepts, M. H. P. Bott, S. Saxov, M. Talwani, and J. Thiede, Eds.: New York: Plenum, 121–131.

Talwani, M., Udintsev, G., Bjorklund, K. R., Caston, V. N. D., Faas, R. W., Kharin, G. N., Morris, D. A., Müller, C., Nilsen, T. H., Van Hinte, J. E., Warnke, D. A., and White, S. M. 1976. Initial Reports of the Deep Sea Drilling Project, Volume 38: Washington, DC: U.S. Government Printing Office.

Theilen, F., Uenzelmann, G., and Gimpel, P. 1984. Sediment distribution at the outer Vøring Plateau from reflection seismic investigations. Sonderforschungsbereich, 313, Sedimentation im Europäischen Nordmeer, 1987.

Thiede, J., Eldholm, O., and Taylor, E. 1989. Variability of Cenozoic Norwegian–Greenland Sea paleoceanography and northern hemisphere paleoclimate. *In* O. Eldholm, J. Thiede, E. Taylor, et al., Proceedings of the Ocean Drilling Program, Scientific Results, Volume 104: College Station, TX: Ocean Drilling Program, 1067–1118.

Torske, T. 1972. Tertiary oblique uplift of western Fennoscandia, crustal warping in connection with rifting and breakup of the Laurasian continent. *Norges Geologisk Undersøkelse* 273:43–48.

Torske, T., and Prestvik, T. 1991. Mesozoic detachment faulting between Greenland and Norway: Inferences from Jan Mayen Fracture Zone system and associated alkalic volcanic rocks. *Geology* 19:481–484.

Townsend, C., and Mann, A. 1989. Late Paleozoic basin development in southern Svalbard: Interacting extensional and compressional events. Abstract. Structural and Tectonic Modelling and its Application to Petroleum Geology, 18–20 October 1989, Stavanger, Norway.

Upton, B. G. J. 1988. History of Tertiary igneous activity in the N Atlantic borderlands. *In* Early Tertiary Volcanism and the Opening of the NE Atlantic, A. C. Morton, and L. M. Parson, Eds.: *Geological Society of London Special Publication 39*, 429–453.

Vågnes, E., Faleide, J. I., and Gudlaugsson, S. T. 1990. Glacial and tectonic uplift of the Barents Sea. Abstract. Post-Cretaceous uplift and sedimentation along the western Fennoscandian Shield. TSGS 7th annual meeting Stavanger, Norway, 3–5 October.

Vogt, P. 1986. Sea-floor topography, sediments and paleoenvironments. *In* The Nordic Seas, B. G. Hurdle, Ed.: New York: Springer-Verlag, 237–386.

Vorren, T. O., Lebesbye, E., and Henriksen, E. 1988. Cenozoic erosjon og sedimentasjon i det sørlige Barentshav. Abstract. 18. Nordic Geological Winter Meeting 1988, Copenhagen, Denmark.

Vorren, T. O., Lebesbye, E., and Larsen, K.-B. 1990. Geometry and genesis of the glacigenic sediments in the southern Barents Sea. *In* Glacimarine Environments Processes and Sediments. J. A. Dowdeswell and J. D. Scourse, Eds.: *Geological Society Special Publication 53*, 269–288.

Vorren, T. O., Lebesbye, E., Andreassen, K., and Larsen, K.-B. 1989. Glacigenic sediments on a passive continental margin as exemplified by the Barents Sea. *Marine Geology* 85:251–272.

Vorren, T. O., Richardsen, G., Knutsen, S.-M., and Henriksen, E. 1990. The western Barents Sea during the Cenozoic. *In* Geologic History of the Polar Oceans: Arctic versus Antarctic, U. Bleil and J. Thiede, Eds.: Dortrecht: Kluwer Academic Publishers, 95–119.

Vorren, T. O., Richardsen, G., Knutsen, S.-M., and Henriksen, E. 1991. Cenozoic erosion and sedimentation in the western Barents Sea. *Marine and Petroleum Geology* 8:317–340.

White, R. S., and McKenzie, D. 1989. Magmatism at rift zones: The generation of volcanic margins and flood basalts. *Journal of Geophysical Research* 94:7685–7729.

White, R. S., Spence, G. D., Fowler, S. R., McKenzie, D. P., Westbrook, G. K., and Bowen, A. N. 1987. Magmatism at rifted continental margins. *Nature* 330:439–444.

Wood, R. J., Edrich, S. P., and Hutchinson, I. 1989. Influence of North Atlantic tectonics on the large scale uplift of the Stappen High and Loppa High, Western Barents Shelf. *In* Extensional Tectonics and Stratigraphy of the North Atlantic Margins, A. J. Tankard and H. R. Balkwill, Eds.: *American Association of Petroleum Geologists Memoir 46*, 559–566.

Ziegler, P. A. 1988. Evolution of the Arctic–North Atlantic and the Western Tethys. *American Association of Petroleum Geologists Memoir 43*, 198.

Ziegler, P. A. 1989. *Evolution of Laurussia: A Study in Late Palaeozoic Plate Tectonics*: Dordrecht: Kluwer Academic Publishers.

III
Late Postrift Evolution

8

Southern Brazil Basin: Sedimentary Processes and Features and Implications for Continental-Rise Evolution

Gilberto A. Mello, Roger D. Flood, Thomas H. Orsi, and Allen Lowrie

In the context of deep-sea sediment studies, the dynamics of sedimentation and surface-sediment erosion are determined by two principal processes: (1) downslope sediment-gravity flows; and (2) transport and dispersion of sediments by geostrophic currents. Both processes shape the physiography of the continental slope and rise (Heezen, Tharp, and Ewing, 1959; Heezen, Hollister, and Ruddiman, 1966), and sculpt the microtopographic features (Hollister, Flood, and McCave, 1978; Kolla et al., 1980; Jacobi and Hayes, 1982). Microtopographic features correspond to *erosional/depositional bedforms*, whose scales range from tens of centimeters (e.g., ripples) to tens or hundreds of kilometers (e.g., furrows and sediment waves). These features are the relict imprints of bottom processes at the seafloor, as they normally preserve internal characteristics and invariably exhibit external structure and orientational parameters inherited from the processes that created them (Heezen and Hollister, 1971; Kennett, 1982; see also Chapters 9, 10, 12, 13). Thus, understanding the origin of sediment bedforms provides a valuable, though indirect, method for interpreting geological and oceanographic processes operating in deep-sea environments.

Since the early 1960s, high frequency (3.5 and 12 kHz) echograms have been the primary data source used in studies of deep-sea sediment bedforms. Detailed geometric studies of echo profiles have helped in understanding the nature of bedforms (Krause, 1962; Flood, 1978, 1980), and through *echo-character* analysis (Laine, Damuth, and Jacobi, 1986), sedimentary processes and their related bedforms have been mapped throughout many ocean basins (see Chapter 12).

The echo-character study of Damuth (1975) defined important aspects of sedimentary processes on deep portions of the northern Brazil continental margin. Damuth and Hayes (1977) extended this study to the eastern and southern sections of the region. They recognized a consistent relationship between reflectivity patterns on echograms and relative abundance of coarse particles (silt/sand) in sediment cores from areas covered by turbidities. Furthermore, they observed that the seafloor on the continental rise returns a multitude of hyperbolic echoes, suggesting that current-created bedforms dominate this area. These studies are integral to our understanding of the major seafloor sedimentary processes of the Brazil continental margin; however, such studies by necessity show only regional variations of echo patterns, without revealing microtopographic features. Detailed study of microtopography is important for interpreting bottom processes and inferring their possible origins, whether from downslope mass movements (Embley, 1975, 1976, 1980; Jacobi, 1976; Embley and Langseth, 1977; Embley and Jacobi, 1977, 1986; Jacobi and Hayes, 1982) or from bottom-current activity (Hollister et al., 1974, 1976; Kolla, Moore, and Curray, 1976; Flood, 1978, 1980; Kolla et al., 1980; McCave and Tucholke, 1986).

The objectives of this study are to investigate in much greater detail the sediment bedforms of the continental slope and rise of the southern Brazil Basin, and, whenever possible, to establish their relationship with associated bottom processes. Due to the presence of several deep water masses and to regional evidence that the southern Brazil continental margin has prograded, the study area is an exceptional region in which to evaluate the role of, and interaction between, alongslope and downslope processes.

REGIONAL SETTING

Physiography

The general configuration of the southern Brazil Basin (Figs. 8-1, 8-2A) results from the interaction of both tectonic–magmatic and tectonic–sedimentary processes that started in the Late Jurassic/Early Cretaceous (Asmus and Guazelli, 1981; Asmus, 1984; Emery and Uchupi, 1984). Igneous and tectonic activity, both related to spreading processes of the South Atlantic, gave rise to the basic regional fabric; this

physiography was gradually modified later by sedimentation and erosion processes (Gorini and Carvalho, 1984). Evaporite deposition during the Aptian (Early Cretaceous) furnished sediments that form the São Paulo Plateau (Leyden et al., 1976; Asmus, 1984). This plateau is a prominent feature of the southern Brazil continental margin, and corresponds to an elongate, north–south trending structure located between the continental slope and rise. The plateau, 1,500–3,000 m deep, exhibits rugged surface topography, which results from intense diapirism (Kumar et al., 1977; Gambôa and Kumar, 1977).

The topography of the continental rise, beyond the limits of São Paulo Plateau, is dictated by depositional, erosional, and volcanic processes. Because of the com-

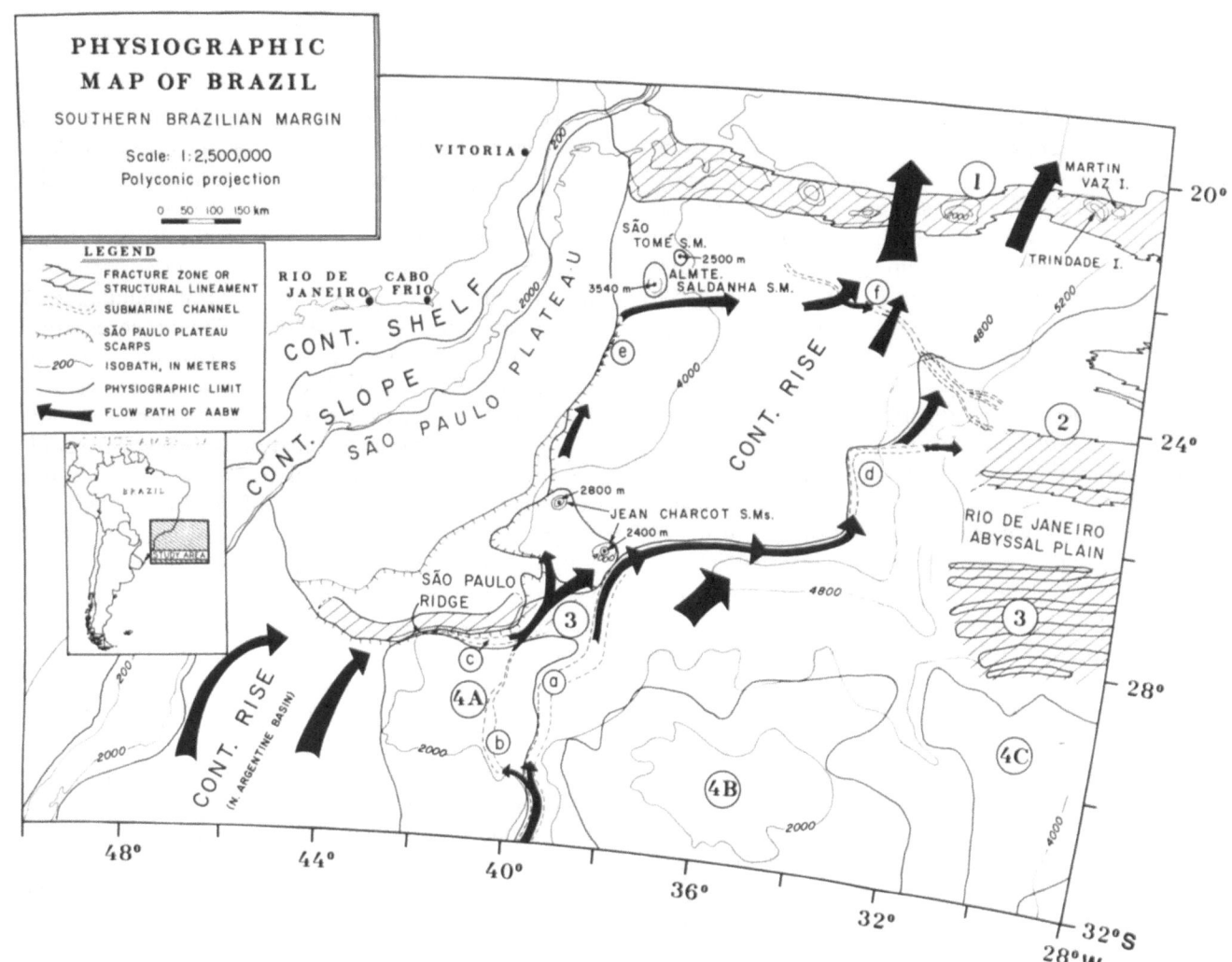

FIGURE 8-1.   Physiography and benthic circulation of southern Brazilian continental margin and adjacent deeps [after: Johnson et al., 1976; Zembruscki (1979); Johnson (1984); and Asmus (1984)]. Major physiographic features are indexed numerically and alphanumerically: (1) Vitória–Trindade Seamount Chain; (2) Rio de Janeiro Lineament; (3) Florianópolis Lineament; and (4A) the western, (4B) central, and (4C) eastern sections of Rio Grande Rise. Letters denote major submarine channels: (a) Vema Channel; (b) secondary branch of Vema Channel; (c) channel at base of São Paulo Ridge; (d) Rio de Janeiro Channel; (e) Guanabara Channel; and (f) Columbia Channel. AABW = Antarctic Bottom Water.

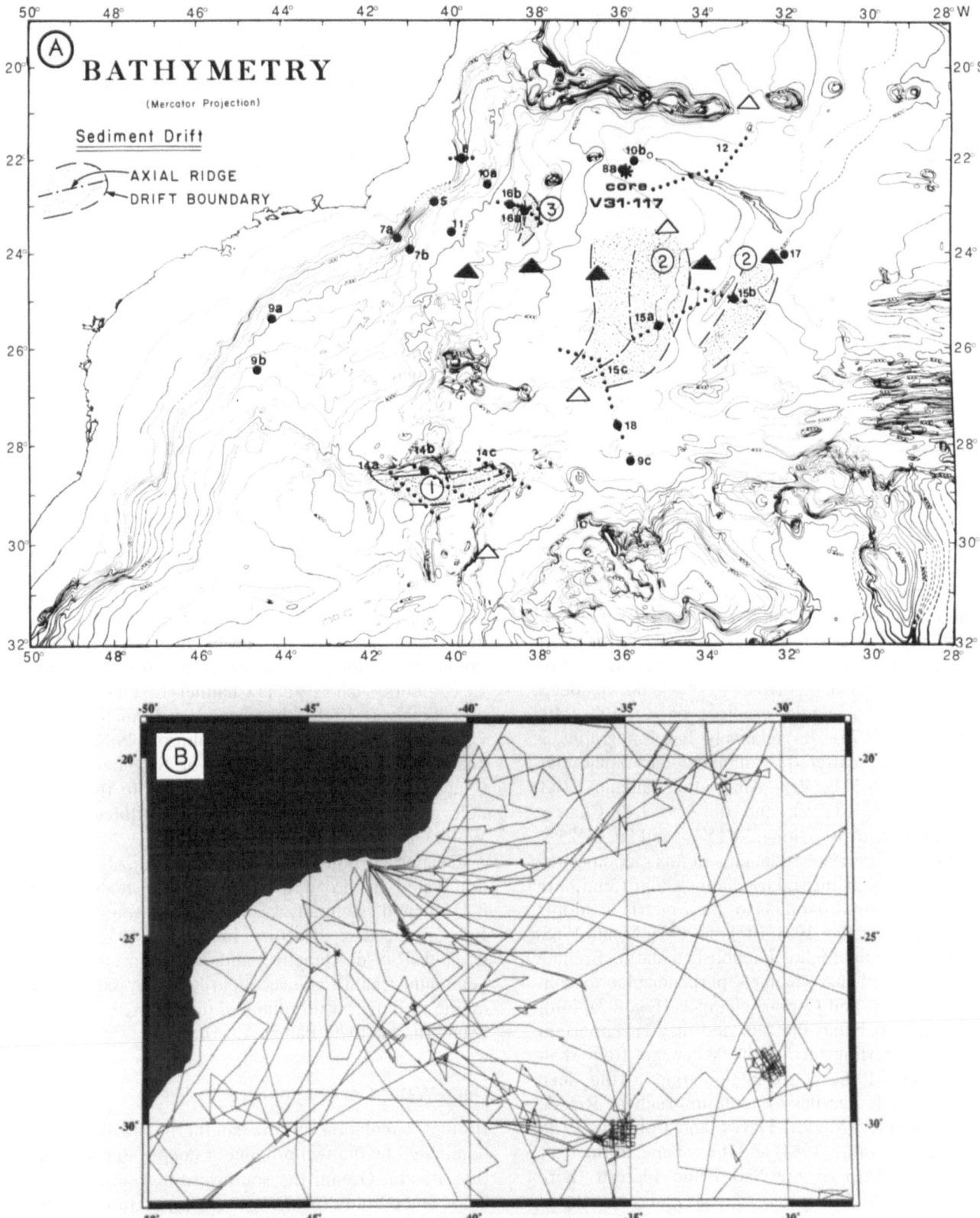

FIGURE 8-2. (A) Bathymetry (contours in meters) and location of sediment drift deposits of southern Brazil Basin (after Cherkis, Fleming, and Brozena, 1989). (1) Vema Drift; (2) Western and Eastern Santos Outer Ridge; (3) Guanabara Drift. Dotted lines and solid circles show locations of seismic and 3.5-kHz echo profiles, respectively, selected from track lines on (B). Numbered dotted lines/solid circles designate figure number of each profile. Locations of hydrographic stations (Figure 8-13) shown for International Geophysical Year (solid triangles) and GEOSECS (open triangles) expeditions. (B) Track lines of 3.5-kHz profiles used to compile map in Figure 8-4.

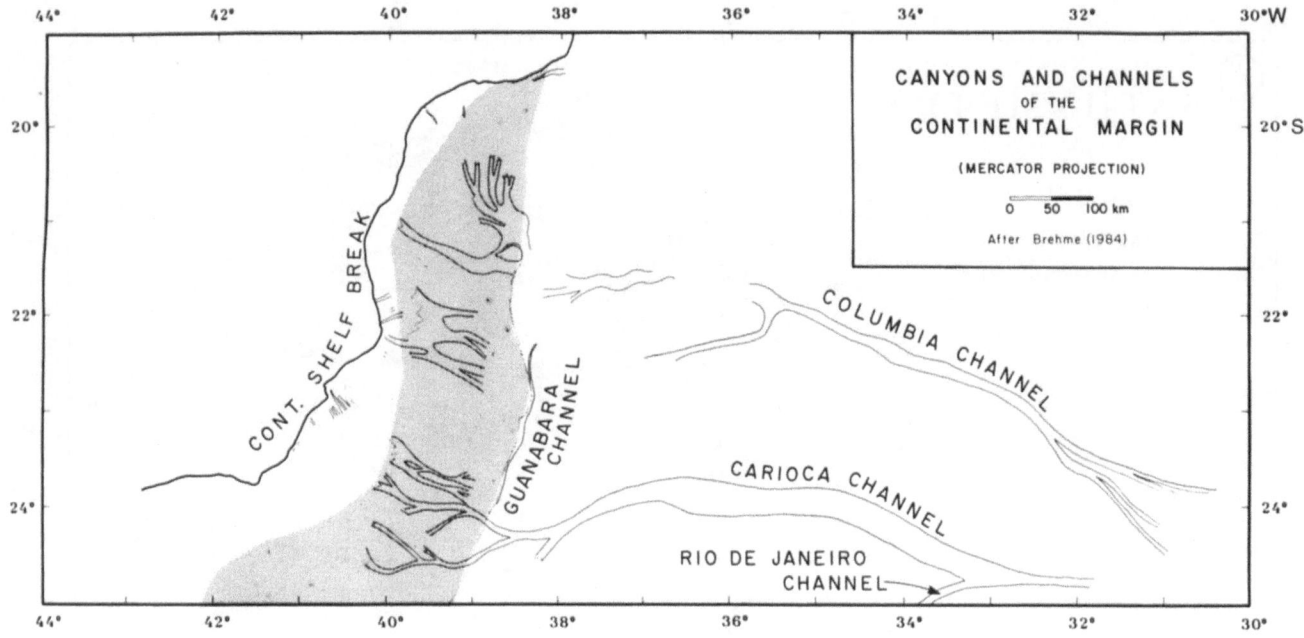

FIGURE 8-3.   Submarine canyons and channels crossing perpendicular and parallel to São Paulo Plateau (shaded region) and continental rise.

plex geomorphologic characteristics of the southern Brazil Basin, the continental rise can be subdivided into two sections, one to the east and one to the south of São Paulo Plateau (Fig. 8-1). The eastern section is broad (> 600 km), from 3,300–4,800 m deep, and possesses a gentle gradient of ~ 1:100–1:400 (Zembruscki, 1979). Its relatively flat topography is interrupted by scattered seamounts: São Tomé (lat. 21°55′S, long. 37°35′W); Almirante Saldanha (lat. 22°25′S, long. 37°35′W); and the two Jean-Charcot seamounts (lat. 26°10′S, long. 39°15′W; and lat. 26°25′S, long. 38°15′W). Surface sediments in this section of the rise are extensively incised by channels that trend both perpendicular and parallel to bathymetric contours (Gorini and Carvalho, 1984; Brehme, 1984; Mello, 1988; Orsi, Lowrie, and Mello, unpublished data). Brehme (1984) suggested that channels perpendicular to contours [Columbia and Carioca channels (Fig. 8-3)] initiate on the São Paulo Plateau, and play an important role in transporting terrigenous sediments from shallow portions of the continental margin to the deep basin. Channels parallel to contours, such as Rio de Janeiro Channel (Moody, Hayes, and Connary, 1979; Gorini and Carvalho, 1984) and the channel at the base of the east scarp of the São Paulo Plateau (here tentatively called *Guanabara Channel*), may be related to scouring by bottom currents flowing along the rise.

The southern section of the continental rise corresponds to the northernmost portion of the Argentine Basin (Palma, 1984; Emery and Uchupi, 1984). This section extends to shallower water (range = 2,200–4,000 m) than the eastern section, but bathymetric

gradients are similar (Zembruscki, 1979). No major downslope channels have been identified in the southern section, though the lower limit of the rise is bounded by numerous channels that parallel bathymetric contours, such as Vema Channel (Le Pichon, Ewing, and Truchan, 1971), the secondary branch of Vema Channel (Johnson et al., 1976; Johnson, 1984), and the channel that follows the base of São Paulo Ridge (Gambôa and Kumar, 1977). We refer to these channels, which are conduits for AABW (Antarctic Bottom Water), as the *Vema Channel Complex*.

Deeper portions of the southern Brazil Basin are dominated by broad tectonic–volcanic features (Fig. 8-1), aligned essentially west–east (from south to north: Rio Grande Rise, structural lineaments of Florianópolis and Rio de Janeiro, and the Vitória–Trindade Seamount Chain); and secondarily by abyssal hills that merge with the western flank of the Mid-Atlantic Ridge (Zembruscki, 1979; Palma, 1984).

Deep-Water Oceanography

Bottom circulation in the southern Brazil Basin is dominated by the two prominent deep water masses of the Atlantic Ocean: the southward-flowing North Atlantic Deep Water (NADW: avg. temperature > 2.0°C; salinity > 34.90‰) and the northward-flowing Antarctic Bottom Water (AABW: avg. temperature < 1.0°C; salinity < 34.90‰) (Wüst, 1933; Fuglister, 1960; see also Chapters 4, 5, 11, 12). The hydrographic configuration of these water masses, calculated by geopotential method (Reid, Nowlin, and Patzert, 1977), shows that higher gradients are located to the west along the continental margin, and decrease progressively to the east. Reid, Nowlin, and Patzert (1977) used relative geopotential fields for NADW and AABW in this area to demonstrate that preferential flux of NADW is southward with an anticyclonic gyre, whereas flux of AABW is northward with a cyclonic gyre.

Lynn and Reid (1968) and Lynn (1971) studied stability parameters for AABW and NADW in the Atlantic through calculations of density potential (σ), taking pressure referentials at sea level ($\sigma_0 = 0$ dbar) and $\sim 4,000$ m ($\sigma_4 = 4,000$ dbar). The authors showed that the approximate lower portion of NADW lies at a density maximum of 27.90 g/l in the profile of σ_0; this maximum value, compared to the σ_4 profile, coincides with a density value of 45.93 g/l, which corresponds to a continuous, horizontal isopycnal extending through almost the entire western section of the Atlantic Ocean. Lynn and Reid (1968) suggested that most of the mixing between NABW and AABW must take place along this continuous strip of water. Reid, Nowlin, and Patzert (1977) added that mixing along this isopycnal enhances the vertical density gradient between the two water bodies, creating a zone of maximum stability, which is maintained for great horizontal distances.

Johnson (1982) examined the geometry of deep water masses of the equatorial and subtropical Atlantic and suggested that mixing between AABW and NADW might occur over a depth interval of 400–500 m. He also added that the mixing zone is inclined ($\sim 3,200$ m at lat. 30°S, but lowering to 4,000 m at lat. 40°S), which may result from wedging of the two water masses moving in opposite directions. Broecker, Takahashi, and Li (1976) used GEOSECS data to conclude that the NADW/AABW boundary is marked by a pronounced break in the curves of many oceanographic parameters. In the curves of potential temperature and salinity, for instance, the change in gradient occurs near values of 2.0°C and 34.90‰, respectively, which defines a zone that Broecker, Takahashi, and Li (1976) and Broecker and Takahashi (1980, 1981) called the *two degree discontinuity*, or TDD. According to these authors, the water mass at the TDD is comprised of ~ 89 and 11% of waters coming from the north and south, respectively.

Johnson (1985) concluded that the flow of AABW and NADW began in the middle Oligocene, approximately 32 and 36 Ma, respectively. The AABW is generally considered to be the dominant geological agent in the southern Brazil Basin, a consensus supported by mineralogical (Biscaye, 1965), geochemical (Lawrence, 1979), 3.5-kHz echo character (Damuth and Hayes, 1977), and low-frequency seismic reflection (Kumar et al., 1979) studies. As AABW enters the region through Vema Channel (Johnson, 1984) at long. 29–30°W (Fig. 8-1), flow is constricted and accelerated,

creating a powerful erosional/redistributional force (Johnson et al., 1976). Although Hogg et al. (1982) noted that flow decelerates once it enters the region, AABW flow is still of sufficient intensity to rework sediments and create a complex pattern of 3.5-kHz hyperbolic echoes (Damuth and Hayes, 1977; Mello, 1988).

METHODS

We used L-DGO single-channel seismic profiles, 3.5–12-kHz echograms, sediment cores, bottom photographs, and water mass properties (salinity, potential temperature, and dissolved oxygen) in our study of downslope and alongslope sedimentary processes. We interpreted sediment gravity processes in two steps following methods described by Embley (1975, 1976, 1980): first we analyzed the single-channel and 3.5-kHz seismic data (Fig. 8-2B), which aided in identifying areas affected by slumps, slides, debris flows, and turbidity currents; second we examined sediment cores for reliable evidence of mass movement (e.g., slumped sediment sections and gradational turbidites). Genetic and geometric classifications of sediment-gravity processes were adapted from models suggested by Jacobi and Hayes (1982), Cook, Field, and Gardner (1982), and Embley and Jacoby (1986) (see also Chapters 12, 13). We studied current-created bedforms by examining bottom photographs, 3.5–12-kHz echograms, and seismic reflection profiles. These analyses were combined with available water mass properties (Fuglister, 1960; Bainbridge, 1976) and current-meter data (Johnson et al., 1976; Reid, Nowlin, and Patzert, 1977; Hogg et al., 1982; Schmitz and Hogg, 1983) to determine the main flow path of AABW within the study area.

Vertical profiles of hydrocast stations from the IGY and GEOSECS expeditions were examined to investigate the NADW/AABW boundary, which is critical for determining both the circulation pattern and which dominant water mass directly affects the seafloor. Having assumed that the TDD zone is a reliable physical barrier between these water masses, we examined potential temperature, salinity, and dissolved-oxygen curves for vertical changes in near-bottom gradients. Next, we studied the geometry and orientation of current-derived bedforms aided by single-channel and 3.5–12-kHz seismic data, according to methods developed by Flood (1978, 1980), Jacobi and Hayes (1982), and McCave and Tucholke (1986). Echograms (3.5–12-kHz) were particularly useful for examining small scale erosional/depositional furrows, which can be inferred from closely spaced (< 100-m) hyperbolic echoes. We analyzed the shape of these echoes through a computer technique (Program HYPECHO; Flood, 1978) to estimate the orientation of the furrows (com-

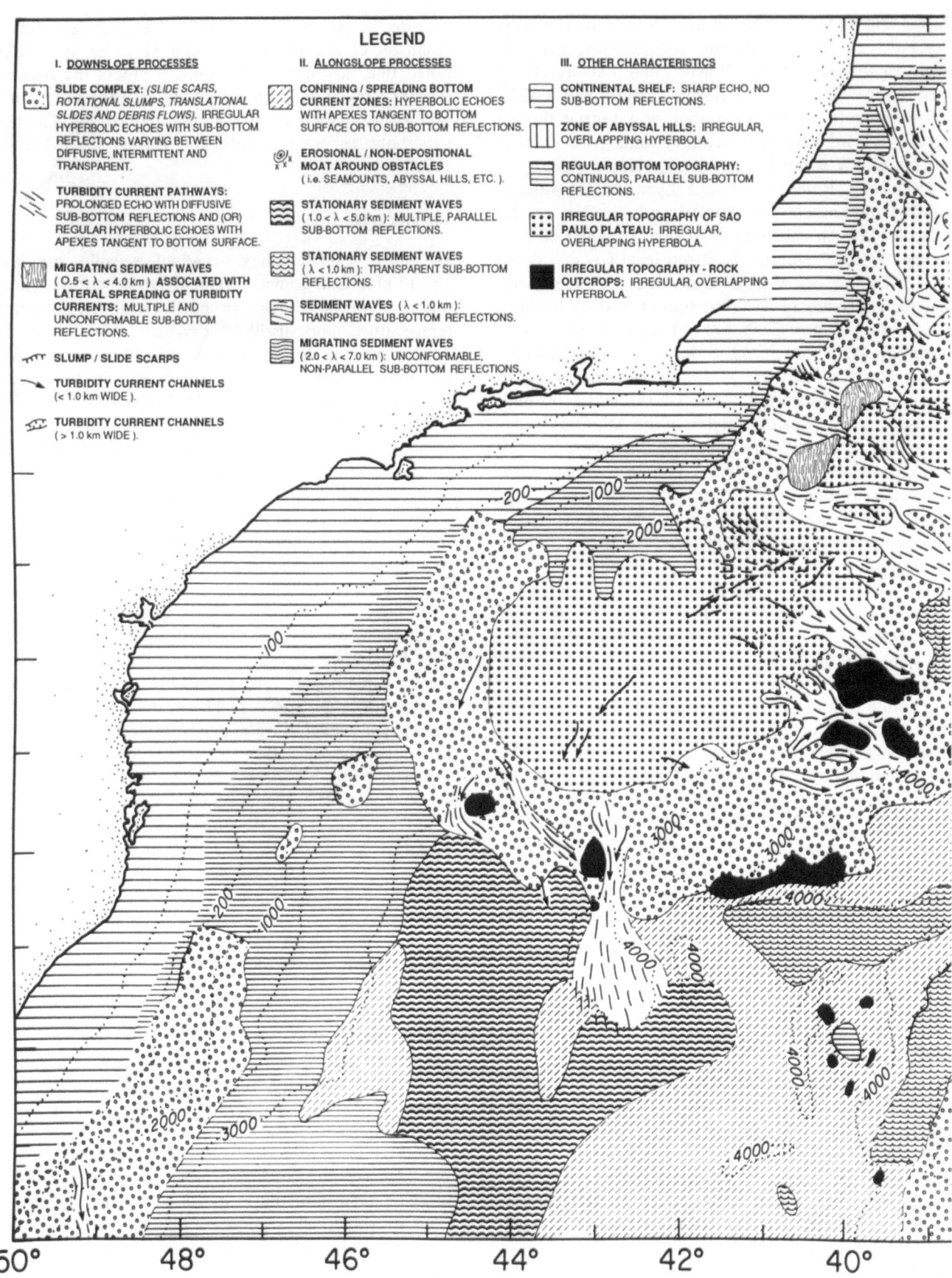

FIGURE 8-4.   Major downslope and alongslope sedimentary processes and microtopography in southern Brazil Basin.

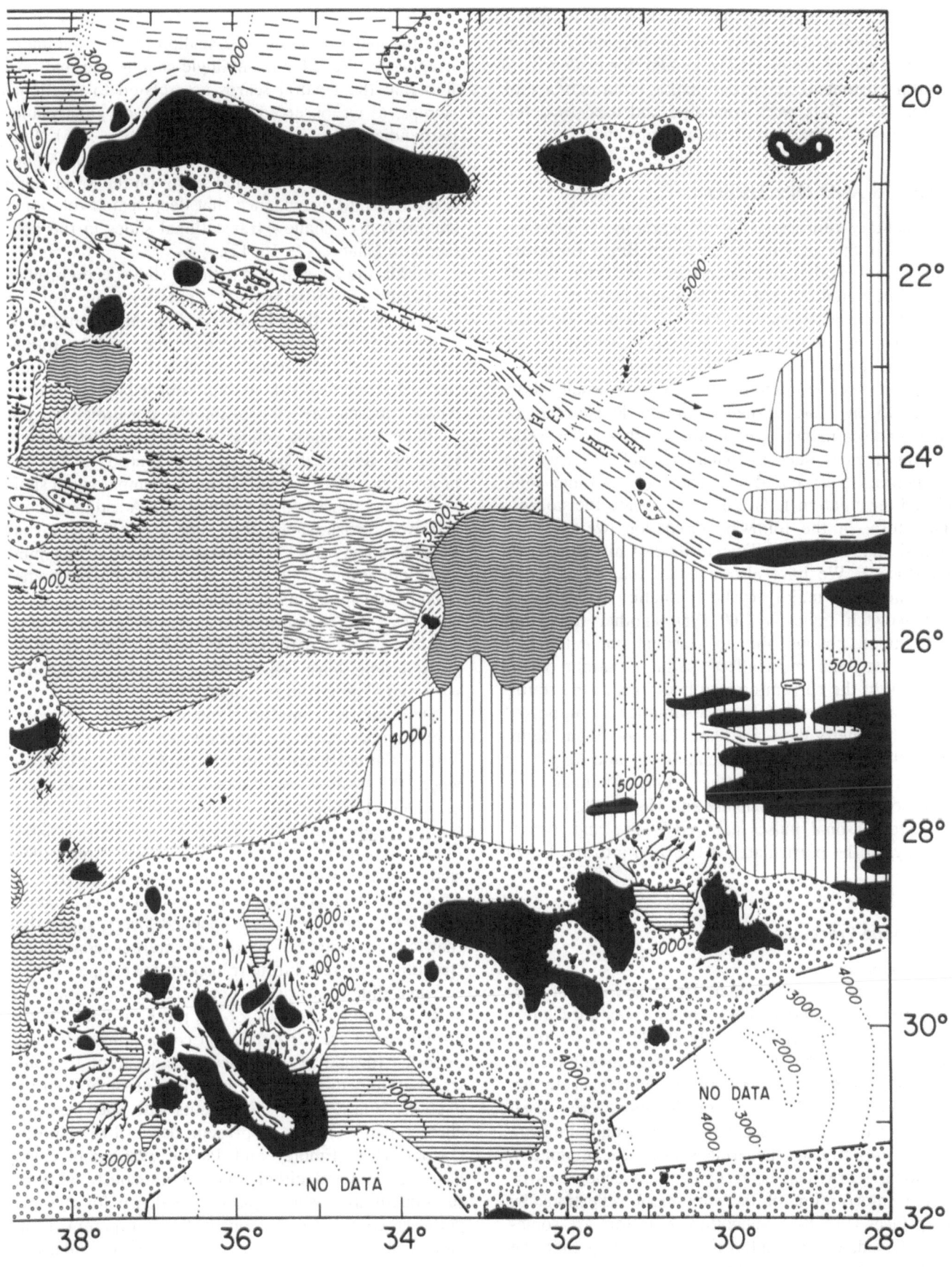

FIGURE 8-4. Continued

pare with Chapters 12, 13). Further, we examined bottom photographs using the method described by Heezen and Hollister (1971), looking for evidences of bottom currents (i.e., current ripples, scour marks, etc). Later, these data were combined and used to determine the flow direction of bottom currents.

RESULTS AND DISCUSSION

Sedimentary processes, microtopography, and associated echo character of the southern Brazil Basin are shown in Fig. 8-4 and summarized in Table 8-1. It should be noted that study of sediment bedforms using high-frequency seismic data (3.5 kHz) is limited by the vertical range of the seismic technique, which images only the upper 50–100 m of the seafloor. However, sediment cores examined in the area are usually short (< 10–15 m) and normally contain Pleistocene or Holocene sediments. Even in areas with very low sedimentation rates, such as Vema Channel (< 1 cm/kyr; Ledbetter, 1979), cored sediments are no older than glacial termination II (127 ka) of the late Pleistocene. Therefore, in the following discussion, we have assumed that surface sediments (< 100 m) are mostly Pleistocene. We have also assumed that, besides the influence of major cyclic climatic events (i.e., Milankovitch cycles: < 100 kyr), another important component, possibly due to seasonal fluctuations of bottom currents, must affect spatial variations occurring within the current-created bedforms in the southern Brazil Basin. This latter assumption is supported by nephelometry data (Eittreim, Thorndike, and Sullivan, 1976; Biscaye and Eittreim, 1977) and long-term current-meter studies in the Vema Channel area (Johnson et al., 1976; Schmitz and Hogg, 1983), which indicate seasonal pulses in AABW flow and, consequently, in the particulate load carried into the Brazil Basin.

Downslope Processes

Slide Complexes

Seismic profiles of the southern Brazil continental margin reveal slides and slumps, which are bounded laterally by distinct scarps, normally 10–80 m (to 150 m) high (Figs. 8-5, 8-6). These scarps occur mostly on the upper continental slope, where gradients typically exceed 1:50 (> 1°). Most scarps on the slope were caused by removal of surface sediments, and usually indicate where translational submarine slides have originated. Erosional scarps can be recognized on echograms by abrupt interruption of the subbottom reflections by scarp faces (Fig. 8-5). Subbottom reflectors are sometimes flexed, but not interrupted, indicating that the scarps are flexural (i.e., monocline flexure; Jacobi and Hayes, 1982). Such erosional and flexural scarps are

generally present in a sequence of downslope steps, forming many offsets of the seafloor, which undulates irregularly. We infer from single-channel seismic profiles that the flexural scarps are related to rotational slumps, as we observed traces of curved glide planes perpendicular to bathymetric trends (Fig. 8-6). Furthermore, progression of flexural scarps on the upper continental slope to erosional scarps on the lower slope (Figs. 8-5, 8-6), is evidence that deep-seated rotational slumps trigger shallow translational slides in the area (see also Chapters 9, 10, 12).

Within the zone of translational slides, echo profiles perpendicular to bathymetric contours show that the seafloor is smooth (Fig. 8-7A), but those parallel to bathymetric contours demonstrate the irregular topography of the slide scars (Fig. 8-7B). These contour-parallel profiles also show numerous hyperbolic echoes whose apexes are approximately tangential to the seafloor. From these echoes we infer that the seafloor is highly dissected by irregularly spaced (< 100 m) grooves, furrows, or small channels (compare with similar features described in Chapter 12). A sediment core retrieved proximal to this slide scar (V14-16: lat. 24°20′S, long. 41°43′W) contains homogeneous mud with undulating silty lamina, which probably originated from lateral spreading of turbidity currents. Thus, some slide scars on the continental slope might also serve as turbidity-current pathways (Table 8-1). Such turbidity currents may have originated or accentuated the furrows and channels in the slide zones.

Slide debris from the slope can eventually reach deeper areas on the continental rise, as seen near the base of the northern portion of the São Paulo Plateau, where the seafloor sometimes rises abruptly some tens of meters, forming irregular mounds (Fig. 8-8A). Such features were probably created by resistant, isolated blocks transported with the slide debris, from either shallow portions of the continental slope or the plateau area. Core V31-117, taken near one of these mounds (Fig. 8-8B), displays steeply inclined layers, suggesting that folding and(or) disruption of internal sedimentary structure occurred during transport downslope.

Some slide zones on the southern Brazil continental slope, and along the base of the northern flank of Rio Grande Rise, are surrounded or covered by wedge-like deposits, formed by acoustically transparent to semi-transparent sediments with intermittent subbottom reflections (Fig 8-9A–C). Sediments with such acoustic characteristics usually represent debris-flow deposits (Jacobi, 1976; Embley and Jacobi, 1986), and could have accumulated either contemporaneously with, or subsequent to, a slide episode. Evidence for this hypothesis comes from acoustically transparent sediments, which almost always fill or follow irregularities of pre-existent relief (Fig. 8-9A). Moreover, the contact between "transparent" deposits and "normal"

TABLE 8-1   Description, distribution, and interpretation of echo types.

Echo Type (Vertical bar = 100 m, Horizontal bar = 2 km)	Symbology (Fig. 8-4)	Echo Description — This Study	Echo Description — Damuth and Hayes (1977)	Physiographic Distribution	Interpretation
		Sharp echo, generally no sub-bottom reflections	IA	Continental shelf	Continental shelf process (not focused in this work)
		Continuous, parallel sub-bottom reflections	IB	Continental shelf edge, slope and rise and top of Rio Grande Rise	Pelagic/hemi-pelagic sedimentation. Surface sediment not disturbed by downslope/alongslope processes
		Irregular hyperbolic echoes with sub-bottom reflections varying between diffusive, intermittent and transparent	IIA, IIIC	Continental shelf edge, slope and rise, São Paulo Plateau and flanks of Rio Grande Rise and major seamounts	Slide scars and slump/slide debris
		Top: Prolonged echoes with diffusive sub-bottom reflections and(or) regular hyperbolic echoes with apexes tangent to bottom surface. Bottom: Channel dimensions (left < 1 km wide; right > 1 km wide)	IIB, IIID	Continental slope, rise and abyssal plain, São Paulo Plateau and flanks of Rio Grande Rise	Turbidity current pathways

Echo Type (Vertical bar = 100 m, Horizontal bar = 2 km)	Symbology (Fig. 8-4)	Echo Description — This Study	Echo Description — Damuth and Hayes (1977)	Physiographic Distribution	Interpretation
		Multiple and unconformable sub-bottom reflections	IIA	São Paulo Plateau	Migrating sediment waves (wavelength: 0.5-4.0 km); lateral spreading of turbidity currents
		Top: Hyperbolic echoes with apexes tangent to bottom surface or sub-bottom reflections. Bottom: Hyperbolic echoes with apexes tangent to bottom surface (cross marks)	IIID, IIIE	Continental rise and abyssal plain, flanks of sediment drift deposits	Zones of confining and (or) spreading of bottom currents; erosional/non-depositional furrows and moats.
		Multiple, parallel sub-bottom reflections	IB, IIIB	Continental rise (northern Argentine Basin)	Current-created stationary sediment waves (wavelength: 1.0-5.0 km)
		Multiple, parallel sub-bottom reflections	IIIB	Continental rise, top of sediment drift deposits and lower western flank of central Rio Grande Rise	Current-created stationary sediment waves (wavelength: <1.0 km)
		Transparent sub-bottom reflections	IA	Continental rise, western section of Santos Outer Ridge	Current-created sediment waves (wavelength: <1.0 km)

TABLE 8-1   Continued

Echo Type — Vertical bar = 100 m, Horizontal bar = 2 km	Symbology (Fig. 8-4)	Echo Description — This Study	Echo Description — Damuth and Hayes (1977)	Physiographic Distribution	Interpretation
		Unconformable, non-parallel sub-bottom reflections	IIIF	Abyssal plain, eastern section of Santos Outer Ridge	Migrating sediment waves (wavelength: 2.0-7.0 km)
		Irregular, overlaping hyperbola	IIIA	São Paulo Plateau	Irregular topography of São Paulo Plateau; possible outcroping diapirs
		Irregular, overlaping hyperbola		Abyssal plain, lower western flank of Mid Atlantic Ridge	Zone of abyssal hills
		Irregular, overlaping hyperbola		Continental slope, rise and abyssal plain; Rio Grande Rise and Vitória-Trindade Seamount Chain	Rock outcrops, sea mounts and structural lineaments

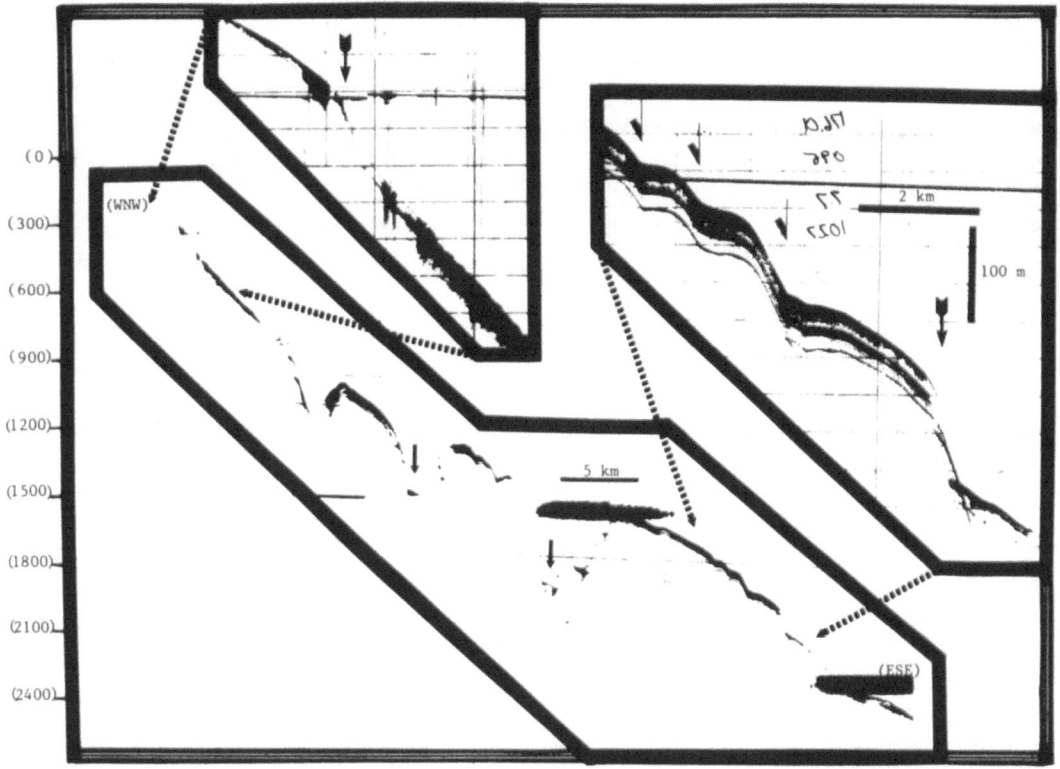

FIGURE 8-5.   Echogram (RC16-11) showing canyons on continental margin and evidences of down-slope processes. Note accumulation of slumped material, marked by inclination of bottom surface, at axes of two canyons (lower two arrows). Detailed portions of profile (top: left and right) show location of erosional (arrows with tails) and flexural (half-arrows) scarps. Vertical scale in meters.

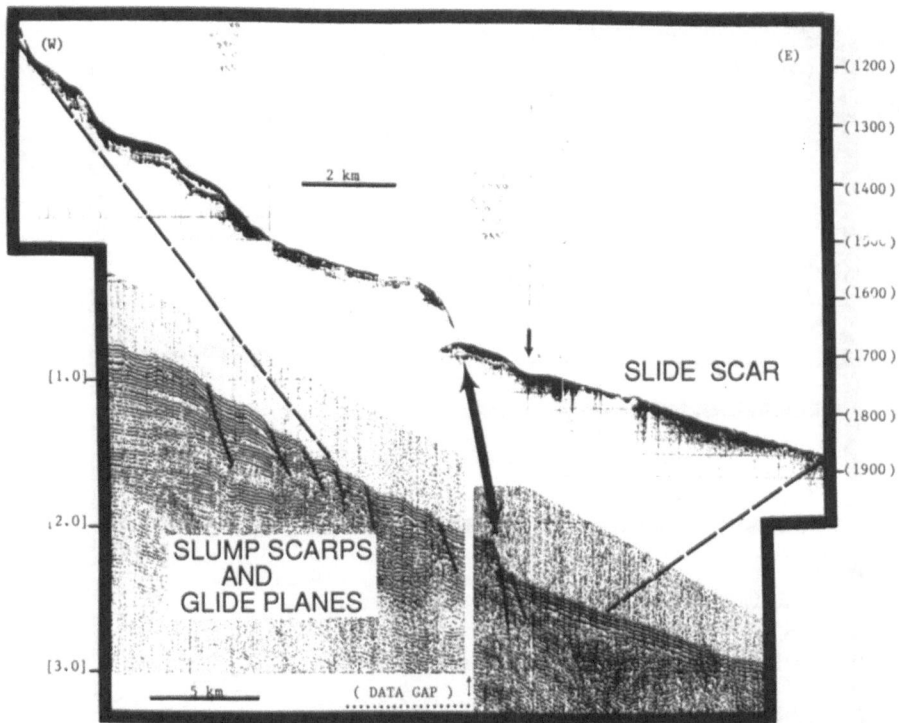

FIGURE 8-6.   Echogram (above) and single-channel profile (below; RC16-11) with evidences of rotational slumps and translational slides on continental slope and base-of-slope near São Paulo Plateau. Double-headed arrow, connecting profiles, identifies major scarp that may be transition between end of rotational slumps and beginning of translational slides. Small arrow on top profile indicates possible head of slide scar. Vertical scales in meters (right), two-way traveltime in seconds (left).

(acoustically laminated) sediments is generally marked by a lateral transition of acoustic facies, in which transparent facies bypass laminated facies and form a lenticular wedge (Fig. 8-9B). On distal parts of the rise, the boundary of these deposits is marked by a terminal head, which rises above the surrounding topography (Fig. 8-9C). The surface of these supposed debris-flow deposits usually exhibits small topographic irregularities, as indicated by irregular hyperbolic echoes, whose apexes extend slightly above the seafloor.

Turbidity–Current Pathways and Turbidites

Turbidity–current pathways are widespread in the southern Brazil Basin, although they are concentrated particularly in the region north of lat. 26°S (Fig. 8-4). The continental rise north of lat. 24°S, however, seems to have been most affected by turbidity currents (contrast with Chapter 9). The seafloor is flat, indicating extensive spreading of turbidities over a gentle bathymetric gradient (see Chapters 12, 13). Here, episodic turbidity currents probably passed north of the Victoria–Trindade Seamounts and along the axis of Columbia Channel. Overall, turbidity–current pathways are observed from shallow portions of the continental slope to deeper regions of the continental rise, the abyssal plain, and in the zone of abyssal hills.

Canyons and valleys carved into the steep slopes of major topographic highs, such as on the northern flank of Rio Grande Rise, also appear to be associated with turbidity–current pathways. They approximately parallel large canyons and channels incised perpendicular to bathymetric contours and commonly surround slide-complex deposits (Fig. 8-4).

The two most common acoustic characteristics diagnostic of turbidity–current pathways are prolonged echoes and hyperbolic echoes with relatively short-spaced apexes (< 10 m) tangential to the seafloor (Table 8-1). We infer from these echo characteristics that the seafloor is intensely dissected by erosional features, probably furrows or small, irregular ripples (Damuth and Hayes, 1977; Damuth, 1980).

In the São Paulo Plateau region, in particular, turbidity–current pathways are mostly related to the complex of channels formed around seafloor irregularities (e.g., salt diapirs and basement highs) (Fig. 8-10A). In deeper regions, such as the Rio de Janeiro Abyssal Plain and south of the southern limit of the São Paulo Plateau, turbidites form planar deposits that fill pre-existing topographic lows.

The spreading of turbidity currents has played an important role in eroding and(or) redistributing surface sediments beyond the base-of-slope and across the

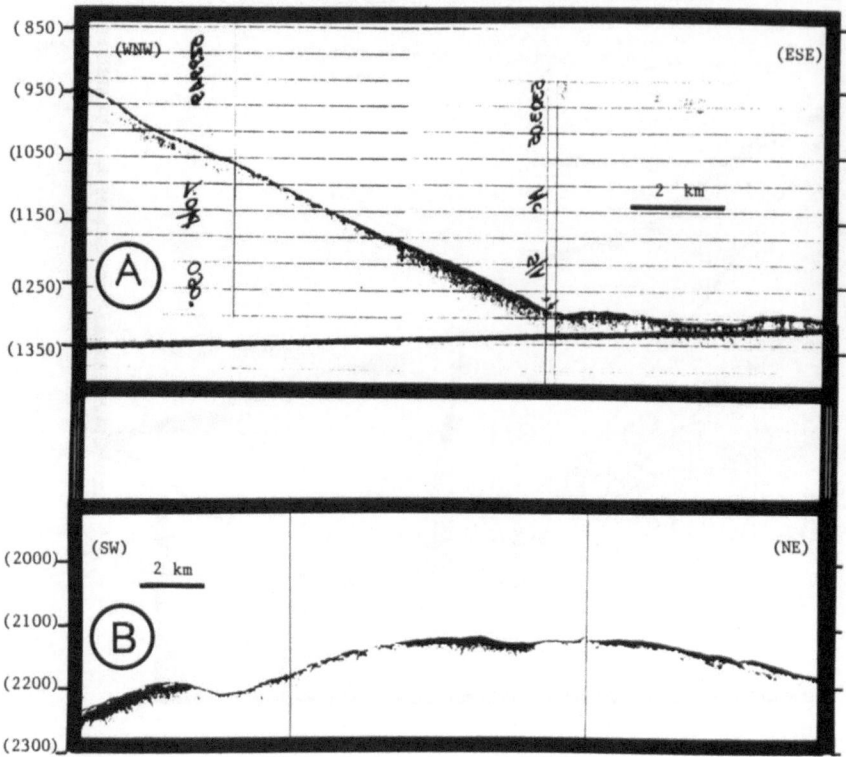

FIGURE 8-7.   Echograms of slide scar zone on continental slope. (A) VM26-05: Downslope profile showing planar seafloor. Note low acoustic penetration (< 50 m), probably caused by scattering of acoustic signals due to seafloor roughness (i.e., microtopography). Vertical scale in meters. (B) VM31-04: Alongslope profile (near profile 7A) showing irregular seafloor with possible furrows, inferred from regularly spaced hyperbolic echoes with apexes tangential to seafloor. Vertical scale in meters.

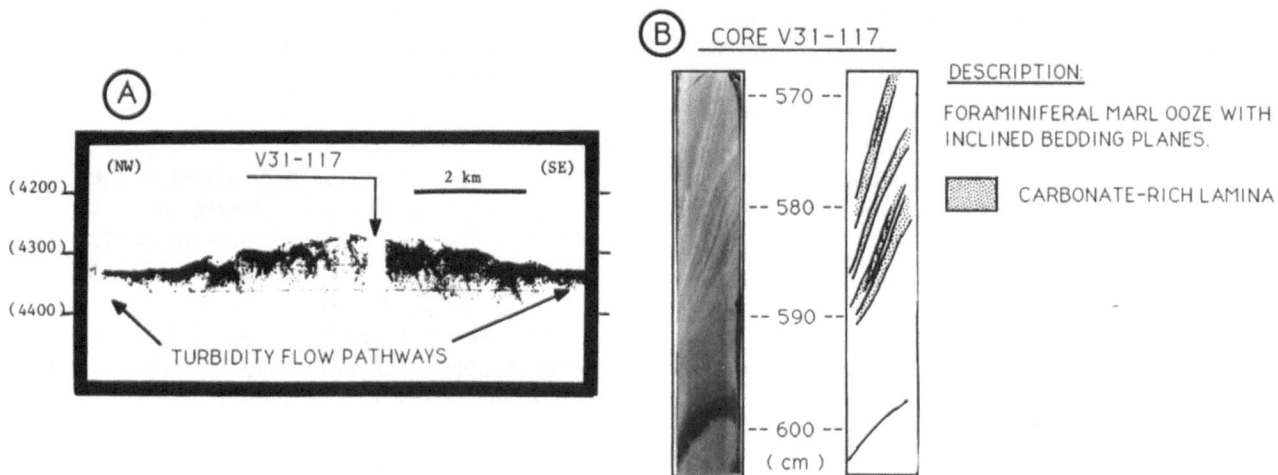

FIGURE 8-8.   Echogram and sediment core on continental rise near axis of Columbia Channel. (A) VM31-05: Profile suggests debris accumulation (blocky?). Note irregular seafloor (irregular hyperbolic echoes) and diffusive subbottom reflections. Vertical scale in meters. (B) Core VM31-117 (water depth 4,307 m; core length 7.60 m) showing inclined bedding, suggestive of slide debris.

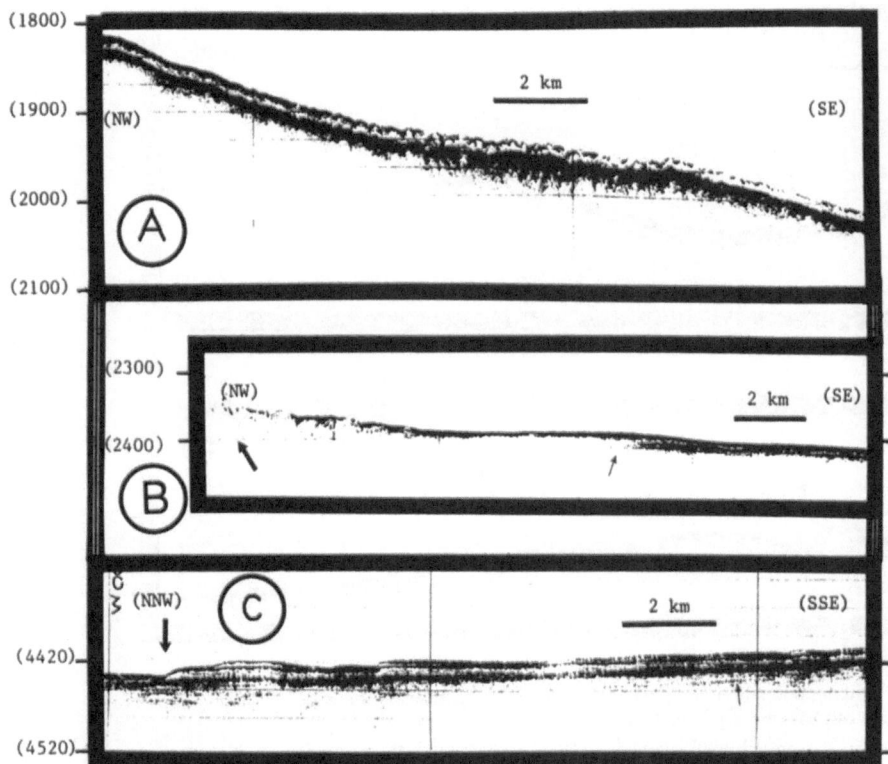

FIGURE 8-9. Echograms showing evidence of possible debris-flow deposits. **(A)** *Glomar Challenger* 72: Lower continental slope at southwestern São Paulo Plateau. Alongslope profile showing acoustically transparent deposit covering ancient slide deposit (hummocky terrain, prolonged subbottom reflections). Vertical scale in meters. **(B)** V31-04: Base of slope at southwestern São Paulo Plateau. Downslope profile showing center of lenticular deposit (thick arrow) and transitional boundary with pre-existing, acoustically laminated deposits (thin arrow). Note slight overlap of transparent sediment over laminated deposit, forming a wedge. Vertical scale in meters. **(C)** *Glomar Challenger* 72: Base of slope at northern flank of central Rio Grande Rise. Note terminal head (thick arrow) of possible debris-flow deposit that reaches abyssal plain, covering pre-existing, acoustically laminated deposits **(thin arrow).** Vertical scale in meters.

continental rise. This is apparent on the upper continental rise east of the São Paulo Plateau, between lat. 22°S and lat. 25°S, where numerous narrow channels with variable axial spacing (< 1–5 km) (Fig. 8-10B) form a complex drainage pattern south of Columbia Channel.

Depositional features, such as upslope–migrating sediment waves observed on the São Paulo Plateau between lat. 23°S and lat. 24°S, are probably related to turbidity–current activity as well. These migrating sediment waves (Fig. 8-11) cover a broad, elongate area of the seafloor, where there are many channels and submarine canyons. It is noteworthy that the morphology of the sediment waves is more irregular away from the axes of these channels, and subbottom reflections are distorted. These evidences suggest the presence of subbottom structures, either allochthonous blocks displaced from the continental slope, or salt diapirs upon which turbidites settled after spreading laterally.

Single-channel seismic profiles reveal a remarkable difference in acoustic facies between the north and south margins of Columbia Channel (Fig. 8-12). South

of the channel, the top layer of sediment (up to 0.5 sec) is acoustically transparent. This layer is in contact with a strong, continuous northward-extending subbottom reflector exposed along the axis of the channel. On the northern margin of the channel, however, this reflector is no longer obvious; instead, the sediment is acoustically transparent (up to 1.0 sec), with the exception of a well-laminated upper layer (< 0.5 sec). Conspicuous acoustic reflections in the laminated layer are parallel and extend for more than 100 km laterally. We attribute this striking difference in acoustic facies to the activity of AABW on its path from south to north, and to the deposition of fine-grained sediments north of Columbia Channel.

Alongslope Processes

Current-created bottom features of the continental rise and deep abyssal plain in the southern Brazil Basin are located primarily in the domain of northward-flowing AABW (compare with Chapters 12, 13). The inferred

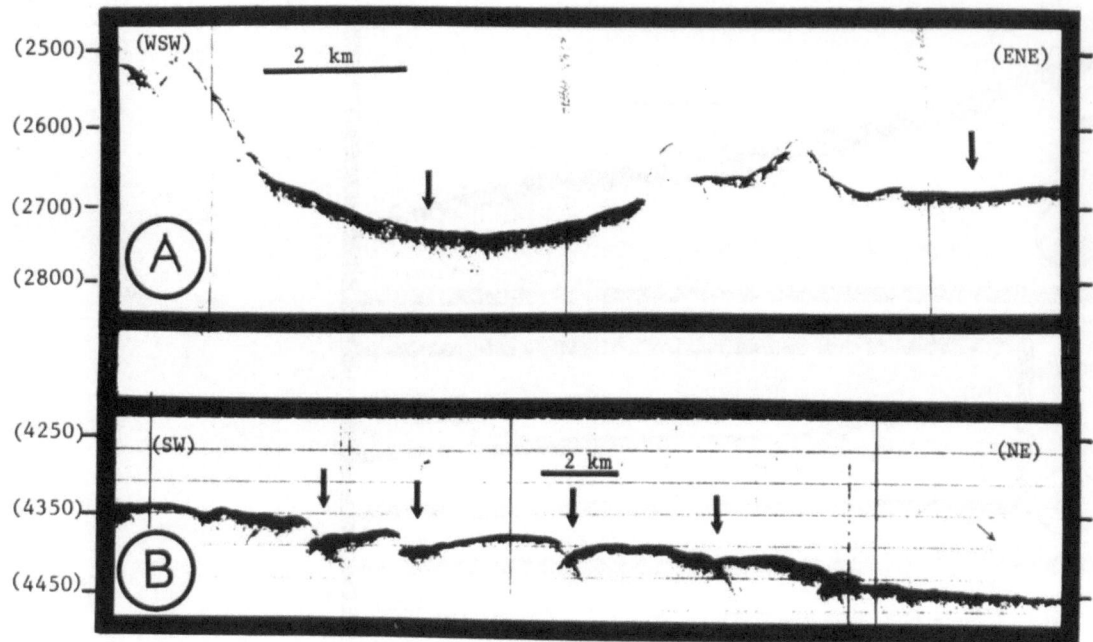

FIGURE 8-10.   Echograms of turbidity flow pathways. (A) V31-05: Northern São Paulo Plateau. Turbidity-current pathways (arrows) flanking possible outcropping salt diapirs. Vertical scale in meters. (B) RC21-05. Continental rise near axis of Columbia Channel (thin arrow). Axes of minor turbidity–current channels (thick arrows) parallel to main channel. Vertical scale in meters.

major path of this abyssal current through the study area (Fig. 8-1) agrees with theoretical directions predicted by geopotential calculations (Reid, Nowlin, and Patzert, 1977). Curves of potential temperature, salinity, and dissolved oxygen (Fig. 8-13) show that the mean TDD zone occurs at ~ 3,400 m. Jones (1984) obtained a similar depth using clay–mineral assemblages of surface sediments in Vema Channel.

Much of the entrained sediments transported (and subsequently deposited) by AABW is derived from high-latitude sources (Biscaye, 1965; Lawrence, 1979), from sediment drifts in the central Argentine Basin (Richardson et al., 1987), and from influx of particulates from turbidity currents (Klaus and Ledbetter, 1987) and benthic storms (Flood and Shor, 1988) originating on the Argentine continental margin.

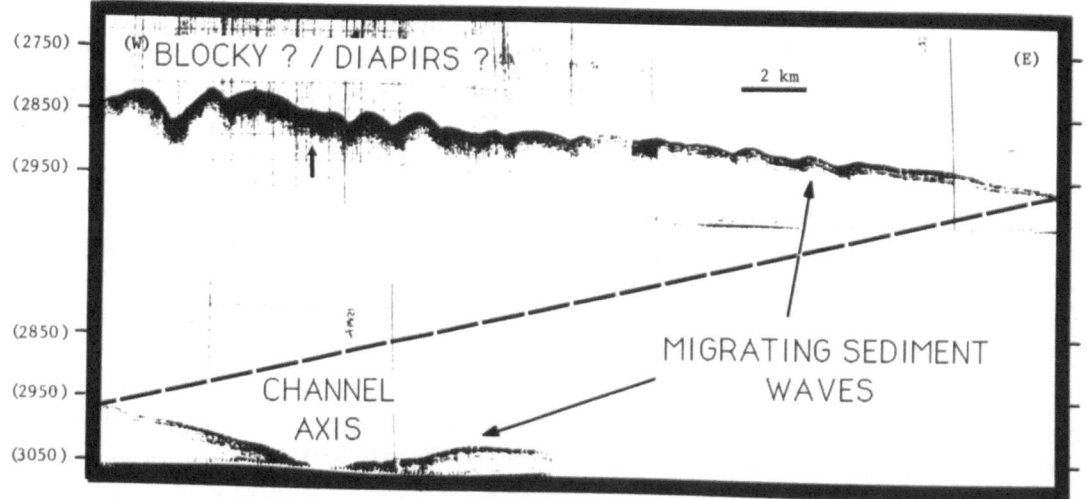

FIGURE 8-11.   Echogram (RC21-05) of migrating sediment waves near axis of one turbidity–current channel on São Paulo Plateau. Note migration upslope (to the west), indicated by inclined focusing echoes at trough of waves. Subbottom reflections appear distorted (short arrow) where sediment waves coalesce with possible allochthonous blocks or piercement diapirs. Vertical scale in meters.

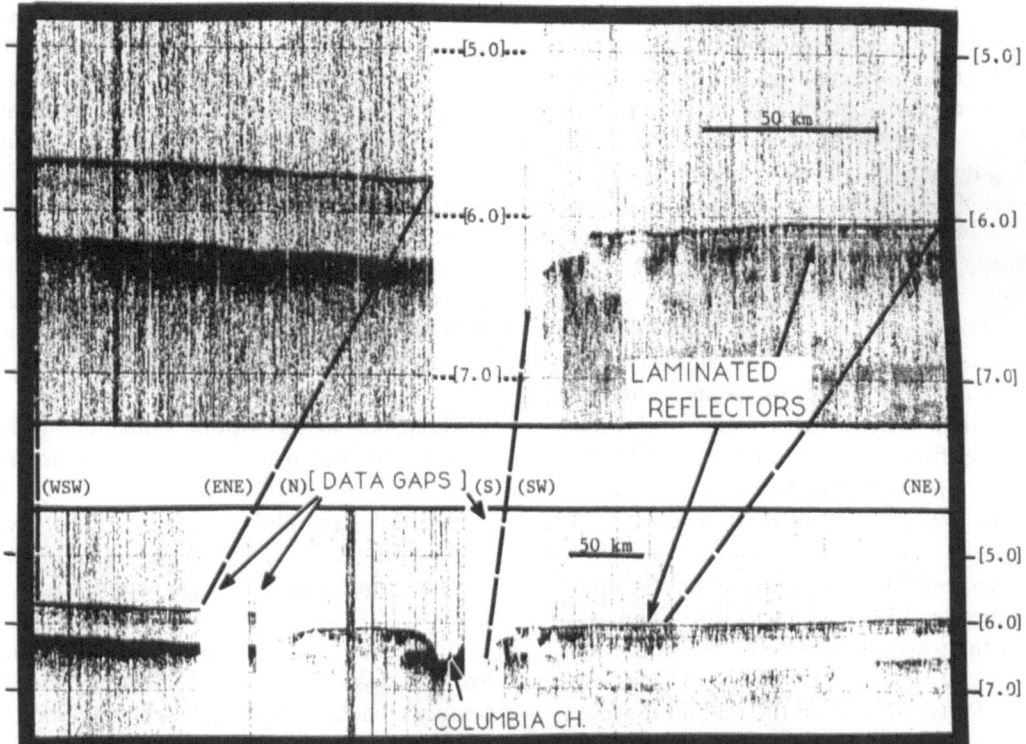

FIGURE 8-12.   Single-channel seismic profile (V24-13) on continental rise crossing axis of Columbia Channel. Top section shows details of selected parts of bottom section, demonstrating remarkable difference in acoustic facies between northeast (right from center) and southwest margins (left from center) of channel. Vertical scale is two-way traveltime in seconds.

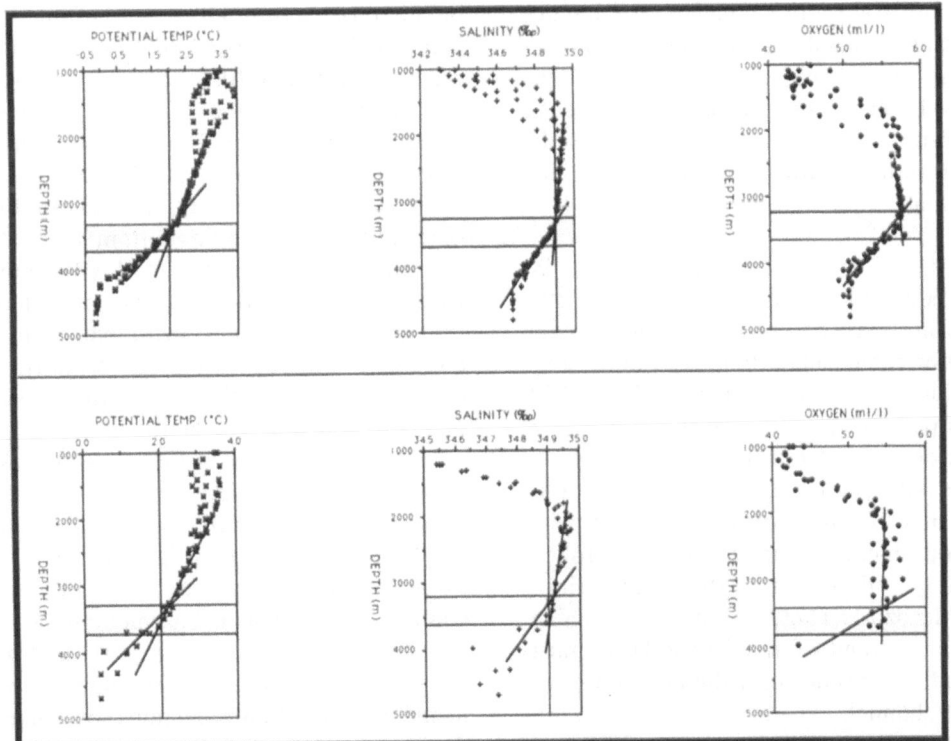

FIGURE 8-13.   Hydrocast stations from GEOSECS (top) and IGY (bottom) expeditions (locations given in Fig. 8-2A). Change in gradients near bottom is emphasized on profiles to show approximate location of two-degree discontinuity (TDD) and mixing zone (shaded area) between NADW (above) and AABW (below) in southern Brazil Basin.

Sediment Drifts and Associated Sediment Waves

Both stationary and migrating sediment waves (Fig. 8-4, Table 8-1) on the continental rise are found associated with sediment drifts, normally on the topmost parts of the drifts. To the south, southeast, and east of the São Paulo Plateau, stationary sediment waves are widespread at 3,400–4,000 m water depth. Migrating sediment waves occur less extensively, although they occupy a wider depth range (3,500–5,200 m).

The geometry of sediment-drift deposits of the southern Brazil Basin is similar to that of drift deposits of the North Atlantic Basin (McCave and Tucholke, 1986), since both are large bodies of sediment that rise from the surrounding seafloor and usually display axial ridges that parallel flow orientation (see also Chapters 6, 9, 12). The three main drift deposits of the southern Brazil Basin (Fig. 8-2A) are, from south to north:

1. *Vema Drift*, located in the Vema Channel Complex, corresponding to a north–northeastern sedimentary extension of the western section of the Rio Grande Rise.
2. *Santos Outer Ridge*, located on the middle continental rise east of the São Paulo Plateau and north–northeast of the Vema Channel Complex, where the deposit is divided into two segments (Western and Eastern Santos Outer Ridge) by the Rio de Janeiro Channel.
3. *Guanabara Drift*, located on the upper continental rise between the axis of the Guanabara Channel and the Almirante Saldanha Seamount. Seismic reflection profiles across these deposits show a relatively thick (> 1,000 m above acoustic basement), acoustically transparent or semitransparent sediment mound, with intermittent, unconformable subbottom reflections.

The Vema Drift, with an east–west axis (long. 42°W to long. 39°W; ≈ 300 km long) at 3,750 m water depth, terminates at the northern end of the Vema Channel Complex. Unconformable subbottom reflectors (Fig. 8-14A–C) are evidence that the structure of the drift must have developed in multiple stages of erosion/sedimentation, probably related to AABW fluctuations. This agrees with studies of AABW fluctuations in the Vema Channel by Johnson et al. (1976), who suggested that bottom-current activity has controlled both the morphology and location of the channel since the end of the Mesozoic. Thus, it seems reasonable that the evolutionary trend of the Vema Drift must have been dictated by episodes of sedimentation/erosion of the Vema Channel.

The Santos Outer Ridge starts at lat. ~ 27°S and has a generally northeast-trending axis. Its northern limit is difficult to discern (Fig. 8-2A), mostly due to the lack of reliable seismic data. Available data, however, suggest that the feature extends to lat. 24°S, where turbidity–current activity related to the Columbia Channel starts to predominate. The Rio de Janeiro Channel, aligned south–north from lat. 26°S to lat. 24°30'S, divides the Santos Outer Ridge into western and eastern sections, each possessing distinctive seismic and topographic characteristics (Fig. 8-15A–C).

The Western Santos Outer Ridge is broad, with a sinuous central ridge (4,000–4,400 m water depth) that approximates the bathymetric trend. The top of the drift exhibits sediment waves with two major acoustic characteristics:

1. Transparent subbottom reflections, from the axis of Rio de Janeiro Channel to approximately long. 35°20'W.
2. Multiple, parallel subbottom reflections, extending from long. 35°20'W toward the middle continental rise, approaching the São Paulo Scarp.

The vertical strip of focussed echoes in the sediment–wave troughs (Fig. 8-15A) suggests that they are stationary. These stationary sediment waves appear to indicate current activity, although their dynamics are not well understood. We think, however, that they mean slow deposition from bottom currents.

Geological evolution of Western Santos Outer Ridge was first studied by Gambôa, Buffler, and Barker (1983) with the aid of both multi- and single-channel seismic data and cores from Deep Sea Drilling Project Site 515 (Fig. 8-15C). The base of this core hole (> 600 m) reached a reflector (0.74 sec) that represents the transition between Oligocene siliceous shale and early Eocene zeolitic shale. This reflector marks an erosional discontinuity (a hiatus of 22 m.y.) at the Oligocene contact, which those authors suggested was the result of a radical change in bottom circulation, probably related to the beginning of AABW flux in the deep Brazil Basin. In addition, they postulated that, after the Eocene–Oligocene erosional event, a new sedimentation regime started to the north of the Vema Channel, when AABW began bringing suspended sediments to construct the continental rise.

The Eastern Santos Outer Ridge (4,600–5,000 m water depth) does not possess a well-marked axial ridge. It is less extensive than the Western Santos Outer Ridge, and ends gradationally northeastward toward the Rio de Janeiro Abyssal Plain, and in gentle contact eastward with basement highs of the abyssal hills (Figs. 8-2A, 8-4). The top of the drift is capped by broad, irregular, migrating sediment waves with unconformable subbottom reflections sometimes truncated at the seafloor (Fig. 8-15B); this is evidence that strong bottom currents have eroded this area (Flood, 1988).

The Guanabara Drift, the smallest of the drift deposits in the southern Brazil Basin (Fig. 8-2A), features

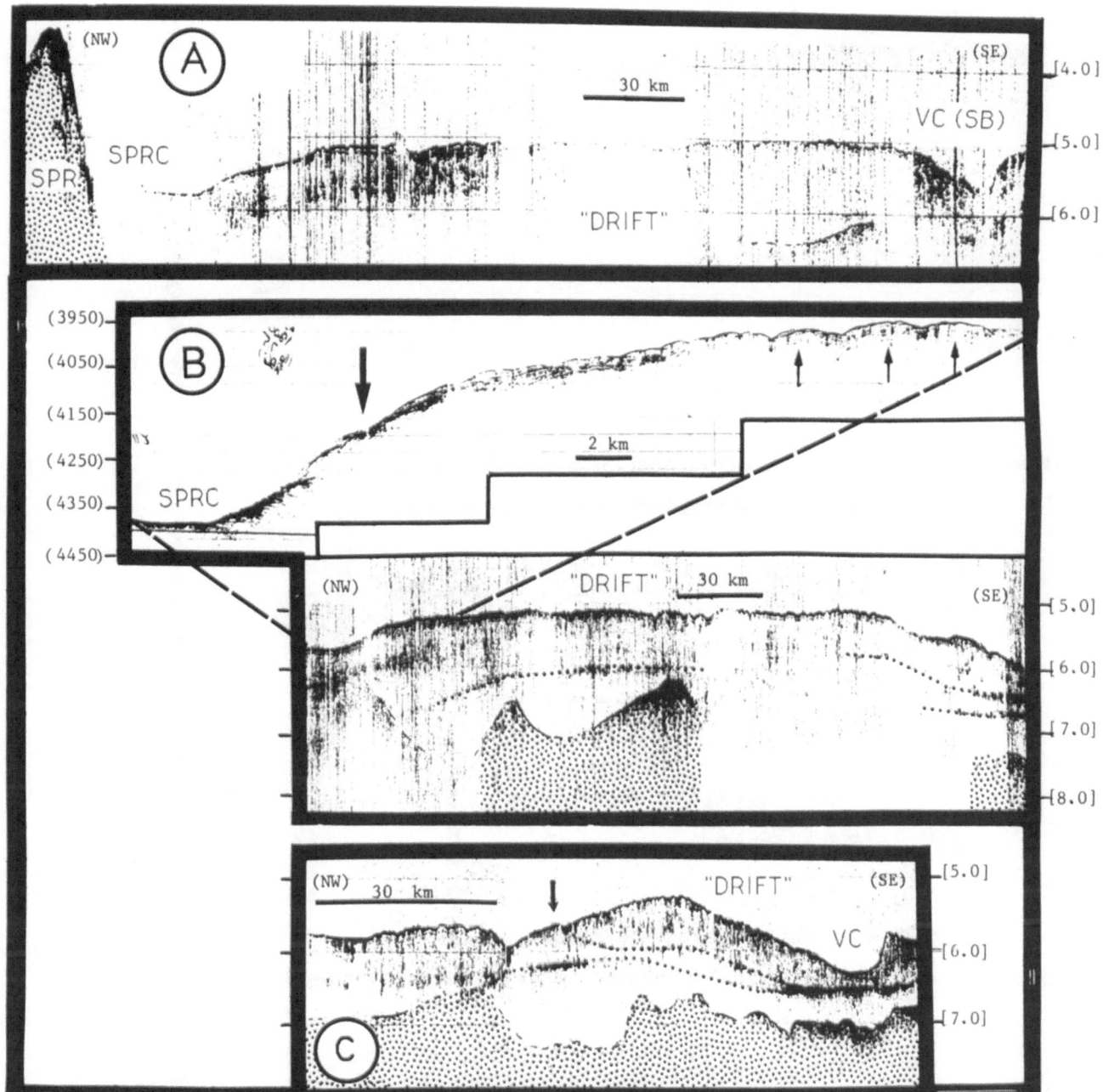

FIGURE 8-14.   Echogram and single-channel seismic profiles across Vema Drift. Acoustic basement stippled on seismic profiles. (A) V24-12: Cross section at western end of drift showing location relative to axes of secondary branch of Vema Channel [VC(SB)] and the channel (SPRC) at base of São Paulo Ridge (SPR). Vertical scale is two-way traveltime in seconds. (B) RC16-10: Echogram (above) showing transition (large arrow) between sediment waves (small arrows) on top and possible furrows (closely spaced hyperbolic echoes) at flank and near axis of São Paulo Ridge Channel (SPRC). Dotted lines on seismic profile (below) emphasize unconformable subbottom reflectors diverging from axis of Vema Channel (SE end of figure). Vertical scale in meters (left); two-way traveltime in seconds (right). (C) V26-06: Unconformable subbottom reflectors (dotted lines) cross almost entire section of drift, and converge toward axis of Vema Channel (VC); this indicates several phases of drift construction. Note apparent ancient ridge where subbottom reflector crops out (arrow). Vertical scale is two-way traveltime in seconds.

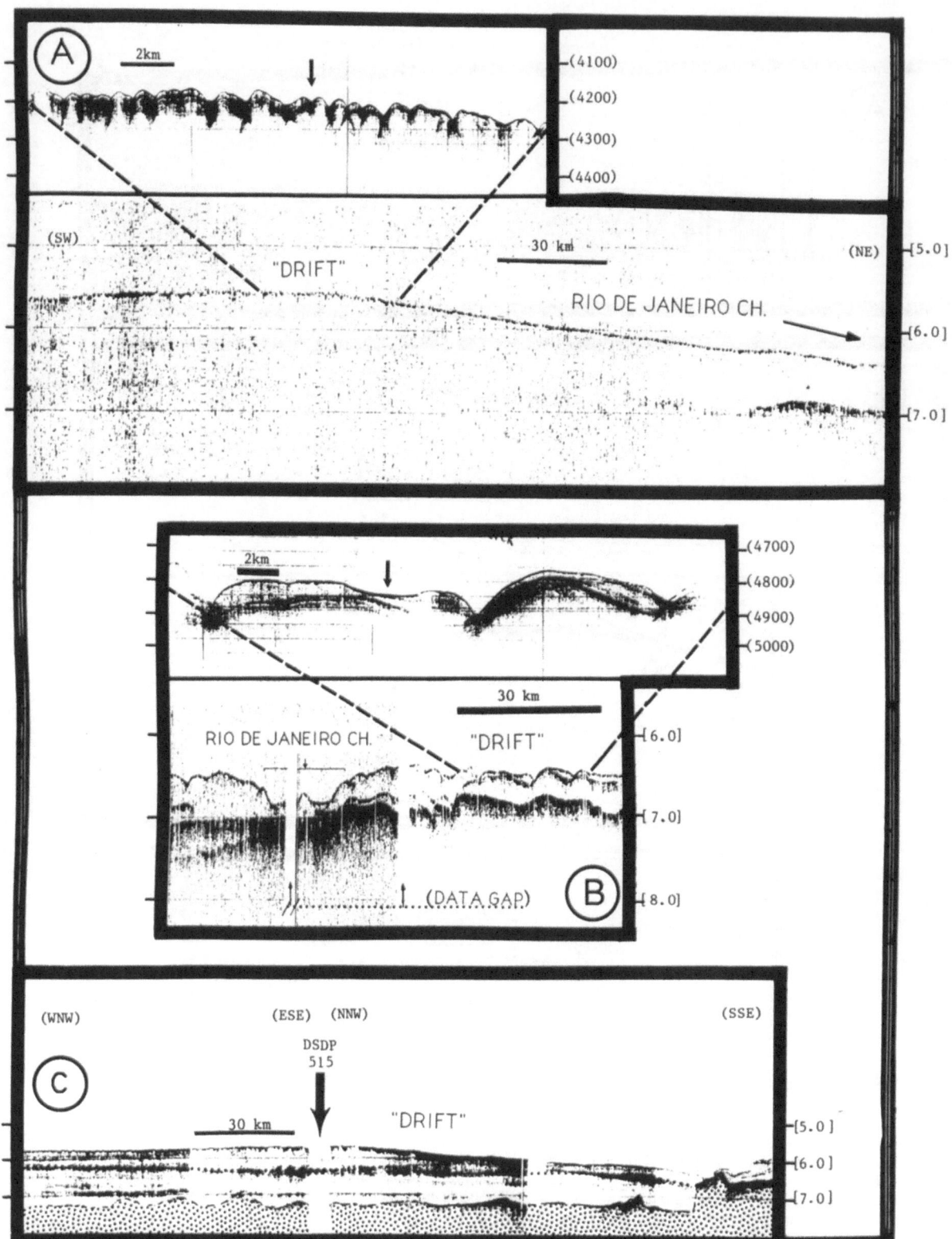

FIGURE 8-15. Echograms and single-channel seismic profiles across Santos Outer Ridge. (A) RC21-05: Stationary sediment waves (top) near axial ridge in western section of drift. Note gradation in sediment reflectivity, from acoustically laminated (left from arrow) to transparent (right from arrow). Also note strong, unconformable subbottom reflector diverging from axis of channel (bottom). Vertical scale in meters and two-way traveltime (seconds). (B) V26-05: Migrating sediment waves (top) in eastern section of drift showing evidence of current erosion (arrow) where subbottom reflector crops out. In bottom profile, note unconformable subbottom reflectors near surface and strong, irregularly undulating reflector near 7.0 sec, diverging from channel axis. Vertical scale in meters and two-way traveltime (seconds). (C) *Glomar Challenger* 72: Location of DSDP Site 515 at axial ridge. Dotted line marks lateral extension of unconformable subbottom reflector (Eocene–Oligocene) contact reached at this site. Acoustic basement is stippled. Vertical scale is two-way traveltime in seconds.

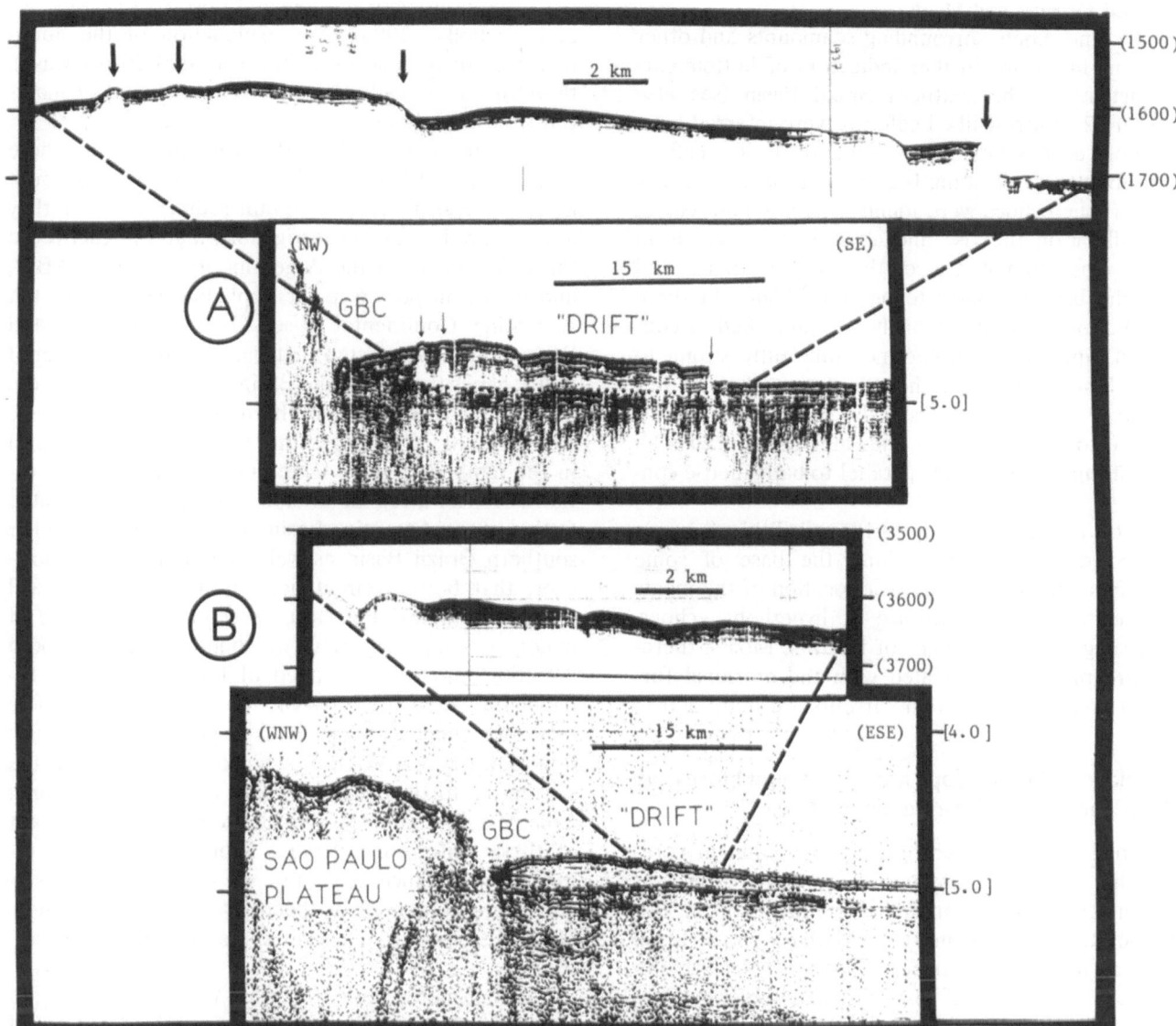

FIGURE 8-16. Echograms and single-channel seismic profiles across Guanabara Drift. Narrow axis of Guanabara Channel (GBC) is confined between drift and São Paulo Plateau scarp. Dotted lines on seismic profiles indicate strong, continuous subbottom reflector marking base of drift. (A) V31-04: Irregular bottom topography at northern end of drift may be caused by piercement structures (diapirs?) (arrows). Vertical scale in meters and two-way traveltime (seconds). (B) RC16-11: Migrating sediment waves (top) decorate surface of drift. Note wedging of drift deposit (bottom) and unconformable subbottom reflectors converging from axial ridge. Vertical scales in meters and two-way traveltime (seconds).

a south–north axial ridge at 3,570 m water depth, which parallels the axis of the Guanabara Channel; the ridge is approximately 5 km from, and rises 75 m above, the channel axis. The Guanabara Drift is approximately 80 km long by 40 km wide, but gradually loses topographic expression to the east and northeast as water depth increases. Single-channel seismic profiles (Fig. 8-16A and B) show that the Guanabara Drift is an asymmetric wedge with subbottom reflections that follow the slopes of the feature, converging laterally from the axial ridge toward the edges of the drift.

Sediment waves, both stationary and migrating, are found on top of the Guanabara Drift. Stationary waves are restricted to the southern portion of the drift, whereas migrating waves predominate in the north and northeastern parts. To the northeast, as the water deepens, these migrating sediment waves gradually lose their characteristics to more subdued surface expression. In this region, the deposit is often disturbed by subbottom structures (possibly diapirs), which sometimes crop out and form scarps with relief of 40 m or greater (Fig. 8-16A).

Erosional Furrows and Moats

Furrows and moats surrounding seamounts and other basement highs are further indicators of bottom-current activity in the southern Brazil Basin (see also Chapter 12). Furrow-like bedforms were inferred from hyperbolic echoes with apexes tangential to the seafloor or to subbottom reflections (Fig. 8-17, Table 8-1). Zones of hyperbolic echoes were found in restricted areas of the seafloor on the rise and abyssal plain, mostly in regions close to the axis of the deep channels and along the base of major topographic highs. In these areas, bottom currents probably are intensified by constriction, and therefore become sufficiently strong to sculpt elongate furrows. The orientation of these furrows, analyzed by variations in the shape of hyperbolic echoes (Flood, 1978, 1980), indicates that the axes of the bedforms are generally parallel to bathymetric contours.

Moats, as shown in Fig. 8-18 and Table 8-1, are localized on the seafloor along the base of some seamounts and abyssal hills. The location of the moats coincides with the occurrence of hyperbolic echoes, which suggests the presence of furrows. Moats, therefore, are probably associated with the erosional furrows in areas where bottom currents are more active.

Interrelation of Downslope and Alongslope Processes: Role in Formation of the Continental Rise

Combined analyses of echograms and sediment cores show that sediments remobilized by gravity processes from the continental margin and the flanks of the Rio Grande Rise reached the deep southern Brazil Basin entrained in turbidity currents. This is particularly noticeable to the north of lat. 26°S, and somewhat less extensively at the southern border of the São Paulo Plateau. Most of the terrigenous material derived from shallower portions of the slope, however, remains on the São Paulo Plateau (Gorini and Carvalho, 1984), which acts as a sediment trap. Part of the material that has bypassed the eastern border of the plateau has accessed the Rio de Janeiro Abyssal Plain through the Columbia and Carioca channels (Brehme, 1984). In the Rio Grande Rise area, turbidites do not appear to have the same areal significance as on the continental margin, although Johnson and Rasmussen (1984) suggested that turbidity–current activity was responsible for the formation of the many canyons cutting the northern flank of the rise.

The presence of significant sediment drifts on the continental rise of the southern Brazil Basin is difficult to explain, because passage of sediment from the continental margin to the deep basin is almost certainly restricted by the São Paulo Plateau. However, sediment drifts can be built by long-distance lateral transport of fine-grain sediments, a principle discussed by many researchers (e.g., Stow and Lovell, 1979; McCave and Tucholke, 1986). Any explanation of the abundance of drift deposits in the southern Brazil Basin, therefore, must call upon a substantial input of material from elsewhere.

We take the example of the Argentine Basin, where Ewing et al. (1971) noticed that drift deposits are built by slow deposition of fine-grain sediments, after they are captured from the terrigenous material (delivered through canyons at the Argentine margin) by AABW, and then transported in the nepheloid layer along the Argentine Continental Rise. Gambôa, Buffler, and Barker (1983) suggested that part of this fine material that escapes the Argentine Basin is deposited along the continental rise in the southern Brazil Basin; Biscaye and Eittreim (1977) noted that the particulate material in the nepheloid layer decreases from lat. 32°S northward. Hence, a regime of slow deposition, comparable to that of the Argentine Basin, must have existed in the southern Brazil Basin as well. It is noteworthy, however, that bottom circulation in the southern Brazil Basin is complex (Fig. 8-1), reflecting the intricacy of major topographic barriers, which locally influence both direction and relative speed of bottom currents. Indeed, the three drift deposits on the continental rise may owe their locations and shapes to the complex flow pattern that AABW acquires once entering the southern Brazil Basin through the Vema Channel Complex. In summary, the location and development of these drifts must depend upon the erosion–sedimentation balance (e.g., rate of deposition inversely related to current flow speed; Davies and Laughton, 1972) resulting from lateral spreading of bottom currents.

Current-meter data from the Vema Channel area (Johnson et al., 1976) reveal mean speed values of bottom currents of 14.7 cm/sec (max. 23.0 cm/sec) and of 6–8 cm/sec (max. 15–19 cm/sec), respectively, at the axes of the main Vema Channel and its secondary branch. It is reasonable to assume, based on our studies of the seismic and bathymetric data, as well as the nephelometry studies of Biscaye and Eittreim (1977), that the progressive lateral loss of bottom-current energy in the Vema Channel Complex might have induced deposition of part of the sediment brought in suspension from the Argentine Basin, thereby contributing to formation of the Vema Drift. Although this current–deposition combination may explain the formation and(or) location of the drift, it is more difficult to interpret how bottom currents contributed to shaping the drift morphology (Figs. 8-2A, 8-14). We suggest, however, that because the seafloor in this area has been intermittently affected by strong bottom currents throughout geologic time, the morphology of the Vema Drift has been modified many times; it exhibits a combination of *plastered* and *double* drifts (Fig. 8-14C),

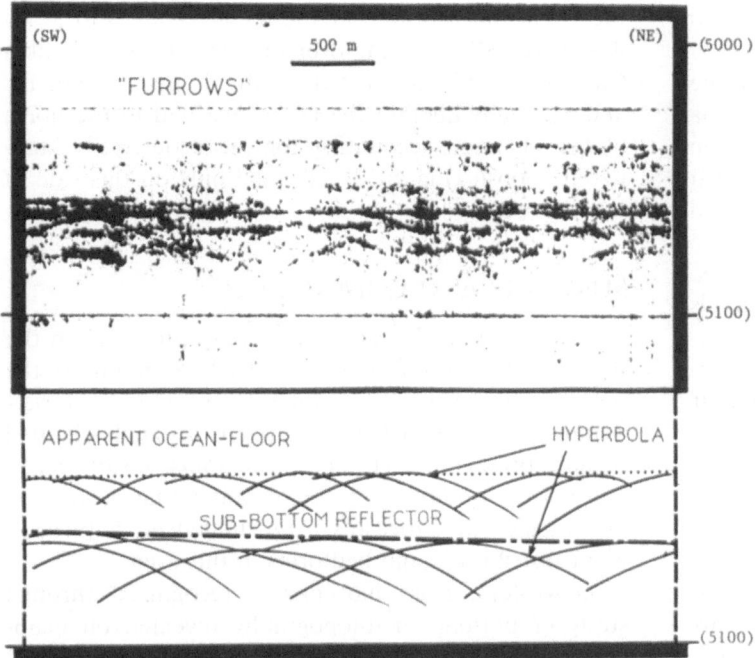

FIGURE 8-17. Echogram (RC21-05) with interpretation of overlapping hyperbolic echoes (furrows). Hyperbolae tangential to subbottom reflector suggest relict furrows formed during a previous erosional episode. Vertical scale in meters.

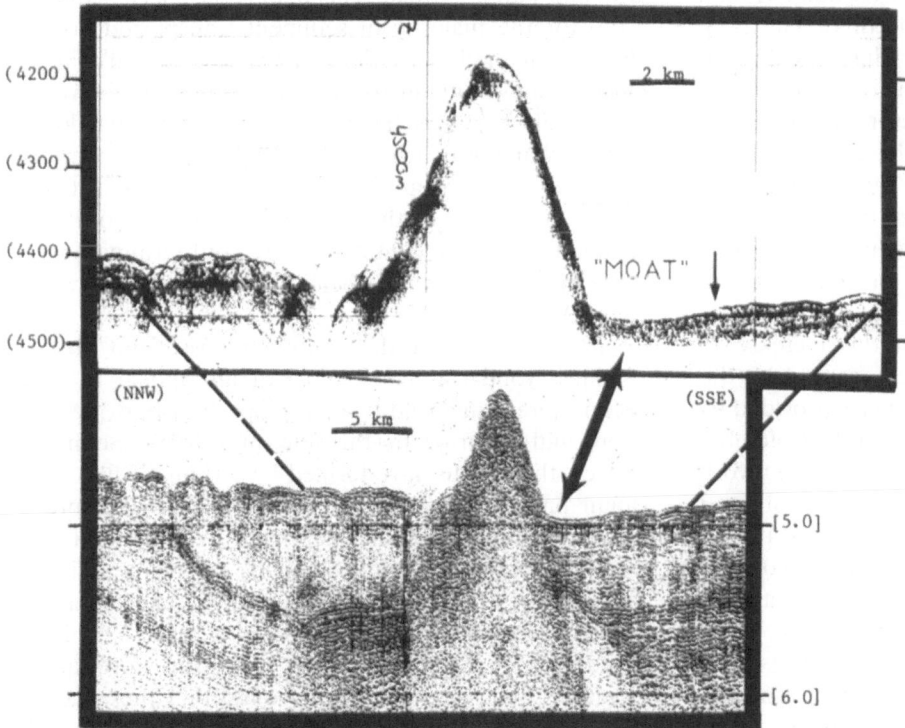

FIGURE 8-18. Echogram and single-channel seismic profile (*Glomar Challenger* 72) showing moat (double-headed arrow) at base of small basement outcrop. Note scouring where subbottom reflection crops out (small arrow). Vertical scales in meters and two-way traveltime (seconds).

according to the terminology suggested by McCave and Tucholke (1986).

Once AABW reaches the southern Brazil Basin, the spreading of bottom currents to the north of the Vema Channel Complex also would favor partial deposition of suspended fine sediments that comprise the drift deposits. Here again, it appears that the geometry and location of the drift deposits are influenced by the balance between sedimentation and erosion. The presence of the Guanabara and Rio de Janeiro channels farther north (Fig. 8-1) indicates that bottom circulation splits into a northern and northeastern component. The northern branch follows the regional contour of the São Paulo Plateau, where the current axis, topographically intensified along the steep eastern scarp of the plateau, carved the Guanabara Channel. Simultaneously, away from the current axis, the Guanabara Drift was built by slow lateral deposition. Such a drift deposit is called a *separated drift* by McCave and Tucholke (1986). The northeastern branch of the current corresponds to the main axis of AABW in the southern Brazil Basin, and is responsible for the creation of the Rio de Janeiro Channel (Figs. 8-1, 8-2A). In this area, the Western and Eastern Santos Outer Ridge are developed on either side of the AABW and, as such, they can be defined as a *double drift*, according to the terminology of McCave and Tucholke (1986).

Drift deposits are not present north of lat. 24°S, primarily due to the large influx of sediment delivered by Columbia Channel, which probably inhibits drift formation. At this latitude, however, Biscaye and Eittreim (1977) reported a considerable increase in the particulate flux transport (8.1 ton/yr); an unusually high figure compared to the average transport found for the area to the north of the Vema Channel (2.5 ton/yr). The authors attributed its magnitude to an error in the methodology used for transport calculation. Alternatively, we propose that this increase in particulate flux may very well be real, caused by the local increment of fine suspended particles provided by periodic pulses of turbidity currents from the Columbia Channel. Such an assumption is supported by the work of Bouma, Rezak, and Chmelik (1969), who showed that terrigenous fine-grained sediments brought in suspension by turbidity currents do not generally concentrate at a level sufficient to transit oceanographic barriers formed between deep water masses. As a result, this fine material is retained along isopycnals until an eventual increase in concentration (= aggregation) forms larger particles, inducing a new phase of deposition. Pierce (1976) noted that fine particles retained along isopycnals may also be laterally transported due to advective effect of currents in abyssal circulation. It is possible,, therefore, that the increased relative particulate flux near the Columbia Channel might result from retention of fine particles in the pycnocline formed at the TDD zone. Northward flux of AABW, crossing over the axis of the Columbia Channel, could transport these fine-grained sediments and gradually liberate the aggregate load to the north of the channel, accounting for the acoustically laminated deposit (Fig. 8-12) observed only on this side of the channel.

SUMMARY AND CONCLUSIONS

We have studied Recent sedimentary processes in the southern Brazil Basin, with special attention to the region between the Rio Grande Rise and the Vitória–Trindade Seamount Chain. Analysis of single-channel seismic profiles, 3.5–12-kHz echograms, piston cores, and bottom photographs demonstrates that downslope and alongslope processes have influenced both sediment distribution and bedforms in this area.

Downslope mass movements, recognized through study of bottom microtopography revealed on echograms and sedimentary structures in cores, were found to have occurred mostly north of lat. 26°S, and on the relatively steep flanks of the Rio Grande Rise. On the continental margin, downslope sediment transport occurs from the continental shelfbreak (≈ 200 m) to ~ 3,200–3,300 m water depth, near the outer scarp of the São Paulo Plateau. This plateau acts as a sediment trap for the majority of sediment transported downslope from shallow portions of the continental margin. Part of the sediment transported downslope, however, eventually bypasses the São Paulo Plateau via downslope channels, such as the Columbia and Carioca channels; these channels form a complex of distributaries on the top of the plateau, extend for more than 600 km across the continental rise, and terminate near the Rio de Janeiro Abyssal Plain ($> 5,000$ m water depth).

Deeper areas of the continental rise and abyssal plain, beyond the São Paulo Plateau, appear to have been more influenced by alongslope transport, deposition, and scouring by the flow of AABW. Sediment waves, thought to have formed by settling of fine sediment particles brought in by AABW, are widespread on the continental rise at depths of $> 3,400$ m. These sediment waves are found overlying extensive bodies of acoustically transparent or semitransparent sediment, which are usually > 100 km long and often exhibit an axial ridge that rises 80–200 m above the surrounding seafloor. Such characteristics suggest that these extensive sediment mounds are drift deposits. These sedimentary features were recognized from the southern end of the study area, near the Vema Channel and the western section of the Rio Grande Rise, to as far north as lat. 24°S, where turbidity currents associated with the Columbia Channel inhibit development of these deposits.

ACKNOWLEDGMENTS

Funding for G.A.M. was graciously supplied by CNPq (Conselho Nacional de Desenvolvimento Cientifico e Tecnologico—Brazil) under project 202576/85. Lamont-Doherty Geological Observatory provided the data and technical support under contract N00014-84-c-0132, Scope SS to R.D.F. Partial funding for T.H.O. was supplied by Planning Systems Inc. (Long Beach, MS) and by an Office of Naval Research–National Defense Science and Engineering Grant (ONR–NDSEG) fellowship. This study benefitted from critical reviews by D. C. Twichell and E. P. Laine. Reviews of earlier versions of the manuscript by M. Kwintner, C. Brenner, and P. Weiss, advice and support by D. V. Kent, and editorial comments on the present version by C. W. Poag are greatly appreciated. Lamont-Doherty Geological Observatory contribution No. 4792.

REFERENCES

Asmus, H. E. 1984. Geologia da Margem Continental Brasileira. *In* Geologia do Brasil, Texto Explicativo do Mapa Geológico do Brasil e da Area Oceânica Adjacente Incluindo Depósitos Minerais, Escala 1:2.500.000, C. Schobbenhaus, D. A. Campus, G. R. Denze, and H. E. Asmus, Eds.: Brasília: D.F.: DNPM, 443–472.

Asmus, H. E., and Guazelli, W. 1981. Descrição sumária das estruturas da Margem Continental Brasileira e das áreas oceânicas adjacentes: Hipóteses sobre o tectonismo causador e implicações para prognósticos de potencial de recursos minerais. *In* Estruturas e Tectonismo da Margem Continental Brasileira e Suas Implicações nos Processos Sedimentares e na Avaliação do Potencial de Recursos Minerais, Série Projeto REMAC, No. 9, H. E. Asmus, Ed.: Rio de Janeiro: PETROBRAS, CENPES, DINTEP, 187–269.

Bainbridge, A. E. 1976. GEOSECS—Atlantic Expedition, V.2—Sections and Profiles, International Decade of Ocean Exploration: Washington, DC: National Science Foundation.

Biscaye, P. E. 1965. Mineralogy and sedimentation of Recent deep-sea clay in the Atlantic Ocean and adjacent seas and oceans. *Geological Society of America Bulletin* 76:803–832.

Biscaye, P. E., and Eittreim, S. L. 1977. Suspended particulate loads and transport in the nepheloid layer of the abyssal Atlantic Ocean. *Marine Geology* 23:155–172.

Bouma, A. H., Rezak, R. R., and Chmelik, R. F. 1969. Sediment transport along ocean density interfaces. *Geological Society of America Abstracts with Programs* 7:259–260.

Brehme, I. 1984. Vales Submarinos Entre o Banco dos Abrolhos e Cabo Frio. M.S. thesis, Universidade Federal do Rio de Janeiro, Rio de Janeiro, R.J., Brazil.

Broecker, W. S., and Takahashi, T. 1980. Hydrography of the Central Atlantic—III. The North Atlantic Deep Water complex. *Deep-Sea Research* 27: 591–613.

Broecker, W. S., and Takahashi, T. 1981. Hydrography of the Central Atlantic—IV. Intermediate waters of Antarctic origin. *Deep-Sea Research* 28A:177–193.

Broecker, W. S., Takahashi, T., and Li, Y. H. 1976. Hydrography of the Central Atlantic—I. The two-degree discontinuity. *Deep-Sea Research* 23:1083–1104.

Cook, M. E., Field, J. V., and Gardner, J. V. 1982. Characteristics of sediments on modern and ancient continental slopes. *In* Sandstone Depositional Environments, P. A. Scholle and D. Spearing, Eds.: *American Association of Petroleum Geologists Memoir 31*, 329–364.

Damuth, J. E. 1975. Echo character of the Western Equatorial Atlantic floor and its relationship to dispersal and distribution of terrigenous sediments. *Marine Geology* 18:17–45.

Damuth, J. E. 1980. Use of high-frequency (3.5–12 kHz) echograms in the study of near-bottom sedimentation process in the deep-sea: a review. *Marine Geology* 38:51–75.

Damuth, J. E., and Hayes, D. E. 1977. Echo character of the east Brazilian continental margin and its relationship to sedimentary processes. *Marine Geology* 24:74–95.

Davies, T. A., and Laughton, A. S. 1972. Sedimentary processes in the North Atlantic. *In* Initial Reports of the Deep-Sea Drilling Project, Volume 12, A. S. Laughton, W. A. Berggren, et al.: Washington, DC: U.S. Government Printing Office, 905–934.

Eittreim, S. L., Thorndike, E. M., and Sullivan, L. 1976. Turbidity distribution in the Atlantic Ocean. *Deep-Sea Research* 23:1115–1127.

Embley, R. W. 1975. Studies of deep-sea sedimentation processes using high-frequency seismic data. Doctoral Thesis, Columbia University, New York.

Embley, R. W. 1976. New evidence for occurrence of debris flow deposits in the deep-sea. *Geology* 4:371–374.

Embley, R. W. 1980. The role of mass transport in the distribution and character of deep-ocean sediments with special reference to the North Atlantic. *Marine Geology* 38:23–50.

Embley, R. W., and Jacobi, R. D. 1977. Distribution and morphology of large submarine sediment slides and slumps on Atlantic continental margins. *Marine Geotechnology* 2:205–228.

Embley, R. W., and Jacobi, R. D. 1986. Mass wasting in the western North Atlantic. *In* The Geology of North America, Volume M. The Western North Atlantic Region, P. R. Vogt and B. E. Tucholke, Eds.: Boulder, CO: Geological Society of America, 470–490.

Embley, R. W., and Langseth, M. G. 1977. Sedimentation processes on the continental rise of northeastern South America. *Marine Geology* 25:279–297.

Emery, K. O., and Uchupi, E. 1984. *The Geology of the Atlantic Ocean*: New York: Springer Verlag.

Ewing, M., Eittreim, S. L., Ewing, J. I., and LePichon, X. 1971. Sediment transport and distribution in the Argentine Basin. Nepheloid layer and processes of sedimentation. *In* Physics and Chemistry of the Oceans, Volume 8, L. H. Ahrens, F. Press, S. K. Runkorn, and H. C. Urey, Eds.: New York: Pergamon, 51–77.

Flood, R. D. 1978. Studies of deep-sea sedimentary microtopography in the North Atlantic Ocean. Doctoral Thesis, Massachusetts Institute of Technology/Woods Hole Oceanographic Institution, Woods Hole, MA.

Flood, R. D. 1980. Deep-sea sedimentary morphology: modelling and interpretation of echo-sounding profiles. *Marine Geology* 38:77–92.

Flood, R. D. 1988. A Lee-wave model for deep sea mud wave activity. *Deep-Sea Research* 35:973–983.

Flood, R. D., and Shor, A. N. 1988. Mud waves in the Argentine Basin and their relationship to regional bottom circulation patterns. *Deep-Sea Research* 35:973–983.

Fuglister, F. C. 1960. *Atlantic Ocean Atlas of Temperature and Salinity Profiles and Data from the International Geophysical Year of 1957–1958. Volume 1*. Woods Hole, MA: Woods Hole Oceanographic Institution.

Gambôa, L. A. P., and Kumar, N. 1977. Synthesis of geological and geophysical data in a 1° square area around Site 356, Leg 39 DSDP. *In* Initial Reports of the Deep-Sea Drilling Project, Volume 39, P. R. Supko, K. Perch-Nielsen, et al.: Washington, DC: U.S. Government Printing Office, 947–957.

Gambôa, L. A. P., Buffler, R. P., and Barker, P. F. 1983. Seismic stratigraphy and geological history of the Rio Grande Gap and Southern Brazil Basin. *In* Initial Reports of the Deep-Sea Drilling Project, Volume 72, P. F. Barker, R. L. Carlson, D. A. Johnson, et al.: Washington, DC: U.S. Government Printing Office, 481–517.

Gorini, M. A., and Carvalho, J. C. 1984. Geologia da Margem Continental Inferior Brasileira e do fundo oceânico adjacente. *In* Geologia do Brasil, Texto Explicativo do Mapa Geológico do Brasil e da Area Oceânica Adjacente Incluindo Depósitos Minerais, Escala 1:2.500.000, C. Schobbenhaus, D. A. Campus, G. R. Denze, and H. E. Asmus, Eds.: Brasília: D.F.: DNPM, 473–489.

Heezen, B. C., and Hollister, C. D. 1971. *The Face of the Deep*: London: Oxford University Press.

Heezen, B. C., Hollister, C. D., and Ruddiman, W. F. 1966. Shaping of the continental rise by deep geostrophic contour currents. *Science* 152:502–508.

Heezen, B. C., Tharp, M., and Ewing, M. 1959. The floor of the oceans, 1: The North Atlantic. *Geological Society of America Special Paper 65*.

Hogg, N., Biscaye, P. E., Gardner, W., and Schmitz, Jr., W. Z. 1982. On the transport and modification of Antarctic Bottom Water in the Vema Channel. *Journal of Marine Research* 40 (supplement): 238–263.

Hollister, C. D., Flood, R. D., Johnson, D. A., Londsdale, P. F., and Southard, J. B. 1974. Abyssal furrows and hyperbolic echo traces on the Bahama Outer Ridge. *Geology* 2:395–400.

Hollister, C. D., Gardner, W. D., Londsdale, P. F., and Spencer, D. W. 1976. New evidence of northward-flowing bottom water along the Hatton Sediment Drift, Eastern North Atlantic. *EOS, Transactions of the American Geophysical Union* S7:261.

Hollister, C. D., Flood, R. D., and McCave, I. N. 1978. Plastering and decorating in the North Atlantic. *Oceanus* 21(4):5–13.

Jacobi, R. D. 1976. Sedimentary slides on the northwestern continental margin of Africa. *Marine Geology* 22:157–173.

Jacobi, R. D., and Hayes, D. E. 1982. Bathymetry, microtopography and reflectivity characteristics of the West African Margin between Sierra Leone and Mauritania. *In* Geology of the Northwest African Continental Margin, U.

von Rad, K. Hinz, M. Sarnthein, and E. Seibold, Eds.: Berlin: Springer-Verlag, 181–212.

Johnson, D. A. 1982. Abyssal teleconnections: Interactive dynamics of the deep ocean circulation. *Palaeogeography, Palaeoclimatology, Palaeoecology* 38:93–128.

Johnson, D. A. 1984. The Vema Channel: Physiography, structure and sediment-current interaction. *Marine Geology* 58:1–34.

Johnson, D. A. 1985. Abyssal teleconnections II. Initiation of Antarctic Bottom Water flow in the southwestern Atlantic. *In* South Atlantic Paleoceanography, K. J. Hsü and H. J. Weissert, Eds.: Cambridge: University Press, 243–281.

Johnson, D. A., and Rasmussen, K. A. 1984. Late Cenozoic turbidite and contourite deposition in the Southern Brazil Basin. *Marine Geology* 58:225–262.

Johnson, D. A., McDowell, S. E., Sullivan, L. G., and Biscaye, P. E. 1976. Abyssal hydrography, nephelometry, currents and benthic boundary layer structure in the Vema Channel. *Journal of Geophysical Research*. 81(33):5771–5786.

Jones, G. A. 1984. Advective transport of clay minerals in the region of the Rio Grande Rise. *Marine Geology* 58:184–212.

Kennett, J. P. 1982. *Marine Geology*: Englewood Cliffs, NJ: Prentice Hall.

Klaus, A., and Ledbetter, M. T. 1987. Mass wasting on the continental margin off Uruguay and northern Argentina revealed by high resolution seismic records (3.5 kHz). *EOS, Transactions of the American Geophysical Union* 68:1335–1336.

Kolla, V., Moore, D. G., and Curray, J. R. 1976. Recent bottom current activity in the deep western Bay of Bengal. *Marine Geology* 21:265–270.

Kolla, V., Eittreim, S. L., Sullivan, L. G., Kostecki, J. A., and Burckle, L. H. 1980. Current-controlled abyssal microtopography and sedimentation in the Mozambique Basin, Southwest Indian Ocean. *Marine Geology* 34:171–206.

Krause, D. C. 1962. Interpretation of echo sounding profiles. *International Hydrographic Review* 39:65–123.

Kumar, N., Gambôa, L. A. P., Schreiber, B. D., and Mascle, J. 1977. Geologic history and origin of São Paulo Plateau (Southeastern Brazilian Margin), comparison with the Angolan Margin, and the early evolution of the Northern South Atlantic. *In* Initial Reports of the Deep-Sea Drilling Project, Volume 39, P. R. Supko, K. Perch-Nielsen, et al.: Washington, DC: U.S. Government Printing Office, 927–945.

Kumar, N., Leyden, R., Carvalho, J., and Francisconi, O. 1979. Sediment Isopachs, Continental Margin of Brazil. *American Association of Petroleum Geologists Special Map Series*, 1 sheet.

Laine, E. P., Damuth, J. E., and Jacobi, R. D. 1986. Surficial sedimentary processes revealed by echo-character mapping in the Western North Atlantic Ocean. *In* The Geology of North America, Volume M. The Western North Atlantic Region, P. R. Vogt and B. E. Tucholke, Eds.: Boulder, CO: Geological Society of America, 427–436.

Lawrence, J. R. 1979. $^{18}O/^{16}O$ of the silicate fraction of Recent sediments used as a provenance indicator in the South Atlantic. *Marine Geology* 33:M1–M7.

Ledbetter, M. T. 1979. Fluctuations of Antarctic Bottom Water velocity in the Verma Channel during the last 160,000 years. *Marine Geology* 33:71–89.

Leyden, R., Asmus, H. E., Zenbruscki, S., and Bryan, G. 1976. South Atlantic diapiric structures. *American Association of Petroleum Geologists Bulletin* 60:196–212.

LePichon, X., Ewing, M., and Truchan, M. 1971. Sediment transport and distribution in the Argentine Basin. *In* Physics and Chemistry of the Earth, Volume 8, L. H. Ahrens, F. Press, S. N. Runcorn, and H. C. Urey, Eds.: New York: Pergamon, 31–48.

Lynn, R. J. 1971. On potential density in the deep South Atlantic Ocean. *Journal of Marine Research* 29:171–177.

Lynn, R. J., and Reid, J. L. 1968. Characteristics and circulation of deep and abyssal waters. *Deep-Sea Research* M15:577–598.

McCave, I. N., and Tucholke, B. E. 1986. Deep current-controlled sedimentation in the Western North Atlantic. *In* The Geology of North America, Volume M. The Western North Atlantic Region, P. R. Vogt and B. E. Tucholke, Eds.: Boulder, CO: Geological Society of America, 451–468.

Mello, G. A. 1988. Processos Sedimentares na Bacia do Brasil: Setor Sudeste-Sul. M.S. thesis, Universidade Federal do Rio de Janeiro, Rio de Janeiro, R.J., Brazil.

Moody, R., Hayes, D. E., and Connary, S. 1979. Bathymetry of the Continental Margin of Brazil. *American Association of Petroleum Geologists Special Map Series*, 1 sheet.

Palma, J. J. C. 1984. Fisiografia da Area Oceânica. *In* Geologia do Brasil, Texto Explicativo do Mapa Geológico do Brasil e da Area Oceânica Adjacente Incluindo Depósitos Minerais, Escala 1:2.500.000, C. Schobbenhaus, D. A. Campus, G. R. Denze and H. E. Asmus, Eds.: Brasília: D.F.: DNPM, 429–441.

Pierce, J. W. 1976. Suspended sediment transport at the shelf break and over the outer margin. *In* Marine Sediment Transport and Environmental Management, D. J. Stanley and D. J. P. Swift, Eds.: New York: Wiley-Interscience, 437–458.

Reid, J. L., Nowlin, W. D., Jr., and Patzert, W. C. 1977. On the characteristics and circulation of the Southwestern Atlantic Ocean. *Journal of Physical Oceanography* 7:62–90.

Richardson, M. J., Biscaye, P. E., Gardner, W. D., and Hogg, N. G. 1987. Suspended particulate matter transport through the Vema Channel. *Marine Geology* 77:171–184.

Schmitz, W. J., Jr., and Hogg, N. G. 1983. Exploratory observations of abyssal currents in the South Atlantic near Vema Channel. *Journal of Marine Research* 41:487–510.

Stow, D. A. V., and Lovell, J. P. B. 1979. Contourites: their recognition in modern and ancient sediments. *Earth Science Reviews* 14:251–291.

Wüst, G. 1933. Bodemwasser und die gliederung der Atlantischen Tiefsee. Wissenschaften Ergebnisse Deutsche D.A.E. "Meteor", 1925–27. Berlin, Bd. VI, 1 Lfg., S. 107.

Zembruscki, S. G. 1979. Geomorfologia da Margem Continental Sul Brasileira e das Bacias Oceânicas Adjacentes. *In* Geomorfologia da Margem Continental Brasileira e das Areas Oceânicas Adjacentes, Série Projeto REMAC, No. 7, H. A. F. Chaves, Ed.: Rio de Janeiro: PETROBRAS, CENPES, DINTEP, 129–177.

9
Southeastern New England Continental Rise: Origin and History of Slide Complexes

Dennis W. O'Leary and Max R. Dobson

It has become clear in the last 10 years that the continental slopes and rises of the Atlantic Ocean are the loci of numerous mass movements, which are among the largest on the planet. The vast extent of debris strewn across the rise and the evidence of recurrent activity indicate that mass movement has been a significant agent in the stratigraphic development of the rise, at least in Pleistocene, and possibly in Holocene time. But while the importance of mass movement as a rise-modifying process has been recognized by many marine geologists (Embley, 1976, 1980; Jacobi, 1984; Vassallo, Jacobi, and Shor, 1984; Tucholke, 1987), the causes and provenance of mass movement, the mode of debris transport, and its relationship to sedimentation processes native to the rise remain poorly understood. In large part, our ignorance of mass movement as a phenomenon affecting continental rises stems from lack of a synoptic view of the seafloor, a view that is now available in the form of GLORIA long-range sidescan sonar image mosaics. However, our ignorance also stems from a tendency to accept unchallenged concepts of mass movement and rise sedimentation, which have been passed through the literature for more than half a century. We are emerging from a period when the role of mass movement in the transfer of sediment was explained speculatively and was seen mainly as a subordinate factor in the generation of turbidity currents.

This paper presents new GLORIA sidescan-sonar data and supporting seismic reflection data, and new interpretations that deal with three issues relevant to mass movement and the geologic evolution of Atlantic continental rises:

1. The sources and causes of mass movement.
2. The mode of debris emplacement and its effect on rise stratigraphy.
3. The rise as an environment of deposition.

We explore these topics by analysis of the origin and constitution of two major slide complexes along the New England margin: the Munson–Nygren slide complex off Georges Bank (O'Leary, 1986a), and the southeast New England slide complex (SENESC), along the continental slope and upper rise between long. 69°30′W and long. 71°45′W, off Massachusetts, Rhode Island, and Connecticut (Fig. 9-1). We focus mainly on the origin and geological history of the SENESC and its place in the stratigraphic development of the southeastern New England continental rise. Details of the SENESC have not been reported elsewhere.

We use the term slide complex to refer to spatially related mass movements dominated by sliding, along with the resulting deposits (Table 9-1). A genetic relationship is inferred. Mass movement is the bulk displacement of a distinctly bounded mass of rock or unconsolidated sediment exposed on a slope or part of a dipping unit, down the slope under the influence of gravity. The mechanical varieties of mass movement (slide and flow) are described and their relationships indicated in Table 9-1. Throughout the text we refer to mass movement deposits generally as debrites. We choose this term because it connotes debris, yet without any implication of origin. The term is described by Stow (1985):

... mixed lithologies with hard pebbles and boulders or soft mudstone clasts set in a muddy matrix; (they) vary in thickness up to several tens of meters. May be disorganized, minimally organized with a basal zone of lensoid (?shear) lamination, a middle zone of high-angle faults and slump folds capped by convolute lamination, and an upper clast-rich zone that can show slight positive grading, water escape pipe and dish structures and some horizontal alignment of clasts.

Core samples and seafloor observations suggest that

214

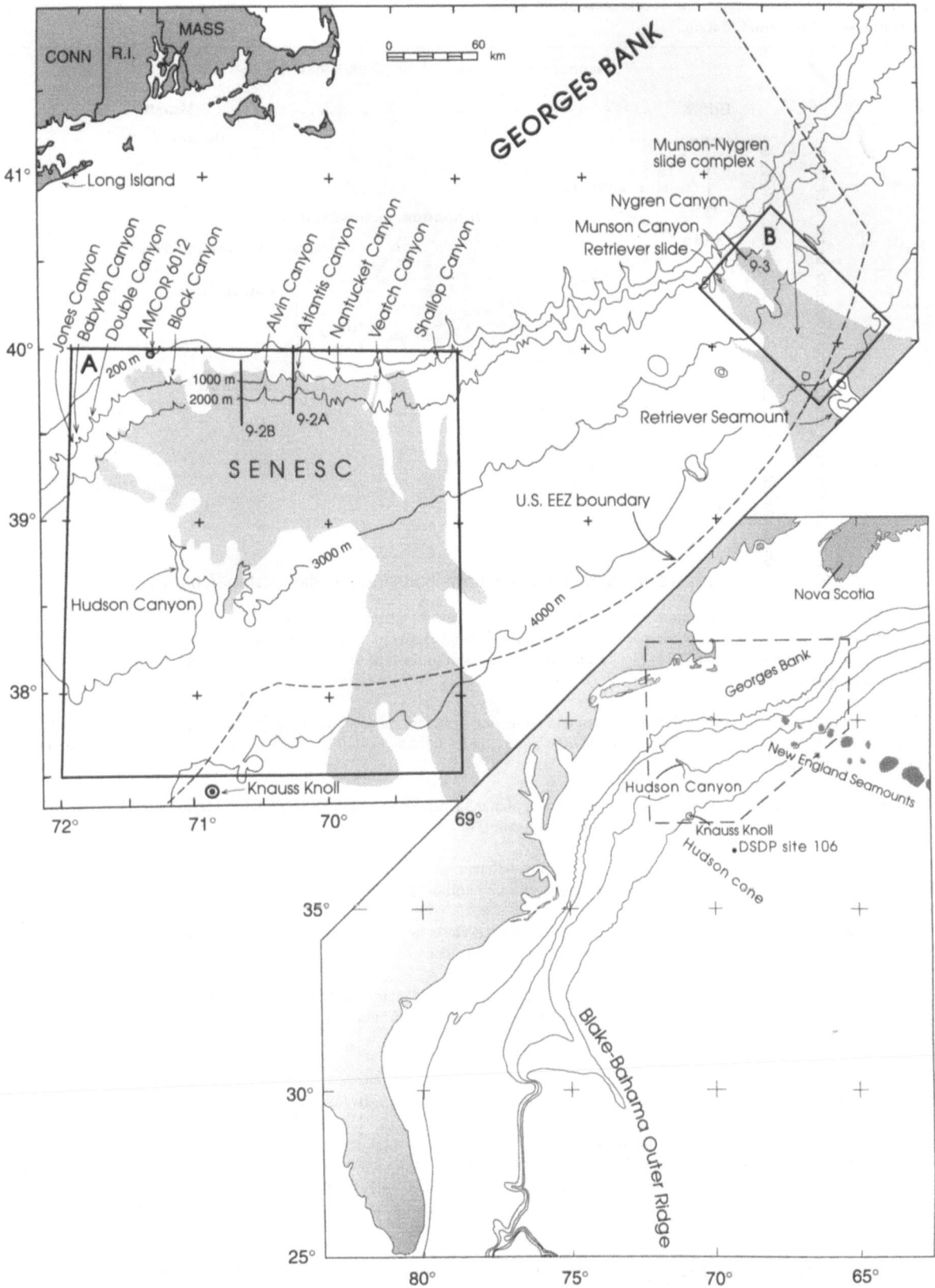

FIGURE 9-1. Locations of SENESC and Munson–Nygren slide complexes off southeast New England continental shelf. Heavy, straight lines are track lines of seismic records illustrated in correspondingly numbered figures.

TABLE 9-1   Mass Movement Types and Terminology and Inferred Relationships as Applied to Mass Movement of the New England Continental Margin.

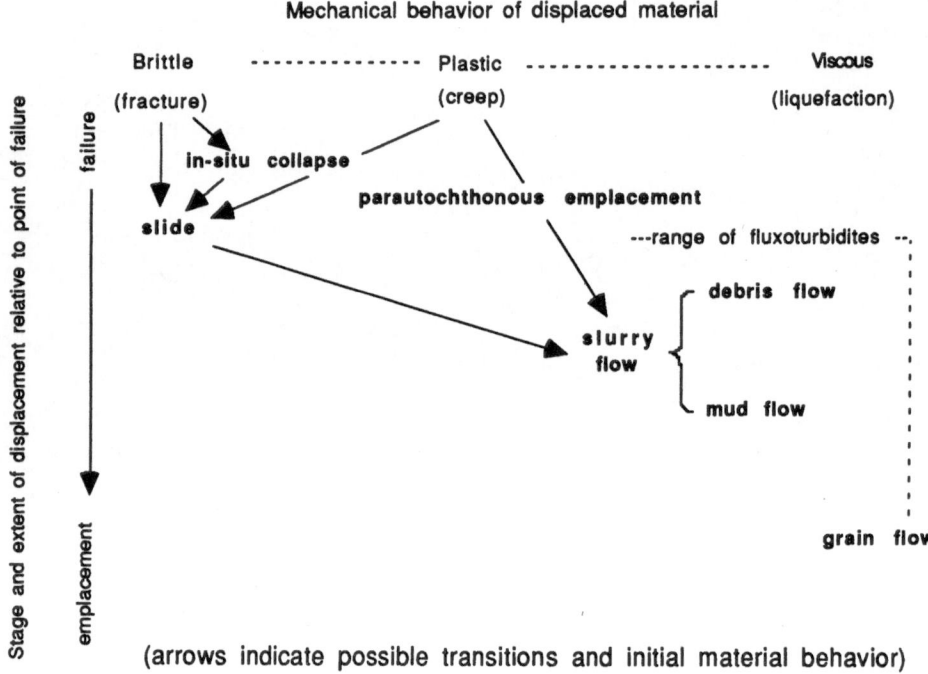

(arrows indicate possible transitions and initial material behavior)

Slide complex—An area of mass movement characterized by multiple and recurrent centers of slope failure, in which initial transport is dominated by sliding. Debris transport is complicated by, and ultimately gives way to, varieties of flow, resulting in broad, coalescent debris fields that include deposits (debrites) ranging from blocky slide rubble to distal grain-flow deposits, and from in situ collapse (lateral spreads) to parautochthonous masses.

Slide—Movement of failed mass governed by sliding friction along one principal shear plane. Type of slide (e.g., slump, slab slide) determined by form and structure of displaced mass and geometry of slide plane. Slides initially exhibit brittle behavior and generally become increasingly fragmented during movement.

In situ collapse—Fragmentation or disaggregation of extensive, planar surficial unit with little or no lateral translation. Involves brittle failure of surficial layer and plastic or viscous deformation (creep, liquefaction) of incompetent substrate.

Slurry flow—Flow (general laminar shear) dominated by viscous behavior of suspension of clay-sized particles in water; dispersive pressure also generated by ordered shear velocity changes as grains approach one another (Carter, 1975).

Debris flow—Slurry flow in which appreciable numbers of suspended clasts are present (Carter, 1975). Typified by poor sorting, high clay content, high viscosity, tendency to gain cohesive strength with loss of velocity.

Grain flow—Flow (general low-viscosity laminar shear) in which suspension is maintained by intergranular impacts that generate dispersive pressure (Carter, 1975). Form very thin tapering units, possibly associated with turbidites. Probably of minor importance in SENESC because they require relatively well sorted, mud-free sediment.

Fluxoturbidite—Confused term mainly of historical interest. General usage presupposes tubidity currents and turbulent flow even though deposits imply laminar flow and even sliding. Term may flag debrites emplaced by slurry or grain flow.

Parautochthonous emplacement—Similar to in situ collapse, but having more disaggregation and displacement; probably involves admixture of mud flow, debris flow, or slide rubble, but translation is minimal compared to flows. Parautochthonous emplacement occurs near or within area of deposition for most debrite, whereas in situ collapse represents head-zone pre-slide failure.

this is a fair description of the mass movement deposits along the upper rise off New England.

GLORIA data indicate that the Neogene and Quaternary history of the continental rise off southern New England is appreciably different from that of the rise off Nova Scotia and south of Hudson Canyon. Accordingly, we do not propose the present study as a model of rise development, but rather as a case history with explanations offered for its distinctive features. Our explanations do bear on universal considerations, however, including seafloor stability, turbidity- and contour-current activity, and the significance of debrites in siliciclastic deposits of the continental rise.

LARGE SLIDE COMPLEXES OF THE ATLANTIC BASIN

During the 1970s and 1980s, marine echo-sounding and coring surveys returned data that indicated that vast sheets of debris coat both the Euro-African and North American Atlantic continental rises (Jacobi, 1976; Embley and Jacobi, 1977; Embley, 1976, 1980; Bugge et al., 1987). Along the U.S. Atlantic rise, the extent of mass movement debris was recognized as enormous. Embley (1980) estimated that at least 40% of the rise, from the New England seamounts to the Blake–Bahama Outer Ridge, is covered with mass movement debris to water depths of 5,400 m. On both sides of the Atlantic, debris generally was recognized as having originated from mass movement of the uppermost 100–300 m of section on the middle to lower continental slopes. Mass movement commonly involved recurrent activity within nested, overlapping, or adjoining areas of sliding; the term "slide complex" could be applied to areas of mass movement spanning hundreds of kilometers along the slope (Jacobi, 1976; Vassallo, Jacobi, and Shor, 1984; Kidd, Hunter, and Simon, 1987; Bugge, Belderson, and Kenyon, 1988; see also Chapters 10, 12, 13) and feeding into stratigraphically complex, but planimetrically unified, debris deposits on the rise.

A late Pleistocene to Holocene age of formation is indicated for these circum-Atlantic slide complexes, as most core samples from the debris fields include Pleistocene and older clasts embedded in debris that is covered by apparently undisturbed hemipelagic sediment tens of centimeters to more than a meter thick. Mass movement seems generally to postdate canyon or channel incision of the Pleistocene section (Kidd, Hunter, and Simon, 1987), and there does not appear to be any particular spatial relationship between debris and previously cut channels (Embley and Jacobi, 1977).

Failure generally has occurred in sections of homoclinal strata that are at least partly indurated and that dip several degrees seaward. The cause of failure and subsequent sliding in the large slide complexes is unknown, but widely guessed at. Earthquake shock and decomposition of clathrate layers are the most frequently cited plausible mechanisms. Rapid deposition along the shelf edge is not likely to have been a triggering mechanism in these cases (Embley and Jacobi, 1977).

RECORD OF THE SOUTHEAST NEW ENGLAND AND MUNSON–NYGREN SLIDE COMPLEXES

The SENESC and the Munson–Nygren slide complex are in many ways typical of the large mass movement complexes described in the preceding text. The SENESC comprises several large slides that originated along a 220-km stretch of continental slope, at water depths of around 750 m and deeper (Fig. 9-1). The slides are indicated by complex, stepped scarps more than 100 m in relief, and by tabular or slab-like depressions within the lower slope and uppermost rise. Each slide seems to have evolved into a debris flow, the different flows having coalesced on the upper rise to form broad, lobate sheets, about 30 m thick, that thin gradually seaward. Debris extends for 200 km and more across the seafloor. The SENESC also includes slope failure and depositional features indirectly related to sliding, but indicative of instability within the complex, such as in situ collapse of upper slope strata, parautochthonous debris on the rise, and debris issued from canyons present within the SENESC.

Knowledge of mass movement comprising the SENESC has been accumulating for nearly a quarter century, beginning with a report by Uchupi (1967) based on a few seismic reflection profiles of the continental slope southeast of Long Island. Despite the fact that he could provide few details, Uchupi drew attention to the great extent of the slide area and its significance to the sedimentary framework of the continental rise. Details of the scarps along the upper slope between long. 70°30′W and long. 71°00′W were illustrated and described by MacIlvaine (1973) based on 3.5-kHz echo-sounder records and seafloor photographs. MacIlvaine (1973), and MacIlvaine and Ross (1979) recognized a sequence of sliding, and MacIlvaine (1973) astutely observed that creep might be involved in the mass movement. Debris-flow deposits of the SENESC were identified and mapped across the rise by means of 3.5-kHz records (Embley and Jacobi, 1977; Embley, 1980; Vassallo, Jacobi, and Shor, 1984; Pratson and Laine, 1989). Vassallo, Jacobi and Shor, (1984) showed a debris field ("sediment slide complexes") 180 km wide and 400 km long, essentially the entire SENESC. Large base-of-slope slide excavations were identified on the basis of high-resolution seismic reflection profiles (O'Leary, 1986b). The slide excavations represent major component slides within the SENESC; they are broadly aligned downslope with tongues of blocky debris mapped by Vassallo, Jacobi, and Shor (1984).

Despite the evidence of extensive debris-flow deposits on the rise between Hudson Canyon and long. 70°00′W, the case for major mass movement remains controversial. Tucholke (1987) opined that this segment of the Atlantic continental slope is an area of minor mass movement, and that the alleged sparsity might be explained either on the basis of low upper-slope gradient, or the character of Pleistocene stratigraphy, or to conditions of Pleistocene deposition.

The Munson–Nygren slide complex consists of one or more large slide-dominated mass movements present on the lower continental slope, at about 2,000 m water depth, between Munson and Nygren Canyons off Georges Bank, and a large slide, the Retriever slide,

TABLE 9-2   Interpretive Key to 3.5-kHz Echo Characteristics of the Seafloor and to GLORIA Image Brightness Patterns.

3.5-kHz Profile Example	3.5-kHz Echo Type	GLORIA Image Brightness Pattern	Seafloor Character	Distribution
	Steep, sharply defined overlapping hyperbolae (type IVA of Pratson and Laine, 1989)	High-contrast bright and dark (slope generated backscatter)	Hard seafloor deeply cut by gullies, canyons	Mostly mid to upper slope
	Distinct (sharp), smooth, flat to broadly undulatory seafloor underlain by: a) transparent substrate b) parallel-layered substrate c) attenuated, pinching or lapping substrate (types ID, IE of Pratson and Laine, 1989)	Uniform and dark (acoustic facies 2, 6 of Kidd et al., 1987)	Smooth, firm seafloor underlain by: a) distal debrite b) hemipelagic layers or contourites; patchy thin debrite; current-scoured seafloor	Middle to lower rise
	Sharp to prolonged echo on markedly uneven (mainly convex forms) seafloor underlain by fuzzy, patchy substrate or pagoda structures (Emery, 1974; Flood, 1980)	Complex pattern in medium tones	Deformed or partly eroded sea-floor possibly underlain by gassy or locally disturbed, intrusive or fractured substrate	Upper rise (swells)
	Prolonged echo on tightly uneven to notched seafloor; slight hyperbolae and focussed echoes; indistinct substrate: disrupted, fuzzy, or mostly transparent	Medium gray, speckled, irregular, with elongate bright or dark tones	Parautochthonous or in situ debris bordered by scarps or channels; borders may be sharp or gradational	Chiefly lower slope, upper rise
	Distinct, sharp, smooth to uneven seafloor underlain by parallel reflections or thin, transparent substrate, abruptly bounded by wispy, deeply nested hyperbolae or long, tailed hyperbolae	Dark, finely figured polygonal areas with sharp margins, indentations surrounded by bright seafloor	Raised inliers or buttes, locally thinly covered by debris, bordered by scarps, collapsed slopes, surrounded by rubble, coarse debris	Chiefly mid to lower slope, uppermost rise

Echo character	Gray tone	Deposit	Setting
Fuzzy, indistinct hyperbolae with random, flat to convex sharp echo segments; irregular, uneven seafloor with local high relief and steep slopes	Generally bright to speckled (acoustic facies 7 of Kidd et al., 1987)	Coarse, proximal debris	Mid slope to upper rise
Distinct sharp to prolonged continuous echo on broadly uneven, mounded seafloor; transparent substrate on strongly reflecting horizon; transparent unit pinches out at scarp edges (type IIB of Pratson and Laine, 1989)	Light to medium uniform gray tone	Homogenized debris caps ("debris flows" of Jacobi, 1984)	Lower slope to upper rise
Generally distinct, sharp echo reflected by uneven mounds separated by notches, slope-breaks; interspersed with indistinct hyperbolic echoes similar to 5 (type IVB of Pratson and Laine, 1989; "blocky" type of Jacobi, 1984)	Light to medium gray tone, local dark or bright streaks or patches (acoustic facies 3 of Kidd et al., 1987)	Coarse debris, varies locally in surface texture and relief; may be covered locally with fine sediment	Upper rise
Dark, prolonged echo with uneven bright gaps and focussed echoes; forms generally rough but low relief surface; underlain by transparent to very weakly layered substrate (type IIB of Pratson and Laine, 1989)	Generally bright to mottled or speckled; may show larger elongated dark patches (acoustic facies 4 of Kidd et al., 1987)	Parauthochthonous or rough debris	Mid upper rise
Sharp to prolonged echo; local small hyperbolae but generally irregular, low-relief seafloor on transparent substrate of uneven thickness that rests on reflection similar in acoustic character to seafloor echo	Generally medium to light gray, mottled areas with relatively distinct borders	Local mounded or stacked debrites ("debris flow" of Jacobi, 1984)	Lower slope to upper rise
Prolonged, slightly fuzzy echo on generally smooth seafloor, grades to transparent or dark borders or very weakly layered substrate; rests on even but weakly reflective horizon (type IIA of Pratson and Laine, 1989)	Generally dark with darker flow forms or dark borders (acoustic facies 5 of Kidd et al., 1987)	Distal debrites or mud flow deposits	Rise

located just west of Munson Canyon (O'Leary, 1992; Fig. 9-1). Debris of both slides merges at 3,000 m water depth, some 65 km downslope from the heads, to form a debris field 50 km wide. The total downslope length of the debris field is at least 260 km. The initial description and map of the Munson–Nygren slide complex was based on high-resolution seismic reflection profiles (O'Leary, 1986a). The full extent of the complex, and the identification of the Retriever slide as a spatially distinct mass movement, was determined on the basis of the GLORIA data set (O'Leary, 1992).

As with the SENESC, the existence of mass movement between Munson and Nygren Canyons is not fully accepted. Danforth and Schwab (1990), in a study based on the same seismic reflection profiles used by O'Leary (1986b), explained stratigraphic truncation (reported by O'Leary as evidence of the Munson–Nygren slide) to be a result of downcutting by a series of canyon systems between Munson and Nygren canyons.

DATA SOURCES AND METHODS

The inconsistent opinions and speculations about mass movement comprised by the SENESC and the Munson–Nygren slide complexes arise partly because of scattered and insufficient data. The opportunity to integrate new, systematically acquired data with published data and to clearly demonstrate the nature of the slide complexes came with the New England leg of the U.S. Atlantic Exclusive Economic Zone (EEZ) GLORIA survey, conducted in March and April of 1987, as a joint effort of the U.S. Geological Survey, the British Institute of Oceanographic Sciences, and the Canadian Geological Survey.

Details of the GLORIA system have been described by Somers, et al. (1978) and Laughton (1981); however, several aspects of this technology must be reviewed, as our survey was conducted in an area that has received detailed profiler coverage (Pratson and Laine, 1989), but little seafloor sampling. The shipboard GLORIA data have a spatial resolution of about 125 m along track and about 50 m across track. Therefore we can provide little comment on individual features of the seafloor smaller than about the size of a football field, except that the brightness of individual pixels is at least partly a function of seafloor roughness on the order of several centimeters (Kidd, Hunter, and Simon, 1987). Furthermore, the GLORIA signal beam is capable of penetrating at least a few tens of centimeters below the seafloor, so that image brightness may also be related to inhomogeneities within the surficial sediment or the roughness of a thinly buried surface (Gardner et al., 1991; see also Chapter 13). We must also bear in mind that GLORIA images provide essentially no infomartion on relief or slopes inclined less than the minimum acoustic incidence angle of a few degrees. This makes interpretation of slope forms (i.e., relief less than about 1:50) particularly difficult, or impossible, without the aid of echo-sounder profiles.

A 3.5-kHz seismic profiler with a calculated resolution of 0.8 m, and a 160-in³ airgun system with a vertical resolution of 25 m, were routinely deployed with GLORIA. Airgun profiles were used to supplement earlier seismic-reflection information on the shallow stratigraphy (O'Leary, 1986b) and to provide information on subseafloor structure in the survey region. The 3.5-kHz profiles were used to determine local relief, areas of outcrop, and depositional features directly linkable to sonar image brightness.

The 3.5-kHz profiler has been widely used to construct echo-character maps (Damuth, 1980; see Chapters 8 and 12). Pratson and Laine (1989) used these characters to establish a 12-fold seafloor classification for a large part of the U.S. Atlantic slope and rise, based on distinct, indistinct, irregular, and hyperbolic echos. For this study, we present a table (Table 9-2) that incorporates both the echo types reported by Pratson and Laine (1989) and the GLORIA sonar facies listed by Kidd, Hunter, and Simon (1987) (compare with Chapters 8, 12, 13).

Additional sidescan sonar image data obtained by the SeaMARC I mid-range system is used in our study. The SeaMARC I was a 30-kHz system that provided a resolution of 2.5 m and a swath width of 5 km. The data were acquired by USGS and Lamont-Doherty Geological Observatory in 1981 under the aegis of the U.S. Bureau of Land Management (O'Leary and McGregor, 1983). Initial interpretation of GLORIA shipboard records and SeaMARC I records indicated unexpectedly complex patterns of debris emplacement and of inferred behavior of surficial sediment on the slope and rise during mass movement. In light of the recognized complexity of the SENESC, we include a modified table of the major types of submarine mass transport, together with suggested criteria for their recognition (Table 9-1). Extent of translation, internal disorganization, and mechanics of movement can be assessed to some degree using geophysical data, although without the confirmation offered by seabed samples, the correct terminology can be difficult to apply. Nevertheless, the terms used in the morphological descriptions that follow derive from those given in Table 9-1.

REGIONAL PHYSIOGRAPHY AND STRATIGRAPHIC SETTING

A geological appraisal of the slide complexes with respect to the evolution of the continental rise is incomplete without knowledge of their stratigraphic and morphologic settings. The following section outlines the important elements of the physiography and stratigraphy that form the geological environment for

TABLE 9-3   Summary of Shallow Stratigraphy.

SENESC	Munson–Nygren Slide Complex
SLOPE SECTION	

Pleistocene

Thin (20–40 m) surficial unit; thin-bedded to massive(?) silt, sand, clay.

Pleistocene

Thick (> 400 m) section of homoclinal to gently offlapping strata up to 30 m thick, form two paraconformable subunits; chiefly gray, compact to plastic silty clay and fine sand; some gassy intervals.

Several packages of foreset beds underlie the shelfbreak and total 450–480 m thick; chiefly sand, silty sand, fine gravel, and clay; locally cemented. Units become parallel-bedded across slope.

Neogene

Chiefly dark gray glauconitic silt and sand, gray clay.

Neogene

Chiefly glauconitic quartz sand and dark glauconitic clay.

Paleogene

Chiefly parallel-bedded to very thinly lensing interval 350–400 m thick; generally conformable with Neogene strata; mainly middle to upper Eocene marl, calcareous claystone, and indurated chalk. Also local thin, calcareous Paleocene and sandy Oligocene.

Cretaceous

Well-stratified section of uncertain thickness. Eroded upper contact; complex subunits are broadly warped and beveled. Chiefly dark bluish- to brownish-gray, fine-grained calcareous sandstone, siltstone, claystone, silty micaceous clay, massive to laminated siltstone; locally glauconitic and generally indurated.

| RISE SECTION | |

Chiefly complex interlensing subunits having strong acoustic contrasts; intervals thin and pinch out to east against an irregular, eroded(?) ascending Paleogene(?) unit, which forms broad arch along Georges Bank; only inferred Pleistocene section carries across to east. Three major subunits:

a. parallel-layered Pleistocene sand, silty clay, loose to deformed to indurated; up to 180 m thick.

b. massive to weakly layered Pliocene compact calcareous to glauconitic silty clay and clay.

c. parallel-bedded upper Miocene dark gray sandstone and siltstone.

Generally uneven, parallel-bedded section up to 300 m thick, continuous with Neogene(?) and Pleistocene slope strata. Broadly unconformable on inferred Paleogene unit. Probably chiefly indurated, locally calcareous and glauconitic sand, silt, and claystone.

each slide complex. The stratigraphy is summarized in Table 9-3.

Continental Slope

Physiography

The SENESC and the Munson–Nygren slide complexes occur in distinctly different physiographic settings. The SENESC is located along an east–west oriented stretch of continental slope characterized by a convex profile and by sparse canyon erosion (Fig. 9-2); the Munson–Nygren slide complex is located low on the concave, extensively canyon-eroded continental slope along Georges Bank (Fig. 9-3). The upper slope shoreward of the SENESC is a gently convex surface,

having an inclination of about 1.4°, that extends from the shelfbreak (about 140 m water depth) to a middle slope water depth of about 750 m. At water depths of 750–1,000 m, the seafloor steepens across a scarp marked by embayments, tabular reentrants and salients, and cliff and terrace intervals several tens to more than 100 m in relief (Fig. 9-2). This break marks the head zone of major slides of the SENESC. From about 1,000 m water depth, the narrow lower slope (≤ 10 km wide) descends, with an average inclination of about 7–8°, down to an abrupt contact with the continental rise between 1,900 and 2,000 m depths (Fig. 9-2A). West of long. 70°30′W, a broad ramp stands between the lower slope and the rise (Fig. 9-2B). The ramp is inclined at about 1.5° and extends from 1,500 m to ~ 2,300 m

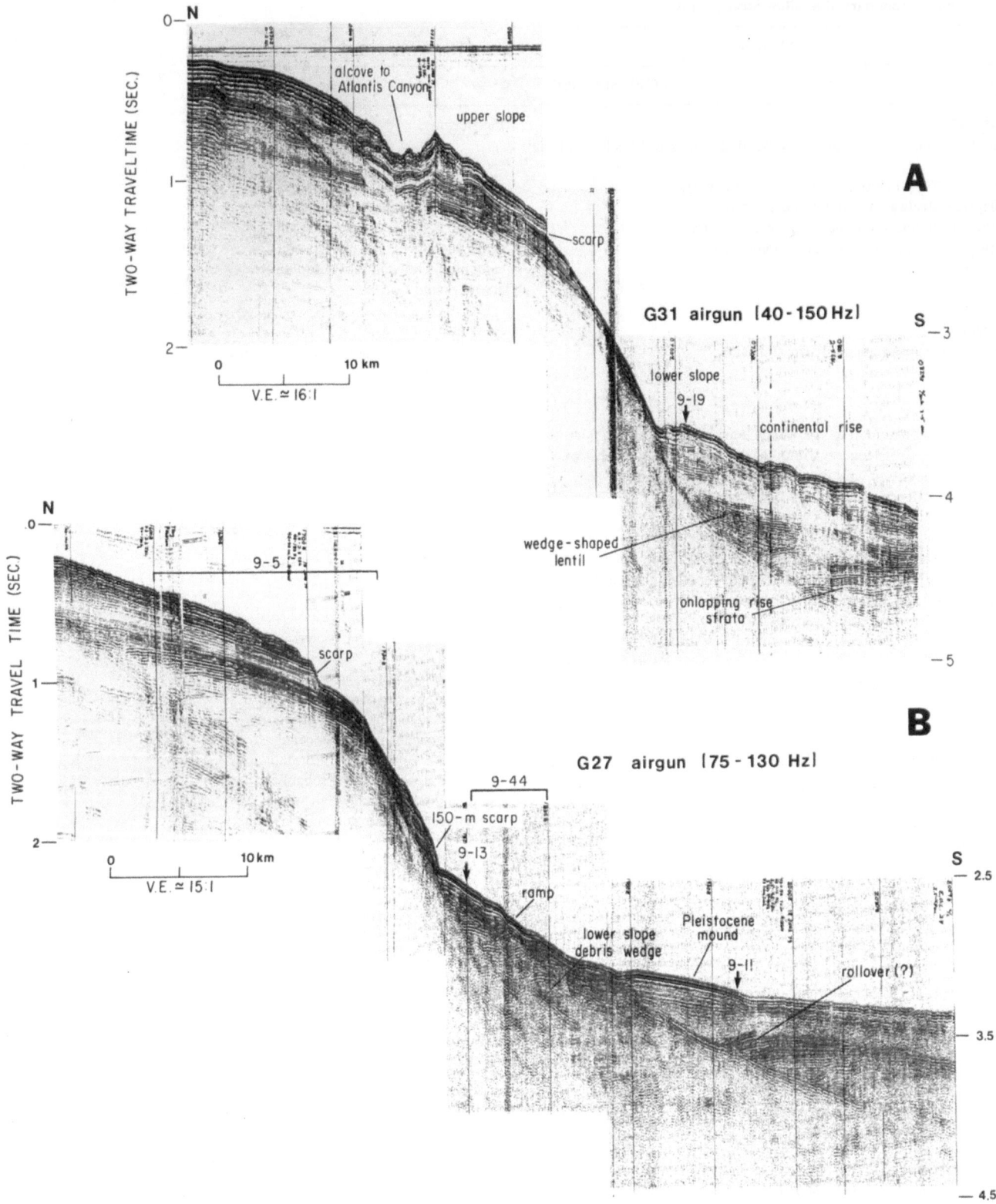

FIGURE 9-2. Typical continental slope profiles along SENESC, as shown by high-resolution seismic reflection records. A. Continental rise units onlapping continental slope. B. Lower slope debris wedge that forms prominent ramp west of 71°15′W. See Figure 9-1 for locations. Numbered arrows indicate track-line crossings of records shown in other figures.

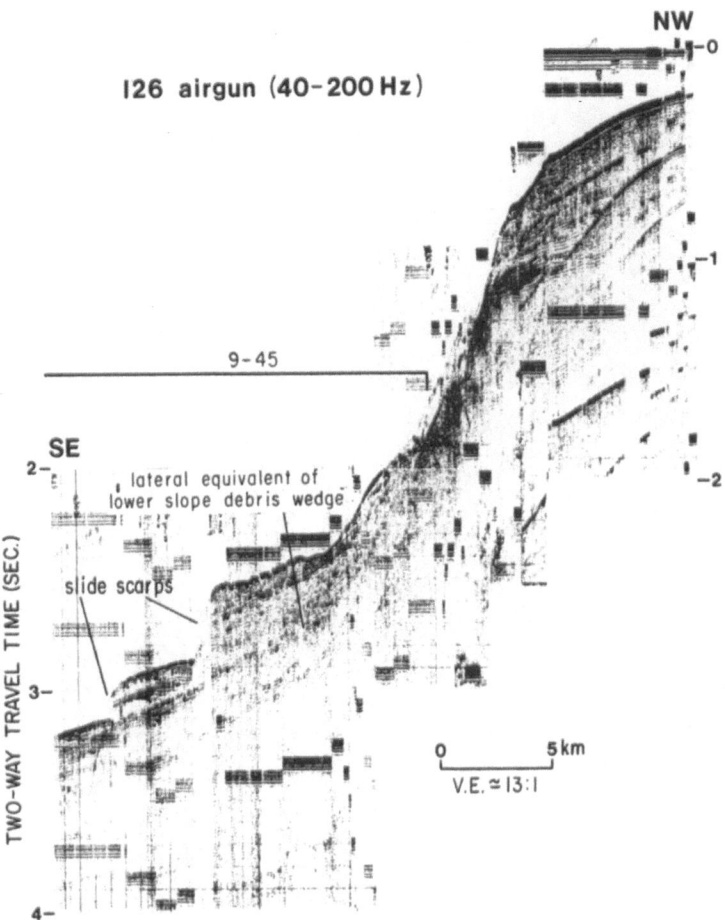

126 airgun (40-200 Hz)

NW

9-45

SE

lateral equivalent of
lower slope debris wedge

slide scarps

TWO-WAY TRAVEL TIME (SEC.)

0   5 km

V.E. ≃ 13:1

FIGURE 9-3.   High-resolution seismic reflection record across continental slope and head of Munson–Nygren slide. See Figures 9-1, 9-45 for location.

depth, a planimetric distance of ~ 17 km. The ramp represents a large base-of-slope accumulation of slide debris, hereafter called the lower slope debris wedge (Fig. 9-4).

Along Georges Bank, the slope falls off from the shelfbreak at a relatively steep angle of 2–3°. Slope inclination decreases with increasing water depth, resulting in a concave profile that fairs onto the rise without stratigraphic interruption. The concave profile is accentuated and complicated by deep, mid-slope erosion related to closely spaced canyons. The large majority of the canyons along Georges Bank head on the upper slope, forming a succession of scored embayments that, in downslope section, give the Georges Bank slope its characteristic steeply concave (~ 9°) and rugged upper to middle slope profile. The Munson–Nygren slide complex occurs below the zone of maximum lateral erosion, where the intact slope is inclined at a little more than 2°.

Whereas the most characteristic feature of the slope along Georges Bank is a deeply etched, pinnate, erosion pattern, the southeast New England slope is notable for sparse and simple canyon incision. Most of the canyons on the southeast New England slope do

not feature the ramose gully pattern so characteristic elsewhere, but instead have simple, chute-like forms with collapsed, embayed walls. The larger canyons include, from west to east, Hudson, Jones, Babylon, Double, Block, Alvin, Atlantis, Nantucket, Shallop, and Veatch (Fig. 9-1). Aside from Hudson, these canyons barely indent the shelf. The relief of these canyons is comparable to the relief of the larger canyons elsewhere on the Atlantic slope (~ 400 m maximum).

Stratigraphy

The Cretaceous interval is broadly and evenly layered on strike, but is disharmonically warped, and has a regionally steeper dip in relation to the largely unconformable superjacent interval. Samples of the upper levels of the Cretaceous interval consist of semiindurated, gray sandstone, siltstone, mudstone, darkgray limey clay, and compact, silty, micaceous clay (Manheim and Hall, 1976; Poag, 1982). Within the SENESC, Upper Cretaceous siltstone is only thinly buried or is exposed in canyon walls (Weed et al., 1974).

A laterally persistent and uniformly thick (~ 400 m) Tertiary unit (see also Chapter 6) generally conforms

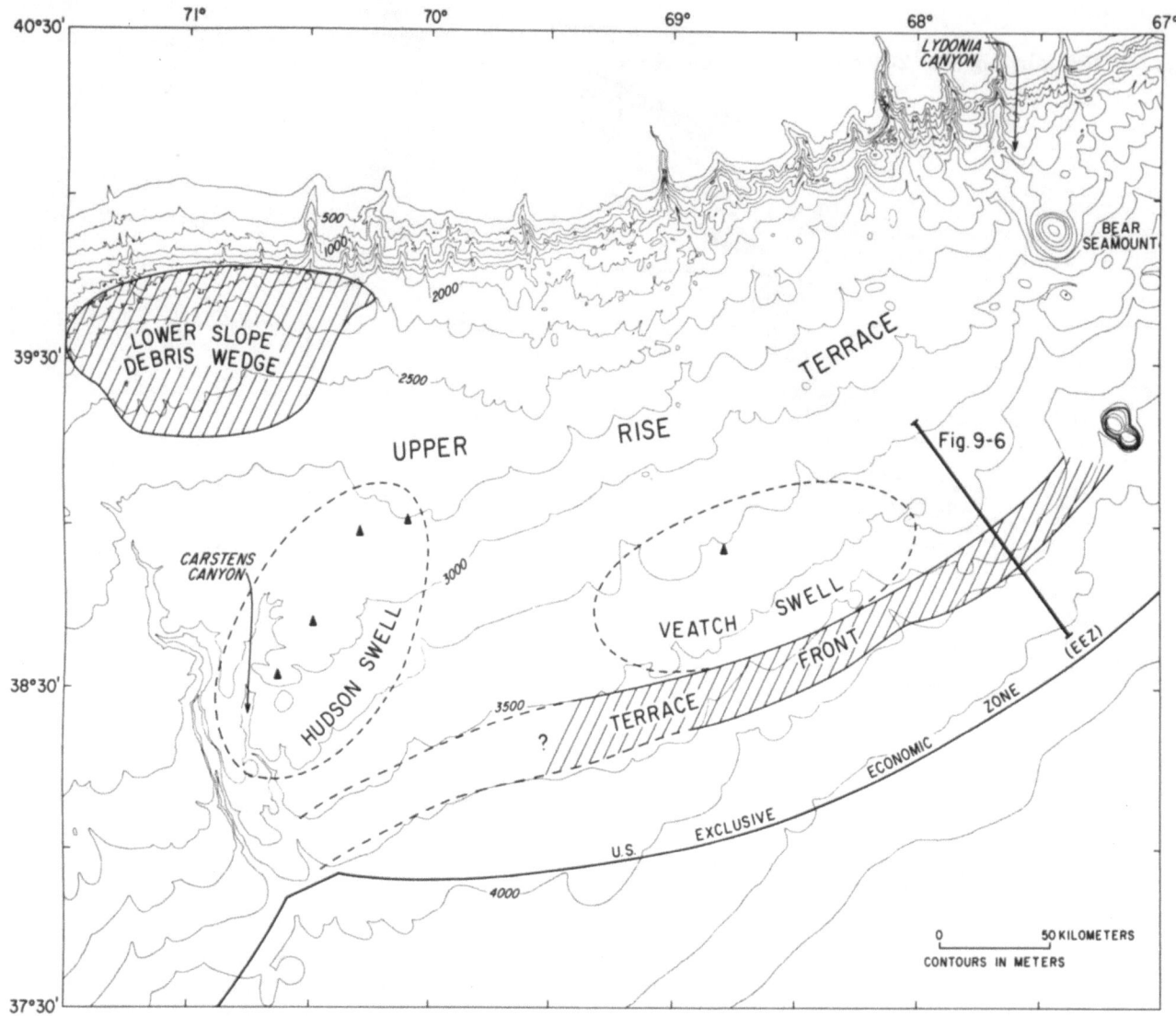

FIGURE 9-4. General bathymetric features of upper rise terrace along southeast New England continental slope, including lower slope ramp (debris wedge), swells and swell crests (triangles), and terrace front.

to the lower slope off southeast New England. The Tertiary unit crops out at water depths that vary from less than 700 m to more than 1,500 m and has been widely sampled (Stetson, 1949; Northrop and Heezen, 1951; Gibson, 1965, 1970). Nearly all samples obtained from the outcropping Tertiary unit consist of middle to upper Eocene chalk or indurated marl. Along Georges Bank, the Tertiary unit is widely exposed in canyon walls. In many places, the unit is gradational, or at least conformable, with the overlying Neogene(?) and Pleistocene interval; elsewhere, the overlying strata have a downlapping contact.

The Neogene(?) and pre-Wisconsinan Pleistocene section is represented upslope from the SENESC by a 425–460-m-thick interval of homoclinal strata. Seaward of the shelfbreak, the strata become progressively bet-

ter defined in seismic profiles across the upper slope. The layers have a distinctive acoustic grading, from sharply bounded, nearly opaque upper parts down to diffuse, relatively thick, transparent lower parts (Fig. 9-5). Each layer is paraconformable with the lower one; the irregular contacts indicate that deformation and(or) erosion occurred between successive episodes of deposition. A great deal of local, fine-scale, acoustic wipe-out in the intervals below distorted reflections implies that intrastratal deformation also has occurred. At Atlantic Margin Coring Project (AMCOR) hole 6012 on the upper slope near Block Canyon (Fig. 9-1), the section is represented by nearly 304 m of mainly sticky-to-plastic, slightly silty, dark-gray clay of Pleistocene age (Hathaway et al., 1979). The sediment is gassy below 90 m. The hole bottomed in dark olive-gray,

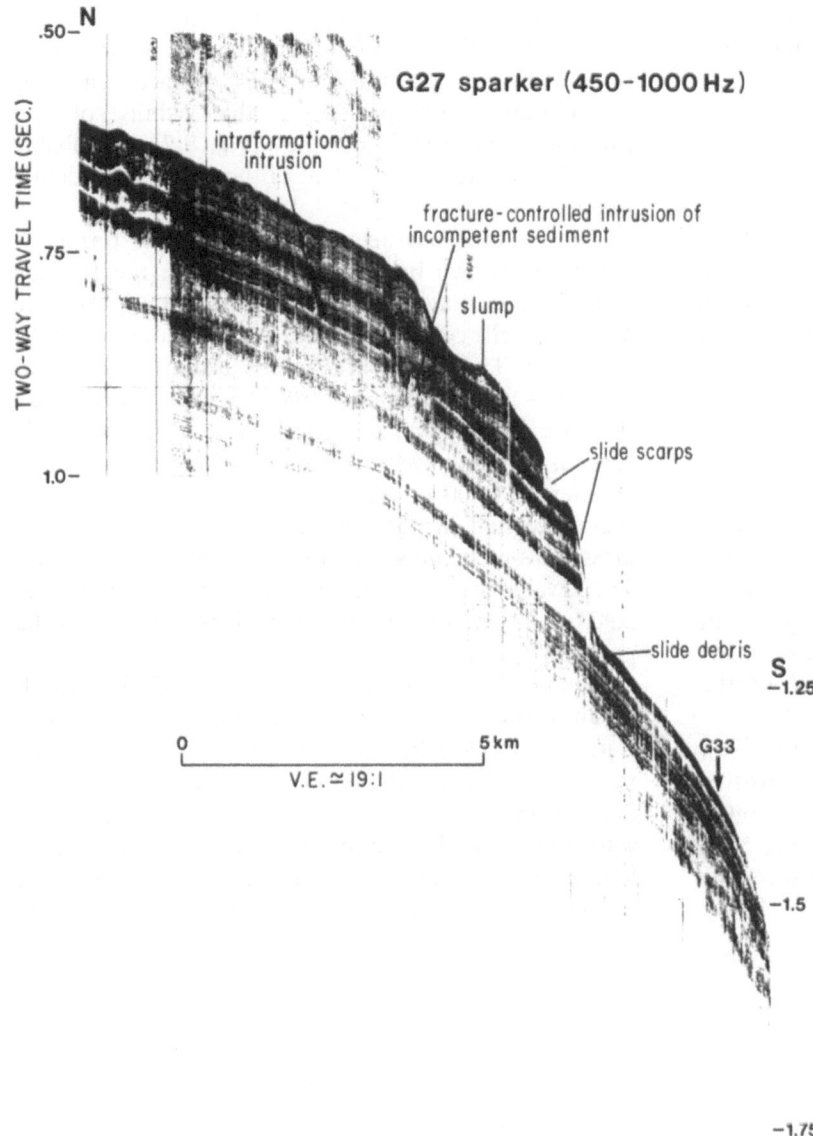

FIGURE 9-5. Complex, stepped, slide scarps at mid-continental slope level of SENESC. Truncated section is Pleistocene. Acoustically faint or transparent layers may be relatively weak intervals composed mainly of clay. See Figure 9-2B for setting.

slightly plastic clay of possibly Miocene age. Within the SENESC, much, or all, of the Pleistocene homoclinal section (as much as 200 m) is truncated along the scarp at mid-slope depths (750–1,000 m; Fig. 9-5). Along Georges Bank, the Pleistocene section is compact and even cemented in places; it supports the deeply gullied seafloor of the upper slope.

The uppermost stratigraphic unit of regional extent on the New England continental margin is a blanket-like deposit, hereafter called the surficial unit, which ranges in thickness from about 20 to 40 m. It is generally conformable with underlying strata, but in places it rests on an eroded surface or drapes scarps and rough-surfaced debrites. We interpret the surficial unit as a regional drape deposit of Wisconsinan age,

derived chiefly from glacial outwash and glaciomarine sedimentation (O'Leary, 1988).

The lower slope debris wedge that underlies the base-of-slope ramp previously described, has a complex stratigraphic relationship with the slope and the rise: its foot is onlapped by strata that make up the continental rise, and its contact with the slope consists of obscure, onlapping, tilted segments (Fig. 9-2B). No clear basal contact is recorded in the high-resolution seismic reflection profiles. The debris wedge consists of several poorly discriminated, acoustically chaotic sub-units, each of which is an olistostrome that ramps up against the slope and grades downdip into relatively evenly layered strata beneath the rise. The uppermost subunit of the wedge is probably chiefly middle

Miocene. Its uneven but well-defined surface is probably equivalent to reflector Merlin of Mountain and Tucholke (1985).

Continental Rise

Physiography

The upper continental rise off southeastern New England is essentially a broad, onlapping terrace that extends along the slope for at least 350 km, and seaward from the slope as far as 175 km (Fig. 9-4). The terrace front descends across a relatively steep (~ 1°) declivity (relative to the the upper rise surface inclination of ~ 0°15') located roughly between the 3,600- and 4,000-m contours (Fig. 9-4). Here, the seafloor steepens gradually from the smooth convex outer margin of the upper rise terrace and descends, becoming increasingly uneven across broad steps and flexures, for a downslope distance of 15–25 km, and for a relief of as much as 550 m (Fig. 9-6). Strata of the upper rise pinch out or thin appreciably as they descend the declivity; strata of the nearly flat lower rise onlap the base of the declivity (Fig. 9-6).

Toward the northeast, in the vicinity of the New England seamounts, the upper rise terrace narrows. The eastern end of the upper rise terrace is obliterated by the wide, eroded floor of Lydonia Canyon (Fig. 9-4). There may be remnants of the terrace surrounding Bear Seamount (Fig. 9-4), but northeast of the

seamounts a terrace-like upper rise is missing; instead, the continental slope forms a generally smooth transition to a more gently dipping rise surface. The flattened upper rise profile in the vicinity of the Munson–Nygren slide complex (Fig. 9-7) is probably caused by a lateral equivalent of the lower slope debris wedge.

The elevated character of the upper rise terrace is enhanced by two broad sediment swells: the Hudson Swell, centered at lat. 38°30'N, long. 71°30'W, and the Veatch Swell, centered at lat. 38°50'N, long. 68°45'W (Fig. 9-4). The swells are areally extensive, being at least 100–150 km long and 50–60 km wide (downslope); they are low, having a vertical expression that rarely exceeds 100 m above an otherwise planar seafloor (i.e., a relief of 1:500 or less). Nevertheless, the influence they have had in ponding or diverting the runout of mass movement debris of the SENESC is impressive. The swells are present at depths in the range of 2,900–3,700 m, and are oriented slightly oblique to the gross bathymetry.

The regional bathymetry suggests that Hudson Swell is an artifact of erosion associated with Carstens Canyon (Fig. 9-4), but this is not truly the case. Airgun records that cross the swell axes display low-angle unconformities associated with units that thicken toward the crests. Low-amplitude waveforms with wavelengths of 1 km pervade the sediments in the region of the swell flanks. These features, and the fact that the

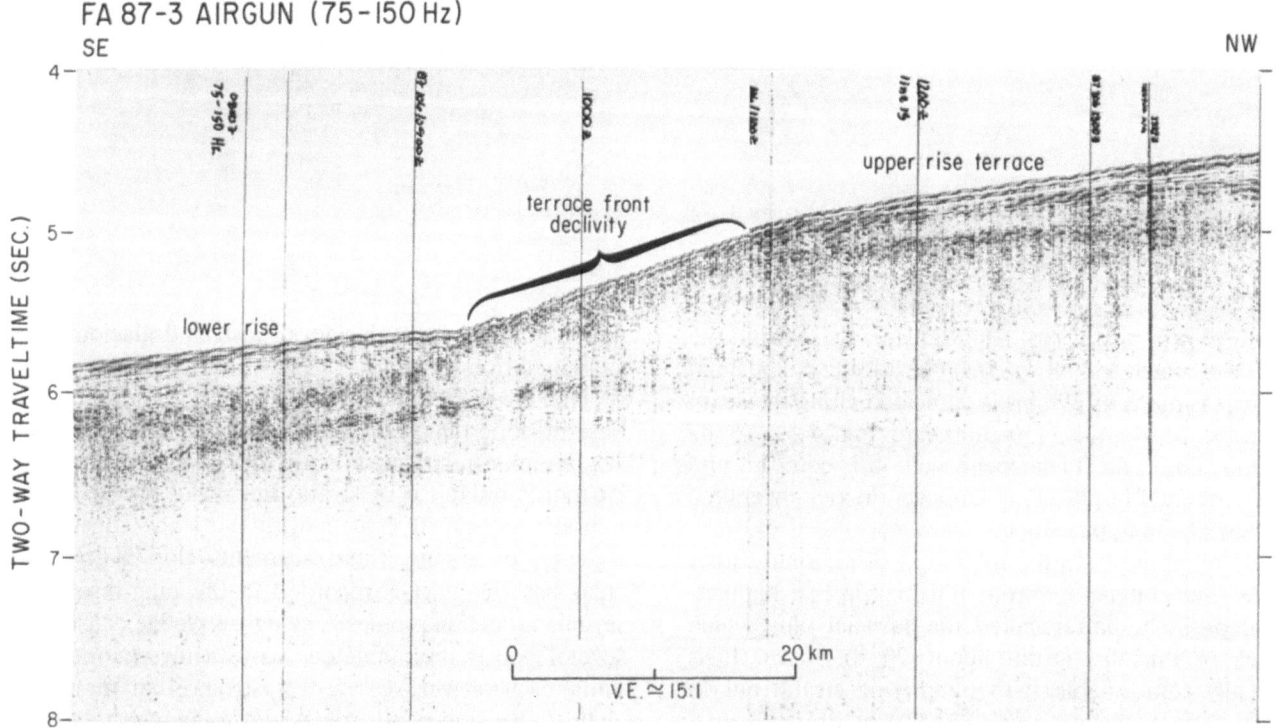

FIGURE 9-6.   Profile of upper rise terrace front shown by high-resolution seismic reflection record. See Figure 9-4 for location.

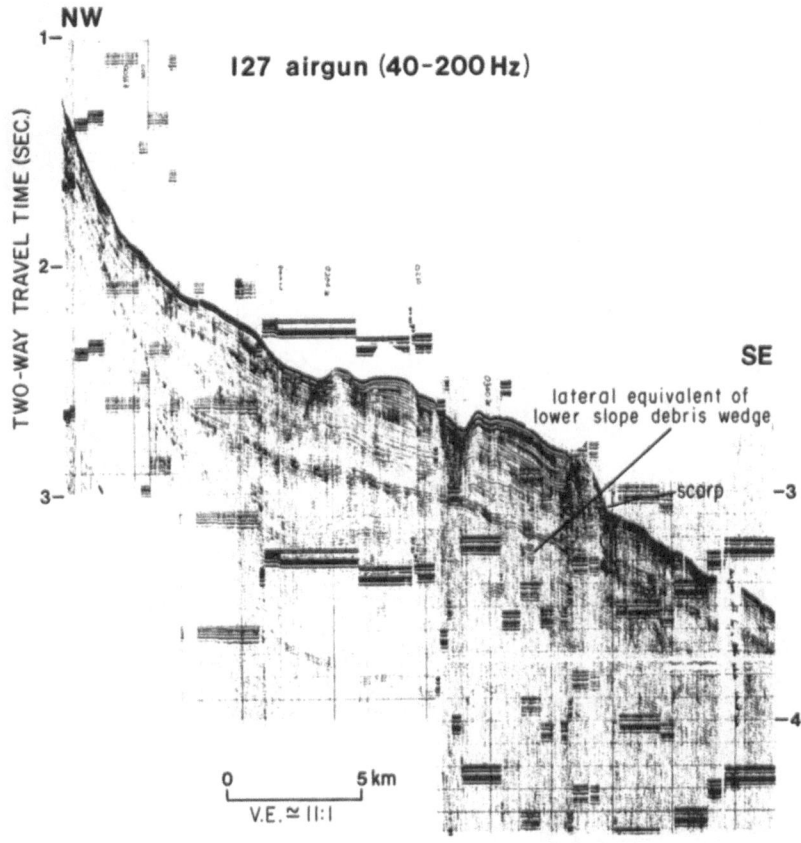

NW

127 airgun (40-200 Hz)

TWO-WAY TRAVEL TIME (SEC.)

SE

lateral equivalent of
lower slope debris wedge

scarp   −3

−4

0   5 km

V.E. ≃ 11:1

FIGURE 9-7.   Upper rise profile shown by high-resolution seismic reflection record across distorted (failed?) Neogene(?) and Pleistocene section adjacent to Munson–Nygren slide. See Figure 9-45 for location.

swells are formed within a narrow 800-m-depth interval, invite comparison with documented sediment drifts along the U.S. Atlantic margin.

The upper rise terrace off southeastern New England appears to be a partly covered extension of the Chesapeake Drift (Fig. 9-8; see also Chapter 6). The Chesapeake Drift is expressed mainly by isopach contours of the upper middle Miocene to upper Pliocene section (Mountain and Tucholke, 1985). Ponding of more than 400 m of upper Pliocene to Pleistocene sediment between the drift crest and the lower slope (Mountain and Tucholke, 1985) has created the terrace-like form of the upper rise. The Hudson and Veatch swells have actually accumulated on top of the ancient and partly eroded Chesapeake Drift.

The Chesapeake Drift was probably itself a bathymetric swell in middle Pliocene time (see Chapter 6). By late Pliocene time the seaward side of the drift was being eroded by currents, chiefly along the present 3,500-m contour (Mountain, 1987a), and the landward side was ponding calcareous and terrigenous mud (Tucholke and Laine, 1982). This depositional and erosional pattern has persisted virtually to the present day.

There can be little doubt that current molding of the upper rise off New England is a recent (Holocene)

and probably ongoing process. Features such as Knauss Knoll (Lowrie and Heezen, 1967; Fig. 9-1), faired rims of upper rise channels (Fig. 9-9), and sediment-filled ripple swales on the upper rise declivity indicate that generally southward-directed currents have had an appreciable effect on deposition of surficial sediment across the southeast New England rise. Furthermore, existence of nearly 100 m of section that constitutes the Hudson and Veatch swells suggests that contour current activity sufficient to form the swells has been a virtual constant during deposition. The only downslope depositional and erosional activities we can surely identify (mass movement and canyon erosion) are simply intrusions or aberrations in what appears to be a bottom-current–dominated depositional regime.

Stratigraphy

The Neogene and Pleistocene stratigraphy of the upper rise off southeast New England consists of at least three, more or less acoustically distinct, slightly discordant units, which onlap the lower slope (Fig. 9-10a, b, c). The lowest unit (a) consists of strongly reflective layers. The middle unit (b) is acoustically transparent or very faintly layered; it attains a maximum thickness of about 200 m close to the slope and thins gradually seaward. It includes in its lower part an onlapping

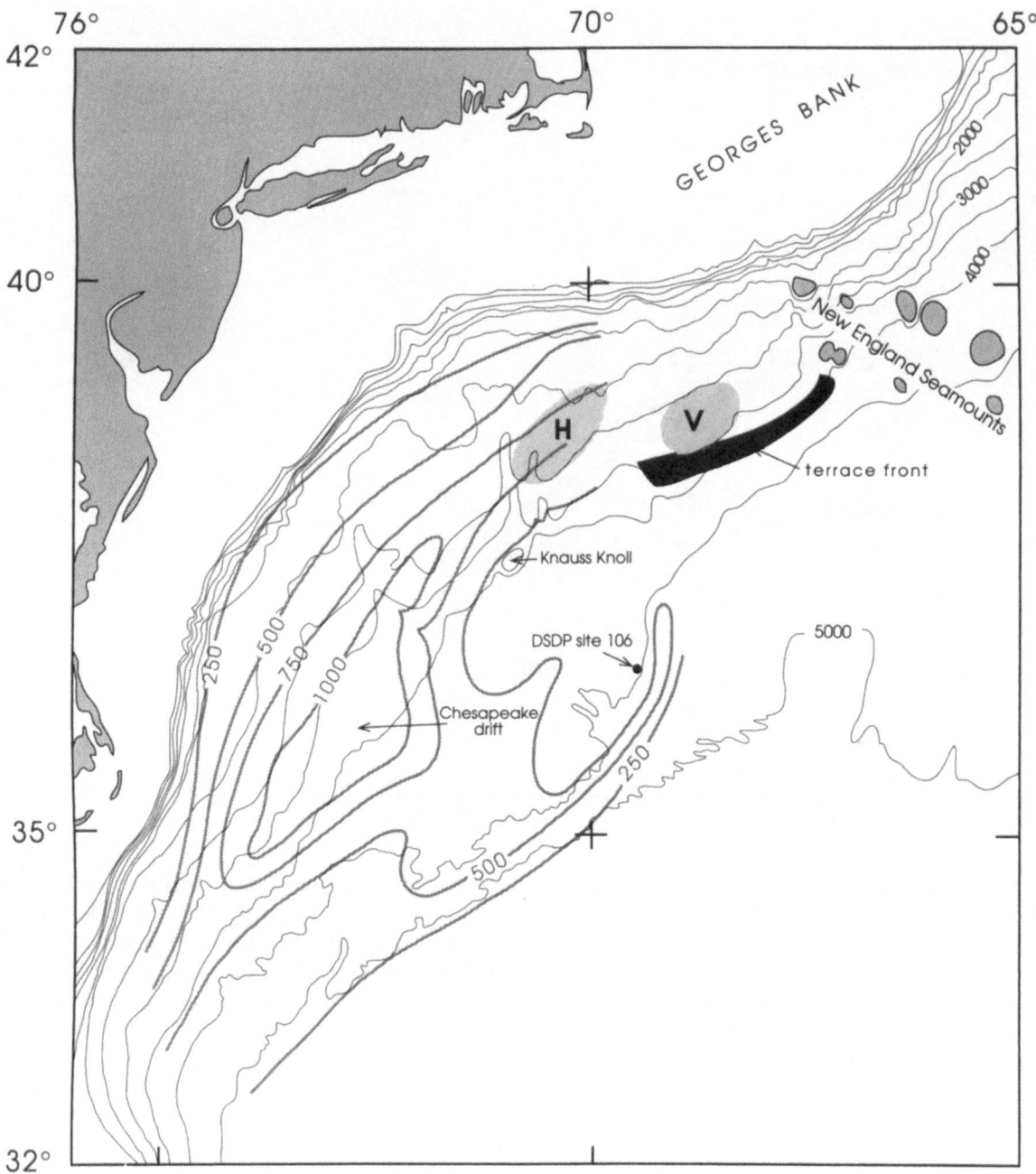

FIGURE 9-8.   Location and structure of Chesapeake Drift shown by upper middle Miocene to upper Pliocene isopachs (between reflectors Merlin and Blue; from Mountain and Tucholke, 1985). Also shown: Hudson (H) and Veatch (V) swells, and upper rise terrace front. Light contours indicate bathymetry in 500-m intervals; heavy contours indicate isopachs in meters.

wedge-shaped lentil (Figs. 9-2A, 9-10) having a distinctive dark acoustic signature. The lentil is probably a local debris deposit of Pliocene age. The uppermost unit (c) is a locally unconformable, highly reflective layered interval, which, along strike, ranges in thickness from 170 to 80 m, and which is correlative with or coextensive with Pleistocene strata of the upper slope.

In contrast to its simple, consistent downdip aspect, the rise section on strike along the southeast New

England slope is remarkably complex. Reflector-bounded intervals of the uppermost reflective layered unit (c) pinch in and out and vary in acoustic amplitude; they are warped, folded, and possibly faulted on the mounded surface of the subjacent transparent or faintly layered unit (b), and the lowermost reflections of the uppermost layered unit fade in and out along strike, suggesting lateral gradation with the underlying transparent unit (Fig. 9-11). Likewise, the uppermost

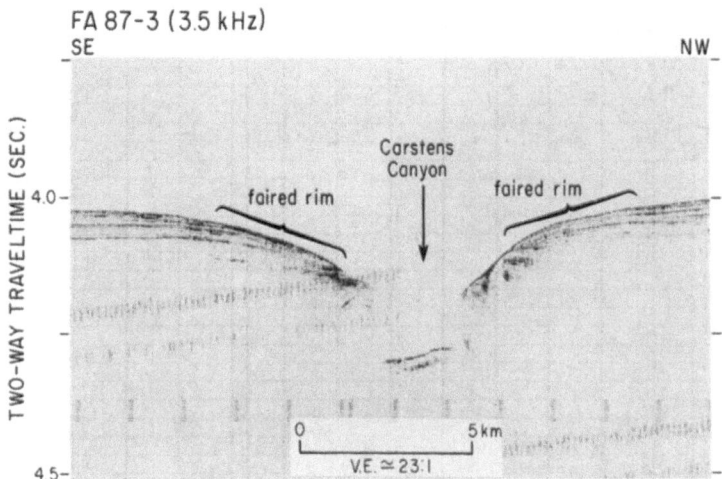

FIGURE 9-9.   Pleistocene strata faired across rims of Carstens Canyon shown by 3.5-kHz record. See Figure 9-14 for location.

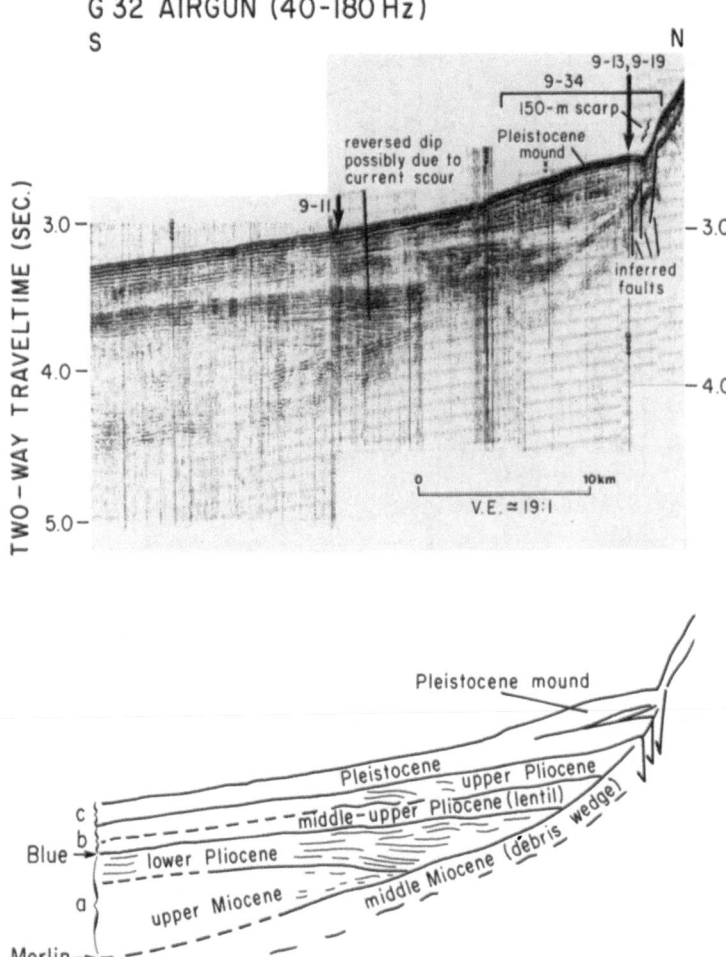

FIGURE 9-10.   Neogene and Pleistocene stratigraphy of upper rise off southeast New England shown by high-resolution seismic reflection record. Line drawing shows inferred chronostratigraphic units and regional horizons Blue and Merlin. See Figure 9-14 for location.

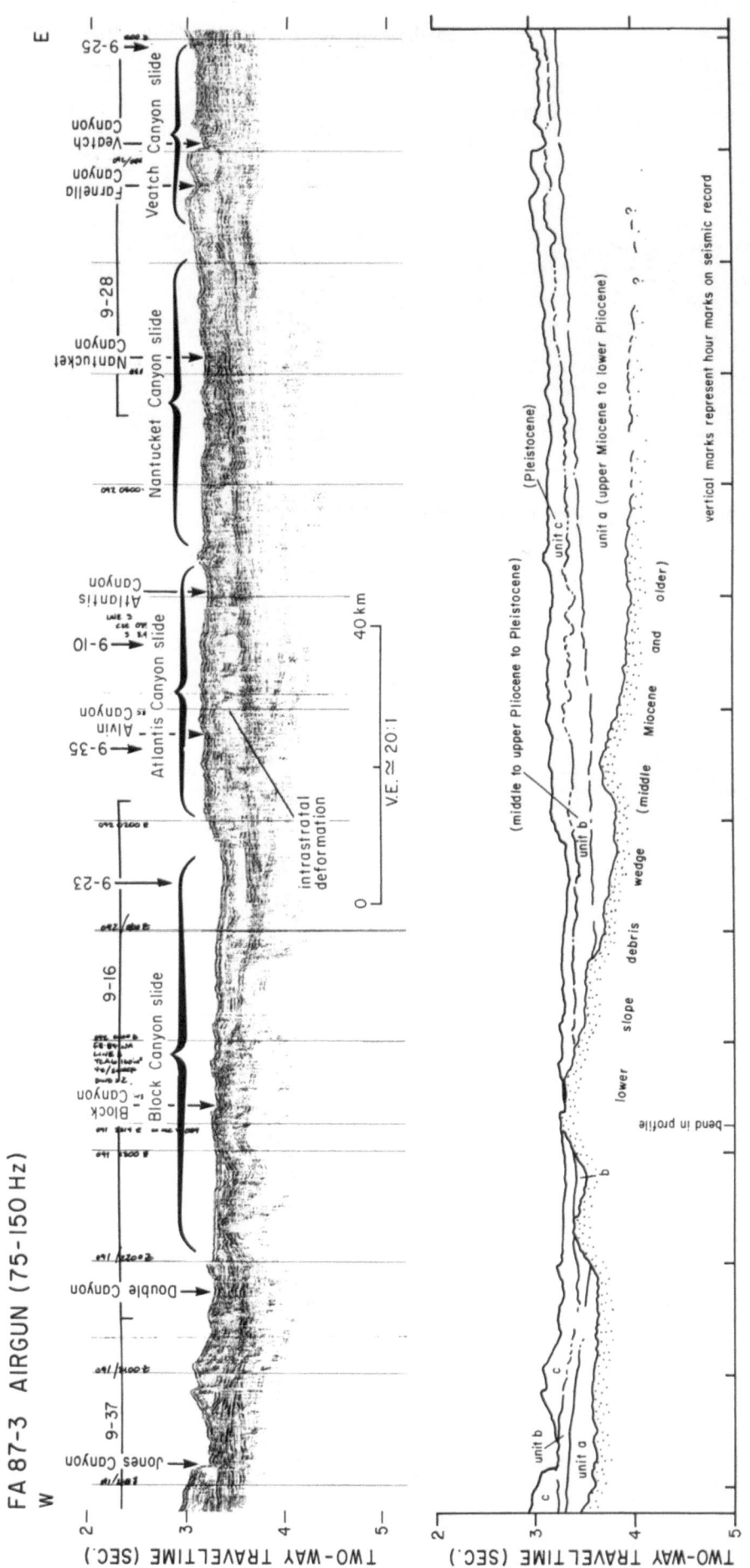

FIGURE 9-11. General stratigraphy and major physiographic features of upper rise along SENESC as shown by high-resolution seismic reflection record. See Figure 9-14 for location.

reflections of the lowest unit are interleaved with the overlying transparent unit and are bowed up, rumpled, and cut out along strike.

Poag (Chapter 6 and unpublished data) traced units at the series level northward from New Jersey to New England, with chronostratigraphic control provided by DSDP holes (Van Hinte et al., 1985; Poag and Mountain, 1987; Wise and Van Hinte, 1987). Based on this work, the onlapping units in our study are probably in the following stratigraphic relationship: the lowest, layered unit (a) is probably lower Pliocene and older; the transparent unit (b) is probably middle to upper Pliocene; the uppermost layered unit (c) is most likely Pleistocene. The DSDP stratigraphic data from the upper rise off New Jersey (Van Hinte et al., 1985; Poag, 1987) imply that the Pleistocene section of the southeastern New England upper rise is probably at least 150 m thick and consists of variably bedded sands to silty clays that range in consistency from gassy and soupy to firm and indurated, and that they include many highly deformed intervals and layers embedded with Eocene and other clasts. The Miocene to Pliocene section probably consists of more uniform, compact, calcareous to glauconitic clay or silty clay.

Additional insight into the slope-onlapping units of the southeast New England rise is provided by comparison with DSDP Site 106 (Hollister, Ewing, and Habib, 1972), located downdip on the lower rise at lat. 36°25′N, long. 69°26′W, at a water depth of 4,500 m (Figs. 9-1, 9-8; see also Chapter 6). Seismic profiles from this site show structural and acoustic features intriguingly similar to those seen in the upper rise off southeast New England. These features include an interval 0.2–0.5 sec thick of laminated reflections that grade down-section into an underlying transparent interval approximately 0.5 sec thick. The surface of the transparent interval is unevenly mounded; it locally breaches the lower part of the superjacent layered interval and, more commonly, cores broad flexures and corrugations of its lower layers. The transparent interval lies on an essentially horizontal, slightly warped interval of parallel, continuous layers.

The upper layered interval at DSDP Site 106 consists of 360 m of Pleistocene sandy to clayey sediment. Some beds are indurated, others are soft and plastic, and highly disturbed beds were penetrated near the base of the interval. The section is probably turbiditic, but no primary structures were seen. Rare plant debris and carbonized wood fragments were found, however. The transparent interval includes pelitic sediment of latest Oligocene through perhaps Pliocene age. About 140 m of Pliocene sediment was penetrated in an interval that spans the upper reflective zone and the underlying transparent zone. The Neogene section consists of chiefly dry, compact to indurated, dark greenish-gray siltry clay containing abundant quartz.

The hole may have penetrated reflector A at a subbottom depth of 1,050 m (Hollister, Ewing, et al., 1972).

SOUTHEAST NEW ENGLAND SLIDE COMPLEX

The southeast New England slide complex incorporates a variety of features that are the results of repeated failures and mass movement in the form of large translational slides and debris flows. Characteristic slide-related landforms include stepped scarps, in situ collapse fields, residual stacks, buttes, inliers, and so forth; depositional features on the rise include debris mounds, shingled debrites, canyon fills, and mud flow deposits. Translational sliding in the SENESC has been complicated by other types of mass movement, including canyon-related mud flows and activity that created the presently enigmatic lower slope debris wedge.

A general chronology of mass movement in the SENESC is demonstrated by cross-cutting landforms and elements of the stratigraphy outlined in the previous section. The sequence of mass movement in the SENESC seems to have been the following:

1. Emplacement of the lower slope debris wedge and recurrent Neogene mass movement.
2. Generation of major slab slides (removal of Pleistocene and older strata).
3. Generation of post-slab slide chutes or canyons (creation of present slope morphology in late Pleistocene).
4. Local subsequent sliding and scarp ravelling (possible Holocene activity).

In the following sections we discuss and analyze these features in chronological order, relate them to rise deposition, and elucidate a concept of mass movement and seafloor instability.

Lower Slope Debris Wedge

The lower slope debris wedge represents the oldest mass movement accumulation in the SENESC and possibly the most chronologically persistent site of accumulation as well. Tilted, step-like reflection segments, and curved internal reflections suggest that the wedge is broken by, and perhaps rests on, listric faults that have apparently propagated near to the seafloor and caused mass movement of surficial sediment (Fig. 9-12). In places, the entire wedge seems to have rotated downdip, producing a 150-m-high scarp in the overlying draped strata where these strata contact the adjoining lower slope (Figs. 9-2B, 9-10). At the foot of the wedge, onlapping Neogene strata exhibit a peculiar downwarping (rollover): reverse dips in these layers suggest a vertical displacement of nearly 90 m (Fig.

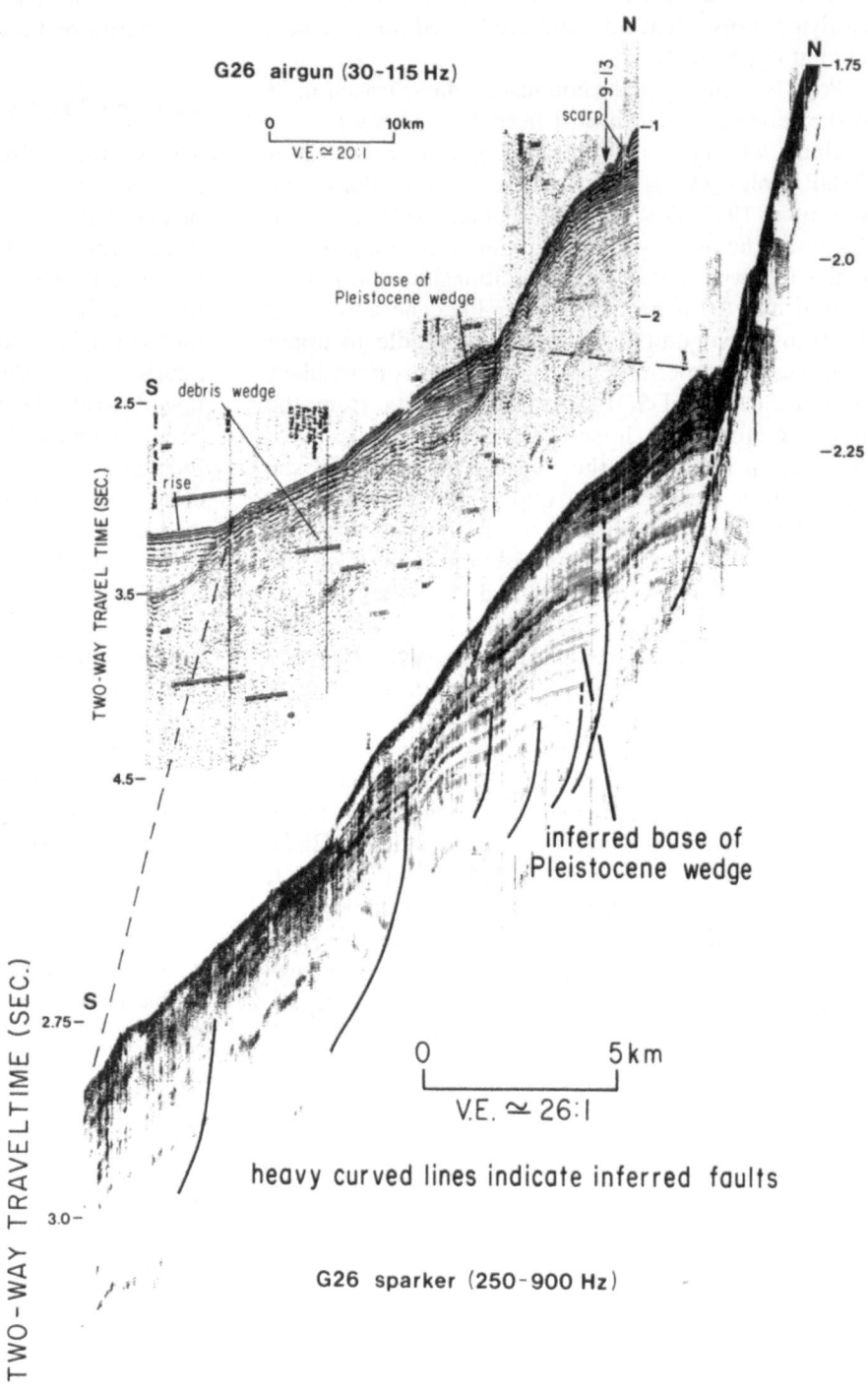

FIGURE 9-12.   Lower slope debris wedge shown by high-resolution seismic reflection record. See Figure 9-14 for location.

9-2B). These distortions may represent postdepositional rotation and seaward translation of the debris wedge, or they may represent current scour.

The upper level of the debris wedge is unconformably overlain by a thick Pleistocene interval that also has the form of a downdip-thinning, downlapping wedge, or mound (Figs. 9-10, 9-12). This Pleistocene package is a prominent feature of the SENESC; its

existence suggests that slide debris accumulated continually in small increments along the base of the slope during the Pleistocene. The Pleistocene accumulation on the debris wedge was subsequently subject to mass movement; a good part of it slid away, exposing the rough surface of the debris wedge that is presently fronted by the rubble-strewn rise (Fig. 9-12). Removal of Pleistocene strata appears to have involved spectac-

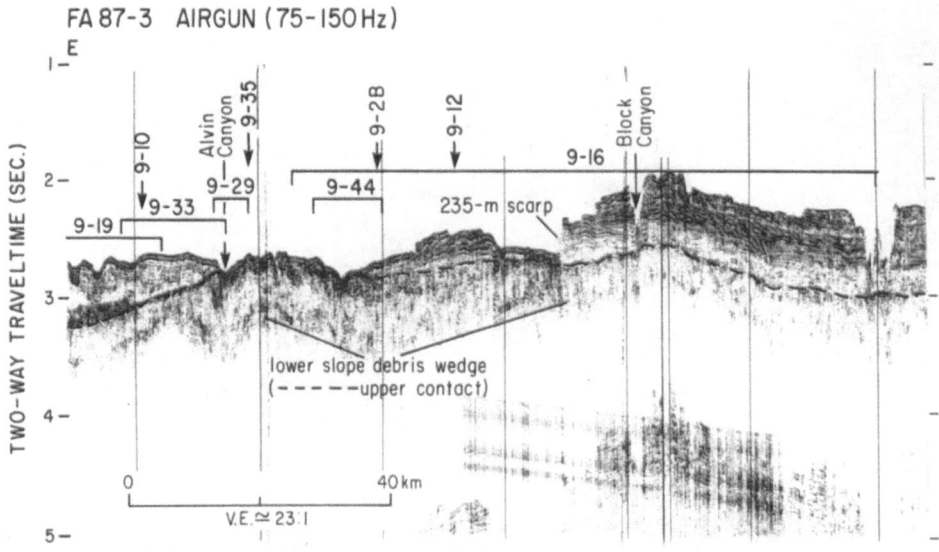

FIGURE 9-13.   Lower slope debris wedge shown along strike by high-resolution seismic reflection record. Here most of Pliocene and Pleistocene section has slid away (Block Canyon slide), leaving remnants and scarps. See Figure 9-14 for location.

ularly large movements, as indicated by a single scarp at least 235 m high (Fig. 9-13).

Major Slides

Major open-slope mass movement of the SENESC seems to have occurred in four more or less separate areas of concentrated and recurrent slide activity. Because sliding was probably the initial mode of mass movement in the head zone of each area, the areas of mass movement are conveniently termed slides. Each slide (or more precisely, group of concentrated and recurrent slides) is named for a major submarine canyon that heads into or passes through the main debris field on the rise. From west to east these are: the Block Canyon, the Atlantis Canyon, the Nantucket Canyon, and the Veatch Canyon slides (Fig. 9-14). The slide zones are distinguished along the lower slope and uppermost rise by broad, scarp-bounded excavations separated by largely intact stretches of the surficial unit. The major slides are difficult to identify otherwise, because their various debris deposits merge without clear stratigraphic distinction lower on the rise, below about the 2,300-m contour; finer designations are impractical because head zones on the slope are not well discriminated in presently available data.

Block Canyon Slide

The westernmost component of the SENESC is the Block Canyon slide (Fig. 9-14). The slide heads in a complex, largely unmapped zone of failure about 70 km long. SeaMARC I images from the eastern end of the zone indicate that slab slides originated on the slope at around 750–1,000 m water depth; GLORIA data suggest that bathymetrically higher levels of fail-

ure exist farther west, near long. 71°00′ W. Along much of the upper slope (~ 600 m water depth) in the Block Canyon region, large patches of the uppermost 10 m of the surficial unit have failed and more or less slid out of place along stratal surfaces (Fig. 9-15).

The main slide excavation on the upper rise–lower slope is bounded laterally by scarps as high as 235 m (Fig. 9-13). Intricate promontories of autochthonous strata riven by narrow cuts, and acoustically dark, isolated stacks occur in water depths as great as 2,300 m (Figs. 9-16, 9-17). The floor of the slide excavation on the rise is relatively smooth, but local culminations having rough, irregular surfaces, probably represent debris- or rubble-covered stumps of autochthonous section. Much of the eastern half of the Block Canyon slide excavation exposes the lower slope debris wedge. The axis of the slide excavation appears to be a broad, concave depression. Block Canyon itself is represented by a shallow depression at the west side of the broadly depressed slide field (Fig. 9-18). Proximal debris has a high uniform acoustic brightness, indicating textural homogeneity, and little in the way of flow pattern. At least two levels of debris are present, the bathymetrically lowest (youngest) of which appears to emanate from chute-like canyons. The debris field extends downslope for only about 100 km before being obstructed by the Hudson Swell (Fig. 9-14). Most of the debris is diverted toward Carstens Canyon and the east side of Hudson Canyon.

Atlantis Canyon Slide

The Atlantis Canyon slide is a 36-km-wide stretch of broadly concave excavations on the lower continental slope that is separated from the Block Canyon slide by a 7-km expanse of elevated seafloor marked by warped

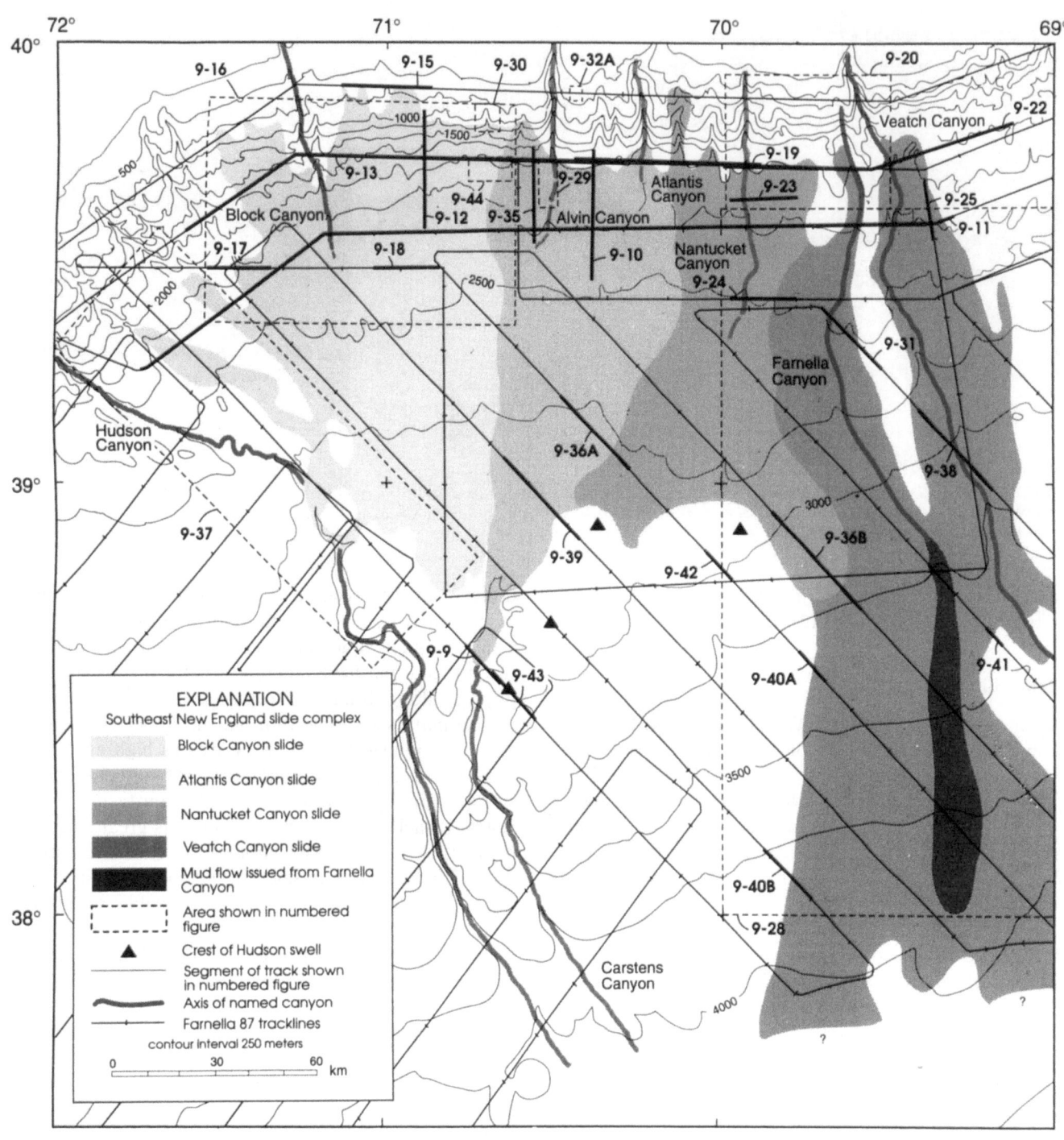

FIGURE 9-14.   Gross bathymetry and planimetric features of SENESC, and locations of areas or track lines of illustrated features. Slide scars and debris deposits, undifferentiated.

and broken remnants of the surficial unit, and capped with patches of debris (Fig. 9-16). The Atlantis Canyon slide is even less clearly distinguished from the Nantucket Canyon slide to the east; between Atlantis Canyon and Nantucket Canyon, the upper slope is deeply cut into broad, rough-topped, butte-like stacks (Fig. 9-19).

The western side of the Atlantis Canyon slide is dominated by open-slope mass movement floored by

the debris wedge; the eastern side of the slide is occupied by Atlantis and Alvin canyons. The two canyons contributed an undetermined amount of debris to the rise, most of it predating deposition of the surficial unit. Atlantis Canyon is the more important debris contributor; on the upper rise it appears to be a nearly flat-floored, rubble-filled trough bordered by steep scarps as high as 65 m (Fig. 9-19), and cut by an axial notch of only about 15-m relief. The floor of the

FA 87-3 (3.5 kHz)

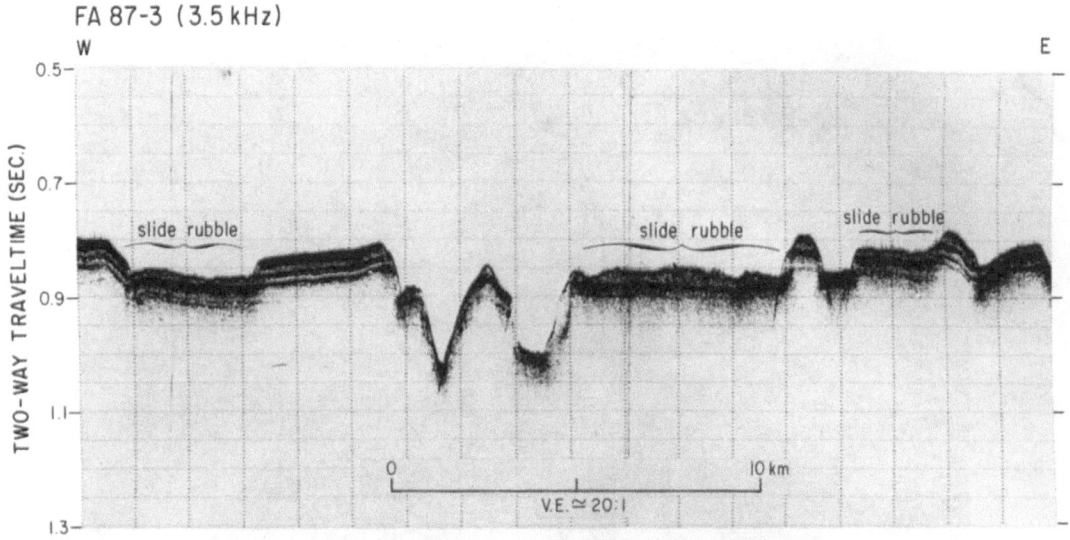

FIGURE 9-15.   Partial removal of Pleistocene surficial unit by sliding along upper continental slope within Block Canyon slide, as shown by 3.5-kHz record. Note contrast in form between cross sections of two canyons near middle of record; compare with Figure 9-21. See Figure 9-14 for location.

canyon or the wall of the axial notch is cut in stratified sediment that apparently predates the slide. Debris of the Atlantis Canyon slide extends downslope for about 100 km. Distal debris is diverted by the Hudson Swell into Carstens Canyon.

Nantucket Canyon Slide

The Nantucket Canyon slide is expressed by debris-floored troughs embayed deeply in the lower slope and upper rise, in contrast to the Block Canyon and Atlantis Canyon slides to the west, which feature widely extended scarps high on the slope. Although the head of the Nantucket Canyon slide seems to be thoroughly cut by deep, narrow debris chutes, as indicated by seismic reflection profiles (O'Leary, 1986b), GLORIA images show a segment of an arcuate scarp zone at about 900 m on the slope (Fig. 9-20).

On the upper rise, at about 2,200 m water depth, the slide excavation is about 35 km wide and is marked by swales that represent individual mass-movement tracks. These are fed by troughs that range in width from 1 to 13 km, separated by low, rough, ridges about 150 m in relief (Fig. 9-19). The troughs have steep, scarp-like walls about 70 m in relief cut into the adjacent ridge flanks, and they are floored and banked with debris (Fig. 9-21). Nantucket Canyon has been swept clean by erosion, unlike the neighboring troughs. There are at least two levels of irregularly mounded debris, the upper of which seems to consist largely of collapsed surficial strata. The surficial unit appears originally to have rested upon the older debris unit. The western part of the slide is morphologically similar to the Atlantis Canyon slide, and is separated from it by a broad, debris-mantled butte (Figs. 9-11, 9-19). The eastern boundary of the Nantucket Canyon slide is marked by a prominent west-tilted slab of relatively

intact strata about 5–8 km across, located in water as deep as 2,250 m (Fig. 9-22).

Within the slide field, Nantucket Canyon is partly filled by proximal debris (Fig. 9-23). At about 2,500 m water depth, Nantucket Canyon is a small notch having about 25-m relief within a trough about 5 km wide having a relief of 10–15 m (Fig. 9-24). The trough is lined and banked with rough debris that extends east, and downslope, forming a broad debrite terrace about 16 km wide and having a total relief of > 50 m. Emplacement of this debrite seems to have involved some lateral thrusting and deformation of the substrate, but the main component of movement was downslope flow, as indicated by faint, bright flow lines in GLORIA image.

Most of the debris of the Nantucket Canyon slide seems to have piled up against the Hudson Swell. Nantucket Canyon and the main slide trough merge into the Atlantis Canyon slide to the west. Westward diversion seems to involve younger and thinner debrites that overrode badly disrupted segments of the surficial unit. To the east, thicker lobes spread out and contact the Veatch Canyon slide, but there appears to have been little mixing. The GLORIA images and the 3.5-kHz records give the impression here of poorly defined, broad, downslope-trending lobes or "shingles" of debris, but a pattern of contacts cannot presently be worked out.

Veatch Canyon Slide

The Veatch Canyon slide consists of two relatively small component slides that head on the slope in well-defined embayments at about 1,800 m water depth. The larger slide is located just east of Veatch Canyon; the smaller is centered at about the same depth, about 10 km west of Veatch Canyon. The main, eastern

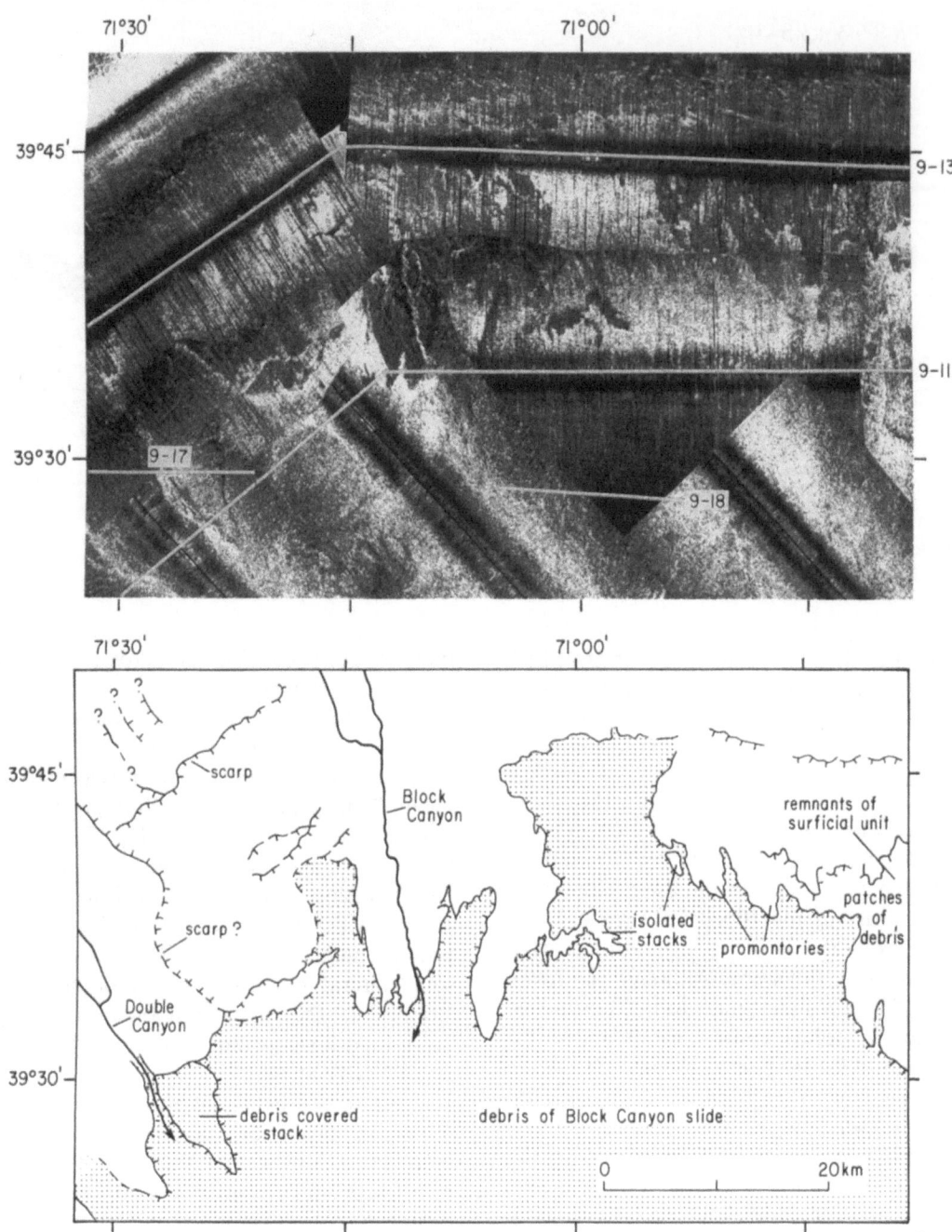

FIGURE 9-16.   GLORIA sidescan sonar mosaic showing details of Block Canyon slide along lower continental slope and upper rise. Very light (bright) areas are slide debris; dark, intricately bordered patches are autochthonous strata and (on east side) lower slope ramp. See Figure 9-14 for location.

excavation of the Veatch Canyon slide is about 20 km wide. It is mantled by local debris deposits of at least two generations. Seismic profile data indicate that the later sliding followed deposition of the 30- to 40-m-thick surficial unit (Fig. 9-25). Debris mounds as thick as 23 m coat remnants of the surficial unit, as well as older debris, and mantle trough walls. The substrate is exposed locally as rough, rounded seafloor protrusions, and the slide excavation is cut by smaller, flat-floored slide troughs. At about 2,000 m water depth, the slide

excavation merges with or cross-cuts Veatch Canyon. The eastern flank of Veatch Canyon is collapsed and the canyon is largely plugged by rubble; the western, intact wall of the canyon forms the western margin of the slide. The wall has a relief as great as 130 m (Fig. 9-26). In contrast, the eastern margin of the slide is a scarp about 40 m high, supported mostly by the surficial unit (Fig. 9-26).

The western slide issued from a graben 3–5 km wide and about 175 m in relief (Fig. 9-27). This was also the

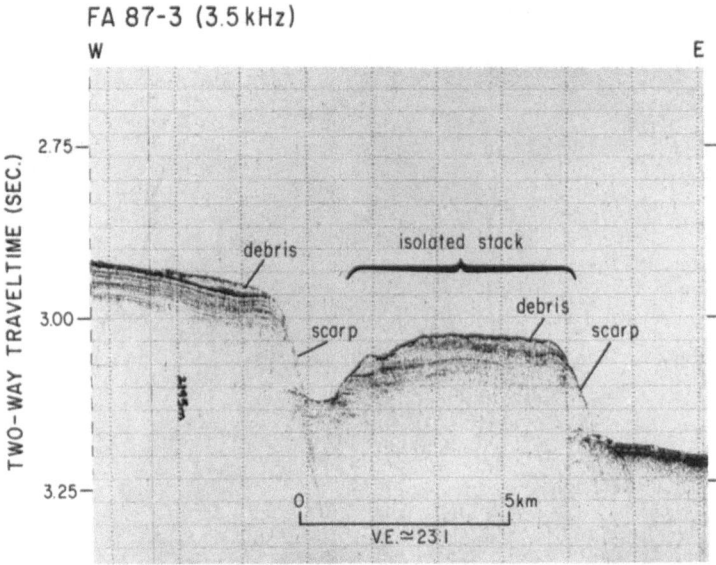

FIGURE 9-17.   Remnant, debris-coated stacks and scarps on west side of Block Canyon slide, as shown by 3.5-kHz record. See Figure 9-14 for location.

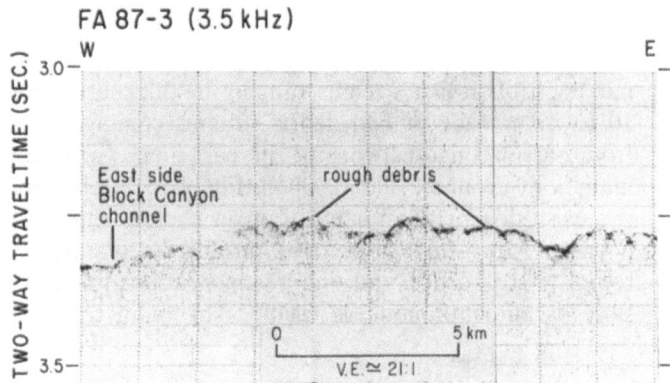

FIGURE 9-18.   Debris field of Block Canyon slide on upper rise. Flat-floored channel of Block Canyon and adjacent rough debris shown by 3.5-kHz record. See Figure 9-14 for location.

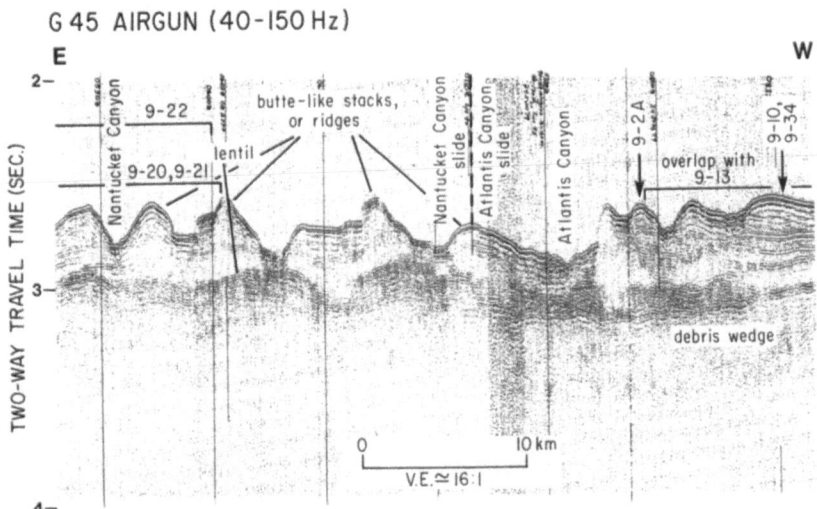

FIGURE 9-19.   Canyon incision along uppermost rise within Atlantis Canyon and Nantucket Canyon slides, as shown by high-resolution seismic reflection record. See Figure 9-14 for location.

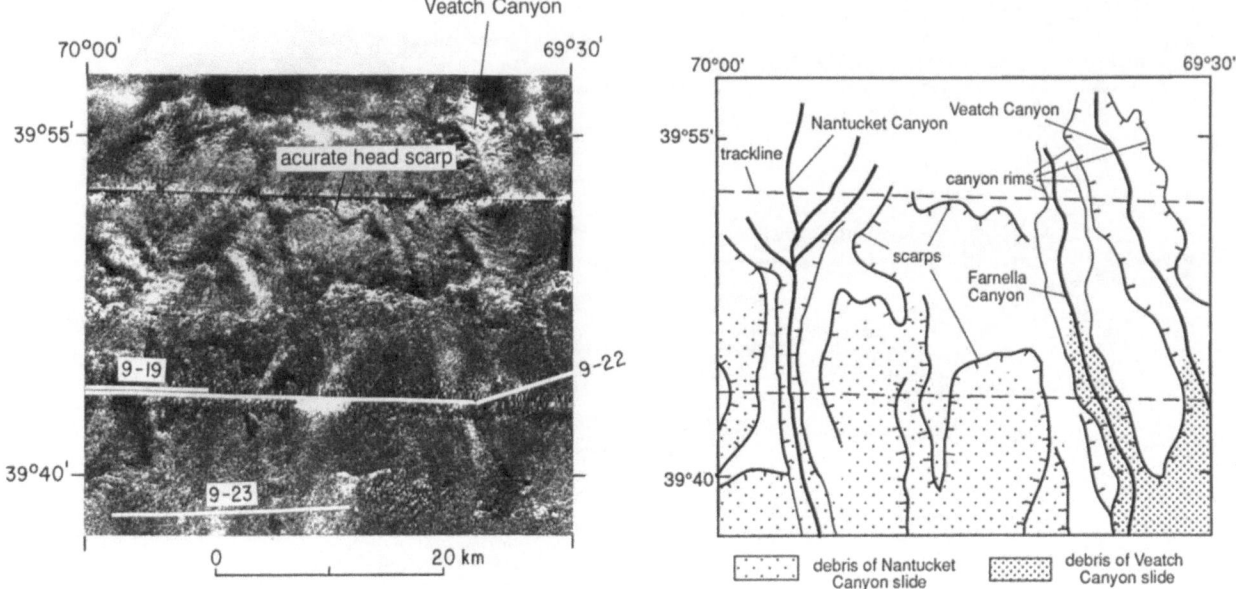

FIGURE 9-20.   GLORIA sidescan sonar mosaic showing arcuate head scarp on east side of Nantucket Canyon slide. Upper reach of Veatch Canyon is also clearly visible. See Figure 9-14 for location.

conduit for an unnamed canyon, hereafter called Farnella Canyon (Fig. 9-14), that extends for about 90 km across the lower slope and rise and controls the distribution of a large debris tongue farther downslope.

The eastern side of the Veatch Canyon slide debris field has pronounced downslope flow lineation and a sharp, smooth contact with the undisturbed seafloor (Fig. 9-28). Apparently debris flowed downslope in narrow, ribbon-like streams ranging in thickness from 10 to 25 m. One or two debris streams coming down from narrow small canyons to the east merge with the main body around 2,900–3,000 m (Fig. 9-28). Debris on the west side of the Veatch Canyon slide is thin and perhaps rocky (in echosounder profiles it features a rough "hard" echo), and underlying stratification, possibly parautochthonous, is visible. The main track of

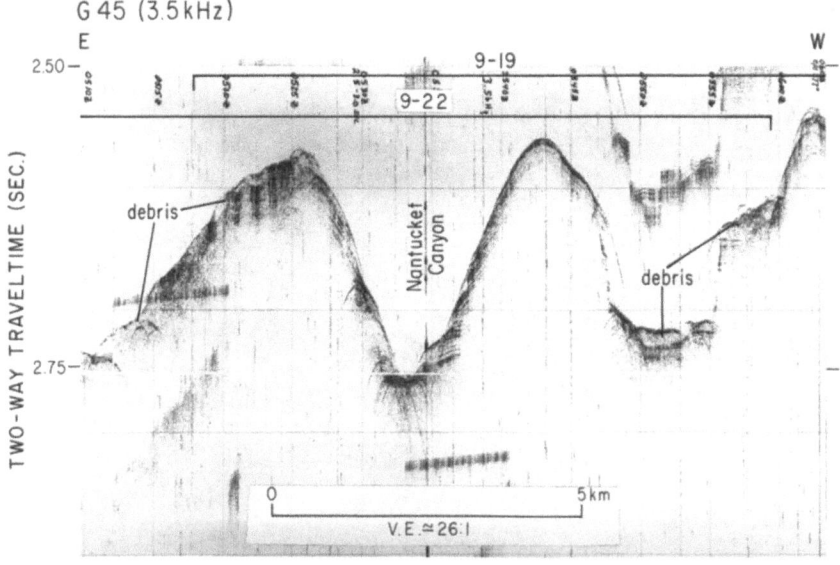

FIGURE 9-21.   Cross section of Nantucket Canyon and adjacent slide troughs, as shown by 3.5-kHz record. Compare smooth, swept appearance of canyon to rough, debris-mounded surfaces of adjacent troughs. Are these troughs canyons that were choked by subsequent mass movement? See Figure 9-19 for setting.

G45 AIRGUN (40-150Hz)

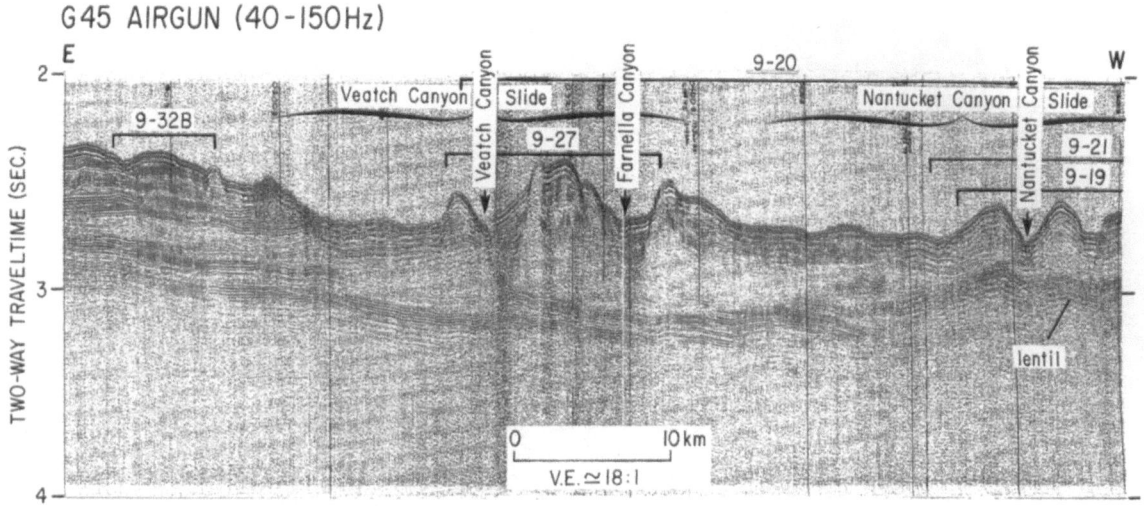

FIGURE 9-22. Profile of Veatch Canyon slide and east side of Nantucket Canyon slide along lower continental slope, as shown by high-resolution seismic reflection record. See Figure 9-14 for location.

the Veatch Canyon slide forms a broad, flattened debrite field, which is at least 200 km long; it extends beyond the survey border.

Other Sources of Debris

We infer that most of the debris in the SENESC was derived directly from open-slope sliding described in the preceding text. Two other sources of debris figure in the total makeup of the complex: flow from canyons, and in situ collapse of surficial strata with little or no transport. At present, the genetic relationships of these sources are unclear, and their significance to rise sedimentation is uncertain.

Flow from Canyons

Some canyons along the southeast New England continental slope contributed substantially to debris deposition on the upper rise, apart from, or in conjunction with, sliding. These canyons are of two types: (1) flat-floored chutes that widen and lose relief as they open on the rise; and (2) narrow, sinuous, round-bottomed, round-banked channels that extend across the rise for

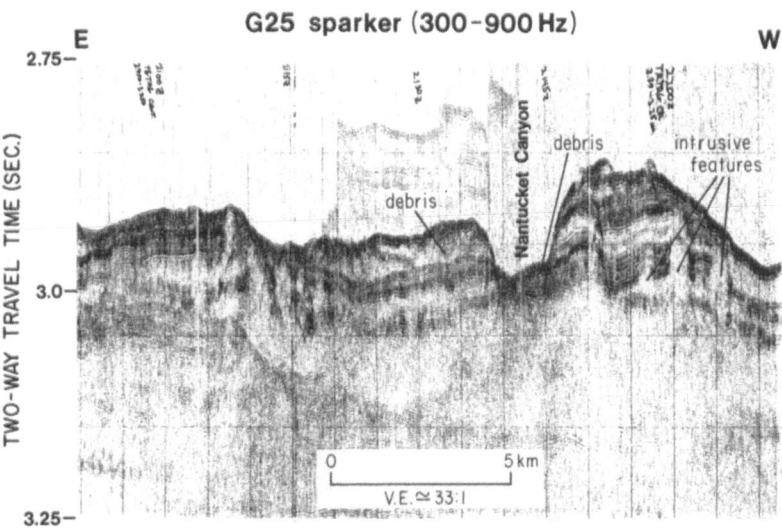

FIGURE 9-23. Cross section of Nantucket Canyon on uppermost rise, as shown by high-resolution seismic reflection record. Canyon is partly filled by debris, and slide excavation to east (left) is thickly banked with debris. Note intrusive forms indicative of intrastratal deformation. See Figure 9-14 for location.

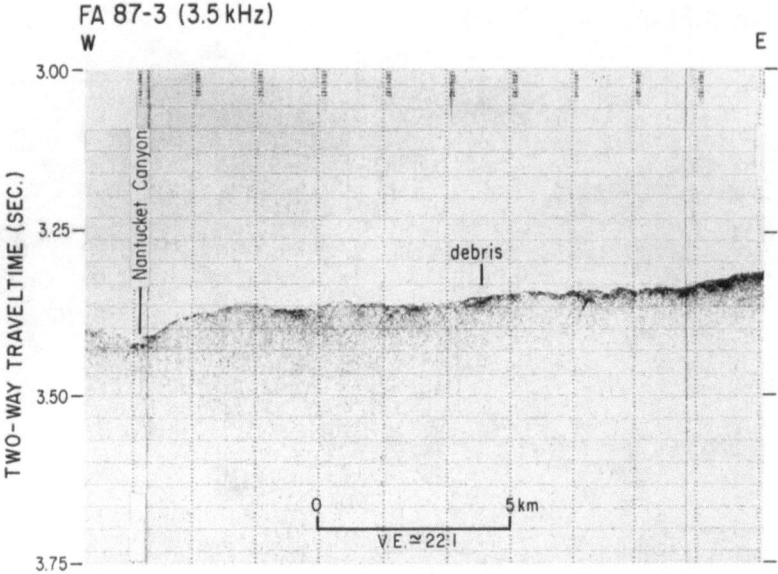

FIGURE 9-24.   Nantucket Canyon is minor notch at east side of shallow depression within debris field on upper rise, as shown by 3.5-kHz record. See Figure 9-14 for location.

tens of kilometers. The canyons lose their identities in the sediment they transferred from the middle and upper slope to the rise.

The canyons are of two generations: (1) those formed prior to deposition of the surficial unit; and (2) those that postdate sliding of the surficial unit. The relationsip of the earlier generation to the open-slope sliding is problematic. For example, it is not clear whether the Atlantis Canyon slide and the Nantucket Canyon slide,

which are canyon-dominated mass movements, were formed by coalescent canyons or whether they originated as open-slope slides low on the slope and were modified by subsequent canyon activity. The early generation of canyons was reactivated to a greater or lesser extent following deposition of the surficial unit. Reactivation may not have contributed much debris. Flow down Atlantis canyon, for example, seems to have simply swept clean a broad path through the debris

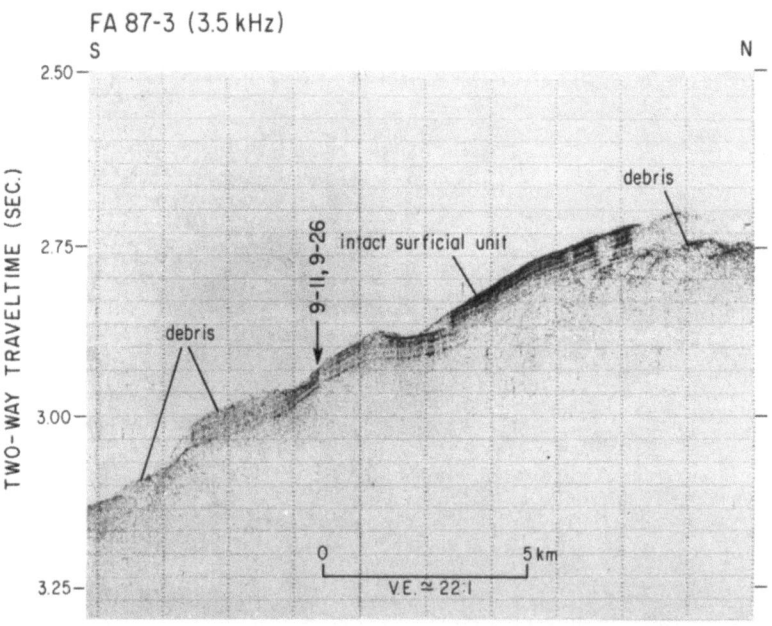

FIGURE 9-25.   Relationship of surficial unit to slide debris along east side of Veatch Canyon slide, as shown by 3.5-kHz record. See Figure 9-14 for location.

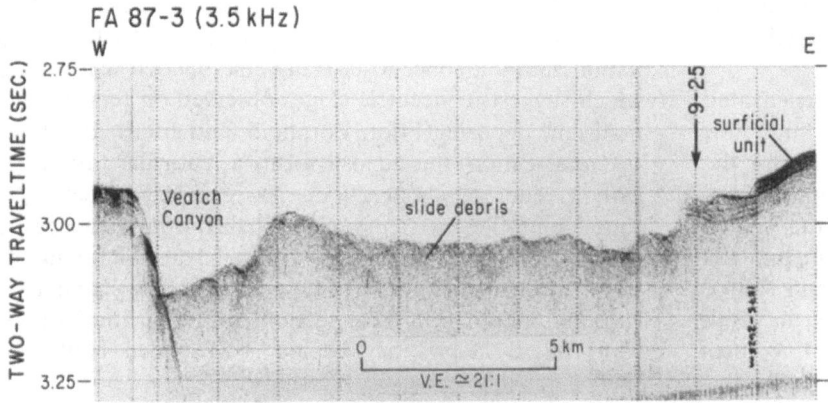

FIGURE 9-26.   Cross section of Veatch Canyon slide along uppermost rise, as shown by 3.5-kHz record. Depression on west side (left) is partly filled channel of Veatch Canyon. See Figure 9-11 for setting.

field. SeaMARC I images of the mouths of Alvin and Atlantis canyons clearly show flow patterns by what seem to be a system of longitudinal bar deposits aligned along the axis of each canyon (Fig. 9-29).

The later generation of canyons originated from point sources along the upper slope, apart from any sliding. The best defined of this generation, to our knowledge, is an unnamed canyon about 20 km west of Alvin Canyon (Fig. 9-14). The unnamed canyon originates at about 900-m water depth from two separate amphitheatric heads having radii of about 350 m (Fig. 9-30). These open into troughs that merge about 2 km farther downslope to form a single incision. At a distance of 2.5 km from the head(s) the canyon has a total relief of about 235 m and a floor width of about 300 m; its rim-to-rim width is nearly 2 km. Erosion has occurred at two stratigraphic levels, resulting in a rough-surfaced bench or lateral terrace about 200 m wide. The upper stratigraphic level, which is cut back to a broadly embayed rim, is about 75 m high. The west side includes the 30-m-thick surficial unit, which is missing on the bathymetrically lower east side (having previously slid away). The "inner gorge" has a relief of about 160 m; its rim (the terrace edge) is also broadly embayed (Fig. 9-30). Clearly, this canyon did not originate by retrogression of the scarp, because the scarp is

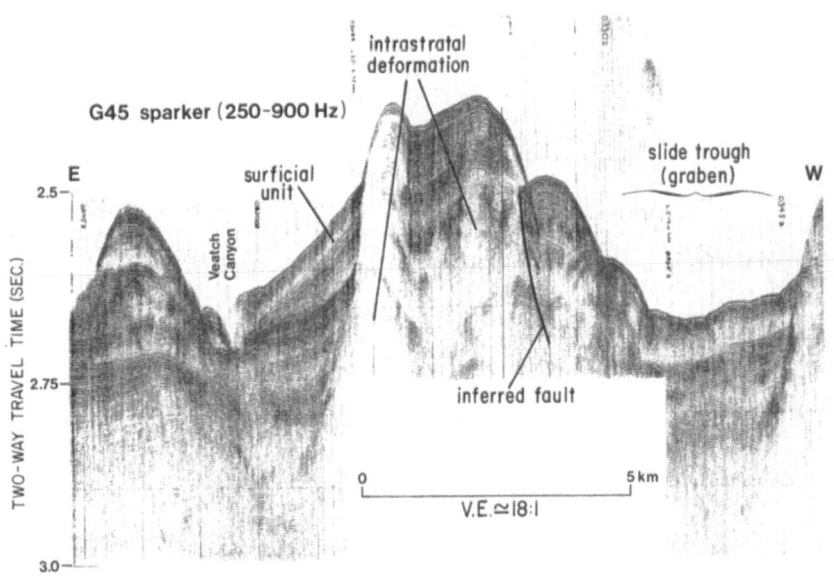

FIGURE 9-27.   Veatch Canyon and slide trough to west along lowermost continental slope, as shown by high-resolution seismic reflection record. Veatch Canyon appears to be cut into surficial unit, which seems to have filled an older, debris-choked trough. Trough to west is partly filled with debris representing different generations of mass movement. There is evidence here of complex intrastratal deformation. See Figure 9-22 for setting.

truncated by canyon walls created by down-cutting; canyon incision distinctly postdates and supercedes the lateral detachment that formed the scarps.

Canyons of the second generation, and reactivated older canyons, have produced debris- or mud-flow deposits that are among the youngest debrites in the SENESC. The most clearly defined example is the deposit issued from Farnella Canyon within the Veatch Canyon slide. The deposit is about 10 km wide and 100 km long (Fig. 9-28); it widens gradually from its source in Farnella Canyon and shows little variation in width for most of its length. Dark-banded lateral flow lines

and low distal backscatter suggest that it is made of largely fine-grained material. Its rounded distal termination shows a concentric, transverse banded texture, which suggests concentric ridges observed on reported mud-flow deposits (Prior, Bornhold, and Johns, 1984). In cross section, the deposit forms a lenticular profile with a relief of about 20 m. As recorded in echo-sounder profiles, it is indistinguishable from adjacent debris. Farnella Canyon appears to be a 90-km-long track for this debris- or mud-flow deposit. The channel is round-bottomed in cross section and has as much as 35-m relief. Debris has coated its rounded banks,

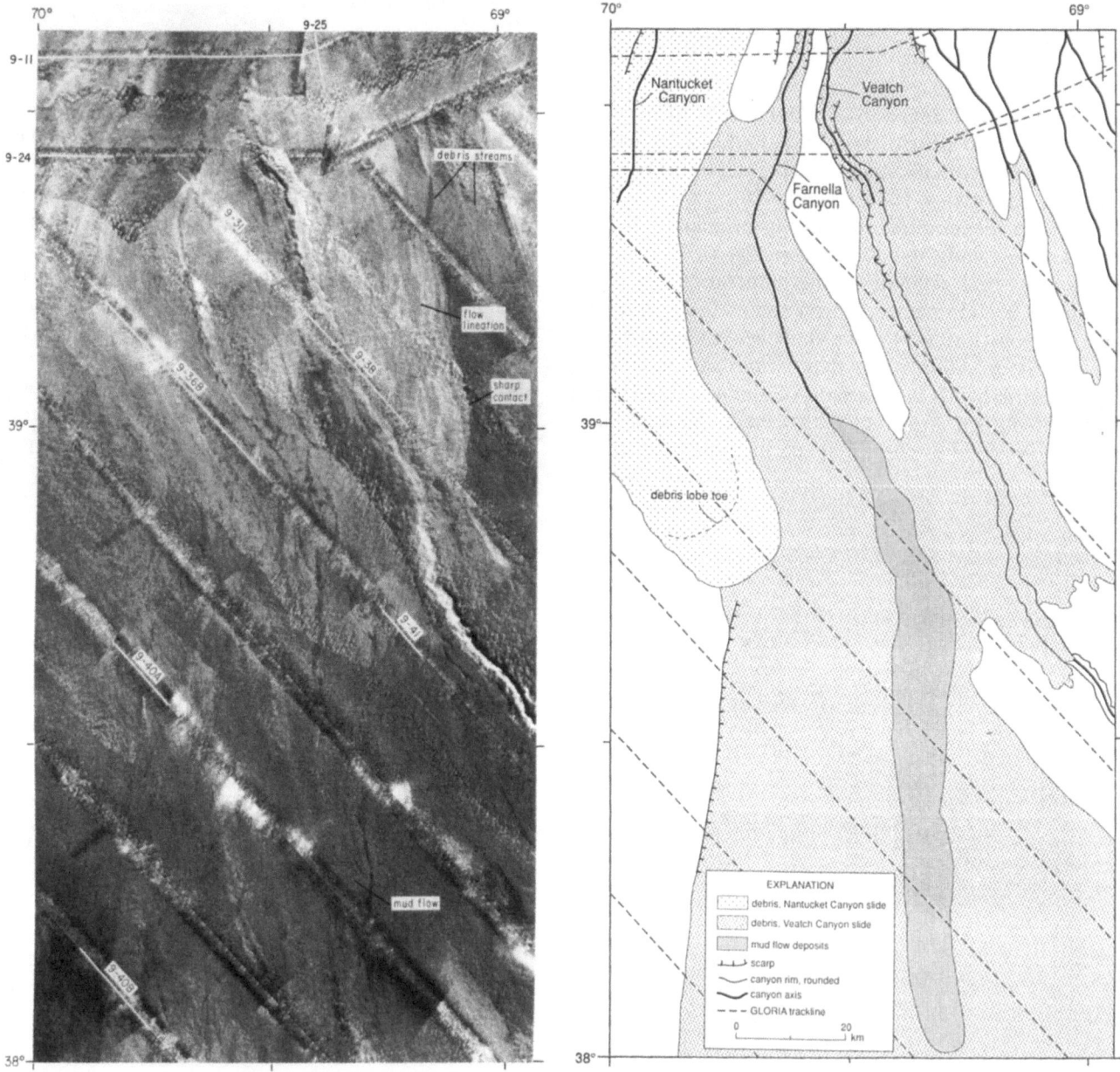

FIGURE 9-28.   GLORIA sidescan sonar mosaic showing features of Veatch Canyon slide and eastern part of Nantucket Canyon slide. See Figure 9-14 for location.

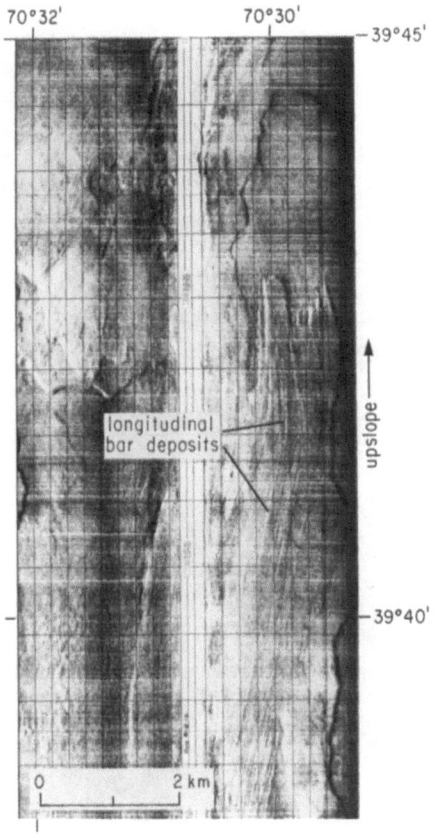

FIGURE 9-29.   SeaMARC I sidescan sonar image showing flow forms oriented along broad floor of Alvin Canyon, where it opens onto rise. See Figure 9-14 for location.

smoothly covering, and apparently grading into, the adjacent intact surficial unit without having formed levees (Fig. 9-31).

Nantucket Canyon seems to have issued a similar deposit 30 km long, 4 km wide, and about 18 m thick, which is occluded by the Hudson Swell. The distal surface is rough with transverse forms. At lat. 39°00′S, Nantucket Canyon forms a nearly flat, hard-bottomed depression about 10 m in relief and 7 km wide, beneath which layered substrate is visible in the 3.5-kHz records.

The chronology and mechanisms of canyon activity cannot be further elaborated here A point we wish to emphasize is that the continental slope has been a primary source of the debrites, whether transported by channelized flow or by open-slope sliding.

In Situ Collapse

In situ collapse of the surficial unit is associated with sliding in the SENESC. Removal of lateral support at the scarp face appears to have been sufficient to bring large tracts of the surficial unit to failure without much subsequent translation. SeaMARC I images of the upper slope show that in situ collapse of the surficial unit (chiefly fragmentation at a scale of meters) was

involved in the formation of polygonal, upper slope scarps (Fig. 9-32A), as well as in the initiation of upper slope canyon amphitheaters. In situ collapse is inferred from echosounder profiles by lateral transition from intact stratification to broken tracts having an uneven surface but no evidence of significant displacement. Along the east side of the Veatch Canyon slide, approximately the uppermost 10 m of section has collapsed, yet only ~ 4 m have actually been removed, presumably by sliding or flowage (Fig. 9-32B). The displaced material exists as a 4-m-thick carpet atop an approximately 13-m-thick sheet of debris within the Veatch Canyon slide excavation.

The widespread occurrence of in situ collapse of the surficial unit along the upper slope suggests endogenous factors, such as earthquake shock, creep, and disruption by intrastratal deformation. Collapse of the surficial layer on the lower slope may be partly accounted for also by imposition of layers of debris derived from farther upslope. An unknown, but possibly significant, component of slope-proximal debris may be in situ. This possibility has important implications for volumetric estimates of allochthonous debris on the rise.

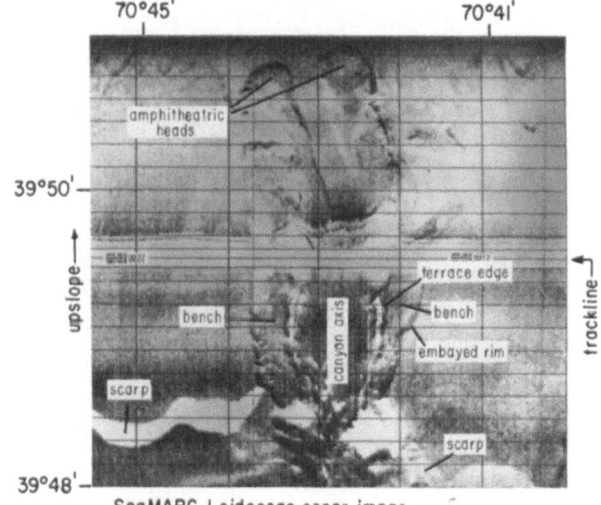

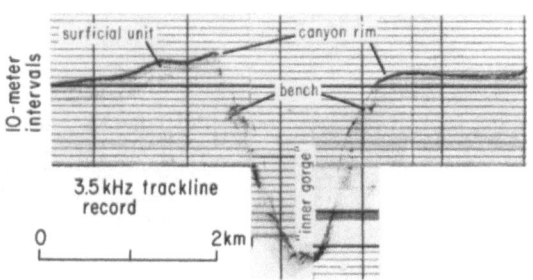

FIGURE 9-30.   SeaMARC I sidescan sonar image showing double-headed canyon that originates at ~ 900-m water depth on the continental slope and cuts through scarp (white band, acoustic shadow facing downslope) of Block Canyon slide. See Figure 9-14 for location.

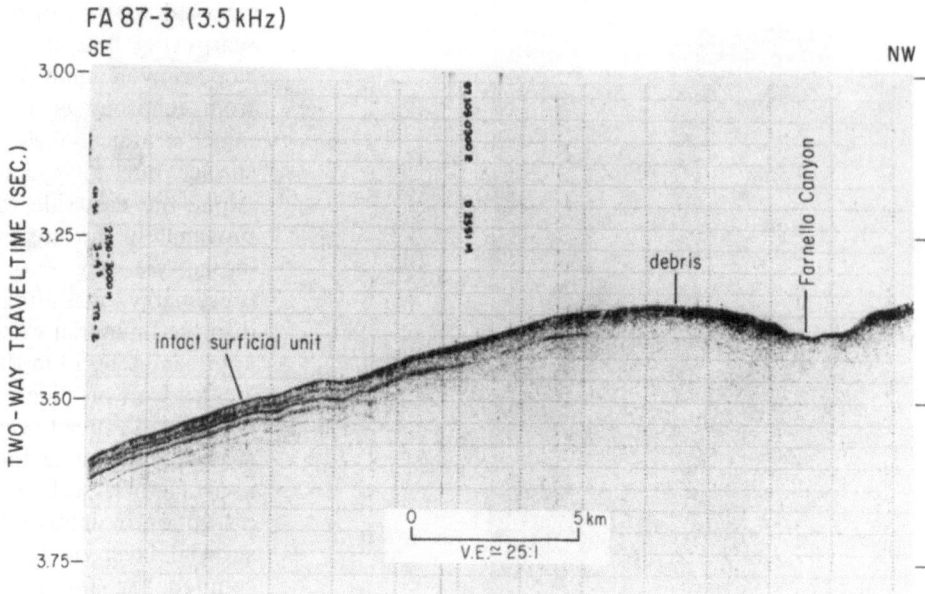

FIGURE 9-31.   Farnella Canyon at mid-rise depth, as shown by 3.5-kHz record. Note gradual transition in echo character with distance from channel, indicating a thinning cover or channel fallout. See Figure 9-14 for location.

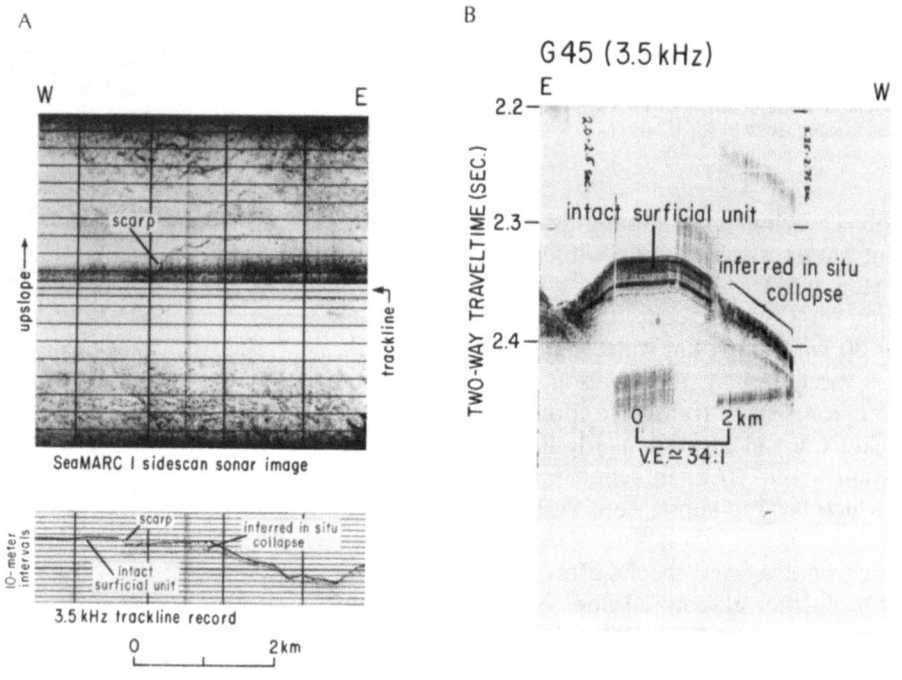

FIGURE 9-32.   Inferred in situ collapse of strata, as shown by sidescan sonar and 3.5-kHz records. A. SeaMARC I sidescan sonar image shows mottled texture indicative of disrupted surficial unit. In situ collapse associated with scarp is shown by 3.5-kHz record along image track line. See Figure 9-14 for location. B. Collapse of surficial unit at east margin of Veatch Canyon slide. Collapse of entire unit seems to have occurred with minimal translation. See Figure 9-22 for setting.

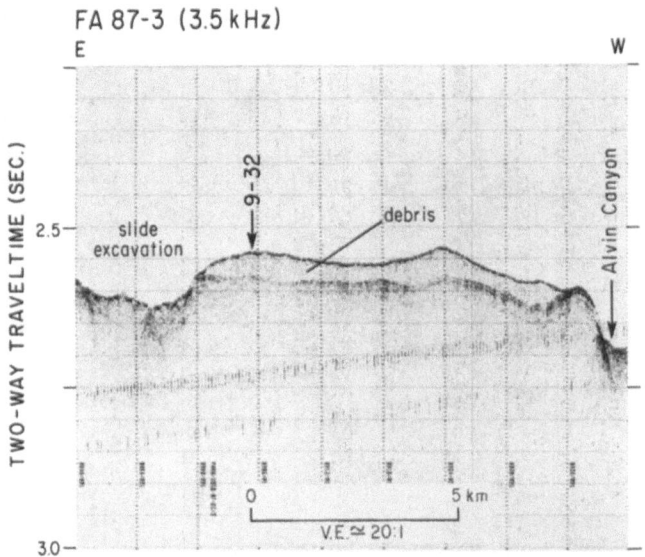

FIGURE 9-33.   Debris deposit banked along lower continental slope in Atlantis Canyon slide, as shown by 3.5-kHz record. Note rough slide excavation on left side. See Figure 9-13 for setting.

Debris Deposits

Debris of the slide complexes deposited on the continental rise is conveniently classified as proximal and distal. Proximal debrites have an upslope contact against the continental slope; distal debrites occur more than half the distance from the lower continental slope to the terminal contact.

Promixal Debrites

The topographically most conspicuous debrites in the SENESC cap the surficial unit on the rise adjacent to the slope. On strike, the slope-proximal debrites form a discontinuous, thickening–thinning unit having an uneven surface, which probably represents numerous penecontemporaneous deposits; individually, these deposits are several kilometers wide (Fig. 9-33). Downdip, the deposits present a distinctive, downslope-thinning, shingle-like profile, the peak of which is separated from the slope by a shallow, commonly flat-floored depression. The peak is actually a broad mound having a maximum thickness of about 35 m, which tapers downslope to a thickness of a few meters, or to pinchout. This change in relief of about 30 m occurs over a downslope distance of 5–10 km (Fig. 9-34). Distal margins are indeterminate; some of the debrites probably thin to layers several tens of centimeters thick over distances of kilometers; others probably have abrupt downslope terminations (Fig. 9-34). The deposits are texturally uniform, consisting of acoustically homogeneous, transparent, rough-surfaced material, which sharply contacts underlying and adjacent strata.

The mounded deposits evidently represent debris that arrived in a completely disaggregated state. Faint scoring on the lowermost slope is evidence of appreciable frictional contact with the seafloor, but not enough to have arrested any of the moving mass on the slope. Thin debris commonly conforms, like a drape, to buried forms on the rise, except that it thickens against the upslope sides of raised, tabular remnants of the surficial unit. This geometry indicates that the moving debris overrode tabular outcrops of subjacent strata, perhaps cutting away some of the upper layers in the process of emplacement (Fig. 9-35). The debris either slid off, or ravelled off, the edges of the low buttes and cuestas upon which it came to rest (Fig. 9-33), resulting in the typically lenticular margins. The mounded, drape-like character of the proximal debris, and its acoustic homogeneity, and the scouring of the adjacent lower slope, imply that debris moved onto the rise as a cohesionless mass having high momentum. The lack of stratification and the presence of convex margins suggest that the mass did not mix or become diluted with seawater, but that it gained frictional strength as it lost momentum.

The proximal debris probably came mainly from detached masses of the surficial unit and from local subsequent slides that occurred on the main slide plane along the lower slope. The mounded deposits appear to be the youngest debrites on the upper rise. However, because these debrites occupy a bathymetrically elevated position, the present data do not provide us with precise knowledge of their relative age; they may be older than bathymetrically lower trough-filling debris that debouched from canyons. Seismic profile data

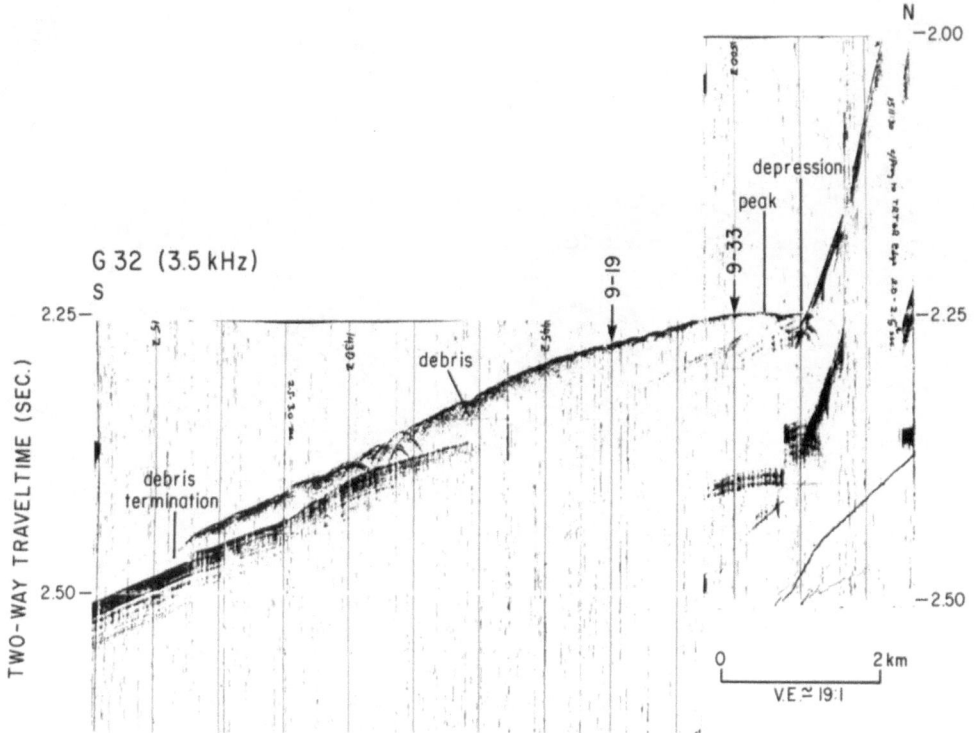

FIGURE 9-34.   Proximal debris mounded upon upper rise, as shown by 3.5-kHz record, clearly postdates surficial unit; compare with Figure 9-33. See Figure 9-10 for setting.

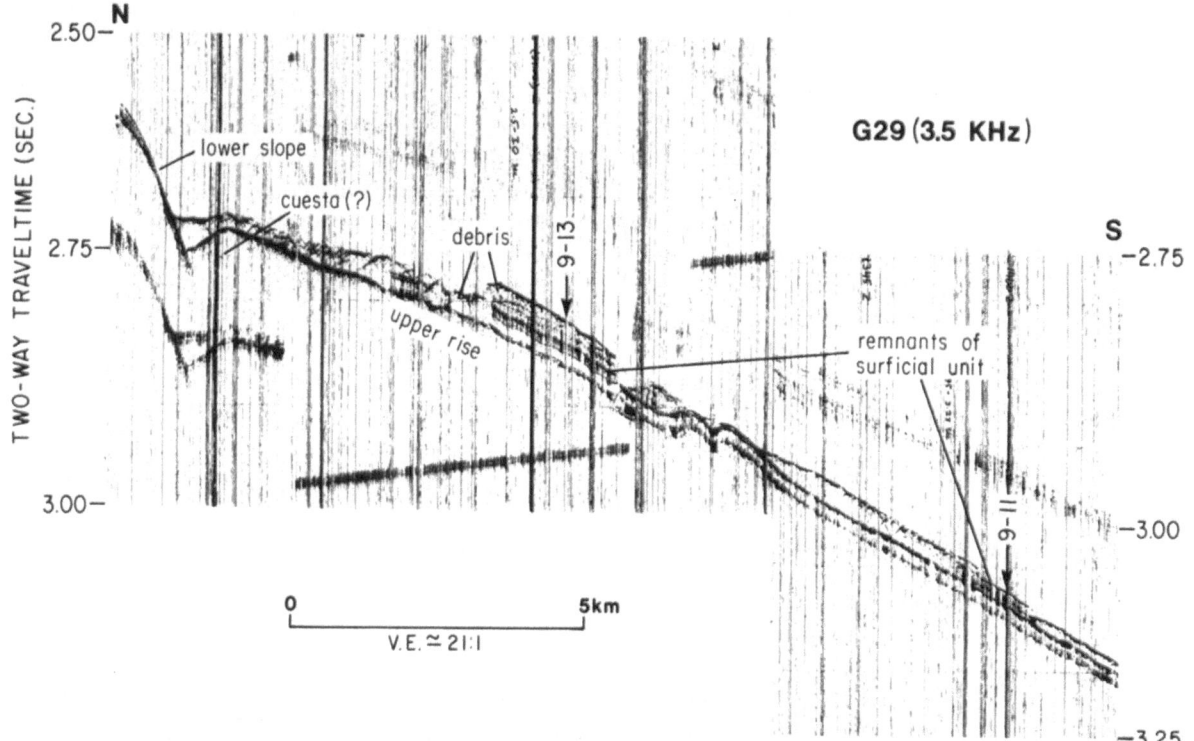

FIGURE 9-35.   Proximal debris, as shown by 3.5-kHz record, banked against lower slope and filling gap created by buried cuesta-like feature. Buried cuesta(?) may represent top of older debrite capped by surficial unit, which was subsequently subjected to mass movement and debris emplacement.

indicate that some of the mounded debris is stacked. Mounded proximal debrites have been observed elsewhere along the U.S. Atlantic upper rise (Robb and Hampson, 1984; Robb, 1986; Farre and Ryan, 1987).

Distal Deposits

Most of the debris within the SENESC is spread out on the rise as remarkably long-travelled, thin debrites, which range in thickness from about 40–50 m to a few meters. Some debrites thin to a feather edge, but others terminate as blunt, lobate fronts, a few meters in relief. The contact profile and the acoustic roughness of each debrite vary little over large stretches of seafloor, but can be distinctly different in adjacent deposits (Fig. 9-36A). This apparently indicates variations in composition or texture expressed at the seafloor. All the debrites are acoustically transparent, so little can be said about their internal makeup. They are typically conformable with the substrate, but the thicker deposits have clearly gouged out some substrate and have ripped up sections several meters thick across wide areas of seafloor, overtopping the autochthonous remnants with debris. It is also clear that some debrites have partly overridden older debrites (Fig. 9-36B), but evidence for this consists mainly of contrasts in surface form; there is little discrimination of buried contacts. However, in many places along the upper rise, the 3.5-kHz profiles show two or three generations of debris separated by laminated strata, which may represent hemipelagic deposition or turbidites. Emplacement of debris has locally ripped out the laminated section.

In general, GLORIA images show the debrites as acoustically bright, indicating a texturally rough surface. Some of the debrites show pronounced flow lines of contrasting tonality extending down the rise for tens of kilometers. Others show a more uniform streaking or wispy texture, aligned along the direction of flow. Some of the debrites blend into intact seafloor; others show a distinct and intricate border. These variations

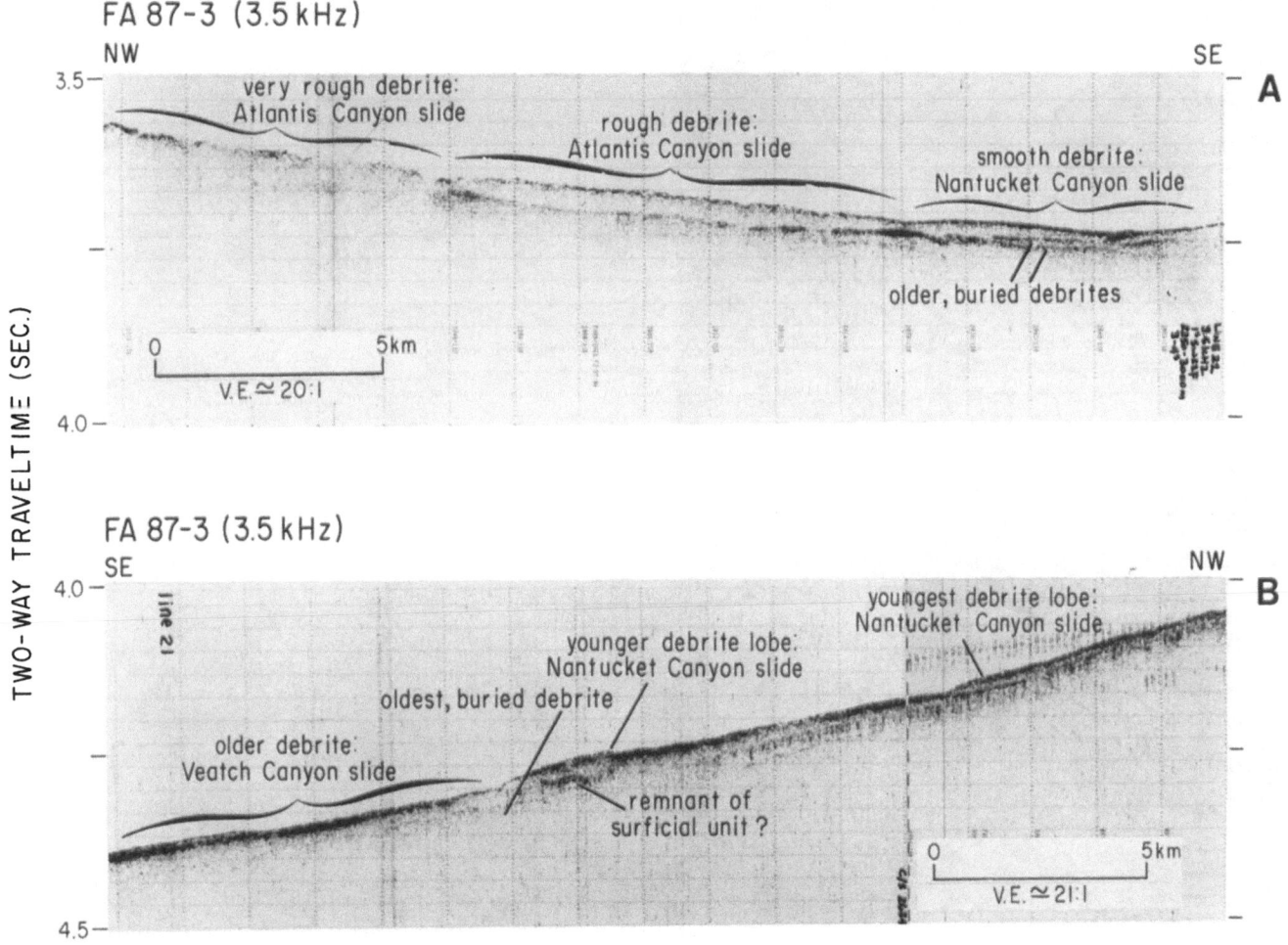

FIGURE 9-36.   Distal debrites, as shown by 3.5-kHz records. A. Possibly two distinct debrites of Atlantis Canyon slide and one of Nantucket Canyon slide. B. Two lobes of Nantucket Canyon slide, lower of which downlaps debris of Veatch Canyon slide. See Figure 9-14 for location.

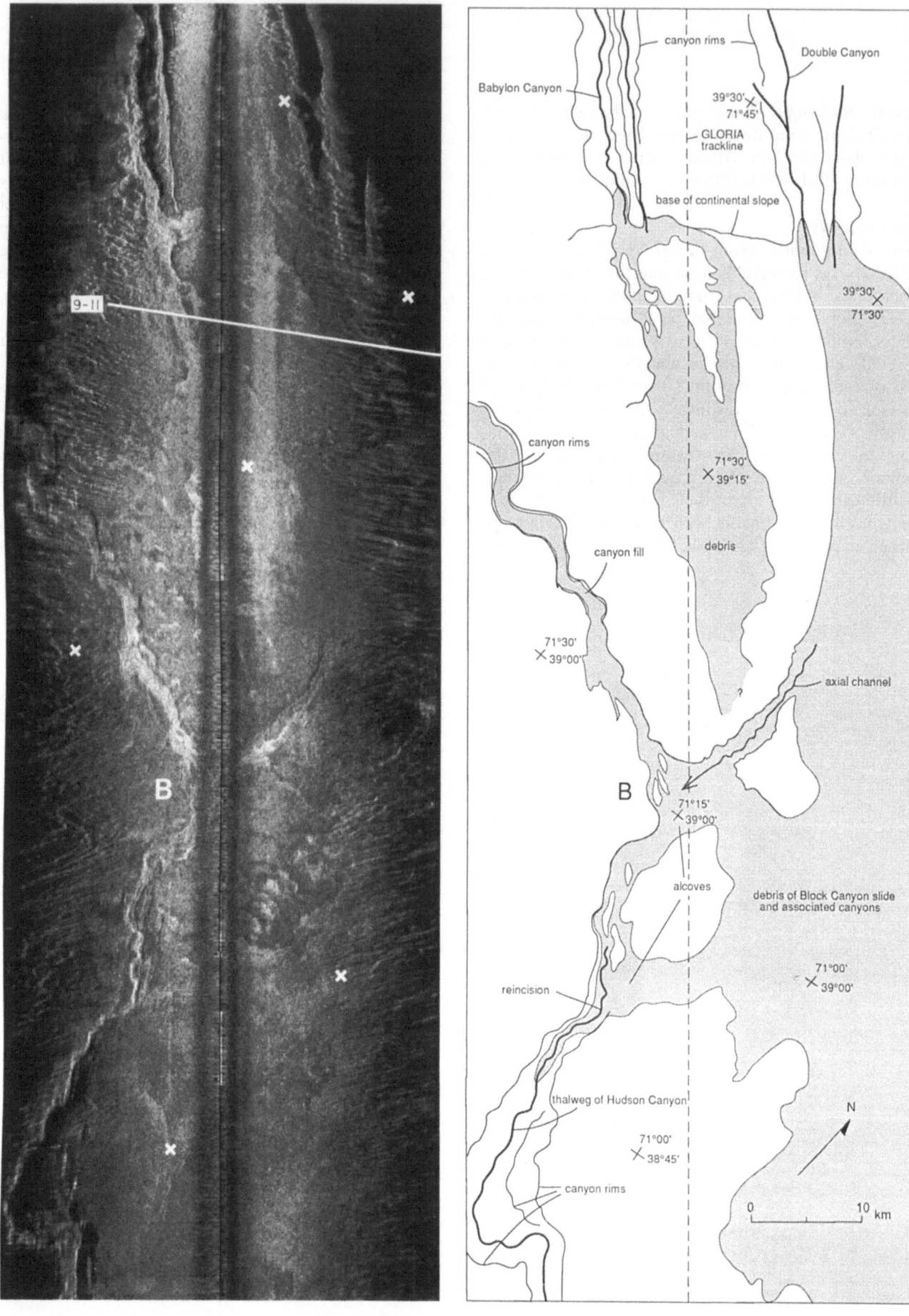

FIGURE 9-37. GLORIA sidescan sonar image showing Hudson Canyon on upper rise. Debris, especially that filling canyon, is light (bright). Debris has entered from east, having originated mainly from chute-like canyons on slope. Note braided pattern, indicative of reincision, in Hudson Canyon below point B.

probably all reflect differences in material properties accentuated during transport.

Relationship of Debris to Older Canyons and to the Swells

Two interesting aspects of the debrites, which bear on their material properties and behavior during emplacement, are their relationships to older canyons and to the swells. One of the most surprising results of this study is the pronounced lack of interaction between early canyon erosion and later mass movement on the rise. Only three canyons, Hudson, Carstens, and Veatch, extend across the southeast New England continental rise (Fig. 9-14). They clearly predate the major phase of mass movement, because each of them is partly plugged by debris.

Debris entered Hudson Canyon from alcoves along its east side in the depth range of 2,300–2,600 m (Fig. 9-37). Based on GLORIA and SEABEAM bathymetry (A. Shor, unpublished data), the debris fill seems to extend as far upchannel as the 2,250-m isobath and as far down channel as the 2,900-m isobath, a channel distance of approximately 50 km. Relief of the open channel above and below those points is at least 100 m, whereas relief of the filled section is as low as 40 m. Most of the debris that fills Hudson Canyon apparently was delivered as a mud flow that originated in Double Canyon (140-m relief) located between Block and Babylon canyons (Fig. 9-1). The widening–narrowing debris path is marked by an axial channel that trends toward a large alcove off Hudson Canyon at lat. 39°00′N (Fig. 9-37). At and below this point of entry, the debris plug in Hudson Canyon appears to have been slightly incised by downchannel flow, which probably also entered Hudson Canyon via the axial mud-flow channel to the east (Fig. 9-37). The feeder channel from the east exhibits a concave section in which debris appears to blur into the adjacent surficial strata. Seismic profiles indicate that Carstens Canyon does not extend

upslope north of lat. 39°00′N, yet within 25 km of its source on the rise, it has a relief as great as 130 m (Fig. 9-9). Debris at least 25 m thick occupies the head and extends for at least 30 km farther down the channel. The debris that fills Veatch Canyon is part of a mass that originally flowed directly downslope across the canyon (Fig. 9-28). Debris extends along the channel of Veatch Canyon for at least 70 km, and attains an estimated thickness of at least 25 m (Fig. 9-38). The maximum thickness of debris in Veatch Canyon is located about 11 km downchannel from the GLORIA line crossing (Fig. 9-28) and is unknown. The termination of the debris along each channel is uncertain; debris probably merges smoothly with the canyon floor. Estimates of fill thickness in the canyons are poor because of poor bathymetric control on the deep channels and because the fill obscures the acoustic floors of the canyons at the line crossings.

Despite the evidence of great mobility, the canyon-filling debris did not sluice down the canyons in the putative manner of a turbidity current. Rather, the debris formed a fixed mass that has remained intact since its emplacement (with the exception of the inferred channel-scoring of the Hudson Canyon fill). The collective features suggest that the debris was transported as a homogeneous, slow-moving mass that was not mixing significantly with seawater, but was close to solidifying upon entering the canyons. GLORIA brightness and acoustic shadow features indicate that debris flow upchannel from the point of debris influx was as readily achieved as the flow downchannel. No samples of the channel fills have been obtained. Perhaps a useful analog of the fills is provided by fills in subaerially exposed channels of the same scale as those of the SENESC, reported by Lajoie and St-Onge (1985).

Contact relationships of the debrites and the swells likewise suggest the grinding halt of a slow-moving mass. Figure 9-39 shows a 35- to 40-m-thick debris lobe

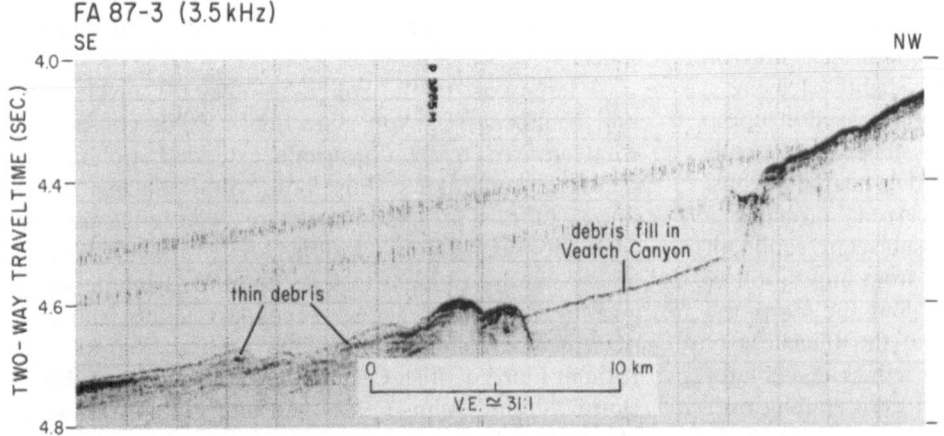

FIGURE 9-38.   Debris nearly filling Veatch Canyon, as shown by 3.5-kHz record. See Figure 9-14 for location.

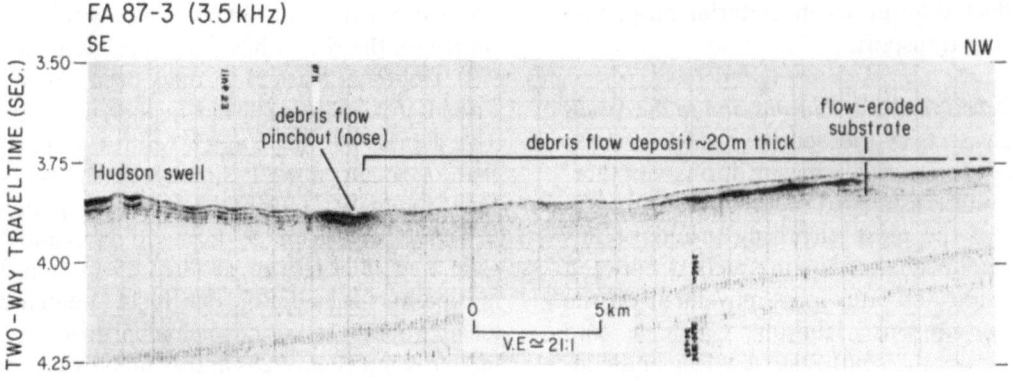

FIGURE 9-39.   Debris of Atlantis Canyon slide has terminated against Hudson Swell, as shown by 3.5-kHz record. Debris seems also to have partly ripped up surficial unit. See Figure 9-14 for location.

of the Atlantis Canyon slide pinching out on the landward side of the Hudson Swell, about 25 m below the swell crest. The debris seems to have flowed as a 10- to 15-m-thick tongue upon the seafloor, but a greater thickness seems to have engendered plowing/mixing of the substrate. Debris flow apparently stopped or was deflected at the base of the 0.5° slope of the swell (Fig. 9-39). Debris of the Nantucket Canyon slide flowed over the crest of the Hudson Swell and downlapped the seaward side.

Parautochthonous Masses

Some of the debris in the SENESC appears to be derived from strata only slightly removed from their site of deposition and with some of the layer-parallel structure preserved. This parautochthonous debris is significant, because it raises the possibility that failure of the rise section may be an important factor in the production of the debrites. Inferred parautochthonous debris is well shown along the rectilinear, north–south oriented western margin of the Veatch Canyon slide (Fig. 9-28). Figure 9-40A shows the margin as a 45-m-high scarp, apparently formed by the collapse and slight downslope shift of the uppermost 45 m of strata. The collapsed section rests on a continuous horizon upon which it has failed and slid slightly, leaving a small gap at the scarp base. The intact section updip from the scarp is capped by 15 m of acoustically homogeneous sediment, perhaps a debrite. The acoustically chaotic mass below the scarp is probably a mixture of emplaced debris and subjacent sediment that has undergone in situ collapse from imposition of the load. With distance downslope along the scarp, the gap and scarp relief diminishes and the thickness of debris increases, ultimately attaining a thickness greater than the adjacent intact strata, and even spilling over onto it (Fig. 9-40B).

The only clearly recorded contact along the eastern border of the Veatch Canyon debris field is shown in

Figure 9-41, where a rough-surfaced deposit terminates as a convex bulge having a relief of about 10 m, against a well-stratified, intact section. The difference in relief between the west and upslope side of the debris field and the east and downslope side, from negative to positive relief, indicates that displacement, however minor, has led to a regionally flatter slope. Failure of this type is well illustrated by example from the continental shelf off northern California (Field and Hall, 1982). Field and Hall make a clear case for a cause by earthquake shock.

The cause of in situ collapse on the rise is unclear. The collapse of large tracts of surficial strata on the rise may be due to earthquakes. It may also be caused by imposition of debris flows (Hutchinson and Bhandari, 1971; Hutchinson, Prior, and Stephens, 1974). The presence of intact substrate below a stratigraphic horizon, and a rectilinear contact along the west side of the Veatch Canyon slide, implies that some material property of the uppermost 40 m of section is involved. The style of failure suggests quick-clay collapse reminiscent of the Turnagain Heights failure (Updike et al., 1988), or the failures described by Field and Hall (1982) and attributed to earthquake shock.

A large part of the confluent Veatch Canyon slide and Nantucket Canyon slide debris fields consists of what appears to be downslope extended, ribbon-like debris or mud flows (Fig. 9-28). Slow moving "mud flows" of this presumed type are reported from the Mississippi delta front (Watkins and Kraft, 1978). Such mud flows are thought to load the underlying deposits and, in some cases, induce secondary failure and displacement. A similar flow lobe on the floor of the Kitimat Fjord, British Columbia, was reported by Prior, Bornhold, and Johns (1984) and interpreted as having obliterated the seafloor and having incorporated the uppermost 4–5 m of strata within the allochthonous mass. An important conclusion of Prior, Bornhold, and

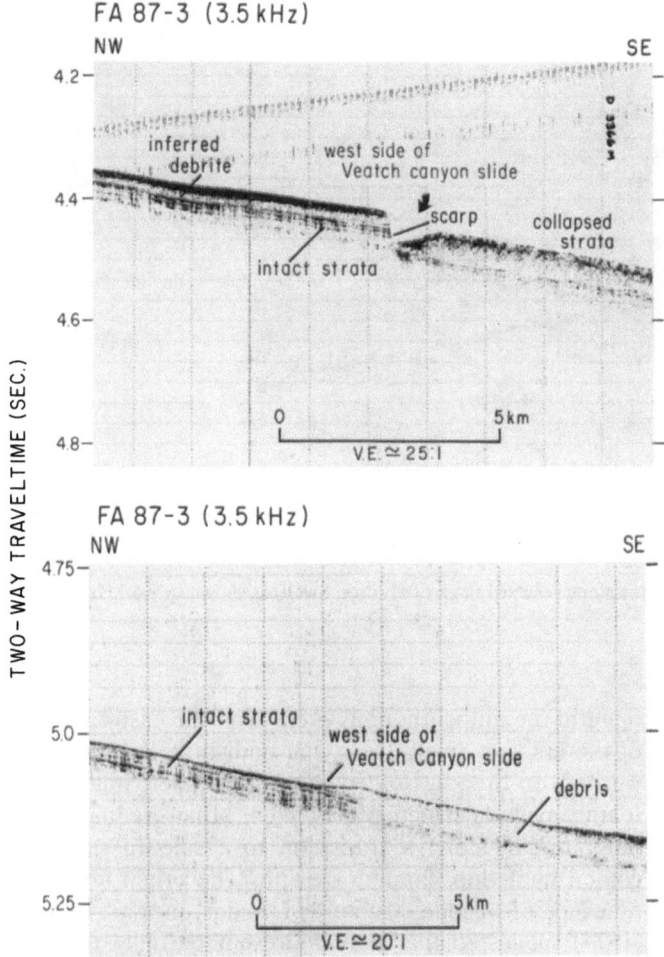

FIGURE 9-40.   Distal west side of Veatch Canyon slide, as shown by 3.5-kHz record. A. Collapse of surficial unit (perhaps due to superposition of debris about 20 m thick) has led to creation of parautochthonous debris. B. Lower on rise, debris along this contact overtops truncated surficial unit. See Figure 9-28 for location.

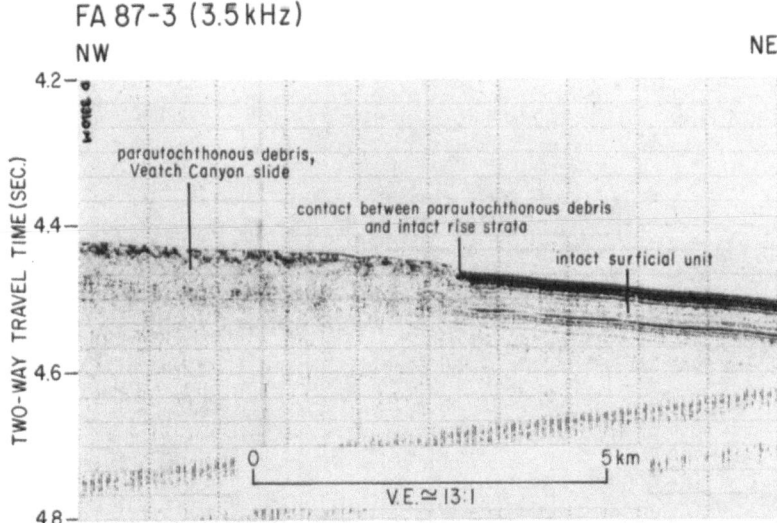

FIGURE 9-41.   Distal east side of Veatch Canyon slide, as shown by 3.5-kHz record. Raised convex surface, patchy parallel subsurface reflections, acoustic focussing, and general acoustic wipeout suggest downslope compression within a parautochthonous mass. See Figure 9-14 for location.

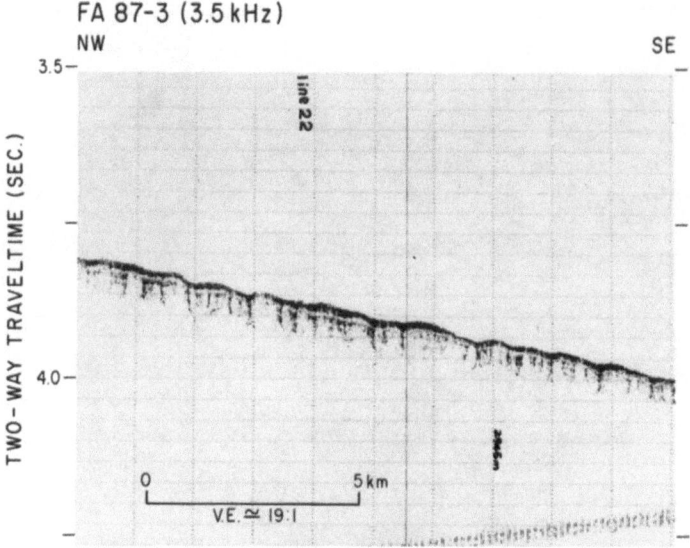

FIGURE 9-42. Pagoda structures along seaward flank of Hudson Swell, as shown by 3.5-kHz record. See Figure 9-14 for location.

Johns (1984), based on samples, is that the high degree of basal erosion, remolding, and mixing, which attends mudflow, produces grain-size distributions that may not be diagnostic of downslope transport. Thus, Prior, Bornhold, and Johns concluded that ancient erosional debris-flow deposits may not be distinguished simply on the basis of textural characteristics, but by structural features, such as flow banding, shearing, and lack of bedding.

A final piece of evidence bearing on parautochthonous debris is the presence of pagoda structures devel-

oped in the autochthonous section on the eastern flank of the Hudson Swell. Here, the undisturbed seafloor is warped or broadly folded, and the underlying strata feature pagoda structures (Fig. 9-42) that gradually die out laterally along a transition to a smooth seafloor coated by a thin, lapping debrite. The origin and significance of pagoda structures remains unclear. Emery (1974) suggested that they represent partially thawed clathrate (see also Chapter 5); Flood (1980) suggested that they represent acoustic focussing phenomena associated with sediment-filled troughs. The possible pres-

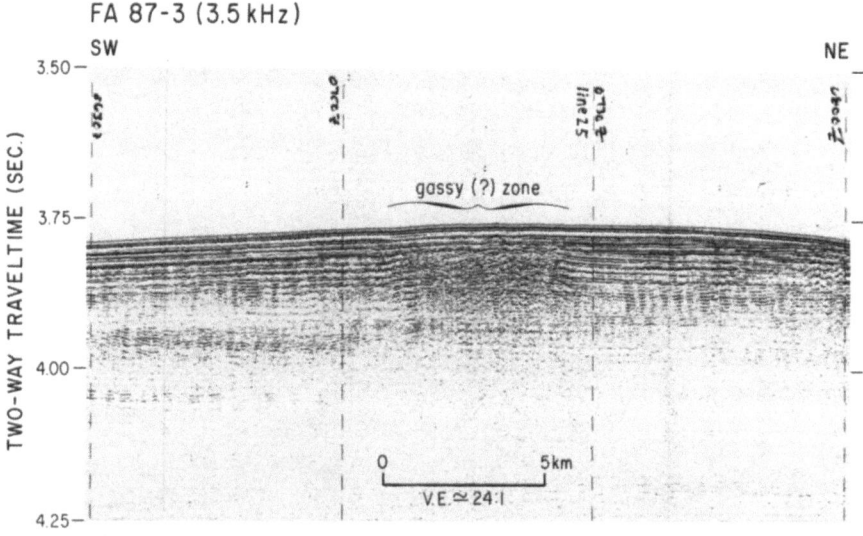

FIGURE 9-43. Fuzzy "adumbration" or disruptive shadowing of upwarped strata beneath crest of Hudson Swell, as shown by high-resolution seismic reflection record; this characteristic suggests gas, perhaps derived from upper Pliocene–lower Pleistocene interval, and possibly of clathrate origin. See Figure 9-14 for location.

ence of partly decomposed clathrate along the flanks of the swells (Tucholke, Bryan, and Ewing, 1977) is interesting, as it could provide appropriate destabilizing conditions required for stratally confined failure. We cannot determine whether or not a gassy substrate is a factor here, though seismic profile data suggest that gassy substrate does underlie the Hudson Swell (Fig. 9-43). Possible parautochthonous sections are also present east of Carstens Canyon below 3,000 m and east of the Munson–Nygren slide.

Subsequent Slides

Subsequent mass movement features in the SENESC include debris tongues or rubble trails that extend from the bases of the larger scarps, and small slides located along the surface of the lower slope ramp in 2,000 m of water, where it is largely stripped of Pleistocene strata between the Atlantis and Block Canyon slides. The small slides are likely the most recent in the SENESC. One of the best examples of these young slides has the shape of an inverted teardrop, similar to the "disintegrating soil slips" reported by Kesseli (1943) from California, and similar to features described by Prior and Coleman (1978) from the Mississippi delta. This young slide is 800 m wide and 4,100 m long; the arcuate head scarp has 10 m or less relief; the toe forms a thin, tapered tail of rubble (Fig. 9-44). A gravity corer dropped in the center of the slide returned about 3 cm of hard gray Pleistocene clay matrix set with small, angular fragments of dark gray, nearly black, Santonian siltstone, white Eocene marl, and light gray Pleistocene clay. (P. C. Valentine, private communication, 1982). The presence of small slides involving reactivated material on the 1.5° lower slope ramp suggests to us that the lower slope debris wedge was unstable at least as recently as late Pleistocene.

MUNSON–NYGREN SLIDE COMPLEX

The Munson–Nygren slide and the Retriever slide were identified and described as a single entity by O'Leary (1986a). The GLORIA data make clear, however, that mass movement on the west side of Munson Canyon constitutes a slide separate from the larger Munson–Nygren slide to the east. The slide west of Munson Canyon is referred to as the Retriever slide, as it sheds debris directly downslope toward Retriever Seamount (Figs. 9-1, 9-45). The Retriever slide and the Munson–Nygren slide share basic stratigraphic and morphologic similarities, and their debris forms a single confluent field on the rise. Their origins must be closely linked, and therefore they are grouped as a slide complex, despite their spatially distinct heads. For convenience, both slides are jointly referred to as the Munson–Nygren slide complex.

Each slide issues from an amphitheatric head about 7 km in diameter and with sufficient relief to be clearly expressed by bathymetric contours in the 1:1,000,000 NOS Northeastern United States Regional Map (National Oceanic and Atmospheric Administration, 1985; Fig. 9-46). The scarps and scars of the Munson–Nygren and Retriever slides are rugged slopes, not unlike the

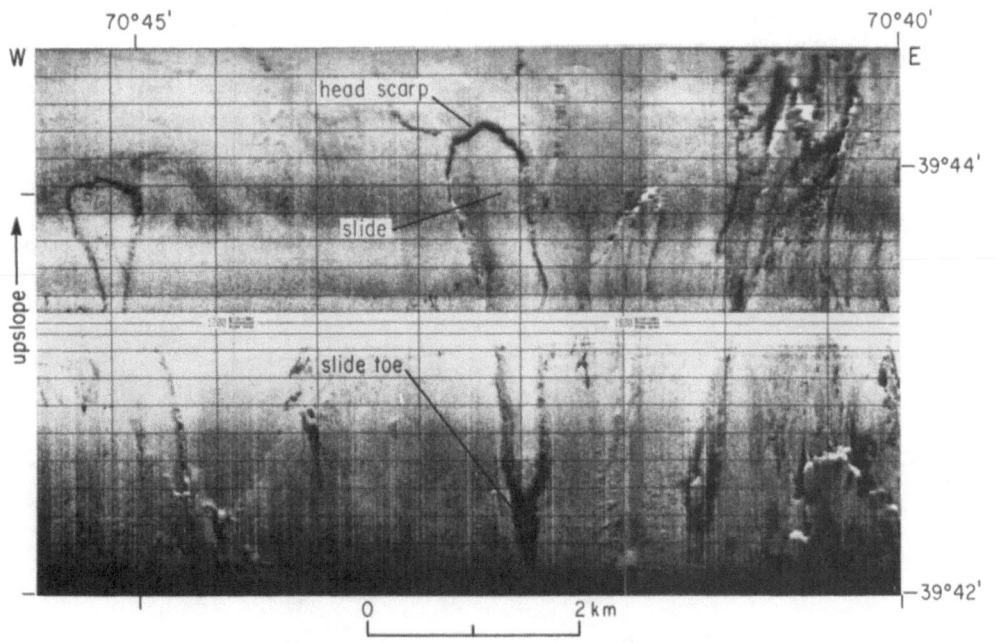

FIGURE 9-44.   SeaMARC I sidescan sonar image showing small, subsequent slides, including "tear drop" slides, formed on lower slope ramp, east side of Block Canyon slide. See Figure 9-14 for location.

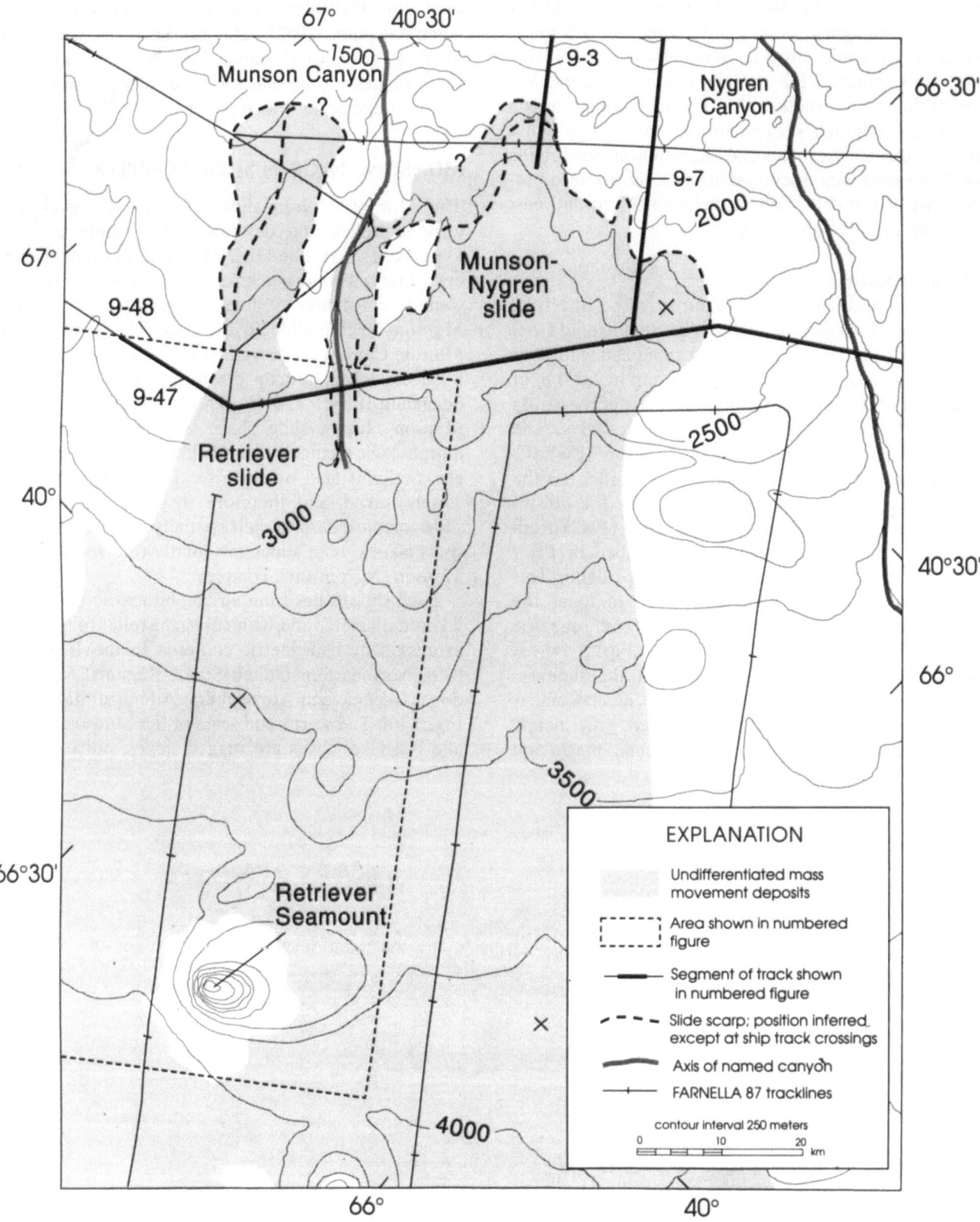

FIGURE 9.45.   Gross bathymetry and planimetric features of Munson–Nygren slide complex, and locations of areas or track lines of illustrated features. Slide scars and debris deposits, undifferentiated.

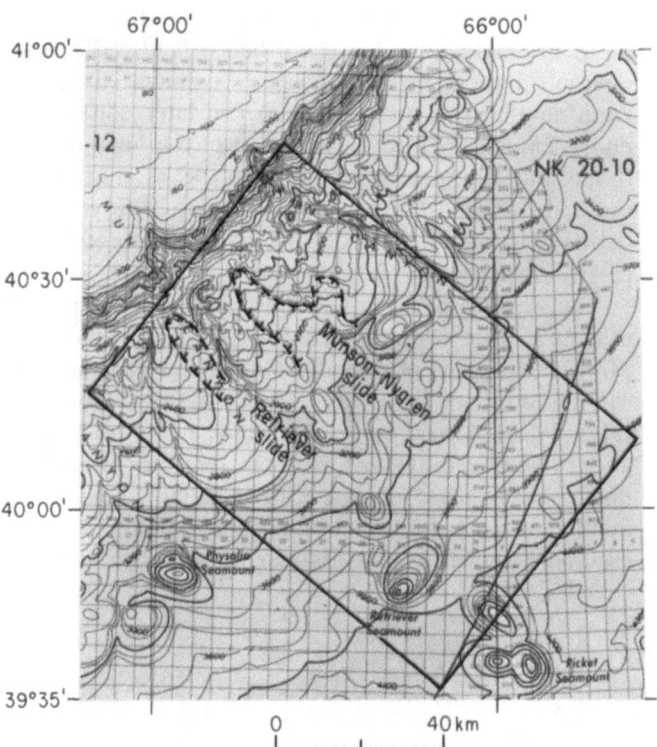

FIGURE 9-46. Part of NOS Northeastern United States Regional Map (NOAA, 1985) showing bathymetric expression of Munson–Nygren and Retriever slides. Dark straight lines mark border of Figure 9-45.

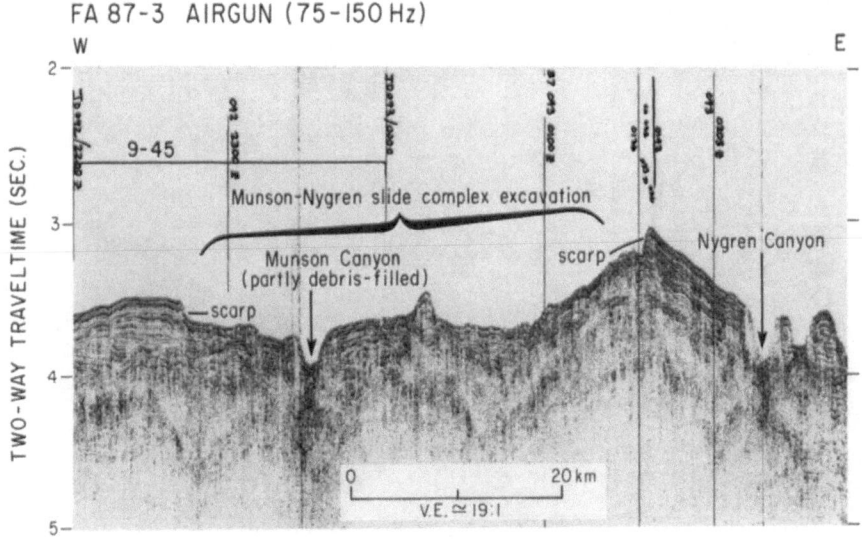

FIGURE 9-47. Cross section of Munson–Nygren slide complex near its head on lower slope, as shown by high-resolution seismic reflection record. See Figure 9-45 for location.

surrounding canyon terrain, and in marked contrast to the tabular stacks and scarps of the SENESC (Fig. 9-47). Failure occurred as deep as 300 m in the section (Fig. 9-3). The debris field is different from that of the SENESC also; missing here are abrupt contrasts in brightness or acoustic roughness that discriminate individual debris lobes. On echosounder profiles, the debris field exhibits a downslope transition of seismic facies from rough to smooth, like that defined by Jacobi (1984). A few flat, raised reflections indicate slabs (olistoliths?) 25–75 m in relief on the seafloor. The debris field gradient is also generally uniform, without the declivities that indicate imbricate deposits.

Mass movement at the two sites has been recurrent. The oldest mass movement postdates formation of Munson Canyon; the youngest mass movement postdates the surficial sedimentary unit. The earlier episode excavated the seafloor by as much as 50 m, creating a lateral border scarp along the distal eastern margin of the debris field. The depressed debris field was subsequently draped by hemipelagic sediment (possibly equivalent to the surficial unit). The later episode of debris emplacement overrode most of the hemipelagic drape, but did not extend debris to the lateral bounding scarp formed by the earlier episode. The younger debris field shows slight declivities, which suggest successive debris lobes.

The western part of the debris field is a composite of debris from Munson Canyon, Retriever slide, the Munson–Nygren slide, and an unnamed canyon that extends down to the east slide of Physalia Seamount (Figs. 9-46, 9-48). Strong contrasts in surface roughness and pronounced downslope flow lines in GLORIA images mark these disparate canyon and mass-movement debris contributions. Despite pronounced local variations shown in echosounder profiles, the debrites

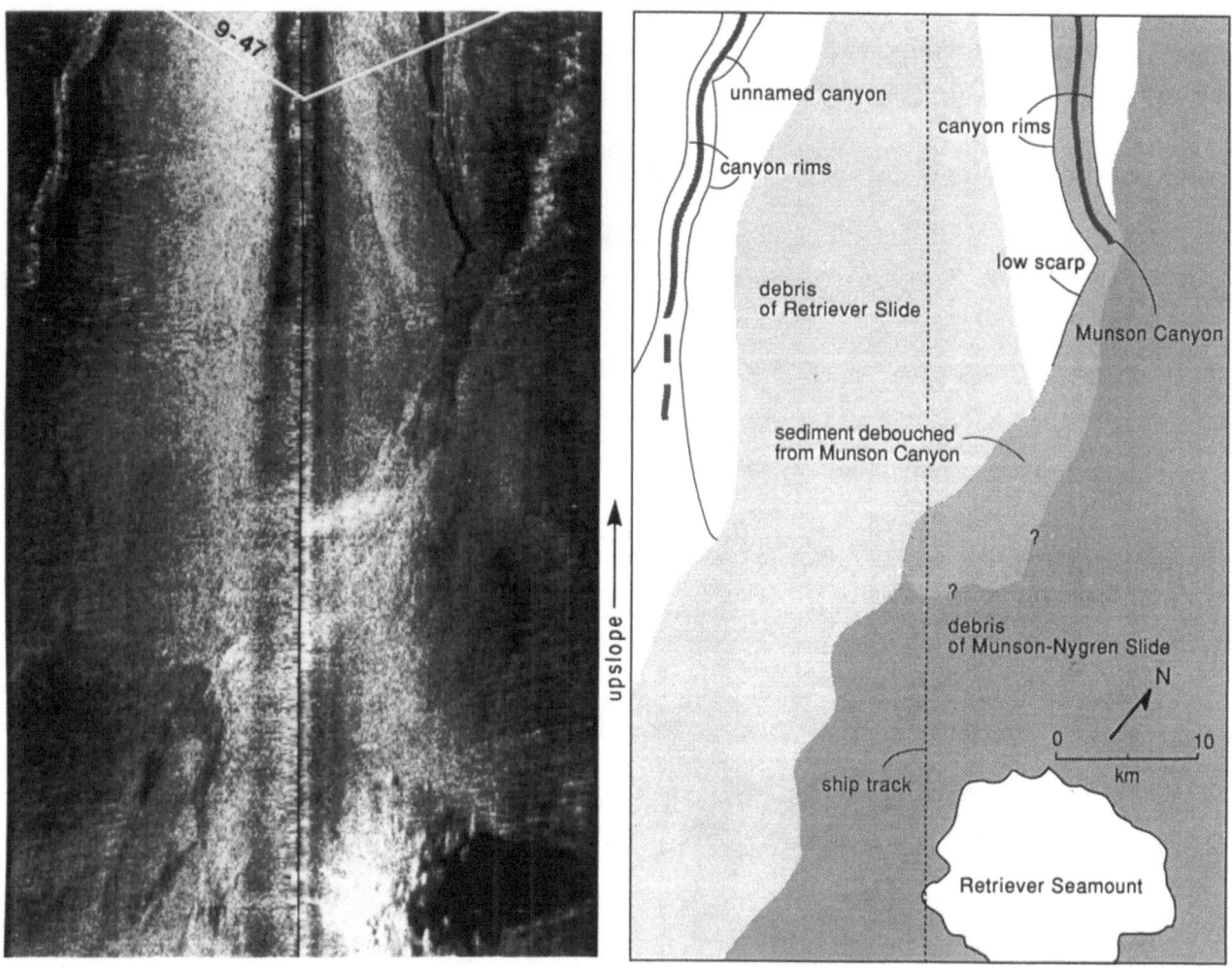

FIGURE 9-48.   GLORIA sidescan sonar image and interpretative map showing complex relationship between Munson–Nygren slide and Retriever slide. Sediment debouched from Munson Canyon trends across Retriever slide debris field, but appears to merge with, or perhaps be diverted by, Munson–Nygren slide. See Figure 9-45 for location.

are acoustically similar in GLORIA images, and are difficult to discriminate according to source. Furthermore, the Retriever slide features a rough-surfaced axial channel, about 1 km wide and 15 m in relief, similar to the canyons to the east and west. Both the Retriever slide axial channel and Munson Canyon become lost in their respective deposits on the upper rise. The relative ages of the various debris contributions are unclear, but because of its bathymetrically lower gradient, we infer that the smoother surfaced Munson Canyon field probably postdates the much rougher, steeper, accumulation of the Retriever slide debrite.

The similarity of form between the slide heads and nearby canyons, and among the deposits associated with these various sources, is striking. In fact, the Munson–Nygren and Retriever slides could realistically be considered as unusually wide, lower slope canyons, or perhaps, features transitional between canyons and open-slope slab slides. Mass movement here must have involved extensive fragmentation at the point of failure, resulting in the wholesale movement of unusually large blocks or slabs, some dimensions of which could be measured in kilometers. The resulting debris was size-graded during transport.

Certain features suggest that the Retriever slide was a powerful, and possibly cataclysmic, event. A salient extending off the west side of Retriever Seamount is banked with debris to a height of 150 m above the upslope seafloor, unusual evidence that the Retriever slide was one of great mobility and momentum. The longer, more regularly graded and attenuated debrites, and deeper excavation of the seafloor, suggest that debris flow in this complex was more energetic than in the SENESC.

DISCUSSION

How has Pleistocene mass movement contributed to the geological evolution of the New England continental rise? Activity of the SENESC has increased the overall elevation of the upper rise terrace and has reduced local relief created by earlier processes of canyon erosion and sediment-swell deposition. This activity seems to have been one of a succession of infilling and upbuilding episodes that formed a major component of rise deposition, and that were sporadically active since at least late Pliocene time, following formation of the Chesapeake Drift (Mountain and Tucholke, 1985; see also Chapter 6). It is not clear, however, whether earlier episodes of cross-sediment deposition were of the scale of the SENESC or involved middle and lower slope sliding, canyon activity, and debrites of the type we have described. The structural and morphological characteristics of the lower

slope debris wedge suggest that pre-Pleistocene mass movement was different in style; it was, perhaps, more widespread along the slope also.

Activity of the Munson–Nygren slide complex seems to have produced a different effect, having reduced the local elevation of the upper rise by redistributing sediment to greater depths on the rise. The result more closely resembles the effects of large mass movements on the U.S. margin south of the Chesapeake Drift (Popenoe, Schmuck, and Dillon, 1992) and on the eastern margin of the Atlantic (see Chapter 12). We, therefore, recognize two distinct regional effects resulting from the slide complexes off New England: (1) *accretion* of the upper rise within the SENESC; and (2) essentially unimpeded *redistribution* of upper rise sediment to the lower rise within the Munson–Nygren slide complex.

An important aspect of flow-emplaced debrites is their characteristic occurrence with turbidites in flysch facies. The Neogene and Pleistocene lithosome of the North American Atlantic margin affected by glaciation is recognized as a flysch, or at least flysch-like, assemblage (Stanley, 1970; Reineck and Singh, 1980). The characterization is significant, because it places the debrites in a setting that implies tectonic activity, and because it underlines the possibility that debrites of ancient flysch successions can be misidentified as tectonic melange. The tectonic affinities of melanges have been strongly emphasized in the literature (Raymond and Terranova, 1984); it is less widely recognized that melanges may actually represent upper rise debrites (Jacobi, 1984). Kleist (1974) and Jacobi (1984) interpreted melanges of the Franciscan and Dunnage Formations, of California and Newfoundland, respectively, as being of sedimentary (mass movement) origin rather than purely tectonic shear phenomena. However, the primary settings for these are considered as having been convergent margins. Our interpretations would imply that debrites, as protomelanges, could be emplaced across a passive margin rise (Jacobi, 1984). If such a margin were later converted to a compressional or sheared convergent setting, all of the stratigraphic affinities except gross layering would be obliterated. On the other hand, turbidites and debrites located in a more distal setting stand a better chance of preservation than heavily clast-laden debrites that are closer to the zone of underthrusting and incorporation into the subduction wedge. But in the distal environment, the affinities to mass movement of even a bona fide debrite might be questionable.

Recent studies of marine sedimentation are dominated by interest in turbidites and submarine fan formation; mass movement is seen mainly as an accessory mechanism in the evolution of fans, and as such, notions of mass movement origin and emplacement of debris appear to be designed to support models of turbidity–current generation. The current terminology of marine sedimentology ineluctably favors interpretations based on turbidity currents. For example, Bouma and Coleman (1985) identified thick sandstone units that contain clay and marl pebbles (either strung out or concentrated) as turbidites with "pebble nests." Some of these units of the Peira-Cava turbidite system, France, were originally considered to be "slump" deposits"; they were reinterpreted by Bouma and Coleman (1985) to fit a Mississippi fan analog. Cazzola, Mutti, and Vigna (1985) described sandstone bodies of the Cengio turbidite system, Italy, as units with thicknesses of 5–25 m that are characteristically devoid of basal channeling and that thin out "abruptly" on mudstones. The units are interpreted as "depositional sandstone lobes" (a type of turbidite) because of their nonchannelized geometry. Perhaps coarse-grained, pinching, clast-laden units such as the above, could usefully be reexamined with an eye to large-scale slope failure and mass movement.

In fact, the debate surrounding debrites and turbidites is not new. The term "fluxoturbidite" (Kuenen, 1958) was applied to the odd deposit, sandwiched within a turbidite succession, that appeared to have characteristics of both sliding and flow (Dzulynski, Ksiazkiewicz, and Kuenen, 1959). The sedimentological definition drew upon some of the same issues of origin that concern us, and it evidently encompassed what we recognize as debrites. However, few realized the importance of these issues (Stanley, 1964), and the ambivalence of the term worked against it. The term was dismissed by Walker (1967) because the definition, with its connotation of sliding, did not fit neatly into the mid-1960s canon of turbidity currents, and its application to mass movement was rejected by Stauffer (1967) because it too clearly implied the action of turbid flow. Stauffer's (1967) rejection may have been overly strict, but his description of what he termed grain-flow deposits of the Santa Ynez Mountains, California, sounds very much like what we would expect a debrite of the New England margin to look like in the field. Subsequent classification of mass movement deposits (Carter, 1975; Middleton and Hampton, 1976) retains the useful notion that some fluxoturbidite identifications in the older literature actually refer to grain-flow or debris-flow deposits.

The depositional character and distribution of debrites on the New England rise do not lend themselves readily to any established depositional model. Certainly the classical model of turbidity current evolution, if it is supported at all, must be of peripheral importance here and must refer to a relatively minor proportion of sediment thrown into the water column during debris flow (Jacobi, 1976). Submarine fan mod-

els, likewise, are difficult to support. The existence of lobes related to a distinct delivery system is unclear. Debris issued from canyons either merges with open-slope slide debris, as with Munson Canyon and the Retriever slide, or forms component flows within (and difficult to distinguish from) the main debrite field. With respect to those canyons that might be considered to be part of, or to feed fan lobes of, their own (such as Hudson and Veatch canyons), the debrites clearly represent a separate and distinct mechanism of delivery and deposition. If a fan deposition model for the SENESC is at all valid, it is the ponded lobe model, characterized in part by the prevalance of mass-movement components (Shanmugam and Moiola, 1991).

A good synthesis of rise deposition systems and a model are provided by Locker and Laine (in press). Their model proposes that submarine fans and sediment drifts develop simultaneously as interactive "companion systems", which may be simply adjacent, or overlapping, or transitional. Alternatively, fan systems may form in the presence of contour currents and simply be modified or eroded by them (see also Chapter 6). Assuming that the SENESC debrites are a "fan" (admittedly a weak assumption, but perhaps plausible if we consider the contributions of Farnella and Munson Canyons, etc.) and that the swells are a drift deposit, our data show no evidence of depositional transition. The "adjacent" fan and drift configuration is presumed to form where individual depositional sequences pass continuously from one depositional setting to the other, reflected only by the change in internal seismic reflection character. "Adjacency" implies that alongslope and downslope currents behave in a compartmentalized manner to an extent reflected by seismic facies transitions. Whatever confidence can be placed in seismic facies representing depositional settings can be used to sustain this model. However, it seems to conflict logically with the current-modified option, or to be mechanistically identical with it. The other alternative, overlapping systems preserved due to joint high accumulation rates, is interesting and possible. Our data do not preclude this interpretation. Our GLORIA mosaic is a snapshot in time; because contour currents are presently active, but downslope transfer is not, we might expect the present overlapping configuration to be replaced eventually by the current-modified fan ("truncated fan") configuration.

Pleistocene open-slope sliding along the New England margin occurred in sections inclined at 2–9°, an inclination clearly sufficient to have afforded stable stratification of tens to hundreds of meters of parallel-bedded siliciclastic strata prior to sliding. All samples and observations obtained so far indicate that the slide complexes did not originate because of initial depositional instability, and they are not a phenomenon of the marine environment that existed during periods of stratification. We must look, therefore, to imposed stresses and(or) some process that selectively weakened the ostensibly stable section at depths tens to hundreds of meters below the seafloor. Both of these destabilizing influences suggest tectonism. The recurrence and the localization of failure in the slide complexes suggest epicentrally clustered earthquakes, an obvious destabilizing agent, especially in light of the 1929 Grand Banks earthquake and its well documented catastrophic effect on the slope off Newfoundland (Piper, Shor, and Hughes Clarke, 1988). However, earthquakes capable of destabilizing the slope within the SENESC or the Munson–Nygren slide complex seemingly should have produced conspicuous mass movement along the steep upper slope east of long. 69°30′W and along Georges Bank, where thick Pleistocene foreset deposits are located. Such mass movement has not been recognized. Some material property peculiar to the fine-grained homoclinal strata, or some mechanism to which they are especially susceptible, must have led to reduced strength thresholds, such that minor shocks could be effective, either in precipitating failure or accumulating strain.

The geometry of the scarps and slide planes, the occurrence of in situ collapse, and seismic profile evidence of intrastratal deformation suggest that specific weak strata within the homoclinal section served as slide planes or shear zones. The term "weak" is relative, of course, because strength of fine-grained sediment is a function of confining stress, pore pressure, and cohesion, each of which can vary with time from bed to bed, so it is unlikely that transient stress loads from earthquakes would trigger all of a potentially allochthonous section at once.

Sparse data (Hathaway et al., 1976) and an abundant geotechnical literature, suggest that the weak layers are probably clay-rich. Under constant gravitational shear stress and over a long period of time, clay-rich layers can weaken, eventually attaining a residual strength that is a fraction of the initial consolidation strength (Suklje and Vidmar, 1961; Bjerrum, 1967). Clay layers in this condition, in the presence of moderate (equilibrium) fluid pore pressure, can slip when subject to seismic shock (Osipov, Nikoloaeva, and Sokolov, 1984); layers with a large content of noncohesive silt to clay-size particles also could collapse or even liquefy under earthquake shock (Field and Hall, 1982; Updike et al., 1988), resulting in in situ fragmentation of overlying beds.

An explanation for mass movement along the New England slope is more difficult to establish than a mechanism for failure. Among the factors that must be considered are the involvement of Paleogene strata in the mass movement, the structure and history of the lower slope debris wedge, and virtually complete removal of detached slabs from the continental slope. These factors are not necessarily interrelated. For example, most of the pre-Pleistocene fragmental material in the debrites may have been issued by the canyons rather than by slab slides. On the other hand, both the Pleistocene slab slides and the older debris wedge may owe their origins to mobile subunits. Seismic profile evidence of quasidiapiric deformation in the upper Pliocene to lower Pleistocene interval of the upper rise (O'Leary, 1986b; Mountain, 1987b), and of faulting related to a deeper, Neogene unit (Macurda, 1988), are sources of instability at a scale commensurate with the SENESC. Intrusive deformation in the upper Pliocene to lower Pleistocene interval has been identified as far north as Nova Scotia (Piper and Sparkes, 1987) and as far south as North Carolina (Cashman and Popenoe, 1985). Deep diapiric deformation that disrupts the sea-floor along the lower slope and upper rise off Nova Scotia and the northern end of Georges Bank was reported by Webb (1973) and Jansa and Wade (1975). The diapiric province may extend farther south along the lower slope, and may have been instrumental in destabilization leading to formation of the Munson–Nygren slide complex.

A final point of discussion concerns the canyons. Submarine landform patterns along the New England margin imply that the canyons are fundamentally dewatering phenomena. The idea of spring sapping as a canyon-forming mechanism is not new (Johnson, 1939); it was reevaluated by Robb et al. (1982) following decades of disfavor. The idea has since attracted interest and ad hoc application as more data have become available. Though an essay on canyons of the U.S. Atlantic continental slope is inappropriate here, a brief discussion may help to clarify the often confused relationships between canyon erosion and mass movement along the New England continental slope.

Technically, the canyons are large rills. Each canyon emanates from one or more point sources circumscribed by amphitheatric or cirque-like margins, the morphology of which is dependent on local stratigraphy or structure. The canyons received nearly their total aqueous input from their heads; there is no evidence for graded overland flow into the canyon heads. The canyons did not originate from deposition along the shelf edge, nor did they act as funnels for sediment slumped from presumed outer shelf deltas built during glacial lowstands. Rather, they cut the very deposits that are commonly thought to have facilitated their origin. The heads of the shelf-indenting canyons are partially choked by glacial sediment of late Wisconsinan age, which has been reexcavated. There has been no significant transport activity in the canyons since the close of the Pleistocene. Thus, there are at least two distinct generations or episodes of canyon incision: late or post-Illinoian, and late Wisconsinan.

It is doubtful that sliding was instrumental in the origin of the canyons, as the canyon heads either postdate formation of slide scars, or the heads are bathymetrically higher than the scarps they cut. On the other hand, it is possible that canyon generation led to mass movement by destabilizing the downslope section and by engendering local collapse and debris flows. In fact, subaerial alpine debris-flow tracks are remarkably similar in form to many of the submarine canyons. We suspect that the initial incision made on the slope by canyon outflow was in the form of a narrow fluid debris-flow track, and that the eroding medium was a slurry flow.

Canyon erosion has penetrated thick, indurated, and fractured units perhaps as deep as the Lower Cretaceous (Ryan et al., 1978). It seems unlikely that debris flows could have downcut to such an extent. However, if the slope canyons originated by artesian piping along anticlinal fractures (a-c joints), actual downcutting may never have occurred. Instead, removal of sediment was effected by liquefaction, by remolding and elutriation of noncohesive beds, and by dissolution, fracturing, and slurry-entrainment of brittle, indurated beds. Headward erosion may have occurred, but to an extent limited by depth and volume of hydraulic outflow across the upper slope.

Submarine spring emission could have occurred along the New England continental slope during glacial lowstands. Computer modeling based on hydraulic conductivities and anisotropies shows that with sea-level drop of 100–150 ft, freshwater discharge is possible along the Atlantic continental slope to water depths of about 1,000 m (Meisler, Leahy, and Knobel, 1984). The presence of a relict freshwater lens beneath the outer continental shelf off New England (Hathaway et al., 1979) implies that the requisite conditions did obtain in the late Pleistocene. Two factors greatly complicate any direct assessment of hydraulic sub-seafloor processes based on present or relict geologic features. One factor is the possibility that the southeastern New England shelf was the site of a glacial foreland bulge, and that the shelf edge and upper slope subsided rapidly during early glacial retreat, whereas the nearshore rebounded (Emery and Garrison, 1967; Newman and March, 1968). Tank drawdown experiments showed that this type of differential tilting could produce artesian escape structures in submerged homoclinally layered sediment (Rettger, 1935). The other factor is the probability of huge amounts of glacial meltwater charged into the rebounding nearshore region during ice retreat. Seismic profile studies suggest

that Block Island Sound was occupied by a large freshwater lake in direct contact with truncated updip ends of Cretaceous strata during late Wisconsinan deglaciation (Bertoni, Dowling, and Frankel, 1977). These conditions of differential uplift and subsidence and groundwater influx, could have provided much of the impetus needed for artesian discharge along the southeast New England slope during coastal deglaciation in late Illinoian and late Wisconsinan time.

The presence of fresh groundwater in fine-grained sediment might also be significant, because yield stress in sediment is lowered with lowered salinity (Locat and Demers, 1988). Thus, we suppose that spring emission along permeable layers and fracture-controlled pipes led to remolding of weakly cohesive marine silts in the consolidated upper slope section. At the site of each canyon head, a cold, freshwater mud slurry formed at the seafloor; the slurry coursed downslope from the zone of emission, and cut a channel subsequently directed by fractured strata. Most of the headwall and flanks were subject to collapse, slumping, and ravelling; the rubble was carried off by the dense slurry. It is possible also that artesian emission (horizontally along permeable beds exposed in canyon headwalls) led to areally extensive liquefaction and collapse conducive to some upper slope mass movement. This movement was, perhaps, sufficient to have produced large mudflows, which fed into canyons like Farnella Canyon, and debouched onto the upper rise.

SUMMARY AND CONCLUSIONS

The continental rise, as currently expressed off southeastern New England, was accreted in late Miocene to middle Pliocene time by sediment drift deposition (see also Chapter 6). During the late Pliocene(?) and most of the Pleistocene, downslope deposition across the entire New England margin, including the rise, was heavy. Tens to hundreds of meters of fine-grained siliciclastic strata of glacial outwash or glaciomarine origin were laid down, including psammitic foreset packages along the outer shelf off Georges Bank, thick ponded and homoclinal sections along the lower slope and the near side of the Neogene sediment drift deposit, and a relatively thin current-faired interval across the sediment drift itself. The result of this deposition was an upper rise terrace, capped by one or two sediment swells, which narrows and pinches out to the northeast along Georges Bank.

The upper rise terrace is an important regional physiographic feature, which controls the pattern of downslope erosion and deposition. Puzzling depositional and erosional forms are well explained by the presence of the swells and the upper rise terrace, but are enigmatic otherwise. Among these features are occlusions and diversions of debris deposits away from swell crests, and the uniform relief and incised channels of the few, major canyons that cross more than 200 km of the upper rise and debouch at about the 4,300-m contour. In contrast, the absence of a terrace-like upper rise north of the New England seamounts is reflected by the flared margins and unimpeded flow forms of the Munson–Nygren and Retriever slide deposits and by the absence of etched rise-crossing channels (see Chapters 10, 13).

The upper Pliocene to lower Pleistocene interval beneath the upper rise terrace is structurally weak; it began to flow in latest Pleistocene and perhaps early Holocene time. This deformation removed downdip support from the Pleistocene section on the lower slope, leading to the major mass movement in the SENESC. Sliding and collapse was conditioned by intrastratal deformation, and probably was triggered by epicentrally clustered earthquakes.

Mass movement in the SENESC comprises chiefly open-slope slab slides, a style of mass movement that has cleared large tracts of the middle and lower continental slope of much, or all, of the Pleistocene section. Sheets of homoclinal strata tens of meters thick, a few kilometers across, and several kilometers long (possibly the entire length of the lower slope), were detached and fragmented upon failure, or during the first instant of movement, much in the manner of a snow slide or avalanche. Such failure and transport occurred successively, peeling away deeper layers, down to, and possibly including, Eocene strata. The failed slabs slid and flowed out across the rise, forming debrites 10–30 m thick, which now extend for more than 200 km downslope.

Evidence indicates that contour-following currents are molding the seafloor at present, but they have not modified the debrite morphology at the scale recorded by GLORIA and echosounder profiles. Either such currents were stronger in the past, or there was more suspended sediment available to effect work, or else bottom-current effects become geologically significant at a time scale orders of magnitude longer than that of mass movement.

Three important observations bear on our interpretations of the origin and significance of the mass movements we have described:

1. The depositionally stable continental slope and the uppermost rise are the only source of debris; failure has occurred across the middle to lower slope in thick homoclinal sections of old, partly indurated strata.

2. Failure has been recurrent, with an unknown recurrence interval; the 25–40-m-thick surficial unit that separates two main failure events suggests that failure may recur over time spans of hundreds to thousands of years.

3. Submarine canyons have a surprisingly complex and unexpected relationship with mass movement.

The early generation of rille-like canyons is clearly unrelated to mass movement. The early chute-like canyons are most closely related in time and function to mass movement, but the nature of the relationship is presently unknown; their chute-like forms may be a result of seismically induced wall collapse during mass movement episodes. A generation of post-slide canyons seems totally unrelated genetically to mass movement. We infer that the canyons that presently score the slope are essentially dewatering phenomena rather than retrogressive features; their origin is tied to Pleistocene lowstands and related changes in the shelf hydraulic regime, complicated by compaction dewatering and structure.

We conclude that failure in the New England slide complexes depended on two fundamental factors: (1) the presence of weak layers subject to cumulative strain within the homoclinal section along the lower slope and beneath the upper rise; and (2) recurrent, clustered earthquakes. The actual composition of the inferred weak layers is unknown; we suppose that some of them are clay-rich and lost cohesion under constant gravitational stress punctuated by earthquakes. A mechanism exists, therefore, to create extensive, multiple-bedding-plane slide surfaces or stratal shear zones. More importantly, weak layers beneath the foot of the slope provide the locus for failure necessary to impose unresolved stress on the upslope section, and thereby, promote wholesale sliding.

Finally, we speculate that large-scale, open-slope sliding of old, homoclinal sections, is a normal consequence of ageing and of ongoing (perhaps sporadic and earthquake-punctuated) differential subsidence of the New England continental margin. It is not necessarily a direct consequence of the Pleistocene proglacial depositional regime or of related changes in sea level.

REFERENCES

Bertoni, R., Dowling, J., and Frankel, L. 1977. Freshwater-lake sediments beneath Block Island Sound. *Geology* 5:631–635.

Bjerrum, L. 1967. Progressive failure in slopes of overconsolidated plastic clay and clay shales. *Journal of Soil Mechanics and Foundation Engineering Division, Proceedings of the American Society of Civil Engineers* 93, SM5, Part I:3–49.

Bouma, A. H., and Coleman, J. M. 1985. Peira–Cava turbidite system, France. *In* Submarine Fans and Related Turbidite Systems, A. H. Bouma, W. R. Normark, and N. E. Barnes, Eds.: New York: Springer-Verlag, 179–184.

Bugge, T., Befrig, S., Belderson, R. H., Eidrin, T., Jansen, E., Kenyon, N. H., Holtedahl, H., and Sejrup, H. P. 1987. A giant three-stage submarine slide off Norway. *Geo-Marine Letters* 7:191–198.

Bugge, T., Belderson, R. H., and Kenyon, N. H. 1988. The Storegga Slide. *Philosophical Transactions of the Royal Society A* 325:357–388.

Carter, R. M. 1975. A discussion and classification of subaqueous mass-transport with particular application to grain-flow, slurry-flow, and fluxoturbidites. *Earth-Science Reviews* 11:145–177.

Cashman, K. V., and Popenoe, P. 1985. Slumping and shallow faulting related to the presence of salt on the continental slope and rise off North Carolina. *Marine and Petroleum Geology* 2:260–271.

Cazzola, C., Mutti, E., and Vigna, B. 1985. Cengio turbidite system, Italy. *In* Submarine Fans and Related Turbidite Systems, A. H. Bouma, W. R. Normark, and N. E. Barnes, Eds.: New York: Springer-Verlag, 179–184.

Damuth, J. E. 1980. Use of high-frequency (3.5–12 kHz) echograms in the study of near-bottom sediment processes in the deep-sea: a review. *Marine Geology* 38:51–75.

Danforth, W. W., and Schwab, W. C. 1990. High-resolution seismic stratigraphy of the upper continental rise seaward of Georges Bank. U. S. Geological Survey Miscellaneous Field Studies Map MF-2111, 11 sheets.

Dzulynski, S., Ksiazkiewicz, M., and Kuenen, P. H. 1959. Turbidites in flysch of the Polish Carpathian Mountains. *Geological Society of America Bulletin* 70:1089–1118.

Embley, R. W. 1976. New evidence for occurrence of debris flow deposits in the deep sea. *Geology* 4:371–374.

Embley, R. W. 1980. The role of mass transport in the distribution and character of deep-ocean sediments with special reference to the North Atlantic. *Marine Geology* 38:23–50.

Embley, R. W., and Jacobi, R. D. 1977. Distribution and morphology of large submarine sediment slides and slumps on Atlantic continental margins. *Marine Geotechnology* 2 (Marine Slope Stability): 205–228.

Emery, K. O. 1974. Pagoda structures in marine sediments. *In* Natural Gases in Marine Sediments, I. R. Kaplan, Ed.: New York: Plenum, 309–317.

Emery, K. O., and Garrison, L. E. 1967. Sea levels 7,000 to 20,000 years ago. *Science* 157:684–692.

Farre, J. A., and Ryan, W. B. F. 1987. Surficial geology of the continental margin offshore New Jersey in the vicinity of Deep Sea Drilling Project Sites 612 and 613. *In* Initial Reports of the Deep Sea Drilling Project, Volume 95, C. W. Poag, A. B. Watts, et al.: Washington, DC: U.S. Government Printing Office, 725–759.

Field, M. E., and Hall, R. K. 1982. Sonographs of submarine sediment failure caused by the 1980 earthquake off northern California. *Geo-Marine Letters* 2:135–141.

Flood, R. D. 1980. Deep-sea sedimentary morphology: modelling and interpretation of echo-sounding profiles. *Marine Geology* 38:77–92.

Gardner, J. V., Field, M. E., Lee, H., Edwards, B. E., Masson, D. G., Kenyon, N., and Kidd, R. B. 1991. Ground truthing 6.5-kHz sidescan sonographs: what are we really imaging? *Geophysical Abstracts in Press* (Supplement to EOS, January, 1991): 7–8.

Gibson, T. G. 1965. Eocene and Miocene rocks off the northeastern coast of the United States. *Deep-Sea Research* 12:975–981.

Gibson, T. G. 1970. Late Mesozoic–Cenozoic tectonic aspects of the Atlantic coastal margin. *Geological Society of America Bulletin* 81:1818–1822.

Hathaway, J. C., Schlee, J. S., Poag, C. W., Valentine, P. C., Weed, E. G. A., Bothner, M. H., Kohout, F. A., Manheim, F. T., Schoen, R., and Miller, R. E. 1976. Preliminary report of the 1976 Atlantic margin coring project of the U.S. Geological Survey. *U.S. Geological Survey Open-File Report 76-844*, 218.

Hathaway, J. C., Poag, C. W., Valentine, P. C., Miller, R. E., Schultz, D. M., Manheim, F. T., Kohout, F. A., and Sangrey, D. A. 1979. U.S. Geological Survey core drilling on the Atlantic shelf. *Science* 206:515–527.

Hollister, C. D., Ewing, J. I., Habib, D., Hathaway, J. C., Lancelot, Y., Luterbacher, H. P., Paulus, F. J., Poag, C. W., Wilcoxon, J. A., and Worstell, P. 1972. Site 106: Lower continental rise. *In* Initial Reports of the Deep Sea Drilling Project, Volume 11, C. D. Hollister, J. I. Ewing, et al.: Washington, DC: U.S. Government Printing Office, 313–350.

Hutchinson, J. N., and Bhandari, R. K. 1971. Undrained loading, a fundamental mechanism of mudflows and other mass movements. *Géotechnique* 21:353–358.

Hutchinson, J. N., Prior, D. B., and Stephens, N. 1974. Potentially dangerous surges in an Antrim mudslide. *Quarterly Journal of Engineering Geology* 7:363–376.

Jacobi, R. D. 1976. Sediment slides on the northwestern continental margin of Africa. *Marine Geology* 22:157–173.

Jacobi, R. D. 1984. Modern submarine sediment slides and their implications for melange and the Dunnage Formation in north-central Newfoundland. *In* Melanges: Their Nature, Origin, and Significance, L. A. Raymond, Ed.: *Geological Society of America Special Paper 198*, 81–102.

Jansa, L. F., and Wade, J. A. 1975. Geology of the continental margin off Nova Scotia and Newfoundland. *In* Offshore Geology of Eastern Canada, W. J. M. van der Linden and J. A. Wade, Eds.: *Geological Survey of Canada Paper 74-30*, 51–105.

Kesseli, J. E. 1943. Disintegrating soil slips of the Coast Ranges of central California. *Journal of Geology* 51:342–352.

Kidd, R. B., Hunter, P. M., and Simon, R. W. 1987. Turbidity-current and debris-flow pathways to the Cape Verde Basin: Status of long-range side-scan sonor (GLORIA) surveys. *In* Geology and Geochemistry of Abyssal Plains, P. P. E. Weaver and J. Thomson, Eds.: *Geological Society Special Publication No. 31*, 33–48.

Kleist, J. R. 1974. Deformation by soft-sediment extension in the Coastal Belt, Franciscan Complex. *Geology* 2:501–504.

Kuenen, P. H. 1958. Problems concerning source and transportation of flysch sediments. *Geologie an Mijnbuouw* 20:329–339.

Lajoie, J., and St-Onge, D. A. 1985. Characteristics of two Pleistocene channel-fill deposits and their implication on the interpretation and megasequences in ancient sediments. *Sedimentology* 32:59–67.

Laughton, A. S. 1981. The first decade of GLORIA. *Journal of Geophysical Research* 86:B11,511–B11,534.

Locat, J., and Demers, D. 1988. Viscosity, yield stress, remolded strength, and liquidity index relationships for sensitive clays. *Canadian Geotechnical Journal* 25:799–806.

Locker, S. D., and Laine, E. P. in press. Paleogene–Neogene depositional history of the middle U.S. Atlantic continental rise: mixed turbidite and contourite depositional systems. *Marine Geology*.

Lowrie, A., Jr., and Heezen, B. C. 1967. Knoll and sediment drift near Hudson Canyon. *Science* 157:1552–1553.

MacIlvaine, J. C. 1973. Sedimentary processes on the continental slope off New England. Ph.D. dissertation, Massachusetts Institute of Technology/Woods Hole Oceanographic Institution.

MacIlvaine, J. C., and Ross, D. A. 1979. Sedimentary processes on the continental slope of New England. *Journal of Sedimentary Petrology* 49:563–574.

Macurda, D. B., Jr. 1988. Contourites and volcanics, Georges Bank, New England. *In* Atlas of Seismic Stratigraphy, A. W. Bally, Ed.: *American Association of Petroleum Geologists Studies in Geology No. 27,* 2:84–87.

Manheim, F. T., and Hall, R. E. 1976. Deep evaporitic strata off New York and New Jersey—Evidence from interstitial water chemistry of drill cores. *U.S. Geological Survey Journal of Research* 4:697–702.

Meisler, H., Leahy, P. P., and Knobel, L. K. 1984. Effect of eustatic sea-level changes on saltwater–freshwater in the northern Atlantic Coastal Plain. *U.S. Geological Survey Water-Supply Paper 2255.*

Middleton, G. V., and Hampton, M. A. 1976. Subaqueous sediment transport and deposition by sediment gravity flows. *In* Marine Sediment Transport and Environmental Management, D. J. Stanley and D. J. P. Swift, Eds.: New York: Wiley-Interscience, 197–218.

Mountain, G. S. 1987a. Cenozoic margin construction and destruction offshore New Jersey. *In* Timing and Depositional History of Eustatic Sequences: Constraints on Seismic Stratigraphy, C. A. Ross and D. Haman, Eds.: *Cushman Foundation for Foraminiferal Research Special Publication No. 24,* 57–83.

Mountain, G. S. 1987b. Underway geophysics during Leg 95. *In* Initial Reports of the Deep Sea Drilling Project, Volume 95, C. W. Poag, A. B. Watts, et al.: Washington, DC: U.S. Government Printing Office, 601–631.

Mountain, G. S., and Tucholke, B. E. 1985. Mesozoic and Cenozoic geology of the U.S. Atlantic continental slope and rise. *In* Geologic Evolution of the United States Atlantic Margin, C. W. Poag, Ed.: New York: Van Nostrand Reinhold, 293–341.

National Oceanic and Atmospheric Administration. 1985. Northeastern United States Regional Map. N. O. S. Washington, DC: Scale 1:1,000,000.

Newman, W. S., and March, S. 1968. Littoral of the northeastern United States: late Quaternary warping. *Science* 160:1110–1112.

Northrop, J., and Heezen, B. C. 1951. An outcrop of Eocene sediment on the continental slope. *Journal of Geology* 59:396–399.

O'Leary, D. W. 1986a. The Munson–Nygren slide, a major lower-slope slide off Georges Bank. *Marine Geology* 72:101–114.

O'Leary, D. W. 1986b. Seismic structure and stratigraphy of the New England continental slope and the evidence for slope instability. *U.S. Geological Survey Open-File Report 86-118,* 1–181.

O'Leary, D. W. 1988. Shallow stratigraphy of the New England continental margin. *U.S. Geological Survey Bulletin 1767.*

O'Leary, D. W. 1992. Submarine mass movement, a formative process of passive continental margins: the Munson–Nygren landslide complex and the southeast New England landslide complex. *In* Submarine Landslides: Selected Studies in the U.S. Exclusive Economic Zone, W. C. Schwab, H. J. Lee, and D. C. Twichell, Eds.: *U.S. Geological Survey Bulletin 2002.*

O'Leary, D. W., and McGregor, B. A. 1983. Midrange sidescan-sonor data and high-resolution seismic data from the continental slope and rise off New England between 67°15'W and 70°55'W. *U.S. Geological Survey Open-File Report 83-805.*

Osipov, V. I., Nikoloaeva, S. K., and Sokolov, V. N. 1984. Microstructural changes associated with thixotropic phenomena in clay soils. *Géotechnique* 34:293–303.

Piper, D. J. W., and Sparkes, R. 1987. Proglacial sediment instability features on the Scotian slope at 63°W. *Marine Geology* 76:15–31.

Piper, D. J. W., Shor, A. N., and Hughes Clarke, J. E. 1988. The 1929 Grand Banks earthquake, slump and turbidity current. *Geological Society of America Special Paper 229,* 77–92.

Poag, C. W. 1982. Foraminiferal and seismic stratigraphy, paleoenvironments, and depositional cycles in the Georges Bank basin. *In* Geological Studies of the COST Nos. G-1 and G-2 Wells, United States North Atlantic Outer Continental Shelf, P. A. Scholle and C. R. Wenkam, Eds.: *U.S. Geological Survey Circular 861,* 43–91.

Poag, C. W. 1987. The New Jersey transect: stratigraphic framework and depositional history of a sediment-rich passive margin. *In* Initial Reports of the Deep Sea Drilling Project, Volume 95, C. W. Poag, A. B. Watts, et al.: Washington, DC: U.S. Government Printing Office, 763–817.

Poag, C. W., and Mountain, G. S. 1987. Late Cretaceous and Cenozoic evolution of the New Jersey continental slope and upper rise: an integration of borehole data with seismic reflection profiles. *In* Initial Reports of the Deep Sea Drilling Project, Volume 95, C. W. Poag, A. B. Watts, et al.: Washington, DC: U.S. Government Printing Office, 673–723.

Popenoe, P., Schmuck, E. A., and Dillon, W. P. 1992. The Cape Fear slide complex; slope failure associated with salt diapirism and gas hydrates. *In* Submarine Landslides: Selected Studies in the U.S. Exclusive Economic Zone, W. C. Schwab, H. J. Lee, and D. C. Twichell, Eds.: *U.S. Geological Survey Bulletin 2002*.

Pratson, L. F., and Laine, E. P. 1989. The relative importance of gravity-induced versus current-controlled sedimentation during the Quaternary along the middle U.S. continental margin revealed by 3.5 kHz echo character. *Marine Geology* 89:87–126.

Prior, D. B., and Coleman, J. M. 1978. Disintegrating retrogressive landslides on very-low-angle subaqueous slopes, Mississippi Delta. *Marine Geotechnology* 3:37–60.

Prior, D. B., Bornhold, B. D., and Johns, M. W. 1984. Depositional characteristics of a submarine debris flow. *Journal of Geology* 92:707–727.

Raymond, L. A., and Terranova, T. 1984. The Melange problem—a review. *In* Melanges: Their Nature, Origin, and Significance, L. A. Raymond, Ed.: *Geological Society of America Special Paper 198*, 1–5.

Reineck, H. E., and Singh, I. B. 1980. *Depositional Sedimentary Environments*: Heidelberg: Springer-Verlag.

Rettger, R. E. 1935. Experiments on soft-rock deformation. *American Association of Petroleum Geologists Bulletin* 19:271–292.

Robb, J. M. 1986. Submarine landslide deposits and avalanche impact features at the base of the east coast continental slope. *Society of Economic Paleontologists and Mineralogists Annual Midyear Meeting Abstracts* 3:95–96.

Robb, J. M., and Hampson, J. C., Jr., 1984. Mounds of sediment along the base of the continental slope of the Mid-Atlantic United States. *Geological Society of America Abstracts with Programs* 16:59.

Robb, J. M., O'Leary, D. W., Booth, J. S., and Kohout, F. A. 1982. Submarine spring sapping as a geomorphic agent on the east coast continental slope. *Geological Society of America Abstracts with Programs* 14:600.

Ryan, W. B. F., Cita, M. B., Miller, E. L., Hanselman, D., Nesteroff, W. D., Hecker, B., and Nibblelink, M. 1978. Bedrock geology in New England submarine canyons. *Oceanologica Acta* 1:233–254.

Shanmugam, G., and Moiola, R. J. 1991. Types of submarine fan lobes: models and implications. *American Association of Petroleum Geologists Bulletin* 75:156–179.

Somers, M. L., Carson, R. M., Revie, J. A., Edge, R. H., Barrow, B. J., and Andrews, A. G. 1978. GLORIA II—an improved long range sidescan sonar. *Oceanology International* 78:16–24.

Stanley, D. J. 1964. Nonturbidites in flysch-type sequences: their significance in basin studies. *Geological Society of America Special Paper 76*, 155–156.

Stanley, D. J. 1970. Flyschoid sedimentation on the outer Atlantic margin off northeast North America. *Geological Association of Canada Special Paper 7*, 179–210.

Stauffer, P. H. 1967. Grain-flow deposits and their implications, Santa Ynez Mountains, California. *Journal of Sedimentary Petrology* 37:487–508.

Stetson, H. C. 1949. The sediments and stratigraphy of East Coast Continental Margin—Georges Bank to Norfolk Canyon. *Massachusetts Institute of Technology and Woods Hole Oceanographic Institution Papers in Physical Oceanography and Meteorology 11*, 1–60.

Stow, D. A. V. 1985. Deep-sea clastics: where are we and where are we going? *In* Sedimentology: Recent Developments and Applied Aspects, P. J. Benchley and B. P. J. Williams, Eds.: *Geological Society Special Publication No. 18*, 67–93.

Suklje, L., and Vidmar, S. 1961. A landslide due to long term creep. *Proceedings of the Fifth International Conference on Soil Mechanics and Foundation Engineering*, Volume 11, Divisions 33-7: Paris: Dunod, 727–735.

Tucholke, B. E. 1987. Submarine geology. *In* The Marine Environment of the U.S. Atlantic Continental Slope and Rise, J. D. Milliman and W. R. Wright, Eds.: Boston/Woods Hole: Jones and Bartlett, 56–113.

Tucholke, B. E., and Laine, E. P. 1982. Neogene and Quaternary development of the lower continental rise off the central U.S. east coast. *In* Studies in Continental Margin Geology, J. S. Watkins and C. L. Drake, Eds.: *American Association of Petroleum Geologists Memoir 34*, 295–305.

Tucholke, B. E., Bryan, G. M., and Ewing, J. I. 1977. Gas-hydrate horizons detected in seismic-profiler data from the western North Atlantic. *American Association of Petroleum Geologists Bulletin* 61:698–707.

Uchupi, E. 1967. Slumping on the continental margin southeast of Long Island, New York. *Deep-Sea Research* 14:635–639.

Updike, R. G., Egan, J. A., Moriwaki, Y., Idriss, I. M., and Moses, T. L. 1988. A model for earthquake-induced translatory landslides in Quaternary sediments. *Geological Society of America Bulletin* 100:783–792.

Van Hinte, J. E., Wide, S. W., Jr., Biart, B. N. M., Covington, J. M., Dunn, D. A., Haggerty, J. A., Johns, M. W., Meyers, P. A., Moullade, M. R., and Muza, J. P. 1985. Deep-sea drilling on the upper continental rise off New Jersey, DSDP Sites 604 and 605. *Geology* 13:397–400.

Vassallo, K., Jacobi, R. D., and Shor, A. N. 1984. Echo character, microphysiography, and geologic hazards. *In* Eastern North American Continental Margin and Adjacent Ocean Floor, 34° to 41°N and 68° to 78°W, Atlas 4, Ocean Margin Drilling Program Regional Atlas Series, J. I. Ewing and P. D. Rabinowitz, Eds.: Woods Hole: Marine Science International, 1–31.

Veatch, A. C., and Smith, P. A. 1939. Atlantic submarine valleys of the United States and the Congo submarine valley. *Geological Society of America Special Paper No. 7*, 1–101.

Walker, R. G. 1967. Turbidite sedimentary structures and their relationship to proximal and distal depositional environments. *Journal of Sedimentary Petrology* 37:25–43.

Watkins, D. J., and Kraft, L. M., Jr. 1978. Stability of continental shelf and slope off Louisiana and Texas: geotechnical aspects. *In* Framework, Facies, and Oil-trapping Characteristics of the Upper Continental Margin, A. H. Bouma, G. T. Moore, and J. M. Coleman, Eds.: *American Association of Petroleum Geologists Studies in Geology No. 7*, 267–286.

Webb, G. W. 1973. Salt structures east of Nova Scotia. *In* Earth Science Symposium on Offshore Eastern Canada,

P. J. Hood, Ed.: *Geological Survey of Canada Paper 71-23*, 197–218.

Weed, E. G. A., Minard, J. P., Perry, W. J., Jr., Rhodehammel, E. C., and Robbins, E. I. 1974. Generalized pre-Pleistocene geologic map of the northern United States Atlantic continental margin. *U.S. Geological Survey Map I-861*, 1–8, 2 sheets.

Wise, S. W., Jr., and Van Hinte, J. E. 1987. Mesozoic–Cenozoic depositional environments revealed by Deep Sea Drilling Project Leg 93 on the continental rise off the eastern United States: cruise summary. *In* Initial Reports of the Deep Sea Drilling Project, Volume 93, J. E. Van Hinte, S. W. Wise, Jr., et al.: Washington DC: U.S. Government Printing Office, (2):1367–1423.

10

Western Nova Scotia Continental Rise: Relative Importance of Mass Wasting and Deep Boundary-Current Activity

John E. Hughes Clarke, Dennis W. O'Leary, and David J. W. Piper

The late Quaternary development of the continental rise off Nova Scotia (Fig. 10-1) has been variously attributed either to predominantly alongslope sediment transport (Hollister and Heezen, 1972) or to predominantly downslope sediment transport (Stanley et al., 1972). The conflicting attributions are based in part on differing interpretation of similar sedimentary facies. The HEBBLE project (Nowell and Hollister, 1985) clearly demonstrated the presence of modern intense deep-boundary-current activity on the continental rise off western Nova Scotia. It was also apparent that the seafloor in the vicinity of the HEBBLE site had been subjected to downslope processes in the past (Shor and Lonsdale, 1981; McCave, 1985). Thus, while the project demonstrated modern deep-water sediment transport, the question remained as to whether this process had been important in the past.

Significantly, the intensity of deep water-mass circulation in the North Atlantic is believed to have increased from the late Pleistocene to the Holocene (Streeter and Shackleton, 1979), while at the same time, the flux of sediment from the continental slope has decreased by at least an order of magnitude (Stow, 1979; Hill, 1984). Therefore, the relative importance of downslope sediment transport and deep boundary currents on Pleistocene rise sedimentology may have been significantly different than at present.

Damuth, Tucholke, and Coffin (1979), Damuth, Tucholke, and Shor (1981), and Swift (1985) have mapped the variations of 3.5-kHz echo character across the continental rise off western Nova Scotia, and have recognized the signature of both alongslope and downslope processes. Ambiguity exists, however, in the definition of sedimentary facies types based on echo character (Damuth, 1980), and the wide spacing of the seismic profiles precludes confident mapping of the orientation of echo-character boundaries.

By using a large-swath sidescan sonar (GLORIA) we have been able to map acoustic backscatter without the necessity of line to line extrapolation (Fig. 10-2). In this manner, both the distribution of acoustic facies and the trend of facies boundaries can be mapped to the limit of the system resolution. We thus are able to delineate the spatial distribution of sediment across the Nova Scotian rise and to infer local depositional environments. We use these new data to investigate the relative importance of mass wasting and deep-boundary-current activity on the geomorphology of the continental rise off western Nova Scotia.

METHODS

This survey was undertaken using the Geologic Long Range Inclined Azdic* (GLORIA) sidescan-sonar imaging system (Somers et al., 1978) designed and built at Great Britain's Institute of Oceanographic Sciences (see also Chapters 9, 13). Most of the ship tracks were oriented at an angle oblique to the inclination of the continental rise in order to enhance our ability to recognize and trace downslope-oriented features and to obtain, along with the sonar records, seismic reflection profiles that provide stratigraphic continuity from

---

*Use of tradenames within this report is for purposes of identification only and does not imply endorsement by the U.S. Geological Survey or Bedford Institute of Oceanography.

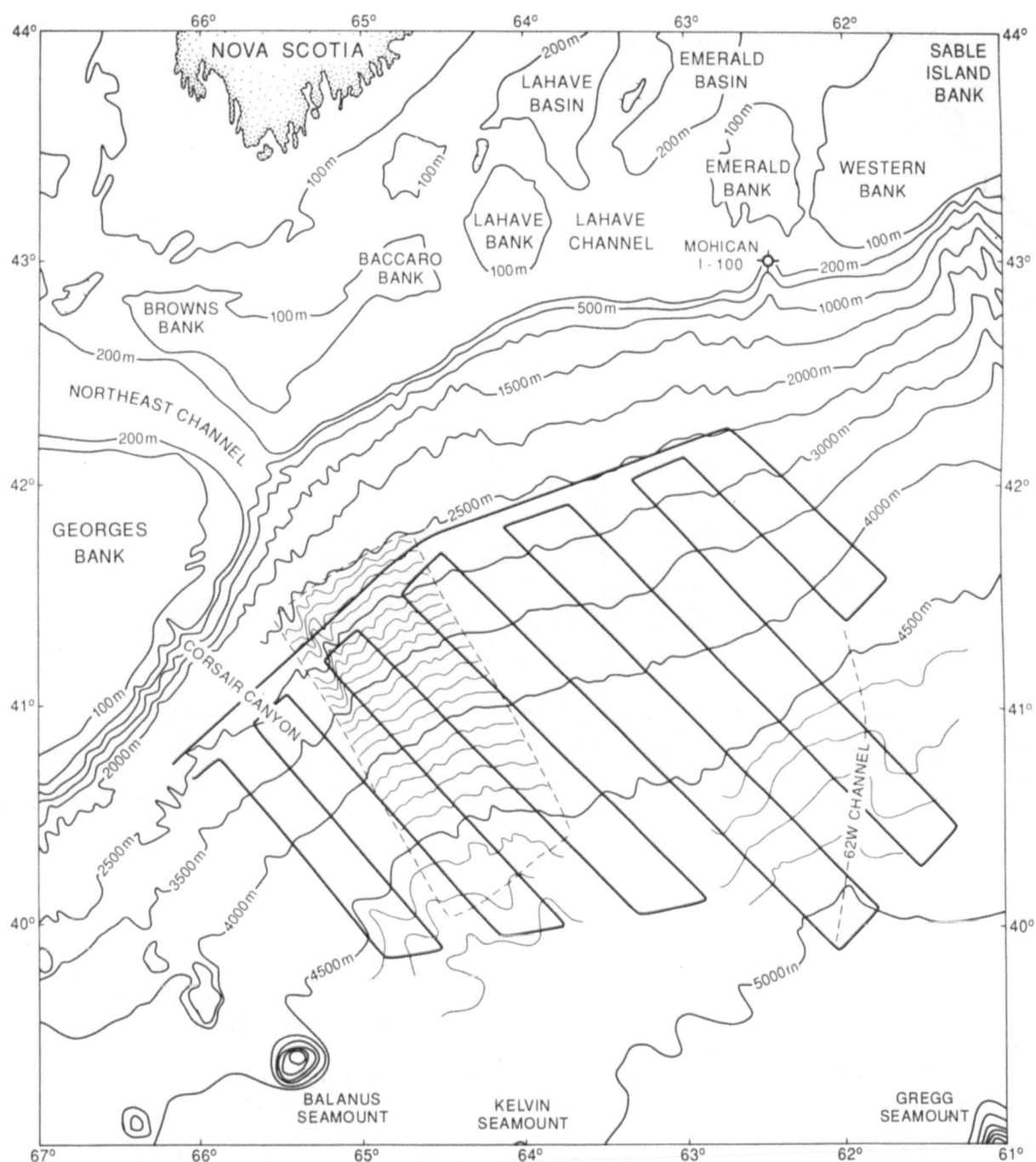

FIGURE 10-1.   Regional bathymetry (corrected meters) based on Shor (1984); area of detailed
contours (within dotted line) based on data collected on *Baffin* cruise 87-039. Heavy lines indicate
GLORIA survey tracklines.

the lower slope to the lower rise. The lines were
spaced to provide 10–50% overlap of image swaths.
The four longest northwest-oriented tracklines were
extended to include the HEBBLE site (Fig. 10-1).
Although the orientation of an elongate target relative
to the fish track creates significant aliasing, the cross-
line links and the echo-sounder profiles acquired in
conjunction with the scanned data indicate that the

survey did not fail to record accurately features oriented along bathymetric contours. Hence we are confident that survey design (Fig. 10-1) did not produce a
biased view of the rise.

At the operating frequency employed by GLORIA
(6.3–6.7 kHz), there will be subbottom penetration of
up to a few meters (see also Chapter 13). The thickness
of Holocene sediment across the Nova Scotian rise

FIGURE 10-2.   Mosaic of GLORIA digital sidescan-sonar images covering 70,000 km² of Western Nova Scotian Rise. Data fitted to Albers equal area projection with grid lines at 30-min spacing. Light regions indicate high backscatter.

rarely exceeds 1.5 m, so GLORIA did image, in part, the distribution of upper Pleistocene sediment.

The digital GLORIA sonar images were corrected for geometry and radiometry in the manner described by Chavez (1986), and the resulting images were mosaicked to generate geographically registered images (Fig. 10-2) of the ocean floor. Archived 3.5-kHz data and 32 piston cores across the survey region, together with seismic and bathymetric data collected on the GLORIA cruise, have been used to complement the sidescan-sonar images. Twenty percent of the study region was bathymetrically surveyed at a 2-km line spacing by the Canadian Hydrographic Service (Fig. 10-1), and these data also were incorporated into the final data set.

GEOLOGIC INTERPRETATION

The region of interest, herein termed the Western Nova Scotian Rise (WNSR), extends latitudinally from 65°W to a channel at 62°W [the boundary between the

downslope and alongslope facies associations of Swift (1985)], which has been termed the 62°W channel by Driscoll, Tucholke, and McCave, (1985). Bathymetrically, the region extends from the base of the slope, at between 2,500 and 3,000 m, to the western part of the Sohm Abyssal Plain in about 5,000 m of water (Fig. 10-1).

The most striking feature of the Western Nova Scotian Rise, as depicted by GLORIA, is the predominance of downslope-oriented landforms (Figs. 10-2, 10-3). We describe these landforms below, as divided on the basis of proposed genesis.

Channels

Unlike channels developed on the Georges Bank rise to the west, which are discrete and lie between regions of hemipelagic or fine overbank deposition, the channels developed on the WNSR are commonly intricately interconnected over wide areas, with no clear definition of an axial channel or of interchannel regions (Fig.

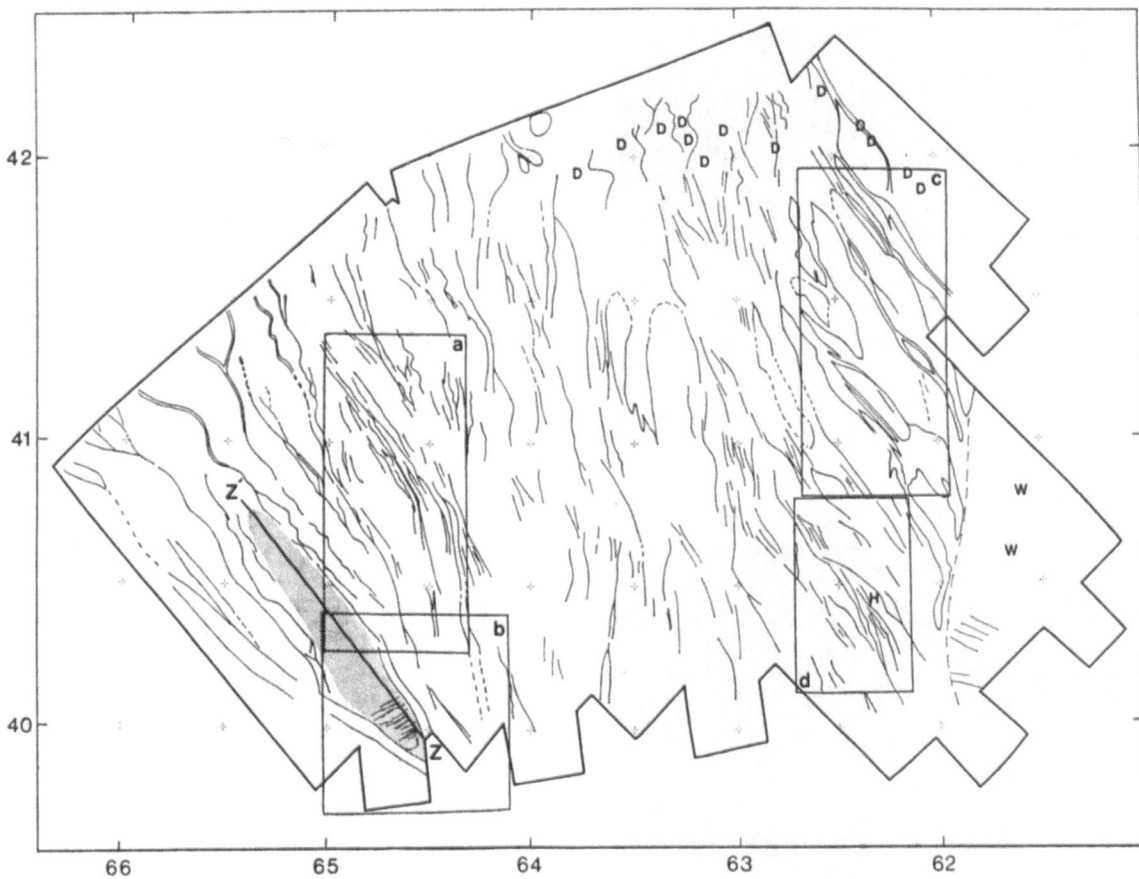

FIGURE 10-3. Line drawing of principal downslope-trending features within GLORIA survey area. Boxes a, b, c, d indicate areas illustrated in subsequent figures; H denotes position of HEBBLE site; D indicates location of diapirs recognized from seismic reflection profiles; W denotes location of long-wavelength bathymetric swell recognized by Swift (1985); line Z-Z' denotes location of seismic profile shown in Figure 10-9; stippled region indicates extent of slide deposit discussed in text.

10-4). Bathymetrically, the channel nets are represented by broad shallow depressions (Fig. 10-5); the sonar images indicate that several channels may occupy a single broad bathymetric depression. When imaged from opposing directions of view, the regions of high and low backscatter remain in the same location, indicating that the expression of the channels in the sonar images is a result of texture differences rather than topography. In water depths less than 4,500 m, topographic levees are not present.

Channels range in width from 200 to 1,000 m, with relief that rarely exceeds 20 m. The channels are commonly discontinuous, extending for distances of 5–100 km. The channels both merge and bifurcate, but no superposition or cross-cutting of channels was observed. Singular channels are rare in water depths of less than 4,500 m, suggesting that there are no preferred paths for downslope sediment transport. In water depths greater than 4,500 m, the channels converge to a few larger channels, which have relief of as much as 50 m and widths of 3–5 km (Figs. 10-4, 10-5, 10-6, 10-7). Piston cores reveal that the channels are floored

by either graded fine to medium sand beds (core H88010-31, Fig. 10-7, 90 cm thick) or graded coarse sand to gravel beds (core V7-68, Fig. 10-5, 189 cm thick; core H88010-30, Fig. 10-7, 156 cm thick). These coarser siliciclastic sediments are overlain by a 1–2-m drape of Holocene hemipelagic sediment. Piston cores elsewhere on the Nova Scotian rise commonly penetrate mud and thin sand laminae (Stanley et al., 1972). The floors of the larger channels, in water depths greater than 4,500 m, commonly appear acoustically smooth (dark) on GLORIA images (Figs. 10-4, 10-6).

Slides

An elevated bathymetric feature located immediately upstream of the junction of two lower-rise channels (Fig. 10-7) appears in GLORIA images (Fig. 10-6) to consist of a series of asymmetric, contour-parallel ridges. Seismic reflection profiles across this feature (Fig. 10-8) reveal a series of discrete blocks of stratified sediment overlying a conspicuous reflector at 100–150-m subbottom. Piston cores recovered from this feature

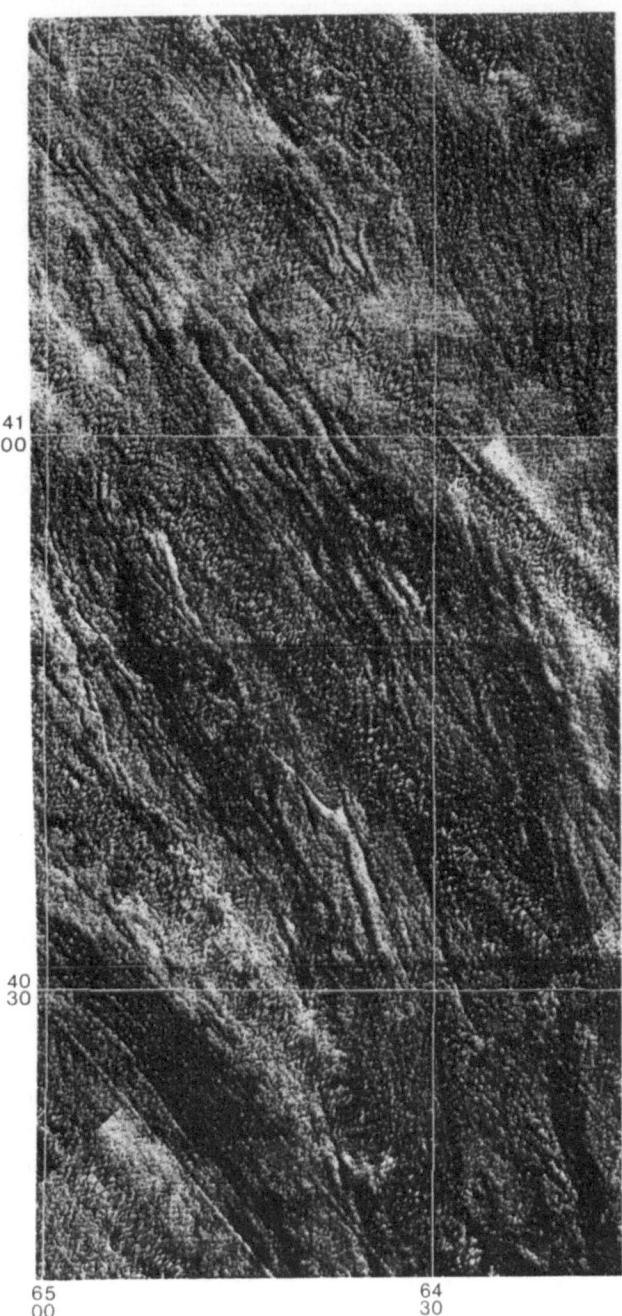

FIGURE 10-4. Enlarged portion of GLORIA mosaic showing complex pattern of rise channels south of Northeast Channel (box a, Fig. 10-3). Bright region along ship track at lat. 41°00′N, long. 64°30′W is an artifact generated by deep scattering layer and is not an acoustic feature of seafloor.

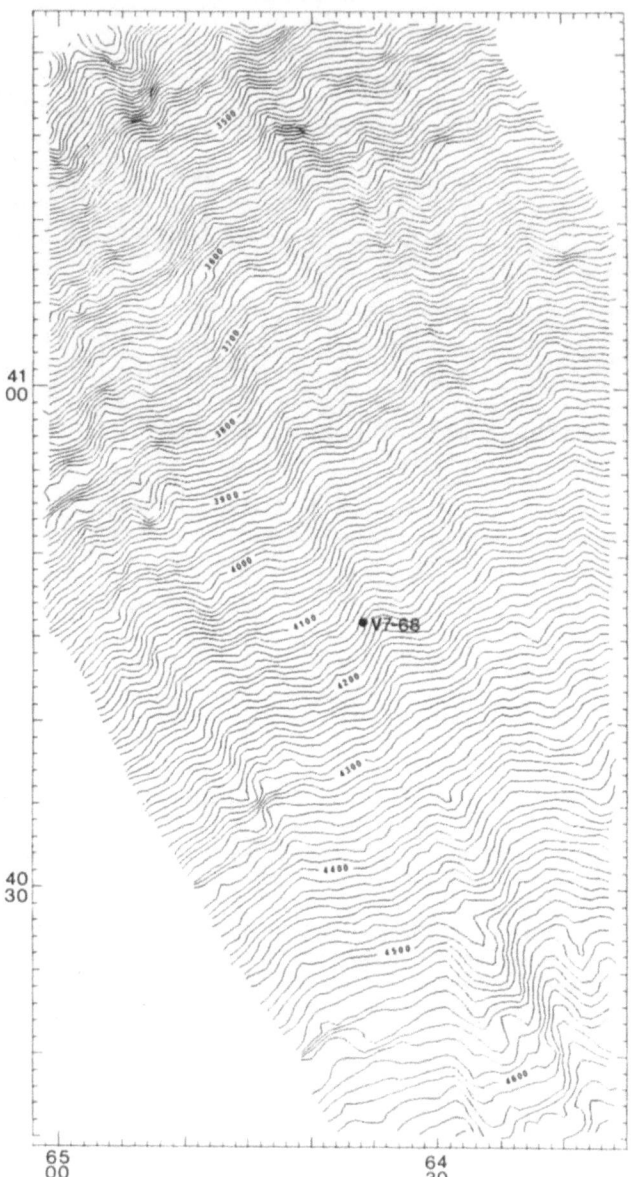

FIGURE 10-5. Bathymetry of same area as Figure 10-4 contoured at 10-m intervals. Depths corrected for sound velocity using synoptic XBT data. Data acquired using conventional echo sounder along 2-km spaced tracks oriented at 075°T. In these water depths the footprint of sounder precludes recognition of features with a dimension of less than 1.5 km. Position of core Vema 7-68 is indicated.

(cores H88010-33, 34, Fig. 10-7) contain interbedded muds and clean sands. Although core recovery was poor because of the presence of thin (1–5 cm) sand beds, we observed no evidence for deformation on the scale of a piston core (∼ 1 m).

Similar, but more subdued transverse ridges, parallel to those within the downslope end of the elevated feature, extend for several tens of kilometers upslope.

Strong hyperbolae on 3.5-kHz records and a zone of near-surface discontinuous reflections (Fig. 10-9, inset) can be traced 90 km upslope. The upslope boundary of the feature is expressed as an abrupt topographic step at 3,500-m water depth. Upslope from the topographic step, the sediment appears evenly stratified and undisturbed (Fig. 10-9). The lateral extent of the disturbed zone is defined (on the basis of 3.5-kHz records, seismic-reflection profiles, and GLORIA images) to be an elongate zone, 90 km long and as much as 30 km wide (stippled area, Fig. 10-3).

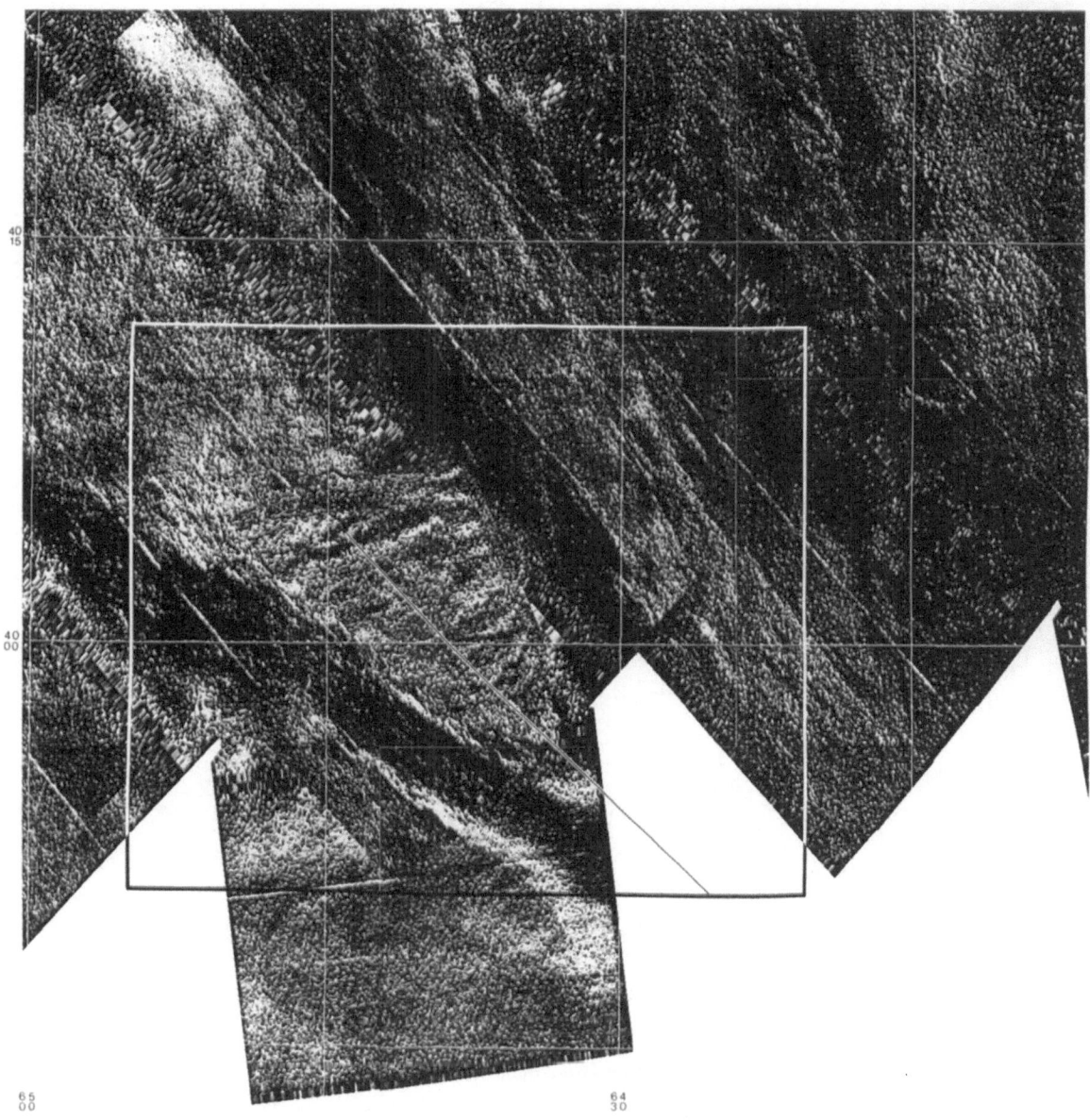

FIGURE 10-6. Enlarged portion of GLORIA mosaic across inferred slide toe and two lower-rise channels (box b, Fig. 10-3). Heavy rectangular outline indicates location of detailed bathymetric map (Fig. 10-7); highlighted trackline segment indicates position of seismic reflection profile shown in Figure 10-8 (top).

The entire feature, extending from 3,500- to 4,700-m water depth is tentatively interpreted to be a complex bedding-plane slide (see also Chapter 9). The rough topography and the absence of seismically identifiable drapes in 3.5-kHz records are evidence for a late Pleistocene age. The transverse ridges within the slide toe resemble intraslide compressional ridges described by Field and Hall (1982), but on a larger scale. Unlike most other known slides, which have their sources on the steeper continental slope (see Chapters 9, 12, 13), this slide was initiated and moved on gradients in the range 0.01–0.015 (0.6–0.85° slopes). The gradient from 3,500 to 4,500 m is remarkably constant, but there is a

marked change in gradient across the toe of the slide. Beyond the toe of the slide (downslope), the seafloor gradients decrease to 0.005 (0.3°).

Debris Flows

Throughout the WNSR, 3.5-kHz records indicate weakly stratified sediment; the echo type (IIA to IIB; Damuth, 1980) suggests a regional preponderance of silty or sandy mud. Piston cores have recovered finely interbedded sands and muds (Hollister and Heezen, 1972; Stanley et al., 1972). In addition, 3.5-kHz records across the WNSR reveal multiple instances of debris-

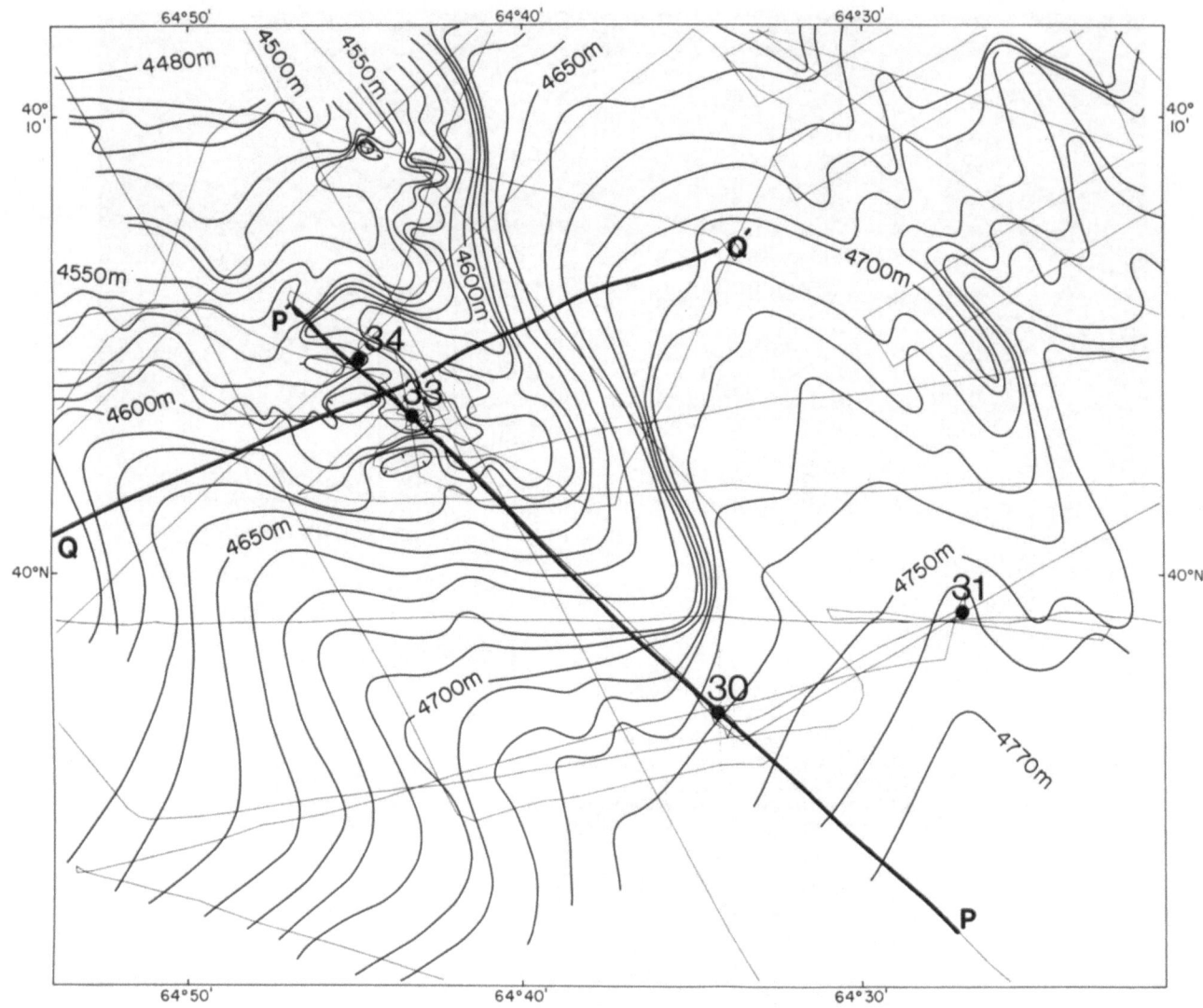

FIGURE 10-7. Detailed bathymetry (Carter corrected meters) over inferred slide toe in 4,700-m water depth. Ship tracks indicated by thin lines. Locations of piston cores, discussed in text, shown by small numbered circles. Q-Q' and P-P' are profiles shown in Figure 10-8.

flow facies. Because the echo character of the surrounding sediment is similar, the spatial boundaries of these debris-flow deposits are hard to define using 3.5-kHz records alone. Using GLORIA, however, the distribution of these debris-flow deposits may be examined by mapping zones of increased acoustic backscatter (Figs. 10-2, 10-10, 10-11).

At about lat. 62°30′W, a number of reflective streams of sediment cross the rise (Figs. 10-10, 10-11). These sediment streams emanate from the lower slope to the north, where reflective sediment appears to cover the entire slope. Downslope, the reflective sediment converges on a series of streams, which ramify on the lower rise and terminate in discrete lobate deposits in water depths of 4,800–5,000 m (Figs. 10-10, 10-11).

Two piston cores obtained from within these features (Fig. 10-11, nos. 27, 29) consist of mud clast

conglomerates underlying about 1 m of Holocene hemipelagic ooze. The reflective sediment streams are thus interpreted to represent upper Pleistocene muddy debris-flow deposits. The stream-like aspect of the debris-flow deposits is a result of restriction into shallow prexisting channels cut into stratified sediments [Fig. 10-12 (top)] or shallow depressions within terrain of similar acoustic character [Fig. 10-12 (bottom)].

The differing intensities of acoustic backscatter from adjacent sediment streams may indicate relative ages. Cross-cutting relationships can be observed locally (Fig. 10-10), but are inadequate to confidently define a relative stratigraphy. These debris-flow deposits lie downslope of, and may be distal continuations of, upper Pleistocene debris-flow deposits, which have been mapped on the slope immediately to the north (Shor and Piper, 1989).

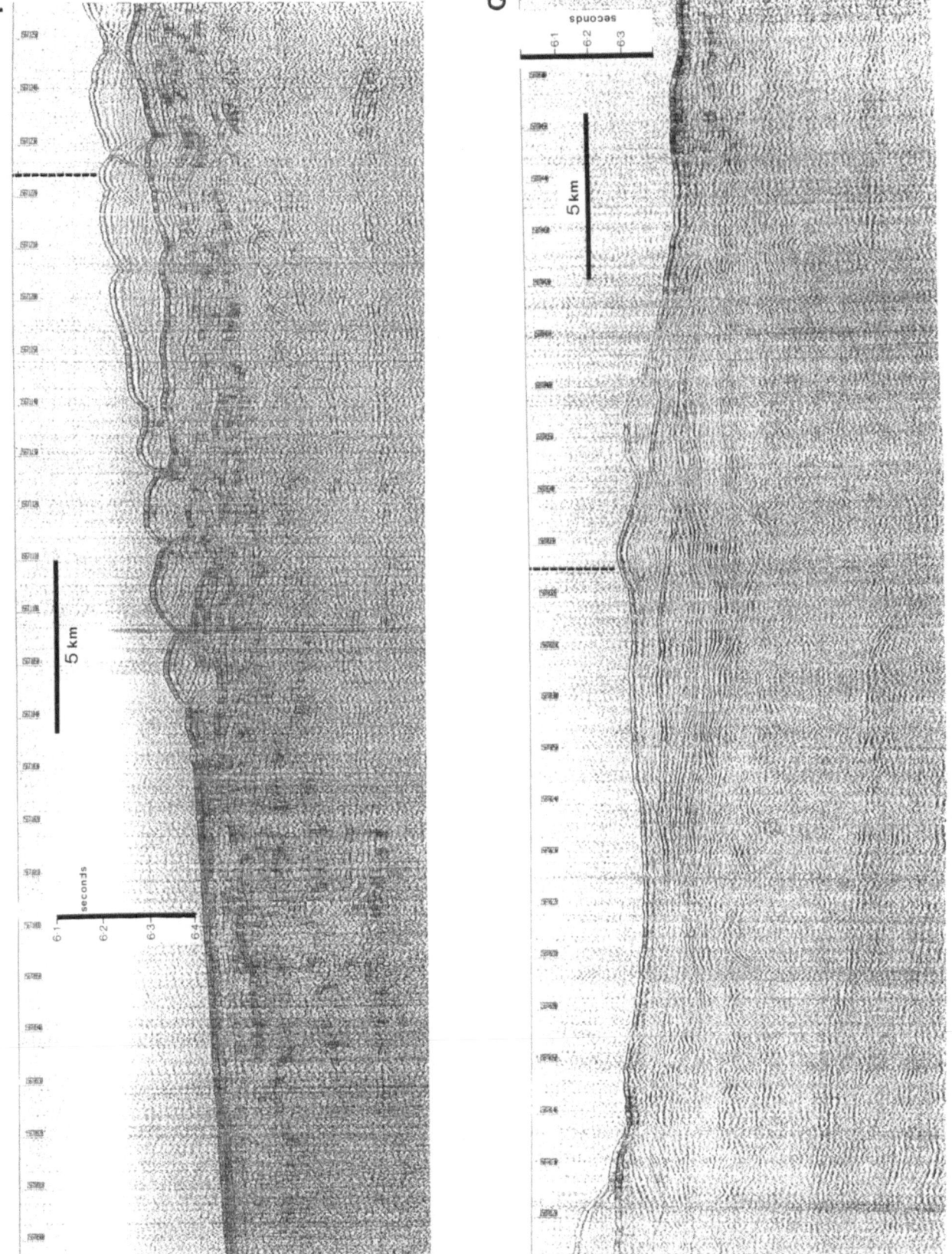

FIGURE 10-8. Two 40-in³ airgun seismic-reflection profiles across slide toe (see Fig. 10-7) illustrating discrete blocks of stratified sediment.

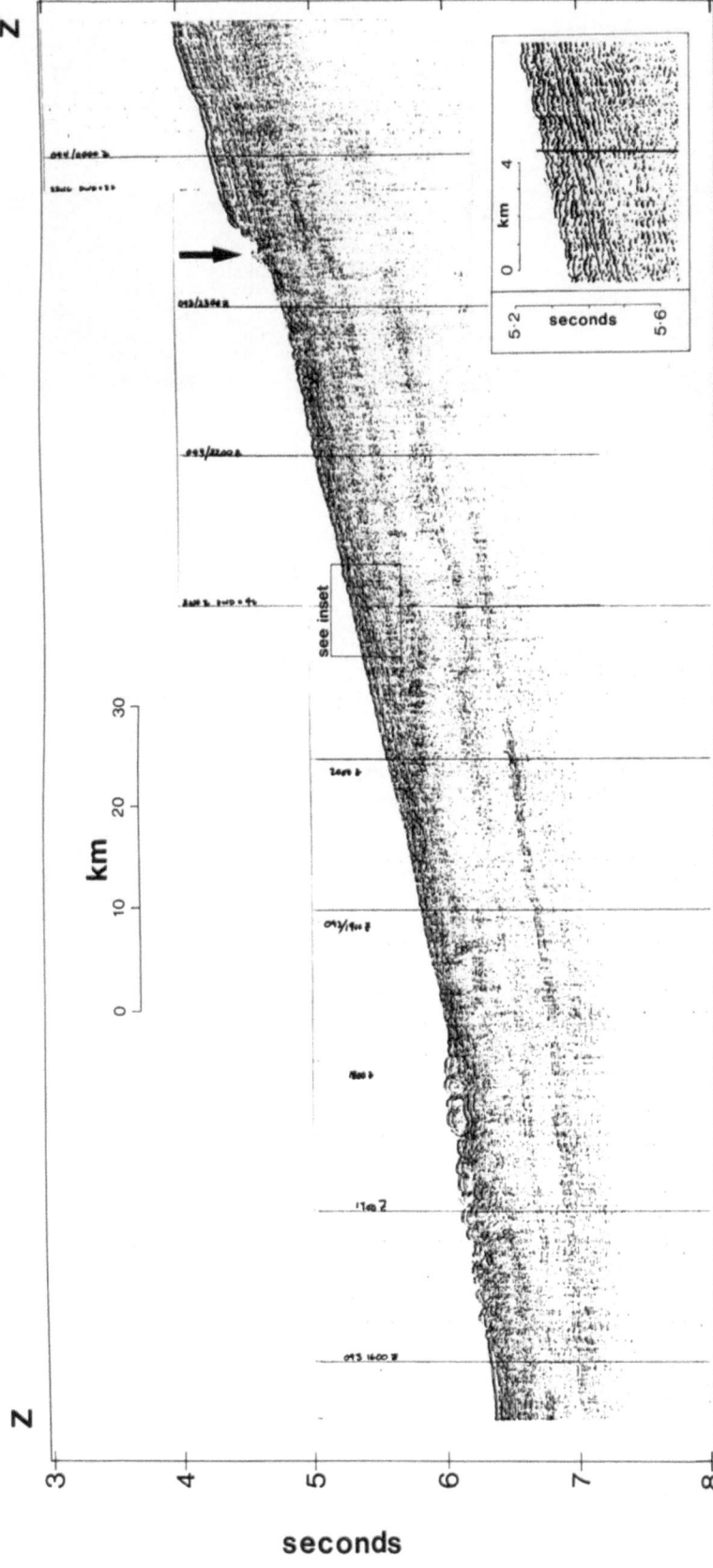

FIGURE 10-9. 160-in³ airgun seismic-reflection profile along line Z-Z′ of Figure 10-3. Inset shows nature of discontinuous convex reflections on rise downslope of scarp at 3,500 m (arrow).

FIGURE 10-10.   Enlarged portion of GLORIA mosaic illustrating stream-like debris-flow deposits (box c, Fig. 10-3).

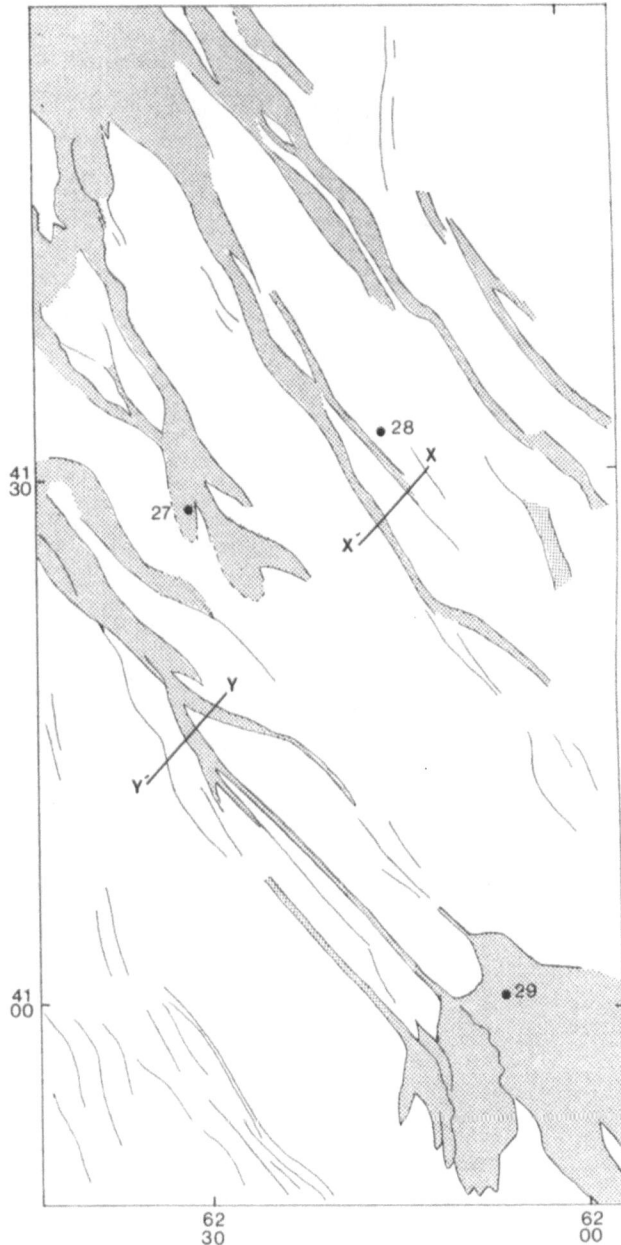

FIGURE 10-11.   Line drawing indicating spatial extent of stream-like debris-flow deposits (stipple pattern). Positions of piston cores 27, 28, and 29 (discussed in text) and 3.5-kHz lines shown in Figure 10-12 (X-X', Y-Y') are indicated.

Deep Boundary-Current Activity

Large scale morphologic features, indicative of the modern flow of water across the WNSR between 4,000 and 5,000 m (the zone of activity of the Cold Filament Current; Weatherly and Kelly, 1985), are conspicuously absent from the GLORIA image mosaic. The along-slope Holocene facies boundaries recognized by Driscoll, Tucholke, and McCave (1985) were not detected using GLORIA (see also Chapter 9).

The HEBBLE site (Hollister and McCave, 1984) lies among a series of narrow lower-rise channels (Figs.

10-3, 10-13). Thin debris-flow deposits, as recognized on 3.5-kHz records (Swift, 1985) and in box cores (McCave, 1985), extend through the HEBBLE site. The channel floors exhibit lower (darker) acoustic backscatter on GLORIA images than does the surrounding seafloor. No coherent evidence of deep boundary-current activity is visible from GLORIA, although the patchiness of regions of higher (brighter) acoustic backscatter may reflect the distribution of fields of small-scale abyssal bedforms (Flood and Lonsdale, 1981).

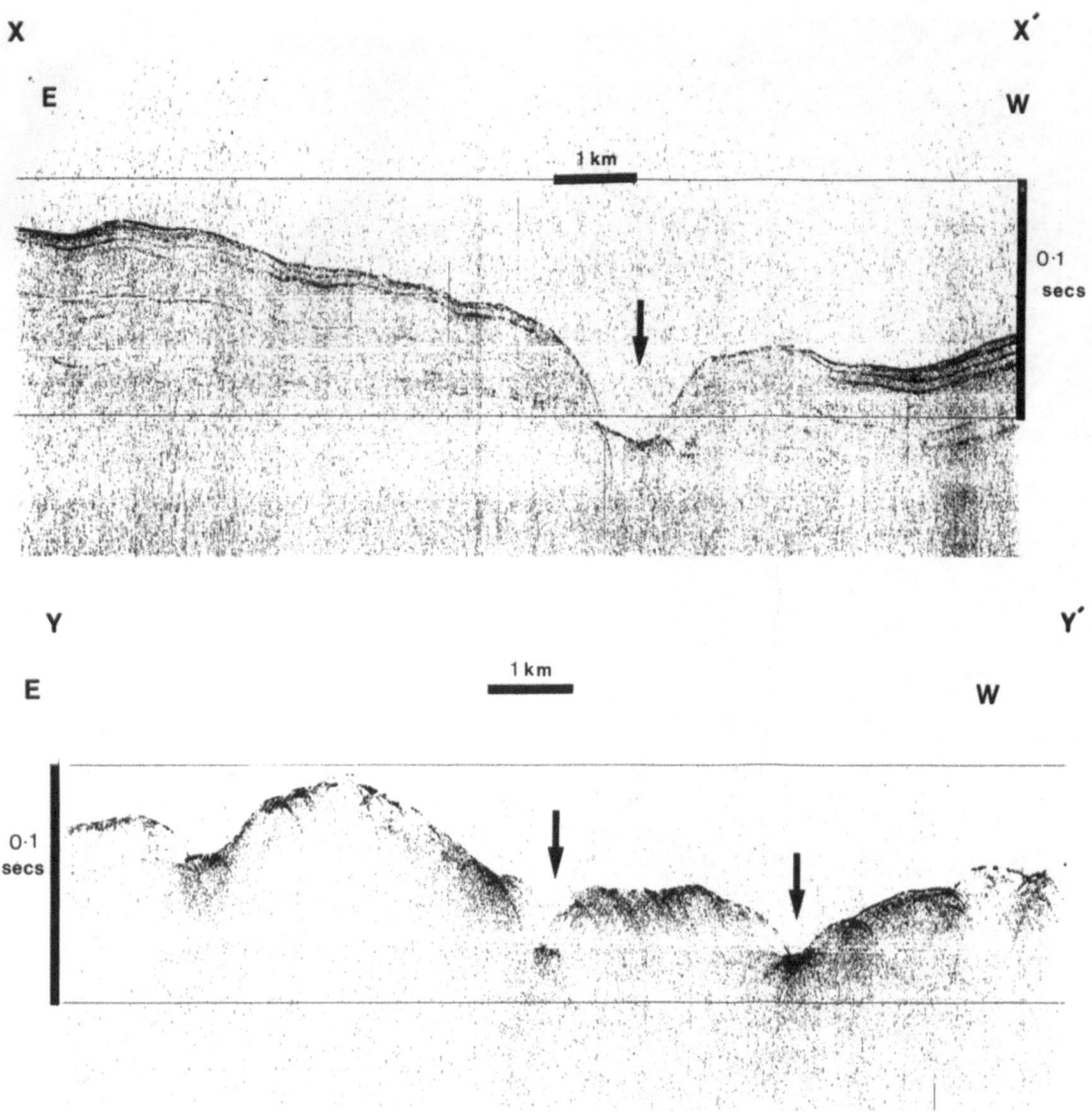

FIGURE 10-12.   3.5-kHz profiles across stream-like debris-flow deposits illustrating their confinement within rise channels (arrows). The location of profiles are shown on Figure 10-11.

Shor and Lonsdale (1981) measured the relief of the rise channels that pass through the HEBBLE site as 3–20 m. In the 12,000 years since the last glaciation, only about 1 m of sediment has been deposited (Stanley et al., 1972; McCave, 1985); thus the topographic expression in the HEBBLE region could not be explained by Holocene sedimentation.

That fraction of the Holocene section that has actively been reworked (foraminiferal sands) comprises less than 25% of the total Holocene record (McCave, 1985). It is thus apparent that the dominant control on continental rise morphology in the vicinity of the HEBBLE site has been quite different from the deep-current flow that is now prevalent. The GLORIA images indicate that late Pleistocene debris flows and

turbidity currents have been the major geomorphic agents (see Chapters 9, 12, 13).

The 3.5-kHz subbottom records of the lower rise east of lat. 62°00′ W (area W, Fig. 10-3) exhibit seafloor undulations with wavelengths of 1–3 km and amplitudes of 10–30 m. Swift (1985) suggested that these undulations may represent abyssal bedform fields of unknown orientation. Our GLORIA data (Fig. 10-2), however, show no coherent pattern of backscatter at the proposed wavelengths, although the rise appears to be faintly mottled. GLORIA is capable of depicting bedforms of the cited dimensions (Roberts and Kidd, 1979; Kidd and Roberts, 1982; Gardner and Kidd, 1987; see also Chapter 13). The changes in slope exhibited by these bedforms is, however, minute. It has been

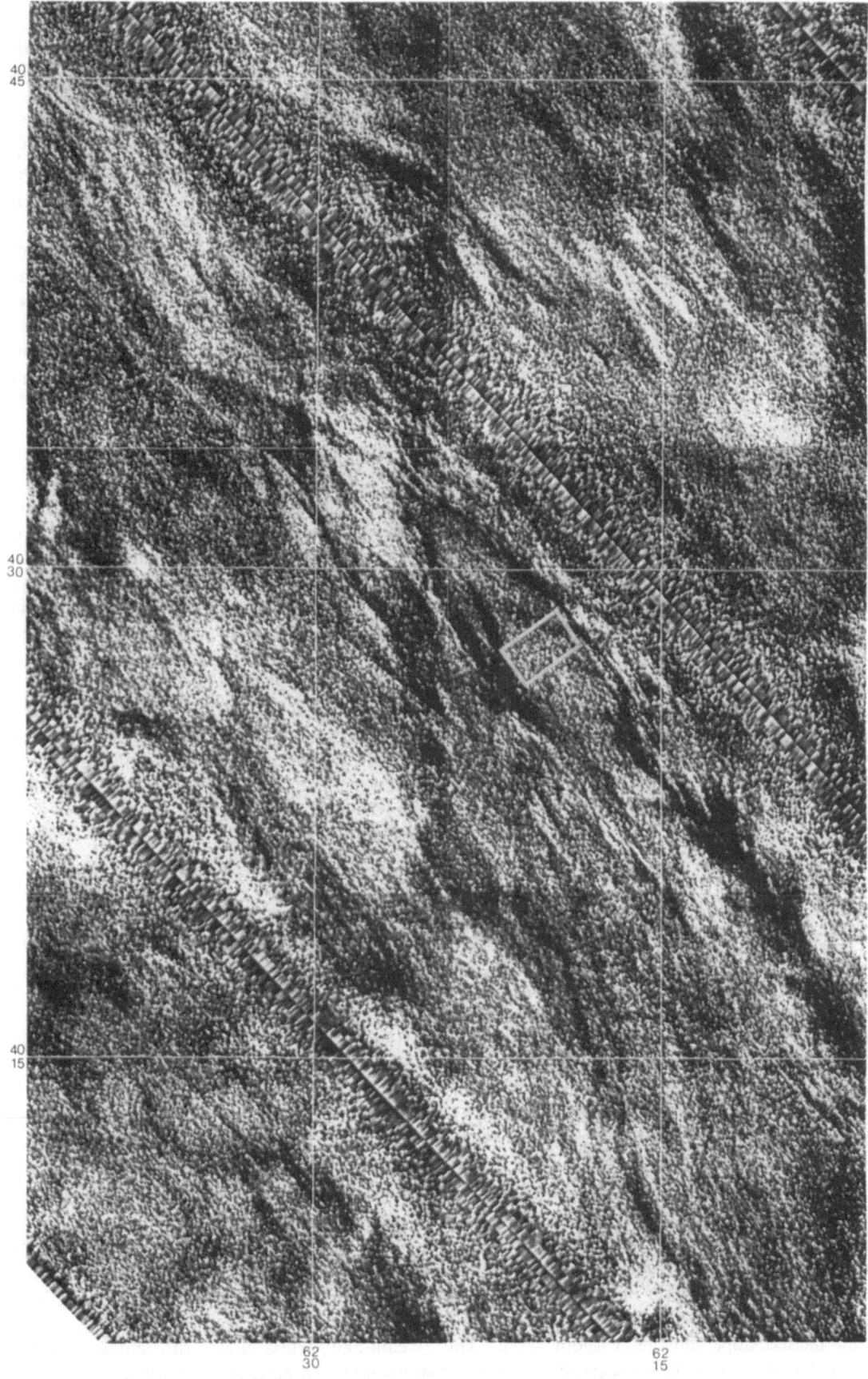

FIGURE 10-13. Detailed GLORIA mosaic in vicinity of HEBBLE meadow (outlined rectangle) showing expression of shallow channels and spatial distribution of reflective patches of seafloor (box d, Fig. 10-3).

suggested that for mid-range grazing angles, texturally induced changes in backscatter on GLORIA images are often of far greater importance than topographically induced changes (Somers and Mitchell, 1988). Thus the appearance or nonappearance of abyssal bedforms may be as much a function of textural difference between crests and troughs as of topographic relief. That there is no good representation of these known bathymetric features on the GLORIA images implies that there are no textural contrasts. This would suggest that the bedforms are inactive and have been draped by a uniform veneer.

Sedimentary Ridge Province

The sedimentary ridge province (Uchupi, Ballard, and Ellis, 1977) is located beneath the lower slope and upper rise between water depths of 2,000 and 3,000 m. Salt diapirs, identified on seismic reflection profiles (denoted by D, Fig. 10-3), are present beneath the lower slope east of long. 64°W; diapir crests rise to within 500 m of the modern seafloor. The sedimentary ridge province has been a major control on the thickness of late Tertiary and Pleistocene sediments (Jansa and Wade, 1974). Little evidence for the presence of diapirs can be interpreted from the GLORIA images, however. Their apparent absence implies that diapiric influence on late Pleistocene and Holocene sedimentation has been minor. The sinuosity of nearby channels (Fig. 10-3) may be evidence that diapirs influence the location of channels by modifying local seafloor gradients (see Chapter 5).

DISCUSSION

The spatial distribution of areas of contrasting backscatter apparent in the GLORIA image mosaic demonstrates a predominant downslope orientation indicative of gravity-driven processes. On the WNSR, however, despite the strong downslope-oriented grain, singular incised or leveed channels, common on the rise both to the east and west, are rarely developed. On the other hand, we observed a multitude of shallow channels with no (or poorly developed) levees.

Channels that cross the continental rise south of Georges Bank are clearly fed by discrete slope canyons (Emery and Uchupi, 1984). The upper sections of the canyons are dendritic in nature (Veatch and Smith, 1939; Scanlon, 1984), and thus presumably collect and focus turbidity currents that are initiated on the slope. Bathymetric data from between lat. 65°00′ and 62°00′W (Shor, 1984) reveal few incised slope canyons, without which turbidity currents would tend to be unfocussed. East of lat. 62°00′W, slope canyons are again common. We infer that the widespread development of many minor channels (rather than a few major channels) on

the rise between lat. 65°00′W and 62°00′W, reflects the poor development of canyons on the adjacent continental slope and the consequent poor focussing of turbidity currents.

The presence or absence of well-developed slope canyons and channels on the Nova Scotian margin can be explained in part by both the deep structure of the continental margin and the glacial history of the shelf. Jansa and Wade (1975) have recognized growth faults parallel to the shelf edge, which were active in the late Tertiary. They infer that down-to-basin movement on these faults, a result of sediment loading and deep shale and salt flowage, has resulted in a bathymetrically deeper shelfbreak and lower regional slope gradients between lat. 62°00′W and 65°00′W. As a result, the continental shelf from Emerald Bank to Brown Bank (Fig. 10-1) remained submergent throughout Pleistocene sea-level fluctuations.

The principal sites of glaciomarine activity during lowstands of sea level were the submergent sections of the shelfbreak, which include Northeast Channel (Torphy and Ziegler, 1956) and La Have Channel [informal name adopted by Swift (1985) for the saddle between Emerald and La Have Banks]. In contrast, Georges Bank and Western Bank both would have been emergent (King and Fader, 1986). Glacial ice probably did not cross Georges Bank (Pratt and Schlee, 1969) and apart from a period in the early Wisconsinan (Boyd, Scott, and Douma, 1988), glacial ice rarely covered Western Bank or Sable Island Bank (King and Fader, 1986).

With the ice margin north of these emergent, outer shelf banks, fine-grained glaciomarine sedimentation was concentrated along the submerged shelf-edge lows (Swift, 1985; Mosher et al., 1989). Because of the low slope gradients, the potential erosive power of turbidity currents was reduced, and the frequency of mass wasting was lowered. Hill (1984) has shown that these conditions resulted in high sedimentation rates on a low-relief slope, whose rare, poorly developed channels terminated in water depths of less than 2,000 m. As a result, a thick Pleistocene sedimentary section has developed on the continental slope (> 1,280 m in the Mohican I-100 well; Fig. 10-1; Jansa and Wade, 1974). In contrast, on the emergent bank tops, subaerial glacial outwash streams supplied coarse sediment to shelf-edge delta systems (Maclaren, 1988). The steeper gradients on the adjacent slopes and the influx of coarse sediment during lowstands of sea level provided conditions for erosion and maintenance of canyons (Piper, Normark, and Sparkes, 1987).

The stream-like debris flows, the slides, and the lower-rise channels all appear to have been active last in the late Pleistocene (and may have been generated then; see also Chapters 9, 12, 13). Apart from a few thin turbiditic muds within the Holocene sediment

(McCave, 1985; piston cores examined in this study), mass wasting appears to have been inactive during the last 12,000 years. Thus, the features visible on GLORIA images predominantly reflect the distribution of late Pleistocene sediments that underlie the veneer of Holocene hemipelagic ooze (Stanley et al., 1972). The effect of modern, intensified, deep boundary currents on the thin Holocene drape does not appear to be noticeable in the backscatter character of the Western Nova Scotian Rise, as recorded by GLORIA.

The presence or absence of contourite sediment in piston cores (Heezen and Hollister, 1972; Stow, 1979) or diagnostic magnetic fabrics (Flood, Kent, and Shor, 1985) can be used to indicate whether or not deep boundary currents were present in the late Pleistocene. Such criteria cannot, however, be used to determine the influence of deep boundary currents on gross morphology of the rise. The expression of late Pleistocene sedimentation, with a predominance of downslope-trending landforms, as revealed by our GLORIA survey, is an indication that deep boundary currents were not an important agent in controlling the morphology of the rise at that time.

SUMMARY AND CONCLUSIONS

A GLORIA survey of the continental rise off western Nova Scotia clearly demonstrates the predominance of downslope-oriented landforms of modern and late Pleistocene age. Holocene hemipelagic ooze as thick as 1.5 m coats the region, but it does not appear to have any significant influence on the spatial pattern of acoustic backscatter, even though it has been eroded locally by deep boundary currents. Coring indicates that the major downslope-oriented features date from the late Pleistocene and comprise primarily multiple shallow channels and stream-like muddy debris flow deposits. The large number and widespread distribution of channels on the rise indicate that there are no preferred sediment pathways across the continental rise east of Georges Bank and west of Mohican Channel (contrast with Chapters 12, 13).

The morphological character and the upper Pleistocene sedimentology of the rise can be related to the interaction between Pleistocene glacial activity on the shelf and the combined effect of outer shelf bathymetry and continental slope gradient. Because these controls have resulted in an aggraded slope, only rarely incised by canyons, sediment transport across the rise is poorly focussed, and lacks local fan development; as a result the rise appears broadly convex. Modern boundary-current activity has been incapable of remolding the rise (at a scale visible from GLORIA) in the 12,000 years since the last major phase of downslope sediment transport.

In general, despite modern alongslope remobilization of deep-sea sediment on the continental rise off western Nova Scotia, the major control on the large-scale late Pleistocene evolution of the rise has been that of turbidity currents, debris flows, and slides. Deep boundary-current activity was of minor importance in the development of the late Pleistocene morphology of the WNSR.

We draw three specific conclusions from this study:

1. The landform and deposition pattern on the Western Nova Scotian Rise as revealed by GLORIA is predominantly the expression of late Pleistocene processes; the pattern does not reflect processes that have been active through the Holocene. This conclusion is, in part, biased by the resolution of the imaging technique we used. GLORIA responds to the dominant acoustic characteristics of the upper few meters of the sedimentary section, which on the Western Nova Scotian Rise includes mainly upper Pleistocene sediments. On the other hand, studies of surficial sediment using box cores or cameras are biased toward modern or Holocene sedimentary processes.
2. Even in the absence of significant downslope activity, modern deep boundary currents have been incapable of remolding the seafloor at a scale detectable by GLORIA during a period of approximately 12,000 years.
3. The low physiographic relief of the WNSR is not a product of current-controlled sediment redistribution, but a result of the interaction between late Tertiary faulting on the continental margin and Pleistocene glaciations on the Scotian Shelf.

ACKNOWLEDGMENTS

The 1987 GLORIA survey was cooperatively conducted by the U.S. Geological Survey and the Canadian Geological Survey. The assistance of the Institute of Oceanographic Sciences, U.K., whose technicians were responsible for operating the GLORIA system and the seismic profiler, and of the Canadian Hydrographic Survey, is greatly appreciated. Archived core data were kindly made available by the Lamont-Doherty Geological Observatory, New York. The first author was in receipt of an Isaac Walton Killam Scholarship and an NSERC visiting fellowship while this research was undertaken. The GLORIA investigation was partly funded by the Canadian Frontier Geoscience Program; subsequent investigation of sediment instability was funded by the Canadian Program for Energy Research and Development, Stability of Continental Slopes project. We thank James M. Robb and C. Wylie Poag for constructive reviews of the manuscript.

REFERENCES

Boyd, R., Scott, D. B., and Douma, M. 1988. Glacial tunnel valleys and Quaternary history of the outer Scotian shelf. *Nature* 333:61–64.

Chavez, P. S. 1986. Processing techniques for digital sonar images from GLORIA. *Photogrammetric Engineering and Remote Sensing* 52:1135–1145.

Damuth, J. E. 1980. Use of high frequency (3.5–12 kHz) echograms in the study of near bottom sedimentation processes in the deep-sea; a review. *Marine Geology* 38:51–75.

Damuth, J. E., Tucholke, B. E., and Coffin, M. F. 1979. Bottom processes on the Nova Scotian continental rise revealed by 3.5 kHz echo character. *EOS, Transactions of the American Geophysical Union* 60:855.

Damuth, J. E., Tucholke, B. E., and Shor, A. N. 1981. Bathymetry and near bottom sedimentation processes off the Nova Scotian continental rise. *EOS, Transactions of the American Geophysical Union* 62:892.

Driscoll, M. L., Tucholke, B. E., and McCave, I. N. 1985. Seafloor zonation in sediment texture on the Nova Scotian lower continental rise. *Marine Geology* 66:25–41.

Emery, K. O., and Uchupi, E. 1984. *The Geology of the Atlantic Ocean*: New York: Springer-Verlag.

Field, M. E., and Hall, R. K. 1982. Sonographs of submarine sediment failure caused by the 1980 earthquake off northern California. *Geo-Marine Letters* 2:135–141.

Flood, R. D., and Lonsdale, P. 1981. Longitudinal triangular ripple distribution patterns in the Nova Scotian continental rise HEBBLE area. *EOS, Transactions of the American Geophysical Union* 62:892.

Flood, R. D., Kent, D. V., and Shor, A. N. 1985. The magnetic fabric of surficial deep-sea sediments in the HEBBLE area. *Marine Geology* 66:149–168.

Gardner, J. V., and Kidd, R. B. 1987. Sedimentary processes on the northwestern Iberian continental margin viewed by long-range sidescan sonar and seismic data. *Journal of Sedimentary Petrology* 57:397–407.

Hill, P. D. 1984. Sedimentary facies of the Nova Scotian upper and middle continental slope, offshore eastern Canada. *Sedimentology* 31:293–309.

Hollister, C. D., and Heezen, B. C. 1972. Geological effects of ocean bottom circulation in the N.W. Atlantic. *In* Studies in Physical Oceanography, Volume 1, A. L. Gordon, Ed.: New York: Gordon and Breach, 37–66.

Hollister, C. D., and McCave, I. N. 1984. Sedimentation under deep-sea storms. *Nature* 309:220–225.

Jansa, L. F., and Wade, J. A. 1975. Geology of the continental margin off Nova Scotia and Newfoundland. *In* Offshore Geology of Eastern Canada, Volume 2, W. J. M. van der Linden and J. A. Wade, Eds.: *Geological Survey of Canada Paper 74-30*, 55–105.

Johnson, D. 1939. The Origin of Submarine Canyons, a Critical Review of Hypotheses: New York: Columbia University Press.

Kidd, R. B., and Roberts, D. G. 1982. Long-range sidescan-sonar studies of large-scale sedimentary features in the North Atlantic. *Bulletin of the Institut Geologique du Bassin d'Aquitaine* 31:11–29.

King, L. H., and Fader, G. B. J. 1986. Wisconsinan glaciation of the Atlantic continental shelf of southeast Canada. *Geological Survey of Canada Bulletin 363*, 72.

McCave, I. N. 1985. Sedimentology and stratigraphy of box-cores from the HEBBLE site on the Nova Scotian continental rise. *Marine Geology* 66:59–90.

Mclaren, S. A. 1988. Quaternary seismic stratigraphy and sedimentation of the Sable Island sand body, Sable Island Bank, outer Scotian Shelf. Technical Report 11, Centre for Marine Geology, Dalhousie University, Halifax.

Mosher, D. C., Piper, D. J. W., Vilks, G. V., Aksu, A. E., and Fader, G. B. 1989. Evidence for Wisconsinan glaciations in the Verrill Canyon area, Scotian Slope. *Quaternary Research* 31:27–40.

Nowell, A. R. M., and Hollister, C. D., Eds. 1985. Deep ocean sediment transport—preliminary results of high energy benthic boundary layer experiment. *Marine Geology* 66:1–409.

Piper, D. J. W., Normark, W. R., and Sparkes, R. 1987. Late Cenozoic stratigraphy of the Central Scotian Slope, Eastern Canada. *Bulletin of Canadian Petroleum Geology* 35:1–11.

Pratt, R. M., and Schlee, J. S. 1969. Glaciation on the continental margin off New England. *Geological Society of America Bulletin* 80:2335–2342.

Roberts, D. G. and Kidd, R. B. 1979. Abyssal sediment wave fields on Feni Ridge, Rockall Trough: Long-range sonar studies. *Marine Geology* 33:175–191.

Scanlon, K. M. 1984. The continental slope off New England: A long-range sidescan-sonar perspective. *Geo-Marine Letters* 4:1–4.

Shor, A. N. 1984. Bathymetry. *In* Ocean Margin Drilling Program Regional Data Synthesis Series Atlas 2, Eastern Northern American Continental Margin and Adjacent Ocean Floor. 39° to 46°N and 54° to 64°W, A. N. Shor and E. Uchupi, Eds.: Woods Hole, MA: Marine Science International, Sheet 1.

Shor, A. N., and Lonsdale, P. 1981. HEBBLE site characterization. Downslope processes on the Nova Scotian continental rise. *EOS, Transactions of the American Geophysical Union* 62:892.

Shor, A. N., and Piper, D. J. W. 1989. A large Late Pleistocene blocky debris flow on the central Scotian Slope. *Geo-Marine Letters* 9:153–160.

Somers, M. L., Carson, R. M., Revie, J. A., Edge, R. H., Barrow, B. J., and Andrews, A. G. 1978. GLORIA II—an improved long-range sidescan sonar. Proceedings IEE/IERE Subconference Offshore Instrument Committee (Oceanology International, 1978): London: B.P.S. Publications Ltd., 16–24.

Somers, M. L., and Mitchell, N. C. 1988. Quantitative backscatter measurements with a long range side-scan sonar. *EOS, Transactions of the American Geophysical Union* 69:1440.

Stanley, D. J., Swift, D. J. P., Silverberg, N., James, N. P., and Sutton, R. G. 1972. Late Quaternary progradation and sand spillover on the outer continental margin off Nova Scotia, Southeast Canada. *Smithsonian Contributions to the Earth Sciences* 8:88.

Stow, D. A. V. 1979. Distinguishing between fine-grained

turbidites and contourites on the deep-water margin off Nova Scotia. *Sedimentology* 26:371–387.

Streeter, S. S., and Schackleton, N. J. 1979. Paleocirculation of the deep North Atlantic: 150,000 year record of benthic foraminifera and oxygen 18. *Science* 203:168–171.

Swift, S. A. 1985. Late Pleistocene sedimentation on the continental slope and rise off Western Nova Scotia. *Geological Society of America Bulletin* 96:832–481.

Torphy, S. R., and Ziegler, J. M. 1956. Submarine topography of Eastern Channel, Gulf of Maine. *Journal of Geology* 65:433–441.

Uchupi, E., Ballard, R. D., and Ellis, J. P. 1977. Continental slope and upper rise off Western Nova Scotia and Georges Bank. *American Association of Petroleum Geologists Bulletin* 61:1483–1492.

Weatherly, G. L., and Kelley, E. A., Jr. 1985. Storm and flow reversals at the HEBBLE site. *Marine Geology* 66:205–218.

11

Guinea and Ivory Coast – Ghana Transform Margins: Combined Effects of Synrift Structure and Postrift Bottom Currents on Evolution and Morphology

Jean Mascle, Christophe Basile, Frances Westall, and Sergio Rossi

The present-day morphology of passive margins is directly influenced by two prevailing factors: (a) the structural framework inherited from their rifting evolution; and (b) the postrift sedimention controlled by paleoclimate and paleoceanography. In many areas, the postrift sedimentary wedge is thick and has totally buried the synrift structural features. As a result, the margin morphology in these thickly sedimented areas is chiefly controlled by oceanographic and climatic characteristics (see Chapters 4–6). Transform margins, in contrast, represent a special case, because tectonic control continues to exert a strong influence on their morphologies even after a long period of geological time (see also Chapter 1).

The morphology of transform continental margins can be generally easily distinguished from that of other divergent margins (see also Chapter 7):

1. They are laterally continuous with major, extinct, oceanic fractures zones.
2. They often display sharp transitions (steep slopes) between shallow continental shelves or plateaus and deep oceanic rises or abyssal plains.
3. They are often characterized by elongated marginal ridges located at the junction between extinct fracture zones and continental platforms (Blarez, 1986; Mascle and Blarez, 1987).

In some areas, these morphological characteristics, especially the presence of long, linear, steep slopes, have provided favorable conditions for bottom currents to control sedimentation and seafloor morphology.

In this chapter, we briefly report the results of bathymetric and seismic-reflection surveys conducted on two segments of the transform margins of the West African borderland, off southern Guinea and the Ivory Coast–Ghana (Fig. 11-1). On the southern Guinea margin, the combination of prevailing structural and oceanographic–hydrologic controls (Fig. 11-2) has led to the formation of large morphostructures, particularly well developed along the continental slope and rise (Mascle, Blarez, and Marinho, 1988; Mascle and Auroux, 1989). Topographically induced erosion is an important control of seafloor morphology in both areas.

STRUCTURAL AND GEOLOGICAL BACKGROUND

The equatorial Atlantic, situated at the junction between two oceanic regions of different ages (the Jurassic North Atlantic and the Cretaceous South Atlantic), appears to be dominated mainly by east–west oriented structural trends, known as the Equatorial Fracture Zone Belt (Heezen et al., 1964; Fig. 11-1). This system of active fracture zones presently offsets the Mid-Atlantic Ridge (MAR) left-laterally by about 2,800 km between lat. 10°N and the equator. This offset closely matches margin offsets around the bulge of Africa and off eastern Brazil at the same latitudes (Le Pichon and Hayes, 1971).

The Equatorial Fracture Zone Belt is bounded to the north by the Doldrums Fracture Zone (~ 08°N; Leclere-Vanhoeve, 1988) and to the south by the Romanche Fracture Zone, which offsets the MAR left-laterally by about 950 km (and represents probably the world's largest ocean rift–rift transform fault; Emery et al., 1975; Gorini, 1977; Chermak, 1979; Figs. 11-1, 11-2). The Doldrums Fracture Zone appears to extend eastward to the northern flank of the Sierra Leone

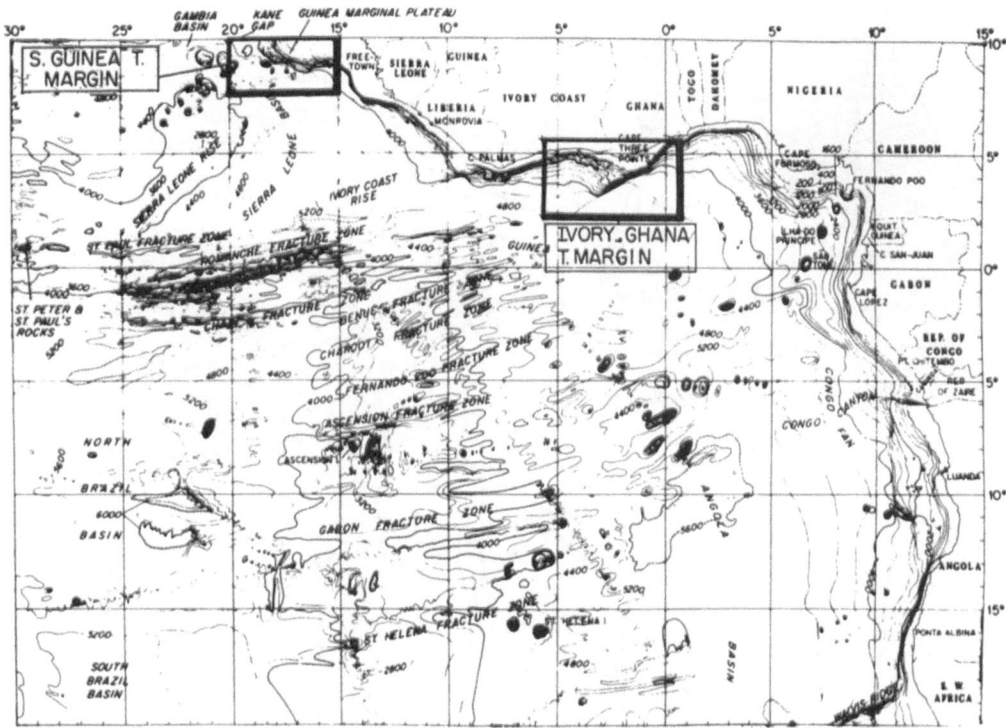

FIGURE 11-1. General bathymetric map of eastern Equatorial Atlantic (after Emery et al., 1975) showing main fracture zones that offset Mid-Atlantic Ridge near Equator, and two transform margin segments discussed in text; Ivory Coast–Ghana transform margin is aligned with largest of equatorial fracture zones, the Romanche Fracture Zone.

Rise (Jones, 1987). Farther east, the fracture zone may connect with another east–west trending fracture, the Guinea Fracture Zone (Krause, 1964). The latter is believed to follow the southern Guinea margin and seems to have been an important structural control on the complex, steep, continental slope facing the deep Sierra Leone abyssal plain (Jones and Mgbatogu, 1982; Mascle, Marinho, and Wannesson, 1986; Mascle, Blarez, and Marinho, 1988). In short, the structural framework of the southern Guinea margin segment can be regarded as the direct heritage of a two-phase rifting history:

1. The Guinea margin was created in the Jurassic as a divergent margin in response to the opening of the central Atlantic (Klitgord and Schouten, 1986).
2. Later, during the Cretaceous, the margin, and particularly its east–west trending southern slope, was reactivated in response to the dominantly oblique transform opening of the equatorial Atlantic (Mascle, Blarez, and Marinho, 1988).

The resulting structural framework consists of a series of slope scarps, narrow, elongated, slope-parallel basins, and protruding volcanic features, draped or covered by current-molded sediments (Jones and Mgbatogu, 1982; Marinho, 1985; Mascle, Blarez, and

Marinho, 1988; Westall, Rossi, and Mascle, in press; Fig. 11-3).

Off the Ivory Coast and Ghana, the transform margin (Fig. 11-4) appears to have resulted from successive shearing between the two rifting continents (Africa and South America; Blarez, 1986; Mascle and Blarez, 1987; Mascle, Blarez, and Marinho, 1988) and it developed between a thinned continental crust (African) and a newly created oceanic crust (in the Cretaceous). Both the Ivory Coast–Ghana and the Guinea margins were later subjected to subsequent thermal cooling and subsidence.

Regardless of the details of geological history along these two separate segments, the most conspicuous structural features are strongly east–west oriented margin segments. Both display steep, linear slopes in rapid transition from the shallow continental shelf and plateau to the adjacent continental rise or the abyssal plain. The continental slope off Ghana, for example, has an average width of about 30 km, which is about one-quarter of the usual slope width at other divergent margins (see Chapters 4–6).

MAIN OCEANOGRAPHIC CHARACTERISTICS

The equatorial Atlantic region appears to be a key zone for the interchange of deep water, intermediate water, and surface water circulation. Bottom water

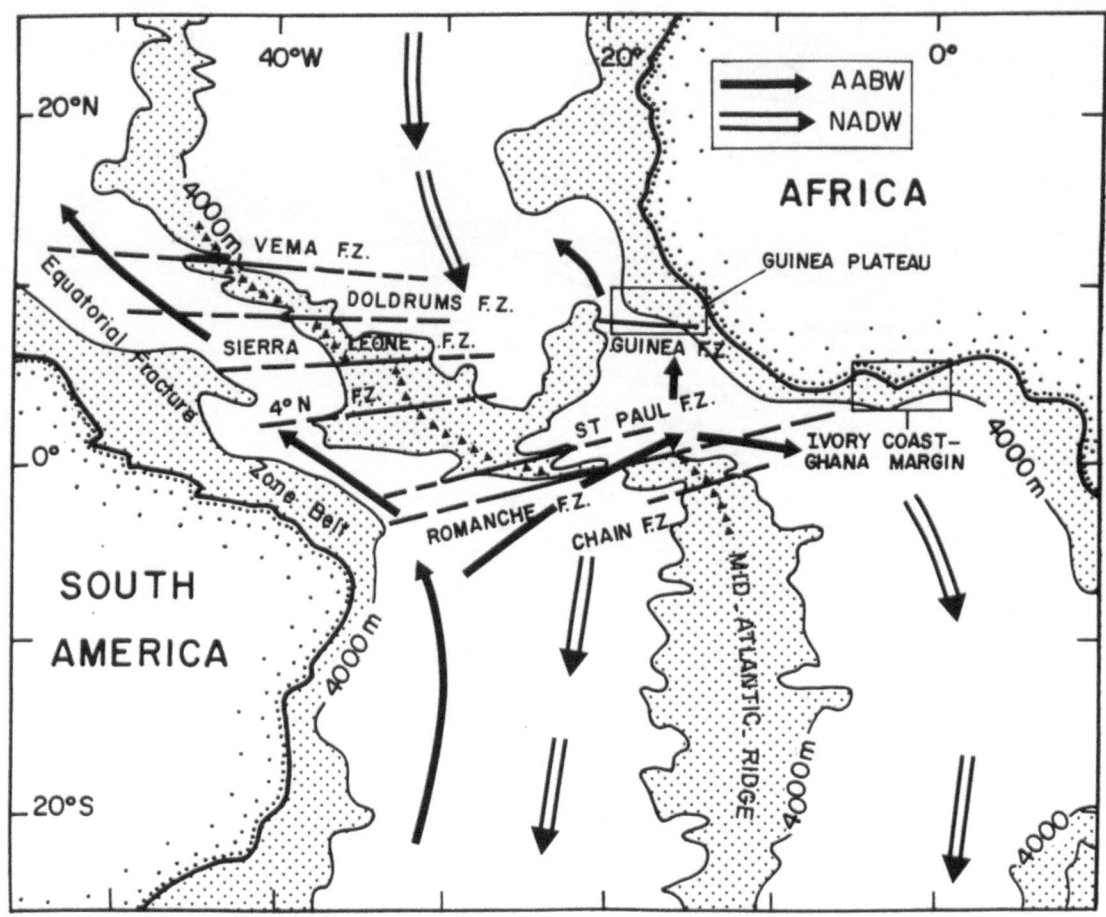

FIGURE 11-2.   Schematic flow diagram for modern AABW and NADW within equatorial area
(adapted from Westall, Rossi, and Mascle, in press). NADW originates in the North Atlantic and flows
southward, passing across the Sierra Leone abyssal plain into the Guinea–Angola Basin; parts of it pass
across the Equatorial Fracture Zone to move into the Brazil Basin. AABW moves northward from
Antarctica as a western boundary current into the Brazil Basin, passing through the deep Romanche
Fracture Zone in the Sierra Leone Basin, and flows north along the west African margin. Here, it mixes
with NADW and flows east and south along the Gulf of Guinea margin.

originating in the deep South Atlantic (Antarctic Bottom Water; AABW) moves from the western South Atlantic basin through the deep trough of the Romanche Fracture Zone into the Sierre Leone Basin to the east (Fig. 11-2). The Romanche Fracture Zone is a particularly important conduit for AABW flow into the eastern Atlantic (Warren, 1981; Mantyla and Reid, 1983). In the Sierra Leone Basin, AABW mixes with the southward moving deep and intermediate water mass, the North Atlantic Deep Water (NADW; Wüst, 1935). The mixed water mass then flows north, washing the foot of the Guinea Rise at depth below 4,250 m (Mienert, 1986), as well as east and southwards into the Guinea Basin, along the foot of the Ivory Coast–Ghana transform margin.

At intermediate depths, the southward and eastward moving NADW sweeps the southern Guinea margin in counterflow to the mixed AABW/NADW at depth (Westall, Rossi, and Mascle, in press; Rossi

et al., unpublished data). Along the Ivory Coast–Ghana margin, however, both NADW and AABW flow southeastward.

CONSEQUENCES FOR MORPHOLOGY OF THE CONTINENTAL SLOPE AND RISE

Southern Guinea Margin

The southwestern and southern slopes of the southwestern segment of this margin display contrasting morphologies (Figs. 11-3, 11-5, 11-7). The western segment, where structural and bottom-current influences are particularly obvious, includes a large, mid-slope basin located between two scarps. The eastern segment is characterized by strong magmatic influences. Detailed multibeam bathymetric and seismic reflection surveys allow us to draw the following conclusions

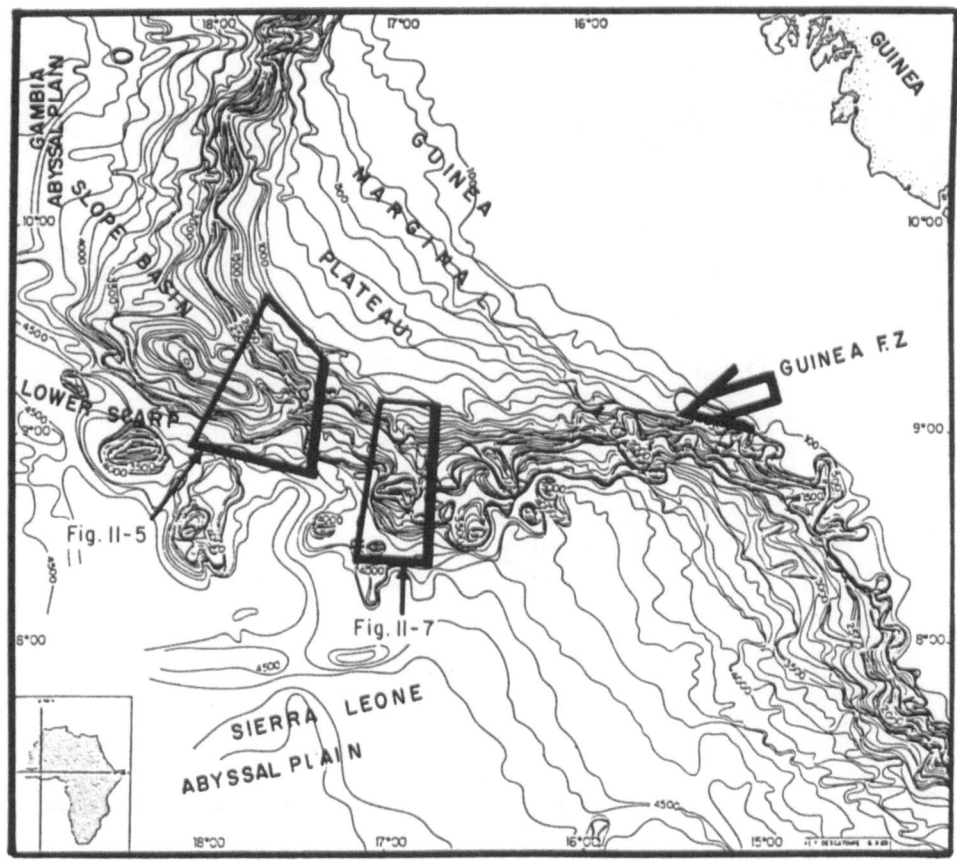

FIGURE 11-3. Bathymetry of Guinea continental margin; contour interval 100 m (from Marinho, 1985). Note sharp difference between western area, facing Gambia abyssal plain, and steep southern slope facing Sierra Leone abyssal plain. Numerous volcanoes are emplaced, mid-slope, along southern segment; they are believed to be connected with formation of Guinea Fracture Zone. The two boxes show area where detailed surveys were conducted in 1988 (from Mascle and Auroux, 1989).

(Mascle and Auroux 1989; Benkhelil et al., 1989; Tricart et al., 1991).

1. The primary control on the present seafloor morphology is the structural framework, particularly the results of the Cretaceous tectonic reactivation phase. This tectonic control is attested by: (a) The general NW–SE to E–N trend (Figs. 11-3, 11-5) of the steep continental slope. (b) The presence of two subparallel fault-controlled scarps. The upper scarp, at the edge of the marginal plateau, results from a series of normal faults, which cut across pre-Cenomanian sedimentary strata. The lower scarp truncates a slope-parallel marginal ridge, and constitutes a sharp transition to the continental rise of the Sierra Leone Basin. This scarp also seems to have cut Mesozoic strata (Marinho, 1985). (c) Finally a mid-slope or intermediate basin, located between the two scarps (Jones and Mgbatogu, 1982), is an important morphological component of the western sector of the southern Guinea margin. This basin contains gently folded pre-Cenomanian strata.

The overall basin structure is believed to have resulted from successive episodes of block faulting and slight folding during the Cretaceous, when the area was submitted to shear stress accompanying the opening of the South Atlantic (Benkhelil et al., 1989; Tricart et al., 1991).

2. A second, important, but secondary, control on the present-day morphology is the bottom-current control of sediment deposition on this margin. Both the NADW (flowing southeastward) and the AABW (flowing northwestward) move along the Guinea margin. We infer that NADW movement is responsible for several bottom-current–controlled features present along the continental slope, particularly along the upper scarp and within the intermediate basin (between 1,200 and 3,000 m); Westall, Rossi, and Mascle, in press; see also Chapters 4, 5, 8, 12). Along the edge of the Guinea marginal plateau (between 1,200 and 1,800 m water depth), bottom currents are strong enough to prevent sediment deposition as well as to erode the sedimentary edge. Simple piston coring has retrieved Lower Creta-

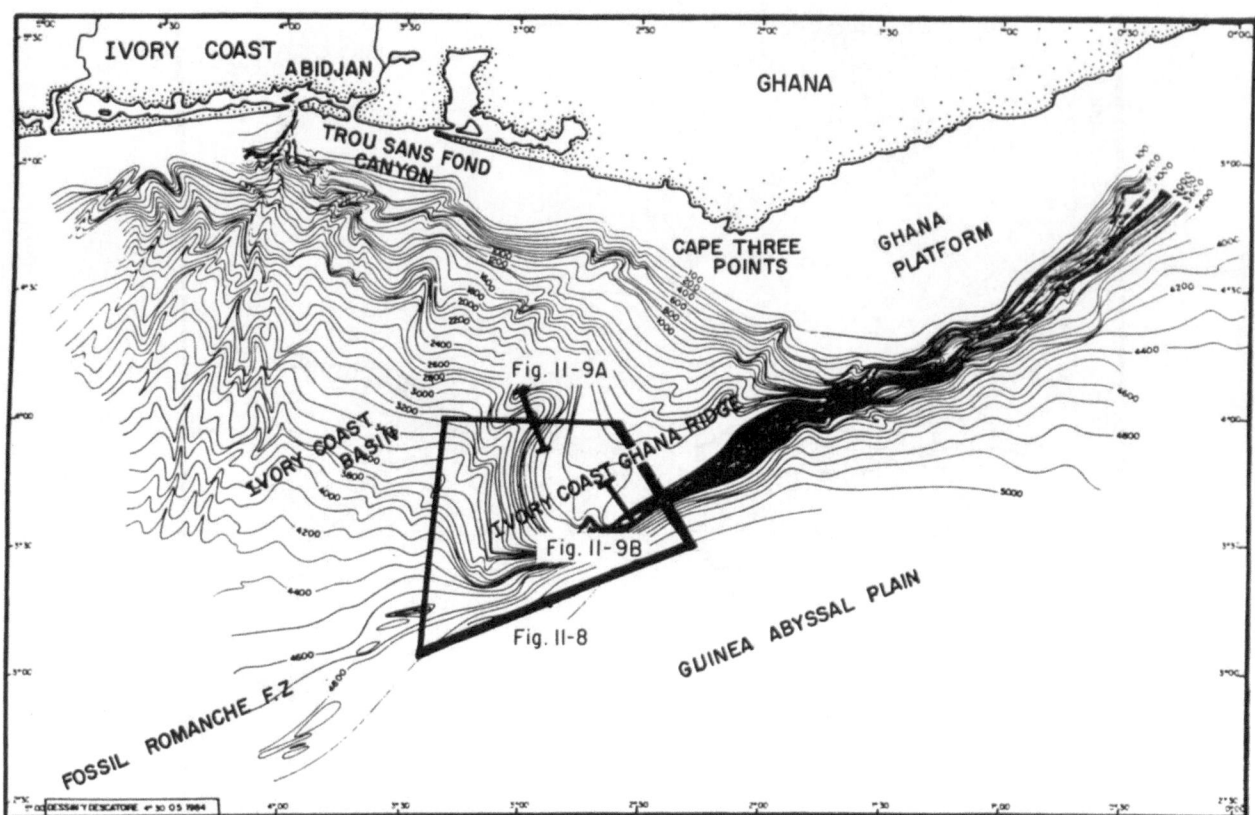

FIGURE 11-4.   Bathymetry of Ivory Coast–Ghana continental margin, contour interval 100 m (from Blarez, 1986). Note very steep Ghana continental slope (off Ghana Platform) that extends southwest-ward as Ivory Coast–Ghana Ridge. Ridge forms southern border of mid-slope Ivory Coast Basin, which represents segment of a divergent margin. Box illustrates area where detailed surveys were conducted in 1988 (from Mascle and Auroux, 1989).

ceous to Cenozoic strata exposed along the edge of the plateau (Fig. 11-6A; Marinho et al., 1984). Con-solidated nonfossiliferous sandstones are exposed along the upper scarp (Marinho, 1985). In the east-ern sector, bottom-current erosion effects also have been detected along the marginal plateau edge. Erosion is indicated by the presence of hardgrounds sampled on the summit of a large, mid-slope vol-cano (Fig. 11-7).

Within the intermediate basin, particularly to the west, unconsolidated sediments are concentrated into large, thick, elongated sediment drifts, whose axes are more or less parallel to the continental slope. These sedimentary drifts are separated by areas of condensed or eroded sediments, which extend into the deeper sedimentary layers (Fig. 11-6B). Those layers, which indicate probable zones of high velocity currents (Westall, Rossi, and Mascle, in press), are evidence of long-term bottom-current activity in this area. The seafloor along the entire slope and rise is covered by current-molded sediment waves of varying sizes (a few

tens of meters to one kilometer), whose wavelengths apparently depend on the slope steepness. These waves are presumed to have resulted from sediment rework-ing and shaping by the eastward-flowing intermediate waters and deep NADW.

Ivory Coast – Ghana Margin

Farther to the southeast, the Ivory Coast–Ghana mar-gin also has developed chiefly as the consequence of shearing processes during the Cretaceous rifting of the equatorial Atlantic. The resulting transform margin is a heritage of a tectonism, which is still active in the Romanche Fracture Zone (Figs. 11-1, 11-2). The main morphological characteristics of this margin segment are: (a) a very steep, sublinear, narrow continental slope off Ghana; (b) a highstanding ridge, known as the Ivory Coast–Ghana Ridge, extending to the southwest as a prolongation of the Ghana slope; and (c) a deep basin, the Ivory Coast Basin, bound to the south by the marginal ridge (Fig. 11-4; Blarez, 1986). SEABEAM and seismic–reflection data collected over the last 10

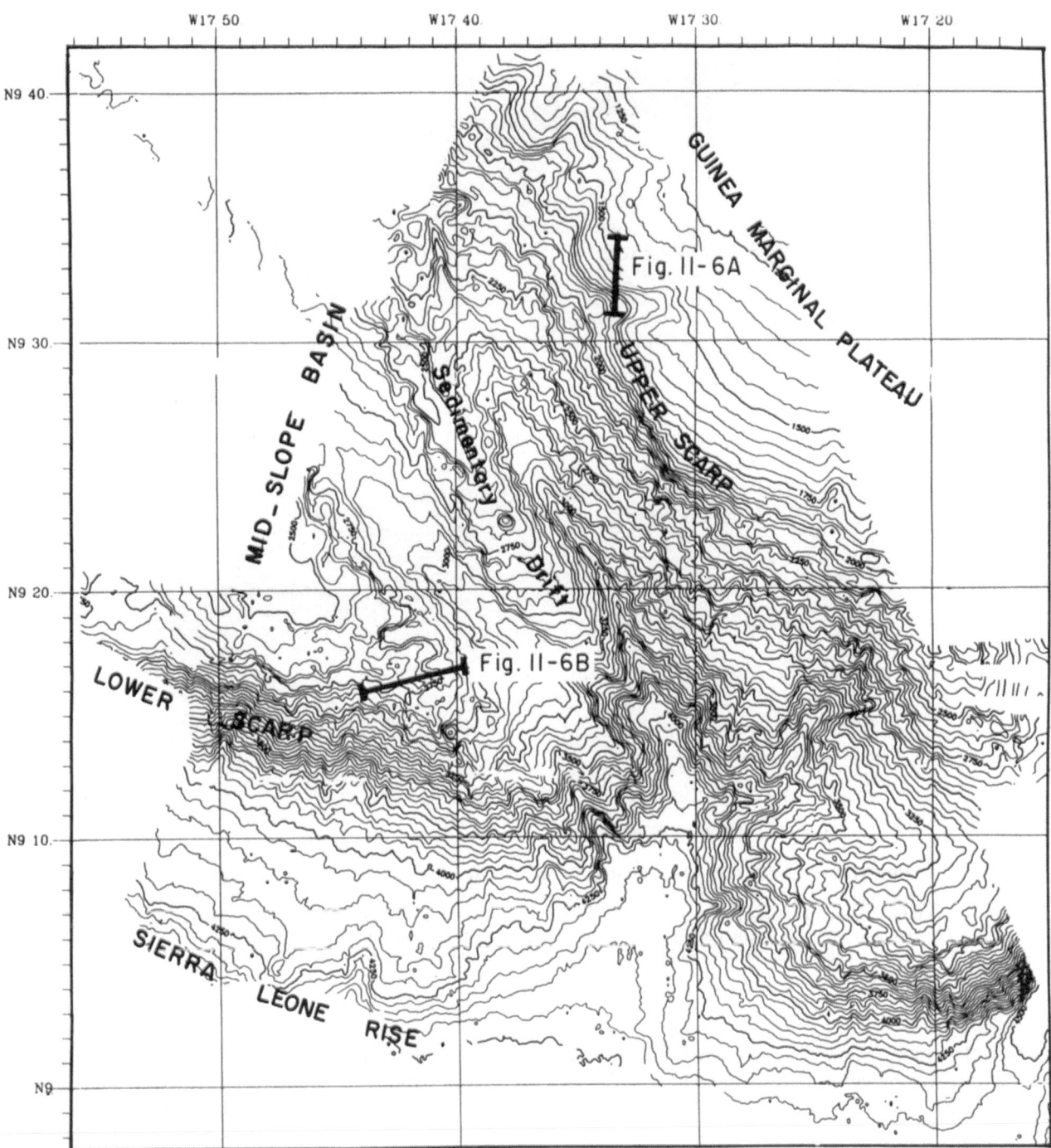

FIGURE 11-5. Detailed SEABEAM map of part of Guinea continental margin (contour interval 50 m). Upper scarp (1,700–2,500 m) bounds Guinea marginal plateau. Lower scarp (3,000–4,000 m) faces narrow continental rise (4,000–5,000 m). Mid-slope valley has developed between scarps; valley is a major conduit for bottom water flow. Large sediment drifts are present within basin, the largest of which extends NW out of study area and attains length of 60 km.

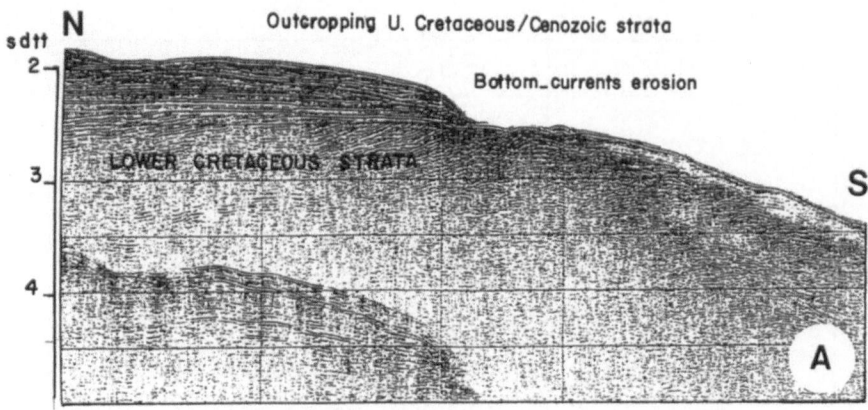

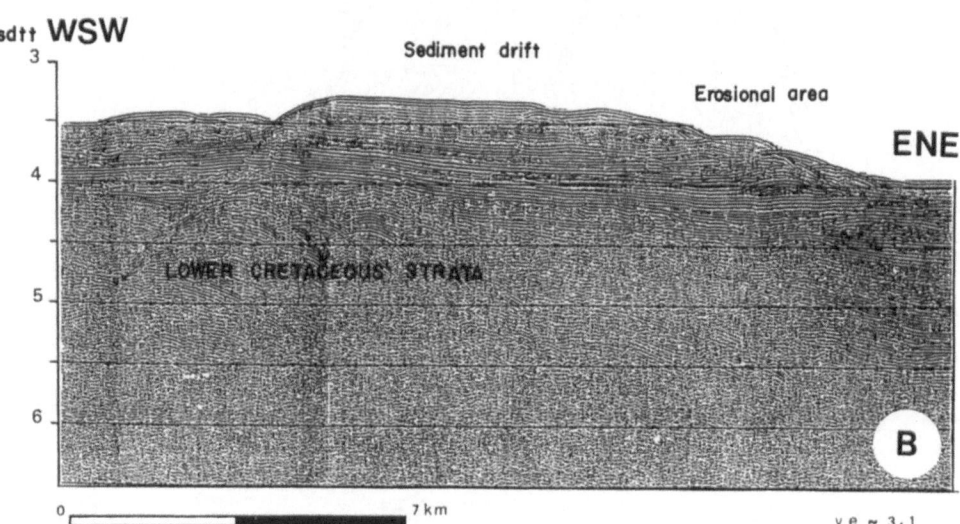

FIGURE 11-6.   A. Seismic profile illustrating sedimentary wedge along edge of Guinea marginal plateau; flat-lying sediments covering most of plateau progressively thin toward southern border; this bottom-current–induced phenomenon has allowed us to sample Upper Cretaceous to Cenozoic sedimentary layers. B. Section of seismic profile across intermediate basin. Sedimentary bedforms and erosional features believed to result from NADW bottom-current activity are particularly well developed in this area.

years (Fig. 11-8) have allowed us to analyze the geological structures of this area. Blarez and Mascle (1988), Basile (1990), Basile, Brun, and Mascle (*in press*) have presented detailed structural studies of this margin, which can be summarized as follows:

1. The present morphology of the margin is a consequence of the shearing that prevailed during the rifting of Africa and South America in Cretaceous time (contrast with Chapters 2, 5). During this Cretaceous interval, megastructures, such as the steep Ghana slope and the marginal ridge, were created. The linear (locally "en echelon") Ghana continental slope, for example, is directly cut into the African basement and its Paleozoic cover, as demonstrated by dredging and seafloor photography (Mascle, 1976; Blarez et al., 1987). The Ivory Coast–Ghana Ridge,

probably a thick pile of deformed Lower Cretaceous sandstones, corresponds to a major shear zone (Basile, 1990). To the north, this ridge is bounded by a series of parallel, strike-slip fault lineaments, whereas its steep southern slope resulted from gravitational readjustments between a highstanding continental crust and a newly created deep oceanic crust (Basile, Brun, and Mascle in press). After active tectonism, the ridge, underwent vertical uplift before cooling and subsiding (Mascle and Blarez, 1987; Basile, 1990; Basile, Brun, and Mascle in press).

The deep Ivory Coast Basin represents an independent divergent margin segment, and at depth is cut into a series of half-grabens and horsts. However, its southern boundary, near the marginal ridge, has undergone shearing stress. This locally reacti-

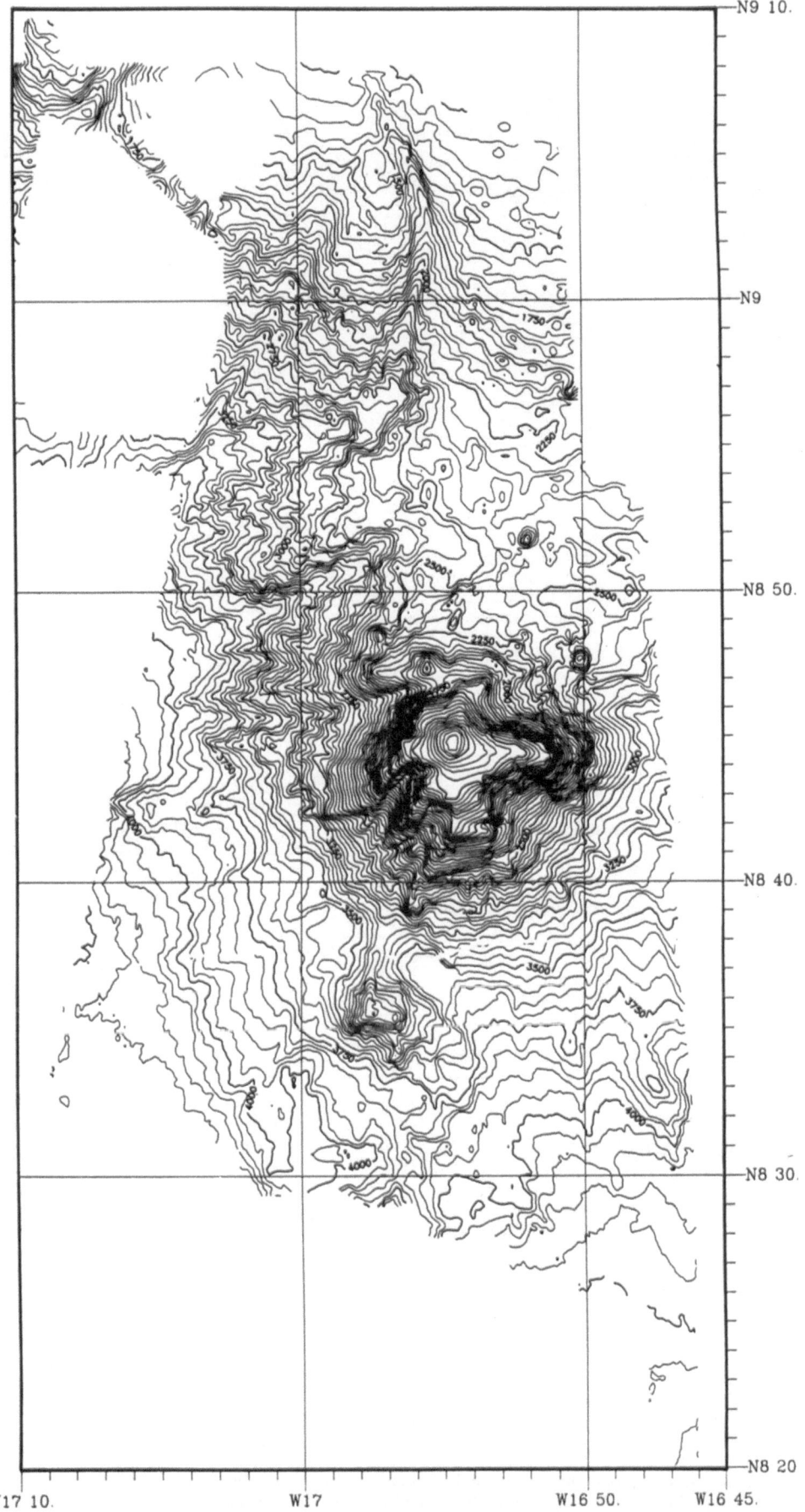

FIGURE 11-7. Bathymetric map of second southern Guinea margin study area. Note 20-km-wide volcano emplaced on mid-slope (2,500–3,500 m). The 2,500-m-high volcano seems to reflect two structural trends, N–S and E–W, which are prevailing trends of margin. Eroded volcano summit stands at −840 m below sea surface, and is characterized by hardgrounds, indicating erosion of sediment cover.

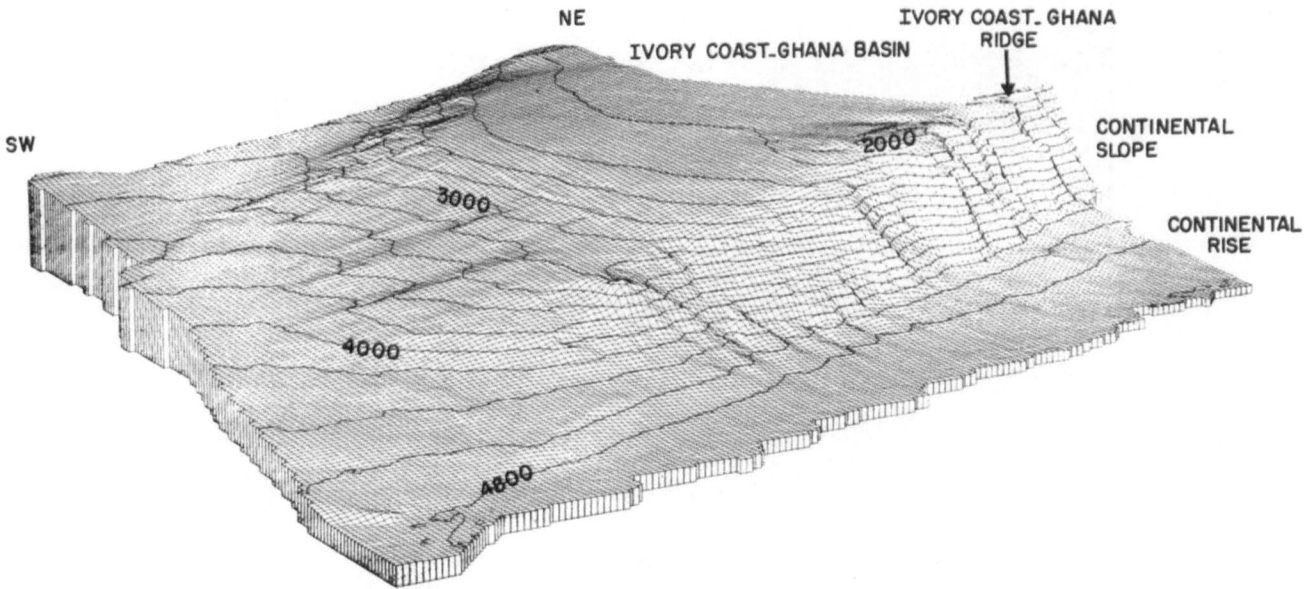

FIGURE 11-8. Three-dimensional representation of morphology of Ivory Coast–Ghana marginal ridge based on SEABEAM mapping of area shown in Figure 11-4. Marginal ridge appears as uplifted southern border of Ivory Coast Basin. Its southern slope, trending NE–SW, is a continuation of very steep continental slope that forms southern border of shallow Ghana Platform.

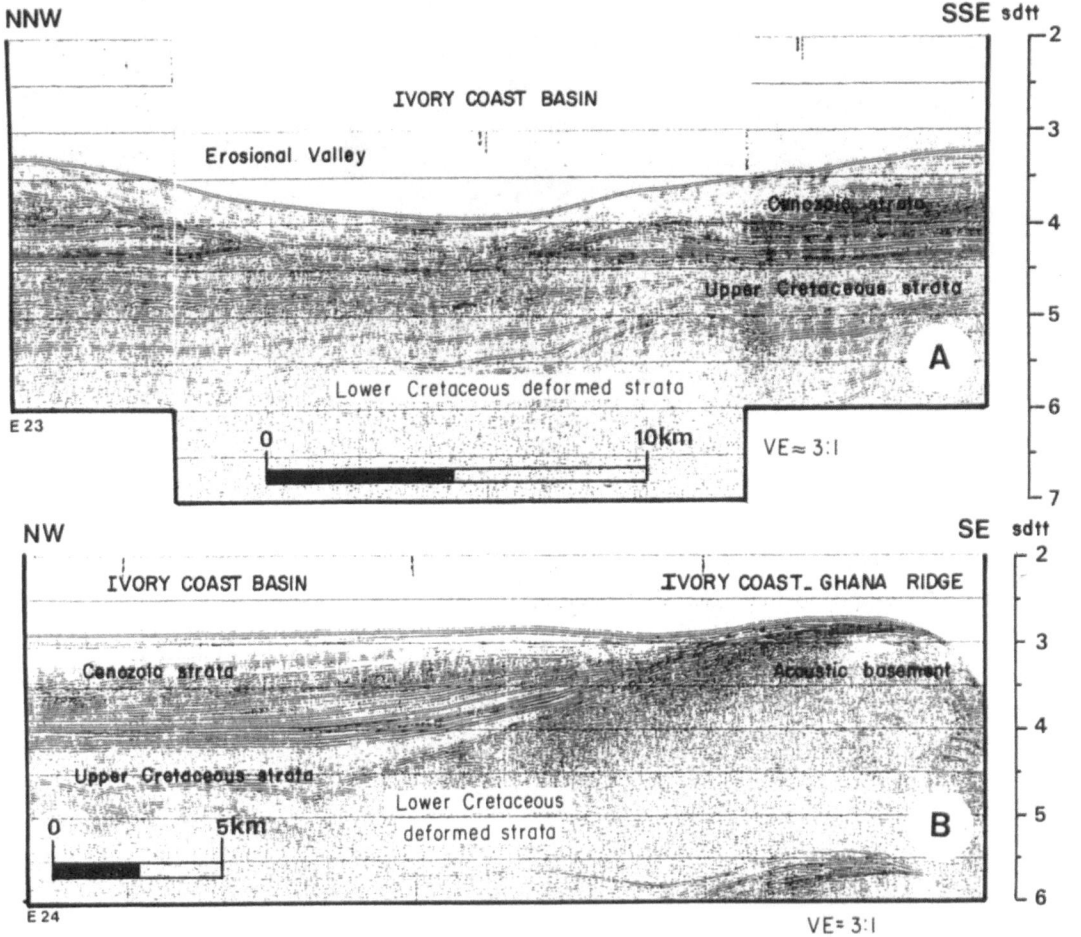

FIGURE 11-9. A. Section of seismic profile showing broad, buried erosional valley cutting across older sedimentary cover of Ivory Coast Basin; valley presumed to be of late Oligocene age; erosion affects Lower Cretaceous strata. B. Section of seismic profile across Ivory Coast–Ghana marginal ridge. Note thin, flat-lying sedimentary cover capping ridge summit. We believe that subaerial erosion was active during thermal uplift of ridge in Cretaceous time; subsequent erosion was dominated by deep bottom currents.

vated and rotated previous extensional structures and slightly folded parts of its Cretaceous cover (Basile, 1990). Most of these features, though, are thickly covered by postrift sedimentary strata, and do not significantly contribute to the present-day morphology. A notable exception is an important normal fault system, located approximately at the transition between the thinned continental crust of the basin and the oceanic crust. This fault system, slightly reactivated by shearing, was the site of a large erosional valley later on in the Oligocene (Fig. 11-9A).

2. Bottom currents within the eastward flowing NADW and AABW appear to have been an important factor controlling the morphology of the Ivory Coast margin. The reworking of the sedimentary cover is, however, less spectacular in this region than along the Guinea margin. The most significant result of the bottom-current activity consists of zones of non-deposition or erosion along the continental slope. Basement rocks (off the Ghana Platform) and Lower Cretaceous sandstones (along the southern slope of the Ivory Coast Ridge) are directly exposed on the seafloor, and have been sampled by dredging (Blarez et al., 1987; Mascle and Auroux, 1989). Thin, flat-lying sediments, resting unconformably above deformed strata on the summit of the Ivory Coast Ridge, are clear evidence of restricted deposition and erosion (Fig. 11-9B). Local occurrences of sediment waves were detected along the narrow continental rise. Sediment erosion and reworking by bottom currents have taken place along the steep, lower southern slope; evidence for such influences has not been detected within the Ivory Coast Basin itself, however. Sedimentary features are present within the basin, but formed during the large-scale Cenozoic erosional episode, as documented on many other parts of the African margin.

CONCLUSIONS

The southern Guinea and Ivory Coast–Ghana transform margins clearly illustrate the combined effects of synrift transform tectonic evolution and oceanographic influences on their present-day morphology. Transform rifting produced major structures, such as very steep, block-faulted slopes (including mid-slope basins), and megashear-zone ridges. Slope steepness and orientation strongly influence deep and bottom-current flow direction, as well as the intensity of flow. This has resulted in a strong bottom-current control of the sediment cover of this margin and has led to erosion and reduced sediment deposition beneath high velocity cores (such that those along very steep slopes or in channels in the mid-slope basins), as well as on the summits of high-standing features such as volcanoes and marginal ridges. Under the influence of the bottom currents, sediment deposition of the Guinea margin has been concentrated into large slope-parallel drifts, the surfaces of which are molded into wave-like structures.

On the Ivory Coast–Ghana margin, although buried erosional features are clearly visible in the basin behind the marginal fracture ridge (the Ivory Coast–Ghana Ridge), the smooth surface of the subsequent sedimentary fill indicates reduced mid-depth and bottom-current velocities in comparison to those of the Guinea margin segment.

ACKNOWLEDGMENTS

Data presented in this paper were collected in 1988 during the EQUAMARGE II cruise of the R. V. J. CHARCOT. IFREMER-GENAVIR provided ship time and most of the technical assistance. INSU-CNRS provided financial support for data processing. Contribution No. 577 of the GEMCO (URA-CNRS 718).

REFERENCES

Basile, Ch. 1990. Analyse structurale et modélisation analogique d'une marge transformante: Exemple de la Marge de Côte d'Ivoire–Ghana. *Mémoires et Documents du Centre Armoricain d'Etude Structurale des Socles 39*, 220.

Basile, Ch., Brun, J. P., and Mascle, J. in press. Structure et formation de la marge transformante de Côte d'Ivoire–Ghana. *Bulletin de la Société Géologique de France.*

Benkhelil, J., Mascle, J., Villeneuve, M., Tricart, P., Auroux, C., Basile, Ch., Ciais, G., and the Equamarge II Group. 1989. La marge transformante sud-guinéenne: premiers résultats de la campagne Equamarge II. *Comptes Rendue Academic Science, Paris* 308(II): 655–661.

Blarez, E. 1986. La marge continentale de Côte d'Ivoire–Ghana: structure et évolution d'une marge transformante. Theses, Université Pierre et Marie Curie, Paris.

Blarez, E., and Mascle, J. 1988. Shallow structure and evolution of the Ivory Coast and Ghana transform margin. *Marine and Petroleum Geolology* 5:54–64.

Blarez, E., Mascle, J., Affaton, P., Robert, C. R., Herbin, J. P., and Mascle, G. 1987. Géologie de la pente continentale ivoiro–ghanéenne: résultats de dragages. *Bulletin de la Société Géologique de France* 8:877–885.

Chermak, A. 1979. A structural study of the Romanche Fracture Zone based on geophysical data. M.S. Thesis, University of Miami.

Emery, K. O., Uchupi, E., Phillips, J., Bowin, C., and Mascle, J. 1975. Continental margin of western Africa: Angola to Sierra Leone. *American Association of Petroleum Geologists Bulletin* 59:2209–2265.

Gorini, M. A. 1977. The tectonic fabric of the equatorial Atlantic and adjoining continental margins: Gulf of Guinea to northeastern Brazil. Ph.D. Thesis, Columbia University, New York.

Heezen, B. C., Bunce, E. T., Hersey, J. B., and Tharp, M. 1964. Chain and Romanche fracture zones. *Deep-Sea Research* 11:30–33.

Jones, E. J. W. 1987. Fracture zones in the equatorial Atlantic and the breakup of western Pangea. *Geology* 15:533–536.

Jones, E. J. W., and Mgbatogu, C. C. S. 1982. The structure and evolution of the West African continental margin off Guinea, Guinea Bissau and Sierra Leone. *In* The Ocean Floor, R. A. Scrutton and M. Talwani, Eds.: London: Wiley, 165–202.

Klitgord, K. D., and Schouten, H. 1986. Plate kinematics of the Central Atlantic. *In* The Geology of North America, Volume M, The Western North Atlantic Region, P. R. Vogt and B. E. Tucholke, Eds.: Boulder, CO: Geological Society of America, 351–377.

Krause, M. C., 1964. Guinea fracture zone in the equatorial Atlantic. *Science* 146:57–59.

Le Pichon, X., and Hayes, D. E. 1971. Marginal offset, fracture zones and the early opening of the South Atlantic. *Journal of Geophysical Research* 76:6283–6293.

Leclere-Vanhoeve, A. 1988. Interpretation des données Seasat dans l'Atlantique Sud: implications sur l'evolution du domaine Caraïbe. Thèse de Doctorat, Université de Bretagne Occidentale.

Mantyla, A. W., and Reid, J. L. 1983. Abyssal characteristics of the World Ocean waters. *Deep-Sea Research* 30:805–833.

Marinho, M. O. 1985. Le plateau marginal de Guinée—transition entre Atlantique central et Atlantique équatorial. Thèse 3ème Cycle, Université Pierre et Marie Curie, Paris.

Marinho, M., Mascle, J., Moullade, M., Robert, Ch., Saint Marc, P., and le Groupe Equamarge 1984. Biostratigraphie de la marge guinéenne: résultats préliminaires de la campagne EQUAMARGE I. *Geologie Mediterraneenne* 11:59–65.

Mascle, J. 1976. Le golfe de Guinée: un exemple d'évolution de marges anlantiques en cisaillement. *Mémoire Societe Geologique Francais, N. S. LV*, 128:104.

Mascle, J., and Auroux, Ch. 1989. Les Marges Continentales transformantes Quest-Africaines Guinée, Côte d'Ivoire, Ghana et la zone de fracture de la Romanche. *Campagnes Océnographiques Francaises* 8:150.

Mascle, J., and Blarez, E. 1987. Evidence for transform margin evolution from the Ivory Coast–Ghana continental margin. *Nature* 326:378–381.

Mascle, J., Blarez, E., and Marinho, M. 1988. The shallow structures of the Guinea and Ivory Coast–Ghana transform margins: their bearing on the Equatorial Atlantic Mesozoic evolution. *Tectonophysics*, 155:193–209.

Mascle, J., Marinho, M., and Wannesson, J. 1986. The structure of the Guinean continental margin: implications for the connection between the central and the South Atlantic oceans. *Geologische Rundschau* 75:57–70.

Mienert, J. 1986, Akutostratigraphic im äquatorielem östatlantik: zur entwicklung der tieffenwasserzirkulation der letzen 3.5 Millionen jahre. "Meteor" forschung. Ergebnisse, C, 40:19–86.

Tricart, P., Mascle, J., Basile, Ch., Benkhelil, J., Ciais, G., and Villeneuve, M. 1991. La tectonique d'inversion médio-crétacée de la marge sudguinéenne (campagne EQUAMARGE II). *Bulletin Societe Geologique Francais* 162:91–99.

Warren, B. A. 1981. Deep circulation of the World Ocean. *In* Evolution of Physical Oceanography, B. A. Warren and C. Wunsch, Eds.,: Cambridge, MA: MIT Press, 6–41.

Westall, F., Rossi, S., and Mascle, J. in press. Current-controlled sedimentation in the Equatorial Atlantic: examples from the southern margin of the Guinea plateau and the Romanche fracture zone. *Marine Geology*.

Wüst, G. 1935. Schichtung and Zirkulation des atlantischen Ozeans. Die Stratosphäre des Atlantischen Ozeans. deutschland atlantische Expedition, "Meteor", 1925–1927. *Wissenschaft Ergebnisse* 6(1):2.

12

Northwest African Continental Rise: Effects of Near-Bottom Processes Inferred from High-Resolution Seismic Data

Robert D. Jacobi and Dennis E. Hayes

In this chapter we summarize the near-bottom processes operating along the Northwest African continental rise and slope from latitude 3°N to 35°N (Figs. 12-1, 12-2), and discuss the sedimentological implications of these processes. Our discussions are based primarily on analyses of microphysiography and reflectivity characteristics of the seafloor (hereafter collectively referred to as *echo character*) inferred primarily from 3.5-kHz seismic records, as well as from 12-kHz records, ~ 100-Hz records, records from various sidescan systems, cores, and bottom photographs. The relations between 3.5-kHz echo characters and near-bottom processes were summarized by Damuth (1980) and Jacobi and Hayes (1982, their Tables 1 and 2; and Table 12-1; see also Chapter 8). Overviews of the geology along the Northwest African margin can be found in Emery et al. (1975), Uchupi et al. (1976), Sarnthein et al. (1982), and von Rad and Wissman (1982); see also Chapters 2 and 5.

The summary figure (Fig. 12-2) is based on three detailed echo-character maps and GLORIA images (Kidd, Hunter, and Simm, 1987). The northern detailed echo-character map [Madeira Island map, latitudes 29–35°N, Fig. 12-3 (foldout)] was published by Jacobi and Hayes (1984a, b). A new detailed echo-character map of the Canary Islands region (latitudes 23–29°N) was constructed for this chapter [Fig. 12-4 (foldout)], and is based on 3.5-kHz seismic profiles recorded aboard Lamont-Doherty Geological Observatory ships, and on GLORIA surveys (Kidd, Simm, and Searle, 1985; Kidd, Hunter, and Simm, 1987; Searle, 1987). The southern area (the Cape Verde Islands map, latitudes 3–23°N) was published in a generalized

form by Jacobi and Hayes (1982), and the detailed version is included here as Figure 12-5 (foldout).

This chapter consists primarily of: (1) new observations of local echo character and its related near-bottom processes for the Canary Islands map area [Fig. 12-4 (foldout)]; (2) discussion of the regional near-bottom processes and their interactions along the entire Northwest African margin; (3) discussion of unresolved questions concerning echo character and near-bottom processes; and (4) implications for studies of ancient continental margins.

SEDIMENTARY PROCESSES IN THE CANARY ISLANDS MAP AREA

New Echo-Character Map

The echo character of the Canary Islands map area was originally studied by Embley (1975, 1976, 1982) who focused on the gigantic Spanish Saharan Submarine Sediment Slide (4S Slide), and on seafloor reflection hyperbolae around the Saharan Seamounts.

Detailed reexamination of the 3.5-kHz records, combined with GLORIA data in the western map area (Kidd, Simm, and Searle, 1985; Kidd, Hunter, and Simm, 1987; Searle, 1987), have allowed us to construct a more detailed map [Fig. 12-4 (foldout)]. This map reveals new features, and permits new interpretations for some features noted previously. Determining the trends of some microphysiographic province boundaries was difficult, however, because most of the ship tracks are widely spaced dip lines.

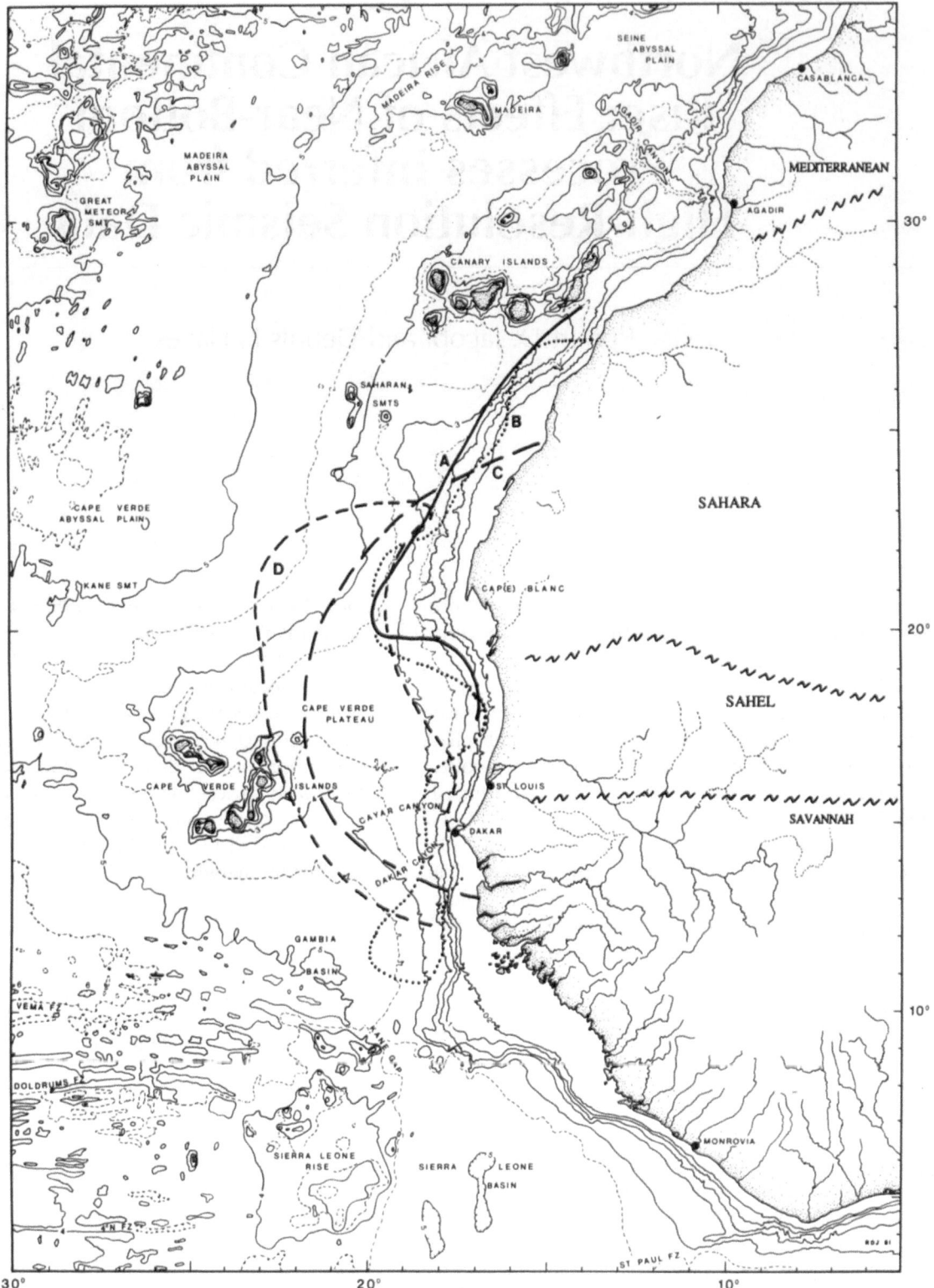

FIGURE 12-1. Generalized bathymetry and physiographic features of Northwest African margin. Primary contours (solid) 1,000 m; dashed contours 500 m. Bathymetry redrawn from GEBCO (Searle, Monahan, and Johnson, 1982). Line A (solid line) marks western limit of strong Pliocene–Pleistocene upwelling and divergence (from Ruddiman, Sarnthein, et al., 1989). Line B (dotted line) encloses nearshore region with greater than 50% biogenic opal (> 6 μm) in surficial sediments (calculated on a carbonate-free basis; from Sarnthein et al., 1982). Line C (long dashes) encloses approximate present region of strong northern summer atmospheric haze (from Stabell, 1989). Line D (short dashes) encloses region in which 18,000-yr-old terrigenous sediments around 24 μm have a Q-mode factor loading of > 75 (and are presumably aeolian; from Sarnthein et al., 1982). Selected climate provinces delimited by squiggly lines (from Sarnthein et al., 1982, and Ruddiman, Sarnthein, et al., 1989).

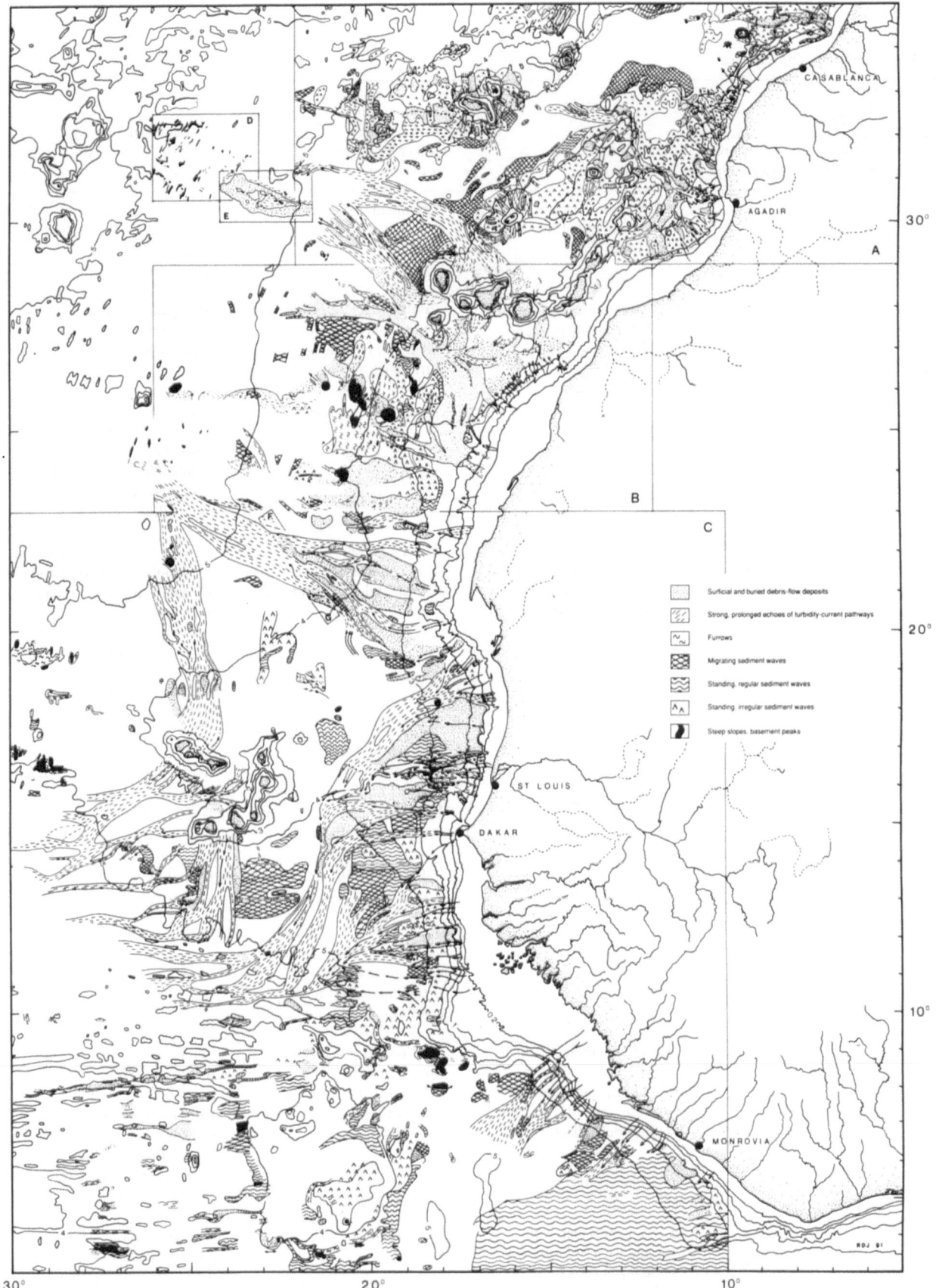

The legend shown in the figure contains the following entries:

- Surficial and buried debris-flow deposits
- Strong, prolonged echoes of turbidity-current pathways
- Furrows
- Migrating sediment waves
- Standing, regular sediment waves
- Standing, irregular sediment waves
- Steep slopes, basement peaks

FIGURE 12-2. Generalized echo character of Northwest African margin. Source for map A (Madeira Island map) is Figure 12-3 (foldout) (Jacobi and Hayes, 1984a). Source for map B (Canary Islands map) is Figure 12-4 (foldout) (this chapter). Source for map C (Cape Verde Islands map) is Figure 12-5 (foldout) (Jacobi and Hayes, 1982). Map D is after a GLORIA image (Searle, 1987). Maps E, F, and G are after GLORIA images from Kidd, Hunter, and Simm (1987). Echo character symbols explained in detail in Table 12-1.

TABLE 12-1   Seafloor Echo Character.*

| Type | | | | | |
This Paper	Symbol on Figure 12-2	Sketch of Echo Type	Description of Echo Type	Physiographic Provinces in Which Echo Type is Common	Possible Origin and/or Associated Deposit
IA	══ ══	━━	Continuous, strong, sharp echo; no subbottom reflectors	Continental shelves	Thin deposits of sand and gravel over consolidated or semi-consolidated sediments
IB	══ ══	☰	Continuous, strong, sharp echo; continuous subbottom reflectors	Continental slope, rise, and abyssal plain; absent in areas of proximal turbidities	In regions of distal turbidites, reflectors commonly are turbidites In pelagic environments, reflectors from variations in density and(or) calcium carbonate content; commonly relatively few sand/silt beds associated with echo
IIA	⫽⫽	▓	Continuous, strong prolonged echo; commonly no subbottom reflectors, except near boundaries of echo type, where discontinuous subbottoms occur	Turbidity-flow pathways (e.g., channels) on continental slope and rise	Echo actually very small hyperbolae, probably result of small, somewhat regular features caused by turbidites (e.g., erosional furrows or flutes) Relatively high percentage of sand/silt beds
IIB	⋰●	IIB	Weak, slightly prolonged echo, commonly forming transparent wedge, elevated with respect to "normal" seafloor	Downslope from slide scar on continental slope and rises, at base of seamounts. Present within and without canyons and channels	Generally related to debris-flow deposits, esp. if slide scar or toe of material is observable
	⠿	Buried IIB	Same as above, but transparent wedge lies below multiple reflectors	Same as above	Buried debris-flow deposit
IIIA		⋀⋀	Large hyperbolae with varying vertex elevations (~ 10–100 m) + λ	Continental slope, canyon walls, abyssal hills, seamounts and fracture-zone topography	
IIIC		IIIC-1 ⩙⩙	Large hyperbolae with varying vertex elevation (> 40 m), some hyperbolae enclose zones of multiple subbottom reflectors	Continental slope and rise, slope and base of seamounts, fracture zones, abyssal hills	"Hummocky material"—slide material near slide scar. Hummocks with internal reflectors probably are slide blocks (olistoliths) or, in some areas, portions of section that did not slide
	■ ■	IIIC-2 ⋀⋀⋀	Large hyperbolae with varying vertex elevation (< 40 m), no subbottom reflectors	Same as above	"Blocky material" slide material downslope from IIIC-1 and upslope from IIB. Apparently material was less rigid than IIIC-1 and more rigid than IIB

TABLE 12-1   Continued

This Paper	Type — Symbol on Figure 12-2	Type — Sketch of Echo Type	Description of Echo Type	Physiographic Provinces in Which Echo Type is Common	Possible Origin and/or Associated Deposit
IIIC	■	IIIC-3	Large hyperbolae, slightly varying vertex elevations, regular-irregular λ. Commonly associated with IIA along channels. Forms areas of raised topography	Along channels on continental rise.	Source unclear: either a result of bedforms along levees (small sediment waves) or slide material that moved down a channel and is presently being reworked
	□ □	Buried IIIC-2/3	Same as IIIC 2 or 3, but transparent wedge lies below multiple reflectors	Same as IIIC 2 or 3	Same as IIIC 2 or 3
IIID					
IIID	/////	IIID-1	Very regular, discrete hyperbolae with vertices tangent to seafloor or a subbottom reflector. Commonly associated with zones of IIA	Zones of IIID-1 occur primarily on continental rise, approximately parallel and adjacent to (or within channel systems)	Source is probable bedforms associated with turbidites
	~ ~ ~	IIID-2	Same as above, except associated with IIIA of seamount or continental slope or IB/IIA of continental rise. Commonly associated with condensed or missing sedimentary section	Zones of IIID-2 are parallel to bathymetric contours on continental rise or occur around seamounts	Source is generally furrows caused by narrow zones of erosion or nondeposition by contour currents. Furrows parallel to current flow
	~ ~	IIID-3	Same as above, except associated with IIB	In areas where both IIB & IIID-1/2 occur; see IIID-1/2	If IIID-3 is caused by contour-current activity, IIID-3 is probably related to ripples or furrows caused by contour current
	■	IIID-4	Same as above, except associated with IIB	Same as IIB; continental slope and rise and at base of seamounts	Sources probably small bumps on top surface of debris-flow deposit, or current furrows
A			Locally planar seafloor	Continental shelf, slope, rise, and abyssal plain	
B	~~~	B-1A	Standing (climbing) sediment waves with regular form (λ and amplitude)	Echoes form zones approximately parallel to bathymetric contours on continental rise; less commonly parallel to channel systems on continental rise; form zones around seamounts	Contour currents and, less commonly, turbidity currents
	⨯⨯⨯	B-1B	Migrating sediment waves with regular form		
	∧ ∧ ∧	B-2A	Standing sediment waves (climbing) sediment waves with irregular form	Same as above	Contour currents and(or) turbidity currents

TABLE 12-1 · Continued

This Paper	Type — Symbol on Figure 12-2	Sketch of Echo Type	Description of Echo Type	Physiographic Provinces in Which Echo Type is Common	Possible Origin and/or Associated Deposit
		B-2B	Migrating sediment waves with irregular form		
C		C-1	Erosional scarp (with no displacement of underlying strata)	Continental slope and rise, seamounts, abyssal hills, and fracture zones	(1) Exposed portion of a slide scar (especially if cross section is open-ended or box-shaped), (2) Steep wall of deep-sea channel
		C-2	Fault scarp (displaced subbottom reflectors occur beneath surface expression of fault)	Same as above	Fault caused by instability of sedimentary sequence because of (1) over-steepened slope caused by (a) contour-current erosion downslope; (b) sediment sliding downslope; (c) high sediment input; (2) tectonic activity in basement
		C-3A	Sharp monoclinal flexure. Note difference in form compared to B-1B and B-2B. C-3A generally has a long gentle slope and a short steep slope	Same as above	Ductile faulting resulting from causes similar to C-2
		C-3B	Relatively steep slope; double parallel lines denote lateral extent of slope	Generally on continental rise	Along channel margins, buried slide scarps, and basement highs that promote "ramps" and "flats"

*After Jacobi and Hayes (1982).

Four principal elements directly affect the architecture of the Canary Islands continental rise: (1) thermohaline contour-following currents (THCC); (2) turbidity currents; (3) morphology of igneous "basement"; and (4) sediment slides. The following section focuses on new echo-character observations concerning these elements.

Zones of Regular, Overlapping Hyperbolae Tangent to the Seafloor

Saharan Seamount Region

Numerous zones of regular, overlapping hyperbolae, tangent to the seafloor, occur in the Saharan Seamount area between latitudes 24°N and 27°N. Embley (1975) and Embley, Rabinowitz, and Jacobi (1978) postulated that these hyperbolae result from "erosional furrows" caused by contour currents. Lonsdale (1978, 1982) performed a deep-tow survey and short-term current-meter study over a hyperbolated province located at lat. 25°55′N, long. 19°50′W (Site L1 on Fig. 12-6). He found that the seafloor lineations giving rise to the hyperbolae are narrow, continuous, and parallel to both the regional bathymetric contours and bottom-water flow. Although Sarnthein and Mienert (1986) believed that the lineations observed on the deep-tow records are narrow sediment waves, the position of shadows on deep-tow records suggests that the lineations are caused by relatively widely spaced seafloor features with negative relief, and are, therefore, furrows (Lonsdale, 1978, 1982).

Extensive studies elsewhere have shown that "erosional furrows" generally are long and straight, but some exhibit "tuning forks" that open in the upflow direction. The furrows are oriented parallel to regional bottom-water flow, which in abyssal areas generally is parallel to regional bathymetric contours (Flood and Hollister, 1974, 1980; Hollister et al., 1974; Embley et al., 1980; Flood, 1981, 1982; Lonsdale, 1978, 1982). The furrows are thought to result initially from "localized abrasion by, or scour around" (Flood, 1982, p. 176) coarse material aligned by helical flow cells in a boundary layer. The initial groove is broadened and

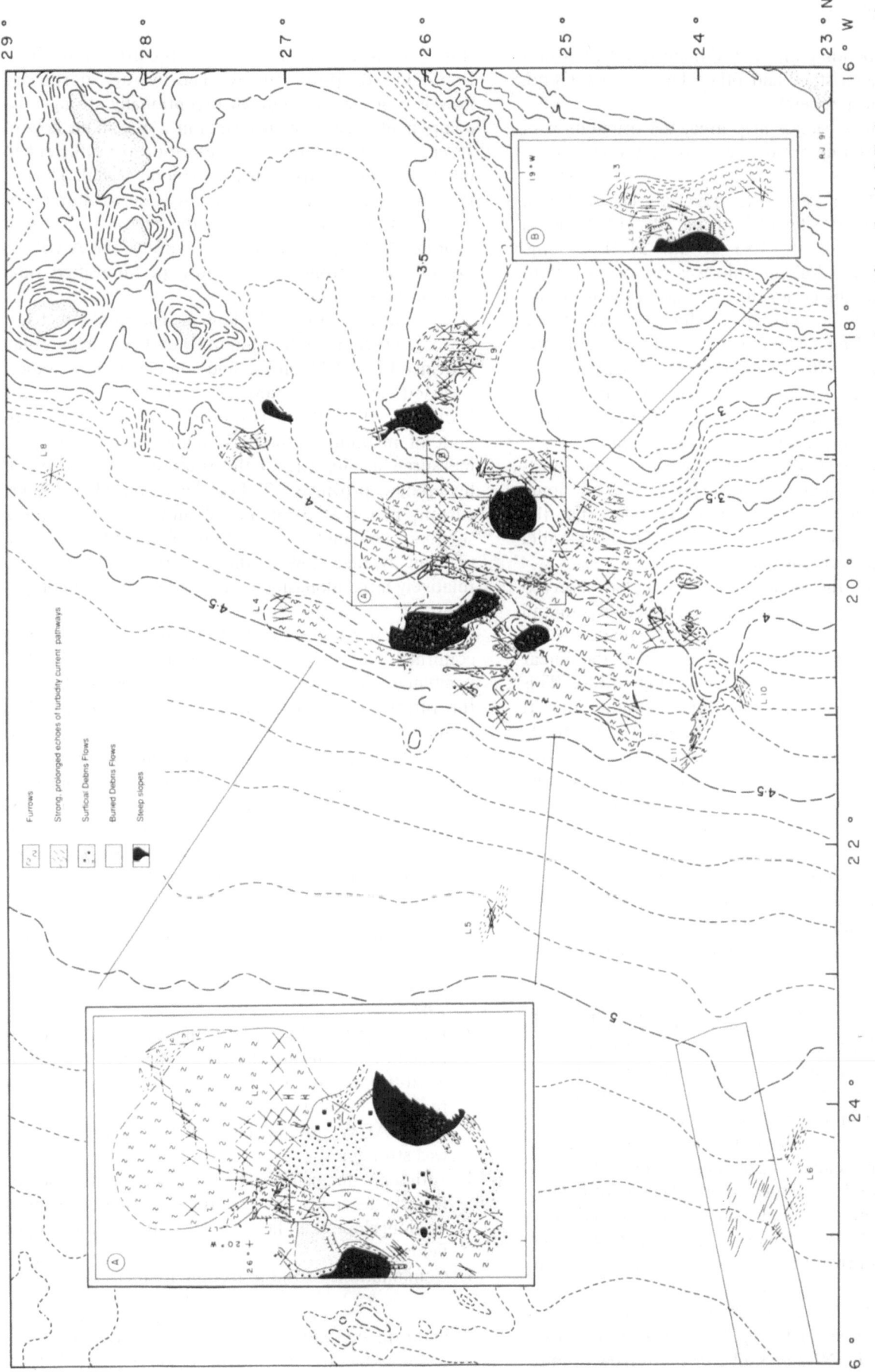

FIGURE 12-6. Lineation (primarily furrow) orientations in the Canary Islands map area. Lineation orientations determined from hyperbolae form (see text for explanation). Each × represents the two potential lineation orientations of a nonunique solution. Dashed portion of ×'s indicate less likely direction, based on a lack of commonality with nearby determinations. Single lines denote unique determinations. Small box labelled L1 in inset A locates deep-tow survey of Lonsdale (1978, 1982). Double-arrow in box denotes average orientation of furrows observed on deep-tow surveys. Lines in rectangular band in southwestern corner indicate sediment lineations observed on a GLORIA image (after Kidd, Hunter, and Simm, 1987). Labelled features discussed in text and compiled in Table 12-2. Primary contours 500 m (heavy contour lines); supplemental contours 100 m. Bathymetry redrawn from GEBCO (Searle, Monahan, and Johnson, 1982).

deepened, depending on the sediment supply, bottom-current characteristics, and other factors (see Flood, 1982, 1983, for a review).

In the Canary Islands map area, the orientation of furrows has been determined by three methods:

1. At three locations (site LS1 at lat. 25°53′N, long. 19°53′W, site LS2 at lat. 25°21.5′N, long. 19°53.5′W; and site LS3 at lat. 25°31′N, long. 19°07′W, Fig. 12-6) the ship was turned in a tight circle. By noting either the bearings of the hyperbolae apices or the bearings at which the hyperbolae flatten out, Embley (1975) and Embley, Rabinowitz, and Jacobi (1978) found that the lineations are nearly parallel to the bathymetric contours; at site LS1 they are parallel to furrows observed on more recent deep-tow images in the same area.
2. Lonsdale's (1978, 1982) deep-tow survey and short-term current-meter study (lat. 25°55′N, long. 19°50′W, site L1 on Fig. 12-6) revealed that the average orientation of furrows (Fig. 12-6) is parallel to the regional bathymetric contours.
3. By analyzing the form of the hyperbolae observed on the 3.5-kHz records in the manner of Bryan and Markl (1966) and Flood (1980), we made > 200 determinations of the angles at which the linear features cross the ship tracks (Fig. 12-6). This method yields two potential furrow orientations. If the lineations are at a high angle to the track, however, then the difference between the two possible trends is small and of little concern. Also, one of the potential orientations can be eliminated if it is not common to one of a second set of determinations located nearby. The reliability of this approach can be judged by the good agreement among all three methods. In the Saharan Seamount area, most determinations yield at least one probable furrow orientation that is approximately parallel to the bathymetric contours (Fig. 12-6).

Most zones of regular, overlapping hyperbolae in the Saharan Seamount area contain seafloor lineations parallel to the regional bathymetric contours, thus suggesting that these provinces delineate areas of accelerated bottom-water flow. The major provinces of hyperbolae are restricted to a band of seafloor depths between approximately 4,500 and 3,200 m [Embley, 1975; Figs. 12-4 (foldout), 12-6]. This depth range straddles the present upper surface of Antarctic Bottom Water (AABW), which is at approximately 3,900 m below sea level (bsl) in the area (e.g., Sarnthein et al., 1982).

Lonsdale's (1978, 1982) 4-day current-meter study at two locations in site L1 (Fig. 12-6) measured bottom-water flow to the NNE, parallel to the furrow orientations deduced from 3.5-kHz records and from Lonsdale's deep-tow survey. The average northward flow is 3–4 cm/sec, but an exceptionally strong tidal component exists, which results in a maximum flow of 18 cm/sec alongslope. Such velocities are sufficient to maintain and construct furrows (Flood, 1982). Thus, if these current-meter records are representative of long-term flow, the furrows in the region of Saharan Seamount site L1 are active, not because the AABW is exceptionally strong, but because of intensified tidal flow in the seamount area.

In contrast to the NNE flow measured at site L1, a 3.5-day current meter moored east of hyperbolae province L2 at 3,495 m bsl, and 5 m above the seafloor, revealed an average flow of 2.4 cm/sec to the SE (Lonsdale, 1978, 1982). However, a NNE tidal component of medium strength is similar in trend to that measured at site L1. In the south-central portion of hyperbolae province L2, Lonsdale moored a 3.3-day current meter at 3,836 m bsl and 50 m above the seafloor. This meter showed an irregular flow pattern with an average flow to the SSE of 1.3 cm/sec and relatively large alongslope and downslope tidal components.

The deeper current meters of Lonsdale (1978, 1982) confirm that a sluggish northward-flowing AABW, combined with a strong tidal flow, is probably keeping the furrows open in the deeper fields of hyperbolae, including the western, deeper portion of L2. However, both bottom-temperature data and current-meter data suggest that the shoaler fields of hyperbolae (including the easternmost portion of L2, and all of L3 and L3A) are at depths presently overlain by sluggish southward-flowing North Atlantic Deep Water (NADW). Possible explanations for the shoaler hyperbolae fields are:

1. The fields are relict features produced by thicker AABW when the flow was stronger during glacial intervals (e.g., Sarnthein et al., 1982; Murray et al., 1986).
2. Eddies and undulations along the transition between the northward-flowing AABW and the southward-flowing NADW permit AABW to periodically reach up to the shoaler depths.
3. The NADW produced the features when its flow was stronger.
4. Although sluggish NADW flows across the shoaler hyperbolae fields, the lineations are actually caused by alongslope tidal flow similar to that of the deeper fields.
5. Combinations of the preceding items.

The principal fields of furrows are restricted along-slope to the Saharan Seamount region [Figs. 12-1, 12-2,

12-4 (foldout), 12-6]; this observation is consistent with the hypothesis that accelerated bottom water and tidal flow around and between the seamounts may have formed the furrows. To the south, where such seafloor restrictions are not apparent, no major fields of hyperbolae are evident [Figs. 12-1, 12-2, 12-4 (foldout), 12-6]. Apparently there is not a one-to-one correlation, however, between hyperbolae fields and seamounts or gateways between seamounts. For example, the large hyperbolae field at approximately lat. 27°N, long. 20°20′W (site L4, Fig. 12-6) is not adjacent to any known seamount; however, the orientations of seafloor lineations are parallel to bathymetric contours and are therefore thought to imply a contour-current origin.

The lack of large turbidity-current pathways in the Saharan Seamount region also promotes the formation of furrows, because the relatively low sediment input allows furrow growth at relatively low flow velocity. In contrast, north of the Saharan Seamount region, hyperbolae provinces did not develop, probably because of turbidity currents that delivered extensive sediment (as evidenced by the turbidity-current pathways), and because of a lack of bathymetric obstructions that might cause the bottom water to accelerate.

Relation to Turbidites

Submarine channels, canyons, and turbidity-current pathways generally exhibit a strong prolonged echo (Damuth, 1980; Jacobi and Hayes, 1982). In the Canary Islands map area, however, a few small zones of sporadic overlapping hyperbolae, tangent to the seafloor, are well removed from the Saharan Seamount provinces (Fig. 12-6). These zones of overlapping hyperbolae are associated spatially with strong prolonged echoes typical of turbidity-current pathways (e.g., provinces L5, L6, L8, portions of province L9 near the prolonged echoes at 25°45′N, 18°15′W, and L10, Fig. 12-6). At province L5, both inferred solutions for hyperbolae-related lineations trend at high angles to the regional bathymetric contours (Fig. 12-6), but are approximately parallel to the assumed path of nearby turbidity currents, marked by strong prolonged echoes. Similarly, at all three determination locations in province L6, one of the possible lineation orientations is colinear with sediment lineations mapped to the north from GLORIA (sidescan) records (Kidd, Simm, and Searle, 1985; Kidd, Hunter, and Simm, 1987). At provinces L8 and L10 both potential trends are downslope, and one trend is approximately parallel to the downslope boundaries of the turbidity-current pathways. At several sites in province L9, the determined lineation trends are oriented 90° to the ship track, and thus are unique solutions. Because these unique solutions are oriented perpendicular to the local slope, they may be related to the nearby turbidity-current pathway. However, neither

potential trend of nearby nonunique solutions in province L9 is oriented downslope. Therefore, some of the furrows in province L9 (Fig. 12-6) may be related instead to thermohaline currents (NADW).

Sporadic, overlapping hyperbolae apparently can result from features related to turbidity currents, as previously proposed for the Cape Verde Islands map area (Jacobi and Hayes, 1982). Such features must be fairly long and straight to return long-tailed hyperbolae; they may be similar to chutes observed in shallow fan systems (Prior and Bornhold, 1989), or similar to sand ribbons with adjacent positive relief (Piper et al., 1985).

Relation to Debris-Flow Deposits

In several areas in the Canary Islands map area, as well as in the Cape Verde map area (Jacobi and Hayes, 1982), transparent debris-flow wedges are associated atypically with areas of regular overlapping hyperbolae, tangent to the seafloor. In some cases, it is clear that a thin debris-flow deposit has partially buried a furrowed seafloor. For example, 3.5-kHz echograms recorded immediately east of site L1 (Fig. 12-6) display only partial hyperbolae immediately adjacent to a debris-flow deposit. Presumably, the missing portion of each hyperbola represents the part of a furrow that is buried by the debris-flow deposit. Lonsdale's (1978, 1982) deep-tow records from the same area confirm such an interpretation.

A different relationship is defined by elevated sediment wedges, which display a single reflector of regular overlapping hyperbolae, lying adjacent to deeper seafloor, which displays multiple reflectors and no hyperbolae (e.g., site L7, Fig. 12-6). The presence of a nose (or toe), separating the sediment wedge from the "undisturbed" seafloor, suggests that the wedge is a debris-flow deposit. The fact that hyperbolae are restricted to the wedge sediments suggests either that: (1) the debris-flow deposit had sufficiently low cohesion for a bottom current to erode or mold the debris; or (2) the regular hyperbolae are the result of regularly spaced folds constructed during flow of the debris. The hypothesis of low sediment cohesion appears more tenable for site L7 (Fig. 12-6) because the orientations of the lineations are parallel to the furrow trends in deep-tow survey L1, and because the lineations are at a high angle to the assumed boundary of the deposit.

In other areas, both debris-flow deposits and the surrounding seafloor exhibit regular, overlapping hyperbolae. Possible explanations include: (1) a very thin debris-flow deposit only partially filled the furrows, because the hyperbolae are definitely tangent to the upper surface of the sediment wedge; and (2) the debris-flow deposit was later furrowed by bottom currents.

Sediment Waves on the Lower Continental Rise

Orientations

There are three possible orientations of sediment-wave provinces on the lower continental rise: (1) parallel to the regional bathymetric contours (contour-current influence); (2) parallel to downslope channel/levee complexes; or (3) parallel to the nearly buried abyssal hills. Orientations of sediment wave crestlines (see Jacobi, Rabinowitz, and Embley, 1975, for method) allow discrimination among these possible origins. We calculated crestline orientations for 74 localities; 27 determinations yielded unique solutions; the remainder have two potential solutions. GLORIA images (Kidd, Simm, and Searle, 1985; Kidd, Hunter, and Simm, 1987; Searle, 1987) complement the crestline orientation study. There are at least two distinct populations of orientations. One trend is approximately parallel to the contours; from other studies (see Table 12-1) and map patterns, we judge this trend to be primarily caused by thermohaline contour currents. On the upper continental rise and slope, the geometry of some sediment waves with alongslope crestlines suggests that these sediment waves are slump or surface-creep folds [e.g., Figs. 12-4 (foldout), 12-7, Table 12-1; Jacobi and Hayes, 1982]. The second prominent set of crestlines trends downslope; on the lower continental rise, these sediment-wave provinces are adjacent to turbidity-current pathways. Their orientation and distribution suggest that these lineaments are sediment waves built by turbidity currents. This may be the first area where such differentiation is clearly possible. Sediment waves that trend downslope adjacent to slide scars are judged to be related to slide events. As discussed later on, basement features also influence some sediment-wave provinces.

Relation to Thermohaline Contour Currents

It has been well established that under appropriate sedimentological and hydrological conditions, thermohaline contour currents (THCC) construct migrating sediment waves with crestlines parallel or oblique to the bathymetric contours (Flood, 1988; Flood and Shor, 1988). In the Canary Islands map area (Fig. 12-7), migrating sediment waves are restricted generally to depths below the approximate present upper limit of AABW at 3,900 m (Sarnthein et al., 1982). The two largest migrating sediment-wave provinces in the Canary Islands map area (provinces W32 and W37) have sediment-wave crestlines oriented generally parallel to the bathymetric contours (Fig. 12-7), and the provinces are deeper than the AABW upper surface. The sediment waves are, therefore, probably related to AABW flow. Smaller migrating sediment-wave provinces W27 and W34A also may have formed by AABW flow, but crestline orientations are not avail-

able to confirm this supposition. The downslope orientations of sediment-wave crestlines in the easternmost part of W34B suggest that they are more likely related to turbidity currents than to THCC. The shoaler parts of province W35A are shoaler than the present upper surface of AABW. These sediment waves may have been formed when AABW had a shoaler upper surface, possibly during glacial times (Sarnthein et al., 1982). Alternatively, the sediment waves may have been generated by the North Atlantic Deep Water (NADW), which flows south above the AABW. The orientations of migrating waves in the southeastern part of W35 (W35C) suggest northward bottom-water flow there, produced either by NADW curving to the north around a local bathymetric promontory located to the west (Fig. 12-7), or by AABW flowing generally north. The low number of crestline orientations in W35 precludes a definitive resolution. To the south, current-meter data of Lonsdale (1978, 1982) suggest a generally southward flow (NADW) in the shoalest depths of W35A.

Relation to Turbidity-Current Pathways

At sites SW4 and SW5 (Fig. 12-7), "sediment lineations" observed on GLORIA records (Kidd, Simm, and Searle, 1985; Kidd, Hunter, and Simm, 1987) trend approximately downslope parallel to lineation orientations calculated from hyperbolae in turbidity-current pathways (Fig. 12-6). Kidd, Hunter, and Simm (1987) suggested that these sediment lineations are the result of narrow ribbons of "overbank" deposits, which buried pelagic sediment. We concur that the presence (or absence) of turbidity-current pathways and their turbidites accounts for the relatively wide features (ca. 10–20 km width), such as those at SW6 (Fig. 12-7). If the zone of relatively low reflectivity (observed on the GLORIA image) is extended along strike to our ship track, then it correlates with a zone lacking turbidity-current features [Fig. 12-4 (foldout)]. However, sediment waves on 3.5-kHz echograms recorded from site SW5 have the same spacing, orientation, and location as the narrow alternations (1–2 km) of "bright and dark" reflectivity observed on the GLORIA image. Because the sediment waves also exhibit some echo prolongation (suggesting turbidity-current activity), the GLORIA sediment lineations probably are the result of both sediment-wave relief and lithofacies differences between troughs and crests caused by the relative importance of turbidites and pelagites.

Two lines of evidence suggest that these sediment waves were not constructed by a THCC, with turbidity currents merely flowing over the waves and ponding turbidites in the troughs. First, the orientation of the sediment-wave crestlines is at a much higher angle to the bathymetric contours than generally is found on continental rises for THCC-generated waves (Jacobi,

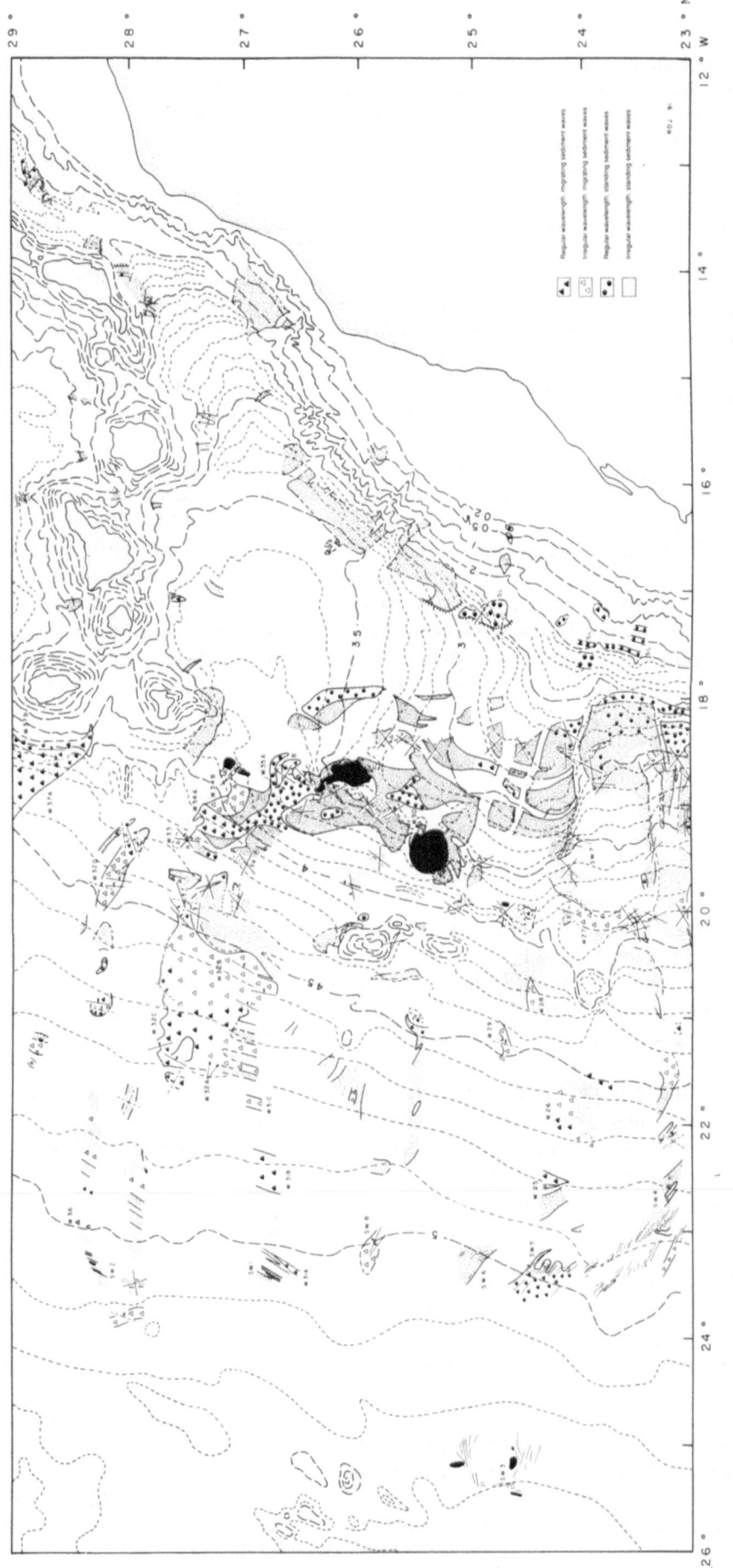

FIGURE 12-7. Sediment-wave provinces and crestline orientations in the Canary Islands map area. Distribution of provinces from Figure 12-4 (foldout). Orientation of sediment-wave crestlines determined following method described by Jacobi, Rabinowitz, and Embley (1976). Potential crestline orientations indicated by ×s, or where unique solutions were possible, by single lines. Migration direction indicated by arrow. Parallel lines in southwest (areas SW3, SW4, SW5) indicate crests and troughs of sediment waves observed on reinterpreted GLORIA records of Kidd, Hunter, and Simm (1987). Labelled features discussed in text and compiled in Table 12-3. Primary contours 500 m (heavy contour lines); supplemental contours 100 m. Bathymetry redrawn from GEBCO (Searle, Monahan, and Johnson, 1982).

Rabinowitz, and Embley, 1975; Flood, 1988). Second, the migration direction of sediment waves in both provinces SW5 and SW8 is to the northeast, which is not compatible with a northward flow of AABW. Thus, either there is an unrecognized southward-flowing bottom current that has constructed sediment waves with crestlines abnormally highly oblique to the bathymetric contours, or the waves were constructed by turbidity currents. We conclude that in the region of SW5 and SW8, (Fig. 12-7) sediment waves and GLORIA "sediment lineations" are both more likely related to turbidity-current pathways.

If the TCP sediment waves are longitudinal waves, then the observation that these waves and associated sediment lineations are generally parallel, linear, and oriented downslope, suggests that on a large scale, turbidity currents do not fan out on the lower fan (continental rise), but generally continue flowing downslope parallel to the regionally linear TCPs. This relationship is consistent with a "flow-stripping" model, in which portions of turbidity currents above the height of the channel walls do not follow channel meanders (e.g., Normark and Piper, 1985). If the sediment waves are some type of TCP-generated "lee wave" (Flood, 1988), then their parallel, linear crestline orientations again argue for a flow-stripping model.

Strong Prolonged Echoes

Elevated Sediment Wedges (Sand Lobes?) along Turbidity-Current Pathways

Strong prolonged echoes (SPE) typical of turbidity-current pathways are thought to indicate short-wavelength bottom roughness and a relatively high sand content (e.g., Damuth, 1975, 1980). On the Laurentian Fan, for example, such echoes correspond to "gravel waves" with wavelengths of ~ 50–100 m, which were constructed during the passage of a turbidity current (Piper, Stow, and Normark, 1985). On the lower continental rise of the Canary Islands map area, the prolonged echo covers large, relatively smooth areas that may include shallow, broad channels. However, near site T1 (Fig. 12-8), slightly elevated regions return strong prolonged echoes, whereas the lower areas return multiple-reflector echoes [Figs. 12-4 (foldout), 12-8]. The boundary between the echo types is relatively sharp and steep, similar to that found at the nose of some debris-flow deposits. At site T1, several narrow (~ 1 km) bands of elevated, strong prolonged echoes occur immediately south of a ~ 34-km-wide zone of elevated prolonged echoes. This wide zone includes both channels and subdued (flat) levees. The boundary of the wide zone also coincides with an approximately downslope sediment lineation imaged by GLORIA

(Kidd, Hunter, and Simm, 1987). The orientation and the probable sandy nature of the narrow bands suggest that they may be similar to "sand ribbons" observed on sidescan surveys elsewhere (Belderson et al., 1972; Piper et al., 1985; Pirmez et al., 1989, 1990). The wide zone is quite broad for a ribbon, however, and is more on the scale of a "sand lobe." Thus, site T1 provides a tantalizing hint of sand ribbons and sand lobes on this continental rise, and implies that, at least here, they are closely associated. Alternatively, the sand ribbon/lobe may actually comprise a debris-flow deposit over which more recent turbidity currents have flowed.

At site D7 (Fig. 12-8), a sediment wedge returning a strong prolonged echo forms a clear "toe", which buries the adjacent region of multiple subbottom reflectors. The sediment wedge can be traced continuously upslope, beneath and alongside the 4S Slide as far as site D8 (Fig. 12-8). The southern (upslope) portion appears to have a ponded or feathered pinchout, whereas at site D7, it forms positive local relief. If the echo is continuous from site D8 to site D7, then it may document the transition from "normal" turbidite deposition upslope to sand lobe deposition downslope.

Turbidites Buried by Debris-Flow Deposits

At several sediment slide locations, debris-flow deposits form relatively transparent sediment wedges over a strong prolonged echo. On the lower portion of the 4S Slide, the strong prolonged echo can be traced continuously upflow underneath the debris-flow deposit from site D7 to site D8 (Fig. 12-8). Between sites D5 and D8, debris-flow deposition only partly covers the turbidites in the pathway. The debris flow and turbidity current may have been unrelated, however, because debris flows sometimes move over and through older channels [Fig. 12-4 (foldout), and also observed on GLORIA records (Kidd, Simm, and Searle, 1985; Kidd, Hunter, and Simm, 1987; Searle, 1987); see also Chapter 13]. However, it is also possible that a turbidity current was generated by the 4S Slide; this turbidity current could have outrun the slower moving debris flow, only to be ultimately buried by the debris flow. A similar scenario has been envisioned for the Grand Banks Slump (Piper et al., 1985).

The narrow Saharan Slide debris-flow deposit, extending from site D9 to site D10 (Fig. 12-8), also overlies sediments returning a strong prolonged echo, and may represent another coupled debris-flow–turbidity-current event. In contrast, at site D4 (Fig. 12-8), the debris-flow deposit that forms toes over the strong prolonged echo, probably had a source to south, whereas the turbidity current probably had a source to the east. Thus, at this site, the two deposits are in close proximity, but represent different events and sources.

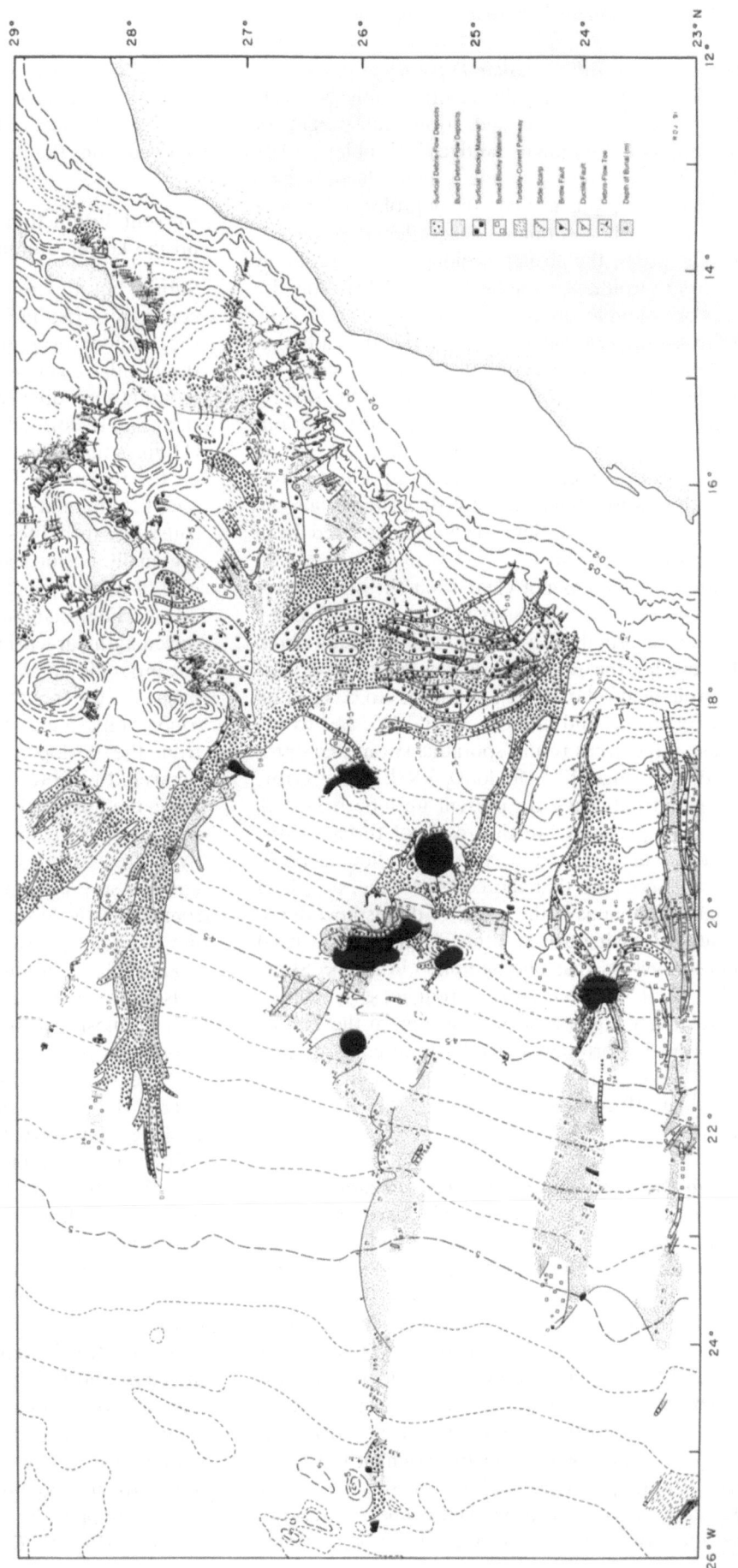

FIGURE 12-8. Sediment slides in the Canary Islands map area. Labelled features discussed in text. Primary contours 500 m (heavy contour lines); supplemental contours 100 m. Bathymetry redrawn from GEBCO (Searle, Monahan, and Johnson, 1982).

Upper Surfaces of Sediment-Slide Deposits
Debris-flow deposits typically return relatively weak, slightly prolonged echoes and form transparent or semitransparent "sediment wedges" (Embley, 1976; Jacobi, 1976). However, in the Canary Islands map area, some echograms display strong prolonged echoes adjacent to clearly identifiable debris-flow deposits.

In some areas, the strong, prolonged echo probably results from turbidites extending across debris-flow deposits. For example, in the 4S Slide at site D1 (Fig. 12-8) the strong prolonged echo is confined to a bathymetric low, whereas the adjacent higher seafloor exhibits the normal debris-flow echoes. On several adjacent ship tracks, we observed the same relationship. Thus, we are fairly confident that the strong prolonged echoes at site D1 are the result of turbidites, which either were more recently deposited than the debris-flow deposit, or (less likely) are covered by a thin debris-flow deposit.

In other areas, it is not clear what the strong prolonged echo signifies (e.g., sites D2 and D3, Fig. 12-8). Echograms recorded both upflow and downflow from site D2 display the usual debris-flow echo. Thus, the strong prolonged echo at site D2 appears to be isolated in terms of source and outflow. At site D3, and to the northeast, large areas return sporadic, strong, semiprolonged echoes [Figs. 12-4 (foldout), 12-8]. Because turbidity-current pathways are present upslope from D3, these echoes probably are the result of turbidity-current activity in the area of the debris-flow deposit. Finally, it is possible that the echo in some, or all, of the enigmatic regions is the result of debris-flow deposition alone—having arisen either from small surficial roughness (such as that photographed in the 4S Slide; Embley, 1975), or from surface roughness combined with a relatively high proportion of sand in the debris-flow deposit. Unfortunately, we have no cores or bottom photographs in the areas of anomalous echoes (e.g., Embley, 1975).

Relation to Thermohaline Contour-Current Features
Strong prolonged echoes commonly have been observed in moats both in the Canary Islands map area and elsewhere (e.g., Jacobi and Hayes, 1982; Damuth, Jacobi, and Hayes, 1983). In the Canary Islands map area, strong prolonged echoes also appear to be associated with furrows. At lat. 26°30′N, long. 20°30′W (south of L4, Fig. 12-6), a zone of prolonged echoes occurs on strike between furrow zones to the north and south. Furrows in the northern zone at L4 are oriented alongslope, and are associated with sediment waves, as are the strong, prolonged echoes at site L8. It is possible that the area of prolonged echoes represents furrows that are too small to produce hyperbolae. Such

a situation has been documented in the western Atlantic (Tucholke, 1979; Flood, 1982). Alternatively, the echo character at site L8 may result from turbidites derived from the nearby seamounts.

Current Interactions with Igneous Basement Ridges on the Lower Continental Rise

Subtle, but apparently common, features on the lower continental rise include local "ramps", or "steps," on the seafloor overlying basement ridges; the crests of these ridges generally parallel the axis of the Mid-Atlantic spreading center. We believe that many of the ramps were formed by two complementary processes: (1) the actual slope of a ramp is the result of deposition on the regionally downslope side of a basement ridge; and (2) the flat seafloor upslope from the ramp is caused by turbidites ponded in a trough located on the regionally upslope side of the basement high (abyssal hill). Ramps typically exhibit various types of sediment waves. For example, standing waves with irregular wavelengths are primarily drape "folds," whereas migrating sediment waves and regular standing waves are probably the result of current interactions with the topographic highs. Strong prolonged echoes returned from the upslope flat seafloor result from turbidites deposited from turbidity currents that were redirected along the upslope side of the ridge. The sediment-wave provinces overlying the basement ridges remain relatively free of turbidites. As deposition progresses, an upslope series of ramps develops, in which each ramp represents the downslope part of a buried ridge, and the flat seafloor is the turbidite pond in the upslope trough. This relationship demonstrates that the abyssal hills are long enough to locally dam turbidites.

The most convincing evidence for the ramp/flat hypothesis comes from GLORIA images (Kidd, Simm, and Searle, 1985; Kidd, Hunter, and Simm, 1987; Searle, 1987). At site SW3 (Fig. 12-7), immediately upslope from NNE-trending basement ridges, the turbidite-related sediment lineations change from approximately downslope to ridge-parallel orientations [Fig. 12-4 (foldout)]. GLORIA data show that a maximum dam length in the area is ~ 55 km, but ridges as short as 10 km also exhibit parallel sediment lineations. Shorter ridges display lineations with oblique orientations.

Variously oriented ship tracks reveal that the trends of ramps and sediment-wave provinces commonly parallel seafloor spreading magnetic lineations. On the Canary Islands map, where sediment-wave fields are restricted to regions underlain by basement highs, we drew province boundaries parallel to the magnetic lineations mapped by Hayes and Rabinowitz (1975). At site SW1 (Fig. 12-7), the ship's course change allowed us to verify that the orientation of basement features in the immediate area is parallel to the trend of magnetic lineations.

As the burial depth of basement ridges increases, their effect on sediment dispersal is lessened. For this reason, basement-controlled sediment features are not prevalent on the upper continental rise. However, if sediment waves build up over a basement high, then they can prolong the influence of the basement ridge long after the ridge is deeply buried (Kidd, Hunter, and Simm, 1987).

In the region of SW4 (Fig. 12-7), a GLORIA image clearly displays an intermediate stage in the burial history of basement ridges where downslope and along-ridge features meet (Fig. 7 of Kidd, Hunter, and Simm, 1987). Northwest-trending sediment lineations clearly cross a broken, linear, wide, low-reflectivity zone, which trends north. The low-reflectivity zone crosses our ship track at a basement-influenced ramp that had relatively little influence on turbidite deposition [Fig. 12-4 (foldout)]. Apparently, as relief on the basement features diminishes during burial, downslope processes begin to cross the features to form cross-ridge trends. On the Canary Islands map [Fig. 12-4 (foldout)], we have interpreted the ramps to trend parallel to basement, whereas the sediment waves and recognizable channels trend downslope. These cross trends demonstrate that interpreting trends for sediment-wave provinces involves a high degree of uncertainty.

In addition to the basement fabric of abyssal hills, fracture-zone ridges can influence deposition patterns (see Chapter 11), especially if the fracture zones are oblique to the regional slope. However, on the lower continental rise in the Canary Islands map area [Fig. 12-4 (foldout)], there is little evidence of such a relationship. We found only one sediment-wave province that may be oriented parallel to a fracture zone that was identified by Hayes and Rabinowitz (1975). In general, fracture-zone relief appears to be small in this area, or perhaps the ridges have been more completely buried here than elsewhere. Fracture-zone ridges also are not apparent in the GLORIA data taken to the west of the Canary Islands map (Kidd, Hunter, and Simm, 1987; Searle, 1987).

Sediment Slides and Slumps

On small-scale maps it is common to represent even major slide complexes as a single unit (e.g., Fig. 12-2). This "lumping" can be misleading; for example, the 4S Slide (especially between sites D1 and D2 of Fig. 12-8) exhibits a number of seismic facies arranged in bands extending down the slope. The presence of multiple bands of blocky and debris-flow sediment in the D1–D2 region suggests that the entire area may not have slid

away at once, but moved in several discrete events. Alternatively, the hyperbolae could be from a debris flow that moved over a rough bottom associated with levees.

Relationships at the toe of the 4S Slide near D11 (Fig. 12-8) support the multi-event hypothesis; there, at least two debris-flow deposits are piled above the sandy turbidite in the turbidity-current pathway. The number of buried slide-scar "channels" in the upper area of the Saharan Slide (e.g., between D1 and D2, Fig. 12-8) also is evidence that the slide complex was formed by several events. Canyons feeding the 4S Slide also contain slide material, which suggests that some of the canyons also contributed material for the 4S Slide.

The downslope portion of the 4S Slide also exhibits multiple seismic facies. In the lower area (D5, D6; Fig. 12-8), small lobes diverge from the main deposit, and near its probable termination (D11, Fig. 12-8), the 4S Slide appears to have split into several narrow lobes (~ 1.5–7 km wide). Similar features are present at the deep-tow survey site (Site L1, Fig. 12-6). There the debris-flow deposit terminates in extremely narrow (~ 0.7–0.15 km), thin tongues that extend downslope for at least 8–10 km [Figs. 12-4 (foldout), 12-6, 12-8; Lonsdale, 1978, 1982].

We observed no major rotated slump blocks at the surface in the Canary Islands map area. However, large fields of standing waves (regular and irregular), and even some "migrating sediment waves," probably originated as surface creep; we infer this interpretation from the sediment-wave morphology, crestline orientation, and location relative to other slump/slide features (Jacobi and Hayes, 1982; Table 12-1). These fields are restricted to the continental slope, as are most ductile and brittle slump faults not directly related to canyon walls. In the Canary Islands map [Fig. 12-4 (foldout)], a large field of probable surface-creep folds (between lat. 23–24°N, long. 17W–18°30′W, Fig. 12-7) is located along slopes with the same gradient as those that failed in the 4S Slide. The upslope boundary of these surface-creep sediment waves is a ductile slump fault (D12, Fig. 12-8), or a possible brittle slump fault in the south. At site D13 [Figs. 12-4 (foldout), 12-8], the field of "migrating sediment waves" immediately upslope from the 4S Slide probably is composed of surface-creep/ductile slump folds, as inferred from their morphology and crestline orientation.

In the southeastern part of the Canary Islands map, a surface scarp (site D14, Fig. 12-8), was suggested to extend more than 2 km below the seafloor as a fault, and to represent a part of the Atlantis Fracture Zone (Rona, 1970; Rona and Fleming, 1973). Embley (1975) and Embley and Jacobi (1977a) concluded from 3.5-kHz data that this scarp marked a slump fault, and that hummocky terrain with parallel reflectors downslope from the scarp represented deformed, rotated slump blocks (the Atlantis Slump). Similar explanations have been proposed for other sets of similar echo-character features. The postulated typical characteristics of shallow sediment slumps or areas of downslope creep are: (1) an upslope scarp that forms the upslope boundary; (2) a body of deformed parallel reflectors that generally are above a subbottom transparent or hummocky zone; and (3) a downslope bathymetric bulge or "compression" toe.

An alternative explanation for such echo characteristics is that most represent buried slide complexes. This alternative was proposed for some subbottom transparent zones by Embley (1980), and for both transparent and hyperbolated subbottom zones by Jacobi and Hayes (1982) (see also Chapter 13). In this explanation, the scarp is considered to be a relict erosional slide scar, not a slump fault. The subbottom hyperbolae represent buried blocky slide material (Jacobi, 1976), and transparent wedges represent buried debris-flow deposits. In many locations, such transparent zones smooth the underlying topography by partially filling bathymetric lows and thinning over bathymetric highs, like surficial debris-flow deposits. The compression toe invariably lies directly above an abrupt pinchout of a relatively thick transparent layer; this relationship suggests that a drape fold lies above the toe of a debris-flow deposit. Several buried debris-flow deposits feather out, rather than abruptly pinching out, and these deposits lack a compression toe. The buried slide hypothesis provides a more tenable explanation for most of the subbottom transparent and hummocky zones because: (1) multiple subbottom transparent zones commonly are observed; (2) the feathered-out terminations do not have a compression bulge; and (3) the transparent zones smooth the underlying topography.

By documenting the location and depth of these anomalous subbottom reflectors, we have been able to show that the Canary Islands map area contains a number of extremely large buried sediment slides, that some areas have undergone repeated sediment sliding, and that in some areas different sediment slides apparently moved at approximately the same time [Figs. 12-4 (foldout), 12-8; see also Chapter 13].

The largest buried sediment slides are in the southern and central map areas [Fig. 12-4 (foldout)]. The 4S Slide can no longer be considered an anomalously large slide—other regions both in the south and central areas of the Canary Islands map area contain large buried sediment slides that reach nearly to the abyssal plain, and farther downslope than the 4S Slide. The Canary Slide, newly described and named by Masson et al. (Chapter 13), reaches into the Madeira Abyssal Plain at long. 24°W (its upper part is shown in Fig. 12-8 at lat. 28–29°N, long. 19–20°W).

REGIONAL SEDIMENTARY PROCESSES OF THE NORTHWEST AFRICAN CONTINENTAL RISE: INTERACTIONS AND IMPLICATIONS

Four primary processes play prominent, interactive roles along the Northwest African margin and directly affect the architecture of the continental rise: (1) thermohaline contour-following currents (THCC); (2) turbidity currents; (3) igneous basement and piercement structures; and (4) sediment slides.

Thermohaline Contour-Following Currents

Two echo types commonly are regarded as characteristic of thermohaline contour-following currents (THCC): (1) regular, overlapping hyperbolae, tangent to the seafloor; and (2) migrating sediment waves. Additional echoes associated with THCC in some areas include strong prolonged echoes and sediment drifts (e.g., Embley, Rabinowitz, and Jacobi, 1978). Most research that includes a THCC component along the Northwest African margin has focused on the effects of Antarctic Bottom Water (AABW) flow. However, North Atlantic Deep Water (NADW) also affects the continental rise; it flows south above the AABW in the depth range of 3,900–1,500 m (e.g., Johnson, 1982; Sarnthein et al., 1982; Murray et al., 1986). At least six distinct water masses are present above the NADW, but none presently reaches the depths of the continental rise.

Antarctic Bottom Water

The present AABW flow originally was presumed to be sluggish in the eastern North Atlantic Basin, as inferred from dynamic oceanographic calculations (e.g., Wüst, 1955), bottom temperatures and other physical/chemical properties (e.g., Heezen and Hollister, 1971), and bottom photographs (Lowrie, Jahn, and Egloff, 1970; Young and Hollister, 1975; Embley, 1975; Jacobi, Rabinowitz, and Embley, 1975). Sediment waves were thought to be either relict features or to have formed by very slow flow (Jacobi, Rabinowitz, and Embley, 1975). The preferred interpretation was that AABW flow formed the sediment waves during accelerated flow characteristic of glacial intervals.

On the basis of potential bottom-temperature isotherms (Worthington and Wright, 1970), the upper surface of AABW was thought to slope gradually down from about 3,750 m near the equator to about 5,000 m near lat. 39°N. More recent data suggest that the upper AABW surface slopes more gently, from about 3,900–3,800 m at the Cape Verde Islands to about 3,900–4,000 m near lat. 35°N (Lutze, 1980; Johnson, 1982; Sarnthein et al., 1982). The upper surface is actually a transition zone several hundred meters thick (Johnson, 1982).

Sluggish AABW crosses from the western Atlantic through the equatorial fracture zones, including the Romanche and Chain fracture zones (e.g., Heezen et al., 1964; Heezen and Tharp, 1965; Mantyla and Reid, 1983) and the Vema Fracture Zone at lat. 10–11°N (Heezen, Gerard, and Tharp, 1964; Heezen and Hollister, 1971; Eittreim, Biscaye, and Jacobs, 1983; Mantyla and Reid, 1983). From the distribution of sediment waves and the relation of seamounts to moats and locally thickened sedimentary sections, Jacobi and Hayes (1982) inferred that AABW flowed north along the eastern margin of the Sierra Leone Basin and then

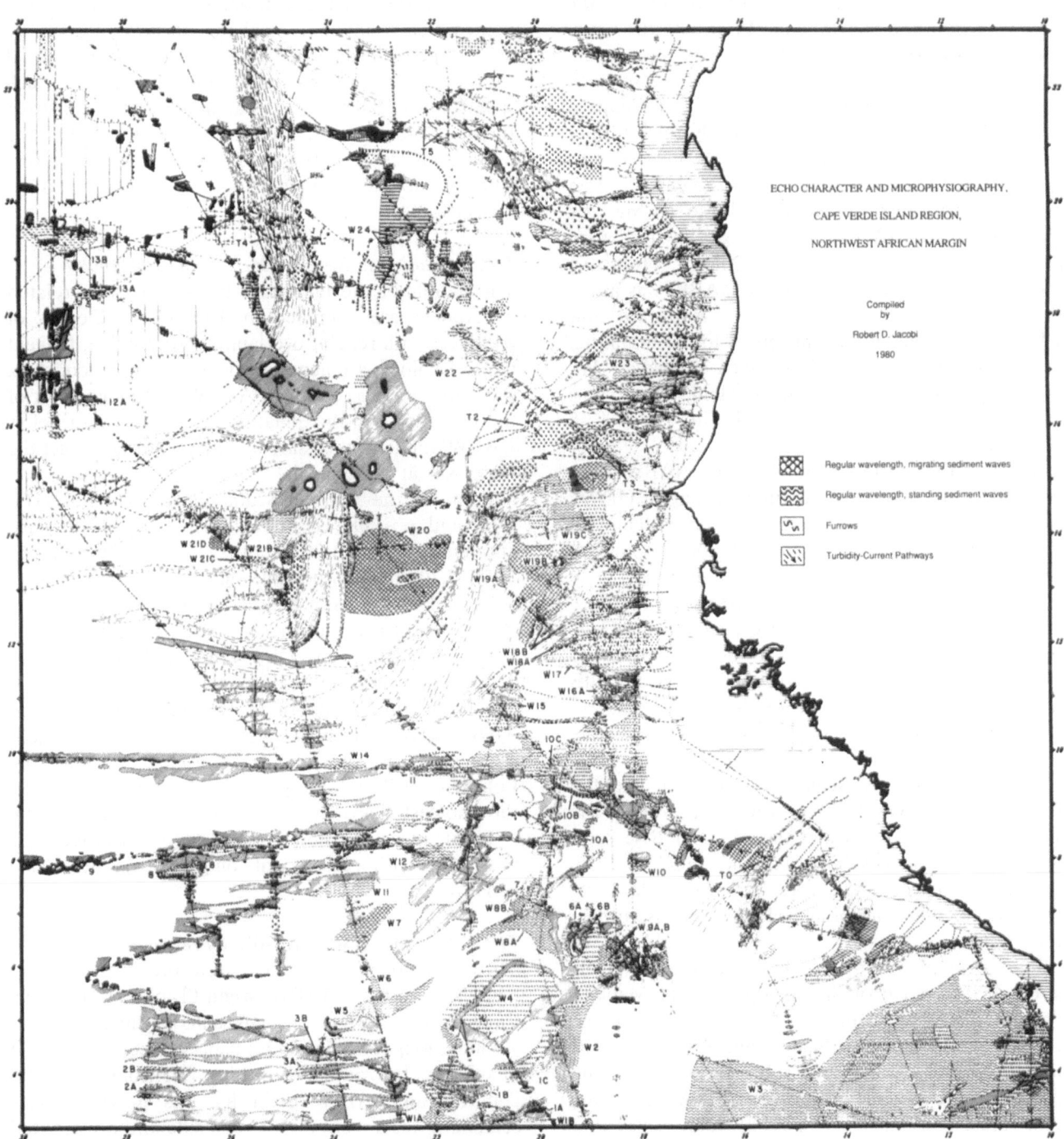

FIGURE 12-9.   Cape Verde Islands map area. Labels with prefix "T" refer to turbidity-current pathways, those with prefix "W" refer to sediment-wave provinces, and those without a letter prefix refer to furrow (hyperbolae) provinces. For other symbol explanations see Figure 12-5 (foldout) (a generalized version of this map appeared in Jacobi and Hayes, 1982).

south along the eastern margin of the Sierra Leone Rise, to form a counterclockwise gyre in the Sierra Leone Basin. The presence of sediment waves buried by turbidites indicates that circulation in the basin was more vigorous in the past. Furrows along the margins of the Kane Gap [Figs. 12-1, 12-5 (foldout)], as well as the distribution of locally extensive zones of nondeposition and(or) erosion, and of sediment drifts, indicate that bottom water generally flowed north through the Kane Gap from the Sierra Leone Basin into the Gambia Basin (Hobart, Bunce, and Slater, 1975; Jacobi and Hayes, 1982). A detailed seismic survey and core study of the Kane Gap area has refined the circulation history (Mienert, 1985, 1986). Relatively thick (upper surface ~ 4,200 m) and vigorous AABW flowed north through the Kane Gap from 3.7 to 2.2 Ma; NADW flowed south through the gap from 1.6 to 1.4 Ma, and relatively thin AABW has flowed sluggishly northward through the gap since 1.4 Ma. During Pleistocene glacial times (1.6 Ma to 10 ka), the AABW was about 100 m thicker than at present. The southward flow of NADW also may have contributed to the formation of the sediment-wave provinces such as that at W2 (Fig. 12-9) and the sediment drifts along the western margin of the Sierra Leone Basin [Fig. 12-5 foldout)].

North of the Sierra Leone Rise, the AABW flows generally north along the continental margin (Jacobi and Hayes, 1982). The distribution of bottom temperatures, echo character, and the orientation of sediment waves with respect to bathymetric contours, allow us to construct a more detailed path for the northward flow. The AABW is deflected to the west around the Cape Verde Islands and then continues flowing northward through, and to the west of, the Saharan Seamounts and around the western side of the Canary Islands. AABW flow splits north of the Canary Islands; one branch flows northeast along the continental margin south and east of the Seine Abyssal Plain, and the other branch curves back to the west around the Madeira Island pedestal and the Madeira Rise sediment drift (Jacobi, Rabinowitz, and Embley, 1975; Embley, Rabinowitz, and Jacobi, 1978; Lonsdale, 1978, 1982; Jacobi, 1982; Jacobi and Hayes, 1982; Sarnthein et al., 1982).

Short-term current meters deployed in a sediment-wave province north of the Agadir Canyon (Madeira Island map area, W41, Fig. 12-10) and in a hyperbolae province in the Saharan Seamounts (Canary Islands map area, H16D, Figs. 12-6, Table 12-2) confirm the general picture of sluggish northward flow (Lonsdale, 1978, 1982). Lonsdale's (1978, 1982) studies show that the current is still active with an alongslope flow of 3–4 cm/sec to the ENE and NE in the sediment-wave province and 5–6 cm/sec (9 cm/sec max) north of the sediment-wave province. If the lee-wave model of Flood

(1988) is correct, and if Lonsdale's (1978, 1982) data are representative of long-term flow, then the average velocity in the wave field is sufficient to produce the sediment-wave morphology observed on the 3.5-kHz records (Flood, private communication, 1991).

A 4-day current-meter study in hyperbolae field H16D in the Canary Islands map area (Fig. 12-6, Table 12-2) measured a 3–4-cm/sec average flow to the NNE and a superimposed, exceptionally strong, tidal component, which results in a maximum speed of 18 cm/sec alongslope (Lonsdale, 1978, 1982). This flow is parallel to the erosional furrow orientations in the area. Such speeds are sufficient both to maintain and to construct furrows (Flood, 1982; see also Chapter 13). On the Madeira Rise, Saunders (1987) recorded near-bottom current velocities as high as 30 cm/sec over a furrow province.

THCC Relation to Regular Overlapping Hyperbolae

The depths of the important fields of current-related hyperbolae are compiled in Table 12-2. Depths of provinces with regular overlapping hyperbolae, tangent to the seafloor, are generally deeper than, or equal to, the upper surface depth of AABW. Although most of the hyperbolae in the Saharan Seamounts occur below AABW, the flow necessary to maintain or construct the furrows is at present tidal.

Some of the Saharan Seamount areas (fields H15, H17, and possibly H18; Table 12-2, Fig. 12-6) have anomalously shallow depths, in the range of 3,600–3,200 m. It is possible that these shoaler hyperbolae fields also were caused by AABW flow, if: (1) AABW was thicker in the past (e.g., Sarnthein et al., 1982); or (2) eddies episodically bring AABW to shoaler depths, especially in regions of obstructions; or (3) uncertainty in the depth of the upper boundary of AABW allows it to be shoaler. Although each of these alternatives is possible, current-meter studies by Lonsdale (1978, 1982) suggest that NADW flows across the shoalest fields in the Saharan Seamount region. Lonsdale measured slow (1.3–2.4 cm/sec average) SSE to SE flow with a large tidal component on either side of H17.

THCC Relation to Sediment-Wave Provinces

The depths of most sediment-wave provinces are also consistent with an AABW origin (Table 12-3). In the Cape Verde map area, a small seamount located at lat. 13°30′N, long. 19°30′W in a major sediment-wave field near Dakar [province W19B, Figs. 12-5 (foldout), 12-9] confirms an AABW origin for this province. This seamount has a moat on the west side and preferential sedimentation on the east side (profile OO in Fig. 13-12 of Jacobi and Hayes, 1982). This relationship is consistent with northward flow of bottom water, and is not easily reconciled with an east-to-west flowing tur-

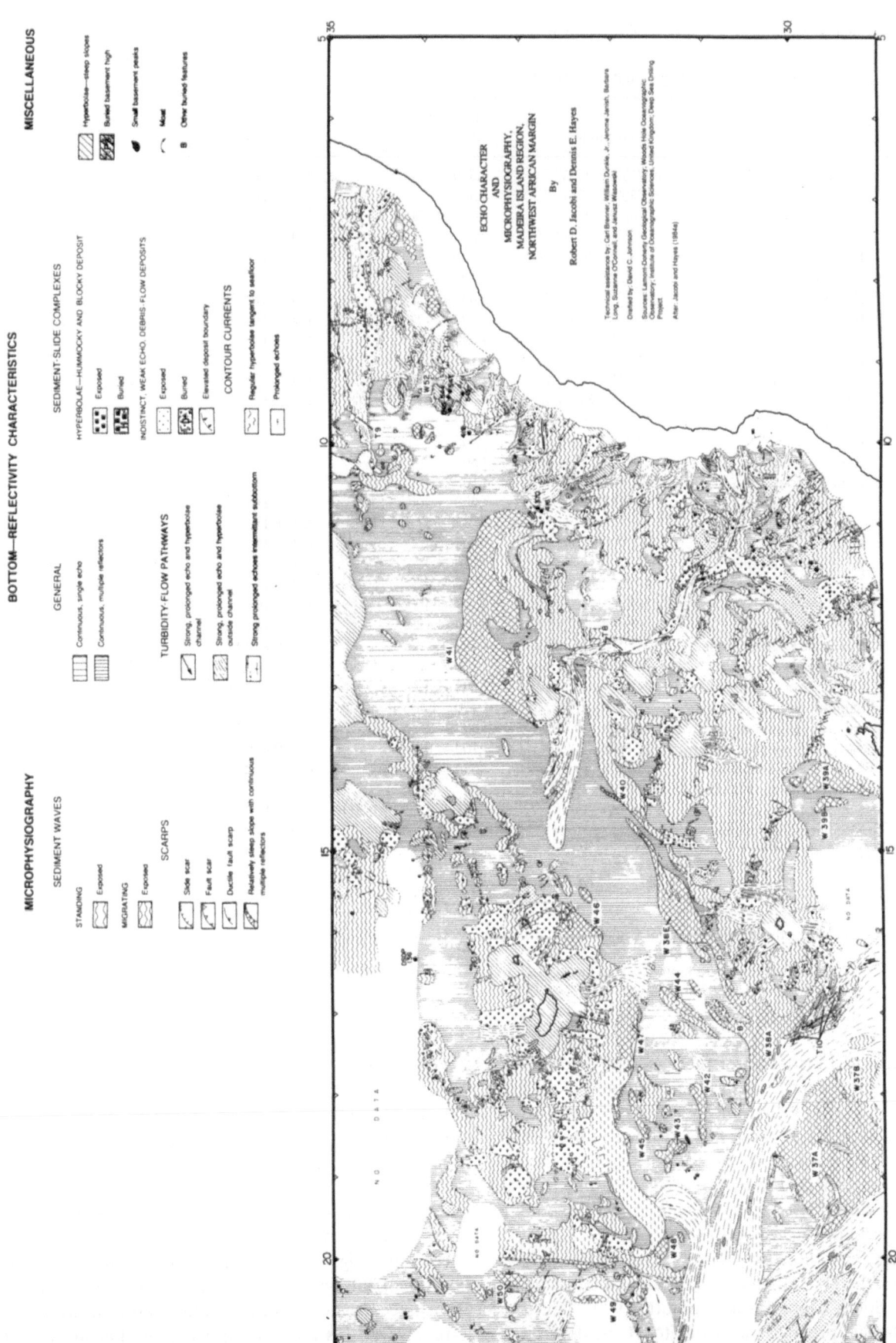

LEGEND

MICROPHYSIOGRAPHY

SEDIMENT WAVES

STANDING

⬚ Exposed

MIGRATING

⬚ Exposed

SCARPS

⬚ Slide scar

⬚ Fault scar

⬚ Ductile fault scarp

⬚ Relatively steep slope with continuous multiple reflections

BOTTOM—REFLECTIVITY CHARACTERISTICS

GENERAL

⬚ Continuous, single echo

⬚ Continuous, multiple reflections

TURBIDITY-FLOW PATHWAYS

⬚ Strong, prolonged echo and hyperbolae channel

⬚ Strong, prolonged echo and hyperbolae outside channel

⬚ Strong prolonged echoes intermittant subduction

SEDIMENT-SLIDE COMPLEXES

HYPERBOLAE—HUMMOCKY AND BLOCKY DEPOSIT

⬚ Exposed

⬚ Buried

INDISTINCT, WEAK ECHO, DEBRIS FLOW DEPOSITS

⬚ Exposed

⬚ Buried

⬚ Elevated deposit boundary

CONTOUR CURRENTS

⬚ Regular hyperbolae tangent to seafloor

⬚ Prolonged echoes

MISCELLANEOUS

⬚ Hyperbolae—steep slopes

⬚ Buried basement high

◣ Small basement peaks

⌒ Moat

B Other buried features

ECHO CHARACTER
AND
MICROPHYSIOGRAPHY,
MADEIRA ISLAND REGION,
NORTHWEST AFRICAN MARGIN

By

Robert D. Jacobi and Dennis E. Hayes

Technical assistance by: Carl Brenner, William Durkle, Jr., Jerome Jarrett, Barbara Long, Suzanne O'Connell, and Januaz Wasowski

Drafted by: David C. Johnson

Sources: Lamont-Doherty Geological Observatory; Woods Hole Oceanographic Observatory; Institute of Oceanographic Sciences, United Kingdom; Deep Sea Drilling Project

After Jacobi and Hayes (1984a)

FIGURE 12-10. Madeira Island map area. Labelled provinces discussed in text and listed in Tables 12-2 and 12-3. Labels with the prefix "T" refer to turbidity-current pathways, and those with a "W" prefix refer to sediment-wave provinces. For Canary Islands map area, see Figure 12-6 for lineations, Figure 12-7 for sediment waves, and Figure 12-8 for sediment slides.

TABLE 12-2   Depth Below Sea Level of Hyperbolae (Furrow) Provinces.*

Site	Locality	Coordinates	Depth (in meters) Max.–Min., or Avg.	References[†]
H1(a)	S Sierra Leone Rise	3°15′N, 20°W	4,500	J, J + H
(b)	S Sierra Leone Rise	3°30′N, 21°W	4,300	J, J + H
(c)	S Sierra Leone Rise	4°N, 20°W	4,300–4,200	J, J + H
H2(a)	Fracture Zone	3°65′N, 27°20′W	4,000	J, J + H
(b)	Fracture Zone	4°N, 27°20′W	4,000	J, J + H
H3(a)	Fracture Zone	4°N, 24°30′W	4,200	J, J + H
(b)	Fracture Zone	4°20′N, 24°30′W	4,000	J, J + H
H4	NW Sierra Leone Rise	5°N, 21°30′W	4,000	J, J + H
H5	Fracture Zone	5°30′N, 28°W	~ 4,050	J, J + H
H6(a)	E Sierra Leone Rise	6°20′N, 19°30′W	4,500	J, J + H
(b)	E Sierra Leone Rise	6°20′N, 19°05′W	4,500–4,400	J, J + H
H7	NW Sierra Leone Rise	7°30′N, 20°30′W	3,910–3,850	J, J + H
H8	Fracture Zone	7°50′N, 26°50′W	4,900	J, J + H
H9	Fracture Zone	8°N, 28°30′W	5,000	J, J + H
H10(a)	Kane Gap area	8°20′N, 19°10′W	4,500	J, J + H
(b)	Kane Gap area	9°10′N, 19°W	4,500–4,400	J, J + H
(c)	Kane Gap area	9°60′N, 19°50′W	4,500–4,400	J, J + H
			N: 4,600	J, J + H
H11	Fracture Zone	9°50′N, 22°10′W	5,000–4,500	J, J + H
H12(a)	FZ + Abyssal Hills	16°30′N, 29°W	5,200–4,950	J, J + H
(b)	FZ + Abyssal Hills	17°N, 29°50′W	5,000–4,900	J, J + H
H13(a)	FZ + Abyssal Hills	18°20′N, 28°45′W	4,500–4,400	J, J + H
(b)	FZ + Abyssal Hills	19°20′N, 29°30′W	4,900–4,500	J, J + H
H14(a)	SW Saharan Seamounts	24°35′N, 21°10′W	4,400– >	E, ERJ, JHa
(b)	SE Saharan Seamounts	24°30′N, 19°50′W	3,800	E, ERJ, JHa
H15	SE Saharan Seamounts	24°50′N, 19°30′W	3,600–3,400	E, ERJ, JHa
H16(a)	Saharan Seamounts (W)	25°25′N, 21°W	4,470–(4,400) – > "b"	E, ERJ, JHa
(b)	Saharan Seamounts (central)	25°45′N, 20°40′W	4,400	E, ERJ, JHa
(c)	Saharan Seamounts (central)	25°30″N, 20°W	4,100–4,000	E, ERJ, JHa
(d)	Saharan Seamounts (E)	25°55′N, 19°50′W	4,040–3,850	E, ERJ, JHa, L
H17	Saharan Seamounts (E)	25°20′N, 19°10′W	3,300–3,200	E, ERJ, JHa, L
H18	Saharan Seamounts	26°20′N, 19°40′W	4,050– > 3,800–3,500	E, ERJ, JHa, L
H19	N Saharan Seamounts	27°N, 20°10′W	4,450–4,280	E, ERJ, JHa, L
H20	SE Madeira Rise	32°N, 19°W	4,750–4,000	E, ERJ, JHb
H21	W Madeira Rise	32°25′N, 19°50′W	4,450–4,200	E, ERJ, JHb
H22	W Madeira Rise	33°N, 19°50′W	4,375–4,200	E, ERJ, JHb

*Depths based on bathymetry of GEBCO (Searle, Monahan, and Johnson, 1982) integrated with echo character from reports as referenced.

[†]E = Embley (1975); ERJ = Embley, Rabinowitz, and Jacobi (1978); J = Jacobi (1982); J + H = Jacobi and Hayes (1982); JHa = this report; JHb = Jacobi and Hayes (1984a); L = Lonsdale (1978, 1982).

bidity current. Furthermore, there are no known turbidity-current pathways near the seamount. We are, therefore, confident that these major sediment-wave provinces near Dakar were caused directly by THCC. However, the position of these fields appears to be governed in part by the availability of sediment from downslope turbidity currents, because most large fields are near turbidity-current pathways.

Turbidity currents also can construct sediment-wave fields (e.g., the Delgada Fan, Normark and Gutmacher, 1985; the South China Basin, Damuth, 1979). Thus, for sediment-wave provinces adjacent to TCPs, depth of a sediment-wave field alone does not discriminate between a THCC and a TCP origin. The orientation of crestlines appears to be the prime discriminator. In the Canary Islands map area, for example, sediment waves

generated by turbidity currents have crestlines oriented approximately downslope, whereas the usual orientation of THCC-generated sediment waves is either alongslope or somewhat oblique to the slope. In some of the major provinces in the Cape Verde Islands map area (W19B, W19C, W20 on Fig. 12-9), sediment waves are oriented parallel to regional bathymetric contours and normal to the general downslope flow of turbidity currents [Fig. 12-5 (foldout)]. The orientation and character of the waves do not appear to change at greater distances away from the channels and turbidity-current pathways; this relationship suggests that most of these sediment waves were built by along slope currents. The distribution and crestline orientations of the sediment waves indicate that the major fields along the Northwest African margin were most likely constructed by

TABLE 12-3   Depth Below Sea Level of Sediment-Wave Provinces.*

Site	Locality	Coordinates	Depth (in meters) Max.–Min., or Avg.[†]
		Cape Verde Islands map	
W1a	S Sierra Leone Rise +FZ	3°20′N, 22°W	4,300–4,100(E)– 4,000(W)
b	S Sierra Leone Rise +FZ	3°10′N, 21°W	4,650–4,300
W2	SE Sierra Leone Rise	4°30′N, 19°20′W	4,900–4,000
W3	Sierra Leone Basin	4°N, 14°W	5,000– ~ 3,000
W4	Top Sierra Leone Rise	5°30′N, 20°30′W	2,700–2,500
W5	W Sierra Leone Rise +FZ	5°15′N, 23°15′W	4,100
W6	W Sierra Leone Rise +FZ	5°45′N, 23°25′W	4,250–4,200
W7	W Sierra Leone Rise +FZ	6°30′N, 23°35′W	4,250–4,200
W8a	E Sierra Leone Rise	6°45′N, 20°W	4,000–3,820(E)– 3,750(W)
b	E Sierra Leone Rise	7°15′N, 20°20′W	3,800–3,750
W9a	E Sierra Leone Rise	6°20′N, 8°45′W	4,750–4,700
b	E Sierra Leone Rise	6°20′N, 18°10′W	4,750
W10	SMT, N Sierra L. Basin	7°50′N, 18°10′W	4,750–4,700
W11	Top of FZ ridge/SMT	7°30′N, 23°40′W	4,000–3,750
W12	W Sierra Leone Rise +FZ	8°N, 22°10′W	4,550–4,250
W13	FZ	9°50′N, 20°50′W	5,040–4,500
W14	FZ	9°50′N, 23°40′W	5,050–4,950
W15	Cont. Rise S Dakar	11°N, 20°30′W	5,020–4,900(?)
W16a	Cont. Rise S Dakar	11°05′N, 18°45′W	4,400(+)–4,200
b	Cont. Rise S Dakar	11°05′N, 18°30′W	4,200–4,100
W17	Cont. Rise S Dakar	11°30′N, 19°10′W	4,800(?)–4,000 or possibly 3,500
W18a	Cont. Rise S Dakar	12°N, 19°45′W	4,650
b	Cont. Rise S Dakar	12°30′N, 19°45′W	4,700
W19a	Cont. Rise W Dakar	13°15′N, 20°45′W	4,600–4,500
b	Cont. Rise W Dakar	13°15′N, 19°50′W	4,700–4,100
c	Cont. Rise W Dakar	14°N, 19°40′W	4,400–3,600
d	Cont. Rise W Dakar	14°15′N, 20°35′W	4,200
e	Cont. Rise W Dakar	14°55′N, 20°15′W	4,000–3,600
f	Cont. Rise W Dakar	15°N, 19°W	3,500–3,100
W20	SE Cape Verde Isds.	13°20′N, 23°W	4,800–4,250
W21a	SW Cape Verde Isds.	13°20′N, 25°W	4,650(TC?)
b	SW Cape Verde Isds.	13°40′N, 25°35′W	4,600–4,480
c	SW Cape Verde Isds.	13°40′N, 25°35′W	4,850–4,800
d	SW Cape Verde Isds.	13°55′N, 26°10′W	4,850–4,650
W22	S Cape Verde Plateau	17°10′N, 21°W	3,700–3,200
W23	Cont. Rise N Dakar	17°N, 18°25′W	3,700–2,700(gen) 1,500(min)
W24	N Cape Verde Isds.	19°10′N, 22°45′W	3,900–3,800
		Canary Islands Map	
W25	SW Saharan Smts.	24°20′N, 22°30′W	4,800
W26a	SW Saharan Smts.	24°15′N, 22°W	4,700–4,650
b	SW Saharan Smts.	24°15′N, 21°45′W	4,600
W27	S Saharan Smts.	23°50′N, 20°W	4,500–4,400
W28	S Saharan Smts.	24°25′N, 20°50′W	4,300–4,250
W29	SW Saharan Smts.	24°25′N, 21°10′W	4,500–4,400
W30	E Sharan Smts.	25°30′N, 18°55′W	3,300–3,100
W31a	W Saharan Smts.	26°35′N, 23°25′W	4,980–4,960
b	WNW Saharan Smts.	26°50′N, 22°30′W	4,900–4,860
c	NW Saharan Smts.	26°50′N, 21°50′W	4,860–4,780
W32a	NW Saharan Smts.	26°55′N, 21°20′W	4,840–4,620
b	NNW Saharan Smts.	26°55′N, 20°40′W	4,580–4,450
c	NNW Saharan Smts.	27°40′N, 21°W	4,730–4,400

TABLE 12-3 (continued)   Depth Below Sea Level of Sediment-Wave Provinces.*

Site	Locality	Coordinates	Depth (in meters) Max.–Min., or Avg.[†]
		Cape Verde Islands map	
W33	N Saharan Smts.	27°30′N, 19°20′W	4,300–4,220
W34a	N Saharan Smts.	27°10′N, 19°10′W	4,150–4,050
b	N Saharan Smts.	27°N, 19°W	4,050–3,850
W35a	N Saharan Smts.	26°35′N, 18°50′W	3,700–3,680
b	N Saharan Smts.	26°35′N, 18°40′W	3,700
W36	W Canary Isds.	28°30′N, 22°55′W	4,920
		Madeira Island Map	
W37a	W Canary Isds.	29°30′N, 18°40′W	4,650–4,200
b	W Canary Isds.	29°20′N, 17°50′W	4,200–4,000–3,800(SL)
W38a	Cont. Rise S Agadir Cyn.	30°15′N, 17°15′W	4,580–3,800(T)
b	Cont. Rise S Agadir Cyn.	30°45′N, 17°05′W	4,450–4,420
c	Cont. Rise S Agadir Cyn.	30°40′N, 16°55′W	4,350–4,275
d	Cont. Rise S Agadir Cyn.	30°50′N, 16°20′W	4,330–4,250
e	Cont. Rise S Agadir Cyn.	31°15′N, 15°40′W	4,420–4,310
W39a	Cont. Rise S Agadir Cyn.	29°30′N, 14°10′W	3,690–3,615
b	Cont. Rise S Agadir Cyn.	29°30′N, 14°30′W	3,575–3,550
W40	Cont. Rise S Agadir Cyn.	31°55′N, 14°20′W	4,330–4,220
W41	Cont. Rise N Agadir Cyn.	33°20′N, 12°W	4,440–4,000
W42	Basin, S. Madeira	31°, 18°15′W	4,680–4,600(T?)
W43	Basin, S. Madeira, SMT	31°20′N, 18°55′W	4,720–4,685
W44	Basin, S. Madeira	31°15′N, 16°55′W	4,460(T?)
W45	Basin S. Madeira	31°50′N, 18°10′W	4,660–4,550
W46	E Madeira	32°30′N, 16°W	4,400–4,100(T? at upper)
W47	S Madeira	31°50′N, 17°10′W	4,500–4,300
W48	S Madeira Rise	31°30′N, 19°50′W	4,740–4,700
W49	W Madeira Rise	32°10′N, 20°25′W	4,800–4,760
W50	W Madeira Rise	33°05′N, 20°10′W	5,000–4,860
W51	W Madeira Rise	33°N, 21°50′W	~ 5,250
W52	Cont. Rise, N Agadir Cyn.	33°55′N, 9°05′W	3,800–3,600

*Depths based on bathymetry of GEBCO (Searle, Monahan, and Johnson, 1982) Integrated with echo character from this report.
[†]SL = probable slump origin; T = probable turbidity–current-generated sediment waves.

AABW. However, long narrow fields adjacent to turbidity-current pathways, especially those that extend up the upper continental rise, significantly above the upper surface of AABW (e.g., field W37C), may have been constructed by turbidity currents.

A few sediment-wave provinces are too shallow to have been formed by the present AABW [W3, W4, W19F, W22, W23 (Fig. 12-9); W30, W35A, B (Fig. 12-7, Table 12-3)]. Some of these are most likely the result of NADW flow, whereas others may be related either to TCP or possibly to eddies from seamount/tidal interactions. Provinces W35A, B and W30 (Fig. 12-7) in the Saharan Seamount area are not adjacent to TCPs. Field W30 (Fig. 12-7) is immediately upslope from the furrow province H17, (which encompasses lineation site L3; Fig. 12-6), that at present is subject to NADW and tidal flow. Both provinces W23 and W19F (Dakar area) are at depths consistent with a NADW origin, but both also lie between TCPs. The irregular

standing waves between W19E and W19F are approximately at the depth of the upper surface of AABW. The upslope progression of waveforms thus may be related to the transition between AABW and the overlying NADW. We have no evidence that bears on the origin of province W23. Province W22 is well removed from present TCPs and is too shallow to be influenced by AABW; these sediment waves are probably the result of accelerated NADW.

Province W3 (Fig. 12-9) consists of several subprovinces. The deeper portions of wave field W3 were most likely caused by AABW (Jacobi and Hayes, 1982), but the shoaler portions probably resulted from a combination of slump (creep) folds and TCP-produced waves [Fig. 12-5 (foldout); Jacobi and Hayes, 1982]. Because the northwestern margin of the wave field is buried by more recent, planar-bedded, distal turbidites [Fig. 12-5 (foldout); Jacobi and Hayes, 1982], these sediment waves are presumed to be relict, and not related to present TCPs. However, the exposed, migrating sediment waves in the northwestern portion of W3 may be related to the adjacent TCPs. Migrating waves at a similar depth to the northwest (~ lat. 8°N, long. 16°W) appear to be the result of TCP flow, because they form matching levees on opposite sides of a TCP (profile UU in Fig. 13 of Jacobi and Hayes, 1982).

Province W4 (Fig. 12-9) is clearly above AABW, and may have been formed by NADW in combination with eddies generated by Sierra Leone Rise obstructions. Because these sediment waves have "double crestlines" on 3.5-kHz records, the crestlines are believed to be crescent-shaped (Jacobi and Hayes, 1982). Limited cores from DSDP Site 366 in province W4 suggest that construction of the waves began either in mid-Miocene time, according to sedimentation rates, or in Eocene time, according to winnowed sediments (Lancelot, Seibold, et al., 1977). Echograms suggest that the mid-Miocene date is more likely. Province W2 [Figs. 12-5 (foldout), 12-9] probably was affected when NADW flowed south through Kane Gap (1.6–1.4 Ma; Mienert, 1985, 1986). However, the province appears to be on strike with moats and sediment drifts adjacent to seamounts and partially buried basement highs [Figs. 12-5 (foldout)]; these relationships suggest longer times of southward flow. We, therefore, believe that sediment waves in province W2 were generated not only by NADW, but also by the southward flow of an AABW gyre.

Turbidity Currents

Turbidity-Current Pathways

Turbidity-current pathways (TCPs) are defined as the channel and adjacent seafloor that returns a strong, prolonged echo (Jacobi and Hayes, 1982). Generally, this echo type is thought to indicate a high proportion of coarse-grained sediment and short-wavelength seafloor roughness (Damuth, 1975, 1980). The seafloor roughness in the Grand Banks Slide area, for example, consists of gravel waves, which were imaged by Sea MARC I (Piper et al., 1985). On the Northwest African upper continental rise, TCPs generally are restricted to channels and channel walls of channel/levee complexes, whereas the relatively well-developed levees exhibit sediment waves with few, if any, strong to semiprolonged subbottom reflectors (e.g., site T0 on Fig. 12-9, and profile UU in Fig. 13 of Jacobi and Hayes, 1982). On the lower continental rise, channels have very low relief and levees are very subdued or nonexistent (the "lower submarine fan"). The TCPs widen downslope (~ 50–100 + km) and extend to the abyssal plains; laterally they extend well beyond the subdued levees. Sediment waves associated with some

major channel systems also appear to extend to the abyssal plains [e.g., Fig. 12-4 (foldout); see also Chapter 13]. That the bulk of the sandy turbidites (recognized as strong prolonged echoes) stretch from the lower fan/middle fan boundary to the abyssal plain is consistent with evidence from cores raised from the abyssal plains that contain multiple sand layers of variable thickness (e.g., Piper, Stow, and Normark, 1985; Weaver, Searle, and Kuijpers, 1986; Weaver and Rothwell, 1987).

Depositional / Sand Lobes

Do strong prolonged echoes represent depositional lobes? Mutti and Normark (1987) noted that "lobe" has disparate meanings for modern turbidity systems and ancient turbidites. In modern settings, "lobe" is primarily a morphologic term. The wide zones of strong prolonged echoes and cross-channel morphology on the Northwest African margin are similar to those identified as a midfan lobe on the Monterey Fan (Normark et al., 1985), as a depositional lobe on the Navy Fan (Normark and Piper, 1985), and as fan lobes on the Mississippi Fan (e.g., O'Connell et al., 1985). Because echograms commonly show no subbottom reflectors below the strong prolonged echo, we cannot determine from echograms alone whether these echoes indicate thick-bedded sands (typical of ancient lobe deposits), or thin-bedded sands (typical of ancient lobe fringe deposits). The strong prolonged echo may well be common to both types of deposits; it appears that both thick- and thin-bedded sandy turbidites occur in regions mapped as lobes or lobe fringe ("sheet") sands (Normark et al., 1985; Normark and Piper, 1985; O'Connell et al., 1985; Piper, Stow, and Normark, 1985). Limited core data from the Northwest African margin show that both thick (1–6 + m) and thin (~ 10 cm) turbidites are present (Sarnthein and Diester-Haass, 1977; Sarnthein, 1978; unpublished data). Other methods, including analysis of long cores, must be used to discriminate thick from thin beds. The extensive distribution of some strong prolonged echoes suggests that these echoes represent sheet sands of unknown thickness (sheet sands in ancient examples are relatively thin). On the Northwest African margin (Fig. 12-2), these depositional lobe sands (in modern turbidity-system terminology) do not form fans and do not coalesce to form continuous deposition alongslope on the lower fan and continental rise; rather, they form fairly restricted pathways.

On the Canary Islands map, we have recognized narrow (~ 1 km) elevated ribbons that return a strong prolonged echo [site T1 on Figs. 12-4 (foldout), 12-8]. Adjacent to the narrow bodies is a wide deposit (~ 34 km) with similar characteristics that spans the entire TCP. These deposits are most likely to be either debris-flow deposits over which more recent turbidity currents have flowed, or thick sandy turbidites—perhaps sand ribbons and(or) a sand lobe. The position of these echoes on the lower continental rise and lower fan, significantly removed (> 300 km) from the middle fan/lower fan boundary, is anomalous with respect to fan models.

Submarine "Fans": Form and Morphologic Zonation

As emphasized by Mutti and Normark (1987), and by several other authors (Bouma, Normark, and Barnes, 1985), many submarine "fans" are not fan-shaped. Figures 12-2, 12-3 (foldout), 12-4 (foldout), and 12-5 (foldout) show that on the Northwest African margin, no major turbidity-current system is fan-shaped. Only the small systems in the area of site T0 (Fig. 12-9) exhibit partial fan shapes.

In terms of morphologic zonation, if middle fans are discriminated from lower fans by channel/levee complexes with prominent relief, then middle fans generally extend only relatively short distances onto the continental rise [Figs. 12-2, 12-5 (foldout), 12-4 (foldout), 12-5 (foldout)]. The lower fan has an immense downslope length compared to that of the middle fan in most of the major turbidity current systems on the Northwest African margin, except for those south of about lat. 10°N.

Meandering Channels

As drawn in Figures 12-2, 12-3 (foldout), 12-4 (foldout), 12-5 (foldout), flow lines in the turbidity-current pathways are relatively straight. This linearity results from the extreme difficulty in mapping, or even in recognizing (from 3.5-kHz records alone), meanders of the kind imaged by sidescan records (e.g., Damuth et al., 1983, 1988), because multiple, closely spaced, well-navigated, ship-track data are rare.

On the Northwest African margin, each detailed map [Figs. 12-3 (foldout), 12-4 (foldout), 12-5 (foldout)] has a few areas where sufficient data indicate that most major TCPs probably include meandering channels (see also Chapter 13). On the Cape Verde Islands map [Fig. 12-5 (foldout)], meandering channels probably occur in the turbidity-current pathways of: (1) the Cayar Canyon system (site T2 on Fig. 12-9); (2) the system extending north from the Cape Verde Islands (site T4 on Fig. 12-9); and (3) the system extending from the Cape Blanc area (site T5 on Fig. 12-9). The Cayar Canyon also meanders at site T3 [Figs. 12-5 (foldout), 12-9]. On the Canary Islands map [Fig. 12-4 (foldout)], buried meanders are evident beneath the 4S Slide at sites T6 and T7 near lat. 25°15′, long. 17°30′W on (Fig. 12-8). On the Madeira Island map, sites T8, T9, and T10 (Fig. 12-10) all exhibit a meandering main channel. Although most areas on the rise where data are sufficiently closely spaced appear to have meandering channels (even areas that would be considered lower fan),

GLORIA data (Kidd, Hunter, and Simm, 1987) immediately west of the Madeira map (Fig. 12-2) reveal that buried channel walls on the lowermost rise do not exhibit distinctive meanders. There, relatively straight channels occur downslope from the meandering channels observed on the Canary Islands map; this change in morphology is consistent with studies that show channels become straighter as the slope decreases (see Damuth et al., 1988, for a review).

Single-channel seismic data that display multiple, active channels with similar depths and relief across a single channel/levee system have imaged either a meandering channel, or a braided channel system. Such a set of channels is observed at site T0 of Figure 12-9 (profile UU on Fig. 13 of Jacobi and Hayes, 1982). Asymmetrical cross sections of channels imaged at site T0 imply that at least two of these channels are probably parts of a meander.

Channel Avulsion versus Braided Channels

Because we have few sidescan data, and because the TCPs reveal little subbottom stratigraphy, it is generally not possible to determine whether the marked variability in number of channels between successive ship crossings of the TCP is the result of local tight meanders, recent channel avulsion, or braided channels. However, at site T4 on the Cape Verde map, and at sites T10 and T11 on the Madeira map (Fig. 12-10), the geometries do not easily fit a meander interpretation. If avulsion were the only active process at these three sites, then avulsion must have occurred in the very recent past (at least twice at site T4, and five times at site T11), because there is no discernible hemipelagic sediment cover over the prolonged echo. Although avulsion coupled with meanders could explain the observed pattern, braided channels also could be the cause, especially at sites T4 and T11. If the channels are braided, then compared to the meandering channels, they have higher slopes, a higher proportion of bed-load material, and carry more episodic turbidity currents ("flashy", e.g., Damuth et al., 1988).

Interactions Between Bottom Currents and Igneous Basement / Piercement Structures

The interactions of near-bottom processes with elevated basement features are both obvious and subtle. Obvious relationships include: (1) derivation of both turbidity currents and sediment slides from seamounts; (2) diversion of turbidity currents, debris flows, and contour currents by seamounts and plateaus; and (3) restriction and acceration of THCCs by seamount groups.

Subtle, but apparently common, interactions on the lower continental rise, occur between long, narrow "basement" highs and bottom currents. The "basement" highs include abyssal hills formed at spreading centers and along fracture zones, and salt/mud/igneous diapirs [e.g., Rona, 1969; Embley and Jacobi, 1977b; Lancelot and Embley, 1977; Hinz, Dostmann, and Fritsch, 1982; Jacobi and Hayes, 1982; Fig. 12-3 (foldout)]. Basement highs localize sediment waves of various types and divert turbidity currents into troughs upslope from the highs. These interactions result in seafloor ramps and flats on the lower continental rise (see earlier discussion of Canary Islands map).

Sediment waves and piercement structures that are localized on "basement ridges" [e.g., Embley and Jacobi, 1977b; see Fig. 12-3 (foldout)] can prolong the influence of buried "basement" ridges. For example, on the Cape Verde Islands map, piercement structures dam sediments and promote the formation of sedimentwaves (e.g., in the region of W24, Fig. 12-9).

On the lower continental rise in the Madeira and Canary Islands maps [Figs. 12-4 (foldout), 12-5 (foldout)], there is little direct effect of fracture-zone ridges on near-bottom processes. Fracture zones are more influential south of the Canary Islands map. In a transition zone near lat. ~ 16–22°N, long. 22–28°W, both fracture-zone and abyssal-hill morphology are important influences on near-bottom processes [Fig. 12-3 (foldout)]. On the lower continental rise south of the Cape Verde Islands, fracture-zone ridges appear to have diverted and controlled the path of most turbidity currents [Figs. 12-2, 12-3 (foldout)]. For example, turbidity currents flowing south from the Cape Verde Islands have swung west parallel to a fracture-zone ridge at lat. 12°N, long. 24–28°W [Figs. 12-2, 12-5 (foldout)].

Sediment Slides

Spatial Distribution
Mass wasting is an important downslope process that significantly affects the continental rise on the Northwest African margin (see also Chapters 3–7, 9, 10, 13). As detailed in Figures 12-3 (foldout), 12-4 (foldout), and 12-5 (foldout), and summarized in Figure 12-2, it is obvious that sediment slides have affected much of the upper continental rise. In some areas, such as the slope and upper rise north of Dakar, the entire seafloor comprises a series of slides [Fig. 12-5 (foldout); Seibold and Hinz, 1974; Jacobi, 1976; Jacobi and Hayes, 1982].

Few slides occur along the slope and rise south of lat. 11°N, and south of 7°N the relatively few slides present are short (measured downslope), because the slope and rise are narrow. North of the Canary Islands, we also observed only relatively small slides. Slides are concentrated primarily along the margin between lat. 15 and ~ 29°N. Much of this region is now bordered by the Sahara Desert, which limits fluvial sediment input. The region of most extensive sliding matches the region of high aeolian input (Koopman, 1980; Fig. 7 of Sarnthein et al., 1982; Fig. 12-1). The area of maximum sliding also borders the region of intense upwelling (e.g., Sarnthein et al., 1982; Fig. 12-1), which results in high biogenic gas concentrations in the associated sediments (Mienert and Schultheiss, 1989).

The lower continental rise clearly does not exhibit as many slides as the upper rise. However, several major slides extend well onto the lower continental rise, nearly reaching the Madeira/Cape Verde abyssal plains (Fig. 12-2; see also Chapter 13). These massive slides extend significantly farther downslope than the base-of-slope (traditionally viewed as the depositional site for slide material), and even beyond the middle fan region, where Shanmugam and Moiola (1985) and Shanmugam, Damuth, and Moiola (1985) believed they generally ended.

Island groups, seamounts, and basement highs, including fracture-zone ridges and abyssal hills, all are sources for sediment slides that flow from steep slopes onto the lower rise and to the abyssal plain. On the Cape Verde Islands map, many of the equatorial fracture zones display multiple semitransparent deposits along much of the fracture-zone floor [Fig. 12-5 (foldout)]. Although Jacobi and Hayes (1982) suggested that the deposits might be debris-flow deposits reworked by THCC, cores recovered by the Ocean Drilling Program (ODP) contain sediment-slide material that does not display evidence of THCC reworking (Sarnthein, private communication, 1987). The great number of slides along the fracture zones may suggest that the slides record relatively frequent earthquake activity related to minor plate adjustments (Jacobi, 1976).

Temporal Distribution and Triggering Mechanism
The main debris-flow deposit in the 4S Slide was active about 16–17 ka, according to data from a core that penetrated sediment below the debris-flow deposit (at ~ lat. 26°30'N, long. 17°35'W; Embley, 1982). However, smaller debris flows apparently were active as recently as 1–2 ka, as indicated by the pelagic cap over debris-flow deposits found in a core taken at ~ 25°30'N, 18°20'W (Embley, 1982). Based on approximate sedimentation rates, buried sediment slides observed on 3.5-kHz echograms in the Canary Islands map area (Fig. 12-8) span several hundred thousand years. Continuous coring by the Ocean Drilling Program permits us to assemble a longer record of slope failures. For example, data from Leg 108 (Sites 657–668, Ruddiman, Sarnthein, et al., 1989) show a concentration of slide events that began about 2.8 Ma and peaked at about 2.3 Ma, with a secondary mode at about 1.2 Ma and tertiary modes at about 3.8 and 0.6 Ma. Elsewhere in the Atlantic, a similar concentration of dates (Wisconsinan or older) has been found (e.g.,

on the Feni Drift; Flood, Hollister, and Lonsdale, 1979), but with a significant number of dates ranging into the present (e.g., on the U.S.–Canadian continental margin, Embley, 1980, 1982; Embley and Jacobi, 1977a, 1986; Piper, et al., 1985; and on the Norwegian continental margin, Jansen et al., 1987).

The high number of slide events of Wisconsinan age (and older glacial intervals) generally has been attributed to high sedimentation rates during glacial lowstands (e.g., Embley and Jacobi, 1977a, 1986). Traditionally, the high sediment input has been ascribed elsewhere to river systems that reached the shelfbreak during lowered sea level. However, during the glacial intervals beginning about 3 Ma (and including the Wisconsinan at 18 ka), West Africa was so arid that fluvial discharge into the Atlantic was almost nonexistent from approximately lat. 30°N to the southernmost part of the mapped area (Pastouret et al., 1978; Sarnthein et al., 1982; Stein et al., 1989; Tiedemann, Sarnthein, and Stein, 1989). Rather, the increased terrigenous (siliciclastic) input observed along much of the Northwest African margin was due primarily to vastly increased aeolian transport (e.g., Koopman, 1980; Fig. 21 of Sarnthein et al., 1982; Fig. 12-1). Estimates vary for the age of onset of the aeolian input in different latitudes, and include 4.6–4.3 Ma (Tiedemann, Sarnthein, and Stein, 1989), 3.8 Ma (Pokras and Ruddiman, 1989), 3.2 Ma (Sarnthein et al., 1982; Stabell, 1989), or 3.1 Ma (Stein et al., 1989); aeolian input intensified at about 2.4 Ma (e.g., Pokras and Ruddiman, 1989; Stabell, 1989; Stein et al., 1989), and again at approximately 0.8 Ma (Tiedemann, Sarnthein, and Stein, 1989). During the past ~ 0.8 Ma, some cores show a significant decrease in the aeolian input in interglacial and warmer intervals (Pokras and Ruddiman, 1989; Ruddiman and Janacek, 1989; Stein et al., 1989). In a core immediately upslope from the 4S slide, Thiede, Suess, and Muller (1982) found a significant increase in the aeolian(?) quartz accumulation rate during the (glacial) oxygen isotope stage 2 (~ 27 to ~ 13 ka), which brackets the age estimate for the main 4S Slide event. Additional terrigenous input included aeolian-sand turbidity currents generated from sand dunes that migrated to the exposed shelfbreak in Pleistocene and late Pliocene time (~ 2.6–1.9 Ma) (Sarnthein and Diester-Haass, 1977; Sarnthein, 1978; Faugeres et al., 1989).

Approximately concurrent with the increased aeolian input was an increase in coastal upwelling that began about 3.2–2.9 Ma (Diester-Haass and Chamley, 1982; Sarnthein et al., 1982; Tiedemann, Sarnthein, and Stein, 1989). The upwelling resulted in a significant increase in biogenic input, which is evidenced in the sediments by biogenic opal and organic matter (including organic carbon and biogenic gas). In the core upslope from the 4S Slide, the biogenic input also increased dramatically during the glacial oxygen-isotope stage 2 (Thiede, Suess, and Muller, 1982). Thus, although river input was minimal during glacial times, increased sediment input by other mechanisms may have promoted more frequent sediment slides.

Post-Wisconsinan slope failures indicate that additional trigger mechanisms for sediment slides should be considered. The increased biogenic silica concentration typical of sediments below the upwelling zone results in dramatically low shear strength values (Mienert and Schultheiss, 1989), perhaps because of underconsolidation caused by interlocking microfossils (e.g., Keller and Bennet, 1973; Mayer, 1982). Beneath the zone of upwelling, high biogenic gas contents also are observed in sediments deeper than about 25 m below the seafloor (Mienert and Schultheiss, 1989). Biogenic gas has been suggested to be a strong factor in destabilizing slopes both elsewhere (e.g., Carpenter, 1981; Prior and Coleman, 1982; see discussion in Embley and Jacobi, 1977a, 1986) and along the Northwest African margin (von Rad and Wissman, 1982). For example, if the gas hydrate (clathrate)/free gas boundary shifts upward because of a rising sea level, free gas would form in previously hydrated sediment; this free gas might cause overpressured sediment. Consistent with the biogenic gas hypothesis, is the high organic carbon content measured at ODP Site 657 in a buried Pleistocene sediment slide, which presently lies outside the area of upwelling and high organic carbon content (Stein et al., 1989). Although high organic carbon itself appears not to affect shear strength in these sediments (Mienert and Schultheiss, 1989), the high organic content may indicate high gas/clathrate content.

The combination of high biogenic silica and gas contents in the sediment below the zone of upwelling appears to make them particularly susceptible to slope failure compared to regions north and south of the upwelling zone. This susceptibility may allow relatively minor additional loadings, such as earthquakes or added sediment to trigger slope failure. Rare earthquakes occur along passive margins such as the event that triggered the Grand Banks Slide (e.g., Piper et al., 1985; see discussion in Embley and Jacobi, 1986). Some earthquakes along the Northwest African margin may be related to plate readjustment along fracture zones, or to volcanism. Volcanism and associated flows may themselves trigger local sediment slides, and extensive buried ash layers may provide slide planes. Possible evidence for a volcanic/earthquake trigger mechanism is found at the seamount located at lat. ~ 26°N, long. 20°20′W. There, slope failures occurred simultaneously on opposite sides of the seamount three different times (as evidenced by the same sequence of buried and surficial sediment slides; Fig. 12-8).

In deeper water, other factors may contribute to slope failures. At the foraminiferal lysocline (about

4,300 m on the Sierra Leone Rise), the rigidity of the sediment (as measured by *p*-wave velocities) decreases significantly because of a breakdown of the foraminifers (Mienert, Curry, and Sarnthein, 1988). von Rad and Wissman (1982) suggested that local undercutting by increased bottom current activity along the Northwest African margin also could lead to an oversteepened, unstable slope.

Sediment Slumps

Few major slump blocks are present at the surface along the Northwest African margin. However, large fields of standing waves (regular and irregular), and even some "migrating waves" probably originated as surface creep, as indicated by their morphology, orientation, and location (e.g., Jacobi and Hayes, 1982). In all three detailed maps [Figs. 12-3 (foldout), 12-4 (foldout), 12-5 (foldout)], the surface-creep fields not directly related to canyon/channel systems are restricted to the continental slope, as are most of the ductile and brittle slump faults. Ductile and brittle slump faults, and slump folds, also are observed commonly along channel/canyon margins on the continental rise [e.g., along the Agadir Canyon at ~ lat. 30°30′N–32°30′N, long. 10–13°W in the Madeira Island map; Fig. 12-3 (foldout)].

UNRESOLVED QUESTIONS

How Do Thermohaline Contour Currents Interact with Turbidity Currents?

On the continental rise in the Madeira Island map area [Fig. 12-3 (foldout)], the depth-range and width of sediment-wave provinces increase near TCPs (W37, W38 on Fig. 12-10). Jacobi, Rabinowitz, and Embley (1975) and Jacobi and Hayes (1984b) suggested that this increase might indicate interaction between THCCs and turbidity currents. Possible types of interactions include:

1. THCCs could redirect the flow of relatively dense turbidity currents into alignment with the THCC.
2. THCCs could redirect or entrain only the relatively dilute turbidity-current "clouds".
3. Material recently deposited by the turbidity current could be entrained by the THCC.

At present we have only indirect evidence to help discriminate among these possible interactions. Except for province SW8 in the Canary Islands map (Fig. 12-7), sediment-wave provinces along TCPs generally are more extensive on the down-THCC-flow side (e.g.,

W38A, Fig. 12-10). Because the turbidity currents have flowed westward, the asymmetry could be evidence of either a Coriolis effect, or any of alternatives 1–3. Province W38A is so narrow but extensive along the margin of the TCP, that its map pattern resembles that of a levee. However, the most distinctive province is W41. This field is not adjacent to any known major TCP, and so is probably not directly related to turbidity currents. It is, however, about 20 km north of the presumed intersection between the TCP from Agadir Canyon and the north-flowing AABW; this relationship suggests that the sediment waves were built here in part because turbidity currents delivered sufficient material for the THCC to sculpt and mold. The position of province W41 also is governed by flow instability caused by nearby basement highs and changes in slope orientation [Fig. 12-3 (foldout)].

That sediment-wave province W41 (Fig. 12-10) is not directly adjacent to the TCP, and that, in fact, there is no wide sediment-wave province adjacent to the canyon, suggests that the sediment-wave material was entrained by the THCC directly from turbidity currents. At least part of the coarser material apparently was deposited in the TCP (as evidenced by the echo type), so we suggest that the THCC may have entrained the dilute, relatively slower portions of turbidity currents as they intersected the THCCs. Generally, sediment-wave provinces do not return strong prolonged echoes indicative of sandy turbidites, but in some provinces adjacent to TCPs (e.g., at lat. 28°25′N, long. 18°30′W in the W37A province of the Canary Islands map area), we have observed such echoes, especially in troughs. This relationship suggests that the sandier turbidites can be diverted in some cases.

Is the Distribution of Erosional Furrows Consistent with Their Proposed Origin?

As discussed by Flood (1982), the maintenance of a furrow depends on the balance between fine sediment deposition during low-velocity flow and erosion during high-velocity flow. To construct a furrow, the sedimentation rate must be relatively low, and the velocity relatively high. These conditions are met in both localities where we mapped furrows. The sedimentation rate on the Madeira Rise has been low (1.5 cm/1000 yr) for the past 225,000 yr (Embley, Rabinowitz, and Jacobi, 1978). This low rate is expected, because there are no TCPs on the Madeira Rise, and only one major TCP is present east of the rise extending south from Madeira. For similar reasons, the sedimentation rate in the Saharan Seamount area is also low (Embley, Rabinowitz, and Jacobi, 1978). Additionally, both hyperbolae zones occur where high-speed bottom-water flow could be expected, and has been measured (in the

Saharan Seamounts, Lonsdale, 1978, 1982; and on the Madeira Rise, Saunders, 1987).

The necessity of relatively low sediment input for the development of furrows is demonstrated south of Madeira [Figs. 12-2, 12-3 (foldout)]. There, where the TCP intersects the approximate level of the AABW upper surface, a sediment-wave province has developed in the down-AABW-flow direction (W47 on Fig. 12-10). However, farther down-AABW-flow, in the same depth range as the sediment-wave province, a major furrow province (H20) is present. Apparently the furrows begin where the sedimentation rate has decreased following deposition in the upcurrent sediment-wave province. The speed of AABW probably also increases as flow is deflected around the rise.

Why Are Furrow-Generated Hyperbolae Not Observed Along Subbottom Reflectors?

Regular, overlapping hyperbolae are tangent only to the seafloor along the Northwest African margin, suggesting that furrows occur only at the seafloor surface (e.g., profile 2 in Fig. 4 of Embley, 1978). The lack of hyperbolae tangent to subbottom horizons does not entirely preclude the possibility of buried furrows, because buried furrows might not return hyperbolae. However, the fact that hyperbolae tangent to subbottom reflectors are observed elsewhere on echograms (see Chapter 8) suggests that burial processes generally do not eliminate the elements necessary for hyperbolae generation.

If furrows are present only at the seafloor along the Northwest African margin, then ancient rates of sediment input relative to the flow velocity must have been significantly higher than in the recent past, when the surficial furrows formed. Although this ancient, relatively high sedimentation rate may have characterized the Northwestern African margin as a whole (Koopman, 1980; Sarnthein et al., 1982), cores raised from the Madeira Rise (Embley, Rabinowitz, and Jacobi, 1978) do not support such a contention. According to dated horizons in these cores, the deposition rate on the Madeira Rise during the past 17,000 yr was higher (~ 2.3 cm/1000 yr) than at any other time in the past 225,000 yr. Between 17 and 73 ka, the sedimentation rate was half that of any other time interval recorded in the cores (0.71 cm/1000 yr). These core data are more consistent with the hypothesis that the furrows are actually relict features, and that they are merely maintained by present bottom-water flow. If this suggestion is correct, then the furrows formed during the last glacial interval; the increased speed of bottom water at that time apparently more than compensated for any increase in sedimentation rate associated with a lowered sea level. The main source for sediment on the Madeira Rise is Madeira, which has no shelf, so sedimentation rates probably did not in-

crease significantly because of an eustatic fall. The proposed time of furrow formation is consistent with the observation that several sediment slides, which have flowed over the furrows, appear to have no new surficial furrows [e.g., ~ SW of Madeira, Fig. 12-3 (foldout); Saharan Seamount area, Fig. 12-4 (foldout), including site L1 and to the south, Fig. 12-6].

The relict-furrow hypothesis is perplexing, however, because the combined tidal and AABW velocities in the Saharan Seamount region (Lonsdale, 1978, 1982) and in the Madeira Rise area (Saunders, 1987) are presently sufficient to construct or maintain furrows (Flood, 1982). Either the current-meter data are not representative, and the present-day velocities are not sufficient to erode furrows, or the debris-flow deposits covering the furrows are too young to have developed furrows. If these current-meter studies were representative, and if the past sediment input were either less than, or equal to, the present rate, then the surficial furrows should have older (buried) counterparts along subbottom reflectors. However, we observed none on the 3.5-kHz records. Our tentative conclusion is that the furrows are presently maintained (and perhaps in some areas, constructed) by accelerated tidal flow, but that in glacial times, accelerated THCCs constructed many of the furrow provinces.

IMPLICATIONS FOR STUDIES OF ANCIENT CONTINENTAL RISES

Sediment-slide deposition is not restricted to sites on the upper rise (at the base-of-slope) or to the middle fan [Figs. 12-2, 12-3 (foldout), 12-4 (foldout), 12-5 (foldout)]. Large sediment slides, such as those in the Canary Islands map area, continue well out onto the lower rise (lower fan), and, in fact, the buried slide deposits in the central western portion of the Canary Islands map nearly reach the Madeira Abyssal Plain (Fig. 12-2). Masson et al. (Chapter 13) present evidence that some slides actually reach the Madeira Abyssal Plain. In the western Atlantic some slides also impinge on the abyssal plain (Embley and Jacobi, 1977a, 1986; see also Chapters 8, 9, 10). It is clear that a single ancient slide complex cannot be used as a reliable indicator of base-of-slope paleodeposition.

Paleoslope determinations (and inferences regarding the orientation of the continental margin) are commonly based on analysis of sediment-slide folds and(or) on flow directions inferred from turbidites. On the Northwest African margin, the great number of seamounts, island group pedestals, and plateaus creates local slopes that can be directed up to 180° away from the regional slope. Because these local slopes can be quite extensive, downslope processes acting over broad areas may leave trails that indicate neither the regional slope nor the continentward direction.

The 4S Slide is an excellent example of a sediment slide with a downslope flow that paralleled the continental margin for 200 km. The slide near the head of the Agadir Canyon also flowed parallel to the continental margin for 100 km. Potentially more confusing to the future geologist are slides that actually have flowed toward the continent from seamounts and plateaus. Many sediment slides have flowed down the eastern flanks of the Saharan Seamounts toward the continent. Other examples include turbidity currents that flowed parallel to the continental margin down the same path as the 4S Slide, and turbidity currents that flowed 700 km north from the Cape Verde Islands parallel to the continental margin.

The Northwest African continental margin poses additional potential problems for the interpretation of ancient TCPs and continental rises. Although TCPs generally are oriented downslope, in some areas the whole pathway appears to trend obliquely to the bathymetric contours (e.g., lat. 24–25°N, long. 23–24°W). Possible explanations include:

1. The flow did not respond instantly to changes in slope direction (a momentum factor).
2. THCCs influenced the direction of the downslope flows.
3. Insufficient data exist to correctly establish the trend of the flow or the bathymetric contours.

If the deduced flow-trend is correct, then this TCP is oriented about 20–40° away from the downslope direction for a distance of > 150 km.

On the lower continental rise, basement ridges (abyssal hills or piercement structures) can form barriers to downslope turbidity currents. GLORIA sidescan data (e.g., Kidd, Simm, and Searle, 1985; Kidd, Hunter, and Simm, 1987) from west of the Canary Islands map show numerous examples of redirected flow; some such flows extend over distances of 60 km (see also Chapter 13).

The fact that channels meander across submarine fans obviously poses a serious problem for determining regional flow (and regional slopes) from limited paleoflow indicators in ancient channel deposits. However, flow indicators outside channels may provide a fairly reliable regional paleoflow direction. For example, the trends of sediment lineations and crestlines of turbidity-current–generated sediment waves along the TCPs in the Canary Islands map are fairly linear, and may be typical of middle/lower fans and lower continental rises.

SUMMARY

Complex interactions among near-bottom processes have formed the present continental rise off Northwest Africa. The three dominant processes are turbidity currents, sediment slides (debris flows) and thermohaline contour-following currents.

North of the Sierra Leone Rise, TCPs generally extend 700 km from the base of the continental slope onto the abyssal plains. In this same region, TCPs with minor levees and shallow channels (the lower fan) dominate the submarine fans; they generally comprise approximately 75% of the total length of the TCPs on the rise. The middle fan, consisting of levees with sediment waves and more pronounced channels, is located near the base-of-slope. We have recognized possible sand lobes on the lower fan 300–500 km downslope from the middle fan. South of the Sierra Leone Rise, the TCPs are dramatically shorter and the lower fan is roughly equal in length to the middle fan.

Sediment slides along the Northwest African margin represent an extremely important downslope transport process. The larger slides are more than 1,000 km long, and stretch from the continental (or insular) slope to the far reaches of the lower rise; some impinge upon the abyssal plains. The large slides clearly are not restricted to the base-of-slope or to any single part of the conventional submarine fan system. Although slides were more active during the last glacial interval, there is evidence that slide activity has continued into the recent past. Factors that may contribute to the high number of slope failures along the Northwest African margin include: (1) increased aeolian sediment input during glacial intervals; (2) increased upwelling that results in elevated biogenic silica and gas contents; and (3) rare earthquakes.

Contour-current activity along this margin remains enigmatic. There is evidence that AABW has swept north along the continental rise since the Miocene. This current has resulted in sediment-wave provinces, furrow provinces, and differential sedimentation around seamounts. However, a number of fundamental questions remain unresolved. Is AABW presently sufficiently strong to form furrows? Short-term current-meter measurements in the furrow provinces in the Saharan Seamount area show that flow speeds are sufficient to maintain furrows (at least locally), but the dominant flow component is tidal. The question remains, were the furrows originally constructed during the glacial intervals by accelerated AABW flow, or is the province being constructed at present by primarily tidal flows that accelerate around seamounts?

With respect to sediment waves, a current-meter study shows that slow AABW speeds measured in an area of migrating waves (Madeira Island map area) are sufficient to cause the degree of wave migration observed between the two uppermost acoustic reflectors, if we assume that Flood's (1988) lee-wave model is correct. In the Canary Islands map area, sediment waves generated by turbidity currents can be distinguished on the lower continental rise from those generated by THCCs. The turbidity-current waves have approximately downslope-oriented crestlines, whereas THCC waves have approximately alongslope (or oblique)–oriented crestlines. The parallel nature of TCP-generated sediment waves and associated sediment lineations suggests that on a large scale, turbidity currents do not "fan-out" on the lower fan (continental rise). The linearity of the crestlines and sediment lineations also is consistent with a flow-stripping model (e.g., Normark and Piper, 1985).

At any one time, different processes and interactions have dominated at different downslope and alongslope positions on the rise. Because the rise is building out away from the continent, the vertical succession of process-related features, observed at any location on the upper rise, should generally mimic the lateral succession of features from the abyssal plain toward the continent.

Remaining questions include:

1. What specifically is the interaction between distal turbidity currents and contour currents? Do the contour currents actually divert dilute turbidity currents into the path of the THCCs, or do the THCCs merely rework sediment already deposited by the downslope flows? Careful consideration of map patterns suggests that in some areas, the THCCs actually have diverted portions of the turbidity currents, especially those that entrain finer-grained material.

2. Why are furrow hyperbolae tangent only to the seafloor on the Northwestern African margin? If the absence of hyperbolae from subbottom reflectors truly indicates that furrows were not present at the time the deeper horizons were formed, then the balance between sediment input and flow velocity must have been recently tipped in favor of higher velocities for the first time in ~ 1 m.y.

ACKNOWLEDGMENTS

We thank E. Uchupi and C. W. Poag for their thoughtful and constructive reviews of the manuscript. Discussions with Roger Flood were always very useful. The late David Johnson drafted Figs. 12-3 (foldout) and 12-5 (foldout), and Ana Maria Alvarez drafted Fig. 12-4 (foldout). Juliet Malin provided excellent technical support and Lew Milholland provided superior photographic services. Early portions of this research were supported by NSF grant OCE 83-15430 and ONR contracts N00014-80-C-0098 and N00014-83-C-0132. This is Lamont-Doherty Geological Observatory Contribution No. 4888.

REFERENCES

Belderson, R. H., Kenyon, N. H., Stride, A. H., and Stubbs, A. R. 1972. *Sonographs of the Sea Floor*. New York: Elsevier.

Bouma, A. H., Normark, W. R., and Barnes, N. E. 1985. *Submarine Fans and Related Turbidite Systems*. New York: Springer-Verlag.

Bryan, G. M., and Markl, R. G. 1966. *Microtopography of the Blake–Bahama Region*. Technical Report 8, CU-8-66-NObsr 85077, Lamont-Doherty Geological Observatory, Columbia University, New York.

Carpenter, G. 1981. Coincident slump-clathrate complexes on the U.S. continental slope. *Geo-Marine Letters* 1:29–32.

Damuth, J. E. 1975. Echo character of the western equatorial Atlantic floor and its relationship to the dispersal and distribution of terrigenous sediments. *Marine Geology* 18:17–45.

Damuth, J. E. 1979. Migrating sediment waves created by turbidity currents in the northern South China Basin. *Geology* 7:520–523.

Damuth, J. E. 1980. Use of high-frequency (3.5–12 kHz) echograms in the study of near-bottom sedimentation processes in the deep-sea: A review. *Marine Geology* 38:51–75.

Damuth, J. E., Jacobi, R. D., and Hayes, D. E. 1983. Sedimentation processes in the Northwest Pacific Basin revealed by echo-character mapping studies. *Geological Society of America Bulletin* 94:381–395.

Damuth, J. E., Folld, R. E., Kowsmann, R. O., Belderson, R. H., and Gorini, M. A. 1988. Anatomy and growth pattern of Amazon deep-sea fan as revealed by long-range side-scan sonar (GLORIA) and high-resolution seismic studies. *American Association of Petroleum Geologists Bulletin* 72:885–991.

Damuth, J. E., Kolla, V., Flood, R. D., Kowsmann, R. O., Monteiro, M. C., Gorini, M. A., Palma, J., and Belderson, R. H. 1983. Distributary channel meandering and bifurcation patterns on the Amazon deep-sea fan as revealed by long-range side-scan sonar (GLORIA). *Geology* 11:94–99.

Deister-Haass, L., and Chamley, H. 1982. Oligocene and post-Oligocene history of sedimentation and climate off Northwest Africa (DSDP Site 369). *In* Geology of Northwest African Continental Margin, U. von Rad, K. Hinz, M. Sarnthein, and E. Seibold, Eds.: New York: Springer-Verlag, 529–544.

Eittreim, S., Biscaye, P. E., and Jacobs, S. S. 1983. Bottom-water observations in the Vema fracture zone. *Journal of Geophysical Research* 88:2609–2614.

Embley, R. W. 1975. Studies of deep-sea sedimentation processes using high frequency seismic data. Ph.D. Thesis, Columbia University, New York, unpublished.

Embley, R. W. 1976. New evidence for occurrence of debris flow deposits in the deep sea. *Geology* 4:371–374.

Embley, R. W. 1980. The role of mass transport in the distribution and character of deep-ocean sediments with special reference to the North Atlantic. *Marine Geology* 38:23–50.

Embley, R. W. 1982. Anatomy of some Atlantic margin sediment slides and some comments on ages and mechanisms. *In* Marine Slides and Other Mass Movements, S. Saxov and J. K. Nieuwenhuis, Eds.: New York: Plenum, 182–213.

Embley, R. W., and Jacobi, R. D. 1977a. Distribution and morphology of large submarine sediment slides and slumps on Atlantic continental margins. *Marine Geotechnology* 2:205–228.

Embley, R. W., and Jacobi, R. D. 1977b. Exotic middle Miocene sediment from Cape Verde Rise and its relation to piercement structures. *American Association of Petroleum Geologists Bulletin* 61:2004–2009.

Embley, R. W., and Jacobi, R. D. 1986. Mass wasting in the western North Atlantic. *In* The Geology of North America, Volume M, the Western North Atlantic Region, P. R. Vogt and B. E. Tucholke, Eds.: Boulder, CO: Geological Society of America, 479–490.

Embley, R. W., Rabinowitz, P. D., and Jacobi, R. D. 1978. Hyperbolic echo zone in the eastern Atlantic and the structure of the southern Madeira Rise. *Earth and Planetary Science Letters* 41:419–433.

Embley, R. W., Hoose, P. J., Lonsdale, P., Mayer, L., and Tucholke, B. E. 1980. Furrowed mud waves on the Bermuda Rise. *Geological Society of America Bulletin* 91:731–740.

Emery, K. O., Uchupi, E., Phillips, J., Bowin, C., and Mascle, J. 1975. Continental margin off Western Africa: Angola to Sierra Leone. *American Association of Petroleum Geologists Bulletin* 59:2209–2265.

Faugeres, J. C., Legigan, P., Maillet, N., Sarthein, M., and Stein, R. 1989. Characteristics and distribution of Neogene turbidites at Site 657 (Leg 108, Cap Blanc Rise, Northwest Africa): variations in turbidite source and continental climate. *In* Proceedings of the Ocean Drilling Program, Scientific Results, Volume 108, W. Ruddiman, M. Sarnthein, et al.: College Station, TX: Ocean Drilling Program, 329–348.

Flood, R. D. 1980. Deep-sea sedimentary morphology: modelling and interpretation of echo-sounding profiles. *Marine Geology* 38:77–92.

Flood, R. D. 1981. Distribution, morphology, and origin of sedimentary furrows in cohesive sediments, Southhampton Water. *Sedimentology* 28:511–529.

Flood, R. D. 1982. Observations, classification, and dynamics of furrows in cohesive sediments. *Bulletin Institute Geologique Bassin d'Aquitaine, Bordeaux* 31:141–149.

Flood, R. D. 1983. Classification of sedimentary furrows and a model for furrow intiation and evaluation. *Geological Society of America Bulletin* 94:630–639.

Flood, R. D. 1988. A lee wave model for deep-sea mudwave activity. *Deep-Sea Research* 35:973–983.

Flood, R. D., and Hollister, C. D. 1974. Current-controlled topography on the continental margin off the eastern United States. *In* The Geology of Continental Margins, C. A. Burk and C. L. Drake, Eds.: New York: Springer-Verlag, 197–206.

Flood, R. D., and Hollister, C. D. 1980. Submersible studies of deep sea furrows and transverse ripples in cohesive sediments. *Marine Geology* 36:M1–M9.

Flood, R. D., and Shor, A. N. 1988. Mud waves in the Argentine Basin and their relationship to regional bottom circulation patterns. *Deep-Sea Research* 35:943–971.

Flood, R. D., Hollister, C. D., and Lonsdale, P. 1979. Disruption of the Feni Sediment Drift by debris flows from Rockall Bank. *Marine Geology* 32:311–334.

Hayes, D. E., and Rabinowitz, P. D. 1975. Mesozoic magnetic lineations and the magnetic quiet zone off Northwest Africa. *Earth and Planetary Science Letters* 28:105–115.

Heezen, B. C., and Hollister, C. D. 1971. *The Face of the Deep*. New York: Oxford University Press.

Heezen, B. C., and Tharp, M. 1965. Tectonic fabric of the Atlantic and Indian oceans and continental drift. *Philosophical Transactions of the Royal Society* 258:90–106.

Heezen, B. C., Gerard, R. D., and Tharp, M. 1964. The Vema fracture zone in the equatorial Atlantic. *Journal of Geophysical Research* 69:733–739.

Heezen, B. C., Bunce, E. T., Hersey, J. B., and Tharp, M. 1964. Chain and Romanche fracture zones. *Deep-Sea Research* 11:11–33.

Hinz, K., Dostmann, H., and Fritsch, J. 1982. The continental margin of Morocco: seismic sequences, structural elements and geologic development. *In* Geology of Northwest African Continental Margin, U. von Rad, K. Hinz, M. Sarnthein, and E. Seibold, Eds.: New York: Springer-Verlag, 34–60.

Hobart, M. A., Bunce, E. T., and Sclater, J. G. 1975. Bottom water flow through the Kane Gap, Sierra Leone Rise, Atlantic Ocean. *Journal of Geophysical Research* 80:5083–5087.

Hollister, C. D., Flood, R. D., Johnson, D. A., Lonsdale, P. F., and Southard, J. B. 1974. Abyssal furrows and hyperbolic echo traces on the Bahama Outer Ridge. *Geology* 2:395–400.

Jacobi, R. D. 1976. Sediment slides on the northwestern continental margin of Africa. *Marine Geology* 22:157–173.

Jacobi, R. D. 1982. Microphysiography of the southeastern North Atlantic and its implications for the distribution of near-bottom processes and related sedimentary facies. *Bulletin Institute Geologique Bassin d'Aquitaine, Bordeaux* 31:31–46.

Jacobi, R. D., and Hayes, D. E. 1982. Bathymetry, microphysiography and reflectivity characteristics of the West African Margin between Sierra Leone and Mauritania. *In* Geology of Northwest African Continental Margin, U. von Rad, K. Hinz, M. Sarnthein, and E. Seibold, Eds.: New York: Springer-Verlag, 182–210.

Jacobi, R. D., and Hayes, D. E. 1984a. Echo character, microphysiography and geologic hazards. *In* Northwest African Continental Margin and Adjacent Ocean Floor off Morocco, Ocean Margin Drilling Program, Regional Atlas Series, D. E. Hayes, P. D. Rabinowitz, and K. Hinz, Eds.: Woods Hole, MA: Marine Science International, Atlas 12:14.

Jacobi, R. D., and Hayes, D. E. 1984b. Echo character and geologic hazards of the Morocco continental margin hazards. *In* Northwest African Continental Margin and Adjacent Ocean Floor off Morocco, Ocean Margin Drilling Program, Regional Atlas Series, D. E. Hayes, P. D. Rabinowitz, and K. Hinz, Eds.: Woods Hole, MA: Marine Science International, Atlas 12:14a.

Jacobi, R. D., Rabinowitz, P. D., and Embley, R. W. 1975. Sediment waves on the Moroccan Continental Rise. *Marine Geology* 19:61–67.

Jansen, E., Befring, S., Bugge, T., Eidvin, T., Holtedahl, H., and Sejrup, H. P. 1987. Large submarine slides on the Norwegian continental margin: sediments, transport and timing. *Marine Geology* 78:77–107.

Johnson, D. A. 1982. Abyssal tectonics: interactive dynamics of the deep ocean circulation. *Palaeogeography, Palaeoclimatology, Palaeoecology* 38:93–128.

Keller, G. H., and Bennett, R. H. 1973. Sediment mass physical properties—Panama Basin and northeastern equatorial Pacific. *In* T. H. van Andel, G. R. Health, et al., Initial Reports of the Deep Sea Drilling Project, Volume 16: Washington, DC: United States Government Printing Office, 499–512.

Kidd, R. B., Simm, R. W., and Searle, R. C. 1985. Sonar acoustic facies and sediment distribution on an area of the deep ocean floor. *Marine and Petroleum Geology* 2:210–221.

Kidd, R. B., Hunter, P. M., and Simm, R. W. 1987. Turbidity-current and debris-flow pathways to the Cape Verde Basin: status of long-range side scan sonar (GLORIA) surveys. *In* Geology and Geochemistry of Abyssal Plains, Geological Society Special Publication 31, P. P. E. Weaver and J. Thomson, Eds.: Boston, MA: Blackwell Scientific Publications, 33–48.

Koopman, B. 1980. Quantitative determination of silt sized biogenic silica in Atlantic deep-sea sediments. International Association of Sedimentologists, Abstracts 1st European Regional Meeting. Bochum: (International Association of Sedimentologists) 30–33.

Lancelot, Y., and Embley, R. W. 1977. Piercement structures in deep oceans. *American Association of Petroleum Geologists Bulletin* 61:1991–2000.

Lancelot, Y., Seibold, E., Dean, W. E., Jansa, L. F., Eremeev, V., Gardner, J., Cepek, P., Krasheninnikov, V., Pflaumann, U., Johnson, D., Rankin, J. G., and Trabant, P. 1977. Initial Reports of the Deep Sea Drilling Project, Volume 41: Washington, DC: U.S. Government Printing Office.

Lonsdale, P. 1978, Bedforms and benthic boundary layer in the North Atlantic: A cruise report of INDOMED Leg 11. *Scripps Institute of Oceanography, SIO Reference* 78–30, 15.

Lonsdale, P. 1982. Sediment drifts of the Northeast Atlantic and their relationship to the observed abyssal currents. *Bulletin Institute Geologique Bassin d'Aquitaine, Bordeaux* 31:141–149.

Lowrie, A., Jahn, W., and Egloff, J. 1970. Bottom current activity in the Cape Verde and Canaries Basin. *Eastern Atlantic Transamerican Geophysical Union* 51:336.

Lutze, G. F. 1980. Depth distribution of benthonic foraminifera on the continental margin off northwest Africa. *Meteor Forschungs. Ergebnisse, Reihe C* 33:31–80.

Mantyla, A. W., and Reid, J. L. 1983. Abyssal characteristics of the World Ocean waters. *Deep-Sea Research* 30:805–833.

Mayer, L. A. 1982. Physical properties of sediment recovered on Deep Sea Drilling Project Leg 68 with the hydraulic piston corer. *In* Initial Reports of the Deep Sea Drilling Project, Volume 68, W. L. Prell, J. V. Gardner, et al.: Washington, DC: United States Government Printing Office, 365–382.

Mienert, J. 1985. Akustratigraphie im aquatorialen Ostatlantik: Zur Entwicklung der Tiefenwasserzirkulation der letzten 3,5 Millionen Jahre. Ph.D. thesis, Christian-Albrechts-Universitat, Keil, unpublished.

Mienert, J. 1986. Akustratigraphie im aquatorialen Ostatlantik: Zur Entwicklung der Tiefenwasserzirkulation der letzten 3,5 Millionen Jahre. *Geologisch-Palaontologisches Institut der Universitat Kiel* 40, 19–86.

Mienert, J., and Schultheiss, P. 1989. Physical properties of sedimentary environments in oceanic high (Site 658) and low (Site 659) productivity zones. *In* Proceedings of the Ocean Drilling Program, Scientific Results, Volume 108, W. Ruddiman, M. Sarnthein, et al.: College Station, TX: Ocean Drilling Program, 397–406.

Mienert, J., Curry, W. B., and Sarnthein, M. 1988. Sono-stratigraphic records from equatorial Atlantic deep-sea carbonates: paleoceanographic and climatic relationships. *Marine Geology* 83:9–20.

Murray, J. W., Weston, J. F., Haddon, C. A., and Powell, A. D. J. 1986. Miocene to Recent bottom water masses of the north-east Atlantic: an analysis of benthic foraminifera. *In* North Atlantic Palaeoceanography, Geological Society Special Publication 21, C. P. Summerhayes and N. J. Shackleton, Eds.: Boston, MA: Blackwell Scientific Publications, 219–230

Mutti, E., and Normark, W. R. 1987. Comparing examples of modern and ancient turbidite systems: problems and concepts. *In* Marine Clastic Sedimentology, J. K. Leggett and G. G. Zuffa, Eds.: London: Graham and Trotman, 1–38.

Normark, W. R., and Piper, D. J. W. 1985. Navy Fan, Pacific Ocean. *In* Submarine Fans and Related Turbidite Systems, A. H. Bouma, W. R. Normark, and N. E. Barnes, Eds.: New York: Springer-Verlag, 87–97.

Normark, W. R., and Gutmacher, C. E. 1985. Delgada Fan, Pacific Ocean. *In* Submarine Fans and Related Turbidite Systems, A. H. Bouma, W. R. Normark, and N. E. Barnes, Eds.: New York: Springer-Verlag, 59–64.

Normark, W. R., Gutmacher, C. E., Chase, T. E., and Wilde, P. 1985. Monterey Fan, Pacific Ocean. *In* Submarine Fans and Related Turbidite Systems, A. H. Bouma, W. R. Normark, and N. E. Barnes, Eds.: New York: Springer-Verlag, 79–87.

O'Connell, S., Sterling, C. E., Bouma, A. H., Coleman, J. M., Cremer, M., Droz, L., Meyer-Wright, A. A., Normark, W. R., Pickering, K. T., Stow, D. A. V., and DSDP Leg Shipboard Scientists. 1985. Drilling results on the Lower Mississippi Fan. *In* Submarine Fans and Related Turbidite Systems, A. H. Bouma, W. R. Normark, and N. E. Barnes, Eds.: New York: Springer-Verlag, 291–298.

Pastouret, L., Chamley, H., Delibrias, G., Duplessy, J. C., and Thiede, J. 1978. Late Quaternary climate changes in western tropical Africa deduced from deep-sea sedimentation off Niger Delta. *Oceanologica A* 1/2:217–232.

Piper, D. J., Stow, D. A. V., and Normark, W. R. 1985. Laurentian Fan. Atlantic Ocean. *In* Submarine Fans and Related Turbidite Systems, A. H. Bouma, W. R. Normark, and N. E. Barnes, Eds.: New York: Springer-Verlag, 137–143.

Piper, D. J., Shor, A. N., Farre, J. A., O'Connell, S., and Jacobi, R. D. 1985. Sediment slides and turbidity currents on the Laurentian Fan: Sidescan sonar investigations near the epicenter of the 1929 Grand Banks earthquake. *Geology* 13:538–541.

Pirmez, C., Breen, N. A., Flood, R., Jacobi, R., Ladd, J. W., O'Connell, S., and Westbrook, G. 1989. Interaction of plate convergence and deep sea fan sedimentation: preliminary interpretations of GLORIA mosaic of the northern Colombian continental margin. *Transactions of the*

American Geophysical Union 70:1345.

Pirmez, C., Breen, N. A., Flood, R. D., O'Connell, S., Jacobi, R., Ladd, J. W., Westbrook, G., Franco, J. V., Garzon, M., and Arias-Isaza, F. A., 1990. GLORIA mosaic of the Magdelena deep-sea fan, northern Colombian convergent margin. American Association of Petroleum Geologists 1990 Convention, Official Program, 153.

Pokras, E. M., and Ruddiman, W. F. 1989. Evolution of south Sahara/Sahelian aridity based on freshwater diatoms (genus *Melosira*) and opal phytoliths: Sites 662 and 664. *In* Proceedings of the Ocean Drilling Program, Scientific Results, Volume 108, W. Ruddiman, M. Sarnthein, et al.: College Station, TX: Ocean Drilling Program, 143–148.

Prior, D. B., and Bornhold, B. D. 1989. Submarine sedimentation on a developing Holocene fan delta. *Sedimentology* 36:1053–1076.

Prior, D. B., and Coleman, J. M. 1982. Active slides and flows in unconsolidated marine sediments on slopes of the Mississippi Delta. *In* Marine Slides and Other Mass Movements, S. Saxov and J. K. Nieuwenhuis, Eds.: New York: Plenum Press, 21–50.

Rona, P. A. 1969. Possible salt domes in the deep Atlantic off North-West Africa. *Nature* 224:141–143.

Rona, P. A. 1970. Comparison of continental margins of eastern North America at Cape Hatteras and Northwestern Africa at Cap Blanc. *America Association of Petroleum Geologists Bulletin* 54:129–157.

Rona, P. A., and Fleming, H. S. 1973. Mesozoic plate motions in the eastern central North Atlantic. *Marine Geology* 14:239–252.

Ruddiman, W. F., and Janecek, T. R. 1989. Pliocene–Pliestocene biogenic and terrigenous fluxes at equatorial Atlantic Sites 662, 663 and 664. *In* Proceedings of the Ocean Drilling Program, Scientific Results, Volume 108, W. Ruddiman, M. Sarnthein, et al.: College Station, TX: Ocean Drilling Program, 211–240.

Ruddiman, W. F., Sarnthein, M., Baldauf, J., Backman, J., Bloemondal, J., Curry, W., Farrimond, P., Faugeres, J. C., Janacek, T., Katsura, Y., Manivit, H., Mazzullo, J., Mienert, J., Pokras, E., Raymo, M., Schultheiss, P., Stein, R., Tauxe, L., Valet, J.-P., Weaver, P. P. E., and Yasuda, H. 1989. Proceedings of the Ocean Drilling Program, Scientific Results, Volume 108: College Station, TX: Ocean Drilling Program.

Sarnthein, M. 1978. Neogene sand layers off northwest Africa: composition and source environment. *In* Initial reports of the Deep Sea Drilling Project, Volume 41, J. Gardner, H. James, et al.: Washington, DC: United States Government Printing Office, 939–959.

Sarnthein, M., and Diester-Haass, L. 1977. Eolian-sand turbidites. *Journal of Sedimentary Petrology* 47:868–890.

Sarnthein, M., and Mienert, J. 1986. Sediment waves in the eastern equatorial Atlantic: sediment record during late glacial and interglacial times. *In* North Atlantic Palaeoceanography, Geological Society Special Publication 21, C. P. Summerhayes and N. J. Schakleton, Eds.: Boston, MA: Blackwell Scientific Publications, 119–130.

Sarnthein, M., Theide, J., Pflaumann, U., Erlenkeuser, Futterer, D., Koopmann, B., Lange, H., and Seibold, E. 1982. Atmospheric and oceanic circulation patterns off Northwest Africa during the past 25 million years. *In* Geology of Northwest African Continental Margin, U. von Rad, K. Hinz, M. Arnthein, and E. Seibold, Eds: New York: Springer-Verlag, 545–604.

Saunders, P. M. 1987. Currents, dispersion and light transmittance measurements on the Madeira Abyssal Plain. *Institute for Oceanographic Sciences Report* 241:1–55.

Searle, R. C. 1987. Regional setting and geophysical characterization of the Great Meteor East area in the Madeira Abyssal Plain. *In* Geology and Geochemistry of Abyssal Plains, Geological Society Special Publication 31, P. P. E. Weaver and J. Thomson, Eds.: Boston, MA: Blackwell Scientific Publications, 49–70.

Searle, R. C., Monahan, D., and Johnson, G. L. 1982, General Bathymetric Chart of the Oceans (GEBCO). Ottawa: Canadian Hydrographic Service, 5.08.

Seibold, E., and Hinz, K. 1974. Continental slope construction and destruction, West Africa. *In* The Geology of Continental Margins, C. A. Burk and C. L. Drake, Eds.: New York: Springer-Verlag, 179–196.

Shanmugam, G., and Moiola, R. J. 1985. Submarine fan models: problems and solutions. *In* Submarine Fans and Related Turbidite Systems, A. H. Bouma, W. R. Normark, and N. E. Barnes, Eds.: New York: Springer-Verlag, 29–34.

Shanmugam, G., Damuth, J. E., and Moiola, R. J. 1985. Is the turbidite facies association scheme valid for interpreting ancient submarine fan environments. *Geology* 13:234–237.

Stabell, B. 1989. Initial diatom record of Sites 657 and 658: on the history of upwelling and continental aridity. *In* Proceedings of the Ocean Drilling Program, Scientific Results, Volume 108, W. Ruddiman, M. Sarnthein, et al.: College Station, TX: Ocean Drilling Program, 149–156.

Stein, R., ten Haven, L., Littke, R., Rullkotter, J., and Welte, D. H. 1989. Accumulation of marine and terrigenous organic carbon at upwelling Site 658 and nonupwelling Sites 657 and 659: implications for the reconstruction of paleoenvironments in the eastern subtropical Atlantic through late Cenozoic times. *In* Proceedings of the Ocean Drilling Program, Scientific Results, Volume 108, W. Ruddiman, M. Arnthein, et al.: College Station, TX: Ocean Drilling Program, 361–385.

Thiede, J., Suess, E., and Muller, P. J. 1982. Late Quaternary fluxes of major sediment components to the sea floor at the northwest African continental slope. *In* Geology of Northwest African Continental Margin, U. von Rad, K. Hinz, M. S. Sarnthein, and E. Seibold, Eds.: New York: Springer-Verlag, 605–631.

Tiedemann, R., Sarnthein, M., and Stein, R. 1989. Climate changes in the western Sahara: aeolo-marine sediment record of the last 8 million years (Sites 657–661). *In* Proceedings of the Ocean Drilling Program, Scientific Results, Volume 108, W. Ruddiman, M. Sarnthein, et al.: College Station, TX: Ocean Drilling Program, 241–272.

Tucholke, B. E. 1979. Furrows and focused echoes on the Blake Outer Ridge. *Marine Geology* 31:M13–M20.

Uchupi, E., Emery, K. O., Bowin, C., and Phillips, J. 1976. Continental margin off Western Africa: Senegal to Portugal. *American Association of Petroleum Geologists Bulletin* 60:809–878.

von Rad, U., and Wissman, G. 1982. Cretaceous–Cenozoic history of the West Saharan continental margin (NW Africa): Development, destruction and gravitational sedimentation. *In* Geology of Northwest African Continental Margin, U. von Rad, K. Hinz, M. Sarnthein, and E. Seibold, Eds.: New York: Springer-Verlag, 106–131.

Weaver, P. P. E., Searle, R. C., and Kuijpers, A. 1986. Turbidite deposition and the origin of the Madeira Abyssal Plain. *In* North Atlantic Palaeoceanography, Geological Society Special Publication 21, C. P. Summerhayes and N. J. Shackleton, Eds.: Boston, MA: Blackwell Scientific Publications, 131–144.

Weaver, P. P. E., and Rothwell, R. G. 1987. Sedimentation on the Madeira Abyssal Plain over the last 300,000 years. *In* Geology and Geochemistry of Abyssal Plains, Geological Society Special Publication 31, P. P. E. Weaver and J. Thomson, Eds.: Boston, MA: Blackwell Scientific Publications, 71–86.

Worthington, L., and Wright, W. 1970. *North Atlantic Ocean Atlas*. Woods Hole, MA: Woods Hole Oceanographic Institution.

Wüst, G. 1955. Stromgeschwindigheiten in Teifen und Bodenwasser des Atlantischen Ozenas auf Grung dynamischer Berechung des deutschen atlantischen Expedition 1925–1927. *Deep-Sea Research Supplement* 3:373–397.

Young, R. A., and Hollister, C. D. 1975. Quaternary sedimentation on the northwest African continental rise. *Journal of Geology* 82:675–680.

13
Saharan Continental Rise: Facies Distribution and Sediment Slides

Douglas G. Masson, Robert B. Kidd, James V. Gardner, Quintin J. Huggett, and Philip P. E. Weaver

Northwest Africa has been a key area in the development of current understanding of continental margin sedimentation. Sediment facies maps, based on 3.5-kHz profiling, were produced for large areas of this margin in the late 1970s. These maps demonstrated a complex interplay between downslope and alongslope sedimentary processes (Embley, 1976, 1980, 1982; Jacobi, 1976; Embley and Jacobi, 1977; Jacobi and Hayes, 1982). Large-scale slope failure, generating both turbidity currents and debris flows, was recognized as an important sedimentary process. This was superimposed on a background of hemipelagic and contour-current sedimentation.

This chapter summarizes our recent work in an area west and northwest of the Canary Islands (Fig. 13-1), based on 6.5-kHz long-range (GLORIA) sidescan sonar data, 3.5-kHz profiling, and sediment coring. The sediment core evidence is presented only in preliminary form, as the data have not yet been studied in detail. We discuss the criteria for the recognition of sediment facies in the area, and how we might use the resulting facies map as an aid in understanding sedimentary processes. In particular, we examine the importance of debris-flow and turbidity-current deposits and their relationship to regional sedimentation patterns. We also redefine the Saharan Slide and describe a new slide, the Canary Slide, which originated on the western slopes of the Canary Islands.

STUDY AREA

The area of the Northwest African continental margin covered by our study extends from lat. 27 to 32°N and from long. 18 to 26°W (Fig. 13-1). East of our study area, the continental slope is relatively smooth, having only a few large canyons, but it is interrupted by a number of seamounts and volcanic islands, most of which lie landward of the 4,000-m contour. The dominant physiographic feature is the Canary Islands chain, a rugged series of active volcanic islands reaching 3,000 m above sea level. The steep slopes of the western islands, Hierro and La Palma, give way to a more gently sloping lower continental slope below 4,000 m water depth and to a smooth, exceptionally broad, gently sloping, continental rise below 4,500 m. Gradients range from about 1° between 3,500 and 4,000 m on the lower slope, to 0.1° below 5,000 m, some 600 km farther west on the lower rise. The Madeira Abyssal Plain, below 5,400 m, marks the western extremity of the study area.

PREVIOUS WORK

Sediment-Facies Mapping

A sediment-facies map based on 3.5-kHz profiles, and covering the area from lat. 23 to 35°N on the Northwest African continental margin, was presented by Embley, Rabinowitz, and Jacobi (1978, their Fig. 1) and Embley (1982, his Fig. 1; see also Chapter 12). The main facies recognized are summarized in Table 13-1. These authors demonstrated the feasibility of mapping facies provinces using acoustic techniques, and were able to describe major geological features, such as the Saharan Slide, in unprecedented detail.

Saharan Slide

Embley (1976) first recognized this huge sediment slide, which originated on the Northwest African margin south of the Canary Islands, in water depths of about 2,000 m. He described debris-flows, which had flowed

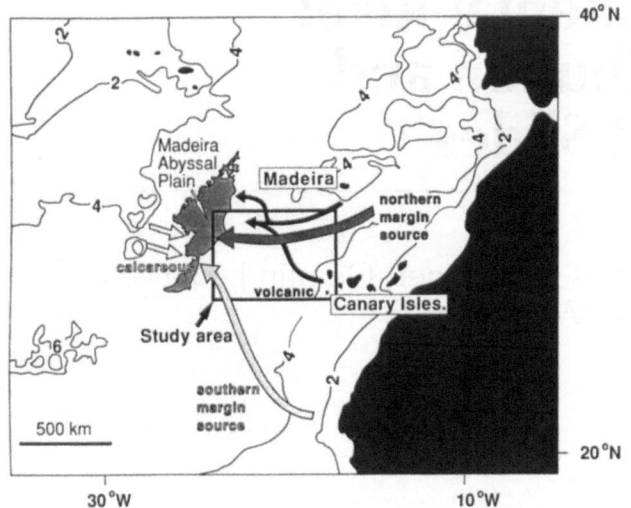

FIGURE 13-1. Location map showing sediment-transport paths to the Madeira Abyssal Plain (from Weaver et al., in press). Abyssal plain defined by the 5,400-m bathymetric contour (stippled area). Other bathymetry in kilometers. Box labelled "study area" locates Figures 13-3 and 13-6.

were postulated as part of the slide complex on the upper continental slope.

Simm and Kidd (1983) and Kidd, Simm, and Searle (1985) examined the area west of the Canaries and mapped debris-flow deposits on the lower continental rise. These they believed were downslope continuations of the flow deposits originally mapped by Embley. They were able to show that debris flows had reached to the very edge of the Madeira Abyssal Plain at long. 24°W, more than 1,000 km from the source of the Saharan Slide. Using a single GLORIA traverse through the debris-flow area, they were also able to demonstrate the potential of this 6.5-kHz sidescan sonar as a tool for mapping the extent of the debris-flow deposits.

Madeira Abyssal Plain

The Madeira Abyssal Plain, lying just to the west of the Saharan Rise, is one of the most intensely studied areas of the world's ocean floor. The evolution of the plain, which is known in detail for the last 730,000 years (Fig. 13-2), has a considerable bearing on our study of the Saharan Rise, because most of the turbidites within the plain must have traversed the Saharan Rise in order to get there. A review of the geology and geochemistry of the Madeira Abyssal Plain was presented by Weaver, Thomson, and Jarvis (1989).

The late Quaternary sedimentary succession of the Madeira Abyssal Plain consists of thick turbidites separated by thin pelagic units, which allow dating of each turbidite. Some of the turbidites are large, reaching 5 m in thickness and 120 km³ in volume. Most are fine grained, although some have basal sands > 1 m thick near the eastern margin of the plain. Turbidites can be classed, according to their composition and source area, as organic-rich, volcanic, or calcareous. These are derived from the West African Slope, the Canary and Madeira Islands, and seamounts to the west of the plain, respectively. Turbidites from the West African Slope can have source areas either north or south of the Canaries; these source areas are differentiated on

north and then northwest, constrained by the topography, before flowing out across the lower slope and upper rise west of the island of Hierro, in total a distance of 700 km. Embley (1976, 1980, 1982) recognized two distinct provinces within the slide. The slide scar, which lies to the southeast of our study area, has a characteristic irregular seabed relief, with small scarps, hyperbolic reflectors, and no persistent subbottom reflections on 3.5-kHz profiles. The slide deposits, which Embley mapped west of the Canaries (extending to 4,800 m), are typically seen on 3.5-kHz profiles as acoustically transparent lenses that onlap the more stratified continental-rise sediments. The thickness of these lenses of slide material exceeds 20 m in places, although nowhere was it believed to exceed 50 m. The total volume of displaced material was estimated at 1,100 km³. The age of the main part of the slide, though poorly constrained, was estimated as 16–17 ka, although smaller and significantly younger slide events

TABLE 13-1   Major Acoustic Facies and their Geological Interpretation as Proposed for the Northwest African Margin by Embley, Rabinowitz, and Jacobi, 1978, and Embley, 1982.

Acoustic Facies	Geological Interpretation
Acoustically transparent lenses	Debris-flow deposits
Waves	Sediment waves
Continuous flat subbottom reflectors, > 50-m penetration	Ponded turbidites
Continuous flat subbottom reflectors, < 20-m penetration	Turbidite pathways on slope and rise
Continuous subbottom reflectors, > 30-m penetration	Hemipelagic continental slope and rise sediments

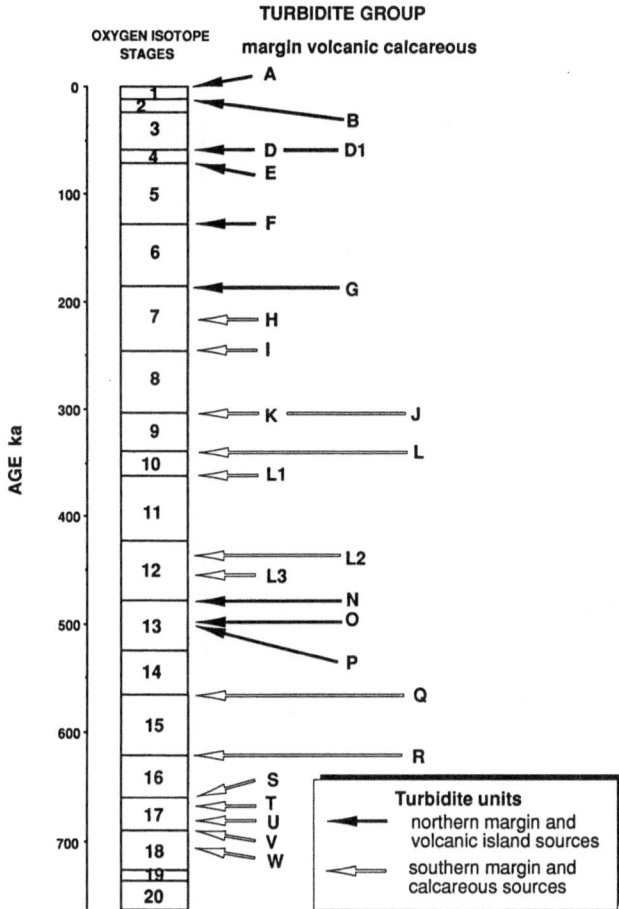

TURBIDITE GROUP

FIGURE 13-2. Summary of turbidite stratigraphy of Madeira Abyssal Plain (from Weaver et al., in press). Turbidites (lettered arrows) are classified in terms of their origins: West African continental margin, volcanic islands of Canaries and Madeira, or seamounts west of abyssal plain. African margin turbidites can be subdivided into those from north of Canaries and those from south.

the basis of their entry points into the plain, which can be either via its northern or southern subbasins (Weaver et al., in press). Turbidite emplacement is infrequent, and appears to have been related to climate and (or) sea-level change; typically, one turbidite is emplaced at each glacial–interglacial and interglacial–glacial transition (Weaver and Kuijpers, 1983).

Simm et al. (1991) attempted to correlate the well-known turbidite sequence of the Madeira Abyssal Plain with that seen in five short gravity cores taken on the adjacent lower Saharan Rise. In general, a first order correlation was possible, although the sequence is much thinner on the rise than on the abyssal plain, and not all the turbidites were seen in both environments. Some difficulty was found in establishing a satisfactory correlation between the sand units in the channel and interchannel areas. Also, the completeness of the rise sections could not be established due to bioturbation and to core disturbance in the short gravity cores. Simm et al. concluded, however, that in principle, with

better quality cores (such as those collected during our March, 1991, cruise), a satisfactory correlation between rise and plain should be attainable. It should be possible to use the detailed knowledge of the evolution of the abyssal plain to establish a framework for the evolution of the rise.

Bottom Currents on the Slope and Rise

Analysis of modern deep-water geostrophic circulation in the Canary Basin indicates that weak bottom currents predominate (Embley, Rabinowitz, and Jacobi, 1978; Saunders, 1987; Dickson, 1990). However, Saunders (1987, p. 26) noted that rare records of near-bottom current velocities of up to 30 cm/sec have also been obtained, and that these "probably exceed the local threshold for sediment movement." Two of the three highest recorded values were over the western flank of the Madeira Rise, in an area where Embley (1976) mapped a zone of possible abyssal sediment furrows (see also Chapter 12). Thus, although photographs obtained by Embley showed a tranquil seabed, it remains possible that transport of sediment by bottom currents occurs episodically. Lonsdale (1982) made current measurements between 3,500 and 4,000 m on the continental slope at lat. 26°N, just south of the Canaries. He measured alongslope currents of up to 18 cm/sec at a depth of 3,950 m, in an area where deep-towed sidescan sonar records showed erosional furrows trending along slope, parallel to the current direction. Farther up slope, however, both bedforms and alongslope currents were absent. Lonsdale (1982, p. 148) concluded that contour currents "have had a significant geologic role" in the development of the Canary Rise area.

DATA COLLECTION AND INTERPRETATION TECHNIQUES

Data Collection

The data on which this study is based were collected on a number of cruises between 1981 and 1991 (Fig. 13-3). The mosaic of GLORIA long-range side-scan sonar data was built up over three cruises in 1981, 1985, and 1990. Sediment cores were obtained on two cruises in 1982 and 1991. All of the above cruises, except the earliest GLORIA cruise, collected 3.5-kHz data.

Echo-Character Mapping Using 3.5-kHz Profiles

The mapping of seabed acoustic facies using 3.5-kHz profiling is a well established technique [see Damuth (1980) for a review of early studies]. In the present study, we established seven echo types, whose charac-

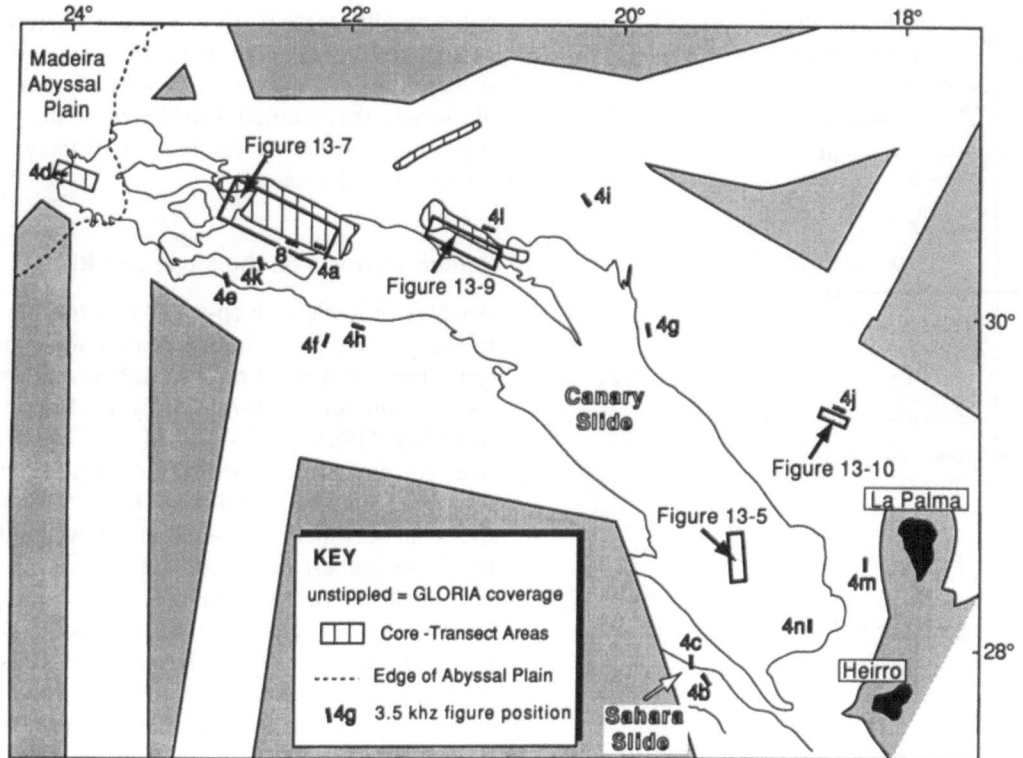

FIGURE 13-3.   Summary of sidescan sonar data in study area. Outlines of Saharan and Canary slides, taken from Figure 13-6, locate context of illustrated sonographs and profiles. Individual core locations cannot be shown at this scale, but are mostly within areas marked by hachuring. Most of area is covered densely by 3.5-kHz profiles. Unfilled boxes locate GLORIA sonographs shown in Figures 13-5, 13-7, 13-9, and 13-10. Short stippled lines locate 3.5-kHz profiles shown in Figures 13-4 and 13-8.

teristics are summarized in Table 13-2 and Figure 13-4 (see Kidd, Simm, and Searle, 1985; see also Chapters 8, 12). Most of the echo types are easy to distinguish, although some grade into others rather than having sharp boundaries. One area of difficulty is the separation of basement outcrop and slide source-area echo types on the island slopes west of the Canaries (Fig. 13-4m, n). In general, basement outcrop gives a sharper echo than slide source areas, is rougher (with larger hyperbolic echos) and, sometimes, has a thin sediment cover. However, it is clearly possible that slides have cut down to basement on parts of the upper island slope, and in this situation, slide source-area echo could pass gradually upslope into basement echo.

A second potential area of confusion is associated with the definition of the irregular and regular penetration echo types (Fig. 13-4f–i). This is clearly a subjective definition, because the two types are part of a continuous series, which extends from virtually no penetration to > 50-m penetration. In practice, there is rarely a sharp boundary between the regular and irregular echo types.

GLORIA Interpretation

GLORIA is a long-range sidescan sonar operating at a frequency of 6.5 kHz (Somers et al., 1978; Somers and

Searle, 1984). The GLORIA swath width of up to 45 km and a tow speed of 8–10 kts allow large areas of seafloor to be mapped relatively quickly. Most of the area discussed in this chapter has been surveyed using GLORIA.

Acoustic facies based on GLORIA data are defined using a combination of backscatter intensity and pattern. It has long been recognized that in sediment-covered areas of low topographic relief, variations in GLORIA acoustic facies might be interpreted in terms of sediment facies (e.g., Kenyon, Belderson, and Stride, 1975; Kidd, 1982; Kidd and Roberts, 1982; Simm and Kidd, 1983; Gardner and Kidd, 1983; Roberts, Kidd, and Masson, 1984). High-resolution 3.5-kHz profiling has been used in many areas covered by GLORIA to support this basic principle, but attempts to calibrate GLORIA images directly using sediment cores have been relatively few (Simm and Kidd, 1983; Kidd, Simm, and Searle, 1985; Gardner et al., 1991; Kenyon, in press). One problem in correlating GLORIA data with cores is the difficulty in making the jump in scale between a core of (at most) a few tens of centimeters diameter and the GLORIA footprint, which is the size of a football field. A second difficulty is our lack of understanding of the degree of penetration into seabed sediments of low grazing-angle energy at the GLORIA wavelength. This problem has been addressed by

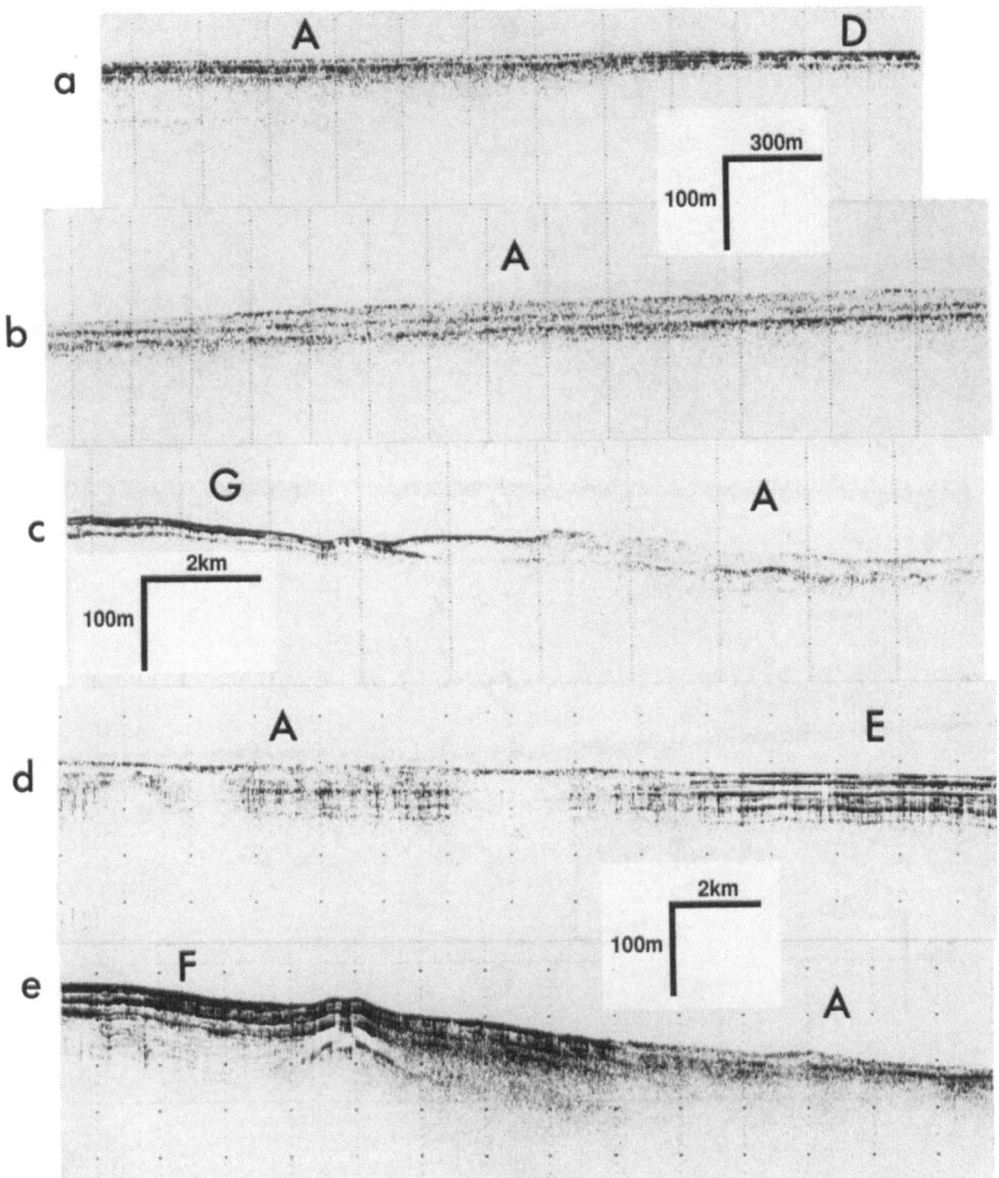

FIGURE 13-4.   3.5-kHz profiles showing examples of major echo types recognized on Saharan rise and slope (letters indicate echo type from Table 13-2). Profiles located on Figure 13-3. (a) Thin debris-flow deposit (echo type A) on Canary Slide onlapping turbidite/hemipelagic sediments (D). (b) Thick, possibly stacked, debris-flow deposits (A) on Saharan Slide. (c) Thick Saharan Slide debris-flow deposit (A) onlapping sediment waves (G). (d) Distal Canary Slide debris-flow deposit (A) onlapping fine-grained turbidites (E) on Madeira Abyssal Plain. Note prolonged echo and masking of underlying layered sequence generated by debris-flow deposit even though the deposit itself is too thin to resolve. (e) Thin Canary Slide debris-flow deposit (A) onlapping pelagic sediments (F). (f–g) Pelagic/hemipelagic sediments (E) on continental rise. (h–i) Turbidite/hemipelagic sediments (D) on continental rise. Note continual variation in penetration from f to i and thus subjective nature of E/D subdivision. (j) Sediment waves (G). (k–i) Channel complexes cut into turbide/hemipelagic sediments (D). Thin veneer of debris-flow deposit (A) has partly buried channel complex on left side of profile k. (m) Volcanic basement outcrop (C) on continental slope west of Canaries. (n) Sediment-slide source area (B).

Huggett and Somers (1988), Gardner et al., (1991) and Huggett et al. (in press), who suggested that several meters of penetration of GLORIA energy is possible, and who cautioned that the observed acoustic facies pattern may represent subsurface rather than seabed sediment-facies patterns. This is confirmed by (and is actually advantageous to) our study, as our data clearly show that GLORIA can detect debris-flow deposits at least 3 m beneath the seabed.

The GLORIA interpretation scheme used here is a modified version of that employed by Kidd, Simm, and Searle (1985) (Table 13-3). As with 3.5-kHz profile interpretation, many facies changes are marked by gradational rather than distinct boundaries; this is es-

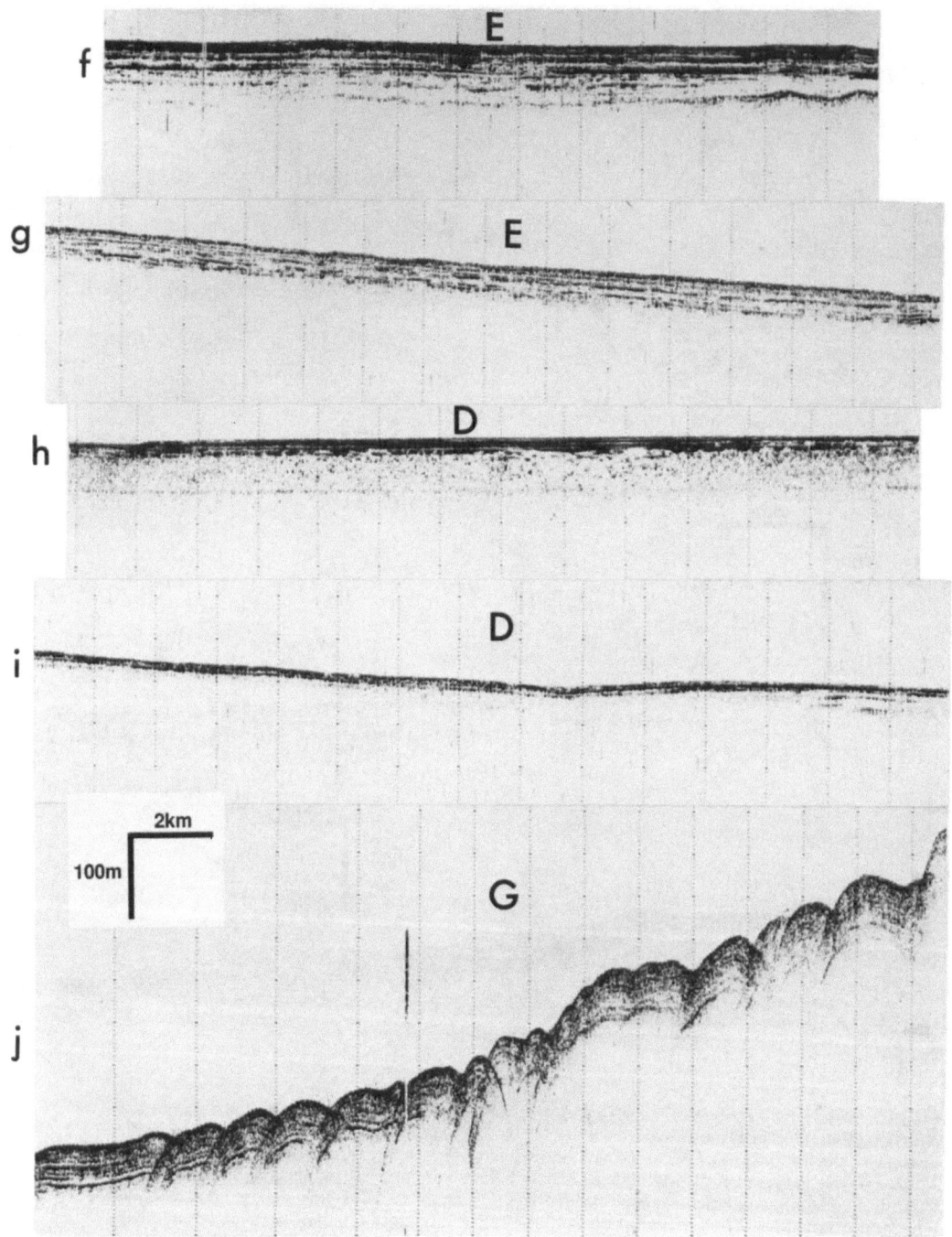

FIGURE 13-4.   Continued

pecially true for boundaries between the acoustic facies interpreted as turbidity-current pathways, hemipelagic sediments, and sediment-wave fields. In the case of turbidites and hemipelagic sediments, it is important to keep in mind that GLORIA may penetrate and receive composite returns from the top few meters of sediment. Because it also appears that the proportion of sand in the top few meters of sediment can be a contributing factor in determining GLORIA backscatter levels (Kenyon, in press), variable proportions of interbedded sandy turbidites and hemipelagic sedi-

ments could, in theory, give a continuously variable range of acoustic returns. Our observations support this theory, because although areas of distinct turbidite and hemipelagic sedimentation can be defined on the basis of the GLORIA data, other areas produce equivocal returns of intermediate character.

Sediment Cores

Gravity, box, kasten, and piston cores have all been collected from various parts of the study area. Full

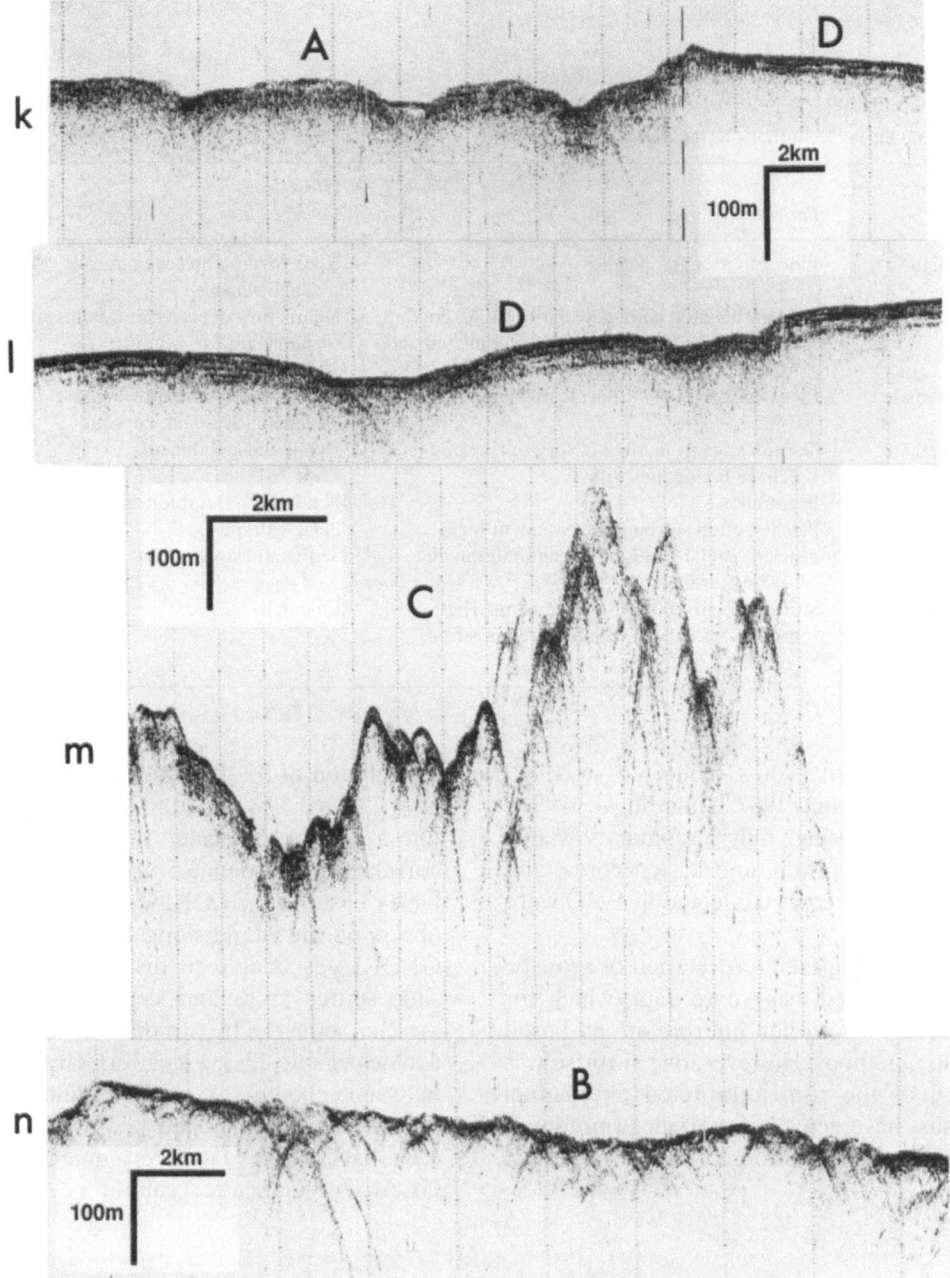

FIGURE 13-4. Continued

TABLE 13-2   Echo-Character Definitions Used in this Chapter Based on 3.5-kHz Profiles.

Facies	Echo Character	Geological Interpretation	Comments
A	Acoustically transparent, lens-shaped units, with prolonged echo. Onlaps other facies	Debris-flow deposits	Prolonged echo may identify this facies in areas where transparent unit thickness too small to be resolved
B	Prolonged echo, no penetration. Small scale rough seabed, many hyperbolae	Sediment-slide source area	Not always easy to distinguish from basement outcrop on island slopes
C	Sharp to prolonged echo, no penetration. Large scale rough seabed, large hyperbolae	Basement outcrop	Seamounts and volcanic island slopes
D	Parallel bedded, typically 10–30-m irregular penetration.	Interbedded turbidites, hemipelagics and pelagics	Grades from almost no penetration where sandy turbidites occur near seabed up to 50 m penetration in pelagic-dominated areas
E	Parallel bedded, continuous penetration > 50 m	Fine-grained turbidites on abyssal plain or pelagics on slope	Effectively flat-lying hemipelagic and pelagic drape on slope and rise
F	Irregularly undulating, penetration > 50 m	Pelagic drape on slope	Confined to abyssal hills
G	Regular undulating, often with marked asymmetry, penetration > 50 m	Sediment waves	

TABLE 13-3   Acoustic Facies Definitions Based on GLORIA Data.

Facies	Backscatter Level	Pattern	Geological Interpretation	3.5-kHz Equivalent
1	Very high	Linear to irregular patches	Rock outcrop on seamounts or volcanic island slopes	C
2	High	Strongly lineated perpendicular to slope, grading downslope to lobate sheets, often lineated and usually with well-defined boundaries	Debris-flow deposits and sediment-slide source areas	A and B
3	Intermediate	Lineated with variable backscatter lineations	Turbidity-current pathways and channel-overbank deposits	D
4	Intermediate to low	None or vaguely mottled or lineated. Poorly defined boundaries with 3	Hemipelagic sediments	E
5	Low	Featureless	Ponded distal turbidites on abyssal plain	E
6	Very low	Patches, often surrounding basement highs	Pelagic drape	F
7	Low to intermediate	distinct, regular bands of low and intermediate backscatter intensity	Sediment waves	G
8	Low to high	Straight or curved narrow lineaments. High backscatter or high with low central stripe	Channels	Topographic lows on profiles

descriptions of the gravity cores (collected in 1982) have been published (Simm, 1987; Simm et al., 1991), but at the time of writing, only preliminary visual descriptions of the box, piston, and kasten cores, and p-wave logs of the piston cores (collected in 1991) were available.

Where possible, geological interpretation of acoustic facies has been calibrated using core data, which for the most part, confirms earlier interpretations based purely on acoustic methods. However, one surprise was that some areas of the rise, interpreted as channel overbank deposits, have actually experienced predominantly hemipelagic sedimentation for at least the last half million years.

Correlation of 3.5-kHz and GLORIA Data

On a first-order scale, there is a good one-to-one correlation between 3.5-kHz echo types and acoustic facies based on GLORIA (Table 13-3). An exception occurs on the island slope west of the Canaries, where GLORIA is unable to distinguish between sediment-slide source areas and slide deposits, although these are characterized by two distinct echo types on 3.5-kHz profiles (Table 13-3, Fig. 13-4). Gradational boundaries between echo types (e.g., D, E, and F over much of the rise) tend to be matched by gradual changes in GLORIA backscatter. However, much subtle variation in GLORIA backscatter cannot be correlated in detail

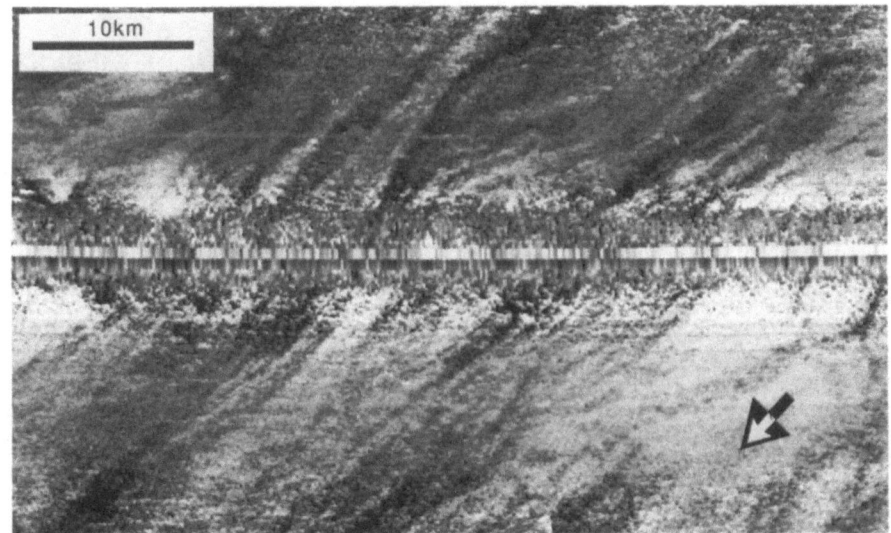

FIGURE 13-5.   GLORIA sonograph showing the strong lineated fabric of upper part of Canary Slide. Arrow shows debris-flow direction. Sonograph located on Figure 13-3.

with 3.5-kHz echo-type changes, although equally subtle changes in regularity and depth of penetration of the 3.5-kHz signal are frequently seen across these areas of subtle GLORIA backscatter variation. This is not surprising, given the different principles involved in collection of the two data types, their different resolutions, and the different seabed characters they measure.

Identification of Debris-Flow Deposits

The criteria used in the identification of debris-flow deposits deserve special discussion, because they are a crucial part of our study (Table 13-3). On the continental slope, where thick (> 5 m) debris-flow deposits occur, acoustically transparent wedges of sediment are clearly seen on 3.5-kHz profiles, as originally described by Embley (1976; see also Chapter 12). These correlate well with areas of high GLORIA backscatter, which have strong downslope-trending lineations (Fig. 13-5). On the continental rise, however, it is frequently difficult to pick the edges of the bodies of transparent material seen on 3.5-kHz data. This is because they thin gradually toward their edges, often to the point where they can no longer be resolved by the 3.5-kHz system (Fig. 13-4a, d). Experience has shown, however,

that the prolonged echo associated with the transparent echo character can usually still be recognized in these regions of thin (< 2–3 m?) debris-flow deposits, and that the edge of the area of prolonged echo correlates well with the edge of the lobate high-backscatter areas seen on GLORIA. Using these criteria, we have been able to map the extent of debris-flow deposits over our entire study area. Only over one small part, along its northwest margin on the lowermost rise, is there any significant doubt as to the location of the edge of the debris-flow deposit. Here, the deposit is buried beneath as much as 3 m of turbidites, which, to the north, contain a thick sand (Weaver and Rothwell, 1987). The result is that the GLORIA backscatter changes little, if at all, across the boundary in this area, whereas the 3.5-kHz echo type changes gradually, but defines no clear boundary.

DISTRIBUTION OF SEDIMENT FACIES

Debris-Flow Deposits and Source Areas

A large part of the area surveyed using GLORIA is covered by a debris-flow deposit or deposits buried beneath 50 cm to 3 m of hemipelagic sediments and

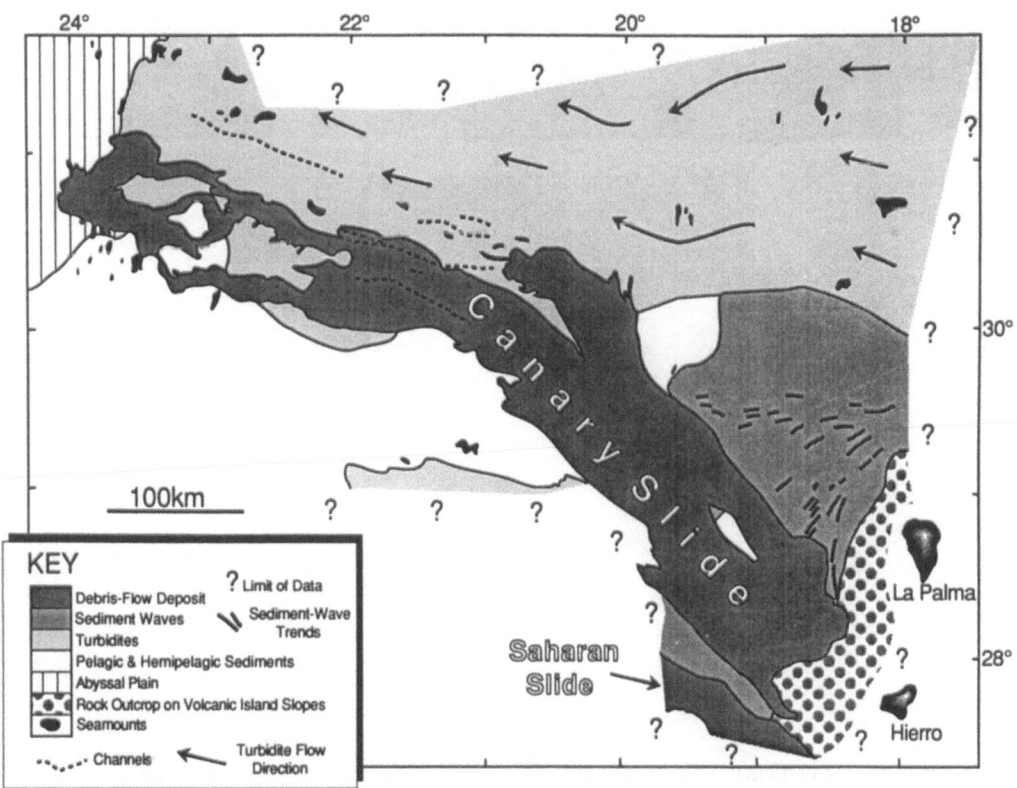

FIGURE 13-6. Facies map based on GLORIA and 3.5-kHz profile data. Area of figure located in Figure 13-1.

turbidites. Previous authors (Kidd and Simm, 1983; Kidd, Simm, and Searle, 1985; Simm et al., 1991) ascribed all the debris-flow deposits west of the Canaries to the single large Saharan Slide event, which originated on the West African continental slope south of lat. 26°N (Embley, 1976, 1980, 1982; see also Chapter 12). However, the complete GLORIA coverage of the area now available clearly shows that a significant proportion of these debris-flow deposits originated, as a distinct event, on the western slopes of the Canaries between 4,000 and 4,500 m water depth (Fig. 13-6). This we now call the Canary Slide.

Our GLORIA data west of Hierro indicates that the Saharan Slide, after passing through the topographic gap south of Hierro, flowed due west, directly downslope. It formed only a narrow (< 25 km wide) but relatively thick (20–50 m) debris-flow tongue on the lower slope and upper rise, as mapped by Embley (1976) and by Jacobi and Hayes (Chapter 12). The Saharan Slide is not seen on a north–south GLORIA line at long. 22°40′W; this confirms that it does not reach the lower rise.

In contrast, the Canary Slide produced a relatively broad (60–100 km wide) but thin (usually < 20 m thick) debris-flow sheet. This sheet covers an area approaching 40,000 km^2, and, with an average thickness of 10 m, has a volume of about 400 km^3. It was derived from a source area of some 6,000 km^2 on the lower slope west of the Canaries, and flowed 600 km to the eastern edge of the Madeira Abyssal Plain. Gradients vary between 1° in the source area to effectively 0° at the edge of the abyssal plain. The removal of up to 65 m of sediment over the entire source area is required to give the calculated volume of the Canary Slide deposit; the

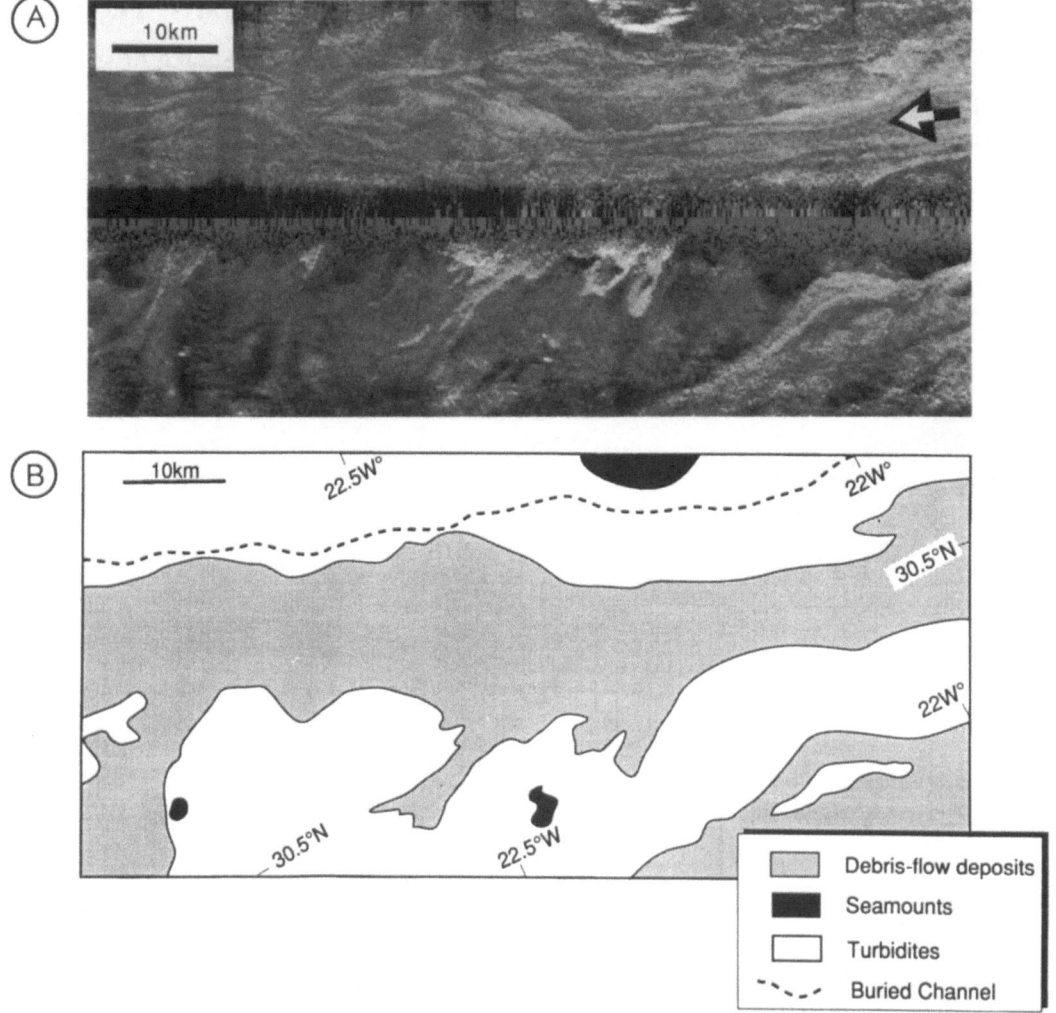

FIGURE 13-7.   (A) GLORIA sonograph from lower part of Canary Slide showing debris-flow deposits burying area of turbidites. Light tones are areas of high backscatter. Arrow shows flow direction of debris. Sonograph located on Figure 13-3. (B) Interpretation of Figure 13-7(A).

exact thickness depends on how much sediment was incorporated into the flow as it crossed the rise.

The Canary Slide has a complex lobate outline, which appears to have been strongly influenced by even the gentlest topography, particularly at its distal end (Fig. 13-6). West of long. 22°W, areas of various facies are isolated as "inliers" within the slide. Although some inliers clearly exist due to their significant relief (tens of meters), many others have an almost unmeasurable topographic expression of 3.5-kHz profiles, and even with our dense profile coverage we cannot be sure that all the inliers have positive relief. Some may exist only because they lie in the lee of an area of positive relief. It is, however, clearly demonstrated by both sidescan and profile data that the slide pathway was able to take advantage of subtle topographic lows. For instance, we see several examples where narrow tongues of debris extend away from the main flow deposit, following partially buried channels (Figs. 13-7, 13-8).

A number of lineaments on the surface of the debris-flow deposit correlate with partially filled channels seen on profile data (Figs. 13-3k, 13-5). Some of these are continuous for distances of up to 100 km (Fig. 13-6). A coring transect over such a feature near lat. 30°40′N, long. 22°20′W shows the presence of a debris-flow deposit buried beneath a thin layer of hemipelagic sediments, both in and on the banks of the channel. The presence of an apparently continuous stratigraphic section above the debris-flow deposit indicates that the channel cannot have been re-excavated following flow emplacement, but that the channel relief must have at least partially survived that emplacement. Why the channels were not completely filled by the flow is not clear, but it may be because the orientation of the channels is generally parallel to that of the flow. Alternatively, postemplacement compaction of the debris-flow deposit may have partially regenerated the channel relief, particularly if the flow deposit is thicker in the channel than on its banks.

As previously noted, it is not possible to distinguish areas of debris-flow deposition from sediment-slide

source areas on the basis of GLORIA data alone, although the source areas tend to have a stronger linear fabric than the depositional area downslope (Fig. 13-5). Somewhat surprisingly, we see no clear scarps at the head of the Canary Slide source area, even though the GLORIA tracks, aligned alongslope, are in the ideal orientation for the detection of such features. This contrasts with the Saharan Slide source area on the West African margin, where distinct scarps as high as 80 m were described by Embley (1976, 1980, 1982; see also Chapters 9, 10, 12). The upper part of the Canary Slide also contains a large "island" of undisturbed slope sediments, centered on lat. 28°45′N, long. 19°15′W (Fig. 13-6). Profiles suggest that this is an in situ block, protected by its position on a basement rise, rather than an allochthonous block, such as those which characterize failures on the flanks of some other volcanic islands (e.g., Hawaii; Moore et al., 1989).

Turbidity-Current Pathways and Channel-Overbank Deposits

The broad valley between the islands of Madeira and the Canaries is interpreted as an important pathway through which turbidites from these volcanic islands, and from the West African margin north of the Canaries, have reached the Madeira Abyssal Plain (Fig. 13-6). The acoustic data (Tables 13-2, 13-3) suggest both a sense of the sediment-flow direction and the presence of sand in the top few meters of sediment. Aspects of this interpretation are reinforced by the presence of volcanic sands in cores from the area and by source direction and provenance studies of megaturbidites on the Madeira Abyssal Plain (Weaver et al., in press). We also note that almost identical GLORIA backscatter patterns, observed on the continental margin 500 km farther south, also have been correlated with an area of sandy turbidites (Kidd, Hunter, and Simm, 1987).

Abundant channels and partially filled channel remnants are seen on 3.5-kHz profiles from the turbidite

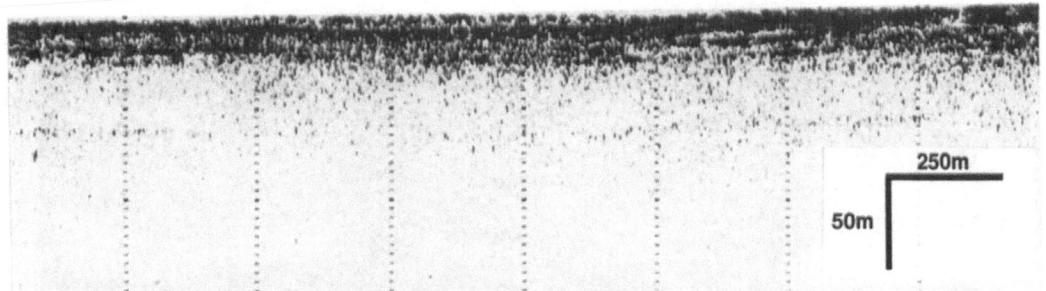

FIGURE 13-8.   3.5-kHz profile showing a tongue of debris-flow deposit, about 5 m thick, filling a shallow channel. Profile located on Figure 13-3.

pathway (Figs. 13-3k, l), although only a few of the more prominent channels can be traced over significant distances on GLORIA data (Fig. 13-9). The larger channels are as deep as 50 m and > 1 km wide. Overall, the discontinuous nature of the channels, and the fact that many are partially or completely filled, strongly suggests that this is a relict channel system. This interpretation is supported by detailed coring around one of the better defined (and thus probably most recently active) channels at lat. 30°30′N, long. 21°05′W. Here, with the exception of a single coarse volcanic sand deposit (which probably correlates with turbidite *b* on the adjacent abyssal plain; Weaver and Rothwell, 1987; Weaver et al., in press), we see no evidence of turbidity-current activity in the upper 2 m of sediments, which a preliminary study suggests represents the last 500,000 years (note that the very low sedimentation rate is the result of carbonate dissolution). This does not, of course, prove conclusively that other parts of the channel system have not been active during this period.

A second area of intermediate GLORIA backscatter with downslope-trending lineations, corresponding to an area of irregular 3.5-kHz penetration, is seen on the continental slope at the southern edge of our study area around lat. 29°N, long. 20–22°W (Fig. 13-6). Although we have not sampled it, cores taken some 100 km to the south contain a mixed volcaniclastic and terrigenous turbidite sequence (Embley, 1982). The position of this area of turbidites is evidence that they are probably derived from the southern slopes of the Canaries and from the African continental margin via the topographic gap just to the south of Hierro.

Hemipelagic and Pelagic Sediments

Much of the lower continental slope and rise to the south of the Canary Slide, characterized by intermediate to low GLORIA backscatter and continuous 50-m 3.5-kHz penetration (Fig. 13-4f, g), is interpreted as an area of hemipelagic sedimentation. The sidescan data show little organized fabric in the area, although vague linear bands of varying backscatter intensity, oriented NNE, may indicate some remnant effect due to the buried basement topography. Areas of pelagic drape occur either in association with abyssal hills or where abyssal hills are sediment covered but still maintain a vestigial relief.

Where hemipelagic sediments have been cored, they consist of interbedded oozes, marls, and clays. Detailed study on the Madeira Abyssal Plain has shown that the more clay-rich units were deposited during glacial intervals and the more calcareous units during interglacials. This relates to climate-driven fluctuations in the level of the carbonate compensation depth (Weaver and Rothwell, 1987; Weaver, Thomson, and Jarvis, 1989).

Abyssal Plain Turbidites

The Madeira Abyssal Plain is a completely featureless low-backscatter area on GLORIA data, and is characterized by continuous 50- to 70-m 3.5-kHz penetration (Fig. 13-4d). However, at its eastern edge, its boundary with the adjacent lower rise hemipelagic and turbidite sediments is gradational; it cannot be defined using the acoustic data. On Figure 13-5, the edge of the abyssal plain is drawn along the 5,400-m contour, following Weaver et al. (in press).

Sediment Waves

A large sediment-wave field occurs on the continental slope northwest of the Canaries in water depths of 3,900–4,600 m (Fig. 13-6). The waves reach 50-m amplitude and 1-km wavelength. Their orientation varies from parallel to almost perpendicular to the regional slope (Fig. 13-6). GLORIA detects the waves only in the center of the field, where amplitudes are greatest (Fig. 13-10). The boundaries of the wave field are,

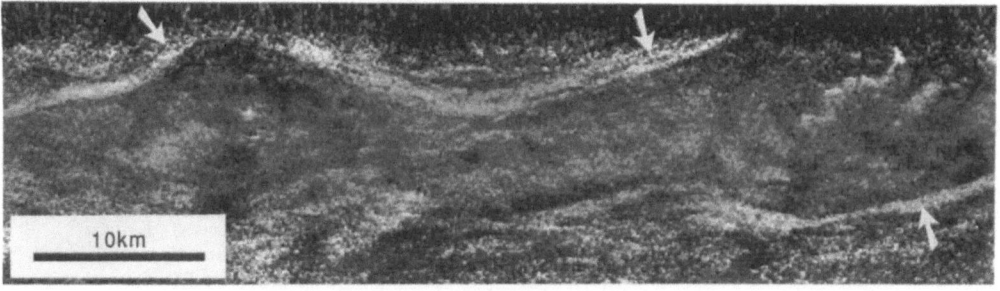

FIGURE 13-9.   GLORIA sonograph showing sinuous channels (arrows) crossing continental rise. Sonograph located on Figure 13-3.

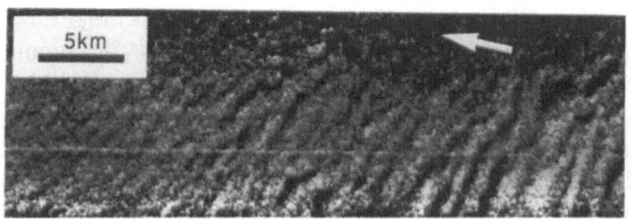

FIGURE 13-10. GLORIA sonograph showing sediment waves on continental slope west of Canaries. Arrow points downslope.

therefore, defined by interpolation between 3.5-kHz profiles.

Around the head of the Canary Slide, there are some features, identified from sonographs as waves, which could be slump folds. The "waves" in this area are parallel to the slope and are more irregular in plan view than those seen farther north; the occurrence of the adjacent slide attests to slope instability. However, 3.5-kHz profiles in this area are mostly subparallel to the wave trend, and are poorly oriented for the purpose of distinguishing slump folds from sediment waves.

Rock Outcrops

Two types of rock outcrop can be defined in the study area. Irregular patches of very high GLORIA backscatter, which occur landward of the 4,000-m contour, mark rock outcrops on the steep island and seamount slopes. Linear to irregular high-backscatter patches, on the rise and abyssal plain, mark outcrops of basaltic rocks on abyssal hills, which protrude through the sediment cover. The linear features trend about 020°, matching the trend of the oceanic magnetic lineations in this area (Searle, 1987). Abyssal hill outcrops are usually small in area, and are rarely > 200 m high.

DISCUSSION

The observed sediment-facies pattern west of the Canaries is dominated by four facies types, which we take as evidence of interaction between four major sedimentary processes: (1) debris flows; (2) turbidity currents; (3) contour currents; and (4) hemipelagic/pelagic sedimentation. The factors controlling these processes are largely independent, but their interaction determines the overall sedimentary environment. In the following section, we discuss the four main facies, and the processes that create them, before describing their interactions.

Debris Flows

Canary Slide

The Canary Slide is one of a number of slides that have occurred on the Northwest African continental margin

between lat. 6° and 32°N (Embley and Jacobi, 1977; Jacobi and Hayes, 1982; see also Chapter 12). The Canary Slide is unusual in that all of the other known slides originated on the upper slope, in water depths less than 2,500 m and on slopes ranging from 1.5 to 6.5°, whereas the source area for the Canary Slide is below 4,000 m and on a slope as gentle as 1°. However, this gentle slope is not atypical of submarine slope failures in general (see Chapter 9); they have been reported from slopes as low as 0.25° (Field et al., 1982; Prior and Coleman, 1982). The morphologies of the debris-flow deposits associated with the Canary and Saharan slides differ primarily in that the former is thinner and spread over a wider area. The Saharan Slide also stopped on a steeper slope than the Canary Slide. This clearly suggests differences in the character of the two flows, presumably resulting from differences in the entrained sediments. The overall impression is that the Canary Slide was an extremely fluid flow, able to follow the subtlest topographic lows, whereas the Saharan Slide appears to have been notably less fluid.

Age of the Canary Slide and Possible Related Events

Analysis of the age of the Canary Slide, based on cores collected in March, 1991, is at an early stage. However, a core that penetrated the slide deposit at lat. 30°48′N, long. 24°01′W, shows that the slide is the same age as turbidite *b* of Weaver and Rothwell (1987), a volcaniclastic-rich turbidite almost certainly derived from the Canary slope and dated at 11–14 ka. Simm et al. (1991) suggested that the debris-flow deposit and turbidite *b* are related features, derived from a single slope failure; we now believe this to be highly probable.

The Saharan Slide was dated at 16–17 ka by Embley (1982), who used conventional radiocarbon dating of bulk samples; this age is slightly older than that of the Canary Slide. However, bioturbation is intense within pelagic sediments in this area, and the sample dated by Embley could have been contaminated, perhaps by material from the underlying foraminiferal-rich turbidite. If so, the error in his date could be much greater than his quoted value of ±475 yr. It is possible (but not proven) that the ages of the two slides are not significantly different; if so, a common age may point to a common cause.

Possible Triggering Mechanisms

Causative mechanisms of submarine slides are notoriously difficult to determine. Some of the more obvious factors, such as underconsolidation due to rapid sedimentation, and the presence of gas hydrates or free gas within the sediment (Karlsrud and Edgers, 1982) are unlikely to apply to the area affected by the Canary Slide. These factors may be more relevant to the Saharan Slide, however, which originated beneath an area of upwelling and high productivity on the continental slope (see also Chapters 9, 10, 12). Megaturbidite emplacement on the Madeira Abyssal Plain has been linked to periods of climate change and sea-level change, although the actual triggering mechanism is not known (Weaver and Kuijpers, 1983); our preliminary dating of the Canary Slide places it within the last deglaciation, an indication that it, too, correlates with a sea-level change. Earthquakes are frequently cited as triggers of sediment failures (e.g., Field et al., 1982; see reviews by Embley, 1982 and Bugge, Belderson, and Kenyon, 1988; see also Chapters 9, 10, 12), and this is perhaps the most likely trigger for the Saharan Slide (Embley, 1982).

Comparisons with Other Oceanic Island Slides

It has recently been recognized that sediment slides can play a major role in the geologic evolution of volcanic islands. The most spectacular examples of these are seen off Hawaii (Lipman et al., 1988; Moore et al., 1989), but it has been suggested that they are characteristic of many other volcanic islands, including the Canaries (Holcomb and Searle, 1990). The slides can be quite large (up to 5,000 km^3) and characteristically involve volcanic basement rocks as well as sediment cover; typical thickness of the mobilized section off Hawaii is 400–2,000 m (Moore et al., 1989). The Canary Slide is different from these Hawaiian-type slides, as it removed only a thin veneer of sediment. It is indistinguishable from many other superficial sediment slides that are present all along the Northwest African margin, and it does not appear to have any special relationship with the volcanic Canary Islands. Indeed, we see no evidence for Hawaiian-type giant slides involving basement rocks around the western part of the Canaries.

Turbidity-Current Pathways

Weaver et al. (in press) have shown that, during the last 180,000 yr, seven megaturbidites reached the Madeira Abyssal Plain using the broad turbidity-current pathway that extends westward from the valley between Madeira and the Canaries. Prior to that, three volcanigenic turbidites reached the abyssal plain by this route at about 500 ka; otherwise the pathway was apparently inactive, at least in terms of deposition on the abyssal plain, at least as far back as 730 ka, the limit of the record so far obtained on the plain. The evidence from the plain for a late Quaternary period of turbidity-current activity fits well with the geophysical data from the continental rise. These data suggest both a flow fabric in the turbidity-current pathway and the presence of sandy sediments in the top few meters of sediment.

Cores from the turbidity-current pathway, however, show a surprisingly variable record of turbidity-current activity. Although some contain several sandy turbidites in the upper 3–4 m (Embley et al., 1978; Simm et al., 1991, and unpublished data), others, particularly along the southern margin of the pathway, contain no turbidites at all. All of the turbidite sands sampled to date are volcaniclastic, indicating a volcanic island source. With the exception of turbidite a, the most recent sequence of turbidity currents derived from the West African continental margin (Fig. 13-2) left no deposits as they crossed the rise. Weaver (unpublished data) has also shown that these African-margin turbidity currents were not erosive as they crossed the rise, because coccolith assemblages in megaturbidites on the Madeira Abyssal Plain contain no significant excess of surface material picked up from the rise. Autosuspension, a possible mechanism that could prevent deposition on the slope and rise, is believed to be particularly effective in fine-grained flows of the type that characterize Madeira Abyssal Plain turbidites (Pantin, 1979). Thus, it is an acceptable thesis that turbidity currents can cross the lower slope and rise without leaving a record.

Important questions concerning the turbidity-current pathway remain unanswered. One concerns the correlation between geophysical and core data, and can be summarized as "what are the geophysical data imaging?" It is an important question, because we cannot, as yet, see a correlation between the variable sequence of turbidites and hemipelagic deposits sampled in the cores and variations in the GLORIA backscatter within the turbidity-current pathway. Clearly, if such a correlation could be established, we would have a very powerful tool for quantifying three-dimensional variations in the sedimentation pattern. At present, however, we have neither enough cores to assess the scale of lateral variation in the sediment sequence, nor an adequate understanding of the subseabed penetration of the GLORIA signal.

The origin of the channel system on the lower rise also remains unexplained. The apparent relict nature of the channels suggests that they predate the most recent turbidity-current influx. In any case, it seems unlikely that channels could be cut, or even be maintained, by turbidity currents in equilibrium with the seafloor, as most of the more recent turbidity currents appear to have been. However, if the channels are older, they must predate the known history of the Madeira Abyssal Plain, because no influx of coarse sediment, which we might expect to be associated with cutting of the channel system, has yet been detected on the plain.

Contour Currents

Our observation of sediment waves over a depth range of 3,900–4,600 m west of the Canaries fits well with regional observations that suggest a northeasterly flowing bottom current at this depth (Lonsdale, 1982). The internal structure of the waves, which indicates upslope migration, and their orientation, parallel to subparallel to the slope, are both typical of sediment waves associated with contour currents (e.g., Embley and Langseth, 1977; see also Chapter 12).

Hemipelagic / Pelagic Sedimentation

Our core data show that, in terms of volume, hemipelagic and pelagic sedimentation have been the primary contributors to sedimentation on the Saharan continental rise during the last 730,000 yr. However, these processes have received little attention, having been largely regarded as "background" sedimentation.

Factors Governing Facies Distributions

The regional picture of sedimentation that emerges from our study, is one of long periods of background hemipelagic sedimentation, modified by contour currents on the slope, and occasionally interrupted by catastrophic turbidity-current and debris-flow events. This picture was first painted for the Madeira Abyssal Plain, and is also true for the rise and lower slope, where we have as yet uncovered little evidence for any greater frequency of downslope sediment transport. However, many turbidity currents travel across the rise without leaving any record, and turbidites form a much smaller percentage of the sediment record on the rise than they do on the abyssal plain.

It is important to realize that catastropic events are actually rare on this margin; only one major turbidite every 30,000 yr (on average) has been deposited during the last 730,000 yr. Debris flows as large as the Canary and Saharan slides are probably even rarer; furthermore, the Canary Slide is the only slide in this region known to have travelled across the entire rise. When they occur, however, turbidity currents and debris flows are much more effective than slow pelagic/hemipelagic deposition in structuring the sedimentary environment of the slope and rise. A good example of this effectiveness is the truncation of the sediment-wave field on the Canary Island slope achieved by both the Canary Slide and the turbidity-current pathway.

The distribution patterns of turbidites and hemipelagic and pelagic sediments in the study area are controlled almost entirely by topography. Turbidity currents from the African continental margin are channelled to the Madeira Abyssal Plain from the south and northeast, because the Canary Islands and the Saharan Seamounts are topographic barriers to flows derived directly from the east (see Chapter 12). Hemipelagic sedimentation predominates in the topographic shadows, except for a narrow zone downslope from the topographic gap between Heirro and the

Saharan Seamounts and, of course, for the area affected by the Canary Slide. Because of the linear nature of the Canary Island chain, volcaniclastic turbidity currents from the islands will initially flow either north or south before being channelled to the west through the broad valley between the Canaries and Madeira, or through the gap south of Hierro.

Possible causes of turbidity currents in the region have been discussed in a series of papers by Weaver and Kuijpers (1983), Weaver and Rothwell (1987), Weaver, Thomson, and Jarvis (1989) and Weaver et al. (in press). The precise triggering mechanism remains elusive, but earthquakes would appear to be most likely. If so, this may imply a relationship between eustatic change and earthquake activity.

SUMMARY AND CONCLUSIONS

This study demonstrates the potential for regional mapping of sediment facies using long-range sidescan sonar in conjunction with traditional techniques, such as 3.5-kHz profiling and sediment coring. We have been able to map, in considerable detail, the distribution of turbidites, debris-flow deposits, contourites, and hemipelagic sediments on the continental rise and slope west of the Canaries. Our revision of the distribution of sediment-slide deposits shows the existence of a large, previously unrecognized slope failure west of the Canaries. This we name the Canary Slide; its source is on the lower slopes of the islands of Hierro and La Palma at about 4,000 m. Debris-flow deposits associated with the slide extend for 600 km to the eastern edge of the Madeira Abyssal Plain.

Our study confirms that the broad valley between Madeira and the Canaries is an important turbidity-current pathway to the Madeira Abyssal Plain. However, whereas turbidity currents from the volcanic islands leave a record on the continental rise in the form of volcaniclastic sands, we have found no trace (on the rise) of those derived from the Northwest African margin. We postulate that these are in a state of autosuspension as they cross the rise. Relict channels within the turbidity-current pathway suggest a history extending back beyond the known 730,000-yr history of the abyssal plain.

ACKNOWLEDGMENTS

The data used in this chapter were collected on a number of cruises of the R.R.S. *Discovery*, R.R.S. *Charles Darwin*, and M.V. *Farnella*, and we gratefully acknowledge the contributions of their respective masters and crews. We thank J. B. Wilson and D. C. Twichell for their constructive reviews of an earlier version of this chapter.

REFERENCES

Bugge, T., Belderson, R. H., and Kenyon, N. H. 1988. The Storegga Slide. *Philosophical Transactions of the Royal Society London*, Series A 325:357–388.

Damuth, J. E. 1980. Use of high-frequency (3.5–12 kHz) echograms in the study of near-bottom sedimentation processes in the deep-sea: a review. *Marine Geology* 38: 51–75.

Dickson, R. R. 1990. World Ocean Circulation Experiment: flow statistics from long-term current-meter moorings: The global dataset in January 1989. World Meteorological Organization World Climate Research Programme, WCRP-30, 1–35 and appendices.

Embley, R. W. 1976. New evidence for the occurrence of debris flow deposits in the deep sea. *Geology* 4:371–374.

Embley, R. W. 1980. The role of mass transport in the distribution and character of deep ocean sediments with special reference to the North Atlantic. *Marine Geology* 38:23–50.

Embley, R. W. 1982. Anatomy of some Atlantic margin sediment slides and some comments on ages and mechanisms. *In* Marine Slides and Other Mass Movements, S. Saxov and J. K. Nieuwenhuis, Eds.: New York: Plenum Press, 189–214.

Embley, R. W., and Jacobi, R. D. 1977. Distribution and morphology of large submarine sediment slides and slumps on Atlantic continental margins. *Marine Geotechnology* 2:205–228.

Embley, R. W., and Langseth, M. G. 1977. Sedimentation processes on the continental rise of northeastern South America. *Marine Geology* 25:279–297.

Embley, R. W., Rabinowitz, P. D., and Jacobi, R. D. 1978. Hyperbolic echo zones in the eastern Atlantic and the structure of the southern Madeira Rise. *Earth and Planetary Science Letters* 41:419–433.

Field, M. E., Gardner, J. V., Jennings, A. E., and Edwards, B. E. 1982. Earthquake induced sediment failures on a 0.25° slope, Klamath River Delta, California. *Geology* 10: 542–546.

Gardner, J. V., and Kidd, R. B. 1983. Sedimentary processes on the Iberian continental margin viewed by long-range sidescan sonar. Part 1: Gulf of Cadiz. *Oceanologica Acta* 6:245–254.

Gardner, J. V., Field, M. E., Lee, H., Edwards, B. E., Masson, D. G., Kenyon, N. H., and Kidd, R. B. 1991. Groundtruthing 6.5 kHz sidescan sonographs: What are we really imaging? *Journal of Geophysical Research* 96:5955–5974.

Holcomb, R. T., and Searle, R. C. 1990. Abundance of giant landslides from oceanic volcanoes. *EOS, Transactions of the American Geophysical Union* 71:1578.

Huggett, Q. J., and Somers, M. L. 1988. Possibilities of using the GLORIA system for manganese nodule assessment. *Marine Geophysical Researches* 9:255–264.

Huggett, Q. J., Somers, M. L., Cooper, A. K., and Stubbs, A. R. in press. Interference fringes on GLORIA sidescan sonar images from the Bering Sea and their importance. *Marine Geophysical Researches*.

Jacobi, R. D. 1976. Sediment slides on the northwestern continental margin of Africa. *Marine Geology* 22:157–173.

Jacobi, R. D., and Hayes, D. E. 1982. Bathymetry, microphysiography and reflectivity characteristics of the West African margin between Sierra Leone and Mauritania. *In* Geology of the Northwest African Margin, U. von Rad, K. Hinz, M. Sarnthein, and E. Seibold, Eds.: Heidelberg: Springer-Verlag, 182–212.

Karlsrud, K., and Edgers, L. 1982. Some aspects of submarine slope stability. *In* Marine Slides and Other Mass Movements, S. Saxov and J. K. Nieuwenhuis, Eds.: New York: Plenum Press, 61–81.

Kenyon, N. H., Belderson, R. H., and Stride, A. H. 1975. Plan views of active faults and other features on the lower Nile Cone. *Geological Society of America Bulletin* 86: 1733–1739.

Kenyon, N. H. in press. Possible geological causes of backscatter variation on GLORIA sonographs from the Mississippi and De Soto Fans, Gulf of Mexico. *Marine Geophysical Researches*.

Kidd, R. B. 1982. Long-range sidescan sonar studies of sediment slides and the effects of slope mass movement on abyssal plain sedimentation. *In* Marine Slides and Other Mass Movements, S. Saxov and J. K. Nieuwenhuis, Eds.: New York: Plenum Press, 289–303.

Kidd, R. B., and Roberts, D. G. 1982. Long-range sidescan sonar studies of large-scale sedimentary features in the North Atlantic. *Bulletin Institut Geologie Bassin d'Aquitaine, Bordeaux*, 31:11–29.

Kidd, R. B., Hunter, P. M., and Simm, R. W. 1987. Turbidity-current and debris-flow pathways to the Cape Verde Basin: Status of long-rang sidescan sonar (GLORIA) studies. *In* Geology and Geochemistry of Abyssal Plains, P. P. E. Weaver and J. Thomson, Eds. *Geological Society of London Special Publication* 31:33–48.

Kidd, R. B., Simm, R. W., and Searle, R. C. 1985. Sonar acoustic facies and sediment distribution on an area of the deep ocean floor. *Marine and Petroleum Geology* 2:210–221.

Lipman, P. W., Normark, W. R., Moore, J. G., Wilson, J. B., and Gutmacher, C. E. 1988. The giant Alika debris slide, Mauna Loa, Hawaii. *Journal of Geophysical Research* 93:4279–4299.

Lonsdale, P. 1982. Sediment drifts in the Northeast Atlantic and their relationship to the observed abyssal currents. *Bulletin Institut Geologie Bassin d'Aquitaine, Bordeaux* 31:141–150.

Moore, J. G., Clague, D. A., Holcomb, R. T., Lipman, P. W., Normark, W. R., and Torresan, M. E. 1989. Prodigious submarine landslides on the Hawaiian Ridge. *Journal of Geophysical Research* 94:17,465–17,484.

Pantin, H. R. 1979. Interaction between velocity and effective density in turbidity flow: phase-plane analysis, with criteria for autosuspension. *Marine Geology* 31:59–99.

Prior, D. B., and Coleman, J. M. 1982. Active slides and flows in unconsolidated sediment on the slopes of the Mississippi Delta. *In* Marine Slides and Other Mass Movements, S. Saxov and J. K. Nieuwenhuis, Eds.: New York: Plenum Press, 21–49.

Roberts, D. G., Kidd, R. B., and Masson, D. G. 1984. Long-range sonar observations of sedimentary facies patterns in and around the Amirante Passage, western Indian

Ocean. *Journal of the Geological Society of London* 141:975–984.

Saunders, P. M. 1987. Currents, dispersion and light transmittance measurements on the Madeira Abyssal Plain: Final Report, March 1987. *Institute of Oceanographic Sciences Report 241*, 1–55.

Searle, R. C. 1987. Regional setting and geophysical characterisation of the Great Meteor East area in the Madeira Abyssal Plain. *In* Geology and Geochemistry of Abyssal Plains, P. P. E. Weaver and J. Thomson, Eds. *Geological Society of London Special Publication* 31:49–70.

Simm, R. W. 1987. Late Quaternary sedimentation on the continental rise off Western Sahara. Ph.D. thesis, University of London, unpublished.

Simm, R. W., and Kidd, R. B. 1983. Submarine debris flow deposits detected by long-range side-scan sonar 1000 km from source. *Geo-Marine Letters* 3:13–16.

Simm, R. W., Weaver, P. P. E., Kidd, R. B., and Jones, E. J. W. 1991. Late Quaternary mass movement on the lower continental rise and abyssal plain off Western Sahara. *Sedimentology* 38:27–40.

Somers, M. L., and Searle, R. C. 1984. GLORIA sounds out the seabed. *New Scientist* 104:12–15.

Somers, M. L., Carson, R. M., Revie, J. A., Edge, R. H., Barrow, B. J., and Andrews, A. G. 1978. GLORIA II—an improved long range sidescan sonar. *In* Oceanology International 78, Technical Session J. London: BPS Exhibitions, 16–24.

Weaver, P. P. E., and Kuijpers, A. 1983. Climatic control of turbidite deposition on the Madeira Abyssal Plain. *Nature* 306:360–363.

Weaver, P. P. E., and Rothwell, R. G. 1987. Sedimentation on the Madeira Abyssal Plain over the last 300,000 years. *In* Geology and Geochemistry of Abyssal Plains, P. P. E. Weaver and J. Thomson, Eds. *Geological Society of London Special Publication* 31:131–143.

Weaver, P. P. E., Thomson, J., and Jarvis, I. 1989. The geology and geochemistry of the Madeira Abyssal Plain sediments: a review. Advances in Underwater Technology, Ocean Science and Offshore Engineering—Disposal of Radioactive Waste in Seabed Sediments, Volume 18: London: Graham and Trotman, 51–78.

Weaver, P. P. E., Rothwell, R. G., Ebbing, J., Gunn, D. E., and Hunter, P. M. in press. Correlation, frequency of emplacement and source directions of megaturbidites on the Madeira Abyssal Plain. *Marine Geology*.

IV
Synthesis

14
Geologic Evolution of Continental Rises: A Circum-Atlantic Perspective

C. Wylie Poag and Pierre Charles de Graciansky

... we know the contours and the nature of the surface soil covered by the North Atlantic for a distance of 1,700 miles from east to west as well as we know that of any part of the dry land. It is a prodigious plain—one of the widest and most even plains in the world. If the sea were drained off, you might drive a wagon all the way from ... Ireland to ... Newfoundland.

Thomas Henry Huxley, 1868

Marine scientists have made quantum leaps toward comprehending the formation, structure, depositional history, and physiography of deep ocean basins during the 124 years following Huxley's presumptuous boast (from the famous lecture "On a Piece of Chalk," delivered in 1868 to a group of working men of Norwich during a meeting of the British Association for the Advancement of Science; Huxley, 1967). Most of this knowledge has been gained in the last 40 years (after World War II), following development of sophisticated hydrographic and geophysical surveying instruments, piston corers, deep-sea cameras and video systems, oceanic drilling technology, and deep-diving submersibles. As demonstrated in this volume, some segments of the Atlantic continental rises have been studied in relatively fine detail (Fig. 14-1). But when one considers the immense Atlantic domain ($\sim 75,000,000$ km^2), only a small percentage of the bathyal and abyssal seafloor is geologically well known.

Physiographically, Atlantic-type continental rises typically constitute gently seaward-sloping parts of the ocean floor (average declination 1:100 to 1:700) between the steep continental slopes (average declination 1:40) and the virtually flat, level, abyssal plains (average declination 1:1,000; Fig. 14-2). Atlantic rises range in width from ~ 100 to 1,000 km and in depth from $\sim 2,000$ to 6,000 m, and cover $\sim 21,000,000$ km^2 (25% of the total Atlantic Ocean floor; Kennett, 1982).

Geologically, Atlantic continental rises are seaward-thinning sedimentary wedges (Fig. 14-2), which contain about 26% ($\sim 18,000,000$ km^3) of the total volume of bathyal and abyssal sediment in the Atlantic (Emery and Uchupi, 1984) and $\sim 15\%$ of all sediment in the Atlantic. Approximately 80% of all continental-rise sediment resides in the Atlantic. Maximum thickness is ~ 12 km in various basins along North America, Northwest Africa, and northern Brazil (Fig. 14-1; Emery and Uchupi, 1984). Elsewhere, the rises are generally less than 6 km thick. Atlantic rises contain largely siliciclastic detritus derived from surrounding continental highlands, but also include sediments swept from bordering carbonate platforms or submergent plateaus and ridges, detritus displaced en masse from continental shelf edges and slopes, and carbonate and siliceous particles derived in situ from shell-secreting oceanic organisms.

There is no unanimity, however, regarding the recognition of continental-rise provinces on several of the Atlantic margin segments (to which the diverse opinions of our contributors attest). Emery and Uchupi (1984), in fact, proposed a new physiographic province in the Atlantic Ocean, called "abyssal aprons," to accommodate sediments deposited by geostrophic bottom currents (mainly sediment drifts, or contourites, formerly considered as part of the rise). This may be a conceptually appealing categorization, but its definition rests not on physiography or depositional style, but on sedimentary genesis. We, therefore, continue to include many sediment drifts (especially those of the North Atlantic) within our concept of continental rises.

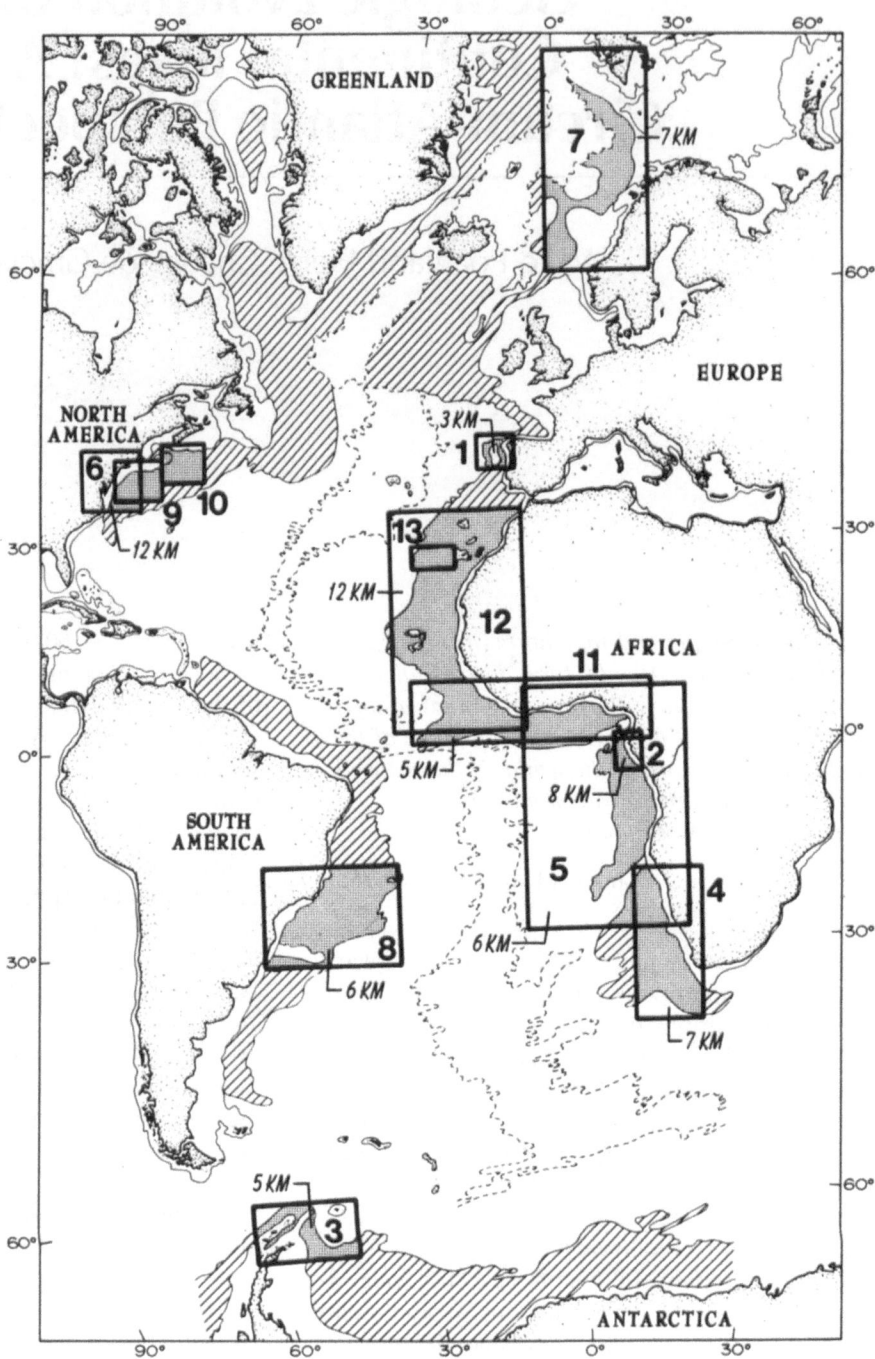

FIGURE 14-1. Location of Atlantic continental rises (diagonal hachures) and segments studied by contributors to this volume (stippled patterns; bold numerals are chapter numbers). Estimated maximum thickness of rise sediments in each study area given in kilometers (thickness data from Emery and Uchupi, 1984).

EARLY STAGES OF BASIN FORMATION: PROVIDING ACCOMMODATION SPACE FOR CONTINENTAL-RISE DEPOSITS

Many authors have used seismic reflection and refraction data, geomagnetic and gravity surveys, deep drilling, and analogies with exposed continental rift systems, to construct models that explain the initiation, location, size, and depth of Atlantic marine basins (Montadert et al., 1979; Watts and Steckler, 1979; Graciansky et al., 1985; Diebold, Stoffa, and The LASE Study Group, 1986; Steckler, Watts, and Thorne, 1986; Vogt and Tucholke, 1986; Boillot, Winterer, et al., 1988; Bond and Kominz, 1988; Manspeizer, 1988; Dunbar and Sawyer, 1989; Keen and Beaumont, 1990; Eldholm, 1991). Atlantic prerift and synrift history has

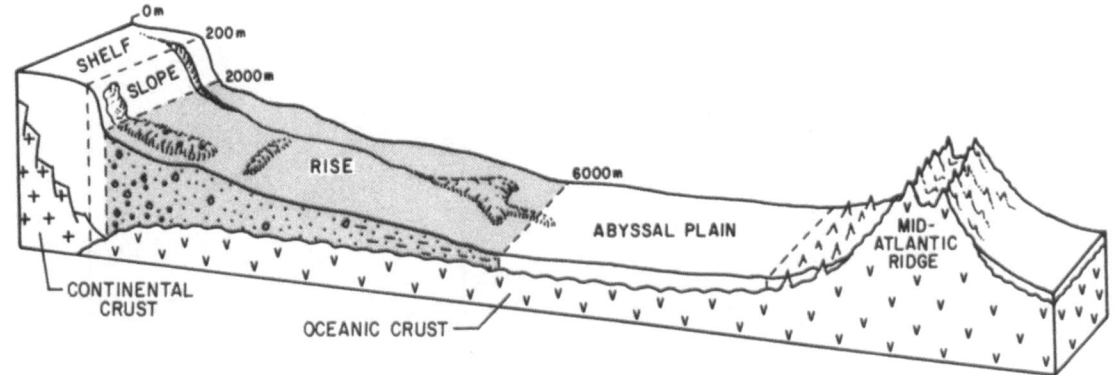

FIGURE 14-2. Major physiographic provinces (not to scale) of a typical sediment-rich Atlantic continental margin. Approximate water depths of province boundaries given in meters.

been most intensely studied along the U.S./Canada margin, the European margin from Rockall Bank to Galicia Bank, and the Norway–Svalbard margin (Fig. 14-1). As Sibuet (Chapter 1), Poag (Chapter 6), and Myhre et al. (Chapter 7) point out, the synrift basins are buried beneath 10–15 km of sediment along North America, and beneath 2–7 km along the Norway–Svalbard margin. In contrast, the sediment-starved European margin allows direct sampling by drilling and dredging, and, in some places, by observation with research submersibles or seafloor photography. Such a thin postrift veneer also allows much finer seismic resolution of prerift and synrift structure and stratigraphy.

Geophysical Models

In the basic model for a "typical" Atlantic-type margin, tensional stresses began to break Pangaea apart along a sublinear axis between North America and Eurasia/Africa sometime in the Late Triassic (~ 230 Ma; Manspeizer, 1988). Rifting was, however, by no means a uniform, synchronous event throughout the entire Atlantic. In some areas, rifting was delayed until the Early Cretaceous (e.g., Galicia; Chapter 1), the Middle Cretaceous (equatorial and southwest Africa; Chapters 2, 4, 5), the Paleocene (Norway; Chapter 7), or even as late as the Pliocene (Bransfield Trough; Chapter 3). In fact, Acosta et al. (Chapter 3) postulate that rifting may still be taking place in parts of the Bransfield Trough. Linear rift grabens and half-grabens produced by this crustal stretching filled with various lacustrine, fluvial, and shallow to deep marine sediments and, in some regions, with thick volcanic flows and volcaniclastic detritus.

Pure and simple shear models (and various modifications thereof) are widely used to explain lithospheric deformation during rifting. Sibuet (Chapter 1), for example, considers the synrift Galicia margin (Fig. 14-1) to have formed by a combination of simple and pure shear, which eventually produced (during the Early Cretaceous) tilted blocks bounded by listric faults (half-grabens; Fig. 14-3). The north–south trending fault blocks of Galicia are clearly delineated on structural maps of the prerift basement surface, and remain conspicuous on SEABEAM maps of the present seafloor west of Galicia.

Drilling during Ocean Drilling Program (ODP) Leg 103 documented the prerift (Tithonian), synrift (Berriasian–early Albian) and postrift (early Albian–Quaternary) stratigraphy, lithologies, and paleoenvironments of the Galicia margin (Boillot, Winterer, et al., 1988). Rifting began with subsidence of a prerift Tithonian carbonate platform (continental crust) accompanied by normal faulting (no tilting) and deposition of Berriasian to lower Hauterivian sediments. A second rifting phase tilted the blocks along listric faults and deposited upper Hauterivian to lower Albian strata.

Sibuet's model postulates the presence of a crustal detachment surface, which extended from the Canadian margin to the western segment of the Galicia margin, where it now marks the base of the rotated blocks (S reflector; Fig. 14-3). Here the block crests are on average 12 km apart and comprise mainly synrift deposits; the horizontal offset along the faults is ~ 4 km; amplitude of magnetic anomalies is ~ 100 nT. Sibuet interprets the structure of this western segment to be the result of simple shear within the brittle crust.

On the other hand, the eastern segment of the Galicia margin displays no detachment surface. Spacing between the block crests is greater (~ 16 km), and the amplitude of magnetic anomalies associated with the blocks is greater (~ 200 nT), but the horizontal offset along faults above synrift deposits is smaller (~ 2 km). Sibuet explains the extension of this eastern segment ($\beta = 3.21$) by a pure shear model applied to the entire lithosphere (both brittle and ductile crust).

NORTH AMERICA

GALICIA

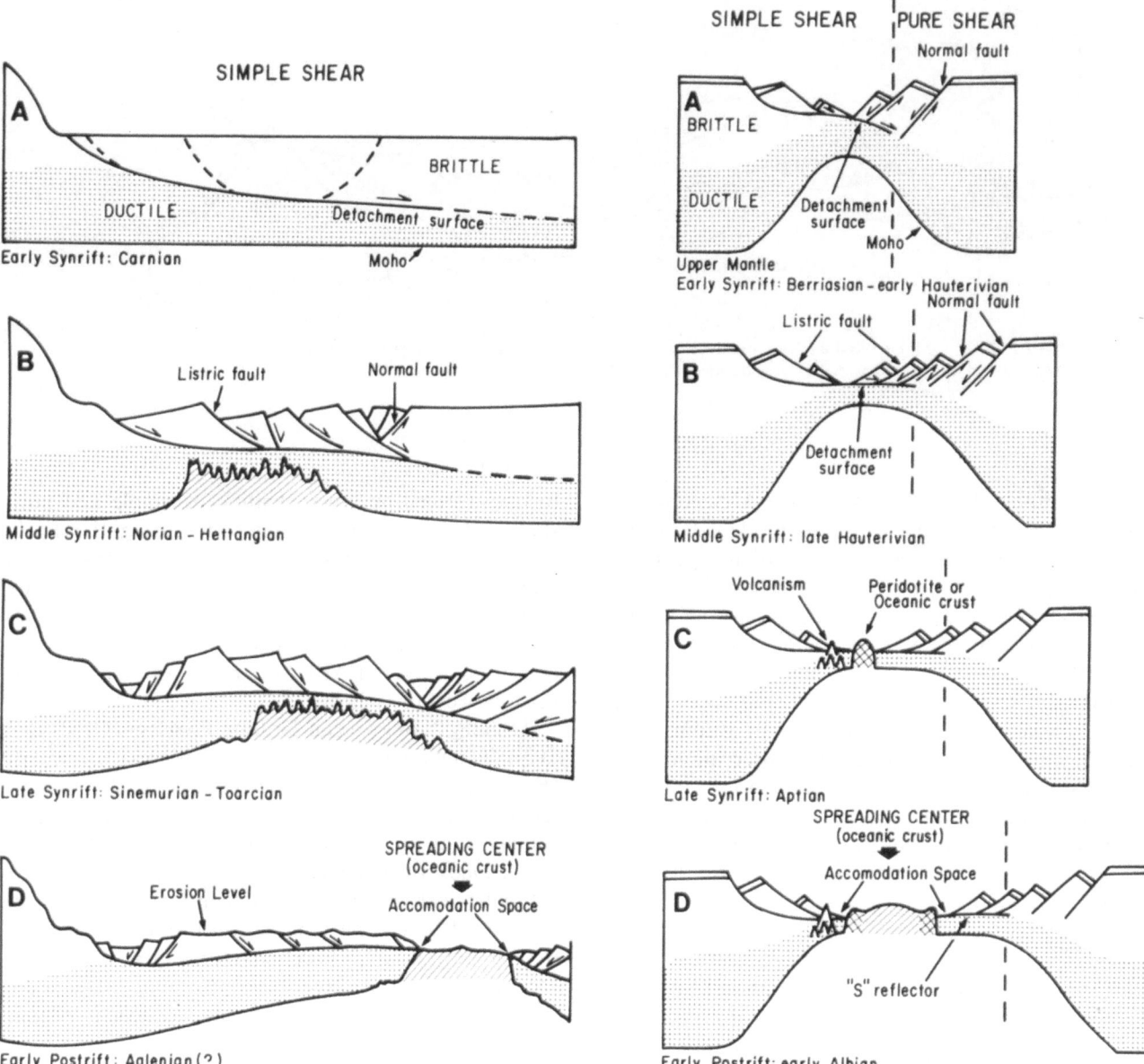

FIGURE 14-3.   Models for crustal evolution (synrift to early postrift) of a sediment-rich margin (North America; Klitgord, Hutchinson, and Schouten, 1986) and a sediment-poor margin (Galicia; Sibuet, Chapter 1).

When rifting ceased, oceanic crust formed along a divergent seafloor spreading axis, the newly formed lithosphere cooled, and the brittle–ductile interface subsided. The subsident oceanic basement provided accommodation space for the accumulation of continental-rise sediments (Fig. 14-3). In this case, however, little siliciclastic detritus found its way from the European continent to the western margin of Galicia, so the continental rise there is atypically thin and narrow. Similar sediment scarcity characterizes parts of the Norway margin (Chapter 7).

On more thickly sedimented divergent Atlantic margins, such as those of North America and most of Africa, similar rifting models have been proposed (Fig. 14-3); initial postrift thermotectonic cooling and subsidence of the oceanic crust also was comparable, though more rapid. On these sediment-rich margins, however, landward flexural bulges and uplifted continental source terrains provided large volumes of siliciclastic detritus to fill the accommodation space in the incipient Atlantic seaway (e.g., Poag and Sevon, 1989). According to current models of postrift basin development (e.g., Keen and Beaumont, 1990), thermotectonic subsidence slowed after ~ 60 m.y.; thereafter, an increasing sediment load (10–12 km thick) provided a major impetus for continued basin subsidence.

The work of Barker and Dalziel (1983), Jeffers, Anderson, and Lawver (1991), Jeffers and Anderson (1990), and Acosta et al. (Chapter 3) reminds us that in some Atlantic regions (the Antarctic, in this case; Fig. 14-1), block-fault rifting and wedgelike synrift deposition began only 3–4 Ma, and may be still taking place. Thus accommodation space for continental-rise sediments is quite limited there.

Chronothermometry and Geothermometry

Along the Gabon–Congo margin, Walgenwitz, Richert, and Charpentier (Chapter 2; Fig. 14-1) use fission-track chronothermometry and fluid-inclusion geothermometry to interpret rift history. They describe an early extensional rift stage (Hauterivian–Barremian; marked by block faulting, graben formation, lacustrine sedimentation), which did not completely separate Brazil from Africa. A late rift stage (Barremian–Aptian; which featured growth faulting), however, marked the successful "breakup" of the São Francisco–Congo craton, and connected the North and South Atlantic with a large rift basin. Rapid uplift and widespread erosion at about 110 Ma (early Albian) immediately preceded the initiation of seafloor spreading in this area. Rapid subsidence during the late Albian provided accommodation space for synchronous filling of the nascent marine basins with 1–2 km of evaporitic, carbonate, and siliciclastic sediments.

Sheared Margins

On still other Atlantic margin segments, early postrift tectonism included mainly lateral crustal shearing along east–west trending transform zones (Guinea–Ivory Coast margin; Chapter 11; Fig. 14-1), or combinations of northwest–southeast shearing and divergent seafloor spreading (Svalbard margin; Chapter 7). In contrast to most divergent margins, sheared margins provided sparse accommodation space for the accumulation of continental-rise sediments.

CONSTRUCTION OF CONTINENTAL RISES: POSTRIFT INTERPLAY OF TECTONISM, EUSTACY, PALEOCLIMATE, AND PALEOCEANOGRAPHY

Sediment-Poor Margins

The early stages of postrift deposition (like the prerift and synrift stages) and initial construction of Atlantic continental rises are best understood in regions of sparse postrift sedimentation, such as the Galicia margin (Chapter 1), the Biscaye–Irish margin (Montadert et al., 1979; Graciansky, Poag, et al., 1987), parts of the Svalbard margin (Chapter 7), and the Antarctic margin (Bransfield Trough; Chapter 3). In the Bransfield Trough, a 30-km-wide depression between the Antarctic Peninsula and Tierra del Fuego (Figs. 14-1, 14-4), an incipient spreading center formed only about 1.6 Ma, so the initial postrift sediments are young (Pleistocene), though relatively thick (~ 500 m). Yet a complex array of sedimentary facies and depositional styles is represented, including siliciclastic turbidites, slumps, debrites, contourites, and draped pelagites. Tectonic relationships also are complex, for contiguous rifting and spreading appear to have taken place since spreading began (~ 1.6 Ma).

In the western half of the Bransfield Trough, patchy debrite aprons and slump deposits form discrete depositional sequences separated by seismic unconformities. Youngest strata are horizontally layered pelagites. Turbidites are also present, having been derived from prograded sedimentary wedges at the mouths of submarine canyons. Some depositional sequences display onlap shifts, which can be interpreted as the result of glacioeustacy. Volcanic dikes and sills intrude sediments near the spreading axis in the northeastern part of the trough.

The spreading axis does not extend, however, into the extreme southwestern part of the trough (Fig. 14-4). The sediments there are thicker (700–800 m), slightly folded, and appear to be part of a continuing rift sequence. These strata comprise alternating units of ash-bearing diatomaceous ooze (interglacial deposits) and turbidites/debrites (glacial deposits). Evidence of paleoceanographic effects on deposition is present in the form of an inferred sediment drift.

The Bransfield spreading axis forms a segmented igneous ridge. The gaps between segments have allowed sediments from the Antarctic continent to reach the northwest side of the trough; minimal sediment appears to have been derived from the South Shetland Islands Ridge.

We see that even on such a geologically young rise, deposition has been strongly influenced by widely variable tectonism, eustacy, and paleoceanography, and by repetitious cycles of glacial and interglacial paleoclimate.

Sediment-Rich Margins

In the thick sedimentary depocenters, such as those of western Africa (Chapters 4, 5) and eastern North America (Chapter 6; Fig. 14-1), any interpretation of early postrift deposition and erosion relies mainly on

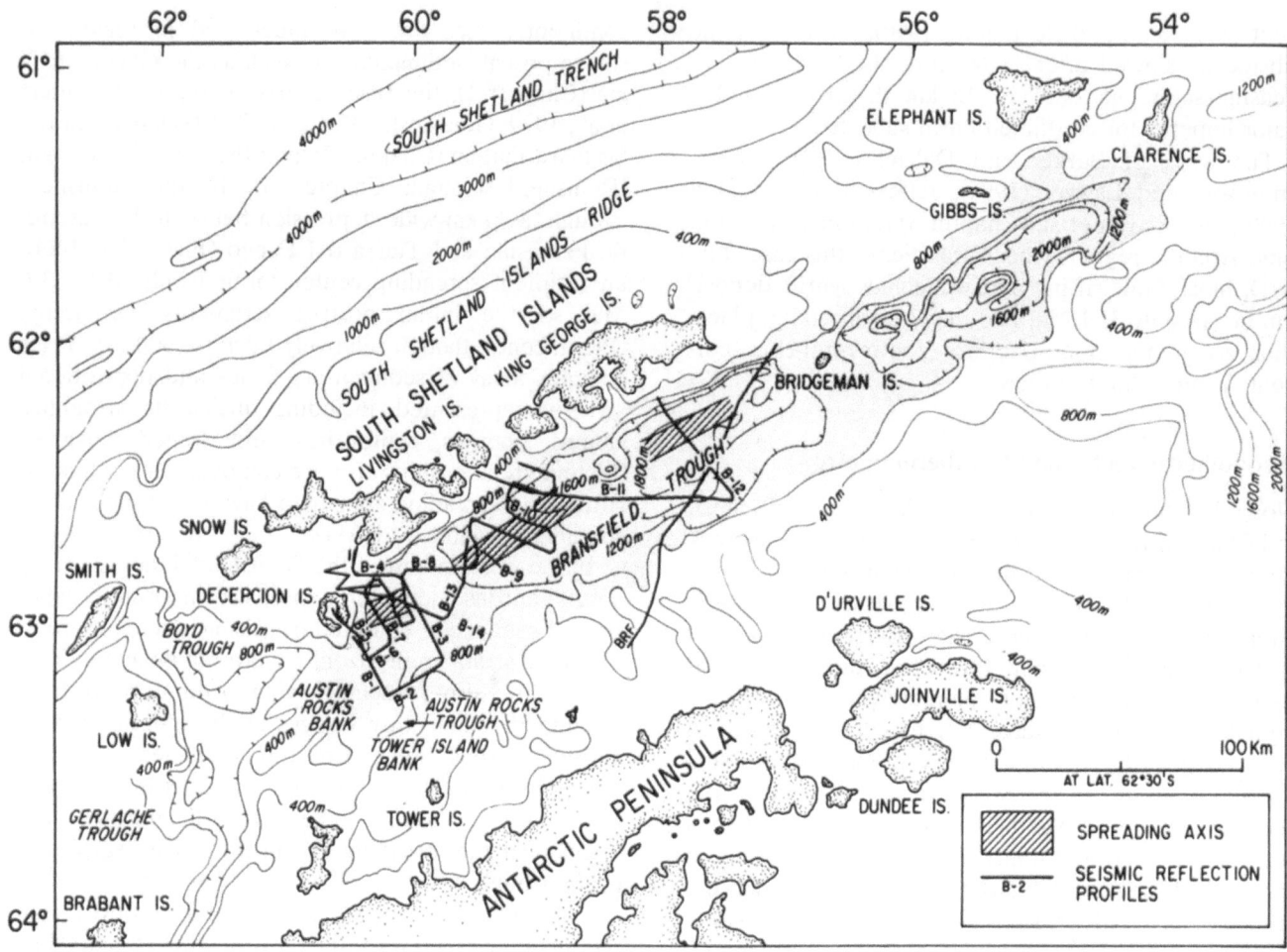

FIGURE 14-4.   Physiography of Bransfield Trough and vicinity (from Acosta et al., Chapter 3).

seismic surveys and magnetic and gravity measurements, supplemented by stratigraphic, paleoenvironmental, and geothermal data from a few scattered boreholes. The oldest postrift strata inferred in this volume are early Middle Jurassic evaporite deposits of the U.S. margin (Chapter 6). By the end of the first major postrift depositional episode [Aalenian(?) allostratigraphic unit], siliciclastic sediments derived from the central Appalachian highlands formed a recognizable continental rise (Fig. 14-5). The rise sediments filled part (the northern part of the Hatteras Basin) of a sinuous trough (500 km × 100 km) on the western side of the narrow proto-Atlantic seaway. On the lower part of this rise, several small submarine fans coalesced into a 180-km-long confluent fan. Distal elements of the fan lapped eastward onto the flank of the spreading ridge and filled narrow swales in the oceanic basement.

Poag (Chapter 6) divided the depositional history of the U.S. Atlantic rise into eight distinct stages on the basis of: (1) variations in the rate of siliciclastic sedi-

ment accumulation; (2) presence/absence of an outer shelf carbonate platform; (3) relative thickness of continental-rise depocenters; (4) relative volume of siliciclastic sediments reaching the rise; and (5) relative abundance of pelagic carbonate on the rise (Fig. 14-6). Deposition during all stages was dominated by downslope sediment gravity flows. Most of the depositional variables were controlled by variations in source-terrain tectonism, paleoclimate, and eustacy, but paleoceanographic regulators (contour currents, coastal upwelling, level of the CCD) significantly affected the relative thickness of depocenters, the presence/absence of carbonate platforms, and the content of pelagic carbonate (especially in the Neogene and Quaternary).

Along the western margin of Africa, postrift deposition was highly anisochronous, and lithofacies and depositional styles varied widely. Northwest Africa was conjugate to the North American margin (Fig. 14-1), so early postrift deposition began simultaneously in the early Middle Jurassic. Continental rise development off Northwest Africa, likewise, paralleled that of North

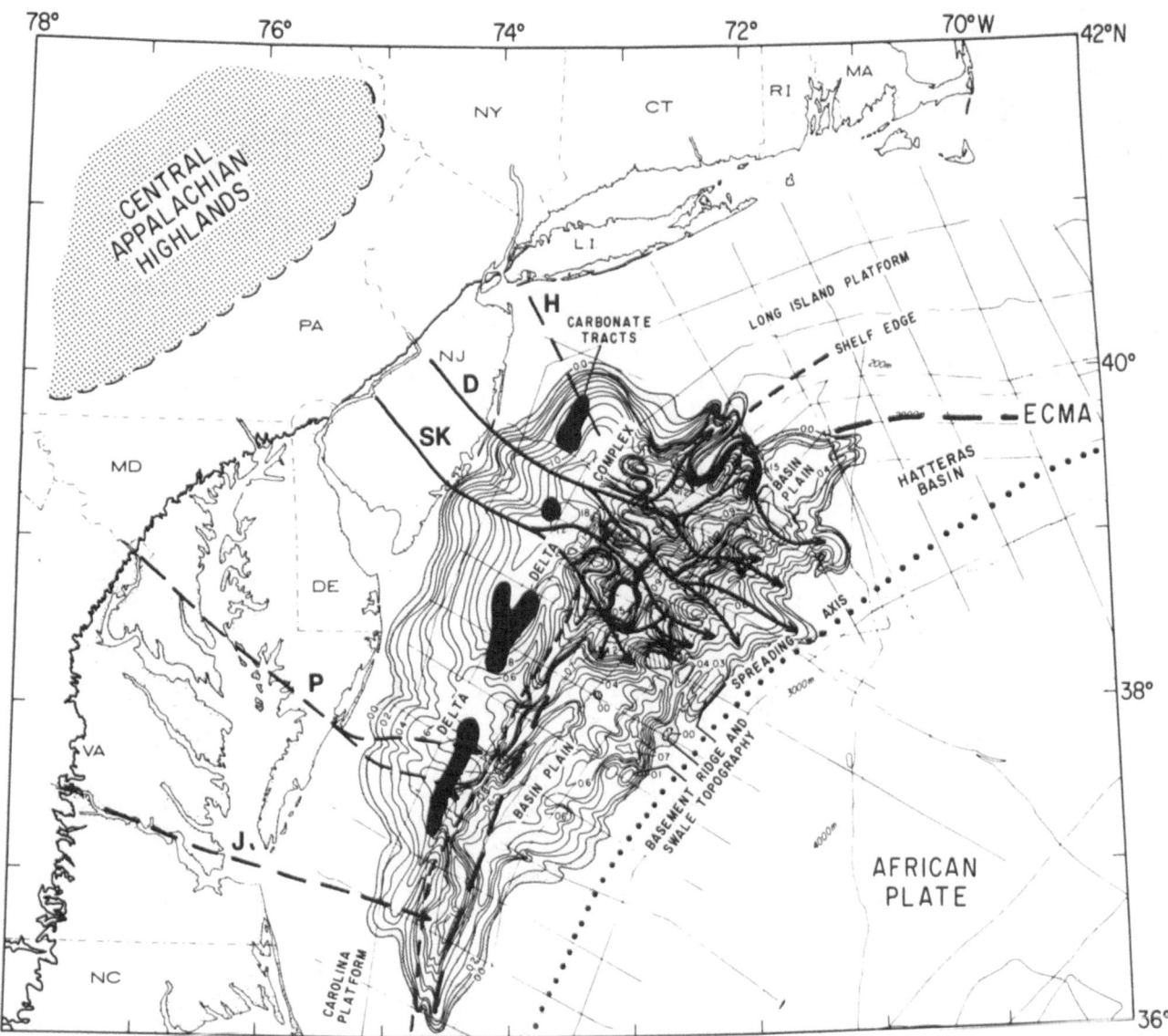

FIGURE 14-5.   Isochron map (contour interval = 0.1 sec two-way traveltime) for initial postrift depositional sequence [Aalenian(?)] of U.S. Middle Atlantic margin. Ancient river systems draining Appalachian highlands indicated by bold arrows (solid arrows = primary dispersal systems; dashed arrows = secondary dispersal systems): H = ancient Hudson River; D = ancient Delaware River; SK = ancient Schuylkill River; P = ancient Potomac River; J = ancient James River; ECMA = East Coast Magnetic Anomaly; NC = North Carolina; DE = Delaware; VA = Virginia; MD = Maryland; PA = Pennsylvania; NJ = New Jersey; LI = Long Island; NY = New York; CT = Connecticut; RI = Rhode Island; MA = Massachusetts (from Poag, Chapter 6).

America through the Jurassic and Cretaceous, but their histories diverged during the Tertiary (Jansa and Wiedmann, 1982).

Along equatorial and Southwest Africa (Chapters 2, 4, 5; Fig. 14-1) postrift deposition began in the Aptian and Albian, 68–74 m.y. later than in the central North Atlantic (Fig. 14-7). On the Gabon–Congo margin, evaporites (500–800 m thick) constitute either the latest synrift (Chapter 5) or earliest postrift (Chapter 2) deposits. Subsequent thermotectonic subsidence was rapid, and 1–2 km of basinal clay and platform carbonates accumulated. Margin subsidence decreased substantially in the Cenomanian; fine-grained siliciclastic sediments continued to accumulate to the present. From fission-track analysis, Walgenwitz, Richert, and Charpentier (Chapter 2) recognize four episodes of geothermal cooling [early Aptian (= early Albian on DNAG scale), late Santonian, middle Eocene, and early Miocene], which signify periods of source-terrain uplift, erosion, and oceanward tilting of the

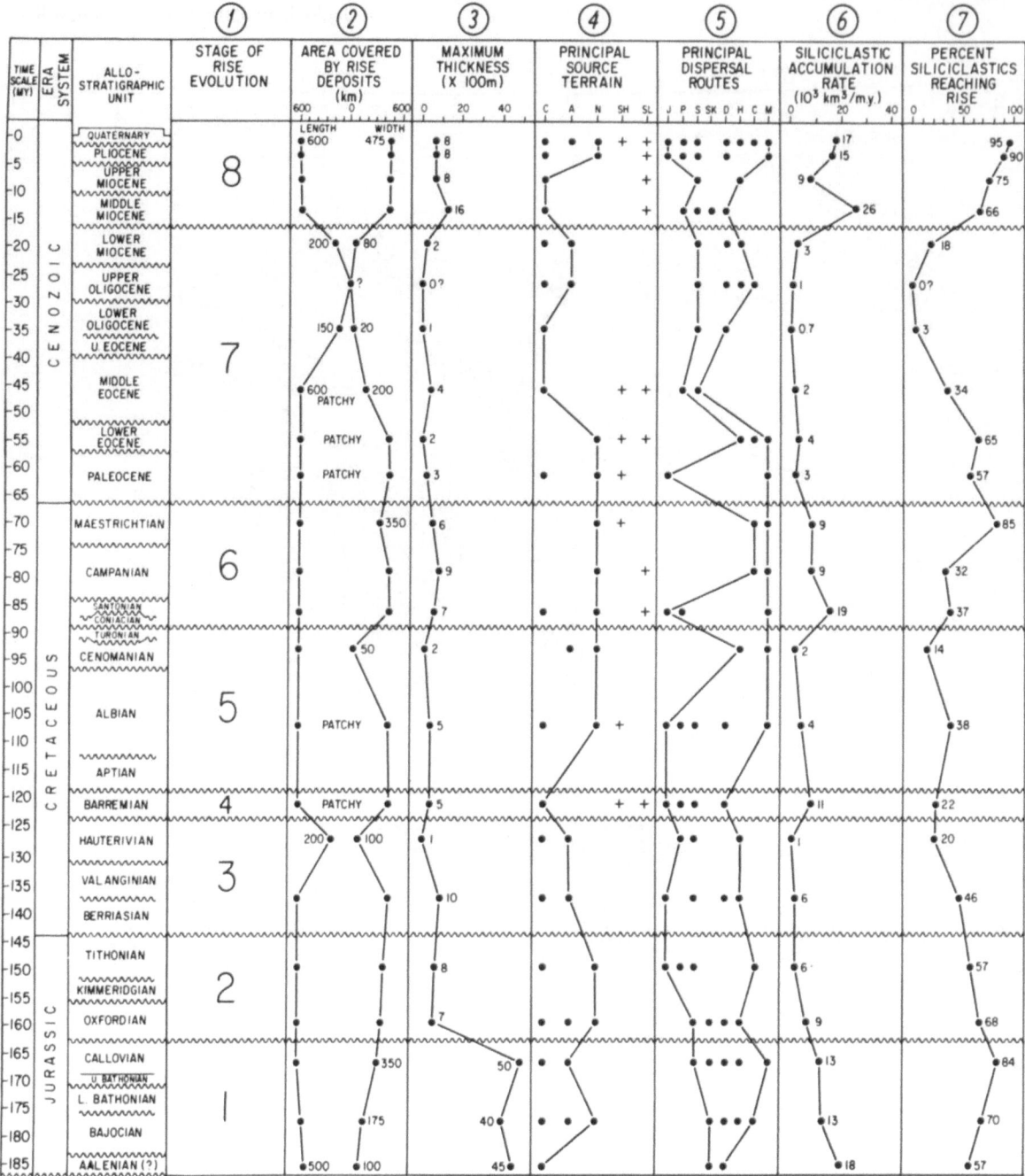

FIGURE 14-6. Key aspects of depositional history of continental rise in northern part of Hatteras Basin, U.S. Middle Atlantic margin. Time scale is that adopted by Geological Society of America for Decade of North American Geology (DNAG) publications (Palmer, 1983). In compiling age data for this synthesis, we have converted ages listed by other authors to the DNAG time scale, wherever possible. Column 1 = stages in sedimentary evolution of continental rise. Column 2 = approximate length and width of area covered by rise deposits, calculated from isochron maps of Poag (Chapter 6). "Patchy" indicates areas of no (or very thin) sediment. Column 3 = maximum thickness of depositional units on continental rise. Column 4 = principal source terrains for siliciclastic sediments, inferred from isochron maps of Poag (Chapter 6). Solid dots indicate principal highland source terrains; solid line traces migration of northernmost principal source terrain through time. C = central Appalachian highlands; A = Adirondack highlands; N = New England Appalachian highlands; crosses indicate secondary lowland source terrains within the offshore basin complex (SH = continental shelf; SL = continental slope). Column 5 = primary dispersal routes for siliciclastic sediments, inferred from isochron maps of Poag (Chapter 6). J = ancient James River; P = ancient Potomac River; S = ancient Susquehanna River; SK = ancient Schuylkill River; D = ancient Delaware River; H = ancient Hudson River; C = ancient Connecticut River; M = ancient river(s) in eastern Massachusetts. Column 6 = approximate volumetric rates of siliciclastic sediment accumulation as calculated from isochron maps of Poag (Chapter 6). Column 7 = percent (by volume) of siliciclastic sediments that reached continental rise (from Poag, Chapter 6).

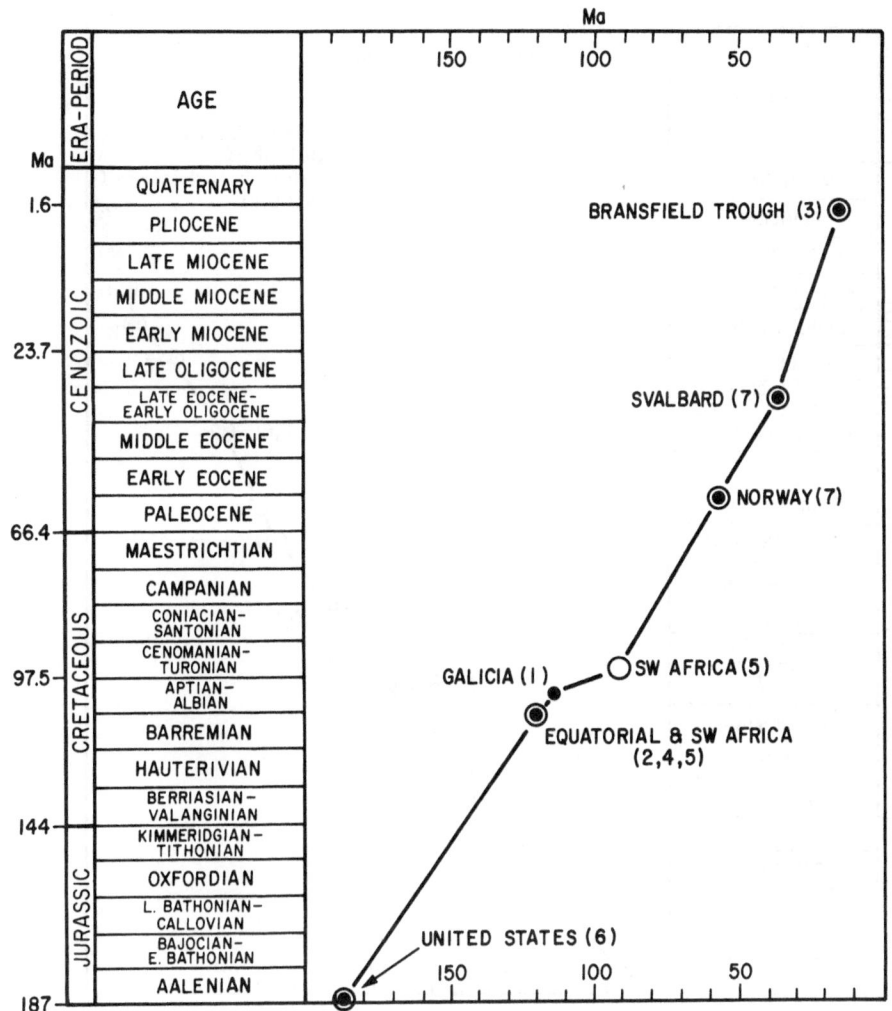

FIGURE 14-7. Time of initiation (DNAG time scale; Palmer, 1983) of seafloor spreading and postrift deposition (solid circles) and first indication of continental-rise development, where known (open circles), in areas studied for this volume (chapter numbers in parentheses).

Gabon–Congo margin (Fig. 14-8). Turbidite deposition, which is particularly associated with the Santonian uplift, continued intermittently through the Maestrichtian. The Eocene and Miocene episodes of uplift and tilting are characterized by deep erosion of channels along the continental slope and increased deposition on the continental rise.

In the Angola Basin and associated basins, evaporite deposition also characterized the Aptian transition from continental rifting to seafloor spreading (Uchupi, Chapter 5; Figs. 14-1, 14-7). This evaporative interval was followed by rapid siliciclastic deposition in the basin and by carbonate production on the shelves. In the southern part of the Angola Basin, sufficient volumes of turbidites and debrites emanated from the Walvis Ridge and from the African continent to form an early continental rise. The alternation of black, green, and red claystones (pelagites and distal turbidites) in the oldest rise sediments attest to periodic

anoxia. Subsequently, three large submarine fans built out by a temporal succession of: (1) volcaniclastic detritus; (2) mixed shallow-water carbonates and volcaniclastics; and (3) muddy pelagic turbidites (rich in biogenic silica) and debrites.

Paleoceanographic influences on these rise deposits were marked. For example, a distinct carbonate decrease in middle Eocene strata is attributable to a shoaling CCD. Middle Eocene to late Oligocene unconformities and condensed deposits are attributable to increased thermohaline bottom circulation. Further evidence of vigorous bottom currents is inferred from the presence of sediment drifts. In addition, the Benguela Current developed in the late Miocene; its associated upwelling enriched the sediments in biogenic silica.

During the Miocene(?) massive loads of siliciclastic sediment depressed the Angola continental shelf and mobilized seaward flow of the basal postrift (or upper

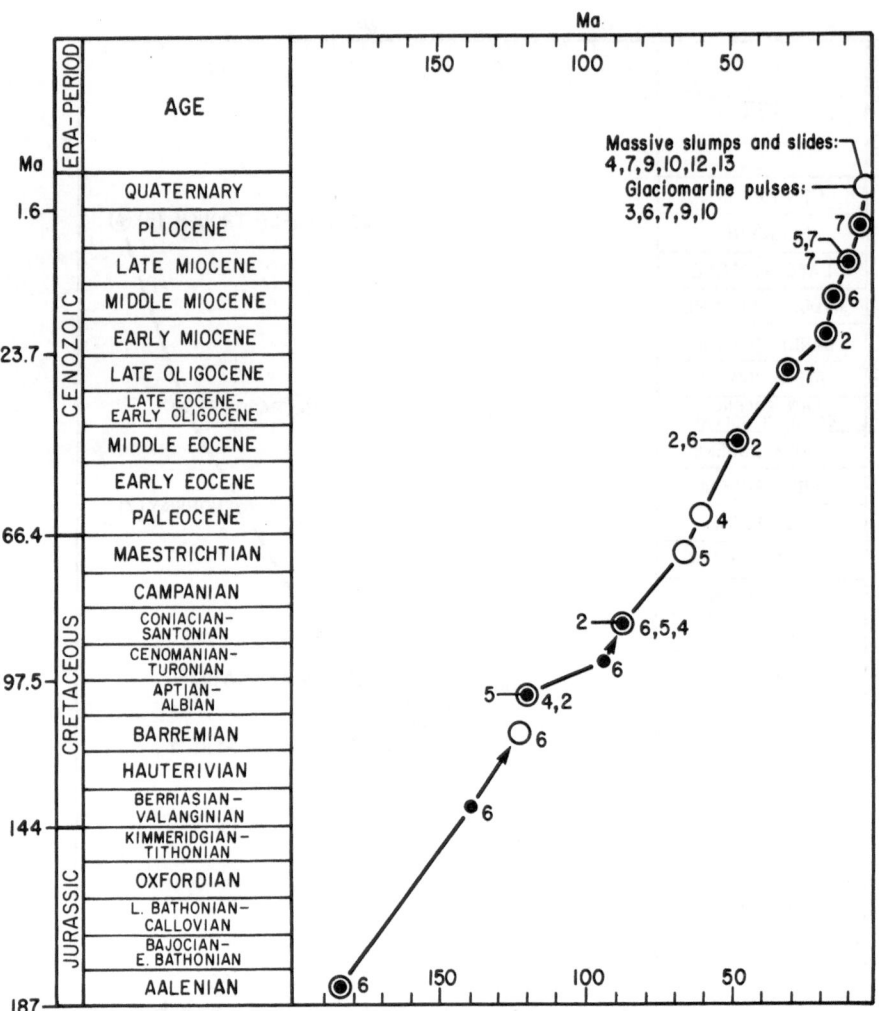

FIGURE 14-8.   Periods of significant source-terrain uplift (solid circles; DNAG time scale; Palmer, 1983) and pulses of siliciclastic deposition on continental rise (open circles) in areas studied for this volume. Bold numerals are chapter numbers. Arrows indicate delayed depositional pulse following uplift of source terrain.

synrift) evaporites to produce a huge diapir field under the slope and rise. The route of the Congo Canyon, incised into the surface of the Congo Fan (probably during glacial lowstands), is now controlled by the distribution of these diapiric barriers.

Off Southwest Africa, Dingle and Robson (Chapter 4; Figs. 14-1, 14-9) contend that no continental rise was present during the late Albian to late Cenomanian interval, though the shelf and slope prograded rapidly. But beginning in the Turonian, a continental rise developed by two different styles of accretion (Figs. 14-7, 14-9). Basement subsided rapidly in the Orange–Luderitz basin during the late Aptian to Maestrichtian, and the resultant accommodation space rapidly filled with sediment. During the Turonian to Oligocene interval, a wide continental rise (~ 200 km wide) was built of allochthonous debris (turbidites, debrites, slumps) derived from the continental slope. The mouth

of the Orange River incised the Cape Canyon during the Oligocene and built a large submarine fan on the continental slope and rise. Paleoclimatic and paleoceanographic effects began to control rise deposition in the late Miocene, with the onset of coastal aridity and vigorous scour by Antarctic Bottom Water. These conditions minimized the dispersal of terrigenous detritus to the rise during the Neogene and Quaternary. Nevertheless, massive slumping continued to transport siliciclastic sediments from the slope to the rise, especially during the Quaternary (Fig. 14-8).

Farther south, in the Columbine–Agulhas region (Fig. 14-9), the continental margin subsided at a significantly slower rate. Late Cretaceous to Oligocene deposition on the rise took the form of large slump blocks. Neogene and Quaternary deposition here, as to the north, was largely limited by the vigor of AABW erosion.

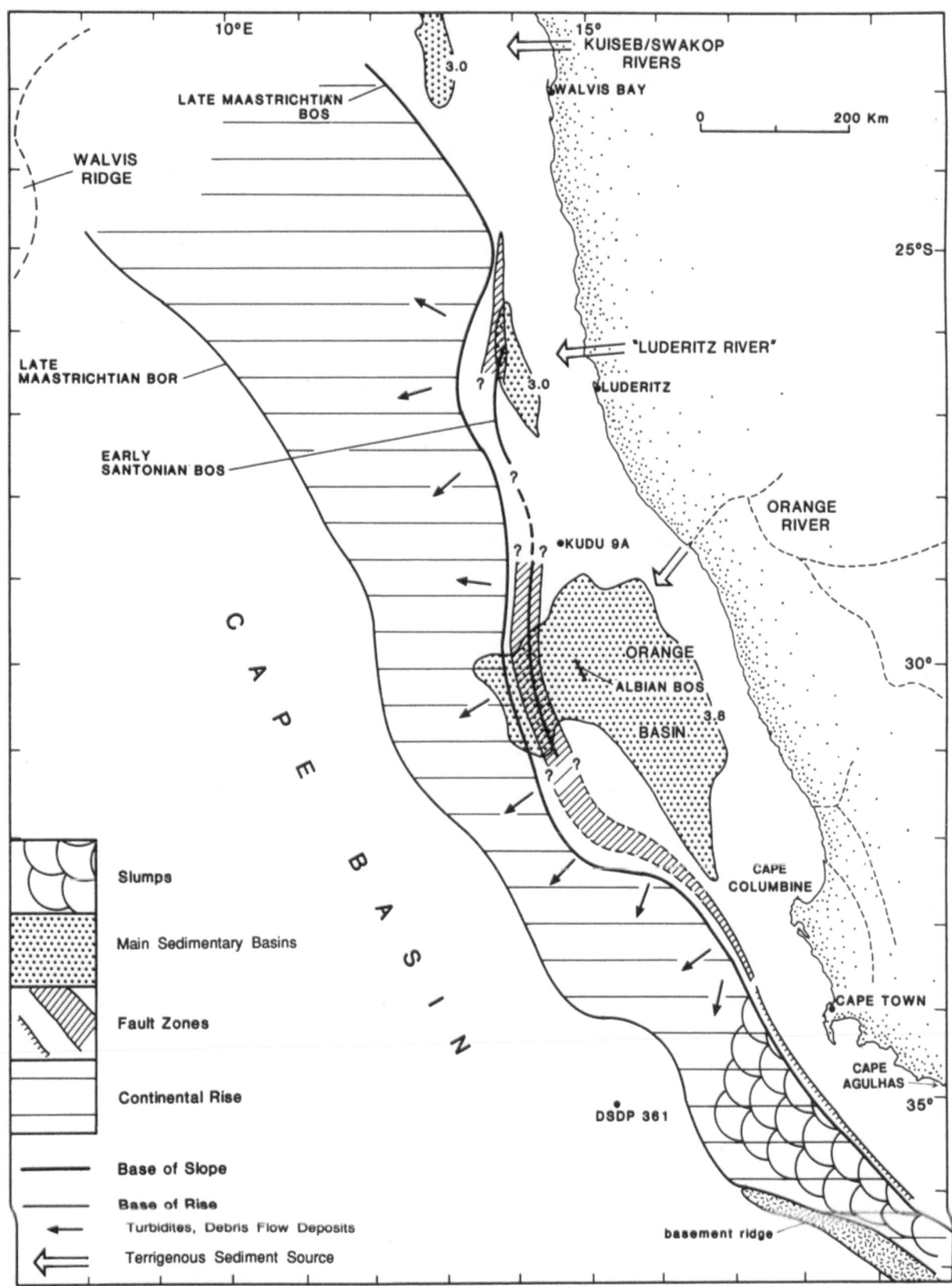

FIGURE 14-9.   Physiography and distribution of early postrift depositional and structural features of Southwest African margin. BOS = base of slope; BOR = base of rise; KUDU and DSDP = borehole sites. Maximum thickness of depocenters given in kilometers (from Dingle and Robson, Chapter 4).

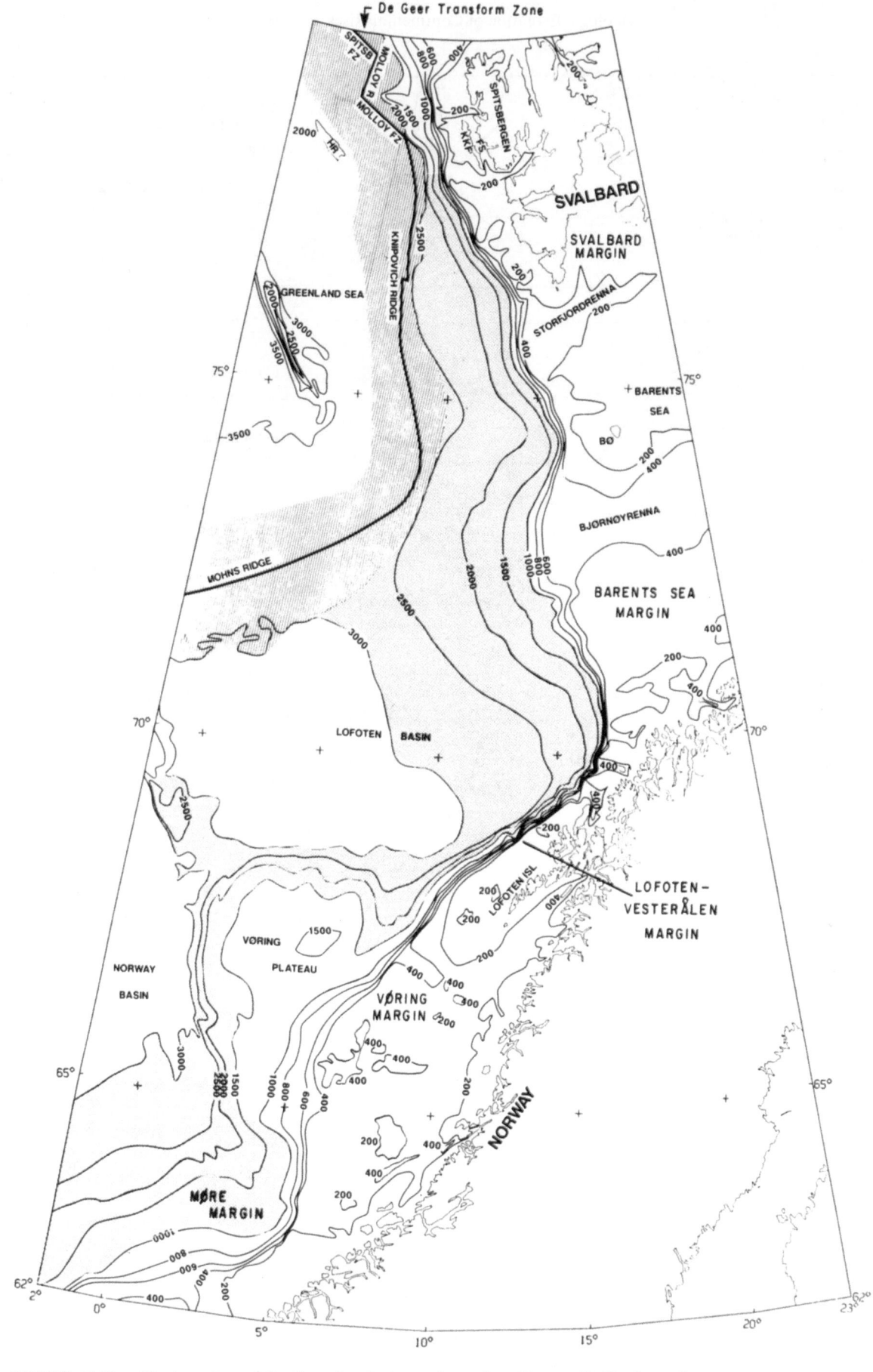

FIGURE 14-10. Physiography and location of geologic provinces along Norway–Svalbard margin (from Myhre et al., Chapter 7). Shaded region between ~ 1,000 and 3,000 m represents continental rise as recognized by Emery and Uchupi (1984); we agree with this interpretation.

On the high-latitude Norway–Svalbard margin, there is disagreement over the distinction of continental rises (Figs. 14-1, 14-10). Emery (1980), Birkenmajer (1981), Vogt et al. (1981), Kennett (1982), and Emery and Uchupi (1984; with whom we agree) recognized linear continental rises on the landward margins of the Norway and Lofoten basins, but Myhre et al. (Chapter 7) include these regions as parts of continental slopes and abyssal plains. Regardless of this nomenclatural problem, we know that postrift sedimentation in those areas began in the Eocene and Oligocene (Fig. 14-7), following a complex history of multiple Paleozoic and Mesozoic rifting, which culminated in a 3-m.y. interval of thoeliitic lava extrusion. The basins of the Norway margin formed as the seafloor diverged from a linear spreading center, but the region off Svalbard formed as a result of regional shearing along the De Geer Transform Zone (Harland, 1969; Chapter 7). Thereafter, structurally undisturbed postrift sediments accumulated on thermotectonically subsiding oceanic crust.

Myhre et al. (Chapter 7) divide the Norway–Svalbard margin into five separate provinces (Møre margin, Vøring margin, Lofoten–Vesterålen margin, Barents Sea margin, and Svalbard margin; Fig. 14-10), each with its individual structural and depositional history. The base of the postrift section in the Møre and Vøring provinces is marked by lower Eocene ash beds (Fig. 14-7). A stratigraphic framework derived from DSDP and ODP drilling can be extrapolated along multichannel seismic profiles (Skogseid and Eldholm, 1989) to infer that lower Eocene strata consist largely of detritus eroded from the antecedent thoeliitic lavas. Middle Eocene to middle Oligocene strata are zeolitic claystones, which also include erosional products from adjacent basement highs. In the Miocene, the shelf began to prograde rapidly as the result of large-scale flexural uplift of the continent and the inner part of the continental shelf. On the Møre margin, progradation culminated with deposition of 2 km of Pliocene clay (now shale), but progradation continued into the Pleistocene on the Vøring margin. Middle Miocene lithofacies are mainly diatomaceous muds and oozes, whereas upper Pliocene and Quaternary deposits are chiefly coarse glacial and interglacial sediments.

A striking feature of the northern Vøring Plateau is a province of piercement diapirs formed by Eocene shales (Caston, 1976; Eldholm, Thiede, and Taylor, 1989). Huge downslope mass movements of sediment characterize the modern continental slope/rise in these two provinces (Fig. 14-8).

On the Lofoten–Vesterålen margin (Fig. 14-10), the low number of seismic profiles and lack of drill sites provide insufficient data to develop a basinwide stratigraphic framework. In general, however, Cenozoic strata of the slope and rise comprise a lower stratified sequence and an upper homogeneous sequence. Seismic profiles show evidence of widespread erosion, slumping, and sliding.

In the Barents Sea province (Fig. 14-10), postrift Cenozoic deposits form the thickest depocenters (5–7 km) of the Norway–Svalbard region. Eocene and early Oligocene volcanism deposited pyroclastic sequences there. Variable uplift along this margin during the Eocene caused both siliciclastic and volcaniclastic wedges to prograde westward.

Since the middle Oligocene (~ 30 Ma), the production and subsidence of oceanic crust throughout the Barents Sea province has accommodated a relatively undeformed wedge of siliciclastic deposits. About half the wedge is believed to be constructed of late Miocene (~ 5.5 Ma) and younger sediments (Fig. 14-8); they are capped by a seaward prograding glacigenic blanket. Large volumes of these sediments have been displaced by mass movement to the slope and rise. The late Miocene acceleration of siliciclastic deposition is attributable to tectonic uplift of eastern source terrains (Fig. 14-8). During the Pliocene and Pleistocene, however, glacial activity amplified the extent and intensity of erosion and deposition.

On the Svalbard margin (Fig. 14-10), postrift depocenters also are relatively thick (4–5 km). Three depositional sequences have been identified here: a Miocene–Oligocene basal sequence (Fig. 14-7); a Pliocene middle sequence (characterized by downslope mass movements); and a Pliocene–Pleistocene upper sequence (containing mainly glacial sediments). Neogene and Quaternary strata predominate in the northern part of the province.

QUATERNARY EROSION AND DEPOSITION: SEAFLOOR FEATURES AND PROCESSES

The enormous data base arising from increasingly rapid, broad-swath imaging techniques is straining our ability to digest and interpret what we see on the seafloor. In fact, new revelations are appearing so rapidly (amply exemplified in this volume) that a fundamental reevaluation of concepts and terminology of deep-ocean sedimentation is taking place (Bouma, Normark, and Barnes, 1985; Normark, Barnes, and Bouma, 1985; Mutti and Normark, 1987). The bias imposed by a given type or scale of imaging (or by a given expert) is clearly manifest in Part II of this volume (Chapters 8–13). Features, for example, that appear to have a certain structure or origin, as determined from two-dimensional seismic profiles or echograms, may require a different interpretation when imaged by GLORIA or TOBI. Actual sampling is requisite for firm documentation, but even core data can confuse the issue if sampling does not sufficiently represent all sedimentary regimes and lithofacies.

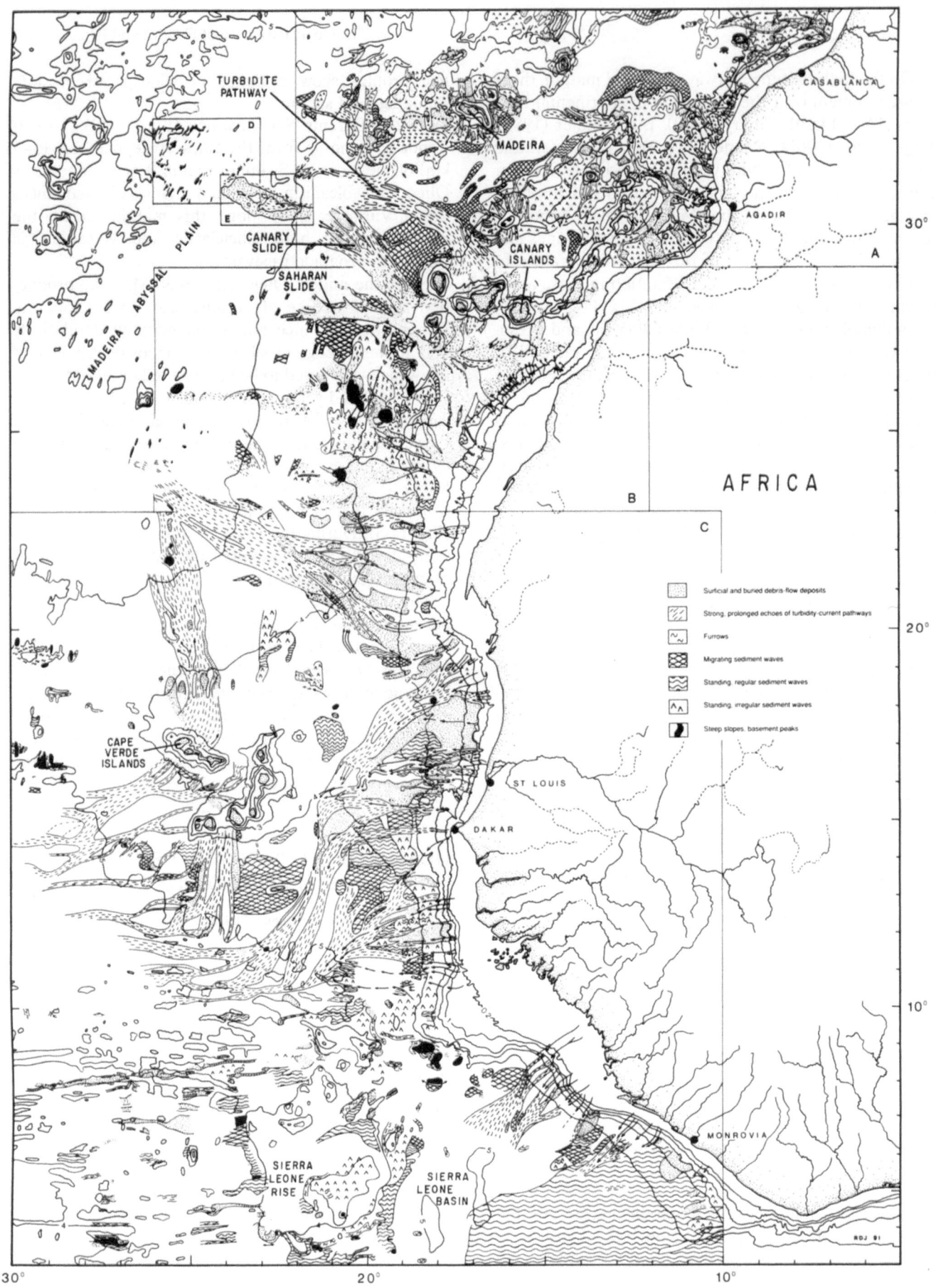

FIGURE 14-11. Physiography and seafloor depositional features of the Northwest African margin (from Jacobi and Hayes, Chapter 12).

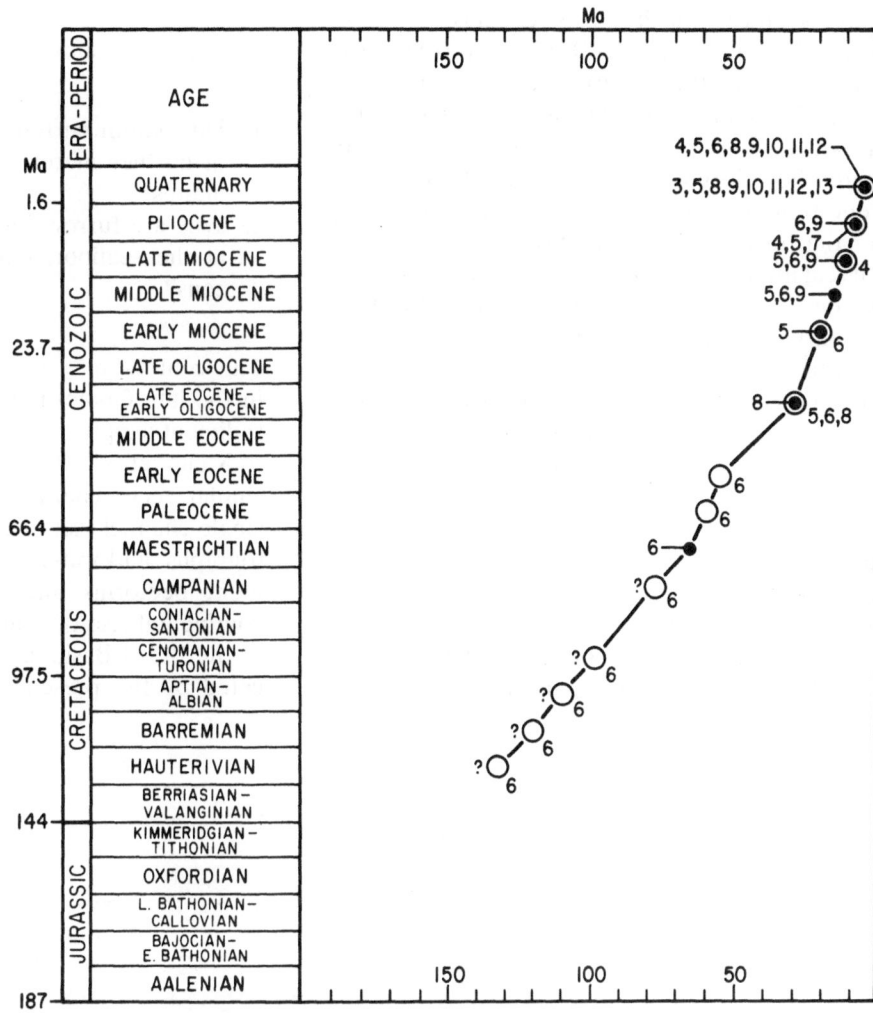

FIGURE 14-12. Periods of bottom-current activity (DNAG time scale; Palmer, 1983) noted in areas studied for this volume. Solid circles = depositional features; open circles – erosional features. Bold numerals are chapter numbers.

Echo-Character Analysis

The most widely applied method of studying Quaternary sedimentary features (bedforms) and processes on continental rises is *echo-character analysis* (Embly, 1975; Damuth, 1980; Laine, Damuth, and Jacobi, 1986). Since the early 1960s, high-frequency (3.5–12 kHz) echograms have been used extensively for this purpose. Correlation of echo-character analysis with current-meter data, sediment cores, and sidescan sonar imagery has been a highly successful method of understanding the distribution and origin of deep-sea bedforms (ripples, furrows, sediment waves). Jacobi and Hayes (Chapter 12), who have made many fundamental contributions to this aspect of continental-rise sedimentology (e.g., Jacobi and Hayes, 1982), analyze herein a huge set of high-resolution seismic data collected along the slope and rise of Northwest Africa (Figs. 14-1, 14-11). In this region, sedimentation is complicated by the presence of seamounts and large islands, which serve as both source terrains and barriers to sediment dispersal. At the heart of Chapter 13 and the other chapters of Part II, is the necessity to distinguish between, and to evaluate the implications of, two primary sedimentary processes: (1) alongslope thermohaline bottom currents (which produce erosional furrows, regional unconformities, ripples, sand waves, and contourite drifts); and (2) downslope sediment gravity flows (which produce turbidites, debrites, slumps, slides, submarine fans, aprons, canyons, and channels).

Massive sediment slides appear to have dominated continental-rise deposition off Northwest Africa in the Quaternary, and some can be traced for more than 1,000 km from the continental slopes to the abyssal plains (Figs. 14-1, 14-11). Thus Jacobi and Hayes assert that the traditional concept of slide deposits collecting at the base of continental slopes is of limited use in interpreting ancient slide deposits.

Contour-current activity along Northwest Africa has been largely attributed to Antarctic Bottom Water (at deeper localities) and North Atlantic Deep Water (at shallower localities), which have swept this area since at least the Miocene (Fig. 14-12). Linear furrows, usually attributed to these currents, are particularly widespread, but a question remains as to whether they were produced under a modern flow regime, or by accelerated flow during Pleistocene glacials. The overriding question raised by Jacobi and Hayes, however, is whether the furrows were really produced by thermohaline circulation at all. Could they have been formed, instead (as hydrographic data indicate), by tide-induced currents that interacted with flow-restricting seamount barriers?

Jacobi and Hayes distinguish sediment waves generated by turbidity currents on the lower continental rise, from those generated by thermohaline currents, by the orientation of wave crest lines (turbidity currents produce downslope orientations in this region; thermohaline currents produce alongslope orientations). These authors show, furthermore, that turbidity currents do not necessarily fan out on the lower continental rise, as some models would require, but may maintain flow parallel to the turbidity-current pathway.

In the opinion of Jacobi and Hayes, two additional important questions remain unanswered:

1. Do contour currents divert active turbidity currents, or do they merely rework previously emplaced turbidites?
2. Why are furrow-generated hyperbolae so prevalent at the seafloor, but absent from subbottom horizons?

This contrast may be evidence of higher flow velocities in the late Quaternary, and(or) of reduced downslope sediment supply.

Mello et al. (Chapter 8; Figs. 14-1, 14-13) used similar echo-character analysis (supplemented by single-channel seismic profiles, cores, bottom photographs, and water mass properties) to examine sediment bedforms and sedimentary processes on the continental rise of southern Brazil. Sediment dispersal in the Brazil Basin, as off Northwest Africa, is complicated by the presence of seamounts; the São Paulo

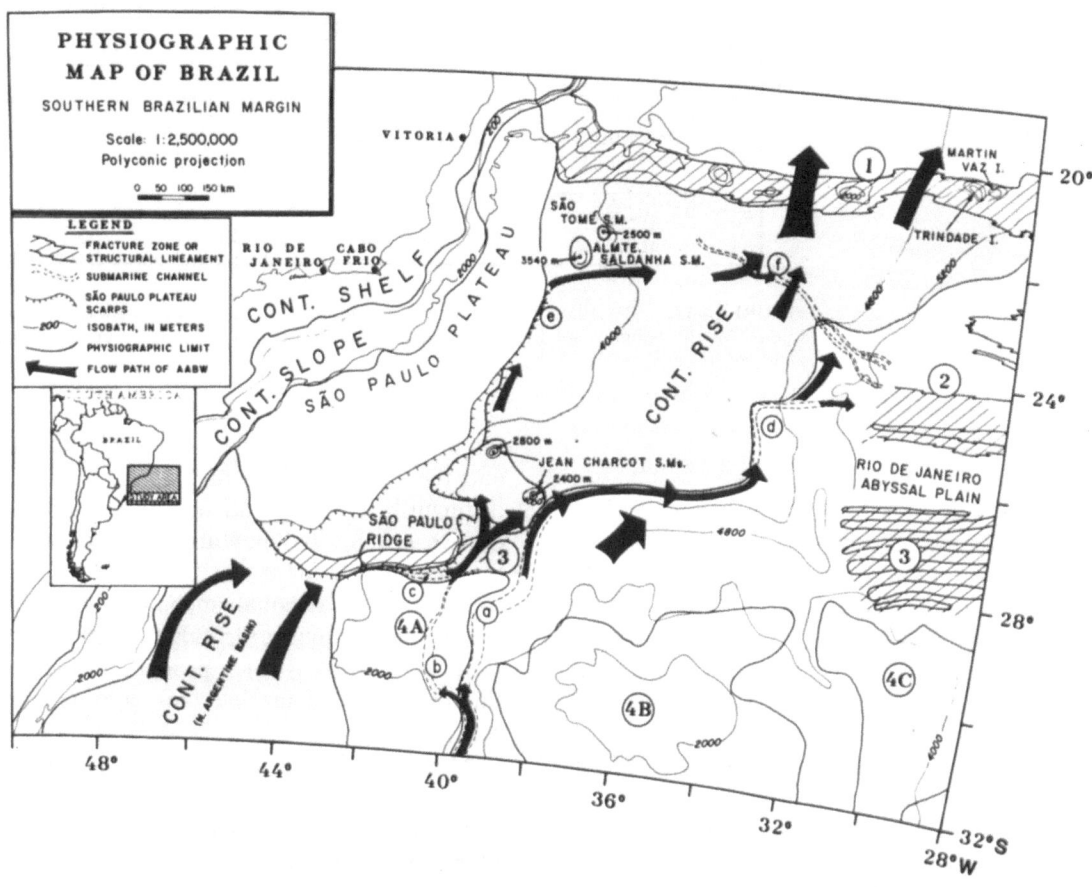

FIGURE 14-13.   Physiographic and oceanographic features of southern part of Brazil Basin. AABW = Antarctic Bottom Water; 1 = Vitoria–Trinidade Seamount Chain; 2 = Rio de Janeiro lineament; 3 = Florianopolis lineament; a = Vema Channel; b = secondary branch of Vema Channel; c = channel at base of São Paulo Ridge; d = Rio de Janeiro Channel; e = Guanabara Channel; f = Columbia Channel; cont. = continental; s.m. = seamount (from Mello et al., Chapter 8).

Plateau and the Rio Grande Rise also stand high above the surrounding seafloor.

In the northern part of the Brazil study area, deep-sea channels incise the rise, both parallel to and perpendicular to bathymetric contours. Mello et al. take these orientations as evidence of alongslope and downslope erosional processes, respectively. The southern part of the study area, however, exhibits only alongslope channels, which are attributed to the northward flow of Antarctic Bottom Water (Figs. 14-1, 14-13), the dominant geological agent in this region (Biscaye, 1965; Lawrence, 1979; Damuth and Hayes, 1977; Kumar et al., 1979).

A long geologic history of drift development in this area (beginning in the Oligocene?) has produced a variety of *plastered*, *double*, and *separated* drifts in the southern part of the Brazil Basin (terminology of McCave and Tucholke, 1986). Mello et al. found such an abundance of drifts to be puzzling, because most terrigenous sediment derived from the continental slope is either trapped on the São Paulo Plateau, or bypasses the plateau and continental rise to reach the Rio de Janeiro Abyssal Plain (Fig. 14-13). The profusion of drifts, thus requires a substantial alternate source of sediment. Mello et al. suggest that suspended fine-grained particles may be derived from the Argentine Basin to the south.

In contrast, drift deposits are absent in the northern part of the Brazil Basin. Apparently, a large influx of sediment through the Columbia Channel (Fig. 14-13) prevents drift formation in this area. Downslope mass movements are common in this area, however, and they also characterize the flanks of the Rio Grande Rise.

Sidescan Sonar Mosaics

With the introduction of advanced sidescan sonar devices, such as Sea MARC, GLORIA, and TOBI, and multibeam bathymetric mapping systems (SEABEAM), linked with digital shipboard data collection and computer-assisted image enhancement, seafloor imaging can now be acquired rapidly in continuous swaths. Mosaics of such images eliminate the data gaps left between seismic survey lines or core stations, and provide three-dimensional perspectives of seabed features (and even of shallow subseabed features).

O'Leary and Dobson (Chapter 9; Fig. 14-1) apply these techniques to explicate large slide complexes on the slope and rise of southeastern New England (Fig. 14-14). Previous studies concluded that this margin accreted since the Miocene by sediment-drift deposition and, presumably, by submarine-fan construction associated with the Hudson Canyon. O'Leary and Dobson demonstrate, however, that beginning in the late Pliocene, massive downslope sediment transport and deposition have built a distinctive terrace on the upper rise; the terrace, in turn, has helped to control patterns of downslope erosion and deposition.

Though thermohaline bottom currents are modifying the physiography of this region at present, O'Leary and Dobson conclude that alongslope effects have been minor compared to those of recurrent slope failure and mass movement (Fig. 14-8). Open-slope slab slides have removed strata (10–30 m thick) from two principal source terrains on the middle continental slope, and have spread it more than 200 km downslope (Fig. 14-14). Much of the Pleistocene section has been excavated from the slope, and beds as old as Eocene are exposed in the slide scars (white Eocene chalk forms conspicuous clasts in Quaternary debrites of the continental rise).

O'Leary and Dobson speculate that these massive slope failures were triggered by epicentrally clustered earthquakes, which acted on sediments preconditioned by interstratal deformation. Such deformation may have been produced, in part, by glacioeustatic falls, which might allow expansion of gas hydrates shallowly buried on the slope and rise. These authors conclude that all Quaternary debrites now on the rise came from failures of the adjacent continental slope, which recurred at 100-yr to 1,000-yr intervals. A generation of post-slide submarine canyons seems to be unrelated genetically to this mass movement.

Without adequate core samples to determine sedimentary and geotechnical properties, and to document contact relationships (particularly at the base of slides) the nature, composition, and origin of the debrites remain speculative. O'Leary and Dobson venture the opinion, however, that during displacement, the debris did not mix thoroughly with seawater; downslope transfer took place as viscous flow, whose motion depended on bulk material properties and momentum. They conclude that mass movement and canyon erosion originated on the continental slope, independent of events taking place on the continental shelf or in continental source terrains. (One might consider, however, the possible effects on slope stability, should a glacioeustatic lowstand move the shoreline to the shelf edge or upper slope.) In the view of O'Leary and Dobson, the sedimentary character, origin, and distribution of continental-rise debrites off New England do not fit established models of deep-sea deposition, but rather, constitute a separate and distinct delivery and deposition system.

A little farther north, off Nova Scotia, Hughes Clarke, O'Leary, and Piper (Chapter 10; Fig. 14-1) also find evidence of extensive mass wasting on the slope and accumulation of debrites on the continental rise. A GLORIA mosaic reveals that downslope-oriented

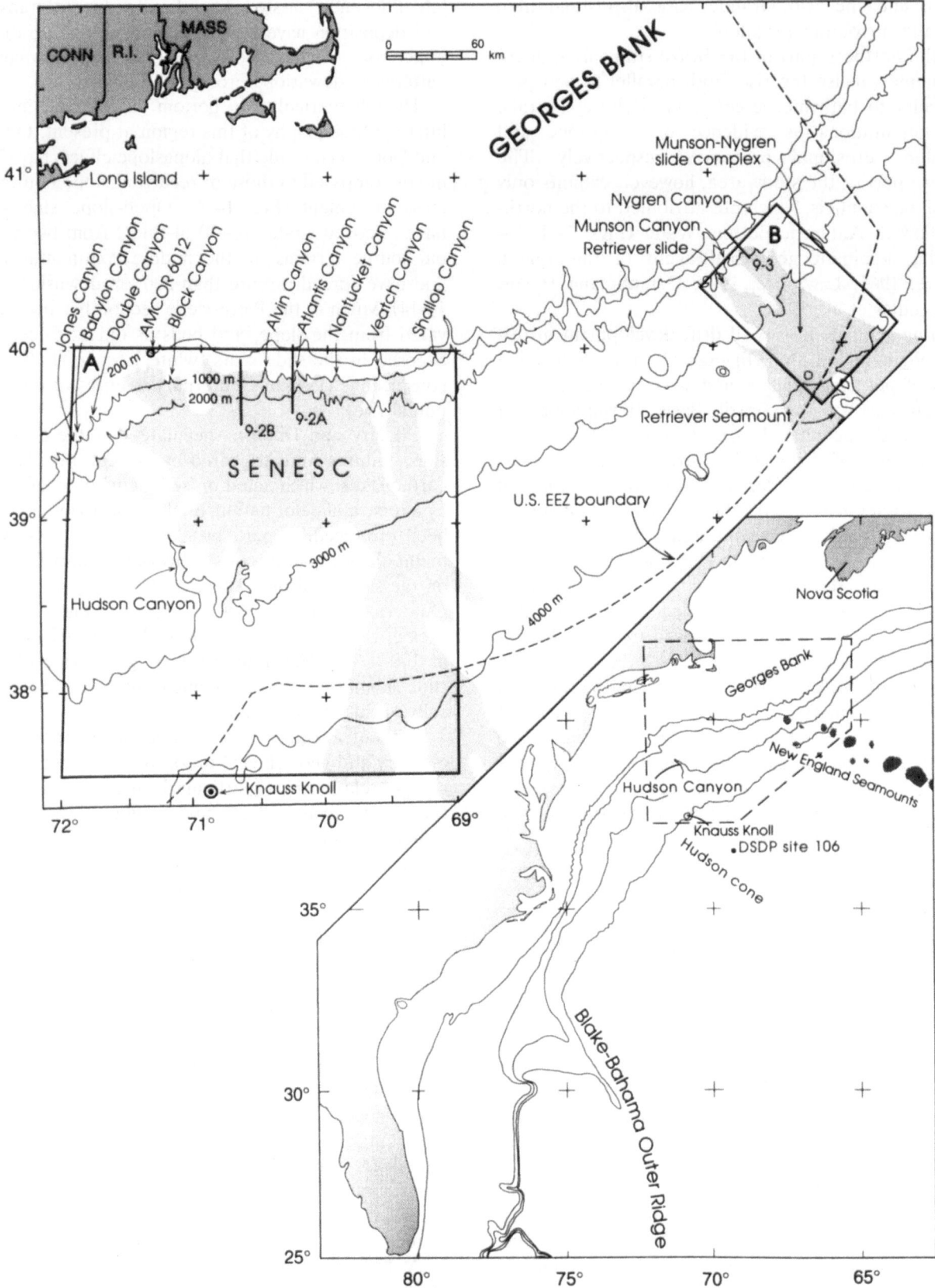

FIGURE 14-14.   Physiography and depositional/erosional features of the southeastern New England margin. EEZ = exclusive economic zone; AMCOR 6012 = Atlantic Margin Coring Project corehole; SENESC = Southeastern New England Slide Complex (from O'Leary and Dobson, Chapter 9).

landforms (channels, slides, debrites) predominate, and that alongslope features are conspicuously lacking, in contrast to earlier interpretations (e.g., Driscoll, Tucholke, and McCave, 1985). Hughes Clarke and his colleagues conclude that the streamlike debrites, slides, and lower rise channels were active in the Pleistocene (and perhaps were generated then), but have been inactive since about 12 ka, when thermohaline deposition increased. Thus, on a short time scale (Holocene), alongslope processes have left a significant imprint in this area, but in the long term (Pleistocene), downslope mechanisms have determined the gross morphological and sedimentary features of this segment of the slope and rise.

Masson et al. (Chapter 13; Figs. 14-1, 14-11) report the first results from an ongoing study, which emphasizes sidescan sonar imaging of the Saharan continental rise, a small portion of the area summarized by Jacobi and Hayes (Chapter 12). Masson et al. used a new deep-towed package (TOBI) to collect 30-kHz sidescan sonar images and 7.5-kHz seismic profiles. They correlate these data with a GLORIA mosaic and a 3.5-kHz echo-character analysis, and with a carefully sited suite of cores.

The results showed that most of the survey area is covered by debrites buried beneath 50 cm to 2 m of turbidites and hemipelagites. A significant proportion of the debrites originated as a single depositional event (Masson et al. name it the Canary Slide), which emanated from the gentle western slope ($\sim 1°$ declivity) between the islands of La Palma and Heirro (Fig. 14-11). The Canary Slide commenced in unusually deep water (4,000–4,500 m depth), and spread more than 600 km onto the eastern fringe of the Madeira Abyssal Plain. On its way, the slide debris took advantage of subtle topographic lows; several narrow debrite tongues can be seen to follow partly buried channels. Surprisingly, no head scarps were observed, which contrasts markedly with the 80-m-high scarps of the nearby Saharan Slide.

Masson et al. also document an important turbidity-current pathway between the island of Madeira and the Canaries (Fig. 14-11), through which volcaniclastic sands have traversed the rise on their way to the Madeira Abyssal Plain. An abundance of open and partly filled channels scar the pathway, some of which are 50 m deep and 1 km wide; few, however, extend for significant distances downslope.

Masson et al. point out some of the difficulties in correlating geophysical data (GLORIA, TOBI) with sedimentological (core) data. For example, though numerous channels are imaged in the pathway by GLORIA, the most recent series of turbidity currents crossed the pathway without any obvious erosional or depositional trace of their passage [10 turbidites were deposited on the Madeira Abyssal Plain in the last 500,000 yr (Weaver et al., in press), but cores show that the last 500,000 yr of deposition in the pathway is represented by hemipelagites]. This relationship implies either that the pathway channels (imaged by GLORIA) were eroded by an older series of slides, or that core coverage is still not sufficiently dense to characterize the entire pathway.

Masson et al. assert that the regional sedimentation regime of the Canaries rise embraced long periods of hemipelagic deposition modified by contour currents and interrupted occasionally by catastrophic slides and turbidite deposition. Like Jacobi and Hayes (Chapter 12), Masson et al. believe that seafloor topography exerts a strong influence on the distribution of sedimentary facies (turbidites, debrites, hemipelagites). The seamounts, in particular (like Galicia Bank and the Bransfield spreading ridge), form effective barriers to coarse sediment transport; hemipelagites predominate in their downslope "shadows."

The trigger(s) for slope failure in this region is not definitely known, but Masson et al. postulate that earthquakes related to eustatic change seem a likely cause (previous workers have demonstrated a strong correlation between turbidite deposition in this area and changes in paleoclimate and sea level).

CONCLUSIONS

Atlantic continental rises are typically gently sloping, seaward-thinning wedges of mainly siliciclastic detritus, which collect at and near the base of continental slopes (Fig. 14-2). Accommodation space for rise deposits forms chiefly through the landward divergence and thermotectonic subsidence of new seafloor from elevated mid-ocean spreading axes (Fig. 14-3). On such divergent margins, rises form thick depocenters (Fig. 14-1), provided continental source terrains are high and paleoclimate is seasonally wet/dry or glacial. Most of the sediment appears to collect there during eustatic lowstands.

In contrast, on margins formed largely by shearing along transform zones, or where paleoclimate has been predominantly arid or ever-wet, or where continental source terrains tend to be low and quiescent, continental-rise wedges are small and thin (Fig. 14-1).

Nonetheless, even at times when fresh supplies of continental detritus are sparse, large volumes of sediment may be transported to continental rises from nearby islands, from submerged seamounts, ridges, and plateaus, and especially, from massive failures of continental slopes and shelf edges (Fig. 14-8).

Regulating agents for continental-rise deposition, such as paleogeography, paleoclimate, cratonic tectonism, seafloor spreading rates, water-mass properties, abyssal circulation, eustacy, and a host of other mechanisms, have varied dramatically during the ~ 187 m.y.

of Atlantic rise evolution. Sediment composition, depositional processes, and seabed morphology of continental rises have varied in response.

On a megascale (millions or tens of millions of years), the most important regulator of rise construction appears to have been the volumetric rate by which siliciclastic detritus reached bathyal and abyssal depths. On most margins, this rate depended heavily on the rates and duration of uplift in bordering continental highlands (Fig. 14-8), and upon the presence of paleoclimates amenable to deep weathering and rapid, long-distance dispersal of the resultant detritus. Eustacy was a secondary regulator, often (but not always) catalyzing sediment delivery to the rise during lowstands. Large-volume, downslope dispersal and deposition mechanisms (sediment gravity flows) pervaded on this megascale, and built rises dominated by slumps, slides, debrites, and turbidites.

On a microscale (hundreds or thousands of years), oceanic current systems have been important sediment dispersal mechanisms; they have eroded, entrained, and deposited fine-grained siliciclastic and biogenic particles, and created rises of dominantly pelagites, hemipelagites, and contourites. The carbonate compensation depth played a secondary role by controlling the volume of pelagic carbonate incorporated into the rise. Megascale and microscale processes have not necessarily been mutually exclusive, however; they often interacted to produce complicated sedimentary facies and seafloor morphology.

The general characteristics and gross interrelationships of these processes and their products are adequately known for the late Quaternary interval, though strongly opposing interpretations exist regarding the implications and relative importance of various processes and features (especially regarding detailed terminology). The geologically extreme aspects of Quaternary paleoceanography and paleoclimate, however, and its geologically brief duration, impede direct extrapolation of its continental-rise construction (processes and products) into the geologic past (especially farther back than the Neogene). Only cautiously general comparisons can be made between deeply buried Jurassic and Cretaceous rise features and those of the Quaternary seafloor.

Computer enhancement of advanced sidescan sonar imagery and multibeam bathymetric data, coupled with high-resolution seismic profiles and precision coring, and tied to improved satellite navigation systems, promise another quantum leap in the explication of modern and late Pleistocene continental rises. To narrow the analytical gap between Quaternary features and geologically older ones, however, will require new imaging devices that provide accurate three-dimensional geometries and increase the resolution of relatively small features at great depths of burial. Above

all, deeper and more densely spaced coring into continental-rise depocenters is imperative.

Modern geological and geophysical investigations, as amply manifest by the contributors to this volume, have firmly established that the bathyal and abyssal seafloor is by no means the simple geologic and physiographic province envisioned by Thomas Huxley. No doubt, Huxley would have been chagrined to learn how badly he underestimated the morphological, structural, and depositional complexity, and the temporal and spatial variability of the Atlantic Ocean's "prodigious plain."

ACKNOWLEDGMENTS

We are grateful to all contributors to this volume for providing such diverse data and stimulating interpretations for Atlantic continental rises, and for allowing us to use selected figures from their chapters. Elazar Uchupi, Peter Popenoe, Richard Dingle, Robert Jacobi, and Annik Myhre kindly reviewed an earlier version of this chapter.

REFERENCES

Barker, P. F., and Dalziel, I. W. D. 1983. Progress in geodynamics in the Scotia Arc region. *In* Geodynamics of the Eastern Pacific Region, Caribbean, and Scotia Arcs, S. J. R. Cabre, Ed. *American Geophysical Union Geodynamics Series* 9:137–170.

Birkenmajer, K. 1981. The geology of Svalbard, the western part of the Barents Sea, and the continental margin of Scandinavia. *In* The Ocean Basins and Margins, Volume 5, The Arctic Ocean, A. E. M. Nairn, M. Churkin, Jr., and F. G. Stehli, Eds.: New York: Plenum Press, 265–330.

Biscaye, P. E. 1965. Mineralogy and sedimentation of Recent deep-sea clay in the Atlantic Ocean and adjacent seas and oceans. *Geological Society of America Bulletin* 76:803–832.

Boillot, G., Winterer, E. L., Meyer, A. W., Applegate, J., Baltuck, M., Bergen, J. A., Comas, M. C., Davies, T. A., Dunham, K., Evans, C. A., Girardeau, J., Goldberg, D., Haggerty, J. A., Jansa, L. F., Johnson, J. A., Kasahara, J., Loreau, J.-P., Luna, E., Moullade, M., Ogg, J. G., Sarti, M., Thurow, J., and Williamson, M. A. 1988. Proceedings of the Ocean Drilling Program, Scientific Results, Volume 103: College Station, TX: Ocean Drilling Program.

Bond, G. C., and Kominz, M. A. 1988. Evolution of thought on passive continental margins from the origin of geosynclinal theory (~ 1860) to the present. *Geological Society of America Bulletin* 100:1909–1956.

Bouma, A. H., Normark, W. R., and Barnes, N. E. 1985. COMFAN: Needs and initial results. *In* Submarine Fans and Related Turbidite Systems, A. H. Bouma, W. R. Normark, and N. E. Barnes, Eds.: New York: Springer-Verlag, 7–14.

Caston, V. N. D. 1976. Tertiary sediments of the Voring Plateau, Norwegian Sea, recovered by Leg 38 of the Deep Sea Drilling Project. *In* Initial Reports of the Deep Sea Drilling Project, Volume 38, M. Talwani, G. Udintsev, et al.: Washington, DC: U.S. Government Printing Office, 761–782.

Damuth, J. E. 1980. Use of high-frequency (3.5–12 kHz) echograms in the study of near-bottom sedimentation processes in the deep sea: A review. *Marine Geology* 38:51–75.

Damuth, J. E., and Hayes, D. E. 1977. Echo character of the east Brazilian continental margin and its relationship to sedimentary processes. *Marine Geology* 24:74–95.

Diebold, J. B., Stoffa, P. L., and the LASE Study Group 1986. A large aperture experiment in the Baltimore Canyon Trough. *In* The Geology of North America, Volume I-2, The Atlantic Continental Margin, U.S., R. E. Sheridan and J. A. Grow, Eds.: Boulder, CO: Geological Society of America, 387–398.

Driscoll, M. L., Tucholke, B. E., and McCave, I. N. 1985. Seafloor zonation in sediment texture on the Nova Scotian lower continental rise. *Marine Geology* 66:25–41.

Dunbar, J. A., and Sawyer, D. S. 1989. Effects of continental heterogeneity on the distribution of extension and shape of rifted continental margins. *Tectonics* 8:1059–1078.

Eldholm, O., Thiede, J., and Taylor, E. 1989. Evolution of the Vøring volcanic margin. *In* Proceedings of the Ocean Drilling Program, Scientific Results, Volume 104, O. Eldholm, J. Thiede, et al.: College Station, TX: Ocean Drilling Program, 1033–1065.

Eldholm, O. 1991. Magmatic–tectonic evolution of a volcanic rifted margin. *In* Evolution of Mesozoic and Cenozoic Continental Margins, A. W. Meyer, T. A. Davies, and S. W. Wise, Eds. *Marine Geology* special issue 102:43–62.

Embly, R. W. 1975. Studies of deep-sea sedimentation processes using high-frequency seismic data. Doctoral Thesis, Columbia University, New York.

Emery, K. O. 1980. Continental margins—Classification and petroleum prospects. *American Association of Petroleum Geologists Bulletin* 64:297–315.

Emery, K. O., and Uchupi, E. 1984. *The Geology of the Atlantic Ocean*: New York: Springer-Verlag.

Graciansky, P. C. de, Poag, C. W., Cunningham, R., Jr., Loubere, P., Masson, D. G., Mazzullo, J. M., Montadert, L., Müller, C., Otsuka, K., Reynolds, L., Sigal, J., Snyder, S., Townsend, H. A., Vaos, S. P., and Waples, D. 1985. Initial Reports of the Deep Sea Drilling Project, Volume 80: Washington, DC: U.S. Government Printing Office.

Harland, W. B. 1969. Contribution of Spitsbergen to understanding of tectonic evolution of North Atlantic region. *In* North Atlantic: Geology and Continental Drift, M. Kay, Ed., *American Association of Petroleum Geologists Memoir 12*, 817–851.

Huxley, T. H. 1967. *On a Piece of Chalk*: New York: Charles Scribner's Sons.

Jacobi, R. D., and Hayes, D. E. 1982. Bathymetry, microtopography, and reflectivity characteristics of the West African margin between Sierra Leone and Mauritania. *In* Geology of the Northwest African Continental Margin, U. von Rad, K. Hinz, M. Sarnthein, and E. Seibold, Eds.: Berlin: Springer-Verlag, 181–212.

Jansa, L. F., and Wiedmann, J. 1982. Mesozoic–Cenozoic development of the eastern North American and Northwest African continental margins: A comparison. *In* Geology of the Northwest African Continental Margin,

U. von Rad, K. Hinz, M. Sarthein, and E. Seibold, Eds.: Berlin: Springer-Verlag, 215–272.

Jeffers, J. D., and Anderson, J. B. 1990. Sequence stratigraphy of the Bransfield Basin, Antarctica: Implications for tectonic history and hydrocarbon potential. *In* Antarctica Exploration Frontier—Hydrocarbon Potential, Geology, and Hazards, Bill St. John, Ed., *American Association of Petroleum Geologists Studies in Geology No. 31*, 13–29.

Jeffers, J. D., Anderson, J. B., and Lawver, L. A. 1991. Evolution of the Bransfield Basin, Antarctic Peninsula. *In* Geologic Evolution of Antarctica, M. R. A. Thomson, J. A. Crame, and J. W. Thomson, Eds.: New York: Cambridge University Press 481–485.

Keen, C. E., and Beaumont, C. 1990. Geodynamics of rifted continental margins, *In* Geology of Canada, No. 2, Geology of the Continental Margin of Eastern Canada, M. J. Keen and G. L. Williams, Eds.: Ottawa: Geological Survey of Canada, 393–472.

Kennett, J. P. 1982. *Marine Geology*: Englewood Cliffs, NJ: Prentice-Hall.

Klitgord, K. D., Hutchinson, D. R., and Schouten, H. 1986. U.S. Atlantic continental margin: Structural and tectonic framework. *In* The Geology of North America, Volume I-2, The Atlantic Continental Margin, U.S., R. E. Sheridan and J. A. Grow, Eds.: Boulder, CO: Geological Society of America, 19–55.

Kumar, N., Leyden, R., Carvalho, J., and Francisconi, O. 1979. Sediment isopachs, continental margin of Brazil. *American Association of Petroleum Geologists Special Map Series*, 1 Sheet.

Laine, E. P., Damuth, J. E., and Jacobi, R. D. 1986. Surficial sedimentary processes revealed by echo-character mapping in the western North Atlantic Ocean. *In* The Geology of North America, Volume M, The Western North Atlantic Region, P. R. Vogt and B. E. Tucholke, Eds.: Boulder, CO: Geological Society of America, 427–436.

Lawrence, J. R. 1979. $^{18}O/^{16}O$ of the silicate fraction of Recent sediments used as a provenance indicator in the South Atlantic. *Marine Geology* 33:M1–M7.

Manspeizer, W. 1988. Triassic–Jurassic rifting and opening of the Atlantic: An overview. *In* Triassic–Jurassic Rifting, W. Manspeizer, Ed.: New York: Elsevier, 41–80.

McCave, I. N., and Tucholke, B. E. 1986. Deep current-controlled sedimentation in the western North Atlantic. *In* The Geology of North America, Volume M, The Western North Atlantic Region, P. R. Vogt and B. E. Tucholke, Eds.: Boulder, CO: Geological Society of America, 451–468.

Montadert, L., de Charpal, O., Roberts, D. G., Guennoc, P., and Sibuet, J.-C. 1979. Northeast Atlantic passive margins: Rifting and subsidence processes. *In* Deep Drilling Results in the Atlantic Ocean: Continental Margins and Paleoenvironments, M. Talwani, W. W. Hay, and W. B. F. Ryan, Eds., Maurice Ewing Series 3: Washington, DC: American Geophysical Union, 164–186.

Mutti, E., and Normark, W. R. 1987. Comparing examples of modern and ancient turbidite systems: Problems and concepts. *In* Marine Clastic Sedimentology, J. K. Leggett and G. G. Zuffa, Eds.: London: Graham and Trotman, 1–38.

Normark, W. R., Barnes, N. E., and Bouma, A. H. 1985. Comments and new directions for deep-sea fan research.

In A. H. Bouma, W. R. Normark, and N. E. Barnes, Eds.: New York: Springer-Verlag, 341–343.

Palmer, A. R., Compiler 1983. The Decade of North American Geology 1983 geologic time scale. *Geology* 11:503–504.

Poag, C. W., and Sevon, W. D. 1989. A record of Appalachian denudation in postrift Mesozoic and Cenozoic sediments of the U.S. Middle Atlantic continental margin. *Geomorphology* 2:119–158.

Roberts, D. G., Schnitker, D., Backman, J., Baldauf, J. G., Desprairies, A., Homrighausen, R., Huddlestun, P., Kaltenback, A. J., Keene, J. B., Krumsiek, K. A. O., Morton, A. C., Murray, J. W., Westburg-Smith, J., and Zimmerman, H. B. 1984. Initial Reports of the Deep Sea Drilling Project, Volume 81: Washington, DC: U.S. Government Printing Office.

Skogseid, J., and Eldholm, O. 1989. Vøring Plateau continental margin: Seismic interpretation, stratigraphy, and vertical movements. *In* Proceedings of the Ocean Drilling Program, Scientific Results, Volume 104, O. Eldholm, J. Thiede, et al.: College Station, TX: Ocean Drilling Program, 993–1030.

Steckler, M. S., Watts, A. B., and Thorne, J. A. 1986. Subsidence and basin modeling at the U.S. Atlantic passive margin. *In* The Geology of North America, Volume I-2, The Atlantic Continental Margin, U.S., R. E. Sheridan and J. A. Grow, Eds.: Boulder, CO: Geological Society of America, 399–416.

Vogt, P. R., Perry, R. K., Feden, R. H., Fleming, H. S., and Cherkis, N. Z. 1981. The Greenland–Norwegian Sea and Iceland environment: Geology and Geophysics. *In* The Ocean Margins and Basins, Volume 5, The Arctic Ocean, A. E. M. Nairn, M. Churkin, Jr., and F. G. Stehli, Eds.: New York: Plenum Press, 493–598.

Vogt, P. R., and Tucholke, B. E., Eds. 1986. The Geology of North America, Volume M, The Western North Atlantic Region: Boulder, CO: Geological Society of America.

Watts, A. B., and Steckler, M. S. 1979. Subsidence and eustacy at the continental margin of eastern North America. *In* Deep Drilling Results in the Atlantic Ocean: Continental Margins and Paleoenvironments, M. Talwani, W. W. Hay, and W. F. B. Ryan, Eds., Maurice Ewing Series 3: Washington, DC: American Geophysical Union, 273–310.

Weaver, P. P. E., Rothwell, R. G., Ebbing, J., Gunn, D. E., and Hunter, P. M. in press. Correlation, frequency of emplacement and source directions of megaturbidites on the Madeira Abyssal Plain. *Marine Geology*.

Index

Boldface numerals indicate illustrations

Additional information of this book

(Geologic Evolution Of Atlantic Continental Rises; 978-1-4684-6502-0) is provided:

http://Extras.Springer.com

Additional information of this book

(Geologic Evolution Of Atlantic Continental Rises; 978-1-4684-6502-0) is provided:

http://Extras.Springer.com